Biomathematical and Biomechanical Modeling of the Circulatory and Ventilatory Systems

Volume 1

For further volumes:
http://www.springer.com/series/10155

Marc Thiriet

Cell and Tissue Organization in the Circulatory and Ventilatory Systems

 Springer

Marc Thiriet
Project-team INRIA-UPMC-CNRS REO
Laboratoire Jacques-Louis Lions, CNRS UMR 7598
Université Pierre et Marie Curie
Place Jussieu 4
75252 Paris Cedex 05
France
Marc.Thiriet@inria.fr

ISBN 978-1-4614-2896-1   ISBN 978-1-4419-9758-6 (eBook)
DOI 10.1007/978-1-4419-9758-6
Springer New York Dordrecht Heidelberg London

Springer is part of Springer Science+Business Media (www.springer.com)

Contents

Introduction

" οψιζ τον αδηλων τα φαινορυενα. [The ob-
servable is the manifestation of the invisible] "
(Anaxagoras).

Volume 1 of this interdisciplinary series "Biomathematical and Biomechanical Modeling of the Circulatory and Ventilatory Systems" aims at presenting cell types of the cardiovascular and respiratory systems as well as major cellular processes that regulate cell fate and can be modeled. These books are conceived with a modeling perspective rather than a purely biological approach. Modeling aims at better understanding, predicting, optimizing, controlling, as well as more easily testing the effects of involved parameter and visualizing. Most often, analytical solutions cannot be used. Numerical procedures are then employed, after basic mathematical and numerical analysis, to check stability and, when possible, solution unicity.

Computational biology tackles biological phenomena using multisample and multivariate data analysis. For example, computational biology aims at pointing out circumstances that trigger a peculiar signaling pathway, among all existing signal transduction axes. Parameters of molecular interactions that occur during a set of chemical reactions that constitute the explored signaling axe must then be estimated. In addition to mathematical modeling and simulation of optimally designed dynamical signaling pathways, computational biology incorporates techniques from optimization, high-performance computing, medical image processing, and data mining and analysis. Bioinformatics, which represents the engineering side of the topics, deals with the production of biological data analysis software. Goals include deciphering the relationships between geno- and phenotype, and correlations between individual genome, nutrition, environment, and life mode, among other health-related factors as well as diseases. Analyses of the human transcriptional and translational landscape (allele-specific expression, differential gene expression according to the context, alternative splicing, small RNA-mediated regulation, RNA editing, etc.) are required to investigate disease-associated traits.

Yet, this book series deals neither with bioinformatics, computational biology, molecular dynamics, structural geno- and proteomics, nor drug development. Instead, it presents the data at molecule, cell, and tissue levels necessary to analyze and represent signal transduction pathways that regulate cell fate and cell's adaptation to its environment. Reaction kinetic models that obey the mass-action law rely on systems of N chemical species that participate in chains of reversible or irreversible, enzymatic and non-enzymatic reactions (receptor or gene binding, molecular complex formation at the cell surface or within the cytosol or nucleoplasm, post-translational modifications, molecular translocation in different subcellular compartments, etc.). Reaction networks can be represented by diagrams; reaction schemes can be depicted using hypergraphs. Temporal dynamics of molecular interaction systems can be modeled using ordinary differential equations on molecule concentrations or continuous-time Markov chains on a number of molecules. Yet, only partial differential equations based on derivatives of concentrations of involved chemical species can express both complex temporal and spatial dynamics of signal transduction pathways. Reaction–transport equations describe the concentration differences that result from production, consumption, and transfer of these chemical species between the various subcellular compartments from the plasma membrane to the cell nucleus, or vice versa, as well as storage and release of molecules. One challenge is to couple models that function at different scales, from the order of the nanosecond to the week and from the order of the nanometer to the centimeter. Different types of models can be applied to distinct scales: e.g., particle methods for tiny structures, partial differential equations for processes at moderate time and space scales, and ordinary differential equations in reduced models at larger scales. Hybrid models incorporate discrete, reactive, moving, deformable objects in a continuum. Another task is to tackle parameters and coefficients via experiments.

Any biological and physiological system can be represented by interactions among molecules, cells, tissues, and organs inside a given organism. *Multiscale modeling* aims at coupling biomathematical models that describe cell and tissue events at nano- and microscales to standard macroscale simulations of any explored physiological process. Modeling of heart functioning illustrates multiscale modeling. In addition, mechanotransduction in the blood circulation is related to the control of the local caliber of the reactive blood vessel wall by mural myocytes. The contraction–relaxation state of these myocytes depends on the experienced stress field and released chemicals from the wetted endothelium. To take into account the reaction of arterial walls to applied hemodynamical stresses, the mathematical model with its set of equations that describes the mechanochemical signaling in vascular cells and paracrine regulation at nano- and microscopic scales can be coupled to that derived from the continuum mechanics theory at the macroscopic scale that governs fluid flow, i.e., the Navier-Stokes equations.[1] This procedure updates the caliber of blood vessel lumen, i.e., the size of the computational domain predicted by the Navier-Stokes equations.

The Navier-Stokes equations arise from balances of physical variables in a homogeneous infinitesimally small material particle (basic control volume). They are straightforwardly exploited at the macroscopic scale to predict the flow behavior of a fluid of given properties in a given segment of a conduit network. The duct wall is either rigid or possesses deformable walls with given rheological features in given dynamical conditions The (*multiphysics modeling*) couples the fluid dynamics to the wall mechanics. Any flow is computed for a selected set of governing dimensionless parameters and boundary conditions.[2] These boundary conditions should take into account input and ouput impedances. Therefore, the *multilevel modeling* aims at coupling models of the flow governing equations at various spatial dimensions (e.g., zero and one- and three-dimensional models). In the future, when appropriate sofware will be available, any multicoupling procedure associated with a multiphysics, multilevel, and multiscale modeling will necessarily use high-performance computing.

> "*Tout organisme, quel qu'il soit, se trouve alors indissolublement lié, non seulement à l'espace qui l'entoure, mais encore au temps qui l'a conduit là et lui donne comme une quatrième dimension. [An organism, whatever it may be, finds itself indissolubly bound, not only to the space which surrounds it, but also to the time which has carried it there, and gives it something like fourth dimension.]* " (F. Jacob) [1]

This series of volumes devoted to Circulatory and Ventilatory Systems in the framework of biomathematical and biomechanical modeling aims at providing basic knowledge and state of the art on the biology and mechanics of blood and air flows. The cardiovascular and respiratory systems are tightly coupled via the supply of oxygen (O_2; οξυς: sharp, bitter;[3] γενεα: begetted, generated) to and removal of carbon dioxide (CO_2; carbo: coal) from the body's cells. Oxygen is not only a nutrient that is used in cellular respiration, but also a component of structural molecules of living organisms, such as carbohydrates, proteins, and lipids. Carbon dioxide is produced during cell respiration. It is an acidic oxide that, in an aqueous solution, converts into

---

[1] Other types of mathematical models target other types of biological processes, such as cell growth, proliferation, migration, and apoptosis, that are observed in certain diseases, including atherosclerosis and cancers, in order to optimize drug design and delivery procedure.

[2] Differential equations are often solved with a set of constraints, — the so-called boundary conditions —, hence the name boundary value problems. A solution to the differential equation obviously satisfies the boundary conditions.

[3] In Greek, οξυσπια means a disorder in which food turns acid.

the anhydride of carbonic acid (H_2CO_3). It is then carried in blood mostly as bicarbonate ions (HCO_3^-) owing to carbonic anhydrase in erythrocytes, but also small fractions that are either dissolved in the plasma or bound to hemoglobin as carbamino compounds. Carbon dioxide is one of the mediators of autoregulation of local blood supply. It also influences blood pH via bicarbonate ions. Last, but not least, it participates in the regulation of air and blood flows by the nervous system.

The Virtual Physiological Human (VPH) project corresponds to a framework program that is aimed at modeling the entire human body, especially its physiology, drawing on recent advances in medical exploration technologies and high-performance computing. The latter, indeed, relies on specific algorithms and coupling platforms to enable integration of various models and rapid computations. The VPH Program's objective is the optimization of medical decision by achieving a better understanding and description of pathophysiological processes, predicting outcomes, and developing and planning new customized treatment procedures based on patient data, in parallel with computer-aided design of drugs and medical devices. Although this project considers the body as a whole made up of many organs rather than a collection of organs, physiological systems that constitute the human body still need to be investigated separately, as their behavior remains to be fully explored before treating their complex interactions with other components. In other words, a new bottom-up integrative research strategy cannot be adequately defined without top-down reductionist approaches.

Although the modeling and simulation of the complex behavior of the physiological systems that govern life in organisms still rely on reductionism, simple resulting models should be representative, i.e., developed after completely understanding the complex reality. In living bodies that are composed of reactive constituents rather than inert elements, phenomena observed at the macroscopic scale that are targeted by biomechanical investigations depend on mechanisms that arise at the microscopic scale. The latter can initiate studies in biomathematics.

Therefore, the first five volumes of this series are devoted to the signaling mediators, cells, and tissues mainly of vascular and ventilatory organs, i.e., to the nano-, micro-, and mesoscopic scales. As biological processes are interdependent, many basic cellular processes are more or less briefly presented, sometimes despite indirect connection with explored phenomena (e.g., protein synthesis), to better understand cell signaling and adaptation as well as tissue remodeling.

In this multidisciplinary context, the three basic natural sciences — biology, chemistry, and physics — interact with mathematics to explain the functioning of physiological flows, such as circulatory and ventilatory streams in their respective pathways. Frameworks are under development to couple biomechanical to biomathematical models of cell signaling and tissue adaptation to better describe the reality, although its complexity still necessitates abstraction. The major objective of the present publication is to present data that will be used to design multiscale models.

> " *Nam quodcumque alias ex se res auget alitque, deminui debet, recreari, cum recipit res. [For any body that augments and feeds others reduces and recreates on receiving them.]* " (Lucretius) [2]

The cell is a heterogeneous medium, even inside the cytosol, nucleus, and cellular organelles. Signaling initiation and the first steps of molecular interactions and transformations of most pathways occur at the cell membrane and cortex. Involved substances can travel between cell compartments and possibly the extracellular compartment to achieve their tasks. Dynamics of a biochemical process can be represented by a set of equations that link time and space variations of concentrations of interacting substances to production and consumption rates.

The complicated physiological system can be analyzed by decomposition into simple parts with identified functions. The combination of these functions allows us to deduce the system functioning based on linear interactions. Deconstruction into parts of physiological systems is necessary in order to understand the complicated behavior of these system compartments, as well as to determine some between-part interactions.

Yet, the cell is a complex system constituted by many components. The features of complex systems are adaptation, self-organization, and emergence. Cells self-organize to operate with optimal performance. The behavior of a complex system is not necessarily predictible from the properties of its elementary constituents, which can non-linearly interact with feedback loops, contributing to system bulk behavior. The organization and bulk behavior of a complex system not only results from the simultaneous activities of its constituents, but also emerges from the sum of the interactions among its constituents. A complex system adapts by changing its organization and possibly its structure in response to environmental stimuli. Yet, a predictive model requires a theory, or at least a framework, that involves relationships.

Models of cell response according to environmental stimuli that treat metabolic and signaling networks can have good predictive potential owing to the limited number of possible states, as cells optimally function in a bounded parameter space (experienced states are defined by a given set of physical and chemical parameters that evolve in known value ranges and feature identified relationships).

The major goal of this book series is to present the basic, exhaustive knowledge necessary to carry out modeling and simulations of behavior and flow pattern in the cardiovascular and ventilatory systems. Multiscale modeling and coupled simulations should be based on a interdisciplinary approach, as physiological conduits have deformable and reactive walls. Macroscopic flow behavior (prediction) should then be coupled to nano- and microsocopic events (corrector scheme of regulated mechanism).

Volume 1 "Cell and Tissue Organization in the Circulatory and Ventilatory Systems" introduces cells (microscopic scale) involved not only in the architecture of the cardiovascular and respiratory systems, but also those conveyed by blood to ensure body homeostasis and defense against pathogens. Volume 1 is also devoted to the basic components of cellular functions, such as mechanotransduction, that enable adaptation to environmental conditions.

Cell fate depends on environment. Volume 2 "Control of Cell Fate in the Circulatory and Ventilatory Systems" starts with command cells — neurons and endocrine cells — that regulate blood circulation and the body's respiration to adapt blood and air flows to the body's needs. Volume 2 details major, controlled stages of the cell life, such as growth, proliferation, migration, and death. Cell activities can be modeled using systems of equations to predict outcomes. Cell processes can be more or less easily described, but a huge number of scenarios can take place. Many models highlight the non-linear dynamics of cell functions.

Volume 3 "Signaling at the Cell Surface in the Circulatory and Ventilatory Systems" focuses on the sensors and receptors that trigger cellular responses to environmental stimuli, particularly mechanical stresses, with cascades of chemical reactions (nanoscopic scale). It actually presents major participants of cell signaling at the cell surface from extracellular ligands (locally released agents for auto- and paracrine regulation, hormones, growth factors, cytokines, chemokines, and constituents of the extracellular matrix, as well as mechanical stresses) that initiate signaling cascades.

Signaling pathways are usually composed of multiple nodes that correspond to major mediators. Intracellular effectors of signaling cascades are characterized in Volume 4 "Intracellular Signaling Mediators in the Circulatory and Ventilatory Systems". Signaling axes trigger the release of substances from intracellular stores that correspond to specific cellular organelles, and the gene expression to synthesize messengers of intra- or autocrine regulation, as well as those of close or remote control (juxta-, para-, and endocrine regulation). Signaling modules and reaction cascades are represented by mathematical models. Primary mediators must be retained in modeling of regulated cellular processes, whereas multiple secondary signaling components are discarded to produce simple, representative modeling and manage the inverse problem. As mathematics deals with abstraction, modeling based on transport equations is preferred to the modeling governed by mass action law that is used by chemists, which features a large number of kinetics coefficients (the values of which are most often unknown).

Volume 5 "Tissue Functioning and Remodeling in the Circulatory and Ventilatory Systems" deals with vascular tissues (blood, heart, blood and lymph vessels) and tissues of the respiratory tract, including interactions between adjoining cells. It is thus devoted to the functioning and remodeling of these tissues, i.e., events with short and long time scales. These regulated processes, especially cell activities involved in adaptation (angiogenesis, blood coagulation, healing, inflammation, mechanotransduction, stress-induced tissue remodeling in response to acute or chronic loadings, etc.) can be described using mathematical models. These models can then be coupled to flow dynamics.

Volume 6 "Circulatory and Ventilatory Conduits in Normal and Pathological Conditions" focuses primarily on macroscopic aspects of the cardiovascular and respiratory systems, acquisition and processing of medical images and physiological signals, as well as diseases of the wall of conduits that disturb blood and air flows. Conversely, local flow disturbances can contribute to trigger pathophysiological processes. Therefore, Volume 6 contains chapters on the anatomy and physiology of the cardiovascular and respiratory systems, and medical signals and images, as well as pathologies of the fluid convection duct network (i.e., heart, blood vessels, and respiratory tract), and their treatment. These diseases — as well as some therapeutic procedures — have been targeted by biomechanical studies. In addition, the development of medical devices incorporates a numerical test stage in addition to experimental procedures.

Volume 7 "Blood and Air Transport in the Circulatory and Ventilatory Systems" addresses the mechanics of air and blood flows in relatively short, curved, deformable conduits that thus convey developing, three-dimensional, time-dependent, mostly laminar flows. Volume 7 takes into account the rheology of blood and deformable walls of respiratory conduits and blood vessels as well as different types of air transport – convection and diffusion – in the respiratory tract. This volume also provides insights into numerical simulations of these types of flows.

Volume 8 contains a set of glossaries, which will aid the reader in rapidly getting information on the parameters and structures that are used in models, as these data arise from diverse scientific disciplines. Certain field-specific vocabulary can indeed limit easy access to this field to researchers of other disciplines. This set of glossaries aims at accessibly explaining the vocabulary and techniques of the 3 involved basic natural sciences (biology, chemistry, and physics) that interact with mathematics to explain the functioning of physiological systems.

Volume 1 is composed of 9 chapters. Chapter 1 provides an introduction to the body's cells and biological tissues. Chapters 2 and 3 give a survey of cell types involved in the vasculature and the respiratory tract, respectively. Chapter 4 describes the cell and its major components, such as the nucleus, which contains the genome, cellular organelles (endoplasmic reticulum, Golgi stack, mitochondria, vesicles, etc.), cytoskeleton, and large molecular complexes immersed in the cytosol. The cell, its nucleus, and organelles are wrapped by typical membranes, which are duplicated in the case of the nucleus and mitochondrion to form envelopes. Mitochondria produce cell energy. Chapter 5 summarizes the current knowledge on protein synthesis, as most signaling mediators are proteins and cell signaling can trigger gene expression. Chapter 6 focuses on the cell cytoskeleton, which is involved in cellular transport, cell division, adhesion, migration, and, last but not least, in cell adaptation to mechanical stresses. Chapters 7 and 8 deal with the cell membrane and its relation to the cell environment. The extracellular matrix is a necessary medium for cell fate, and tissue formation and remodeling due to applied mechanical stresses. Multiple cell processes are primed by activated ion carriers and receptors of the plasma membrane. Numerous cell processes involve the cell cytoskeleton. Chapter 9 deals with molecular transport in the cell in interactions with its environment. Intracellular transfer of cargos uses nanomotors and the tracks of the cytoskeleton.

Common abbreviations such as "a.k.a." ("also known as") and Latin-derived shortened expressions ("e.g." [exempli gratia: for example] and "i.e." [id est: in other words]) are used throughout the text to lighten sentences. Rules adopted for substance aliases as well as alias meaning and other notations are given at the end of this book.

Acknowledgments

These books result from lectures given at Université Pierre et Marie Curie in the framework of prerequisite training of Master "Mathematical Modeling", part of Master of "Mathematics and Applications", Centre de Recherches Mathématiques,[4] and Taida Institute for Mathematical Sciences,[5] the latter two in the framework of agreements with the French National Institute for Research in Computer Science and Control.[6] These lectures mainly aim at introducing students in mathematics to basic knowledge in biology, medicine, rheology, and fluid mechanics in order to conceive, design, implement, and optimize appropriate models of biological systems at various length scales in normal and pathological conditions. These books may also support the elaboration of proposals following suitable calls of granting agencies, in particular ICT calls "Virtual Physiological Human" of the European Commission. The author takes the opportunity to thank the members of ERCIM office (European Consortium of Public Research Institutes) and all of the participant teams of the working group "IM2IM" that yields a proper framework for such proposals. These books have been strongly supported by Springer staff members. The author thanks especially S.K. Heukerott and D. Packer for their help and comments.

The author, an investigator from the French National Center for Scientific Research[7] wishes to acknowledge members of the INRIA-UPMC-CNRS team "REO",[8] and Laboratoire Jacques-Louis Lions,[9] as well as CRM (Y. Bourgault, M. Delfour, A. Fortin, and A. Garon), being a staff member in these research units, and TIMS (I.L. Chern, C.S. Lin, and T.W.H. Sheu), as well

---

[4] CRM: www.crm.umontreal.ca.

[5] TIMS: www.tims.ntu.edu.tw.

[6] Institut National de la Recherche en Informatique et Automatique (INRIA; www.inria.fr).

[7] Centre National de la Recherche Scientifique (CNRS; www.cnrs.fr

[8] www-roc.inria.fr/reo

[9] www.ann.jussieu.fr

as members of the Dept. of Bioengineering and Robotics from Tohoku University (Japan) led by T. Yamaguchi for joint PhD experience and research. The author also acknowledges the patience of his family (Anne, Maud, Julien, Jean, Raphaëlle, Alrik, Matthieu [Matthew], Joanna, Damien, and Frédéric [Fryderyk]). This book is dedicated to the author's mother and Flemish and Prussian grandmothers.[10]

---

[10] They learned me to use a proper grammar and spelling (but partly failed) as well as stimulate the search of the etymology.

1

Cells and Tissues

The body organization can be described according to several length scales from physiological systems, organs, tissues, cells, subcellular structures, to molecules (Table 1.1). A physiological system is commonly constituted of various organs. A body's organ usually contain several tissue types. A biological tissue is often composed of different types of cells. A cell consists of an intracellular medium wrapped by a plasma membrane that encloses a nucleus and diverse types of organelles immersed in the cytosol (Chap. 4).

Cells, although specialized, share a common bulk structure with a genetic code-containing nucleus and a set of organelles. The cell nucleus possesses an envelope (double membrane) that separates the nucleoplasm from the cytoplasm. Cytoplasmic organelles, strictly speaking, also possess either a membrane or an envelope. In addition, large proteic complexes, such as cytoskeletal filaments, spliceosomes, and ribosomes within the cytoplasm, and nucleolus inside the nucleoplasm, correspond to cellular substructures like membrane-bound organelles. Whatever the length scale, a biological structure is endowed of compartments.

Anatomy and histology deal with the macroscopic (organs) and microscopic (tissue and cell structures) scale, respectively. In addition, cells synthesize the components of the interstitium. Any cell aims at surviving, growing, proliferating, and migrating, as well as, for certain cell types, differentiating, before disappearing (Vol. 2 – Chaps. 2. Cell Growth and Proliferation and 4. Cell Survival and Death). Cell replacement and regeneration to which stem and progenitor cells contribute occur in 2 contexts: renewal of the cell population and injury.

Regulated assemblies of cells produce biological tissues. Tissue morphology depends on cell shapes, hence cell cytoskeletons and adhesions that are controlled by the cell response to applied forces. Biological tissues are able to resist stresses and remodel.

Water, proteins, and lipids are the main constituents of the cell. Although the water content is quite high, the deformable cell has been considered as a slightly compressible material.

Table 1.1. Tier architecture of living systems. Levels and sizes (in adults). The right lung is lower and thicker than the left lung due to the liver and heart, respectively, but is globally larger. Lung height is correlated with body's height; lung width with body's height, weight, and sex; heart size with age, weight, and sex [3]). At rest, vital capacity depends on subject's age, sex, and size (VC 1.9–4.4 l in women and 2.4–5.9 l in men; Source: [4]). Albumin is a monomeric, non-glycosylated, water-soluble protein commonly detected in blood plasma (human serum albumin constitutes $\sim$ 60% of human plasma protein). It regulates the blood oncotic (or colloidal osmotic) pressure. It carries molecules of low water solubility (hormones, bile salts, free fatty acids, and cations (Ca^{++}, K^+, and Na^+). It is currently used in experiments on the permeability of blood vessel walls.

Level	Size
Human height	1.40–1.90 m
Macroscopic scale	
Lung height	21–26 cm
Lung thickness	25–32 cm
Heart length	11–15 cm
Heart width	8–9 cm (broadest section)
Heart thickness	~ 6 cm
Microscopic scale	
Epithelial cell	$\sim 30\,\mu$m
Erythrocyte	$\sim 7\,\mu$m
Nanoscopic scale	
Albumin	~ 6 nm
H_2O	~ 0.5 nm

1.1 Cell

The cell is the basic structural and functional unit of the organism (typical size $\mathcal{O}[10\,\mu\text{m}]$). The cell nucleus houses the genetic material, i.e., a set of genes that form a long molecular chain, the deoxyribonucleic acid (DNA). In the extranuclear space, the cytoplasm is made of the cytosol and organelles. An organelle is a specialized subunit within a cell that has a specific function.

Unlike large proteic complexes, cell organelles are membrane-bound structures. Among the organelles, the nucleus and mitochondria are coated by an envelope, i.e., a double membrane that limits a thin space. Single membrane-bound organelles encompass the endoplasmic reticulum, Golgi body, mitochondria, various types of endo- and exosomes, and cilia. Other cellular functional units include, within the nucleoplasm, the nucleolus and, within the cytoplasm, ribosomes, the cytoskeleton with its 3 components — microfilaments, microtubules, and intermediate filaments — and centrioles. The wall of each centriole are usually composed of 9 triplets of microtubules. The stress fibers are components of the actomyosin cytoskeleton that support the cell contraction or relaxation.

The endoplasmic reticulum contributes to the synthesis and folding of new proteins in its rough compartment covered with ribosomes and production of lipids in its smooth compartment. The Golgi body functions in sorting and modification of proteins. Mitochondria are specialized in energy production. Endo- and exosomes carry numerous types of molecules inside the cytosol. Autophagosomes sequester cytoplasmic material and organelles for degradation. Lysosomes are specialized compatments for the breakdown of molecules. Peroxisomes are used for the degradation of hydrogen peroxide. Melanosomes are pigment stores.

Nucleolus is an intranuclear compartment involved in the ribosome generation. In the extranuclear space, ribosomes are sites of translation of the genetic information contained in mature messenger RNAs that is processed to synthesize proteins. Many eukaryotic cells are endowed with a primary cilium. The primary cilium serves as signaling platforms. Primary cilia can act as chemo-, thermo-, and mechanosensor. Some polarized cells possess multiple cilia at their apical (luminal, or wetted) surface in contact with a fluid that are aimed at sweeping this fluid.

1.2 Stem Cells

Stem cells replace dead, fully differentiated cells, especially in tissues characterized by high cell turnover (blood, skin, and gut epithelia) or occasionally for tissue repair and regeneration in response to tissue damage.

Stem cells may thus be employed in therapies that require repair, replacement, and regeneration. Tissue repair and regeneration involve cell interaction over a long time scale to produce suitable cell type, number, and location. The regeneration capacity of stem cells that reside in adult tissues depends on the organ type. In many tissues, stem cells give rise to progenitors (intermediate cells) that have a limited self-renewal potential and elevated probability to undergo differentiation. In terminal differentiation, precursor cells become very specialized and irreversibly lose their ability to self-renew.

During the early stages of embryogenesis, the fertilized egg starts to divide into blastomeres. *Embryonic stem cells* (ESC), derived from *totipotent cells*[1] of preimplantation embryos, are *pluripotent cells* (Fig. 1.1). After gastrulation, they give birth to *multipotent cells* that are irreversibly programmed for a given tissue. Multiple stem cell populations have been discovered from various adult tissues. The number in tissues is less than 1 for 10^4 cells. *Unipotent progenitors* differentiate into a single cell type.

Mesenchymal stem cells localize mainly to the adipose tissue and bone marrow. In the bone marrow, these stem cells participate in the maintenance

---

[1] A single totipotent cell leads to the development of several hundreds of cell types.

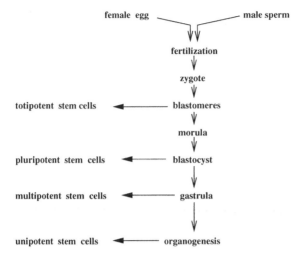

Figure 1.1. Different types of stem cells. Fusion between female and male gametes (fertilization) leads to the zygote and embryogenesis. After fertilization, a set of fast mitoses occurs (cleavage) with formation of blastomeres that build the blastula. In mammals, the blastula is called the blastocyst. The morula corresponds to an organized cell set with external and internal cells. After the mitosis rate has slowed down, blastomeres move (gastrulation), forming the gastrula with 3 germ layers: ecto-, endo-, and mesoderm.

of *hematopoietic stem cells*. They can differentiate in many cell types, such as adipocytes, neurons, osteoblasts, chondroblasts, vascular cells, and cardiomyocytes. They thus contribute to repair of damaged tissues (Vol. 5 – Chap. 11. Tissue Growth, Repair, and Remodeling). They also operate as immunosuppressors (that lower the body's immune response).

In the bone marrow, blood cell formation from a single cell type — the hematopoietic stem cell — maintains the circulating cell pool, because mature blood cells continuously undergo senescence with a given degradation rate. All hematopoietic lineages derive from hematopoietic stem cells that generate both myeloid or lymphoid cells, intermediate progenitors having both lymphoid and myeloid potential.

Tissue repair and regeneration involve cell interactions over a long time scale to produce suitable cell type, number, and location. Relevant mesenchymal progenitor cells can undergo multi-lineage differentiation into mesenchymal tissues, such as bone, cartilage, and adipose tissue.[2]

Regeneration potential relies on tissue-resident stem cells. Tissues with high cell turnover can be restored to their original state after injury, whereas other tissues such as the myocardium can be repaired, but not completely

---

[2] Derivation of mesenchymal stem cells from human embryonic stem cells and their commitment to the chondrogenic lineage can be obtained in a relevant environment using morphogenetic factors from chondrocytes [5].

restored. Regeneration capacity of cardiac stem cells that reside in adult heart is actually small, although they can give rise to the 3 main heart cell types: cardiomyocytes and smooth muscle and endothelial cells.

Self-renewing stem cells constantly produce similar daughter cells or more mature daughter cells with restricted properties. The switch between self-renewal and differentiation is discriminative, as a less primitive cell cannot naturally become a stem cell. Cells indeed differ according to the expressed part of the genome, i.e., its transcriptional and epigenetic regions. During development and adult life, stem cells integrate and react to regulatory signals that maintain self-renewal or prime differentiation, i.e., they change their transcriptional program. Gene expression is in fact determined by the presence of transcriptional regulators (non-coding RNAs, ATP-dependent chromatin-remodeling complexes, chromatin-binding proteins, DNA methyltransferases, histone-modifying enzymes, etc.; Chap. 5). Genes that encode proteins and microRNAs and prime the cell differentiation are repressed in self-renewing stem cells. Although adult stem cells specific to a given tissue are only able to form their tissue of origin, upon stimulation using genes and molecules that define embryonic stem cells, they can return to a more primitive cell state and then transform themselves into other tissue cells (*trans-differentiation*). In addition, fetal and adult hematopoietic stem cells differ in their expression of cell-surface markers and proliferation rate.

In addition to the original quartet of transcription factors[3] (e.g., octamer-binding transcription factor Oct4, Sry-related HMG box Sox2, Krüppel-like factor KLF4, and MyC), relatives of these transcription factors and other transcription factors such as homeobox gene product Nanog,[4] as well as other proteins (i.e., RNA-binding protein Lin28 homolog-A[5] and nuclear receptor NR5a2)[6] can contribute to the reprogramming of somatic cells into pluripotent stem cells, the so-called *induced pluripotent stem cells* (IPSC).

Transcription factors — Oct4, Sox2, KLF4, MyC, and Nanog — coordinately participate in stem cell pluripotency and self-renewal of embryonic stem cells, activating genes that maintain pluripotency and repressing those that are required for differentiation. The transcription factor Sal-like protein SalL4[7] influences the transcription of pluripotency regulators from genes Klf4, MYC, OCT4, SOX2, and NANOG in mouse embryonic stem cells [6]. Nevertheless,

---

[3] A transcription factor is a sequence-specific DNA-binding protein that controls the transcription of genetic information from DNA to messenger RNA. Transcription factors cooperate with coregulators, chromatin remodelers, and histone acetylases, deacetylases, methylases, and kinases in gene transcription (Chap. 5).

[4] In Gaelic, tir na nog means land of the ever young.

[5] The alias Lin stands for abnormal cell lineage in Caenorhabditis elegans (Sect. 2.1.4).

[6] A.k.a. liver receptor homolog LRH1.

[7] A.k.a. zinc finger protein ZNF797.

transcription factor Oct4 may be sufficient to generate pluripotent stem cells from adult mouse neural stem cells [7].[8]

Stem cells are located in specific *niches*, specialized microenvironments, where their proliferation rate is regulated.[9] Stem cell fate (maintenance, survival, self-renewal, and differentiation) depends on many factors: (1) cell-surface molecules of cellular adhesion (Sect. 7.5); (2) secreted substances from stem cells and/or stromal support cells within these niches (e.g., ions such as Ca^{++}, growth factors, Wnt morphogens, etc.); (3) neural and blood inputs; (4) niche stiffness and mechanical stresses; and (5) spatial information.[10] Stem cell niches receive and transmit signals to ensure not only stem cell function and mobilization in response to tissue injury, but also stem cell recruitment (Vol. 2 – Chap. 5. Circadian Clock, Sect. Circadian Rhythm Influence on Stem Cells). Hematopoietic stem cells that reside along the endosteal surface

---

[8] Among the original quartet of transcription factors, Oct4 was supposed to be critical, as it could not be substituted by other transcription factors, whereas KLF, MyC, and Sox2 can be replaced by other factors. However, unrelated transcription factors such as orphan nuclear receptor can replace Oct4 in reprogramming to induced pluripotent stem cells [8].

[9] Epithelial stem cells that regenerate hair follicle and sebaceous glands are detected within the follicular bulge of the outer root sheath of the hair follicle. Basal keratinocytes that repopulate the interfollicular epidermis are observed at the base of epidermis above the basement membrane that separates it from dermis. Gut stem cells reside at the base of intestinal crypts and produce precursor cells that differentiate as they migrate toward intestinal villi [9]. A feedback control between enterocysts and intestinal stem cells involves the Janus kinase–signal transducers and activators of transcription, Jun N-terminal protein kinase, and Notch pathways, as well as epidermal growth factor receptor signaling [10]. Skeletal muscle stem cells are found along myofibers. Neural stem cells are situated in the subventricular zone of the hippocampus and in olfactory bulb, where they are adjacent to endothelial cells [9]. Neuronal cell fate is controlled by neurogenin and NeuroD basic helix–loop–helix (bHLH) proteins that interact directly with the nucleosome-remodeling factor subunits of the switch/sucrose non-fermentable (Swi/SNF) ATP-dependent chromatin-remodeling complex [10]. Nuclear geminin blocks neurogenesis and favors the proliferation of stem cells, as it antagonizes bHLH factors and Swi/SNF components. Carbonic anhydrase-2+ pancreatic cells are progenitors for both endocrine and exocrine pancreas [11].

[10] Interactions between human embryonic stem cells and extra-embryonic endoderm that is mediated by localized secretion of bone morphogenetic protein-2 by the latter and antagonistic growth differentiation factor-3 by the former controls niche size-dependent spatial gradient of receptor-associated proteins SMAD1 [12]. Mediator SMAD1 processes spatial niche information to control self-renewal and differentiation of human embryonic stem cells. Markers of human embryonic stem cells comprise glycolipid antigens SSEA3 and SSEA4, keratan sulfate antigens TRA1-60, TRA1-81, GCTM2, and GCT343, and a set of protein antigens (CD9, CD90 [also known as Thy1], alkaline phosphatase antigens, and class-1 HLA antigens), as well as genes NANOG, OCT4 (also called POU5F1), TDGF, DNMT3B, GABRB3, and GDF3 [13].

of trabecular bone close to osteoblasts and vascular endothelial cells (Vol. 5 – Chap. 2. Hematopoiesis) can leave their niches, enter into the blood circulation, and return to niches.

During embryogenesis, hematopoietic stem cells arise in the *aorta–gonads–mesonephros region*. The first heartbeat in embryos creates pulsatile flow (Sect. 1.9.2) of blood that consists mainly of plasma and primitive red blood cells. Blood flow promotes the formation of hematopoietic stem cells in close connection with the endothelium of embryonic blood vessels from hemangioblasts, particularly in the ventral wall of the dorsal aorta, before the bone marrow appears. Hematopoietic stem cells can sense flow-generated stresses. Normal hemodynamic stress that causes vasodilation and raises blood flow increases the expression of the transcription factor Runx1, a master regulator of hematopoiesis, via nitric oxide [14,15].

Cardiac stem cells are able to generate cardiomyocytes and coronary vessel cells. Human self-renewing, multipotent, SCFR+ cardiac cells (but negative for hematopoietic and endothelial antigens CD34, PECAM1, PTPRc, and VEGFR2) differentiate predominantly into cardiomyocytes and, to a lesser extent, vascular smooth muscle and endothelial cells [16]. Human circulating endothelial progenitors that express CD34, PECAM1, VEGFR2, and SCFR molecules and can migrate to and accumulate in the heart cannot generate a functional myocardium in vivo.

Circulating stem and progenitor cells contribute to angiogenesis and arterial repair after injury. However, the main source of progenitor and stem cells resides in the vascular wall and perivascular region. SCA1+, $SCFR^{low}$, Lin−, $CD34^{low}$ progenitors,[11] also called side population cells, that express ATP-binding cassette transporter ABCg2 have been detected in the tunica media of adult mice aortas [17]. They are able to acquire the phenotype of endothelial cells,[12] when they are stimulated by vascular endothelial growth factor (Vol. 2 – Chap. 3. Growth Factors), and smooth muscle cells[13] in the presence of transforming growth factor-β1 and platelet-derived growth factor-BB, respectively.

1.3 Cellular Differentiation

Three basic categories of cells include germ, somatic, and stem cells. *Germ line cells* give rise to gametes (eggs and sperm). During the body's development, immature *precursor cells* — *stem cells* that can can replicate indefinitely and differentiate into numerous cell types and *progenitor cells* that divide a limited number of times and give rise to a specific cell type — differentiate

---

[11] The aliases SCA and Lin stand for stem-cell antigen and lineage, respectively.

[12] These progenitor cells can produce CD31, vascular endothelium (VE)-cadherin, and von Willebrand factor.

[13] These progenitor cells can synthesize α-smooth muscle actin, calponin, and smooth muscle myosin heavy chain (Chap. 6).

Table 1.2. Stages of the vascular differentiation. Zygocyte, or zygote, is the initial cell formed by male and female gamete cells (from maturation of germ cells). Hemangioblasts, the common blood and vascular precursors, originate from mesodermal progenitors of the embryonic mesoderm and extra-embryonic yolk sac (sac attached to the human embryo that operates as the developmental circulatory system). Hemangioblasts of the yolk sac generate hematopoietic stem cells and, then, angioblasts, or vasoformative cells, that form early blood islands. Mesodermal hemangioblasts give rise to angioblasts that aggregate and join blood islands to create the primitive vascular plexi. Differentiation of arterial and venous endothelial cells leads to the mature vasculature, lymphatic endothelial cells originating from venous endothelial cells.

	Zygocyte	
	Blastocyte	(single-layered blastula)
Ectoderm	Germ layer cells	(three-layered gastrula)
	Mesoderm	Endoderm
	Hemangioblasts	Respiratory tract
	Hematopoietic stem cells Angioblasts	
	Primitive vasculature	
	Differentiated vasculature Arteries and veins	
	Mature blood vasculature Lymphatic vessels	

into mature specialized cell types.[14] Multipotent progenitors form the cells of the 3 germ layers: (1) the mesoderm, from which are derived connective tissue, muscles, and the circulatory system (Table 1.2); (2) the ectoderm (nervous system, skin, etc.); and (3) the endoderm (digestive and respiratory tracts, and endocrine glands).

Cellular differentiation happens during embryo- and fetogenesis (Vol. 5 – Chap. 11. Tissue Growth, Repair, and Remodeling) as well as after birth in children and adults for normal cell turnover and during tissue repair.

Cell differentiation changes the cell's size, shape, membrane polarity, metabolic activity, and responsiveness to signals because of controlled modifications in gene expression. On the other hand, dedifferentiation allows differentiated cell to revert to an earlier developmental stage.

---

[14] *Totipotent cells* (zygote and early embryonic cells) can differentiate into all cell lineages. *Pluripotent stem cells* (e.g., hematopoietic and mesenchymal stem cells) are able to differentiate into many cell types. *Multipotent progenitors* give rise to functional cells.

Table 1.3. Features of transcription factors and microRNAs (Source: [18]).

	Transcription factors	MicroRNAs
Pleiotropy	+	+
Cooperation	Cooperative binding to cognate DNA sequences Cooperative recruitment of transcription cofactors	Cooperative activity
Recognition of site accessibility and binding	Nucleosome coating	RNA-binding proteins Folding of target mRNA sequences Other microRNAs
Regulation	Activation or repression Slower activity (± translocation)	Repression mostly Fast, reversible repression Quick reactivation
Expression	MicroRNA dependency Feedback Post-translational regulation (phosphorylation)	Post-translational regulated maturation RNA editing (adenosine-to-inosine conversion); MicroRNA cofactor modifications

Each differentiated cell type expresses a gene subset of the genome due to a peculiar pattern of gene expression. Growth factors control the switch from one gene expression pattern to another. Distinct cell types differentially retrieve the genetic information encoded in the genome. Transcription factors and microRNAs (Sect. 5.3) form the largest families of gene regulatory factors [18]. Sets of combinatorially expressed transcription factors and microRNAs delineate cell types. Transcription factors and microRNAs share many similar features although they have their specific properties that determine specialized regulatory niches (Table 1.3). Some microRNAs are expressed in a cell- or tissue-specific manner and contribute to the cell identity. MicroRNAs can control the expression of transcriptional regulators. They also regulate alternative splicing during tissue development (Table 1.4; Sect. 5.4).

1.4 Recognition of the Body's Cells – Major Histocompatibility Complex

A large set of proteins on the cell surface such as blood group markers on the erythrocyte surface serve as *antigens*. The superclass of MHC genes actually

Table 1.4. Example of alternative splicing. Intronic and exonic splicing enhancers and silencers determine exon inclusion in or exclusion of the mature transcript. Alternative splicing variants may have distinct functions. Moreover, the splicing modes may depend on cell type due to the restricted presence of regulators that orchestrate splicing decisions. Splicing regulators repress or promote the formation of spliceosomes, often according to the location of specific binding sites within exon (E) or intron (I) sequences.

Pre-mRNA	5′UTR—E1—I1—E2—I2—E3—3′UTR
mRNA variant 1	5′UTR—E1—E2—E3—3′UTR
mRNA variant 2	5′UTR—E1—E3—3′UTR

expresses cell-surface antigens that operate as both *self-antigens* within the body and *non-self-antigens* outside of the organism.[15]

In humans, molecules of the major histocompatibility complex (MHC) are also called human leukocyte antigens (HLA). The immune system uses HLAs to differentiate native cells and foreign organisms. Genes of human leukocyte antigen categories HLA-A to HLA-C as well as HLA-E to HLA-G are class-1 MHC genes, whereas those of categories HLA-DP-A1 and -B1, HLA-DQ-A1 and -B1, and HLA-DR-A and -B1 to -B5 are class-2 MHC genes. Class-1 MHC molecules are present at the surface of almost all cells, whereas class-2 MHC molecules are restricted to antigen-presenting cells, B lymphocytes, and a subset of T cells. Human leukocyte antigens of categories HLA-A to HLA-C and HLA-G have been detected in soluble form in body fluids.[16] Class-1 MHC molecules are homo- and hetero-oligomers.[17] Both classes of MHC proteins in infected or dysfunctional cells are able to initiate an immune response via antigen presentation to T lymphocytes. Class-3 MHC genes encode for immune components (e.g., complement components and cytokines).

Embryonic stem cells produce small amounts of class-1 MHC molecules, but no class-2 MHC molecules [20]. Some embryonic stem cells synthesize H antigens that can induce rejection by the immune system.[18] Expression

---

[15] The immune system comprises 2 main components: (1) humoral immunity that ensures the body's protection is found in humor (cell-free fluid) and involves antibodies against involved antigens or complement and (2) cellular immunity that is defined by activation of cells, such as macrophages, natural killer cells, and antigen-specific cytotoxic T lymphocytes, and release of cytokines in response to an antigen that influences the activity of cells involved in adaptive and innate immune responses.

[16] Soluble HLAs can trigger apoptosis of CD8+ T lymphocytes and natural killer cells [19].

[17] For example, tetrameric complexes of HLAb27 homodimers bind to natural killer (NK) receptors and related immunoreceptors on lymphocytes, monocytes, and natural killer cells [19].

[18] Moreover, fetal antigens that are absent in the thymus (thus not recognized as self-antigens) can be present at the surface of embryonic stem cells.

of class-1 MHC molecules is upregulated in embryoid bodies (aggregates of embryonic stem cells that have begun to differentiate).

The classical function of class-1 MHC molecules at the plasma membrane corresponds to presentation of peptides for recognition by T-cell receptors (Vol. 3 – Chap. 11. Receptors of the Immune System). To maintain self-tolerance, natural killer cells (Sect. 2.2.3.2) express combinations of class-1 MHC inhibitory receptors.

Class-1 MHC molecules also intervene in cell growth and differentiation and organogenesis. Class-1 MHC molecules that are not associated with β2-microglobulin (i.e., class-1 MHC molecules correctly folded) serve as signaling mediators [19].

Class-1 MHC molecules that are involved in signaling to immune cells have many partners that bind to various motifs of MHC molecules. Several class-1 MHC receptors bind to class-1 MHC ligands. Clustering of MHC class-1 molecules enhances their recognition by T-cell receptors. In addition, class-1 MHC molecules are able to prime signaling cascades that control cell fate, such as decisions toward survival and proliferation. Class-1 MHC signals are transduced in combination with integrin signaling (Sect. 7.5.4), in particular using β_4 integrins in endothelial cells. Low concentration of circulating class-1 MHC antibodies leads to cell survival via the focal adhesion kinase pathway (Vol. 4 – Chap. 3. Cytosolic Protein Tyrosine Kinases), whereas a high level enhances amount of fibroblast growth factor (Vol. 2 – Chap. 3. Growth Factors) on the plasma membrane and favors cell proliferation via the mitogen-activated protein kinase pathway [19] (Vol. 4 – Chap. 5. Mitogen-Activated Protein Kinase Modules).

In adaptive immunity, T cells recognize pathogen-derived peptides on the surface of antigen-presenting cells. Activation of T cells is triggered when T-cell receptors recognize the major histocompatibility complex ligand that displays the appropriate antigenic peptide. Within 5 mn of contact, antigenic peptide–TCR complexes form molecular clusters owing to the actin cytoskeleton that brings components from the periphery to the center of the *immunological synapse*. Concomitantly, $\alpha_L\beta_2$ integrins on the surface of T cells tether to intercellular adhesion molecules ICAMs1 on the antigen-presenting cell, in a peripheral ring that surrounds the central TCR cluster. Paired immunoglobulin-like receptor (PIR) expressed by B lymphocytes, mastocytes, and myeloid cells binds to several class-1 MHC molecules and decreases synaptic adaptation and synapse remodeling in response to stimuli.

Among non-classical class-1 MHC molecules, cluster of differentiation molecules CD1e (expressed only in dendritic cells) interacts with phagosome membrane to present large glycolipids for cleavage. CD1-restricted $\gamma\delta$ T lymphocytes recognize pollen-derived phospholipids in allergy [19]. Neonatal Fc receptor that also belongs to non-classical class-1 MHC molecules operates in the passive transfer of maternal immunoglobulin-G to the fetus. Moreover, overexpression of this receptor can prolong the circulatory half life of immunoglobulin-G [19].

Class-1 MHC molecules are able to modulate the function of other cell-surface receptors. Another non-classical MHC molecule, HLAh[19] actually links to transferrin receptor to modulate its affinity for transferrin and participates in erythropoiesis [19] (Vol. 5, Chap. 2. Hematopoiesis).

1.5 Mechanotransduction-Directed Cell Shape

Mechanotransduction in mechanically loaded tissues, such as those of conduit walls of the cardiovascular and ventilatory apparatus, enables the cell to quickly adapt its shape and function to change in loading. The cell senses force exerted to on it by its environment as well as matrix rheological status, transduces (converts) mechanical stimuli into chemical signals, and responds by adapting its shape and fate (survival, growth, differentiation, etc.) [21].

Mechanosensing involves molecule conformational changes following variation in force and/or structure of recognition sites. Mechanotransduction integrates the entire set of mechanical stimuli and triggers chemical reactions and substance fluxes. In addition, mechanoresponse reorganizes relatively quickly the cytoskeleton (Chap. 6) and adjusts the metabolism.

Force sensing relies on detection of local changes in protein conformation that lead to ion channel opening, protein unfolding, modified enzyme kinetics, and variations in molecular interactions following exposure of buried binding site or, conversely, hiding them. Matrix stiffness sensing involves the detection of spacing between matrix constituents and membrane curvature variations.

Mechanotransduction initiates several signaling pathways. Multiple mediators include: (1) at the cell surface, enzymes, adhesion molecules (Sect. 7.5), ion channels (Vol. 3 – Chap. 3. Main Classes of Ion Channels and Pumps), and receptors (Vol. 3, from Chap. 6. Receptors), as well as specialized membrane nanodomains of the plasma membrane (Sects. 7.2.1, 7.2.5, and 7.2.6); (2) at the cell cortex, signaling adaptors and effectors (e.g., monomeric guanosine triphosphatases, heterotrimeric guanine nucleotide-binding (G) proteins, kinases, phosphatases, ubiquitins, etc. [Vol. 4]); (3) in the cytosol, enzymes, scaffolds, carriers such as endosomes (Chap. 9), calcium concentration, and transcription factors; and (4) in the nucleus, nuclear pore carriers (Sect. 9.1.8), enzymes, and the transcriptional and translational machinery (Chap. 5).

1.6 Cell and Tissue Morphology

Morphogenesis describes the regulated spatial distribution of cells, their controlled growth and differentiation, and the localization, shape, structure, and function of the body's tissues and organs during embryo- and fetogenesis as well as after birth.

---

[19] A.k.a. hemochromatosis protein "high iron" (HFe).

Multiple chemical signals and physicochemical processes contribute to morphogenesis. Morphogens and growth factors are soluble messengers that target their cognate receptors to control the cell fate (Vol. 3 – Chaps. 8. Receptor Kinases and 10. Morphogen Receptors). Morphogenesis also involves cell interactions (between them and with the extracellular matrix) and differentiation processes, such as epithelial–mesenchymal and mesenchymal–epithelial transitions.

Tissue morphogenesis can depend on small local asymmetries in localization and activity of cellular proteins, such as myosin-2 and actin, as well as in the extracellular matrix [22]. Numerous molecules cooperate to conduct morphogenesis that results from signal integration.

Segmentation refers to the progressive formation of body's segments. It relies on the coordination of 3 processes [22]: (1) oscillations of gene expression driven by internal clocks that determine the phase of segment formation; (2) generation of a wavefront that determines the competence zone characterized by bistability for the creation a new segment at each cycle; and (3) cell division coupled to posterior growth to generate new precursors for subsequent segments. Two opposing phases are identified in the oscillating behavior: (1) a phase associated with genes of the Notch and fibroblast growth factor (FGF) pathways; and (2) another linked with genes of the Wnt axis (Vol. 3 – Chap. 10. Morphogen Receptors). The wavefront of competence is directed by a FGF gradient and an opposing retinoic acid gradient that originates from different regions with feedbacks.[20]

Biological tissues have an architecture and a functioning that allows resistance and adaptation to stress, respectively. Cell adhesion is a major factor for tissue organization and remodeling. The cortical actin network regulates the cell-surface mechanics, hence it controls the spatial and temporal cell shape and behavior. The tissue-specific extracellular matrix is crucial for tissue development, maintenance, repair, and functioning, as it integrates cells into appropriate structural and functional assembly.[21]

Morphogenesis that develops over medium (minute) to long (hours) time scales is supposed to be governed by viscous effects with an equilibrium due to surface tension, whereas short time scale (seconds) response relies on elasticity [24].

The surface tension (σ_s) in a cell set minimizes the contact area with the environment.[22] Increasing the surface area (A) requires an energy $\delta E =$

---

[20] Fibroblast growth factor stimulates retinoic acid synthesis, but retinoic acid represses FGF signaling, hence, the bistability.

[21] Extracellular matrix dysfunction can result from mutations that reduce synthesis or increase degradation of structural constituents and disturb molecular interactions due to secreted, mutant proteins. Besides, mutant misfolded extracellular matrix proteins (e.g., collagens and matrilin-3) trigger the unfolded protein response following endoplasmic reticulum stress [23].

[22] The surface tension that measures the resistance of an interface to a force has the dimension of force per unit length or energy per unit area (J/m^2). Biological

$\sigma_s\,\delta A.$[23] Contacting cells tend to minimize the total surface area. The least cohesive cell groups surround the other ones. Cell compartments are separated by connective tissues.

The Steinberg differential adhesion hypothesis states that differences in cellular adhesion guide tissue segregation. Tissue surface tension increases linearly with the concentration of cell adhesion molecules. Adhesion strength and specificity rely on types of implicated adhesion molecules and bonds, as homophilic binding is often stronger than heterophilic associations, and adhesion kinetics.

Cellular surface tension results not only from adhesion, but also from cortical tension that opposes membrane flexibility, which is exerted by the actin–myosin contractile cytoskeleton. Cortical tension is exerted by filamentous actin (^{F}actin) and myosin-2 anchored to the plasma membrane. Cellular surface tension thus depends on the density of the cortical actin network as well as the number and dynamics of connections with the plasma membrane.

Cell internalization by another most often corresponds to transient engulfment of apoptotic cells (Vol. 2 – Chap. 4. Cell Survival and Death) by phagocytosis[24] or transcellular migration, but a viable cell can invade another cell during large periods. Cells can internalize into different or same cell type (hetero- or homotypic cell-in-cell structure). Invading cells can move and divide. Unusual cell internalization can carry out a specific function (e.g., thymocyte maturation within thymic nurse cells) [25]. Host cells can protect and promote differentiation and maturation of internalized cells. However, internalized cells can destroy or be degraded by host cells. Various cell types are potential host cells, but invading cells most often are blood cells. In addition, tumor cells can internalize other neoplastic cells and leukocytes.

---

tissues can be assumed to behave like fluids, cells in a tissue like molecules in a liquid despite heterogeneous material, and different cell populations characterized by cohesion of cells with mobility potential like immiscible multiphase fluids.

[23] The physical significance of liquid surface tension is related to the cohesive interactions between molecules in a liquid (Van der Waals forces, hydrogen bonds, ionic interactions). Intermolecular forces within the liquid are, on average, balanced. Molecular interactions that occur with similar molecules disappear with different molecules at any interface. Surface-tension forces arise from imbalance between intermolecular forces. The resulting force tends to bring surface molecules inside the liquid (away from the interface), thus stretching the surface of the interface.

[24] Apoptotic cell phagocytosis is launched by phosphatidylserine that becomes exposed on the outer leaflet of the plasma membrane of apoptotic cells to bind to host-cell receptors, such as brain-specific angiogenesis inhibitor BAI1, stabilin-2, and T-cell immunoglobulin and mucin domain-containing protein TIM4. Phagocytosis is driven by cytoskeletal rearrangements within the host cell that are regulated by small GTPase Rac (Vol. 4 – Chap. 8. Guanosine Triphosphatases and their Regulators). Host receptor activation stimulates RhoG via its guanine nucleotide-exchange factor (GEF) Trio, which in turn hastens Rac via GEF complex 180-kDa protein downstream of CRK (DOCK180, or dedicator of cytokinesis DOCK1) and engulfment and cell motility protein (ElMo).

Table 1.5. Extracellular matrix constituents in the cardiovascular system and regulator kidney (Source: [23]).

Component	Gene	Tissue
Collagen-1	COL1A2	Heart
Collagen-3	COL3A1	Blood vessels
Collagen-4	COL4A1	Kidney, basement membranes
	COL4A3, COL4A4	Kidney, basement membranes
	COL4A5, COL4A6	Kidney, basement membranes
Elastin	ELN	Blood vessels
Fibrillin-1	FBN1	Cardiovascular system
Fibronectin	FN1	Kidney
Laminin	LAMB2	Kidney
Perlecan	HSPG2	Basement membranes

Most of the body's cells are polarized, i.e. they display shape and functional asymmetry. Cell polarity refers to the asymmetric organization of the cell surface, distribution of intracellular organelles, and orientation of the cell cytoskeleton (Sects. 4.2.2, 6.10, 7.7.5, and 9.4). Cells coordinate spatially and temporally their polarity to form polarized tissues. Polarity complexes, signaling cascades, transport pathways, lumen design mechanisms for formation of conduits and hollow organs, and polarity type transitions correspond to tissue building processes [26]. In epithelial–mesenchymal and mesenchymal–epithelial transitions (Sect. 1.13), cells convert between 2 types of cell polarity. Transcription factors control these 2 processes by regulating cell adhesion and polarity complexes.

The extracellular matrix is composed of numerous structural components (elastin, 28 collagen types that provide the matrix scaffold for interaction of extracellular components, hyalectins, proteoglycans, among others; Table 1.5). Matrix–cell interaction relies on functional assemblies of these constituents for signaling and mechanotransduction. The extracellular matrix is an important factor of initiation and maintenance of tissue polarity, as it yields a structural scaffold and conveys positional and differentiation cues. These signals are transduced by various receptors (integrins, dystroglycans, and proteoglycans; Sects. 7.5.4 and 8.1.1) and their effectors of the RHO superfamily of small GTPases (Vol. 4 – Chap. 8. Guanosine Triphosphatases and their Regulators).

During embryo- and fetogenesis, cell differentiation and displacement (Vol. 2 – Chap. 6. Cell Motility) must occur in correct locations at proper times. Spatial and temporal signals that prime changes in gene expression patterns must then be integrated. Migration timing and spatial patterning depend on various agents, such as hormones, morphogens, and microRNAs.

Table 1.6. Tissue organization and development are governed by cell interactions with the environment. Between-cell communications are either direct via cell contacts or indirect via signaling molecules. Cells communicate with other cells via adhesion molecules, growth factors, morphogens, as well as chemoattractants and -repellants.

Inputs	Cell number and functions for tissue formation
	(cell molecular pathways)
	Matrix features
Cell fate	Growth factors
	Cell loadings
	Cell adhesiveness
	Migration potential, chemotactic factors
Growth control	Anchorage to adjacent cells and extracellular matrix
	State of the surrounding extracellular matrix
	(local degradation)
	Nutrient input

Tissue development, organization, and functioning depend on cellular interactions between a given cell and not only its matrix, but also surrounding cells (Table 1.6). These interactions, indeed, occur either by transmission of chemical cues and stresses through the extracellular matrix (Chap. 8) that separates neighboring cells, or, for cells in contact, by stimulated between-cell adhesion complexes and intercellular communications. Signaling pathways that consist of chemical reaction cascades then convert stimuli into cell action, modulating especially the cytoskeleton dynamics (Chap. 6). Among sent signals, *exosomes* secreted into the extracellular environment contain mRNAs and microRNAs. Force transmission to/by the extracellular matrix and cell contacts affects the cell maturation and assembly. The time and space variations in cell interactions regulate the cell fate.

Tissue development and differentiation pathways are activated by growth factors (Vol. 2 – Chap. 3. Growth Factors) and morphogens that stimulate their cognate receptors (Vol. 3). Liganded receptors at the cell surface trigger signaling cascades that stimulate or repress the transcription of development genes.

During organ motions, especially in heart and lungs, which bear large deformations during the cardiac and respiratory cycles, cells undergo *mechanical stresses*: stretch, shear, and possible torsion. Stretch stimulates growth, differentiation, migration, and remodeling. The cytoskeleton transforms from a solid-like to a fluid-like phase (polymerization–depolymerization cycle) in response to transient stretches of relatively small amplitude [27]. Loaded soft cells quickly fluidize and then slowly resolidify (slow relaxation). The closer the initial state from the solid-like state, the greater the fluidization, and the faster the subsequent recovery, except in the case of ATP depletion. The cell ability to fluidize suddenly in response to mechanical stresses and to subsequently resolidify appears to depend much more on stresses themselves than on molecular reactions triggered by stresses.

Arrangement of tissues created by generations of seemingly random cell division depends on cell connections. Imaging techniques do not allow exploration of cell division strategy, but computational models are able to test hypotheses on cell division [28].

1.7 Mechanotransduction during Embryogenesis

Embryo- and fetogenesis rely on the regulation of cell proliferation, differentiation, and migration in the framework of spatially and temporally coordinated changes in the gene expression pattern that is responsible for the mandatory synthesis of soluble morphogens.

In addition, Local and regional variations in mechanical loading can influence tissue pattern formation. Morphogenetic movements can be associated with nuclear translocation of transcription factors. Cells exert traction forces on their matrix via cell–matrix adhesions (Sect. 7.7) and then spread and migrate.

External forces applied to cells by their environment influence cell signaling and functioning. Moreover, forces applied at the cell surface can reorient the cell division plane, thereby influencing tissue formation, as DNA, its proteic scaffolds, and nucleo- and cytoskeleton operate as a single structure [29]. Conversely, internal forces exerted by the contractile actomyosin cytoskeleton trigger cell signaling to regulate cell behavior. Therefore, mechanical forces that are transduced into chemical signals participate in the control of cell and tissue structure and function during embryo- and fetogenesis.

Vasculo- and angiogenesis (Vol. 5 – Chap. 10. Vasculature Growth) are partly based on tensed regions in the extracellular matrix that channel the cell displacement. Cell migration and differentiation depend on spreading configuration, i.e., on cell and matrix rheology and strain and stress fields cells exert on the extracellular matrix, and vice versa.

The mechanical stress field originates from contractile activity of cells in their environment with its given rheological properties. Cells develop contacts onto their environment with a given strength. They contract and pull on the extracellular matrix and adjoining cells to which they are attached. Consequently, cells generate internal mechanical forces that are transmitted to their environment, particularly via adhesion sites (Sect. 7.7), to which their surrounding responds (external reaction force). According to its stiffness that depends on involved types of extracellular matrix components, the external medium favors or impedes tissue deformation.

Mechanical stress regulates the spatial organization of cells and structural rearrangement during development [30]. During their migration, cells stretch and are subjected to external forces generated by cell motion. Resulting tension causes nuclear translocation of transcription factors and cofactors that regulate the expression of cytoskeleton genes. The dynamics of filamentous actin control cell migration. Moreover, signaling by the morphogen Wnt (Vol. 3 – Chap. 10. Morphogen Receptors) that can be triggered by mechanosensors modulates organization of the actin cytoskeleton, hence contractile forces, via the RhoA–RoCK axis. In addition, cells maintain a rate of delivery of proteins using the microtubule cytoskeleton and a rate of retention of proteins in given cell regions such as polarity complexes at cell tips.

Specification of mesenchymal stem cells (Sect. 1.2) to different lineages is influenced by the medium stiffness [30]. In soft matrix that mimics cerebral medium, these cells differentiate into neurons, whereas in intermediate stiffness matrix that mimics striated muscle, they differentiate into myoblasts, and on stiff matrix that mimicks bone, they undergo osteogenesis.

Mechanotransduction regulates differentiation, as cell population extension and motion can activate regulators of differentiation [30]. The RhoA–RoCK pathway that generates cell contraction stimulates stem cell commitment into osteoblast. In fibroblasts, the GTPase-activating protein RhoGAP5 enhances and reduces adipogenesis and myogenesis, respectively. Moreover, spread cells with augmented contractility educe osteogenesis, whereas unspread cells with repressed contractility promote adipogenesis.

Cell proliferation is heightened by tension and dampened by compression, as cytoskeletal tension regulates cell division [30]. Mathematical models can demonstrate this mechanical feedback. Mechanical signals regulate cell proliferation, at least partly, via GTPase RhoA that promotes G1–S transition during the cell division cycle via cyclin–cyclin-dependent kinase complexes (Vol. 2 – Chap. 2. Cell Growth and Proliferation) and its effector RoCK kinase (Vol. 4 – Chap. 4. Cytosolic Protein Ser/Thr Kinases). The RhoA–RoCK axis increases myosin activity by acting on myosin light-chain kinase and phosphatase to favor myosin light-chain phosphorylation. The RhoA–RoCK axis hence supports cell contractility and stretch-dependent proliferation.

Forces generated by blood flow are major determinants of vascular morphogenesis during embryo- and fetogenesis (Vol. 5 – Chap. 10. Vasculature Growth), in addition to the regulation of vessel bore to match blood flow to demand in oxygen and nutrients. Mechanical stress regulates cardiogenesis (Vol. 5 – Chap. 6. Heart Wall), in addition to angiogenesis. Blood flow exerts a stress field inside the vessel lumen as well as within the vessel wall (Sect. 1.9.2). Blood flow mainly exerts: (1) wall shear stress at the interface between blood and vascular endothelium that results from friction and (2) pressure-induced circumferential stretch throughout the wall thickness (in addition to radial and axial stresses). Mechanical forces are thus applied to distinct vascular cell types as well as different cellular components in a given cell. Among these cell components, mechanosensors that are devoted to shear or stretch sensing initiate cell signaling to favor normal blood flow conditions as well as a standard strain–stress status inside the vessel wall. Cyclic mechanical stress is involved in the organization of the primitive vascular plexus into a properly organized vascular tree and development of the outflow tract of the heart [31]. When the heart begins to beat and blood cells enter the blood circulation, the primitive vascular plexus reorganizes into a typical vascular tree. Arteries branch out into smaller arteries, arterioles, and capillaries that merge to form venules, small veins, and veins.

1.8 Biological Tissues

Biological tissues correspond to an organizational level intermediate between cells and organs. Cells create and organize the body's tissues. Biological tissues consist of a collection of connected and/or separated cells embedded into the extracellular matrix. These cells send signals and interact. Their cooperative activity optimize tissue functioning.

1.8.1 Tissue Types

A biological tissue carries out a set of specific functions. Biological tissues can be grouped into 4 basic types [32–34]: epithelial, connective, muscular, and nervous.

Epithelial tissues (type I) are packed cell sheets that cover surfaces of the body and organs. The vascular endothelium is an epithelium that lines the wetted surface of the entire circulatory system. This living tissue reacts to flow forces and regulates cell and substance transport at the blood–wall interface. The respiratory epithelium covers the entire wetted surface of the respiratory tract from the nose and mouth to lung alveoli. However, the constituents of the respiratory epithelium change from upper airways to distal alveoli.

Connective tissues (type II), in which cells are separated in a matrix, bind the different structure components together via adhesion molecules.[25] The mesenchyme, an embryonic and fetal connective tissue, is a precursor for the early formation of blood and blood vessels.[26] Blood is a particular type-II tissue composed of cells suspended in a plasma, which replaces the matrix of connective tissues.

Muscular tissues (type III) are made of specialized elongated cells that contract to produce motion (displacement and deformation), using energy and chemical reactions. *Nervous tissues* (type IV) send electrochemical impulses to regulate the functioning of physiological systems, such as blood circulation and breathing.

---

[25] Connective tissues possess a high content in either fibers or cells. In addition, connective tissues are either loose (poor in fibers, but rich in ground matrix) or dense (rich in fibers, but poor in ground substance).

[26] The mesenchyme also generates connective and muscle tissues. Mesenchyme cells are pluripotent.

Both large blood vessels and airways possess a wall with a trilaminate structure. Each wall layer is made of a composite material. Large blood vessels and airways possess an adventitia that contains a perfusing microvasculature, a middle muscularis layer that directs the luminal caliber, and a surface layer with endothelial and epithelial cells, respectively. Endo- and epithelial cells sense the stress field exerted by the flowing fluid (liquid, i.e., blood, and gas, i.e., humidified and heated air, respectively) transmitted via the glycocalyx (Vol. 5 – Chap. 9. Endothelium) and the airway surface liquid (inner mucus gel and outer periciliary liquid layer; Vol. 5 – Chap. 12. Airway Surface Liquid and Respiratory Mucus), respectively, and regulate the wall response.

Walls of both blood vessels and airways modulate locally the flow resistance to blood and air streams within the lumen as well as regulate the local transfer of leukocytes through it during inflammation, healing, and repair (Vol. 5 – Chap. 11. Tissue Growth, Repair, and Remodeling) as well as mural cell proliferation. In fact, the vascular endothelium and airway epithelium not only serve as barriers at the interface between the wall and fluid, but they also control vascular and airway smooth muscle cell tone (Vol. 5 – Chap. 8. Smooth Muscle Cells), local fluid balance, catabolism and clearance of blood-conveyed and inhaled agents, and attraction and activation of leukocytes in the case of invading pathogens. Vascular endothelial and airway epithelial sensor cells actually synthesize vaso- and bronchomotor tone regulators, inhibitors of cell proliferation (to prevent hyperplasia) and migration (in the absence of inflammation) in normal conditions, and, conversely when stimulated, growth factors, cytokines, chemokines, and adhesion molecules.

1.8.2 Epithelia

Epithelia cover all internal and external body surfaces. Many organs contain interconnected tubular networks that have the same design: a series of duct generations with closed ends (acini, end bud, alveoli, or cyst according to tissue types). In particular, the skin is lined by the epidermis. The inner (luminal) surfaces of all of the hollow organs, anatomical conduits, and glands of the body are also coated by epithelia, whether they open to the atmosphere (e.g., the respiratory tract and sweat glands) or not (e.g., the vascular network and endocrine glands).

Epithelia serve as protective barriers between the organ and its environment. They derive from the ecto- and endoderm. Many cell types are epithelial or have an epithelial-derived origin.

1.8.2.1 Free Epithelial Surface

Epithelia consist of cells tightly packed with tiny intercellular spaces crossed by cell-adhesion junctions. They can constitute a monolayer or stratified structure. All the epithelia have a free surface (without adhesion to cellular or extracellular elements). The free surface of some epithelia possesses microvilli to increase the absorptive surface and cilia to clean the surface.

Connected Epithelial Surface – Basement Membrane

Epithelia, whatever the type, are usually separated from the underlying tissue by a basement membrane, a thin specialized sheet of the extracellular matrix (Sect. 8.2). The basement membrane contacts the basal surface of epithelial cells of monolayered epithelia or cells of the outer layer of pluristratified epithelia (inner layer of epidermis). The basement membrane yields not only a structural support for the epithelium, but also a binding and communication matrix with neighboring tissues.

The basement membrane comprises several components. A dense layer — the *basal lamina*, or *lamina densa* — consists of a network of fine filaments.[27] The basal lamina is connected to the plasma membrane of epithelial cells by filaments of the *lamina lucida*. Anchoring microfibrils tether the basement membrane to reticular (collagenous) fibers of the underlying connective tissue.

1.8.2.2 Epithelium Functions

Epithelia have several functions: (1) protection from mechanical and chemical aggressions, microorganisms, and water loss;[28] (2) sensing; (3) secretion, especially lubricating fluids; (4) absorption and transport of substances; and (5) cleaning. Epithelial cells are held together via occluding junctions — the tight junctions — that create a selectively permeable barrier (Sect. 7.7.3).

1.8.2.3 Epithelium Types

Epithelia are classified according to the number of cell layers and the cell shape. Epithelia made of a single cell layer are called simple. Epithelia arranged in 2 and more cell layers are termed stratified. The outermost layer of a stratified epithelium determines its classification.

Three basic types of epithelia are defined according to the cell shape: (1) *squamous* (pavement), when epithelial cells are flat and thin; (2) *cuboidal*, when epithelial cells have about the same height, width, and depth;[29] and (3) *columnar*, when epithelial cells are elongated in the normal direction to the epithelium (column-shaped).[30]

---

[27] The 2 terms — basement membrane and basal lamina — are often synonyms.

[28] The respiratory epithelium contributes to the immune defense (non-respiratory function), as it produces mucus that contains antimicrobial compounds. The ciliary motion eliminates entrapped particles into the digestive tract.

[29] A monolayer of cuboidal epithelium is thicker than a monolayer of squamous epithelium.

[30] Squamous cells possess elliptical nuclei. Cuboidal cells contain central spherical nuclei. Columnar cells have elongated nuclei usually located near the cell base. Epidermis that carpets the body is a keratinized, stratified, squamous epithelium. Collecting ducts in the kidney medulla are covered by a monolayer of cuboidal epithelium. The small intestine is coated by a monolayer of columnar epithelium. The mouth cavity has a stratified epithelium.

In addition, 2 special types of epithelia exist: pseudostratified and transitional. *Pseudostratified epithelia* such as the respiratory epithelium has the appearance of stratified tissues, because cell nuclei are located at different heights within the epithelium. This feature results from difference in cell height; some small cells do not reach the free surface, whereas tall cells do. *Transitional epithelia* such as the urinary epithelium accomodate to large distension. They are essentially impermeable to salts and water. Their thickness depends on the state of deflation (≥ 6 cell layers) or inflation (2–3 cell layers).

Glandular epithelia contain mucous and serous cells. Mucous cells secrete mucus;[31] serous cells release enzymes.[32] *Ciliated epithelia* generate flow for mucus clearance in the respiratory ducts (as well as the transport of cerebrospinal fluid in the cerebral ventricles). Oriented and coordinated motions of cilia of cell set are favored by flow. Motion orientation is aligned with the streamwise direction of imposed unidirectional flow during tissue development after a first programmed step of rough orientation, allowing cilia to produce directional flow [110].

1.8.2.4 Endothelium and Mesothelium

Mesothelia that cover the narrow pleural and pericardial cavities[33] and endothelia that line the interior (luminal or wetted) surface of the entire circulatory system (cardiac chambers and blood and lymph vessels) are specialized types of epithelia.[34] These thin cell monolayers of squamous epithelia lie on a basement membrane supported or not by a connective tissue.

Mesothelia are constituted by a monolayer of epithelial cells on a basement membrane supported by a connective tissue. Parietal and visceral mesothelia

---

[31] Mucin, the major component of mucus, is secreted from goblet cells of the airway epithelium and mucous cells of submucosal glands.

[32] Lysozyme is an enzyme secreted by serous cells of submucosal glands of the respiratory tract. Secretagogues, such as prostaglandin-F2α(PGF2α), adenosine triphosphate (ATP), and neutrophil elastase (NE), stimulate the secretion of mucin and lysozyme [35]. The relative potency on an equimolar basis is the following: PGF2α ≤ ATP < NE for mucin and ATP ≤ PGF2α < NE for lysozyme secretion. Moreover, an anatomic gradient exist for constitutive and stimulated mucin and lysozyme secretion. Caudal segments of the trachea secrete more mucin and lysozyme than cranial segments.

[33] Mesothelia line several narrow body's cavities that coat organs. In addition to the pleura and pericardium, mesothelium covers peritoneum in the abdominal cavity as well as male (tunica vaginalis testis) and female (tunica serosa uteri) internal reproductive organs.

[34] The corneal endothelium lines the corneal posterior surface (fifth, innermost layer) of the cornea. It governs material transport through this surface and maintains the cornea in a slightly dehydrated state for optical transparency. Transparency can be lost when corneal endothelial cell density is reduced below a critical level of 500 cells/mm^2. Endothelial cell density in a normal human adult cornea ranges between 1,500 and 2,000 cells/mm^2.

coat the chest wall and lungs, respectively. The luminal surface of parietal and visceral mesothelia is covered with microvilli that trap proteins and serosal fluid to provide a frictionless surface for organ sliding. Cell nuclei bulge out along the thin cellular sheet.

The endothelium of the inner surfaces of heart chambers is called the endocardium. Endothelial cells are flat, squamous cells, except in postcapillary, high endothelial venules of lymphatic organs (lymph nodes, but not the spleen).

Three main types of vascular endothelia exist: continuous, discontinuous, and fenestrated endothelia. In continuous endothelia, adjacent endothelial cells are completely sealed together. Moreover, the basement membrane is complete. Consequently, any transfer of materials that is mostly transcellular is strongly controlled. In discontinuous endothelia, apposed endothelial cells are incompletely attached to one another. The basement membrane is incomplete. The vascular permeability thus increases. In fenestrated endothelia, endothelia contain holes, or fenestrations (caliber 70–100 nm). Fenestrated endothelia are exclusively encountered in capillaries.[35] In liver sinusoids,[36] kidney glomeruli,[37] and most endocrine glands, the endothelium is fenestrated. The basement membrane of fenestrated endothelia is complete (e.g., kidney glomeruli) or lacking (e.g., hepatic sinusoids). However, fenestrated capillaries possess fenestral diaphragms (caliber 60–70 nm), except capillaries of the kidney glomerulus.[38]

Endothelial (interfacial) cells produce substances that control the local blood flow via the vasomotor tone that directs the lumen caliber, blood coagulation, flowing cell extravasation, mural cell proliferation, and angiogenesis.

---

[35] Vascular endothelial growth factor is a potent enhancer of endothelial permeability. It primes the opening of endothelial intercellular junctions in capillaries and venules as well as provokes the appearance of fenestrae in venular and capillary endothelia that, normally, are not fenestrated [37].

[36] In the liver, the absence of basement membrane enables blood plasma to enter the spaces of Disse.

[37] Endothelium of the glomerulus have large fenestrae that are not covered by diaphragms, unlike those of other fenestrated capillaries.

[38] Fenestral diaphragm is composed of a meshwork formed by radial fibrils (caliber ~ 7 nm) [38]. These radial fibrils start at the rim and interweave in a central mesh. Plasmalemma vesicle-associated protein (PlVAP or PV1) is a major structural element of fenestral diaphragms [39]. It is an endothelium-specific, single span, type-2 integral membrane glycoprotein that homodimerizes. This protein localizes to sites of transendothelial exchanges, as it also resides in stomatal diaphragms of caveolae. Fenestrated endothelia from various tissues, such as endo- and exocrine pancreas, adrenal cortex, and renal peritubular capillaries, have the same diaphragmatic structure.

1.8.3 Connective Tissue

The connective tissue is a major component of body organs. It derives from the mesenchyme. It is made up of cells immersed into the extracellular matrix. These cells are scattered within the extracellular matrix instead of being attached to one another. They are separated from one another by varying volumes of interstitium.

Cell types of the connective tissue include both resident cells (fibroblasts, fibrocytes, adipocytes, mastocytes,[39] and macrophages) and immigrant cells (e.g., monocytes, lymphocytes, and granulocytes [basophils, eosinophils, and neutrophils]).[40] These wandering cells are mainly involved in immune defense and inflammation, 2 processes that trigger leukocyte migration (Vols. 2 – Chap. 6. Cell Motility and 5 – Chap. 11. Tissue Growth, Repair, and Remodeling).

In addition to cells, the connective tissue matrix contains the ground substance and different types of fibers. In ordinary connective tissue, the ground substance consists of water stabilized by proteoglycans and glycoproteins. Specialized types of connective tissues comprise cartilage, bone,[41] lymphoid tissue (Fig. 1.2; Vol. 5 – Chap. 4. Lymphatic System), and blood. In blood, the matrix is a fluid, the plasma.

1.8.3.1 Connective Tissue Function

The connective tissue has several functions: (1) transport of nutrients and metabolites as well as wastes and catabolites;[42] (2) reservoir of growth factors; (3) inflammation and immunological defense; and (4) structural and mechanical support to enable the maintenance of anatomical shape of tissues and organs and withstand applied forces. Specialized connective tissue types have additional functions. Adipose tissue intervenes in energy storage, brown adipose tissue in the generation of thermal energy (heat; Sect. 1.12), bone marrow in hematopoiesis (formation of blood cells; Vol. 5 – Chap. 2. Hematopoiesis).

---

[39] Mastocytes are also called mast cells. The term "mast cell" is a misnomer, as the noun mast refers to tree fruits such as nuts, hence food. The secretory vesicles of mastocytes were assumed to result from phagocytosis, when they were discovered. Mastocytes are secretory alarm cells.

[40] Neutrophils, eosinophils, and basophils are in fact neutrophilic, acidophilic, and basophilic granulocytes according to the staining properties of their specific granules.

[41] In bones, the ground substance is mineralized, hence solidified by calcium deposits that form typical mineral structures.

[42] Nutrients exit from capillary beds and can travel through the adjoining connective tissue to reach cells.

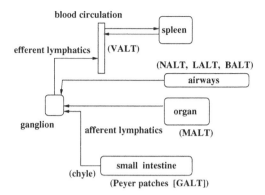

Figure 1.2. The lymphatic system is a compartment of the immune system with primary, secondary, and tertiary lymphoid organs interconnected and connected with other organs via a network of lymphatic vessels, or lymphatics, that carry lymph. Primary lymphoid organs include the thymus and bone marrow. Secondary lymphoid organs comprise lymph nodes with its afferent and efferent lymphatics that bring in and drain out lymph, and lymphoid follicles in spleen. Tertiary lymphoid organs embody lymphoid follicles of the mucosa-associated lymphoid tissue (MALT) with many subcompartments, such as the gut-associated lymphoid tissue (GALT), which includes Peyer's patches, tonsils, adenoids, and localized lymphoid aggregates of the digestive tract, skin-associated lymphoid tissue (SALT), eye-associated lymphoid tissue (EALT), vascular-associated lymphoid tissue (VALT) in arterial walls, and nasal- (NALT), larynx- (LALT), and bronchial-associated (BALT) lymphoid tissues.

1.8.3.2 Connective Tissue Types

Connective tissues are classified into various types according to the relative proportion of cellular and extracellular components. *Ordinary connective tissues* contain all of the basic components of connective tissue. In particular, ordinary fibrous connective tissue forms the outer layer of blood vessels (adventitia). *Special connective tissues* are localized types of connective tissue that include adipose tissue, bone, cartilage, lymphoid tissue (e.g., spleen and lymph nodes), and blood.

Connective tissues can be distinguished as loose (or areolar) or dense, depending on the fiber density (low or high, respectively). *Loose connective tissue* contains abundant ground substance and relatively sparse fibers. It is primarily associated with epithelia. It fills spaces between myofibers, surrounds blood and lymph vessels, exists in the serosal lining membranes of the peritoneal, pleural, and pericardial cavities, in the lamina propria of the digestive and respiratory tracts, etc.

Dense connective tissue is characterized by abundant collagenous fibers arranged in bundles. It is further described as either regular or irregular, according to the fiber orientation. In a dense, regular connective tissue (tendons, ligaments, and aponeuroses), fibers are all aligned in a single direction to confer tensile strength mainly in that direction. Collagenous fibers are packed in dense regular arrays between rows of cells. In an irregular connective tissue, fibers are randomly arranged.

1.8.3.3 Connective Tissue Cells

During embryo- and fetogenesis, mesenchymal stem cells generate the cell types of the connective tissue. In adults, a small number of mesenchymal stem cells can persist in connective tissue.

Undifferentiated, specialized mesenchymal cells — the pericytes — are perivascular cells that coat small blood vessels (arterioles, capillaries, and venules).[43]

In many types of connective tissues, fibroblasts[44] are matrix-secreting cells (Sect. 2.1.4). Fibroblasts actually release all of the types of fibers (collagen and elastin) and carbohydrates of the ground substance.

Like active fibroblasts, relatively inactive fibrocytes and adipocytes are resident cells of connective tissue. Fibrocytes are stimulated by tissue damage and transform into fibroblasts. Fibroblasts synthesize and secrete proteins to repair the damaged tissue. Reticular cells are usually larger than fibrocytes. Adipocytes (Vol. 2 – Chap. 1. Remote Control Cells) can localize to loose connective tissue.

Other cell types can migrate into a connective tissue, such as mono-cytes, macrophages, mastocytes, granulocytes (basophils, eosinophils, and neutrophils), lymphoid cells (lymphocytes and plasmocytes; Chap. 2 and Vol. 5 – Chap. 3. Blood Cells). Macrophages, or histiocytes, are phagocytic cells derived from monocytes.

1.8.3.4 Ground Substance

The extracellular matrix consists of a ground substance and fibers. The ground substance occupies the space between cells and fibers. It consists

---

[43] Pericytes, or adventitial cells, of postcapillary venules dysplay more signs of ac-tivity when they are stimulated by foreign proteins than those of capillaries [40]. They can detach and probably develop into macrophages or plasmocytes. They correspond to smooth muscle cells that reside in walls of larger arterioles and venules, but do not have the structural characteristics of smooth muscle cells. They form a discontinuous coat that attaches to adjacent endothelial cells. The basement membrane that lines the endothelial layer is situated between endothe-lial and perivascular cells as well as around the latter.

[44] The noun fibroblast is a misnomer, as it suggests a precursor cell that differen-tiates into a specialized cell. This interpretation is false because the fibroblast is a differentiated, mature cell, despite its capacity to differentiate into other mes-enchymal cell types. In addition to fibroblasts, chondroblasts and osteoblasts are mature cell types (but not undifferentiated cells that transform into mature cell types) that secrete cartilage and bone materials, respectively.

mainly of proteoglycans (made of chondroitin, dermatan, keratan, and heparan sulfates; Sect. 8.1.1).[45] It has a high water content. Polyanionic carbohydrates of constitutive glycosaminoglycans with their high density of negative charges attracts water and cations to form a hydrated gel. It also contains low-molecular-weight substances, in addition to water and ions. Owing to water-bound molecules, it is easily crossed by low-molecular-weight molecules, such as gases, ions, and small molecules. Large molecules must travel through available spaces of the proteoglycan mesh.

1.8.3.5 Connective Tissue Fibers

Three types of fibers are secreted by connective tissue cells: collagenous, reticular, and elastic fibers. Collagen is a common protein that confers tensile strength with flexibility. Collagen fibers can withstand a strain of 15 to 20% of their resting length. Elastic fibers restore initial tissue configuration after distortion. Elastic fibers can be stretched to about 150% of their unstressed length. Collagen fibers associated with elastic fibers in a tissular matrix serve to prevent overstretching under large mechanical stress.

Fibrocollagenous or fibrous tissue contains a substantial proportion of collagen. Fibrous tissue is thus deformable with great resistance to stretch. *Elastic tissue* is a dense connective tissue that contains predominantly elastic fibers rather than collagen. Elastin fibers confer resiliency.[46]

Collagenous Fibers

Numerous types of collagen have been identified. Type-1 collagen is the most spread type of collagen ($\sim 90\%$ of collagen). Collagen fibers (bore 1–10 μm) are composed of thinner collagen fibrils (caliber 0.2–0.5 μm), themselves constituted of microfibrils, which are assemblies of tropocollagen.

Reticular Fibers – Type-3 Collagen

Reticular fibers are related to collagen fibers, as they are made of type-3 collagen fibrils, sometimes in association with type-4 collagen. They are

---

[45] Proteoglycans are very large macromolecules made of a core protein to which many glycosaminoglycans, or mucopolysaccharides, i.e., long chains of polysaccharides, are attached. These polysaccharides are constructed form disaccharide units. One of the sugars in each disaccharide unit is a hexosamine, or glycosamine. Many of the sugars in glycosaminoglycans have sulfate and carboxyl groups, which makes them highly negatively charged. Hyaluronan, or hyaluronic acid, is the dominant glycosaminoglycan.

[46] Resilience is the ability of a material to return to its original configuration after being loaded, i.e., the property to absorb energy when it is deformed and, when unloaded to restitute the stored energy. The elastic potential energy measures the resilience of a material.

arranged in a mesh-like pattern that yields a support for cells. They form fine networks instead of thick bundles.

In most tissues, these types of collagenous fibers emanate from fibroblasts. However, in hematopoietic and lymphatic tissues, they are synthesized by reticular cells. In the endoneurium of peripheral nerves, they are produced by Schwann cells. In the media of large blood vessels, they are generated by smooth muscle cells.

Reticular connective tissue that consists of reticular cells and a network of reticular fibers is particularly observed in organs characterized by cell movements, such as the spleen and lymph nodes.

Elastic Fibers

Elastic fibers are thinner than collagenous fibers. They yield the ability of biological tissues to cope with stretch. Elastin fibers are interwoven with collagen fibers to limit distension and prevent tearing. They are constituted of elastin and microfibrils, i.e., fibrillar glycoproteins.

Elastins are crosslinked to each other by desmosine and isodesmosine that lodge between elastin molecules. Tension straightens the crosslinked mesh of elastin coils.

Elastin is an important component of blood vessel walls. Dense elastic connective tissue is found in elastic arteries near the cardiac pump that store a fraction of the systolic ejection volume (expelled by heart ventricles) for diastolic restitution. During these processes, elasticity is of paramount importance. In elastic arteries, elastin is constructed by smooth muscle cells of the tunica media. However, vascular smooth muscle cells do not produce microfibrils, but only elastin that form concentric fenestrated sheets between layers of smooth muscle cells.

1.8.4 Muscular Tissue

Myocytes, or muscle cells, create a specialized contractile tissue. They derive from the mesoderm. Muscles produce forces and cause displacements. Myocytes are classified into 3 distinct categories: (1) smooth muscle cells that reside in the inner layers of organ walls (Sects. 2.1.2 and 3.4.7 and Vol. 5 – Chap. 8. Smooth Muscle Cells); (2) skeletal muscle cells that operate in body's motions;[47] and (3) cardiac muscle cells, or cardiomyocytes (Sect. 2.1.5 and Vol. 5 – Chap. 5. Cardiomyocytes).

Cardiomyocytes contract to expel blood from the cardiac ventricles into the arterial trees of the systemic and pulmonary circulations to supply nutrients to cells and ensure gas exchange, respectively. They then relax to suck up

---

[47] Skeletal muscles are attached to the skeleton and control the body's posture and movements. This type of myocytes also lodges in soft tissues, such as the tongue, pharynx, and upper part of the esophagus (visceral striated myocytes).

blood from the draining venous bed into the heart cavities. The contraction of myocardial and smooth muscle cells is unconscious. On the other hand, skeletal muscles can provoke voluntary motions, in addition to the maintenance of the body's posture.

Cardiac and skeletal myocytes are striated because they contain sarcomeres, a set of regular, contractile arrangements of actin and myosin fibers. Skeletal myocytes are organized in parallel myofibers. On the other hand, cardiomyocytes are connected by intercalated discs at various angles. Striated myocytes contract in short, intense bursts and, then, relaxes. Striated myocytes form a multinucleated syncytium owing to the fusion of individual myoblasts during development.

Smooth muscle cells sustain persistent contractions. They do not possess sarcomeres, but a collection of stress actomyosin fibers. Smooth muscle cells pertain to the walls of organs and conduits, such as the components of the digestive tract (esophagus, stomach, and intestines), urinary tract (urethra and bladder), uterus, hair follicles of the skin (arrector pili), respiratory tract, especially the tracheobronchial tree, and blood vessels.

Connective tissue is associated with skeletal muscles. The endomysium with capillaries and small nerves surrounds myocytes. Groups of myocytes are packed into fascicles by the perimysium. Muscles are bounded by the epimysium that continues into the tendon.

1.8.5 Nervous Tissue

The nervous, or neural, tissue constitutes the nervous system, a network of specialized cells — the neurons — that generate, sense, and transmits electrochemical impulses (Vol. 2 – Chap. 1. Remote Control Cells). A neuron possesses 3 parts: (1) a soma, or cell body, that contains the cell nucleus; (2) dendrites that receive and relay the information to the soma; and (3) axon that transmits information from the soma. Neurons are assisted by glial cells.

The nervous system consists of 2 main compartments, central and peripheral. Most often, somas (or somata) with their dendrites aggregate in clusters. Clusters of somas in the central and peripheral nervous systems are named nuclei and ganglia, respectively. In the central nervous system, other regions that contain somas and their myelinated and unmyelinated processes as well as glial cells are termed gray matter. Regions that are crossed by myelinated axons (predominantly) and some unmyelinated axons and possess glial cells are referred to as white matter. The gray matter localizes to the cortex and central region in the brain as well as the spinal cord center (dorsal and ventral horns).

The peripheral nervous system is subdivided into the somatic and autonomic nervous systems. The somatic nervous system coordinates the body's activities under conscious control. The autonomic nervous system include 3 components: the sympathetic, parasympathetic, and enteric systems.

The central nervous system includes the brain, spinal cord, and retina. The peripheral nervous system comprises the cranial and spinal nerves associated with ganglia.[48] Sensory and motor nerves connect the body's organs to the central nervous system. A neuron can have components in both the central and peripheral nervous systems.

The brain encompasses 5 main regions: (1) the *cerebrum*, or *telencephalon*, i.e., the cerebral hemispheres with the cerebral cortex, basal nuclei, and limbic system that operate in motor functions of the body, sensory information processing, awareness, thought, language, learning, and memory;[49] (2) *diencephalon*, or *interbrain*, with the thalamus, hypothalamus, subthalamus, and epiphysis that regulate the activities of the body's organs;[50] (3) *mesencephalon*, or *midbrain*, that include the tectum with the corpora quadrigemina (quadruplet bodies), i.e., the inferior and superior colliculi involved in preliminary visual and auditory processing, and the cerebral peduncles, or crus cerebri, with the tegmentum involved in unconscious homeostatic and reflexive commands;[51] (4) *cerebellum* that serves in the motor control (body's

---

[48] In addition to the basal nuclei, i.e., a group of nuclei in the brain interconnected with the cerebral cortex, thalamus, and brainstem (the term basal nuclei is recommended instead of basal ganglia), 2 major groups of ganglia exist in the peripheral nervous system: (1) dorsal root ganglia, or spinal ganglia, that embody the somas of afferent sensory nerves and (2) autonomic sympathetic and parasympathetic ganglia that incorporate the somas of autonomic nerves. Neurons of any spinal ganglion relay sensory information to the central nervous system. Neurons of autonomic ganglia form the junction between autonomic nerves from the central nervous system and autonomic nerves that innervate target organs. Some of the cranial nerves are exclusively or largely afferent (I, II, and VIII), others are predominantly efferent (III, IV, VI, XI, and XII), and a third category is mixed with both afferent and efferent fibers (V, VII, IX, and X).

[49] The cerebral cortex is divided into frontal, parietal, occipital, and temporal lobes. Cerebral lobes are superficially bounded by various sulci and fissures. The cerebrum contains the anterior olfactory nucleus, i.e., the origin of the olfactory nerve (cranial nerve I).

[50] The *epiphysis* (a.k.a. pineal gland, pineal body, and epiphysis cerebri) is an endocrine gland that produces melatonin. The latter regulates the circadian and seasonal rhythms. The *thalamus* contains many nuclei that process and relay sensory and motor signals to the cerebral cortex. The medial and lateral *geniculate bodies*, prominences on the posteroinferior part of the thalamus, operate as auditory and visual relay centers, respectively. The *hypothalamus* corresponds to the part of the brain that comprises the floor and inferolateral walls of the third ventricle. It processes the autonomic and neuroendocrine signals. The *mamillary bodies* are components of the hypothalmus. The hypothalamus is connected to the *hypophysis*, or pituitary gland, another endocrine gland. The interbrain with ganglion cells of retina transmits visual information to the brain via the *optic nerves* (pair of cranial nerves II).

[51] The *cerebral peduncles*, which are the ventral part of the midbrain, correspond to 2 large bundles of nerve fibers that connect the cerebral cortex to the brainstem. The *tegmentum* represents most of the midbrain. The *tectum*, which is the

equilibrium, posture, and motion) and some cognitive functions;[52] (5) *brainstem* that contains the command centers of the blood circulation and ventilation as well as reflex centers of coughing and sneezing, among others, and can be decomposed into a cranial segment, the *pons*, or *metencephalon*, and a caudal region, the *medulla oblongata*, or medulla.[53] The brainstem and the cerebellum constitute the *hindbrain*, or *rhombencephalon*. The *forebrain*, or *prosencephalon*, includes the telencephalon and diencephalon.

Each brain region has a complicated internal structure. Some compartments, such as the cerebrum and cerebellum, are coated by convoluted or folded layers of cell bodies (cerebral and cerebellar cortex). These compartments as well as other brain regions comprise inner clusters of cell bodies that form deep nuclei.

The central nervous system is surrounded by the protective meninges, a set of connective tissue sheets (dura, arachnoid, and pia mater) that separate the skull and spine column from the nervous tissue. The dura (outer layer) is composed of an external endosteal and internal meningeal layer. Four cranial dural septa divide the cranial cavity mainly into 3 intercommunicating compartments, one sub- and 2 supratentorial.[54] The superficial venous drainage is composed of dural venous sinuses. Arachnoid villi, or granulations, are small protrusions of the arachnoid (middle layer) through the dura mater from which the cerebrospinal fluid enters the blood stream. The dura and pia mater are collectively termed the leptomeninges. The subarachnoid space between the arachnoid and pia mater (inner layer) contains the cerebrospinal fluid as well as the cerebral arteries and their cortical branches. The subarachnoid space contains expansions, the so-called subarachnoid cisterns.

---

dorsal part of the midbrain, contains 4 bumps: the superior and inferior *colliculi*. The superior colliculi participate in visual reflexes, the inferior in auditory signaling. The *oculomotor nerves* (pair of cranial nerves III) emerge from the interpeduncular fossa at the border between the pons and the cerebral peduncles. The *trochlear nerves* (pair of cranial nerves IV) emerge from the midbrain just below to the inferior colliculi.

[52] The cerebellum is attached by the 3 cerebellar peduncles, or pedunculi cerebelli (superior, or brachium conjunctivum, middle, or brachia pontis, and inferior peduncles), to the midbrain, pons, and medulla oblongata, respectively.

[53] The pons contains the nuclei of the vestibulocochlear (cranial nerve VIII), facial (cranial nerve VII), abducens (cranial nerve VI) and trigeminal (cranial nerve V) nerves. The medulla oblongata possesses the nuclei of the hypoglossal (cranial nerve XII), accessory (cranial nerve XI), vagus (cranial nerve X), and glossopharyngeal (cranial nerve IX) nerves.

[54] The falx cerebri, the largest of the dural septa, divides the brain into right and left cerebral hemispheres. The tentorium cerebelli separates the cerebral hemispheres from the cerebellum. It has a small gap called the tentorial incisure, or tentorial notch, to allow the brainstem to cross it. The falx cerebelli divides the cerebellum into right and left hemispheres. The diaphragma sellae encircles the stalk, or infundibulum, of the pituitary, or hypophyseal, gland.

The ventricular network is a set of 4 ventricles (right and left lateral, third, and fourth ventricle) that generates and contains the cerebrospinal fluid (CSF) in the brain ($\sim$ 20% of the CSF volume). It is continuous with the central canal of the spinal cord. Several foramina connect these ventricles, such as right and left interventricular foramina (or foramina of Monro), which link the lateral ventricles to the third ventricle, and the cerebral aqueduct (or aqueduct of Sylvius), a communication path between the third and fourth ventricle. In addition, a median aperture, or foramen of Magendie, and right and left lateral apertures, or foramina of Luschka, drain the cerebrospinal fluid from the fourth ventricle into the principal openings in the subarachnoid space, the cerebellomedullary cistern[55] and superior cistern,[56] respectively.[57]

Cerebrospinal fluid is produced by the choroid plexi in all compartments of the brain ventricular system, except the cerebral aqueduct and posterior and anterior horns of the lateral ventricles. It flows from the lateral ventricles via the interventricular foramina into the third ventricle, and then the fourth ventricle via the cerebral aqueduct in the brainstem. Afterward, it travels into the central canal of the spinal cord or into cisterns of the subarachnoid space via the central foramen of Magendie and 2 lateral foramina of Luschka. Most of the cerebrospinal fluid drains from the subarachnoid space ($\sim$ 80% of the CSF volume) through arachnoid villi. Small amounts of the cerebrospinal fluid flow along the proximal segments of the spinal nerves into vertebral venous plexi or lymphatic vessels.

1.9 Blood – A Flowing Tissue

Blood is a special type of body's tissue that constitutes a specialized compartment of body's fluid. Blood limits its loss by clotting.

1.9.1 Blood Functions and Content

Blood carries out 3 major functions (Chap. 2 and Vol. 3 – Chap. 1. Blood): (1) transport of cells, molecules, and thermal energy throughout the body; (2) regulation of various equilibria, such as water–electrolyte and acid–base balances; and (3) the body's immune defense against foreign bodies.

---

[55] A.k.a. cisterna magna and cerebellopontine angle cistern.

[56] A.k.a. quadrigeminal cistern, cistern of great cerebral vein, and Bichat's canal.

[57] The CSF-filled expansions of the subarachnoid space of the head contain the proximal segment of some cranial nerves and basal cerebral arteries. The other principal cisterns are the pontine cistern, a.k.a. prepontine cistern and cisterna pontis, between the pons and medulla oblongata, the interpeduncular cistern between the cerebral peduncles, and the ambient cistern, or cisterna ambiens along the midbrain that connects the superior cistern to the interpeduncular cistern.

Blood is a concentrated suspension of "cells" (packed cell volume fraction, or hematocrit [Ht], 35–50%)[58] in plasma that contains macromolecules: (1) capsules that contain an hemoglobin solution for oxygen transport — erythrocytes —, or red blood cells (97% of blood cell volume); (2) actual cells — leukocytes — or white blood cells; and (3) cell fragments — platelets —.

Plasma is a Newtonian fluid composed mainly of water (92%), proteins, carbohydrates, mineral ions, various types of messengers such as hormones, gasotransmitters, and dissolved gas, e. g., carbon dioxide (5–10% of CO_2 carried in blood).[59]

1.9.2 Fluid Mechanics –A Cursory Introduction

The main fluid variables, the velocity vector ($\mathbf{v}$) and the stress tensor ($\mathbf{C}$), use the Eulerian formulation. The main wall quantities are the stress tensor and displacement ($\mathbf{u}$). At interfaces, fluid, compliant vessel walls, and possible flowing particles are coupled by constraint continuity. The set of conservation equations is closed by the relationships between the transmural pressure (p)[60] and the cross-sectional area (A; state law). The connected segments of the fluid circuit are decoupled with accurate boundary conditions (BC) most often unknown. Constitutive laws of involved materials depend on the microstructure.

Fluid flows are governed by mass, momentum, and energy balance principles, which are expressed by partial differential equations (PDE). The governing equations of a unsteady flow of an incompressible fluid (mass density [ρ] and dynamic [μ] and kinematic [$\nu = \mu/\rho$] viscosities given in Table 1.7) in an anatomical vessel conveyed with a velocity $\mathbf{v}(\mathbf{x}, t)$ ($\mathbf{x}$: Eulerian position, t: time), are derived from the mass and momentum conservation.[61]

---

[58] Normal values vary with age and sex (at birth: 42–60%; 6–12 months: 33–40%; adult men: 42–52% and women: 35-47%).

[59] The remaining part of CO_2 carried in blood is composed of bicarbonate ions that arise from CO_2 conversion by erythrocytic carbonic anhydrase (70–80%) and carbamino compounds bound to hemoglobin (5–10%).

[60] $p = p_i - p_e$ (p_i: internal pressure), where the external pressure (p_e), the distribution of which is currently supposed to be spatially uniform, is assumed to equal zero. This assumption, which is a good approximation for superficial vessels, becomes questionable when the vessel is embedded in an environment that constraints the vessel. In addition, p_e can depend on time such as in the thorax, where the pressure undergoes cyclic changes. Inflation and deflation of respiratory alveoli are, indeed, induced by nearly oscillatory variations in intrathoracic pressures generated by respiratory muscles, with a more or less strong gradient in the direction of the body height.

[61] Mass and momentum conservation equations are obtained from the analysis of evolution of involved quantities in an infinitesimal control volume, the so-called fluid particle (Vol. 7 – Chap. 1. Hemodynamics). The first basic postulate states that matter is neither destroyed nor created, meaning that any matter finite

Table 1.7. Physical properties of air, water, and blood.

	μ ($\times 10^{-3}$ Pl)	ρ (kg/m^3)	ν ($\times 10^{-6}$ m^2/s)
Air	$19{,}04 \times 10^{-3}$	1.0683	17.82
(37 C; saturated in water vapor)			
	18.1×10^{-3}	1.199	15.1
(20 C; 101.1 kPa)			
Water	1.002	998	1.004
(37 C)			
	0,692	993	0,696
(37 C)			
Water–glycerol	4	1140	3.5
(22 C; 60%–40%)			
Blood	3–4	1055	2.8–3.8
(37 C; Ht = 45%)			
Plasma	1.2	~ 1030	~ 1.2

The equation of mass and momentum conservation of a moving, incompressible fluid (i.e., blood and air during rest breathing) are:

$$\nabla \cdot \mathbf{v} = 0,$$
$$\rho(\partial_t \mathbf{v} + \mathbf{v} \cdot \nabla)\mathbf{v} = \mathbf{f} + \nabla \cdot \mathbf{C}, \tag{1.1}$$

where $\mathbf{C} = -p\mathbf{I} + \mu(\nabla\mathbf{v} + (\nabla\mathbf{v})^T) = -p\mathbf{I} + 2\mu\mathbf{D}$ ($\mathbf{I}$: metric tensor; $\mathbf{D}$: rate of deformation tensor) for a Newtonian fluid. The stress tensor is then expressed by a linear relationship between the extra-stress tensor $\mathbf{T} = 2\mu\mathbf{D}$, i.e., the velocity gradient $\nabla\mathbf{v}$, and pressure. In addition, the fluid viscosity depends on temperature (T), concentration of conveyed particles, and pressure, but neither on time nor any kinematic quantities such as rate of deformation tensor $(\mathbf{D})$.

In isothermal conditions and in the absence of body forces, the equation set (1.1) leads to the simplified form of the Navier-Stokes equation:

$$\rho(\partial_t \mathbf{v} + (\mathbf{v} \cdot \nabla)\mathbf{v}) = -\nabla p_i + \mu\nabla^2\mathbf{v}. \tag{1.2}$$

When the inertia term $(\mathbf{v} \cdot \nabla)\mathbf{v}$ is neglected, i.e., the diffusive term $\mu\nabla^2\mathbf{v}$ is predominant, in the small blood vessels (arterioles) and bronchioles, the

---

volume neither disappears nor becomes infinite. The second basic postulate states that matter is impenetrable; any matter element does not share the same location with another element (*continuity axiom*). The following differential operators are used: (1) the gradient operator $\nabla = (\partial/\partial x_1, \partial/\partial x_2, \partial/\partial x_3)$, (2) the divergence operator $\nabla\cdot$, and (3) the Laplace operator $\nabla^2 = \sum_{i=1}^{3} \partial^2/\partial x_i^2$. The gradient ∇p of scalar p is a vector of component $(\nabla p)_i = \partial_i p$. The gradient $\nabla\mathbf{v}$ of vector $\mathbf{v}$ is a second order tensor with component $(\nabla\mathbf{v})_{ij} = \partial_i v_j$.

momentum conservation equation is called the *Stokes equation*:

$$\partial_t \mathbf{v} + \boldsymbol{\nabla} p_\mathrm{i}/\rho - \nu \boldsymbol{\nabla}^2 \mathbf{v} = 0. \tag{1.3}$$

In stagnant flow regions, in particular when an aneurysm develops in the arterial bed with a configuration that promotes a local flow stagnation, red blood cells can have time to aggregate and form *rouleaux*. These rouleaux can be modeled as purely elastic (Hookean) dumbbells, the length of which change as rouleaux aggregate and fragment under shear. The resulting suspension of isolated erythrocytes and erythrocytic rouleaux oriented in the streamwise direction constitutes an incompressible, non-Newtonian fluid with a shear-thinning,[62] viscoelastic, and thixotropic behavior. The set of equations to solve becomes [41]:

$$\rho \frac{D}{Dt}\mathbf{v} = -\boldsymbol{\nabla} p + \boldsymbol{\nabla} \cdot (2\mu_P \mathbf{D}) + \boldsymbol{\nabla} \cdot \mathbf{E},$$

$$\boldsymbol{\nabla} \cdot \mathbf{v} = 0,$$

$$2\mu_P \mathbf{D} = \mathbf{E} + \tau\big(\frac{D}{Dt}\mathbf{E} - [\boldsymbol{\nabla}\mathbf{v} \cdot \mathbf{E} + \mathbf{E} \cdot (\boldsymbol{\nabla}\mathbf{v})^T]\big),$$

$$\frac{D}{Dt}\hat{N} = -\frac{1}{2}\mathsf{F}(\mathbf{D})(\hat{N} - \hat{N}_{st})(\hat{N} + \hat{N}_{st} - 1), \tag{1.4}$$

where $2\mu_P \mathbf{D}$ and $\mathbf{E}$ represent the viscous and elastic components of the stress tensor, $\mathbf{E}$ being the contribution of the elastic dumbbells to the total Cauchy stress, $\mu_P = \mathrm{Ht}(k_B T + \kappa)\tau$ the polymeric viscosity (Ht: blood cell concentration, k_B: Boltzmann constant, T: absolute temperature, $\tau(\hat{N}, \mathbf{D})$: relaxation time that depends on erythrocyte aggregation and fragmentation rates and relaxation time of a single particle), F the erythrocytic rouleau fragmentation rate, $\hat{N}_{st}(\mathbf{D})$ the value of average aggregate size $\hat{N}$ in a simple shear, steady flow with shear rate $\mathbf{D}$, D/Dt the material derivative defined as

$$\frac{D}{Dt} = \frac{\partial}{\partial t} + (\mathsf{v} \cdot \boldsymbol{\nabla}).$$

In the elastic stress component due to aggregates of all sizes (various aggregated cell number $[N]$), the relaxation time τ_{N_i} for aggregates of size N_i is replaced by an average value $\tau_{\hat{N}}$.

1.9.2.1 Dimensionless Form of Navier-Stokes Equations

The Navier-Stokes equations of a flow of a Newtonian fluid in isothermal conditions in the absence of body forces can be transformed into a classical dimensionless form using appropriate scales for the length ($L^\star$) that can be the boundary layer thickness or the duct hydraulic radius (R_h) for fully-developed

---

[62] The shear rate-dependent viscosity decays with rising shear rate down to a quasi-plateau (μ_∞).

flow in a straight pipe, the time $(T^\star)$ that can the flow cycle (cardiac or breathing) period, the velocity $(V^\star)$ that can be the mean or, better, maximal cross-sectional velocity, and the pressure $(P^\star)$.[63]

The classical formulation of the dimensionless Navier-Stokes equations are hence obtained:

$$\underbrace{\overbrace{\frac{\rho V^\star}{T^\star}\frac{\partial \tilde{\mathbf{v}}}{\partial \tilde{t}}}^{T1}}_{\text{local inertia force}} + \underbrace{\overbrace{\frac{\rho V^{\star 2}}{L^\star}(\tilde{\mathbf{v}}\cdot\tilde{\boldsymbol{\nabla}})\tilde{\mathbf{v}}}^{T2}}_{\text{convective inertia force}} = -\underbrace{\overbrace{\frac{P^\star}{L^\star}\tilde{\boldsymbol{\nabla}}\tilde{p}}^{T3}}_{\text{pressure}} + \underbrace{\overbrace{\frac{\mu V^\star}{L^{\star 2}}\tilde{\boldsymbol{\nabla}}^2\tilde{\mathbf{v}}}^{T4}}_{\text{friction}} . \tag{1.5}$$

The right hand side of equation 1.5 can also be given be in the case of a generalized Newtonian model:

$$\tilde{\boldsymbol{\nabla}}\cdot\left(-\frac{P^\star}{L^\star}\tilde{p}\mathbf{I} + \frac{\mu_\infty V^\star}{L^{\star 2}}\tilde{\mu}\big(\tilde{\boldsymbol{\nabla}}\tilde{\mathbf{v}} + (\tilde{\boldsymbol{\nabla}}\tilde{\mathbf{v}})^T\big)\right), \tag{1.6}$$

where μ_∞ and $\tilde{\mu}$ are the viscosity scale (e.g., maximal shear rate) and dimensionless viscosity (Vol. 7 – Chap. 4. Rheology). The expression 1.6 does not change the conclusion based on the classical formulation.

According to the predominant force type (convective inertial forces or viscous forces) the dimensionless coefficients are displayed in Table 1.8.

The dimensionless Navier-Stokes equations suitable for bends of uniform curvature in a single curvature plane can be found in [42] (also Vol. 7 – Chap. 1. Hemodynamics).

1.9.2.2 Dimensionless flow governing parameters

The formulation of the dimensionless equations depends on the choice of the variable scales ($\bullet^\star$). For example, the length scale is most often the vessel hydraulic radius ($L^\star \equiv R$), the time scale is most often the inverse of the flow pulsation ($T^\star = \omega^{-1}$), and the velocity scale either the mean ($V^\star \equiv \overline{V}_q$) or the peak ($V^\star \equiv \widehat{V}_q$) cross-sectional average velocity (V_q). The dimensionless equations exhibit a set of dimensionless parameters.

The *Reynolds number* Re $= V^\star L^\star/\nu$ is the ratio between convective inertia and viscous effects. When the flow is unsteady, both mean $\overline{\text{Re}} = \text{Re}(\overline{V}_q)$ and peak Reynolds numbers $\widehat{\text{Re}} = \text{Re}(\widehat{V}_q)$, characterize the flow. The study of pipe flow stability relies on the peak Reynolds number computed with the Stokes unsteady boundary layer thickness ($\delta_S = (\nu/\omega)^{1/2}$), i.e., the ratio between the peak Reynolds number and the Stokes number ($\text{Re}_{\delta_S} = \text{Re}/\text{Sto}$).[64]

---

[63] Therefore, $\tilde{t} = t/T^\star$, $\tilde{x} = x/L^\star$, $\tilde{v} = v/V^\star$, $\tilde{p} = p/P^\star$, and the dimensionless differential operators are then expressed by $\tilde{\boldsymbol{\nabla}} = L^\star\boldsymbol{\nabla}$ and $\tilde{\boldsymbol{\nabla}}^2 = L^{\star 2}\boldsymbol{\nabla}^2$.

[64] Both Stokes and Rayleigh boundary layers thickness are $\propto (\nu T)$. The Rayleigh boundary layer deals with a flow over a flat plate that suddenly moves in its own plane, with a constant speed (transient regime). The Stokes boundary layer deals with a harmonic motion of a flat plate in its own plane, with an angular frequency ω (periodic flow). The latter case is more relevant to physiological flows.

Table 1.8. Dimensionless coefficients of the terms of the Navier-Stokes equations in the absence of body forces (LHS, RHS: left and right hand side of equation 1.5; Ti: equation term, $i = 1, \ldots, 4$). C_p is the pressure coefficient. Na, Re, Sto, and St are the Navier, Reynolds, Stokes, and Strouhal numbers. ($T_{\text{diff}}^\star = L^{\star^2}/\nu$: diffusion time scale; $T_{\text{conv}}^\star = 1/\omega$ [ω: flow pulsation]: convection time scale).

Coefficient	Predominant convective inertia (reference coefficient $\dfrac{\rho V^{\star^2}}{L^\star}$)	Predominant friction (reference coefficient $\dfrac{\mu V^\star}{L^{\star 2}}$)
T1 (LHS)	$\dfrac{L^\star}{T^\star V^\star} \equiv$ St	$\dfrac{T_{\text{diff}}^\star}{T_{\text{conv}}^\star} \equiv \text{Sto}^2$
T2 (LHS)	1	$\dfrac{L^\star V^\star}{\nu} \equiv$ Re
T3 (RHS)	$\dfrac{P^\star}{\rho V^{\star^2}} \equiv C_p,$ $C_p \equiv 1$, if $P^\star = \rho V^{\star^2}$	$\dfrac{P^\star}{\mu V^\star / L^\star} \equiv$ Na, $\text{Na} \equiv 1$, if $P^\star = \mu V^\star / L^\star$
T4 (RHS)	$\dfrac{\mu}{\rho L^\star V^\star} \equiv \text{Re}^{-1}$	1

The *Stokes number*[65] Sto $= L^\star(\omega/\nu)^{1/2}$ is here defined as the square root of the ratio between time inertia and viscous effects. It can also be defined as the ratio between the vessel hydraulic radius and the unsteady boundary layer thickness (Sto $= R/\delta_S$), as well as the ratio between the time scale of momentum diffusion when the boundary layer is established and the flow period (Sto$^2 = (R^2/\nu)/T = T_{\text{diff}}/T$).

The *Strouhal number* St $= \omega L^\star/V^\star$ is the ratio between time inertia and convective inertia (St $= \text{Sto}^2/\text{Re}$).

The *Dean number* De $= (R_h/R_c)^{1/2}$Re (R_c: curvature radius of tube axis) for laminar flow in curved vessels is the product of the square root of the vessel curvature ratio by the Reynolds number.[66]

---

[65] The Stokes number is also called Womersley number or Witzig-Womersley number in the literature.

[66] The Dean number is usually calculated in simple bends of constant curvature, such as those used in experiments or as simulation benchmarks. In image-based

The *modulation rate*, or amplitude ratio, is used when the time-dependent component of the fluid flow is a sinusoid of amplitude $V_\sim$ superimposed on a steady one ($\gamma_v = V_\sim/\overline{V}$) to assess the magnitude of the near-wall back flow during the bidirectional phase of the pulsatile flow. The lower the modulation rate, the smaller the retrograde flow.

The *Navier number* $\mathrm{Na} = \dfrac{P^\star}{\mu V^\star/L^\star}$ is here defined as the ratio between the pressure gradient ($\nabla_x p$; normal stress component) and friction ($\mu \nabla_x^2 \mathbf{v}$; tangential stress component).[67]

The *Deborah number* $\mathrm{Deb} = (\tau V^\star/L^\star) = \tau/T_{\mathrm{conv}}$, an inverse Strouhal-like number, is the ratio between the relaxation time or the time constant for polymerization or cell aggregation, and the flow time scale. When $\mathrm{Deb} < 1$, erythrocytes have enough time to aggregate.[68] In the dimensionless form, the inertia term of the momentum conservation equation of the equation set 1.4 has a coefficient that is the product of the Reynolds and Deborah numbers:

$$\mathrm{Re} \times \mathrm{Deb} = \frac{\rho V^\star L^\star}{\mu_\infty} \times \frac{\tau^\star V^\star}{L^\star}.$$

The *Mach number*, or speed index, is the ratio between the cross-sectional average velocity and the speed of propagation of pressure wave ($\mathrm{Ma} = V_q/c$),

---

flow models, the vessel axis varies continually in every direction; consequently, the Dean number is not computed.

[67] The pressure gradient varies non-linearly in the pipe entrance segment, in which the boundary layer grows, and linearly when the flow becomes fully developed, i.e., downstream from the station in which the boundary layer reaches the duct axis in the case of a straight conduit, or, better, from the cross section in which the cross-sectional distribution of the velocity becomes invariant, whatever the tube shape and local curvature.

[68] The Deborah number takes its origin from the biblical song of prophetess Deborah (Judges 5:5) "The mountains flowed (quaked) before the Lord". When the Deborah number is small, i.e. the relaxation time is lower than the observation time (the time scale of the explored process), the material has time to relax and its behavior is rather viscous, exhibiting relaxation of stress and creep, which are features of all biological tissues. A polymer has time to polymerize (to assemble, stream in, flow toward a point: προσρεω; to assemble: θυρσοφορεω). The relaxation time is associated with the polymerization rate. Erythrocytes (deformable capsules that mainly contain hemoglobin metalloproteins) in a plasma with bridging proteins (fibrinogen and globulins) have time to aggregate and form rouleaux. On the other hand, when the Deborah number is great, the material behaves rather elastically. Polymers that flow at high Deborah number becomes oriented in the streamwise direction and stretched. Erythrocytic rouleaux subjected to a shear field split down to isolated red blood cells oriented in the streamwise direction (Vol. 7 – Chap. 4. Rheology). Blood, in the absence of erythrocytic rouleaux after experiencing high shear and shear rate during the ventricular isovolumetric contraction and systolic ejection, can be considered as a Newtonian fluid in large vessels, in which the flow is characterized by a relatively large length scale and small time scale.

Table 1.9. Dimensionless governing parameters of arterial blood flow (cardiac frequency 1 Hz, blood density of 1025 kg/m^3, dynamic viscosity 3.5×10^{-3} Pa.s). Both the Stokes and Strouhal number are important to assess the flow unsteadiness. In steady flow, the pressure gradient balances the friction (and gravity when it has significant effect, as in blood vessels in standing position). In quasi-steady flow the rate of change in boundary conditions is so slow that the momentum has the time to diffuse during the flow period. In unsteady flow, the local inertia effects are preponderant.

Quantity Parameter	Large artery		Small artery	Arteriole
Radius (mm)	12.5		1.25	0.06
	Peak	Mean		
Cross-sectional velocity (m/s)	0.75	0.10	0.30	0.02
Reynolds number	2700	370	110	0.4
Stokes number	7	7	0.7	0.03
Strouhal number	0.02	0.13	0.004	0.003
Flow features	Pulsatile	Quasi-steady		Steady
	Homogeneous fluid			Particle flow
	Navier-Stokes equation			Stokes equation

the wave speed $c = (\rho/([\partial A/A_0]/\partial p))^{1/2}$ depending on the vessel distensibility much more than on fluid compressibility.

Calculated values for the 3 main dimensionless governing parameters are given in Table 1.9, using the values of hemodynamic quantities of Tables 1.10 to 1.13.

1.10 Processing and Clearance of Unwanted Substances

Detoxification aims at removing toxic products from the organism using conjugation and excretion of these molecules from the body to its environment. In addition to exogenous compounds, metabolism can produce harmful substances that are rendered less toxic by reduction and oxidation. Detoxifying enzymes comprise cytochrome-P450 oxidases, UDP-glucuronosyl transferases, and glutathione S-transferases. Two major organs contribute to the processing and clearance of unwanted substances: the liver and kidneys.

1.10.1 Liver

The liver localizes to the right hypochondriac and epigastric regions of the body. This large organ (average length $\sim$ 28 cm and height $\sim$ 16 cm; maximal thickness $\sim$ 8 cm; weight1, 2 − −1, 5 kg [blood component $\sim$ 40 % of

Table 1.10. Values of hemodynamic parameters. (**Part 1**) Peak systolic velocity (m/s) in arteries (MCA: middle cerebral artery) measured by ultra-sound Doppler velocimetry. In the ascending aorta, data are given in 10 healthy young volunteers during supine bicycle and upright treadmill exercise [44].

Artery	Peak systolic velocity (m/s)	Reference
Ascending aorta	1.12	[43]
	0.91 (supine, rest)	[44]
	1.36 (supine, exercise)	[44]
	0.75 (upright, rest)	[44]
	1.39 (upright, exercise)	[44]
Abdominal aorta	0.75	[43]
	0.75	[45]
Femoral artery	0.60	[43]
Carotid artery	0.80	[43]
MCA	0.90 (peak systolic velocity)	[46]
	0.73	[47]
	0.40 (end-diastolic velocity)	[46]
	0.55 (mean)	[46]

Table 1.11. Values of hemodynamic parameters. (**Part 2**) Ratios of blood flow velocities in the internal carotid artery (ICA) to those in the common carotid artery (CCA) for the peak systolic velocity (PSV_{ICA}/PSV_{CCA}; mean and upper and lower values) in 343 healthy subjects [48].

Age range	Women	Men
< 40 y	0.81 (0.48, 1.14)	0.65 (0.32, 0.98)
40–60 y	0.88 (0.36, 1.40)	0.72 (0.39, 1.05)
> 60 y	0.90 (0.36, 1.40)	0.91 (0.27, 1.56)

Table 1.12. Values of hemodynamic parameters. (**Part 3**) Flow rate (ml/s) and artery caliber (mm).

Site	Flow rate (ml/s)	Area (mm^2)	Reference
Aortic orifice	460	550	[49]
Retinal arterioles	0.0002	0.0115	[50]

liver constituents]) is the body's chemical reprocessing plant. The liver receives a dual blood supply from the hepatic artery and portal vein (Vol. 6 – Chap. 1. Anatomy of the Cardiovascular System).[69]

---

[69] The portal vein is formed by the junction of the superior mesenteric and splenic veins. The tributaries of the splenic vein are gastric, left gastroepiploic, pancreatic, and inferior mesenteric vein. The inferior mesenteric vein receives the

Table 1.13. Values of hemodynamic parameters. (**Part 4**) Aorta bore (mm) measured by computed tomography and magnetic resonance imaging. Data measured in 100 normal subjects (CT or MRI) [51]. Stations 1 and 4 on the one hand and station 2 and 3 belong to the same transversal plane.

Station 1	Proximal ascending aorta	36	[51]
Station 2	Distal ascending aorta	35	
Station 3	Proximal descending aorta	26	
Station 4	Middle descending aorta	25	
Station 5	Distal descending aorta	24	
Ascending	Inner caliber (women)	31.0	[52]
aorta	Outer caliber (women)	33.5	
	Inner caliber (men)	33.5	
	Outer caliber (men)	36.0	
wall thickness-to-radius ratio $h/R < 1$			

The liver has a wide range of functions. It is involved in the production of substances necessary for digestion as well as processing of compounds taken up by the digestive tract and conveyed by the portal vein, in addition to detoxification. The catabolism of exogenous compounds generates nutrients and detoxifies poisons.

The liver functions in: (1) glucidic metabolism (glycogenesis, glycolysis, and neoglucogenesis from amino and fatty acids); (2) lipidic metabolism (synthesis and oxydation of fatty acids for energy production, production of lipoproteins for lipid transport, synthesis of cholesterol for production of hormones and biliary acids, and transformation of glucose and proteins into lipids for storage); and (3) protidic metabolism (transformation of amino acids [desamination and transamination]) and synthesis of hepatic and plasmatic proteins, as well as urea).

Other tasks include: (1) storage of vitamins (A, B12, and D) as well as iron; (2) synthesis of some coagulation factors (II, V, VII, IX, and X); and (3) bile secretion for digestion assistance via lipid emulsification, in addition to the metabolism of hormones, drugs, and toxics.

In addition to cell respiration for energy production, oxygen is also aimed at serving in the partial oxidation of organic compounds to create molecules for specialized tasks, such as signaling and toxin removal. Di- and monooxygenases achieve the processing. In the liver, metabolic substances, toxins, and drugs are detoxified by heme-containing members of the cytochrome-P450 set. Cytochrome-P450 monooxygenases are responsible for the catabolism of approximately 75% of pharmaceuticals. During detoxification, the iron com-

---

hemorrhoidal, sigmoid, and left colic veins. The superior mesenteric vein receives the intestinal, ileocolic, right and middle colic, right gastroepiploic, and pancreaticoduodenal veins. The gastric coronary, pyloric, and cystic veins end in the portal vein.

ponent of cytochrome-P450 converts substrates, which are compounds often highly resistant to chemical reaction, to oxygenated products [54].[70]

1.10.2 Kidney

Kidneys are located in the abdominal cavity in the paravertebral gutters. They receive blood from the renal arteries that branch directly from the abdominal aorta. Despite their relatively small size, their blood supply amounts for about 20% of the blood flow rate.

Kidneys are filters of blood that participate in the body's homeostasis. They actually regulate electrolyte fluxes, acid-base balance, and blood volume and pressure.

Kidneys synthesize hormones, such as: (1) calcitriol, the activated form of vitamin-D, that fosters dietary calcium uptake from the digestive tract and reabsorption of calcium in kidneys; (2) erythropoietin that controls erythropoiesis; and (3) renin that hydrolyzes angiotensinogen secreted from the liver into angiotensin-1. Conversely, various hormones control the renal function (aldosterone, antidiuretic hormone, atrial natriuretic peptide, angiotensin-2, among others; Vol. 2 – Chap. 1. Remote Control Cells – Sect. Endocrine System and Hormones).

Kidneys produce urine that permits waste removal, whereas they allow reabsorption of water, electrolytes, glucose, and amino acids (Vol. 2 – Chap. 1. Remote Control Cells – Sect. Nephron Cells). In particular, kidneys excrete urea and uric acid generated from protein and nucleic acid metabolism, respectively. In humans, catabolic P450 iron-containing enzymes of the liver and kidney are involved in drug metabolism.

1.11 Water and Electrolyte Balance

The body's metabolism refers to the sum of activities of constituent tissues in terms of physicochemical changes regulated by the availability, utilization, and disposal of proteins, carbohydrates, lipids, minerals, vitamins, and water, under the control of the endocrine system.[71]

---

[70] Cytochrome-P450 enzymes can oxidize normally unreactive carbon-hydrogen bonds. Their heme iron site binds to O_2, breaks the O–O bond, and reduces one O atom to form water [53]. The other oxygen atom is incorporated into a substrate.

[71] The endocrine system is a network of glands that secretes various types of hormones directly into the blood stream to regulate the activity of the body's tissues and homeostasis (Vol. 2 – Chap. 1. Remote Control Cells – Sect. Endocrine System and Hormones). In addition to the specialized endocrine organs (hypothalamus, epiphysis [or pineal gland], hypophysis [or pituitary gland], thyroid, parathyroid, and adrenal glands, other organs have secondary endocrine functions, such as the heart, adipose tissue, bone marrow, kidney, liver, and organs of the digestive tract and reproductive system.

A set of physiological processes maintain the body's water volume and concentrations and distribution of electrolytes (especially the number of elements per unit volume in intra- and extracellular media), including hydrogen ions (or protons), and, hence, the structure of cells and tissues. The evolution of minerals in dilute aqueous solutions is a much more simple process than any type of metabolism with its more or less huge cascade of chemical reactions.

1.11.1 Functional Compartmentation

Water is the largest constituent of living organisms. The 2 main compartments of body water are the intracellular ($\sim$40% of body weight; 65–75% of the total body water) and extracellular ($\sim$16% of body weight; 15–25% of the total body water) water. Extracellular water is further decomposed into 2 subcompartments: intertitial water ($\sim$11% of body weight) and plasma ($\sim$5% of body weight).

Extracellular fluids carry out 2 major functions: (1) transport of nutrients and wastes and (2) maintenance of the body's homeostasis. Plasma enables the communication between exchange organs, namely, lungs, kidneys, and organs of the digestive tract. In addition, several specialized tissue water exist, such as the cerebrospinal and synovial fluids.

1.11.2 Electrolytes

Motions of freely diffusible water from the compartment of low to the compartment of high concentration across the cell membrane are determined by the concentration of osmotically active electrolytes, i.e., result from the osmotic pressure.[72]

The colloid osmotic pressure of plasma is a part of the osmotic pressure that results from plasmatic proteins. The water exchange between the plasma and the interstitial water is controlled essentially by the capillary hydrostatic pressure and osmotic pressure of plasma colloids, which are almost entirely plasma proteins.[73] However, the colloid osmotic pressure incorporates also the

---

[72] Electrolyte movements initiate water flux between adjacent compartments. The osmotic pressure depends on concentrations of free particles in a solution. It determines the water flux across a semipermeable membrane. The unit of osmotic pressure is the osmol (osm). One equivalent of univalent, ionized or unionized molecule exerts an osmotic pressure of 1 osm. The milliosmolar value can be calculated using the following approximation:

$$\text{mosm} = \text{mg}/100\,\text{ml} \times 10/\text{atomic weight}.$$

[73] Colloid, or dispersed phase, is a substance microscopically dispersed evenly in a continuous phase, or dispersion medium. A (2-phase) colloidal system can be solid, liquid, or gaseous. Examples are plasma (Sol [solid in liquid]) and liquid (e.g., mist and fog) and solid (e.g., smoke and air particulates) aerosols.

Table 1.14. Normal values of electrolytes in plasma. Valence is the maximum number of valence bonds of the given element. Electrolyte concentration can be expressed in terms of weight per unit volume, number of elements (mol) per unit volume, or number of charges (Eq) per unit volume.

	Atomic weight	Valence	Range (mEq/l)	Range (mg/100ml)
Na^+	23	1	135–147	310–340
K^+	39	1	4.1–5.7	16–22
Ca^{++}	40	2	4.5–6.0	9.0–11.5
Mg^{++}	24	2	1.5–3.0	1.8–3.6
Cl^-	35.5	1	98–106	350–375
HCO_3^-			25–31	55–70
HPO_4^{2-}			1.2–3.0	2.4–5.0
SO_4^{2-}			2.5–4.5	4.7

pressure that results from the heterogeneous distribution of diffusible electrolytes. Whenever a colloidal electrolyte localizes to one side of an impermeable membrane, other ionized subtances for which the membrane is freely permeable tend to concentrate in the solution on the opposite side of the membrane (*Donnan effect* or Gibbs-Donnan effect). Therefore, the colloid osmotic pressure of plasma is greater than that due to colloidal material alone.

The control of the volume and, thus, the osmotic pressure of body's liquid is related to a relatively small number of substances. Primary ions that maintain the body's fluid volume and osmotic pressure are sodium (Na^+), potassium (K^+), calcium (Ca^{++}), magnesium (Mg^{++}), chloride (Cl^-), hydrogen phosphate (HPO_4^{2-}), hydrogen carbonate (HCO_3^-), and sulfates (SO_4^{2-}; Table 1.14)). Other involved substances include proteins, glucose, and organic acids in a given amount of water.

Electrolytes have an important physiological role due to the number of charges they possess in solution. Therefore, the measurement unit is given in terms of equivalents (Eq or mEq; Table 1.15), in addition to measurements reported in number of elements present, i.e., number of moles (mol or mmol).

Electrolytes can be transferred across the cell membrane by enzymatic mechanisms. Changes in the concentration of hydrogen ions in the intracellular water may ionize, or, conversely, render osmotically inactive intracellular electrolytes, when electrolytes bind to cellular proteins.

Table 1.15. Distribution of ions (estimated values in mEq/l) in intracellular and intertitial water and plasma in normal conditions. Electrolytes in any biological fluid maintain electroneutrality of the fluid.

	Cell	Intertitium	Plasma
Na^+	14	138	142
K^+	157	5	5
Ca^{++}	10^{-4}	5	5
Mg^{++}	26	3	3
Cl^-	3–9	108	103
HCO_3^-	10	27	27
HPO_4^{2-}	110	2	2
SO_4^{2-}	1	1	1

1.11.3 Hydrogen Ion Control

The body's homeostasis[74] partly relies on hydrogen ion control, or the so-called acid–base balance.[75] The state of biological liquids is normally slightly alkaline ($7.35 \leq pH \leq 7.45$); i.e., the amount of hydroxyl ions (OH^-) is greater than that of hydrogen ions (H^+).[76] Acidosis (or acidemia in blood) and alkalosis (or alkalemia) occurs When pH is lower or greater than the inferior and superior thresholds of the normal alkaline range, respectively.

Biological fluids contain 2 types of anions: (1) *fixed anions*, the state of which does not change with pH (at least in normal conditions; e.g., chloride, sulfate, and organic acids) and (2) *buffer anions* that lose part or all of their negative charges when hydrogen cations accumulate. Buffer anions in biological solutions include bicarbonates, phosphates, plasma proteins, and hemoglobin.

Alkalinization results from elimination of hydrogen ions. The *alkali reserve* is a measure of the content in carbon dioxide of blood released from carbonic

---

[74] C. Bernard (1813–1878) considered that the medium in which humans live is water instead of air, because the circulating blood irrigates every body's tissue and the intertitial fluid bathes every cells. In a broad range of environmental conditions, the homeostatic regulation, which depends, in particular, on the release of hormones, controls the body's temperature, blood concentration of numerous substances, and water volume. Homeostasis principle states that the osmotic pressure and the blood pH can vary only in a small interval to maintain a stable condition.

[75] The term "acid–base balance" is a misnomer, as the physiological process does not deal with acids and bases, but with anions and cations. The body's fluids comprise weak acids and bases due to the presence or absence of hydrogen ions. The fundamental process in acid–base reactions is the transfer of hydrogen ions. Acids and bases are H^+ donators and acceptors, respectively. The dissociation of a weak acid (H^+A^-) is the dissociation of a hydrogen ion and its anion (A^-).

[76] The quantities of H^+ and OH^- are much smaller than those of other cations and anions.

acid and bicarbonate. This concept thus do not incorporate hemoglobin, a major buffer anion, or hydrogen ion acceptor, as well as phosphate and protein buffers.

Hydrogen ion is passed back and forth between bicarbonate ion and hemoglobin during gas transport. In red blood cells, hemoglobin accepts a hydrogen ion from carbonic acid (H_2CO_3) to unite with carbon dioxide (CO_2) and to form bicarbonate. Another erythrocytic buffer is the couple potassium dihydrogen phosphate–potassium monohydrogen phosphate.

Hydrogen and hydroxyl ions exist in the free state in body's liquid in minute amounts. Large quantities of H^+ are buffered and excreted daily.[77]

Several buffers reversibly bind hydrogen ions and impede change in pH. Extracellular buffers include bicarbonate and ammonia; intracellular buffers proteins and phosphate.

Three blood buffers exist in the plasma: (1) bicarbonate–carbonic acid buffer ($HCO_3^- - H_2CO_3$); (2) sodium phosphate–sodium bisphosphate buffer ($Na_2HPO_4 - Na_2H_2PO_4$); and (3) plasma proteins

The *bicarbonate buffer* enables a shift from carbon dioxide (CO_2) toward carbonic acid (H_2CO_3) to hydrogen cation (H^+) and bicarbonate anion (HCO_3^-):

$$HCO_3^- + H^+ \leftrightarrow H_2CO_3 \leftrightarrow CO_2 + H_2O.$$

According to *Le Chatelier's principle*, when a chemical system at equilibrium experiences a change in concentration, temperature, volume, or partial pressure, then the equilibrium shifts to counteract the imposed change and a new equilibrium is established. An acid–base imbalance that overcome the bicarbonate buffer can be compensated in the short term by changing the rate of ventilation, i.e., by varying the blood concentration of carbon dioxide. When the blood pH drops to an abnormally low value, the kidney correct acidosis by excreting hydrogen ions, but this compensation is slower.

Breathing and renal excretion contribute to the regulation of the concentration of hydrogen ions. Lungs under the fast control of the nervous system regulate the amount of exhaled CO_2, hence the concentration of volatile carbonic acid. Kidneys under the slower control of the endocrine system regulate the amount of excreted cations (ions and non-volatile organic [lactic and pyruvic acids] and inorganic [phosphoric and sulfuric acids] acids) available to form bicarbonates.

1.12 Heat Production and Transfer

The internal temperature of the human body is maintained in a small range from either side of 37 C to maintain physiological functions. Biochemical processes, which require chemical energy as well as mechanical energy associated

---

[77] Hemoglobin and oxyhemoglobin that serve as buffers compete for hydrogen ion.

with molecular transport, indeed, depend on the temperature. The body permanently generates heat, or thermal energy. Local heating is controlled by both nervous and endocrine signals.

Blood flow ensures the distribution of thermal energy throughout the entire body. In addition, blood convection through the vasculature causes a heat dissipation due to friction on vessel walls.

Thermal energy is produced in the human body by the metabolism of nutrients. In addition, adipocytes, especially multilocular adipocytes, or brown fat cells (Vol. 2 – Chap. 1. Remote Control Cells), store lipids during the postprandial period. During fasting and cold exposure, these lipids are catabolized and thermal energy is produced. Thermal energy in excess in the body is dissipated to support the body's homeostasis.

In addition, heat can be added (hyperthermia) or removed (hypothermia and cryotherapy [subfreezing temperature]) during diagnostic or therapeutic procedures.[78]

A fraction of the metabolic rate is usually applied to work. The remainder is dissipated as heat based on convection, conduction, radiation, and evaporation of sweat at the skin surface supported by vasodilation of dermal blood vessels and via the respiratory tract. Bioheat transfer refers to the transport of thermal energy in biological systems. The convective heat transfer depends on the features of blood perfusion of the body's tissues and organs. In general, an increase in local temperature causes a rise in local blood flow, and vice versa.

Thermoregulation is a control system that maintain the temperature in various regions of the body near physiological set points, whatever the environmental conditions and metabolic rate.

The conservation of energy, or thermal energy balance over time, within the human body poises the internal production (sources) and loss (sinks) [55]:

$$\Delta E = M - (W + Q_{\text{conv}} + Q_{\text{diff}} + Q_{\text{rad}} + Q_{\text{evap}} + Q_{\text{resp}}) \qquad (1.7)$$

where ΔE is the rate of energy storage in the body, M the metabolic energy production, W the external work, Q_{conv} the surface heat loss by convection, Q_{diff} that by conduction, Q_{rad} that by radiation, and Q_{evap} that by evaporation, and Q_{resp} the respiratory heat loss.

---

[78] The volumetric heating created by an electromagnetic field is governed by the electrical conductivity (G_e [siemens per meter (S/m)]), the imaginary part of the electrical permittivity (ϵ_e [farads per meter (F/m)]; a measure of the ability of a medium to polarize in response to the electric field and transmit this field) and the magnitude of the local electric field ($|\mathbf{e}|$ [V/m]) [55]:

$$q_{\text{T}} = (G_e + \omega \epsilon_e)|\mathbf{e}|^2,$$

where ω is the angular frequency of the field.

Table 1.16. Estimated human metabolic rate in some common activities (Source: [58]).

Activity	Metabolic rate (W/m^2)
Lying	46
Seated relaxed	58
Standing relaxed	70
Low-speed walking	110
Walking	200
Running	550
Swimming	350
Bicycling	290
Teaching	95
Domestic work	100–170

1.12.1 Metabolic Rate

The human body produces thermal energy that is exchanged with the ambient air. The metabolic rate (W/m^2) refers to the heat production by the human body (Table 1.16). The basal metabolic rate is the minimal amount of energy released in an environment of thermal neutrality at rest, several hours after feeding. The total metabolic heat of a body can be calculated by multiplying the metabolic rate with the surface area of the human body.[79]

Approximately 80% of the energy used by the human body is eliminated as heat (Table 1.17). On the other hand, during cold weather, body's activity and insulation is required to maintain an adequate body's temperature.

The human body loses thermal energy to stabilize its temperature using many processes (Sect. 1.12.2). When the body's temperature rises such as during exercise, sweat glands in the skin bring additional moisture to the surface of the skin that is transformed into vapor to remove heat. In addition, breathing discards heat. Conversely, clothing insulates the human body from a cold environment.

1.12.2 Heat Transfer

The generation and transport of thermal energy in living tissue comprises several mechanisms, such as metabolism, conduction (or diffusion), convection,

---

[79] The human body surface area (BSA) can be calculated using the formula of DuBois and DuBois [56, 57]:

$$BSA = 0.007184 w^0.425 H^0.725$$

(w: body's weight [kg]; H: body's height [cm]).

Table 1.17. Energy cost (Source: [59]). Metabolism that maintains concentration gradients on both sides of surfaces is related to body's area rather than mass or volume, hence the unit type.

Activity	Cost $(kJ/(m^2.h))$
Sleeping	150
Sitting	210
Standing	350
Working at a desk	250
Washing, dressing	420
Walking (5 km/h)	590
Running	2510
Swimming	1465
Bicycling	1050

radiation, transport associated with mass transfer, and phase change (e.g., evaporation).[80]

1.12.2.1 Heat Conduction

The transfer of thermal energy[81] by conduction between regions of a biological tissue results from a temperature gradient to reach a thermal equilibrium. The thermal energy is thus the energy of a biomaterial, the constituent particles of which randomly collide with each other and objects in their environment. In biofluids, heat conduction is due to diffusion and collisions of fluid molecules.

The law of heat conduction — the Fourier's law — states that the time rate of transfer of thermal energy (power) through a material is proportional to the temperature gradient (∇T) and the area normal to the transfer direction. The local flux density of thermal energy ($\mathbf{q}_T$) is then given by:

$$\mathbf{q}_T = -G_T \nabla T, \tag{1.8}$$

where the thermal conductivity (G_T) measures the tissue ability to conduct heat.

---

[80] In this framework, evaporation is the change of phase of sweat. The rate of sweat is related to the thermal flux by the latent heat of vaporization (2428 kJ/ kg at body temperature). In addition to sweat, water vapor is lost during breathing.

[81] The thermal energy of a material has the dimension of energy, but, unlike temperature, is not a state variable; its value does depend on the past history of the material.

1.12.2.2 Heat Transfer Parameters

The temperature of a biological tissue rises or decays when thermal energy
is added to or removed from the tissue. The *heat capacity* (C) of a biological
tissue is the ratio between the amount of thermal energy (Q_T) transferred to
or from the tissue and the resulting increase or decrease in temperature (ΔT)
of the biological tissue:

$$C = \frac{Q_T}{\Delta T}.$$

The heat capacity is defined at constant volume or constant pressure:

$$C_V = \left(\frac{\partial Q_T}{\partial T}\right)_V \quad \text{and} \quad C_p = \left(\frac{\partial Q_T}{\partial T}\right)_p. \tag{1.9}$$

The heat capacities at constant volume and at constant pressure are related
by the following expression:

$$C_p - C_V = VT\frac{\alpha_T^2}{\beta_T}, \tag{1.10}$$

where α_T is the *coefficient of thermal expansion* ($\alpha_T = \dfrac{1}{V}\dfrac{\partial V}{\partial T}$) and β_T the
coefficient of isothermal compressibility ($\beta_T = -\dfrac{1}{V}\dfrac{\partial V}{\partial p}$).

The *specific heat capacity*, or thermal capacity, at constant volume or constant pressure is the amount of heat required to change the tissue temperature
per unit mass (joules/[kelvin × kilogram]):

$$c = \frac{C}{m} = \frac{C}{\rho V},$$

(ρ: material density). More precisely, the specific heat capacities are defined
by:

$$c_p = \left(\frac{\partial C}{\partial m}\right)_p; \quad c_V = \left(\frac{\partial C}{\partial m}\right)_V, \tag{1.11}$$

and related by:

$$c_p - c_V = \frac{\alpha_T^2 T}{\rho \beta_T}. \tag{1.12}$$

Values of the specific heat capacity at constant pressure of common fluids
are given in Table 1.18). Parameters derived from the specific heat capacities
include: (1) the *molar heat capacity* (C_{mol}), the mole being the unit for amount
of substance (J/[mol × K]) and (2) the *volumetric heat capacity*, or volume-
specific heat capacity, i.e., the heat capacity per unit volume (J/[m^{-3}.K]).

The *thermal conductivity* (G_T [W/m.K]) leads to the expression of other
quantities. The reciprocal of the thermal conductivity is the *thermal resistiv-
ity* (m.K/W). The *thermal conductance* (W/K) is the quantity of heat that

Table 1.18. Specific heat capacity at constant pressure of air and water (at 25 C [298 K], Source: Wikipedia) and human blood (Source: [60] among others; Ht: hematocrit).

Medium	c_p (J/[kg·K])	Reference
Air (room conditions)	1012	
	963	[59]
Water	4181	
Water	4186	[59]
Water vapor	2009	[59]
Blood	3770	[61]
	3740 (Ht = 30–32.4)	[62]
	3650 (Ht = 41.6–42.4)	[62]
Plasma	3950	[62]
Clotted blood	3500±800	[63]
Biological tissue	3560	[59]

Table 1.19. Thermal conductivity of air and water (Source: Wikipedia) and blood (Sources: [59,63,64]; clot density $1.06–1.10 \times 10^3$ kg/m^{-3}). Thermal conductivity for a given power depends on both a surface area and a temperature gradient.

Material	Thermal conductivity (W/[m·K])
Air (101 kPa, 27 C)	0.024–0.026
Water	0.56–0.63
Biological tissue	0.21
Blood	0.49–0.51
Plasma	0.56–0.58
Clotted blood	0.59±0.11
Aorta	0.44–0.52
Heart	0.46–0.60
Brain	0.50–0.56
Kidney	0.50–0.54
Lung	0.31–0.41
Adipose tissue	0.20–0.22
Skin	0.29–0.32
Atheroma	0.44–0.56

crosses per unit time a plate of given area (A) and thickness (h) with one kelvin difference between its opposite faces ($G_T \times A/h$). The reciprocal of the thermal conductance is the *thermal resistance* (K/W). Values of the thermal conductivity of common fluids are given in Table 1.19).

Table 1.20. Thermal diffusivity of selected media (Sources: Wikipedia, [65]). The blood thermal diffusivity is computed for a density $\rho = 1,055\,\text{kg/m}^3$, a thermal conductivity $G_T = 0.5\,\text{W/(m·K)}$, and a specific heat capacity $c_p = 3770\,\text{J/(kg.K)}$.

Material	Thermal diffusivity (m^2/s)
Air	2.216×10^{-5}
(100 kPa, 300 K)	
Water	1.4×10^{-7}
Blood	1.3×10^{-7}
Aorta	$1.1\text{--}1.2 \times 10^{-7}$
Heart	$1.3\text{--}2.1 \times 10^{-7}$
Lung	$1.1\text{--}1.2 \times 10^{-7}$
Kidney	1.3×10^{-7}

The *thermal diffusivity* in a tissue $(\mathcal{D}_T)$ is defined by:

$$\mathcal{D}_T = \frac{G_T}{\rho c_p}.$$

Values of the thermal diffusivity of common fluids are given in Table 1.20).

The *heat transfer coefficient* (h_T) is the quantity of heat that crosses per unit time a unit area of a given thickness of material under a temperature difference of one kelvin $[\text{W/(m}^2.\text{K)}]$:

$$h_T = q_T / A\Delta T \tag{1.13}$$

Its reciprocal is the *thermal insulance* $(\text{m}^2.\text{K/W})$. The *thermal transmittance* considers the heat transfer in a material due to diffusion, convection, and radiation. The basic relationship used to estimate the heat transfer coefficient is given in the case of heat transfer at the surface by conduction by:

$$h_T = G_T \frac{dT}{dn} / \Delta T, \tag{1.14}$$

where $\mathbf{n}$ is the local normal to the surface.

In biofluids, the heat equation is modeled by a parabolic partial differential equation, the convection–diffusion equation:

$$\rho c_p \frac{DT}{Dt} = \boldsymbol{\nabla} \cdot (G_T \boldsymbol{\nabla} T) + 2\mu \mathbf{D} : \mathbf{D} + S, \tag{1.15}$$

where S is a source term and $D/Dt = \partial/\partial t + (\mathbf{v} \cdot \boldsymbol{\nabla})$ the material derivative.

1.12.2.3 Dimensionless form of heat transfer equations

In the absence of heat source term other than viscous dissipation of the flowing fluid, the dimensionless form of the heat equation using proper scales for the length ($L^\star$; boundary layer thickness or duct hydraulic radius), time ($T^\star$; flow cycle period or inverse of flow pulsation), velocity ($V^\star$; mean or maximal cross-sectional velocity, and pressure ($P^\star$) is expressed as:

$$\underbrace{\rho c_p \frac{\Delta T}{T^\star} \frac{\partial \tilde{T}}{\partial \tilde{t}}}_{T1} + \underbrace{\rho c_p \frac{V^\star}{L^\star} \Delta T (\tilde{\mathbf{v}} \cdot \tilde{\boldsymbol{\nabla}} \tilde{T})}_{T2} = \underbrace{G_T \frac{\Delta T}{L^{\star 2}} \tilde{\boldsymbol{\nabla}} \cdot (\tilde{\boldsymbol{\nabla}} \tilde{T})}_{T3} + \underbrace{2\mu \left(\frac{V^\star}{L^\star}\right)^2 \tilde{\mathbf{D}} : \tilde{\mathbf{D}}}_{T4},$$

(1.16)

with the dimensionless temperature define as $\tilde{T} = \frac{T - T_0}{\Delta T}$ (T_0: body temperature [310 K]) ΔT the process temperature $\Delta T \sim \dot{M} L^\star / G_T$ obtained via a dimensional analysis on fluxes (M: the metabolic rate [$\sim 50\,\mathrm{W/m^2}$ at rest]).

Table 1.21. Dimensionless coefficients of the terms of the heat equation in the absence of source terms (LHS, RHS: left and right hand side of equation 1.16; Ti: equation term, $i = 1, \ldots, 4$). Br, Ec, Pe, Pr, St, Sto are the Brinkman, Eckert, Péclet, Prandtl, Strouhal, and Stokes numbers.

Coefficient	Predominant convection (reference coefficient $\rho c_p V^\star \Delta T / L^\star$)	Predominant diffusion (reference coefficient $G_T \Delta T / L^{\star 2}$)
T1 (LHS)	$L^\star/(V^\star T^\star) \equiv \mathrm{St}$	$L^{\star 2}/(\mathcal{D}_T T^\star) \equiv \mathrm{Pr.Sto}^2$
T2 (LHS)	1	$(L^\star V^\star)/\mathcal{D}_T \equiv Pe$
T3 (RHS)	$\mathcal{D}_T/(L^\star V^\star) \equiv Pe^{-1}$	1
T4 (RHS)	$(\nu/c_p)(V^\star/L^\star)\Delta T^{-1} \equiv \mathrm{PrEc}$	$(\mu/G_T)(V^{\star 2}/\Delta T) \equiv Br$

1.12.2.4 Dimensionless governing parameters

The formulation of the dimensionless equations depends on the choice of the variable scales ($\bullet^\star$). For example, the length scale is most often the vessel hydraulic radius ($L^\star \equiv R$), the time scale is the flow period (T) or, most often, the inverse of the flow pulsation ($T^\star = \omega^{-1}$), and the velocity scale either the mean ($V^\star \equiv \overline{V}_q$) or the peak ($V^\star \equiv \widehat{V}_q$) cross-sectional average velocity.

Some dimensionless parameters associated with fluid flows have been identified (Sect. 1.9.2.2). Additional dimensionless parameters associated with heat and mass transfer can be deduced from similar analyses.

(1) The *Prandtl number* is the ratio of the momentum diffusivity to thermal diffusivity ($Pr = \nu/\mathcal{D}_T$) and the *Péclet number* for thermal diffusion in fluid flows is given by: $Pe_T = Re \times Pr$.[82]

(2) The *Nusselt number* is the ratio of convective to conductive heat transfer, i.e., the ratio of the product of the convective heat transfer coefficient and the characteristic length (e.g., boundary layer thickness or duct hydraulic radius) to the thermal conductivity of the fluid ($Nu = h_T L^\star/G_T$).[83]

(3) The *Brinkman number* ($Br = (\mu V^2)/(G_T \Delta T)$) is related to heat conduction from a wall to a flowing viscous fluid. The Brinkman number is respectively 3.15×10^{-3}, 5.04×10^{-3}, and 4.67×10^{-3} for large and small artery and arteriole, respectively. Hence the viscous dissipation term can be neglected.

(4) The *Graetz number* ($Gr = (D_h/L)Pe$) determines the thermally developing flow entrance length in ducts. It can be defined as the ratio of the product of the mass flow rate per unit volume of a fluid times its specific heat at constant pressure to the product of its thermal conductivity and a pipe length.

(5) In the fluid flows with heat transfer. the *Bejan number* is the dimensionless pressure drop along a channel of length L ($Be = \frac{\Delta p L^2}{\mu \mathcal{D}_T}$).

(6) The *Eckert number* expresses the relation between the flow kinetic energy and enthalpy.

1.13 Epithelial–Mesenchymal and Mesenchymal–Epithelial Transitions

Cellular flexibility with transformations from polarized epithelial cells into migratory mesenchymal cells as well as, conversely, reactivation of epithelial properties in mesenchymal cells, occurs during the body's development. Signal transduction triggers cell differentiation as well as morphological events, such as epithelial–mesenchymal (EMT) and mesenchymal–epithelial (MET) transitions.

Epithelial–mesenchymal transition occurs during embryo- and fetogenesis (e.g., gastrulation, mesoderm development, and neural plate formation) to support tissue remodeling, as well as wound healing and tumor metastasis. Polarized epithelial cells change their morphology and become mobile mesenchymal cells. Epithelial–mesenchymal transition involves the disruption of

---

[82] Similarly, the *Schmidt number* is the ratio of momentum diffusivity to mass diffusivity ($Sc = \nu/\mathcal{D}$) and the Péclet number for mass diffusion in fluid flows is given by: $Pe_m = Re \times Sc$.

[83] Similarly, the *Sherwood number* is the ratio of the convective to diffusive mass transfer ($Sh = h_m L^\star/\mathcal{D}$; h_m: mass transfer coefficient).

polarized adhesion, reorganization of the cytoskeleton, change in composition and organization of the basement membrane, and adoption of motility.

On the other hand, during the mesenchymal–epithelial transition, cells assemble and develop apicobasal polarity, express epithelial-specific proteins, form stable adhesions, and eventually generate conduit lumens.

1.13.1 Epithelial–Mesenchymal Transition

During epithelial–mesenchymal transition, cells lose their intercellular contacts before leaving the epithelial layer. Downregulation of the cell adhesion molecule E-cadherin and upregulation of vimentin are classical EMT features. E-Cadherins are inhibited by several transcriptional repressors that prevent the synthesis of proteins involved in epithelial polarization such as polarity complexes. These transcriptional repressors include E box-binding factors Snail (Snail homolog)[84] and Snai2,[85] basic helix–loop–helix factor Twist (or bHLHa38), transcription factor TcF3,[86] as well as zinc finger E-box-binding homeobox ZEB1 and ZEB2.[87]

Numerous overlapping signals initiate, amplify, and modulate the epithelial–mesenchymal transition. Transforming growth factor-β and hepatocyte growth factor (Vol. 2 – Chap. 3. Growth Factors) as well as morphogens such as Wnt (Vol. 3 – Chap. 10. Morphogen Receptors) can induce epithelial–mesenchymal transition, as they stimulate members of the families of Snail homologs (Snai1–Snai3 and Snai1L1) and zinc finger E-box-binding homeodomain-containing (ZEB) factors (ZEB1–ZEB2) as well as mediators and components of the cytoskeleton and extracellular matrix, in addition to intercellular junctions [26]. Moreover, SNAI family members can induce the coexpression of ZEB transcription factor that further enables additional expression of transforming growth factor-β (positive feedback loop). The transcriptional repressor Snail can also alter molecular transfer, as it can inhibit Rab25 GTPase.

The epithelial–mesenchymal transition is also controlled by a negative feedback loop governed by microRNAs. Some members of microRNA families (miR141, miR200, and miR205) operate in epithelial–mesenchymal transition, as they cooperatively inhibit transcription repressors ZEB1 and ZEB2 [66].[88] Among 207 known microRNAs, miR200s that target ZEB1 and ZEB2 mRNAs can be considered as markers of cells that express E-cadherin, but not or

---

[84] A.k.a. Slug homolog SlugH2.

[85] A.k.a. Slug homolog SlugH1.

[86] A.k.a. E2α, E2a, E12, and E47.

[87] Protein ZEB1 is also called transcription factor TcF8, δ EF1, zinc finger homeobox protein ZFHX1a; ZEB2 SMAD-interacting protein SIP1, or SMADIP1, and ZFX1b, or ZFHX1b. Snail and ZEB1 impede the expression of components of both the Crumbs and Scribble polarity complexes.

[88] Five members of the microRNA-200 family (miR141, miR200a–miR200c, and miR429) and miR205 are downregulated in cells that undergo epithelial–mesenchymal transition in response to transforming growth factor-β [66].

poorly vimentin [67]. Inhibition by microRNAs reduces E-cadherin expression (miR200 inhibition also increases vimentin expression) and induces epithelial–mesenchymal transition. Conversely, microRNA expression in mesenchymal cells initiates mesenchymal–epithelial transition. In cancer cells, expression of miR200 upregulates E-cadherin and reduces their motility.

Alternative splicing contributes to epithelial–mesenchymal transition.[89] Regulated alternative splicing is achieved by interactions between numerous RNA-binding proteins that associate with pre-mRNAs to positively or negatively influence mRNA splicing. The set of developmentally and tissue-restricted splicing factors encompasses not only ubiquitous factors, but also cell-specific splicing factors. Ubiquitous alternating Ser/Arg (SR) repeat-containing splicing factors and heterogeneous nuclear ribonucleoproteins indeed cooperate with tissue-specific regulators of splicing such as tissue-restricted SR repeat splicing factors.[90] The conversion of epithelial to mesenchymal cells relies on the regulation of 2 related, epithelial cell-specific factors — the *epithelial splicing regulatory proteins* ESRP1 and ESRP2 — that intervene in about 100 alternative splicing events [68]. The ESRP proteins intervene in splicing of mRNA precursors that produce proteins involved in intercellular adhesion, polarity, and migration. They induce the expression of epithelial-specific isoforms of fibroblast growth factor receptor-2 as well as of other genes [68]. Epithelial–mesenchymal transition results partly from the repression of epithelial-specific splicing controlled by ESRP proteins.

1.13.2 Mesenchymal–Epithelial Transition

Mesenchymal–epithelial transition is a program component of organo-genesis that also occurs in cancers. Mesenchymal–epithelial transitions happen during somitogenesis that produces the somites, i.e., the precursor cells

---

[89] Alternative splicing of precursors of messenger RNAs is a stage of protein synthesis (Chap. 5).

[90] RNA-binding domain-containing proteins of the heterogeneous nuclear ribonucleoprotein (hnRNP) family and SR superfamily of Arg/Ser domain-containing proteins control alternative splicing. Members of both families repress or promote the formation of spliceosomes, often according to the location of specific binding sites within exon or intron sequences. Tissue-specific regulators of splicing include neuron-specific neurooncological ventral antigens NOVA1 and -2, 100-kDa neural-specific SR-related protein nSR100, neuron- and myocyte-enriched RNA-binding protein RBFox1 and -2 (a.k.a. ataxin-2-binding protein-1 and hexaribonucleotide-binding proteins HRNBP1 and HNRBP2), neural polypyrimidine tract-binding proteins (widely expressed PTBP1 and neural, myoblast, and testis-expressed PTBP2), muscleblind-like proteins (MBNL family), and members of the CUG triplet repeat RNA-binding (CUGBP) Elav-like family (CELF) [68].

of bones and skeletal muscles. Mesenchymal–epithelial transition also elicits kidney development and celomic cavity formation.[91]

Several mechanisms of conduit lumen formation exist that more or less involve mesenchymal–epithelial transition: epithelial sheet rearrangement and folding, membrane repulsion, cavitation, and hollowing [26]. During cavitation, a group of cells proliferate to form a cell mass, inner cells of which undergo apoptosis to create a lumen (Vol. 5 – Chap. 10. Vasculature Growth – Sect. Lumenogenesis). During hollowing, intracellular vesicles are delivered to intercellular regions to generate rudimentary lumens that merge into a single, larger lumen.

Small GTPases of the RHO superfamily — CDC42, Rac, and Rho — (Vol. 4 – Chap. 8. Guanosine Triphosphatases and their Regulators) that control the cytoskeletal dynamics, maintain cadherin-mediated intercellular adhesions and stabilize the epithelial structure [69]. Transcription factor FoxC2 of the Forkhead box class promotes epithelial differentiation [70]. Upregulation of FoxC2 in injured tubular cells activates epithelial cell redifferentiation rather than dedifferentiation during organ repair.

---

[91] A celom is a cavity coated by an epithelium derived from a mesoderm such as the pleuropericardial celom. Celoms in the embryonic mesoderm form the pleural, pericardial, and peritoneal cavities.

2

Cells of the Blood Circulation

Blood supplies oxygen, a mediator for cell energy, and conveys nutrients to tissues and removes waste products of cell metabolism toward lungs and purification organs, in addition to the convection of thermal energy. Moreover, blood transmits messengers such as hormones to target organs.

Blood is involved in body's defense against infection, transporting immune cells and antibodies, and in repair processes after injury. The immune response yields an example of communication between and coordination of different cell types. Immunity illustrates the importance of between-cell signaling to provoke cell proliferation, differentiation, and maturation in dedicated sites, as well as migration to target loci and coordinated functioning to achieve assigned tacks.

Blood circulates from the heart through arteries into arterioles down to capillary beds. Capillaries form extensive networks for molecular exchange between blood and cells. Blood returns to the heart through venules and veins. Blood transport adapts to needs of the body's tissues, especially growing tissues. Capillaries are able to sprout and branch to form new networks. Moreover, arteries and veins can expand and remodel.

Cells related to the cardiovascular system encompass cells of the walls of the heart and blood and lymph vessels that bring needed materials to tissues, as well as those that are conveyed in the blood stream and involved in the regulation of the blood circulation. The flowing blood conveys circulating cells that include 3 main categories of cells — erythrocytes, leukocytes, and thrombocytes — the main function of which is oxygen transport, immunity, and plug formation and activation of blood coagulation factors to limit hemorrhage, respectively. Immune cells circulate in blood flow before scouting tissues in search for possible foreign elements. Nervous cells control the activity of the cardiovascular system according to inputs provided by sensory cells of the vasculature. Even after the final stage of differentiation, cells keep a potential of flexible fate because they can be reprogrammed by a more or less small set of transcription factors.

Any cell interacts with its surrounding medium, as it emits, receives, transmits, stores, and treats information. Many cell stimuli induce signaling cascades that lead to cell responses following molecule synthesis or release from stores (Vol. 3 – Chap. 1. Signal Transduction). Hundreds of mechanical, physical, and chemical stimuli control cell function using a limited repertoire of signaling pathways that prime distinct cell responses. Signal transduction relies on multiple types of ion carriers (Vol. 3 – Chaps. 2–3) and receptors (Vol. 3 – Chaps. 6–11) located in the cell membrane. In addition, the extracellular matrix is required for the assembly of differentiated cells into a functional tissue. The extracellular matrix may also control the fate of stem and progenitor cells.

2.1 Cells of Vasculature Walls

Many types of cells are related to the cardiovascular system. A first set comprises mural cells of: (1) blood vessels, such as vascular endothelial cells and either smooth muscle cells and fibroblasts in the macrocirculation or pericytes in the microcirculation (blood vessel caliber $\mathcal{O}[1]\mu m$–$\mathcal{O}[100]\mu m$); and (2) heart, such as cardiomyocytes, nodal cells, and adipocytes.

Each cell type is characterized by appropriate features. In particular, cardiomyocyte properties include excitability, electrochemical conductibility, and contractility, as well as relaxation and repolarization capacity.

Many tissular characteristics are obtained by message exchanges between different cell types in a given tissue. Blood vessel size, thereby blood pressure, is controlled partly by interactions between vascular endothelial and smooth muscle cells.

2.1.1 Endothelial Cells

Endothelial cells are situated at interfaces between 2 types of connective tissues: (1) circulating blood in the vascular lumen and (2) subendothelial layer of the intima of large vessel walls or surrounding medium of microvessels. The endothelium, a single layer of endothelial cells (Vol. 5 – Chap. 9. Endothelium), is a specialized type of mesenchymally-derived epithelium that lines the interior (wetted) surface of the heart (endocardium) as well as blood and lymphatic vessels.

At the microscopic length scale, endothelial cells build a thin layer with corrugations (extensions toward the vessel lumen) due to the presence of cell nuclei. The wetted surface of the vascular endothelium is covered by a glycocalyx of variable height (50–500 nm) depending on water concentration. The glycocalyx that contains proteoglycans experiences friction forces. Endothelial cells are separated by clefts (length 400–450 nm, caliber 20–25 nm).

Quiescent endothelial cells have an apical primary cilium, a solitary organelle that serves as coordinator of signaling pathways.[1] The primary cilium emanates from the cell surface (length $\mathcal{O}[1]\mu m$) and contains microtubules. It operates in both chemo- and mechanosensation. It, indeed, undergoes bending in response to fluid flow[2] and detects hormones, growth factors, and morphogens (e.g., Hedgehog and Wnt).[3]

Endothelial cells are sensitive to relatively high-magnitude stretch and low-amplitude shear stress and can regulate the vasomotor tone that directs the vessel caliber. Stress distribution within large vessel wall is heterogeneous, as the wall is composed of 3 layers (intima, media, and adventitia) of composite materials. Stress and strain fields bear large temporal and spatial variations, as blood flow is three-dimensional, developing, and unsteady. Mechanical forces are thus applied to distinct vascular cell types as well as different cellular components in a given cell (Fig. 2.1).

Mechanosensors and -transducers include cell-surface receptors, mechanosensitive ion channels, cell-adhesion molecules, and plasmalemmal enzymes, as well as components of the glycocalyx and primary cilium, such as cilia-associated mechanosensors polycystins-1 and -2. The luminal surface of endothelial cells contains shear sensors that are sensible to time-varying, low-magnitude hemodynamic stress. The lateral cell membrane contains stretch sensors that are sensible to time-varying, high-magnitude hemodynamic stress and cell adhesion sites that serve in junctional signaling from and to adjacent cells. Adhesion molecules bind to their counterparts on adjacent cells. Vascular endothelial growth factor receptor (VEGFR) connects laterally to VE-cadherin. Platelet–endothelial cell-adhesion molecule PECAM1 activates Src kinase in response to stress. VE-Cadherin links to VEGFR2 and PECAM1 to facilitate Src–VEGFR2 interaction. Activated VEGFR2 recruits and activates the PI3K–PKB–NOS3 pathway (PI3K: phosphatidylinositol 3-kinase; PKB: protein kinase-B; NOS: nitric oxide synthase). The basal cell membrane also comprises stretch sensors associated with integrin-containing adhesomes that anchor the cell to the basement membrane, a component of the extracellular matrix. Focal adhesions with integrins and their partners (talin, vinculin, etc.) connect the extracellular matrix to actin filaments. Activated integrins excite small GTPase Rho that acts on the actomyosin cytoskeleton. Cells also have ATP pumps and/or cell surface ATP synthase. ATP synthesis, release, and transport in the cell cortex are stimulated by flow. Autocrine regulator ATP activates purinergic receptors.

Applied stresses are transmitted to cytoskeleton filaments, particularly via adhesion molecules and their associated linker proteins. Intracellular strain

---

[1] Various receptors, ion channels, transporters, and signaling effectors localize to the primary cilium or its basal body.

[2] The primary cilium of smooth muscle cells directly interacts with the extracellular matrix.

[3] Lipid accumulation in adipocytes is associated with transient formation of the primary cilium.

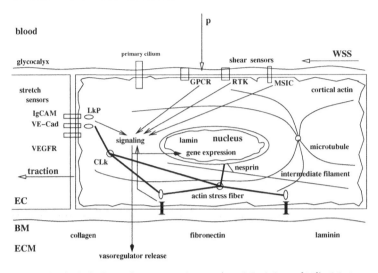

Figure 2.1. Endothelial mechanotransducers (modified from [31]). Major stresses applied to blood vessels include: (1) axial stress induced by longitudinal tension; (2) circumferential stress that results from perpendicularly applied blood pressure; (3) radial compressive stress developed by blood pressure (P) that have much smaller magnitude with respect to axial and circumferential stresses; and (4) wall shear stress (WSS) exerted tangentially by blood flow at the wetted surface of the vessel wall (luminal surface of the endothelium). Mechanotransduction is done via many types of molecules. Mechanosensors and -transducers of endothelial cells (EC) include cell-surface receptors (mainly G-protein-coupled receptors [GPCR] and receptor Tyr kinases [RTK]), mechanosensitive ion channels (MSIC), cell-adhesion molecules (e.g., immunoglobulin-like cell adhesion molecules [IgCAM] and vascular endothelial cadherins [VE-Cad] that can associate laterally with vascular endothelial growth factor receptor [VEGFR]) and their associated linker proteins (LkP), and plasmalemmal enzymes, as well as components of the glycocalyx and primary cilium. The basal cell membrane also comprises stretch sensors associated with integrin (I)-containing adhesomes that anchor the cell to the basement membrane (BM), a component of the extracellular matrix (ECM). Cortical actin, actin stress fibers, microtubules, and intermediate filaments and their crosslinkers (CLk) connect different cell regions and transmit applied forces. Signal transduction triggers gene expression and causes release of regulators of the vasomotor tone.

can induce conformational changes in cytoskeletal elements, such as filaments, crosslinkers, and nanomotors, thereby changing binding affinity. Nesprins-1 and -2 bind to actin filaments, whereas nesprin-3 links to intermediate filaments via plectin. Nesprins interact with lamins of the nuclear cortex via inner nuclear membrane proteins SUn1 and SUn2. Nucleus deformations can modify chromatin conformation, hence modulating access to transcription factors. Activated sensors prime signaling cascades that mainly include kinases and phosphatases. Signal transduction triggers gene expression to synthesize new

Table 2.1. Adrenergic receptors on vascular smooth muscle cells (vSMC) and cardiomyocytes (CMC; Source: Wikipedia).

Type	Action
α1	vSMC contraction
α2	vSMC contraction
β1	CMC contraction
β2	vSMC relaxation

regulators and/or causes release of stored regulators of the vasomotor tone that target adjacent smooth muscle cells.

Endothelial cells secrete substances that control local blood flow by controlling the vasomotor tone (vasoconstriction or -dilation) of smooth muscle cells, blood coagulation (avoiding thrombosis in normal conditions and triggering fibrinolysis of eventual thrombi), molecular transfer into and out of the bloodstream, leukocyte extravasation, smooth muscle cell proliferation, and formation of new blood vessels (angiogenesis). Endothelial cells can construct specialized filters (e.g., renal glomerulus and blood-brain barrier). Endothelial dysfunction is characterized by disturbed nitric oxide production.

Capillary endothelial cells are closely associated with pericytes. The endothelium is continuous or fenestrated (liver sinusoids, kidney glomeruli, and endocrine glands) whether or not large between-cell gaps exist. Endothelial cells of large blood vessels belongs to the tunica intima (inner wall layer), which is isolated from the tunica media (middle wall layer) by a more or less thin subendothelial layer of connective tissue and more or less well defined internal elastic lamina. Endothelial cells are in contact with smooth muscle cells of the subendothelial layer or of the media via gaps through the internal elastic lamina.

2.1.2 Vascular Smooth Muscle Cells

Vascular smooth muscle cells are elongated (diameter <10 µm; length 200 µm; nucleus size 3–4 µm) within the thick wall of large and mid-size blood vessels (Vol. 5 – Chap. 8. Smooth Muscle Cells). Vascular smooth muscle cells appear separated, unlike striated muscle cells, but closely connected. Electrochemical activitation of smooth muscle cells travels from one cell to another by gap junctions. In addition, motor endings of autonomic axons are not directly linked to individual smooth muscle cells. However, vascular smooth muscle cells receive signals primarily from the sympathetic nervous system via adrenergic receptors (adrenoceptors; (Table 2.1).

The media of large blood vessels consists of smooth muscle cells and elastic tissue in varying proportions. Elastic arteries have a higher content of elastic fibers, whereas muscular arteries have a larger proportion of smooth muscle cells. In most vessels, the smooth muscle cells are arranged circumferentially.

Vascular smooth muscle cells are also mechanotransducers. They contracts or relaxes in response to locally applied blood pressure. The vascular smooth muscle tonus regulate the caliber of blood vessels, thereby allowing redistribution of the blood flow to regions that need blood supply due to temporarily enhanced oxygen consumption. Vascular smooth muscle cells ensure not only the regional control of blood flow distribution, but also they preserve a local constant flow rate. The myogenic effect that is responsible for autoregulation (Vol. 6 – Chap. 3. Cardiovascular Physiology) states that vascular smooth muscle cells respond to acute changes in blood pressure. For example, elevated pressure in small resistance arteries triggers smooth muscle cell contraction that narrows the lumen, thereby raising local resistances to flow to keep blood flow constant in the downstream vascular bed. On the other hand, prolonged augmented blood pressure causes wall remodeling (thickening; Vol. 6 – Chap. 7. Vascular Diseases).

Skeletal and smooth muscle can mutually transdifferentiate, each muscle differentiation program can be subverted to the other in the presence of coregulators acting as a molecular switch. Skeletal muscle differentiation is controlled by 4 skeletal muscle-specific transcription factors: myogenic differentiation factor (MyoD), myogenin, MyF5, and myogenic regulatory factor MRF4. These regulatory factors cooperate with myocyte enhancer factor MEF2 to direct the development of skeletal muscles. Myogenic regulatory factors can convert various cell types (including smooth muscle cells) into skeletal muscle cells. Myocardin[4] promotes smooth muscle differentiation [71]. Myocardin represses myogenin in common progenitors of skeletal and smooth muscle lineages. It also activates SMC contractile protein genes. Conversely, myocardin inhibition favors skeletal muscle differentiation.

Smooth muscle cells migrate and accumulate in the intima of artery walls during atherosclerosis and neointimal hyperplasia (Vol. 6 – Chap. 7. Vascular Diseases). Circulating smooth muscle progenitors give rise to cells that produce CD34, $\alpha_5\beta_1$ integrins, receptors VEGFR1 and -2, α-smooth muscle actin, myosin heavy chain, and calponin, but not TIE2 receptor [72]. Although circulating bone marrow-derived progenitor cells are able to differentiate into smooth muscle cells, the main source of invading smooth muscle cells is local [73]. Medial smooth muscle cell proliferation and migration is also observed in response to injury and hypertension [74]. Progenitor cells in the adventitia can serve as an additional source of smooth muscle cells [75]. The adventitia in aortic roots indeed harbor cells that have stem cell markers (e.g., SCA1, SCFR, CD34, and VEGFR2). SCA1+ progenitor cells are able to differentiate into smooth muscle cells upon stimulation by platelet-derived growth factor-BB.

---

[4] Myocardin is a serum response factor coactivator. Serum response factor, which is necessary for SMC differentiation, is a weak transcription activator. It requires cofactors for gene transcription.

2.1.3 Pericytes

Pericytes are cells with contractile properties associated with walls of small blood vessels, such as arterioles, capillaries, and venules (Vol. 5 – Chap. 7. Vessel Wall). Pericyte coverage varies according to the tissue and vessel type. The overall coverage ranges between 11% in myocardial capillaries to about 50% in retinal capillaries. The coverage is more extensive on venules than capillaries. Average pericyte coverage is 84% in venules. Pericytes have prominent nuclei and long processes extending circumferentially and longitudinally around small blood vessels.

Pericytes are closely apposed to endothelial cells (typical endothelium–pericyte distance 100–200 nm).[5] Pericytes are anchored to neighboring endothelial cells by adhesion plaques.

Pericytes have adipogenic, myogenic, osteogenic, and chondrogenic potential according to their locations. Pericytes can in fact differentiate into fibroblasts, adipocytes, smooth muscle cells, or macrophages. They are particularly involved in blood–brain barrier stability and angiogenesis. Pericytes regulate capillary flow.

Pericytes operate in transvascular fluid and material exchange. Over 90% of water crosses venular walls through interendothelial clefts (width ∼20 nm). Protein transport across venular walls is done by diffusion and convection (predominantly owing to the difference between the luminal and intramural pressure)[6] through endothelial clefts and by intracellular flux via vesicular transport. Hydrostatic and oncotic pressure differences are those across the endothelial glycocalyx, i.e., not only between blood and interstitium.[7] The

---

[5] The between-cell thickness depends on the hydrated state of the interstitium. Full hydration can increase the size up to 500 nm.

[6] In the microvasculature, particles are convected by a Stokes flow (Vol. 1 – Chap. 1. Cells and Tissues), in the absence of body forces:

$$\nabla^2 \mathbf{v} = \mu^{-1} \nabla p$$

($\mathbf{v}$: flow velocity; p: pressure; μ: blood dynamic viscosity). Their transport can be modeled by the steady convection–diffusion equation:

$$\mathbf{v} \cdot \nabla c = \mathcal{D} \nabla^2 c$$

(c: concentration, $\mathcal{D}$: diffusivity).

[7] Transport through a porous medium such as the vessel wall can be modeled using the Darcy's law:

$$\mathbf{J}_D = -(\mathcal{P}/\mu)\nabla p,$$

where $\mathbf{J}_D$ is the Darcy flux per unit area (m/s) and $\mathcal{P}$ the material permeability (m^2). Transport through the glycocalyx layer can be assessed by the anisotropic Brinkman equation given in steady state by:

$$\nabla^2 \mathbf{J}_D + \mathbf{P} \cdot \mathbf{J}_D = \mu_{eff}^{-1} \nabla p,$$

rate of filtration from venular lumen to the interstitial space that contains lymphatics depends on the lumen and tissue pressure.

Pericytes that cover endothelial cleft ends create small trapped microdomains and regulate the temporal response of thses trapped domains to changes in endoluminal pressure, hence transient reabsorption kinetics [76]. Trapped microdomains allow quasi-steady equilibration on a time scale that is 2 to 3 orders of magnitude shorter than that of large interstitial spaces.

Pericytes act in angiogenesis (Vol. 5 – Chap. 10. Vasculature Growth). Mature vessel walls such as the rat aorta contain primitive mesenchymal cells that are pericyte progenitor cells [77]. They express precursor cell markers CD34 and TIE2, but neither endothelial (PECAM1 and NOS3) nor smooth muscle (α-smooth muscle actin) cell markers. Upon exposure to serum, these cells lose CD34 expression, but produce α-smooth muscle actin. When embedded in collagen gels, they rapidly migrate in response to platelet-derived growth factor-BB and become dendritic. In the presence of endothelial cells, they transform into pericytes.

2.1.4 Fibroblasts and Fibrocytes

Fibroblasts correspond to mature, differentiated cells (not progenitor cells [footnote 44]). They are the most common resident cells in connective tissue. Fibroblasts can slowly migrate. They derive from primitive mesenchyme. In certain situations, epithelial cells can give rise to fibroblasts (epithelial–mesenchymal transition; Sect. 1.13). Conversely, fibroblasts can give rise to epithelial cells (mesenchymal-epithelial transition). Fibroblasts are morphologically heterogeneous according to their location and activity, as their gene expression pattern is adapted to tissue organization.

Fibroblasts continuously secrete all the components of fibers and ground substance of the extracellular matrix, in particular, glycoproteins, proteoglycans, collagen, and elastin. They thus maintain the structural integrity of connective tissue.

Collagens are delivered by fibroblasts as procollagens that are converted extracellularly into tropocollagens, which self-assemble into fibers. Collagens-3, -4, -7, and -8 reinforce the basement membrane. Reticular fibers made from collagen-3 crosslink to form a fine meshwork (reticulum) that supports cells in certain organs (bone marrow, spleen, liver, etc.).

Fibrocytes are blood-borne cells that produce extracellular matrix. Fibrocytes express class-2 major histocompatibility complex molecules and costimulatory molecules. They migrate to wound sites for tissue repair.

Fibroblasts can be reprogrammed into cells with many properties of pluripotent embryonic stem cell by the ectopic expression (retroviral transduction) of 4 transcription factors: octamer-binding transcription factors

---

where μ_{eff} is an effective viscosity and **P** the Darcy permeability tensor with zero off-diagonal components and diagonal components $P_{ii} = \mathcal{P}_i^{-1}$ ($i = x, y, z$: direction).

Table 2.2. Features of cardiomyocytes and nodal cells. Nodal cells generate and transmit action potentials that cause cardiomyocyte contraction. Cardiomyocytes also conduct action potentials via gap junctions at intercalated discs.

	Nodal cell	Cardiomyocyte
Sarcomere	−	+
T tubule	−	+
Gap junctions	+	+
Self-excitability	+	−
Capillary	Fenestrated	Continuous

Oct3 and -4, Sry-related HMG box Sox2, Krüppel-like factor KLF4, and MyC [78,79].

Induced pluripotent stem cells are similar to pluripotent embryonic stem cells, as they can give rise to all the body's cell types and display genetic and morphologic embryonic stem cell characteristics.[8] The RNA-binding protein Lin28[9] functions with 3 other factors (Oct4, Sox2, and Nanog) to reprogram fibroblasts to pluripotent stem cells [80]. Protein Lin28, a priLet7g-binding protein, selectively blocks the processing of priLet7 microRNAs.[10] Protein Lin28 suppresses miR-mediated differentiation in stem cells.

2.1.5 Cardiomyocytes

Cardiomyocytes are involuntary striated myocytes of the myocardium (middle layer of the cardiac wall; Vol. 5 – Chap. 5. Cardiomyocytes). They contract to propel blood into arteries of the 2 compartments in series of the circulatory system; the right pump expels blood into the pulmonary trunk of the pulmonary circulation and the left ventricle into the aorta of the systemic circulation. Cardiomyocytes are excited by the nodal tissue (Table 2.2) under the control of the autonomic nervous system.

Adjacent cardiomyocytes are connected by intercalated discs. Intercalated discs contain 3 types of cellular junctions: fascia adherens, a type of adherens

---

[8] Fibroblasts, as well as other cell types, such as mesenchymal stem cells, hepatocytes, and gastric epithelial cells, can generate induced pluripotent stem cells. Hepatocytes and gastric epithelial cells belong to epithelial cell populations characterized by cell adhesions supported in particular by E-cadherin (Sect. 7.5).

[9] A.k.a. zinc finger CCHC domain-containing protein ZCCHC1.

[10] Among small, non-coding RNAs, microRNAs post-transcriptionally repress the gene expression. Mature microRNAs are produced from primary microRNAs (pri-miRs) by sequential cleavages by Microprocessor, a small complex made of ribonuclease-3 Drosha and double-stranded RNA-binding protein DGCR8 (a.k.a Pasha), which recognizes the RNA substrate, and, then, by the Dicer complex, which is composed of human immunodeficiency virus transactivating response RNA-binding protein and Argonaute-2, to release pre-miRs and mature miRs, respectively.

junction, macula adherens, or desmosomes, and macula communicans, or gap junctions. Fascia adherens anchor actin and connect sarcomeres for synchronized contraction. Macula adherens or desmosomes bind intermediate filaments for coordinated contraction. Gap junctions allow ion fluxes associated with action potentials for fast traveling of depolarization waves between cardiomyocytes.

Cardiomyocytes are resistant to fatigue. They have a large number of mitochondria that enable continuous aerobic respiration and high content in myoglobins to store oxygen. They are close to capillaries that supply nutrients and fuel. They are not well adapted to hypoxia.

Excitation–contraction coupling that generates heart contraction is done by calcium ions from the extracellular space and intracellular store, i.e., the sarcoplasmic reticulum. Excitation-induced Ca^{++} cycling with influx and outflux via sarcoplasmic reticulum Ca^{++} pumps, ryanodine receptors, and Na^{+}–Ca^{++} exchangers (Vols. 3 – Chap. 3. Main Classes of Ion Channels and Pumps and 5 – Chap. 5. Cardiomyocytes) modulated by β-adrenoceptors governs the strength of cardiomyocyte contraction. Cardiomyocyte contraction requires ATP manufactured by oxidative phosphorylation (cell respiration).

Cardiomyocytes respond to applied strain and stress using mechanosensors. Mechanosensors include mechanosensitive ion channels, cell-surface receptors, integrin-based adhesomes with their receptors, signaling mediators, and actin-linking and -polymerizing modules, dystrophin-associated complex that links actin filaments to matrix, and certain sarcomeric proteins. Mechanotransduction relies on various effectors (Vols. 3 and 4), such as Ras GTPase, mitogen-activated protein kinase, phospholipase-C, protein phosphatase PP3, Janus-activated kinase, nitric oxide, calcium, and microRNAs, to trigger adaptative hypertrophy gene expression by activated transcription factors, such as nuclear factor-κB, nuclear factor of activated T cells, and signal transducer and activator of transcription (Vol. 5 – 5. Cardiomyocytes).

2.1.6 Nodal Cells

Myogenic, self-excitable nodal cells are responsible for the genesis (heart automaticity) and propagation of the electrochemical wave, or action potential, that triggers cardiomyocyte contraction (Vol. 5 – Chap. 6. Heart Wall). Nodal cells differ from cardiomyocytes (Table 2.2).

Isolated nodal cells send impulses and contract rhythmically at a given rate according to their situation along the nodal tissue, hence type, density, and activity mode of ion carriers involved in the genesis and propagation of the action potential. Pacemaker cells in the sinoatrial node determine the heart frequency. The activity of nodal cells is modulated by the autonomic nervous system via its sympathetic and parasympathetic system.

The cardiac action potentials differ significantly in different regions of the nodal tissue and myocardium due to different ionic current characteristics (different types and proportions of involved ion channels). The propagation speed and timing are set up to produce coordinated contraction of different regions of the ventricular myocardium that are depolarized with a more or less great delay. Purkinje fibers and atrio- and ventriculomyocytes are characterized by fast response, whereas sinoatrial and atrioventricular nodal cells have transmembrane potential that rises slowly.

The rate of spontaneous depolarization of sinoatrial nodal cells determines the frequency of propagated action potentials. Multiple ionic currents through Ca_V1 and -3, hyperpolarization-activated, slow and rapid delayed rectifying K^+, Na^+–K^+ pump, and Na^+–Ca^{++} exchangers, in addition to sustained and background transient outward currents regulate depolarization slope (spontaneous frequency) of sinoatrial nodal cells. Minimum membrane potential level due to outward K^+ currents activates inward hyperpolarization-activated current for membrane depolarization that in turn stimulates Ca_V1 and -3 channels. Resulting Ca^{++} influx primes Ca^{++} release by ryanodine receptors that produces a subsarcolemmal Ca^{++} increase that spreads by Ca^{++}-induced Ca^{++} release and amplifies Ca^{++} influx via currents through Na^+–Ca^{++} exchangers for action potential upstroke [81]. The coordinated activity of ryanodine receptors and Na^+–Ca^{++} exchangers explains, at leat partly, the spontaneous frequency of action potentials in sinoatrial nodal cells.

The capillaries in the atrioventricular node and nodal bundles of the heart are fenestrated, whereas the non-specialized myocardium is supplied by continuous capillaries [82]. Fenestrae (size 50 ± 5 nm) are bridged by diaphragms.

2.2 Blood-Circulating and Tissue-Scouting Cells

Hematopoietic stem cells (HSC)[11] that commit into differentiation cascades to form progenitors and then mature blood cells. Multipotent, self-renewing stem cells generate progenitors with a more limited differentiation potential. They also reproduce themselves (self-renewal), hence producing progeny with the same developmental potential (into various cell types) as the original cell. They reside in stem cell niches in the bone marrow. Self-renewal is intrinsic, but stem cell niches influence the type (symmetrical or asymmetrical) of stem cell division.

Hematopoietic stem cells generate red and white blood cells and platelets via various intermediate progenitor cells and 2 developmental lymphoid and myeloid lineages using growth factors and transcription factors involved in differentiation switches (Vol. 5 – Chap. 2. Hematopoiesis).

In opposition to erythrocytes and thrombocytes, the leukocyte compartment comprises a large variety of specialized cell types that share a common

---

[11] Hematopoietic stem cells were the first discovered type of stem cells (1961). They express a set of markers, such as stem cell factor receptor (SCFR; a.k.a. kinase in Tyr [KIT]) and stem cell antigen SCA1, but not lineage (Lin) markers. Fetal and adult HSCs do not behave similarly. Developmental stage-specific transcription factor Sox17 is produced by fetal, but not adult hematopoietic stem cells.

function in innate and adaptive immunity. The innate immune system is the first line of defense against invading pathogens. Detection of foreign microorganisms by the body involves several families of pattern-recognition receptors and a set of signaling molecules to produce factors of immune response.

Components of innate immunity are monocytes and macrophages, granulocytes (neutrophils, eosinophils, and basophils), mastocytes, and dendritic and natural killer cells. On the other hand, cells of adaptive immunity include B lymphocytes and 2 main types of T lymphocytes, CD8+ cytotoxic and CD4+ helper T lymphocytes.

Lymphocytes that originate in the bone marrow mature in peripheral lymphoid organs, such as the thymus and spleen. Double-negative (CD4−, CD8−) DN1 cells migrate from the bone marrow to the thymus to initiate development of T lymphocytes, macrophages, and, to a lesser extent, granulocytes, and natural killer and dendritic cells [83, 84].

The classical model of hematopoiesis states that a hematopoietic stem cell gives rise to progenitor cells unable to self-renew. These progenitors latter generate common lymphoid progenitors (CLP; lymphopoiesis) devoid of myeloid and erythroid potentials and common myelo-erythroid progenitors (CMP; myelopoiesis) lacking lymphoid potential, which give birth to granulocyte–macrophage progenitors (GMP) and megakaryocyte–erythroid progenitors (MEP).

The lymphoid lineage comprises B, T, and NK cells. The remaining leukocyte types as well as erythrocytes and thrombocytes are formed from the myeloid lineage. However, a fraction of macrophages derives from T-cell progenitors. Moreover, a substantial proportion of thymic granulocytes originates from DN1 cells.

Double negative DN1 cells lack B-cell potential, but a substantial proportion of DN1 and DN2 cells (proT category of T-cell progenitors) have both T-cell and myeloid potential that is lost by DN3 cells (preT category of T-cell progenitors) [83, 84] (Fig. 2.2).

2.2.1 Erythrocytes

Erythrocytes or red blood cells (typical size 6–8 μm) constitute the major compartment of blood cells. At usual altitudes, the erythrocyte number is equal to 4 to 5 million in women and 5 to 6 million in men per cubic millimeter (microliter) of blood. Red blood cells are separated from blood plasma by centrifugation. Blood tests include red blood cell count (cell number per unit of blood volume) and hematocrit (blood volume fraction occupied by erythrocytes). They deliver oxygen from lungs to body's tissues.

Erythrocyte is a circulating capsule rather than cell, as the nucleus has been expelled during the reticulocyte stage. This capsule contains a hemoglobin solution. It contains about 270 million hemoglobins, with each carrying 4 heme groups. Vitamin-B12 is needed for the production of hemoglobin. Erythrocytes store about 65% of the total iron contained in the body. Iron circu-

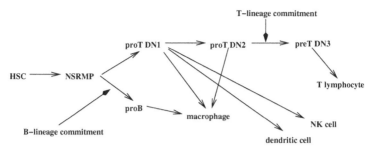

Figure 2.2. Thymus hematopoiesis (Sources: [83, 84]). Non-self-renewing multipotent progenitor (NSRMP) is the thymus seeding cell. In opposition to the classical model of hematopoiesis, the lymphoid lineage has a myeloid potential (HSC: hematopoietic stem cell, a self-renewing multipotent progenitor; proT (preT at more mature stage) and proB: progenitors that produce T and B lymphocytes; DN1, DN2, and DN3: double-negative-1, -2, and -3 cell [DN1 is also called early T-cell progenitors] with respect to CD4+, CD8+ double-positive cell; NK: natural killer cell).

lates in plasma bound to the carrier transferrin. After hemolysis, hemoglobin is bound to plasma haptoglobin.

Erythrocytes have surface glycoproteins that determine blood types. Proteins on the erythrocyte surface include the anion exchanger of the solute carrier class SLC4a1[12] and glycophorins.

Erythrocytes are produced by erythropoiesis in the bone marrow of large bones (liver in embryo), at a rate of about 2 million per second. The production from committed CFUe erythroid progenitors is stimulated by erythropoietin synthesized in the kidney. Erythroid proliferation comprises an early erythropoietin-dependent phase followed by a fibronectin-dependent stage. In late stages of erythropoiesis (when erythropoietin is not involved), erythroblast proliferation (but not differentiation) requires signaling by $\alpha_4\beta_1$ integrins that adhere to fibronectin [85].

Reticulocyte formation from late erythroblasts results from an asymmetrical cell division during which the nucleus is compacted and extruded. Actin filaments form a cortical actin ring around the condensed nucleus, which contracts and pinches the nucleus off. Actin cytoskeleton regulators, GTPases Rac1 and Rac2 (Vol. 4 – Chap. 8. Guanosine Triphosphatases and their Regulators), bind to RhoA effector Diaphanous-2 and promote enucleation (without intervention of Diaphanous-1) [86].

Reticulocytes ($\sim$1% of circulating red blood cells) leave the bone marrow to form mature erythrocytes. Erythrocytes develop from committed stem cells to erythrocytes in about 7 days and live about 120 days. The aging erythrocyte is recognized by phagocytes for processing in the spleen, liver, and bone marrow. A schistocyte is a red blood cell that undergoes cell fragmentation, or a fragmented part of a red blood cell.

---

[12] A.k.a. anion exchanger AE1 and Band-3.

2.2.2 Thrombocytes

Thrombocytes, or platelets, are cell fragments (hence, anuclear cells; size 1.5–3.0 μm) circulating in the blood ($\sim$150,000–400,000/mm^3). They trigger primary hemostasis leading to the formation of blood clots and participate in inflammation. Thrombocytes are produced from megakaryocytes; each megakaryocyte produces 5,000 to 10,000 platelets. Platelets circulate during approximately one week. They are then destroyed in the spleen and liver by Kupffer cells.

Thrombocytes contain RNA, mitochondria, a canalicular system, and several types of granules: (1) lysosomes with hydrolases; (2) dense bodies with ADP, ATP, serotonin, histamine, and calcium, and (3) α-granules with fibrinogen, factor V, vitronectin, thrombospondin, and von Willebrand factor.

Platelets activated by the contact with collagen, thrombin, among others, release the granule content, mainly coagulation factors and platelet activators. Thrombocytes also secrete platelet-derived growth factor, as well as cytokines, chemokines, and other inflammatory mediators.

Thrombocytes adhere to each other via adhesion molecules (integrins) and to endothelial cells. Thrombocytes can contract during aggregation owing to their high concentration of myosin and actin filaments to reinforce the hemostatic plug.

2.2.3 Leukocytes

Two principal populations of leukocytes, or white blood cells, circulates in blood (Table 2.3): (1) granular leukocytes, granulocytes, or polymorphonuclear leukocytes, which include neutrophils, eosinophils, and basophils; and (2) agranular leukocytes or agranulocytes (mononuclear leukocytes), which comprise lymphocytes, monocytes, and macrophages.

Leukocytes are cells of the immune system that defend the body against foreign agents. Leukocytes (4,000–10,000 per microliter of blood) are produced by hematopoietic stem cells in the bone marrow.

2.2.3.1 Granulocytes

Basophils are responsible for antigen response by releasing histamine that causes inflammation. They have bi- or tri-lobed nuclei. Eosinophils are involved in the treatment of parasitic infections as well as in allergic reactions. They have bilobed nuclei. Neutrophils participate in the defense against bacterial and fungal infections. They have a multilobed nucleus (Table 2.3). Neutrophils cannot renew their lysosomes and die after phagocytosis.

2.2.3.2 Lymphocytes

Lymphocytes are small circulating cells with large, eccentric nuclei and small content of cytoplasma. Lymphocytes scout as they move from blood to

Table 2.3. Leukocyte types and main features (Source: Wikipedia).

WBC type	Fraction (%; adult)	Size (μm^3)	Targets	Nucleus	Lifetime
Neutrophil	40–75	10–12	Bacteria, Fungi	Multilobed	6 h–3 days
Eosinophil	1–6	10–12	Parasites, Allergy	Bi-lobed	8–12 days
Basophil	< 1	9–10	Allergy	Bi-/tri-lobed	
Lymphocyte	20–45	7–8		Large	Weeks–years
B			Pathogens		
CD4+ T_H			Bacteria		
CD8+ T_C			Virus, Tumors		
NK			Virus, Tumors		
Monocyte	2–6	14–17	Various	Curved	Months–years

connective tissues and back again to target possible invading microorganisms. They are especially concentrated in lymphoid tissues (spleen, thymus, lymph nodes, Peyer's patches of ileum, and tonsils) that include germinal centers where activated lymphocytes proliferate.

Different types of lymphocytes exist: (1) B lymphocytes (or B cells), and (2) T lymphocytes (or T cells). B lymphocytes develop in niches of the bone marrow of adults, whereas T lymphocytes mature in the thymus. B lymphocytes synthesize antibodies that recognize and bind to foreign molecules for material destruction. The antibodies are either secreted or bound to the lymphocyte membrane. Plasmocytes are differentiated lymphocytes that produce large amounts of antibodies.

Antigen receptor of B lymphocytes corresponds to a combination of immunoglobulin heavy (IgHC) and light (IgLC) chains. Recognition of foreign antigen by mature B lymphocytes triggers clonal expansion and antibody secretion, whereas reactivity to auto-antigens primes cell death to ensure self-tolerance (clonal-selection concept). In addition, an alternative process that affects antigen receptor specificity in immature lymphocytes, which already exhibit a receptor, is initiated in central lymphoid tissues (receptor-selection notion).[13] Antigens are recognized by specific, plasmalemmal, heterodimeric T-cell receptors. T-cell receptors-α and -β confer the antigenic specificity

---

[13] Receptor editing corresponds to a mechanism in which signaling via an autoreactive antigen receptor promotes further receptor-gene rearrangements. This recombination eliminates the receptor gene that confers autoreactivity and replaces it with an non-self-reactive gene. Self-tolerance is thus obtained without eliminating the cell [87].

of $\alpha\beta$T lymphocytes, whereas T-cell receptors-γ and -δ are expressed by $\gamma\delta$T lymphocytes. Polypeptide chains of both B-cell (BCR) and T-cell (TCR) receptors contain a constant domain and a variable region.

T Lymphocytes

T Lymphocytes constitute an infrafamily of the lymphocyte subfamily (of the leukocyte family) that includes developing thymocyte subsets, naive CD4+ and CD8+, TCR$\alpha\beta$+ and TCR$\gamma\delta$+ T cells, and differentiated effector subsets that comprise (Vol. 5 – Chap. 3. Blood Cells): (1) CD4+, helper T cells (T$_H$), with types 1 (T$_{H1}$) and 2 (T$_{H2}$); (2) CD8+, cytotoxic T lymphocytes (T$_C$ or CTL); (3) CD4+, CD25+, regulatory T cells (T$_{Reg}$); and (4) natural killer (NK) cells.

Multipotent progenitors from the bone marrow colonize the thymus and differentiate into various T-cell lineages according to environmental signals that drive specific transcriptional programs. T-cell differentiation into $\alpha\beta$ and $\gamma\delta$ lineage is regulated by $\gamma\delta$-specific transcription factor Sox13 [88]. Commitment into CD4+ and CD8+ lineages is controlled by zinc finger and BTB[14] domain-containing transcription factor ZBTB7b [89].[15] Invariant NKT cells develop owing to zinc finger and BTB domain-containing transcription factor ZBTB16 [90].[16] After development in the thymus, mature T lymphocytes recirculate. Mature CD4+, CD8+ naive and memory T lymphocytes cells reside in the bone marrow.

T lymphocytes form 2 supertribes — $\alpha\beta$ and $\gamma\delta$ — that differ both in function and distribution through the body. $\alpha\beta$L̃ymphocytes are selected in the thymus by engagement of their T-cell receptors with self-peptides presented by major histocompatibility complex molecules. $\gamma\delta$L̃ymphocytes localize to epithelia (mucosa and skin). They are predominant in mucosa of intestine, lungs, and reproductive tract, which are the primary barriers to diverse environmental aggressions. Epidermal $\gamma\delta$l̃ymphocytes produced a plasmalemmal marker of the immunoglobulin superclass, Selection and upkeep of intraepithelial T cell Skint1 encoded by a gene on chromosome-4 [91].

Adhesion- and degranulation-promoting adaptor (ADAP) regulates function of T-lymphocyte integrins, which depends on T-cell receptors, and thereby controls T-lymphocyte adhesion and activation.[17]

---

[14] The BTB domain refers to BrC, Ttk, and BAB. BrC Alias stands for Broad complex, Ttk for Tramtrack, and BAB for bric-à-brac. Molecules BrC and Ttk that are encoded by Drosophila genes are zinc finger proteins.

[15] A.k.a. T-helper-inducing PoZ (for Pox virus and zinc finger)–Krüppel-like factor (THPoK).

[16] A.k.a. promyelocytic leukemia zinc finger protein (PLZF).

[17] Adaptor ADAP also stimulates TCR and CD28 costimulatory receptor to activate nuclear factor-κB. Protein ADAP interacts with plasmalemmal adaptor CARMA1 that couples TCR signaling to NFκB [92]. Proteic complex made of

Memory T Lymphocytes

Immunological memory is done either by circulating antibodies or expanded populations of antigen-specific B and T lymphocytes with effector functions, such as recognition of previously encountered pathogens and reactivity toward allergens and other non-infectious antigens.[18] Memory T lymphocytes are generated for adaptive immunity, as they confer long-lived, antigen-specific protection against repeated invasions of pathogens. Both viral and bacterial infections stimulate Ifnγ+, CD4+, CD8+ T lymphocytes that can differentiate into long-lived, memory T cells. Long-lived, CD8+ memory T-cell population is larger than that of CD4+ memory T lymphocytes. Both CD4+ and CD8+ memory T-cell subsets with distinct functional potentials arise from Ifnγ+ T-cell progenitors [93].

Helper T Lymphocytes

Helper T cells (T$_H$) coordinate adaptive immune defense against specific pathogens. T$_H$ Cells comprise 3 major subtribes: helper type-1 (T$_{H1}$) and type-2 (T$_{H2}$) as well as interleukin-17-secreting helper T cells (T$_{H17}$). T$_{H17}$ Cells are involved in host defense against specific pathogens and induce autoimmunity and tissue inflammation. They then interact with regulatory T cells to prevent tissue inflammation and mediate self-tolerance.

Cytotoxic T Lymphocytes

Cytotoxic T-lymphocyte granule exocytosis and subsequent release of granule content are aimed at eliminating virus-infected and tumor cells by cytotoxic T and natural killer lymphocytes. Cytotoxic granules contain perforin and granulysin that alter cell membranes, as well as granzymes that are serine proteases bound to proteoglycan serglycin. Cytolysis is mediated by secretion at the immunological synapse formed between killer and target cells of granzymes and granulysin that cause cell death by caspase-independent and -dependent pathways [94] (Vol. 2 – Chap. 4. Cell Survival and Death). In addition, killer cells engage plasmalemmal death receptors such as tumor-necrosis factor receptors (e.g., TNFRSF6a) that activate the caspase pathway. Cytotoxic T and natural killer lymphocytes destroy target cells using the same basic mechanisms, but secretion is triggered by distinct receptors (antigen and NK receptors, respectively). Moreover, the expression of cytolytic molecules is constitutive in NK cells, but regulated in cytotoxic T-lymphocytes.

CARMA1, caspase-like protein MALT1, and adaptor BCL10 is required for TCR-dependent activation of IκB kinase and subsequent nuclear translocation of NFκB.

[18] After meeting a known antigen, circulating antibodies bind the corresponding antigen for elimination by phagocytic cells, whereas certain types of immune cells present this antigen and secrete cytokines for expansion of antigen-specific memory B and T lymphocytes.

Natural Killer Lymphocytes

Natural killer cells are effector lymphocytes of the innate immune system with natural cytotoxicity and cytokine production that limit extension of tumors and infections. Therefore, NK cells spread throughout lymphoid and non-lymphoid tissues. Natural killer cells are also regulatory cells that interact with dendritic cells, macrophages, T lymphocytes, and endothelial cells. They are thus able to discriminate target cells from healthy cells using a detection system based on plasmalemmal activating and inhibiting receptors [95]. Activating NK-cell receptors detect alert molecules on cells in distress. NK Cells recognize glycolipid antigens presented by the non-classical major histocompatibility complex-1-like molecule CD1d[19] and subsequently secrete cytokines and chemokines. Distinct NK-cell subsets have been defined according to phenotype, function, and site.

Natural killer cells acquire their features during development in the thymus. They arise from CD4−, CD8− precursors and undergo a series of differentiation stages characterized by the sequential expression of several plasmalemmal markers (cell-adhesion molecule CD24, epican or CD44, and NK1-1 factor). Afterward, natural killer cells navigate to bind transformed or infected cells via integrins and NK receptors that form a lytic synapse. Both receptor types depend on the cytoskeleton, particularly actin filaments with their actin-binding proteins and regulators that switch NK cells from motion to a stable contact with a target cell.

Cell recognition by NK cells via NK-cell receptors not only targets peptide-presenting domain of class-1 MHC molecules (using NK-cell receptors killer cell immunoglobulin receptors [KIR] and NKG2), but also their binding regions (owing to killer cell lectin-like receptor KLRa1, the single member of the human Ly49L family of NK-cell inhibitory receptors) [19].

Regulatory T Lymphocytes

Regulatory T cells are immunosuppressive lymphocytes that maintain tissue tolerance to avoid autoimmunity. Regulatory T cells also hamper excessive immune reactions to avoid or, at least, limit chronic inflammation.[20] However,

---

[19] CD1d is expressed by both CD4+, CD8+ thymocytes and epithelial cells of the thymus.

[20] Regulatory T cells can have several basic modes of immunosuppressive action [96]: (1) release of inhibitory cytokines, such as interleukin-10 and -35 and transforming growth factor-β; (2) cytolysis via granzyme and perforin (e.g., CD4+, CD45Ra+ regulatory T cells stimulated with CD3- and CD46-specific antibodies; CD45R, or CD45, is the type-C protein Tyr phosphatase receptor (PTPRc); CD45Ra is an isoform that contains exon A). (3) apoptosis mediated by cytokine deprivation that depends on CD25 (or IL2 receptor-α); (4) cell inhibition by intracellular transfer of cAMP through gap junctions; (5) immunosuppression via adenosine receptor A_2 stimulated by pericellular generation of adenosine due to concor-

regulatory T cells must preserve useful pathogen-specific immune responses as well as antitumor immunity. CD4+, CD25+, FoxP3+, regulatory T cells develop in the thymus and exhibit a set of T-cell receptors that are specific for self-antigens. Regulatory T cells can also be generated from effector T cells; these cells then usually do not express forkhead box FoxP3 factor.

Regulatory T cells promote early antiviral immune response at a site of infection [97]. Regulatory T cells accumulate, are activated, and proliferate both at the site of infection and in the draining lymph nodes. Regulatory T cells require transcription factor FoxP3 to develop and function. The presence of regulatory T cells in the lymph node prevents influx and activation of effector T, natural killer, and dendritic cells that thus migrate into the site of virus infection.

B Lymphocytes

The phenotype of B lymphocytes is partly defined by its pattern of active and inactive gene expression. During development, progenitor cells divide and differentiate into specific lineages, and daughter cells retain the same activity profile as parent cells.

Immature, IgM^{dull}+ B lymphocytes leave the bone marrow and mature in the spleen. Splenic B lymphocytes include 2 populations of precursors of mature B lymphocytes (Table 2.4): (1) recently immigrated, transitional, type-1 B lymphocytes and (2) transitional type-2 B lymphocytes in primary follicles of the spleen that form from type-1 B lymphocytes [98]. Mature B lymphocytes are generated from both types of transitional B lymphocytes. Induction of B-lymphocyte longevity and maturation requires different signals derived from the B-cell receptor (BCR). B-Cell receptor is composed of membrane-bound immunoglobulin (Ig) and Igα–Igβ heterodimer. Spliced forms of protein Tyr phosphatase PTPRc (or CD45) expressed on the surface of B and T lymphocytes target SRC family kinase Lyn that regulates BCR signaling. Bruton's Tyr kinase (BTK), which is specifically expressed in myeloid cells and B lymphocytes, is activated by SRC family kinases after BCR crosslinking and interacts with numerous ligands to participate in BCR signaling.

Mature IgD+, IgM+ B lymphocytes recirculate and can localize in the bone marrow. Only mature B lymphocytes enter lymphoid follicles of spleen and lymph nodes to participate in the immune response. Two B-cell types are generated in germinal centers: (1) quiescent memory B lymphocytes that

---

dant action of ectoenzymes ectonucleotide pyrophosphatase–phosphodiesterase ENPP1 (or CD39) and ecto-5′-nucleotidase NT5E (or CD73); (6) inhibition of dendritic cell maturation by lymphocyte-activation gene LAG3 (or CD223) that binds to class-2 MHC molecules expressed by immature dendritic cells to suppress immunostimulation of dendritic cells; and (7) production by dendritic cells of immunosuppressor indoleamine dioxygenase due to cytotoxic T-lymphocyte antigen CTLA4 that interacts with costimulatory receptors CD80 and/or CD86.

Table 2.4. Tissue distribution of various types of B lymphocytes (Source: [98]).

Type	Location
Immature	Bone marrow
Transitional type 1	Bone marrow, blood, spleen (red pulp and outer periarteriolar lymphoid sheet close to primary follicle)
Transitional type 2	Spleen (follicle)
Mature	Bone marrow, blood, spleen (follicle), lymph node
Marginal zone	Spleen (specialized area surrounding the white pulp outside the marginal sinus)

wait for re-exposure to antigen to be reactivated and (2) long-lived antibody-secreting plasmocytes (terminally differentiated B lymphocytes) that continuously secrete antibodies, which circulate in blood.

Memory B lymphocytes synthesize plasmalemmal immunoglobulins IgG, IgE, and IgA that allow recognition and reaction to free antigens. Depending on the antigen type, the differentiation of B lymphocytes into plasmocytes usually requires help from antigen-specific, CD4+ T lymphocytes. After immunization, antigen-specific B lymphocytes are, with basophils, the leukocyte populations that can bind substantial amounts of intact antigens.

Activated B lymphocytes migrate to the T-cell zones of local secondary lymphoid tissues and soon move within this secondary lymphoid tissue, either to extrafollicular foci where they grow as *plasmablasts* or to follicles in which they form *germinal centers*. Plasmablasts in the extrafollicular foci mainly form short-lived plasmocytes. In germinal centers, plasmablasts differentiate into either antibody-producing plasmocytes or memory cells. Plasmocytes that leave germinal centers often migrate to distant sites of antibody production where they can have a long lifespan.

B-cell differentiation into antibody-secreting plasmocytes results from a combination of signals provided by T and dendritic cells and pathogens via cytokines and B-cell receptors as well as receptors for innate signals. The pattern of plasmocyte response depends on antigen and sites of immune recognition and antibody production.[21] The bone marrow constitutes a storage site for long-lived plasmocytes that are formed in germinal centers [99].

The bone marrow is the hematopoiesis locus that serves as a primary lymphoid organ, as it yields niches for stem cells for lympho- and myelocytogenesis. The bone marrow also acts as a secondary lymphoid organ that hosts mature lymphocytes in the initiation of B- and T-cell responses. Bone marrow-resident mature B lymphocytes participate in immune responses. A large proportion of mature B lymphocytes occupies a perisinusoidal niche in the bone marrow. In opposition to B lymphocytes in the follicular niche, perisinusoidal

---

[21] Antibodies are producted in the spleen, lymph nodes, lamina propria of the gut, bone marrow, and peritoneal and pleural cavities.

B lymphocytes are activated by microbes in a T-cell-independent manner to generate specific IgMs [100].

Bone marrow-resident dendritic cells are organized into perivascular clusters that are seeded with mature B and T lymphocytes [101]. These dendritic cells promote the survival of recirculating B lymphocytes that are located in the bone marrow, as they produce macrophage migration-inhibitory factor (MIF).[22]

2.2.3.3 Monocytes and Macrophages

Monocytes operate in phagocytosis and antigen presentation to T lymphocytes. Monocytes leave blood to become tissue-resident macrophages. Monocytes are able to replace their lysosomes and have a longer life than that of neutrophils.

Cytokines and chemokines released during infection or in tumor environment recruit macrophages. Upon activation by microbial products or interferon-γ, macrophages can produce cytokines and cytotoxic compounds to destroy invading pathogens or tumor cells. Macrophages are able to digest by-products of dead cells and bacteria.

Macrophages also target cells infected by fungi or parasites. Macrophages hence contain numerous lysosomes for degrading ingested material. Besides, macrophages can accumulate indigestible residue. Macrophages operate as phagocytes[23] in innate, non-specific defense and cell-mediated, specific immunity.

Macrophages are mobile scavengers within connective tissue. They continuously sample their environment by endo- and phagocytosis. They sense invading pathogens using pattern recognition receptors that bind common microbial structures. When tissue damage requires reinforcements, monocytes attracted to a damaged site by chemoattractants cross the vascular endothelium to differentiate into macrophages, thereby increasing the tissue-resident macrophage population. Macrophages are able to cross epithelia to scavenge foreign material on exposed surfaces such as the alveolar wetted wall.

---

[22] Macrophage migration inhibitory factor activates the receptor complex made of a widely expressed, type-2 transmembrane, homotrimeric protein, the class-2 histocompatibility antigen invariant chain (CD74), which serves as the MIF receptor (MIFR), and the cytosolic Tyr kinase activator CD44, which is the indispensable MIFR coactivator, also called homing cell adhesion molecule (HCAM) and epican. Agent MIF initiates the phosphorylation of ERK1 and ERK2 kinases. It also primes the activation of phospholipase-A2, which targets arachidonic acid. Substance MIF triggers the anti-apoptotic phosphatidylinositol 3-kinase–protein kinase-B pathway (Vol. 2 – Chap. 2. Cell Growth and Proliferation). In addition, MIF can associate with soluble CD44.

[23] Phagocytosis include several steps. A phagosome is formed upon ingesting a foreign material (necrotic debris, dust, pathogens, etc.), that fuses with lysosomes to create a phagolysosome in which enzymes act. Waste material is expelled or assimilated.

Table 2.5. Signaling and tasks primed by lipopolysaccharide-activated Toll-like receptors as well as kinases stimulated during TLR4 signaling in response to lipopolysaccharides in activated macrophages (Source: [102]; CamK: calmodulin-dependent kinase; EGFR: epidermal growth factor receptor; MAPK: mitogen-activated protein kinase; NFκB: nuclear factor-κB; PK: protein kinase; TGF: transforming growth factor; TNF: tumor-necrosis factor; TOR: target of rapamycin).

Major mediators	Programmed tasks
Adipocytokine	Actin cytoskeleton dynamics
Caspase	Cell communication
(cleavage of cytoskeletal proteins)	Cell proliferation
EGFR1	Kinase activity
Insulin	(ATMK, ATRK, Aurora,
NFκB–MAPK	CamK2, ChK1, CK1,
PI3K–PKB–TOR	ERK, GSK3, NEK6,
Rho	PKA, PKB, PKD, PLK)
Ca^{++}–CamK2–NFAT	
TGFβ	
TNFα	

Macrophages secrete many substances, such as monokines, enzymes, and complement proteins. Conversely, they have lymphokine receptors for activation. Macrophages not only serve as scavengers and sources of secreted agents, but also as antigen-presenting cells after digesting a pathogen that target helper T lymphocytes. Macrophages integrate antigens into their plasma membrane for exhibition attached to a class-2 MHC molecule. Activated macrophages have a lifespan of a few days.

Recognition of microbial signals by Toll-like receptors (TLR; Vol. 3 – Chap. 11. Receptors of the Immune System) causes reprogramming of gene expression in macrophages (Table 2.5). To avoid excessive inflammation, macrophage activation is controlled by immunosuppressive regulators such as interleukin-10. The activation program is then downregulated (synthesis of inhibitor of nuclear factor-κB and dephosphorylation [inactivation] of mitogen-activated protein kinases by dual specificity phosphatase, etc.)

Numerous macrophages are assigned to strategic regions with high probability of occurrence of microbial invasion or dust accumulation (Table 2.6). Moreover, in certain organs, macrophages can operate as paracrine regulators of tissue-specific cells (e.g., testicular macrophages that interact with Leydig cells).

Despite their cytotoxic potential, macrophages can fail in their struggle against infections or tumors, because certain pathogens or tumors modify macrophage activity characterized by production of interleukin-10 and tumor-necrosis factor rather than interleukin-12 [103]. In addition, antimicrobial and antitumoral activity of macrophages increases when the inhibitor of nuclear

Table 2.6. Macrophages allocated to specific tissues.

Tissue	Macrophage name
Pulmonary alveolus	Alveolar macrophage
Connective tissue	Histiocyte
Nervous tissue	Microglial cell
Bone	Osteoclast
Liver	Küpffer cell
Spleen	Sinusoidal lining cell
Kidney	Mesangial cell

factor-κB kinase IKKβ is inhibited [104, 105]. Inactivation of Nuclear factor-κB reduces production of IL10 and TNF by macrophages, whereas it augments expression of nitric oxide synthase NOS2 and synthesis of IL12. Interleukin-12 mediates recruitment of natural killer cells.

Transcription factors C/EBPα[24] and SPI1[25] cause a transdifferentiation of committed lymphoid precursors into macrophages. Fibroblasts with activation of macrophage-associated genes and extinction of fibroblast-associated genes by combined action of C/EBPα and SPI1, SPI1 alone, or C/EBPβ, acquire a macrophage-like phenotype [106]. The conversion is primarily induced by SPI1, as C/EBPα serves as a modulator of macrophage-specific gene expression.

2.2.4 Dendritic Cells

Dendritic cells of the immune system can have branched projections (dendrites). They derived from hemopoietic progenitors in the bone marrow. Dendritic cells check tissues for possible microbial infection. When an invader is

---

[24] CCAAT/enhancer-binding protein (C/EBP) binds to target DNA regulatory segments. The encoded proteins are crucial for terminal differentiation and maturation of committed hematopoietic progenitors.

[25] Spleen focus forming virus proviral integration proto-oncogene Spi1 encodes the transcription factor SPI1, a.k.a. PU.1, that activates gene expression during myeloid and B-lymphoid cell development. This protein binds to a purine-rich sequence (PU-box) near the promoters of target genes, and regulates their expression in coordination with other transcription factors and cofactors. SPI1 5′UTR regulatory element is an RNA element in the 5′UTR of SPI1 mRNA that inhibits the translation SPI1 transcripts. Several transcription factors have been isolated that regulate differentiation of committed progenitors into given cell lineages. Factor SPI1 commits cells to the myeloid lineage, whereas GATA1 supports erythropoietic and megakaryocytic differentiation. Transcription factor SPI1 is specific to macrophages, dendritic cells, and B cells. Therefore, SPI1 interacts with transcription factors in different cell lineages according to the context. Factor SPI1 is a primary regulator of macrophage genes in mesenchymal cells. It can coopt C/EBPα as a hematopoietic cofactor. In addition, SPI1 upregulates the CSF1R gene.

Table 2.7. Types of dendritic cells (Ifn: interferon; IL: interleukin; TLR: Toll-like receptor; Source: Wikipedia). Dendritic cells express T-cell marker CD4 or myeloid-cell markers class-2 MHC molecules CD36 (a.k.a. thrombospondin receptor, fatty acid translocase, glycoprotein GP3b and GP4, platelet collagen receptor, and scavenger receptor ScaRb3) and CD68 (a.k.a. macrosialin, ScaRd1 and GP110).

Name	Function and secretion	TLR
Myeloid (mDC)	IL12, IfnI mDC1 stimulates T cells mDC2 targets wound infection	TLR2, TLR4
Plasmacytoid (pDC)	Ifnα	TLR7, TLR9

found, innate immune cells are alerted and local dendritic cells capture microbial antigens at the site of infection and transport them to the draining lymph nodes. Dendritic cells process antigens for presentation to other immune cells.

Activated dendritic cells have a lifespan of a few days. These antigen-presenting cells are mainly present in small quantities in tissues in contact with the body's environment, such as skin (Langerhans cells) and the inner tunica of the respiratory and digestive tract. Monocytes can give rise to immature dendritic cells.

Immature dendritic cells are conveyed by blood and permanently search for pathogens in the surrounding environment, using pattern recognition receptors. Immature dendritic cells phagocytoze and degrade pathogens and up-regulate plasmalemmal receptors for T-cell activation. Once activated, they migrate to the lymphoid tissues, where they interact with helper T, killer, and B lymphocytes to prime the adaptive immune response. In particular, dendritic cells stimulated by antigens or allergens favor the differentiation of helper T cells into T_{H1}, T_{H2}, or T_{H17} cells according to cytokine types.[26]

Any antigen-presenting cell (dendritic cell, macrophage, and B lymphocyte) activates a resting helper T lymphocyte when the presented antigen matches the target that specifies the helper T-cell function.[27] The dendritic cells interact with other cells either by direct contact via complementary plasmalemmal receptors or cytokines (Table 2.7).

Categories of dendritic cells have been identified in humans: CD11+, myeloid and CD11−, plasmacytoid dendritic cells. *Plasmacytoid dendritic*

---

[26] A SCFR–PI3K–IL6 signaling pathway promotes differentiation into T_{H2} and T_{H17} cells from dendritic cells exposed to allergens, which express receptor Tyr kinase SCFR [107]. SCFR− dendritic cells elicit T_{H1}-cell differentiation, whereas interleukin-6 prevents T_{H1}-cell differentiation. Stimulated SCFR by its ligand, the stem cell factor, induces IL6 production via phosphoinositide 3-kinase.

[27] Macrophages and B lymphocytes only activate memory T lymphocytes, whereas dendritic cells, the most potent antigen-presenting cells, activate both memory and naive T lymphocytes.

cells (pDC) are relatively poor T-cell stimulators and less efficient for antigen uptake. They activate natural killer cells, macrophages, and CD11c+ dendritic cells. Plasmacytoid dendritic cells that express Toll-like receptors TLR7 and TLR9 sense viral nucleic acids within TLR-containing early endosomes and produce large amounts of interferons-I upon viral infection [108].[28] In addition, plasmacytoid dendritic cells can detect self-nucleic acids in the extracellular medium and respond when cell death occurs in normal tissues during tissue renewal and repair.[29]

2.2.5 Mastocytes

Mastocytes, or mast cells, act as secretory alarm cells of connective tissues (skin and mucosa)[30] that release chemical signals, which diffuse through the extracellular matrix to trigger inflammation. Mastocytes are small, scattered cells with secretory vesicles. These granules contain histamine, heparin, and various other mediators.

Mastocyte promotes vasodilation and tissue reaction such as inflammation in response to IgE-dependent antigens and IgE-dependent immediate hypersensitivity (IgE-associated allergy), as well as in some innate and adaptive immune responses supposed to be independent of immunoglobulin IgE [109].

Mastocytes operate as effectors during innate and adaptive immune responses, as they kill pathogens, degrade endo- or exogenous toxics, and regulate the function of immune cells (dendritic cells, B and T lymphocytes, monocytes, macrophages, and granulocytes) and other cell types (fibroblasts

---

[28] The innate immune system senses viral infection via nucleic acid recognition owing to: (1) endosomal Toll-like receptor TLR7, -8, and -9 and myeloid differentiation primary-response gene MyD88; (2) endosomal TLR3 and Toll–interleukin-1 receptor domain-containing adaptor inducing interferon-β (TRIF); and (3) retinoic acid-inducible gene RIG1. The TLR7/9–MyD88 pathway is exclusively used by plasmacytoid dendritic cells, whereas the TLR8–MyD88 and TLR3–TRIF pathways are primed by myeloid dendritic cells. The RIG1 pathway serves to detect viral RNA in myeloid cells and all non-hematopoietic cells. IfnIs produced by pDCs trigger generation of interleukins IL12, IL15, IL18, and IL23 by mDCs, and stimulate monocyte differentiation into dendritic cells [108]. They also increase the ability of mDCs to present antigens to CD8+ T cells and induce the differentiation of naive T cells into T helper-1 cells. Interferons-I derived from plasmacytoid dendritic cells also stimulate NK cells and favor, in conjunction with IL6, B-cell differentiation into mature antibody-secreting plasmocytes.

[29] This response is usually prevented by: (1) rapid clearance of dying cells by macrophages; (2) degradation of self-nucleic acids by DNases and RNases; (3) compartmentation of TLR7 and TLR9 within the endosomes; and (4) receptor-mediated repression of TLR7- or TLR9-induced interferon responses in plasmacytoid dendritic cells [108].

[30] Mucosa is a tissue lining covered by an epithelium. It lines body cavities exposed to the body's environment. It secretes mucus in certain anatomical tracts.

and neural, epithelial, smooth muscle, and vascular endothelial cells) by secreting various products.[31] However, mastocytes can have anti-inflammatory and immunosuppressive functions.

---

[31] Mastocyte increases the vascular permeability via histamine, provokes bronchoconstriction via leukotriene-C4, and elicits collagen synthesis by fibroblasts via tryptase.

3

Cells of the Ventilatory Tract

The ventilatory apparatus optimizes the surface area for gas exchange between blood and air in the prescribed thoracic volume. The ventilatory system undergoes strong changes in configuration during breathing determined by variations in shape of the thoracic cage caused by the contraction–relaxation cycle of respiratory muscles. These variations in anatomical configuration must maximize changes in lung volume, but minimize changes in flow resistance.

In addition to breathing and related functions (air heating and humidification), the respiratory tract has non-respiratory functions, as it contributes to phonation, acid–base balance,[1] thermoregulation (Sect. 1.12), and immune defense,[2] The ventilatory apparatus have specific metabolic and endocrine functions. It synthesizes surfactant. In endothelial cells of lung-perfusion vessels, angiotensin-converting enzyme predominantly convert the inactive decapeptide angiotensin-1 into the aldosterone-stimulating octapeptide angiotensin-2.

The walls of conducting airways consist of 3 main layers: (1) a *mucosa* composed of the respiratory epithelium and a connective tissue lamina; (2) a *submucosa* with smooth muscle cells; and (3) an *inner tunica* with cartilage in large bronchi. In general, the respiratory epithelium becomes thinner peripherally.

---

[1] The body's respiration influences the blood pH via the blood content of carbon dioxide.

[2] Immunoglobulin-A is secreted by the respiratory epithelium. Moreover, released mucus contains antimicrobial substances, such as the glycoproteic bacteriocide and fungicide lactoferrin, or lactotransferrin, the glycoside hydrolase lysozyme (a.k.a. muramidase and ^{N}acetylmuramide glycanhydrolase), the antibacterial lactoperoxidase, and dual oxidase-1 (DuOx1) and -2 (DuOx2), or thyroid oxidase, which generate reactive oxygen species. In addition, airway epithelial cells produce collectins, such as surfactant proteins-A (SftPa1 and SftPa2) and -D (SftPd), defensins, proteases, and reactive nitrogen species. Ciliary motion removes particles and pathogens entrapped in mucus toward the pharynx.

Many types of cells are related to the ventilatory system. A first set comprises cells of the wall structure of: (1) respiratory conduits, such as epithelial cells, cells of submucosal glands and airway cartilages, smooth muscle cells, and connective tissue cells; and (2) blood and lymph vessels, such as endothelial cells, fibroblasts, and vascular smooth muscle cells in the macrocirculation or pericytes in the microcirculation.

Flowing blood that is strongly connected to inhaled and exhaled air for gas exchange conveys circulating cells (erythrocytes, leukocytes, and thrombocytes). Immune cells require blood circulation for scouting tissues in search for possible foreign elements. Nervous cells control the activity of the ventilatory system according to inputs provided by sensory cells of the respiratory apparatus and vasculature. Even after the final stage of differentiation, cells keep a potential of flexible fate because they can be reprogrammed by a more or less small set of transcription factors.

Many tissular characteristics are obtained by crosstalk (message exchange) between different cell types in a given tissue. The abluminal edge of ventilatory epithelial cells contacts the extracellular matrix that is required for the assembly of differentiated cells into a functional tissue. The extracellular matrix could also control the fate of stem and progenitor cells. The luminal edge is a wetted surface that can transduce received mechanical, physical, and chemical stimuli associated with the air stream.

In humans, 4 cell types compose the respiratory epithelium of proximal ducts of the tracheobronchial tree: ciliated, mucus-secreting goblet, indeterminate, and basal cells. In terminal bronchioles, an additional secretory cell type is interspersed between ciliated cells: the Clara cell. Ciliated cells are observed in the respiratory tract down to respiratory bronchioles.

Pulmonary peptide-producing neuroendocrine cells with dense-cored granules are identified isolated or grouped, often innervated. Neuroendocrine cells are the first cell type to differentiate and mature within primitive airway epithelium (from week 8 of gestation). Submucosal nerves and ganglia are found from week 10 of gestation [111]. At the end of gestation, the population of neuroendocrine cells and neuroepithelial bodies increases in peripheral airways.

Leukocytes are observed in the basal zone of the epithelium in 16-wk-old human fetuses [111]. Mastocytes are found in normal adult airway epithelium, but have not been described in the fetus. Because air is contaminated with infectious agents and more or less toxic particles, the respiratory epithelium is associated with specialized cells of the immune system. Protective functions include the epithelium barrier and mucociliary clearance, cough, resident phagocytes, and secretion of antimicrobial peptides and proteins (lysozyme, lactoferrin, defensins, and cathelicidin) in airway surface fluid [112].[3] The

---

[3] Peptides and proteins of the airway surface fluid can act as microbicides as well as chemotactic factors for cells of the innate and adaptive immune systems. Human defensin-β1 (Defβ1), or β-defensin BD1, is constitutively present, but

Table 3.1. Airway epithelial cells at the interface between flowing air in the airway lumen and the remaining part of the airway wall. Goblet and ciliated cells as well as macrophages collaborate to clean respiratory conduits. The number of basal cells is related to the height of the epithelium. Basal and parabasal cells operate as progenitor cells. Intermediate (parabasal) cells that form a poorly defined layer located just above the basal cells are undifferentiated cells (progenitors) that can transform into any epithelial cell type. They are observed in all airways. Brush cells that are characterized by a brush border of microvilli may resorb periciliary fluid. Clara cells are supposed to be stem cells in the bronchiolar epithelium.

Conduction zone	Respiratory zone
Ciliated cell	Pneumocyte 1
Goblet cell	Pneumocyte 2
Brush cell	Alveolar macrophage
Clara cell	
Basal cell	
Intermediate cell	

respiratory epithelium can be penetrated by lymphocytes and other cells of the immune system. Alveolar septa contains alveolar macrophages that can amplify innate immune responses.

3.1 Respiratory Epithelia

The respiratory epithelium is composed of 3 main types of epithelial cells: goblet cells, ciliated cells, and basal cells (Table 3.1). Basal cells are close to the basement membrane, between the ciliated and goblet cells. They do not extend up to the surface of the cell monolayer. Consequently, the respiratory epithelium has a pseudostratified structure.

As the epithelium structure and cell types that constitute the respiratory epithelium vary along the respiratory tract, the respiratory epithelium cannot be considered unique. The cell composition of airway compartment (set of bronchial generations) is given in Tables 3.2 and 3.3).

The pseudostratified columnar ciliated epithelium lines most of the respiratory conduits. A stratified squamous epithelium covers the epiglottis (except a part of the posteroinferior surface), aryepiglottic folds, and true vocal folds. A stratified columnar ciliated epithelium resides in transitional regions between stratified squamous and ciliated pseudostratified columnar epithelia.

Respiratory epithelia are composed of columnar cells that display a strong apicobasal polarity over most of the surface of the respiratory tract. The

---

Defβ4 (a.k.a. Defβ2 and BD2) and Defβ103, or BD3, are inducible in response to bacterial recognition by Toll-like receptors (Vol. 3 – Chap. 11. Receptors of the Immune System) [113].

Table 3.2. Mean number of airway epithelial cells per unit length (mm) of basement membrane and assessed percentage of basal and parabasal cells among epithelial cells in various territories of the tacheobronchail tree (Source: [114]).

Airway bore (mm)	≥ 4	2–4	0.5–2	≤ 0.5
Total number of epithelial cells	338	271	223	173
Basal cells	31 ± 7	30 ± 7	23 ± 7	6 ± 4
Parabasal cells	7 ± 4	3 ± 2	$1 \pm 2!$	0

Table 3.3. Mean number of airway epithelial cells per unit length (mm) of basement membrane and assessed percentage of goblet and Clara cells among epithelial cells (Source: [115]).

Airway type	Bronchi	Bronchioles	Terminal bronchioles	Respiratory bronchioles
Internal bore (mm)	2–17	0.5–1.8	0.4–0.9	0.2–0.4
Cell number	349 ± 34	284 ± 27	214 ± 20	184 ± 5
Goblet cells	11 ± 6	10 ± 5	$2 \pm 3!$	0
Clara cells	0	$0.4 \pm 1!$	11 ± 3	22 ± 5

apical surface of epithelial cells at the luminal interface regulate the exchange of materials (absorption and secretion). The lateral surfaces of epithelial cells contact adjacent cells by specialized junctions (desmosomes and adherens and tight junctions; Sect. 7.7).

The respiratory epithelium is situated more or less near blood circulation. Functional regions for air conditioning and gas exchange have an epithelium particularly close to blood vessel endothelium.

Epithelial cells can contain keratin filaments. Different cell types express distinct pairs of keratin proteins. In basal cells that are firmly attached to the basement membrane, keratin-5 and -14 can attach to hemidesmosomes.

The percentage of ciliated cells in large and intermediate sized ducts of the tracheobronchial tree (generations G0–G3) of dogs decays from 22 and 25% in the trachea and main bronchi to 8 and 3% in lobar and segmental bronchi, respectively [116].[4] In ciliated cells, cilia become shorter and more sparse in distal respiratory ducts than those in large airways.

---

[4] Mucus velocity is correlated to the percentage of ciliated cells. The fastest mean mucous velocity is detected in the trachea (80 μm/s). Mucus velocity is 38, 55, and 66% lower in main, lobar, and segmental bronchi than that in the trachea [116].

3.1.1 Formation of the Respiratory Tract

The formation of the lower respiratory conduits begins in embryonic week 4. The laryngotracheal tube arises from the foregut[5] The laryngotracheal tube divides distally to form 2 lung buds. The lung bud and the tracheobronchial tree have an endodermal origin. Most of the other tissues of the lower respiratory passages derives from the splanchnic mesoderm.[6]

The lung development can be divided into 4 periods: (1) the *glandular period* (embryonic and fetal weeks 7–17),[7] during which bronchi grow and branch;[8] (2) the *canalicular period* (until fetal week 25 to 28), during which bronchi and bronchioles expand and branch, primitive alveoli appear, and lung tissue is vascularized; (3) the *saccular period* (until ∼ fetal week 36), or terminal sac period, during which primary alveoli and its associated vascularization develop; (4) the *alveolar period* that begins shortly before birth, but the first mature alveoli appear only after birth.

The capillary network develop between terminal sacs, from which alveoli mature. Alveoli continue to be formed during childhood up to year 8. Alveolar maturation and growth continue for another decade, but the alveolus number do not change.

3.1.2 Differentiation of Human Respiratory Epithelium

The respiratory tract is a tubular network arising by branching morphogenesis in which cells bud from an epithelium and organize into a tube, being guided by growth factors and other signaling molecules. Two types of epithelial cells are implicated in budding: (1) tip cells which lead branch outgrowth, and (2) trailing cells which form a tube. The cells with the highest FGF receptor activity take the lead positions. Budding cells are functionally specialized [117]. Notch activity prevents other cells from driving the lead. Trailing cells, thereby, do not require FGF receptors. Nonetheless, they follow the lead cell according to FGF attraction.

Maturation of the respiratory epithelium starts from the proximal airways and progresses distally. The conducting and respiratory segments of the

---

[5] The foregut is the anterior part of the alimentary canal, from the mouth to the duodenum at the entrance of the bile duct. The foregut can be divided into 3 regions. The first segment of the foregut creates the primitive pharynx; the second part the lung bud and oesophagus; the third compartment the stomach and related structures. A laryngotracheal groove appears on the ventral edge of the foregut and creates a respiratory diverticulum that bifurcates into right and left buds.

[6] The branching pattern is regulated by the surrounding mesoderm. The mesoderm surrounding generated trunks prevents branching, whereas the mesoderm surrounding the tips fosters branching. Lobar and segmental bronchi are observed during embryonic week 5 and 6, respectively.

[7] Generated bronchi resemble glandular acini.

[8] The main generations of the bronchial tree are formed at embryonic week 6.

Table 3.4. Timing of differentiation of human respiratory epithelial cells (Sources: [118, 119]).

Gestational age (week)	Cellular event
4	Primitive respiratory epithelial cells
8	Neurosecretory cells
10–12	Presecretory and preciliated cells
12	Mucous glands
14	Neuroepithelial bodies, ciliated cells, prebasal cells
16	Goblet, serous, and basal cells, prepneumocytes-2, preClara cells
24	Pneumocytes-1 and -2
24–26	Clara cells

respiratory tract contain at least 11 epithelial cell types. The approximate timing of differentiation of human respiratory epithelial cells is given in Table 3.4.

3.1.2.1 Development of Ciliated Cells

Cilia are seen in the trachea and main bronchi during gestational week 10. By fetal week 13, cilia can be detected in the distal part of the developing bronchial tree. By week 24, the ciliated epithelium with 2 or more cell layers gradually transforms into a single layer of columnar and then cuboidal cells still covered with cilia [120]. In the most peripheral parts, epithelial cells are flattened and free of cilia; capillaries penetrate between these cells and reach the surface of bronchiolar wall. By term, ciliated cells reach the terminal bronchioles.

3.1.2.2 Development of Intrapulmonary Mucus-Secreting Cells

Mucous glands first appear in month 4 of fetal life. Small buds emerge from basal cells of bronchial epithelium and penetrate into the subepithelial layer. They are usually generated from bottom or lateral wall of epithelial creases.

Goblet Cells

Goblet cells and glands are first detectable in the tracheal epithelium of 13-week fetuses, when both buds and primitive tubular glands are observed [120]. At week 14, the number of tubular glands in proximal bronchi is the greatest; many glands show lateral buds and a lumen. Glands can be identified at week 14 in intrasegmental bronchi as well as following generations, but with a smaller density.

Between week 12 and 24, goblet cells are found only in the most proximal intrasegmental generations, hence spreading much less distally than glands [120]. Goblet cells are not detected downstream from large bronchi at week 22 of gestation [111]. At term, goblet cells are absent in bronchioles. The number of goblet cells peaks at midgestation (30–35% of lining cells) and decays at end of gestation. With increasing gestational age they extend distally down to bronchioles. The proportion of goblet with respect to ciliated cells increases in the first 4 weeks after birth.

Submucosal Glands

Submucosal glands appear in the human trachea at week 10 of gestation [111]. They then develop progressively. They are indeed detected in the carina one week later and in bronchi at month 4 of fetal life. In extrapulmonary bronchi, gland formation rate reaches a peak during the period of week 12 to 14 of fetal life. Gland formation decays afterward and terminates during the middle of gestational week 23.

3.1.3 Development of Airway Smooth Muscle Cells

Airway smooth muscle cells are identified in the trachea and main and lobar bronchi during weeks 6 to 8 [111]. Smooth muscle cells then develop in walls of segmental bronchi and terminal and respiratory bronchioles. Bronchial smooth muscles have a mature structure at birth. However, the amount of bronchial smooth muscle increases immediately after birth.

3.1.3.1 Development of Pulmonary Neuroendocrine Cells

In developing human lung, the first neuroendocrine cells appear at gestational week 8, when all other epithelial cells are still undifferentiated [121]. During terminal stages of development, neuroendocrine cells are observed in small peripheral airways and primitive saccules. By the end of the glandular period, single and groups of neuroendocrine cells reside within the entire length of primitive bronchial epithelium. In postnatal lungs, both isolated neuroendocrine cells and neuroepithelial bodies are concentrated in small peripheral airways.

3.1.4 Cell Turnover

The respiratory epithelium is in continuous renewal; lost cells are replaced by the controlled proliferation and differentiation of stem cells. Because airways consist of a heterogeneous cell population, interactions between the different cell types control morphogenesis, maintain respiratory tissues, and regulate repair after injury. In addition, the development of the respiratory epithelium depends on the extracellular matrix, particularly on distribution and deposition of elastin and collagen.

3.1.5 Nitric Oxide

Nitric oxide, or nitrogen monoxide (Vols. 3 – Chap. 6. Receptors and 4 – Chap. 9. Other Major Signaling Mediators), is a gasous signaling molecule involved in many physiological and pathological processes. This highly reactive radical gas has a lifetime of a few seconds. It is involved in the production of reactive nitrogen intermediates. This messenger diffuses easily across membranes. Among its actions, it causes vasodilatation. In addition, it participates in neurotransmission, like other gasotransmitters, such as hydrogen sulfide and carbon monoxide.

3.1.5.1 Aggresome and NOS2 Inhibition

Aggresomes form in response to stress. Aggresomes that sequester inducible nitric oxide synthase (NOS2) impede nitric oxide production. In particular, human bronchial epithelial cells use aggresomes to inactivate cytokine-inducible NOS2 that, unlike other constitutive NOS isoforms (NOS1 and NOS3), can synthesize relatively large amount of nitric oxide [122]. Inflammation of the distal respiratory tract can be detected by exhaled nitric oxide concentration.

3.1.5.2 Nitric Oxide and Mitochondrial Function

Nitric oxide regulates cellular respiration and mitochondrial genesis. It binds to complexes-1, -2, and -4 of the mitochondrial respiratory chain (Sect. 4.7.5) and impedes their activities. On the other hand, the NO–cGMP-dependent mitochondrial genesis is associated with augmented oxygen consumption and ATP content. Attenuated mitochondrial function, related to a decreased number of mitochondria and/or impaired mitochondrial activity, labels pulmonary hypertension.

Nitric oxide also operates as a vasodilator (Vol. 5 – Chaps. 8. Smooth Muscle Cells and 9. Endothelium), once released from pulmonary artery endothelial cells. In idiopathic pulmonary arterial hypertension, NO production is reduced (but not in pulmonary arterial hypertension with known etiology). Mitochondria have a lower oxygen consumption. Energy production then depends mainly on glycolysis [123].[9]

---

[9] Aerobic glycolysis that depends less on oxygen, yields only 2 ATP per glucose, whereas the complete oxidation produces 38 ATP per glucose. In addition, aerobic glycolysis minimizes exposure to reactive oxygen species, as it diminishes oxidative metabolism.

3.2 Cell Growth and Differentiation of the Respiratory Epithelium

Respiratory epithelium is constantly, very slowly renewed and can be efficiently repaired after injury owing to stem and diversely committed progenitor cells within the respiratory mucosa as well as dedifferentiation of epithelial cells [124]. Cells are lost and replaced by controlled proliferation and differentiation of stem cells.

Interactions between different cell types of airway walls regulate: (1) morphogenesis during the lung development; (2) maintenance of tissues; and (3) repair. Intercellular interactions involve: (1) constituents of desmosomes and tight and gap junctions, as well as other cell adhesion constituents such as integrins; (2) intercellular communication particularly via auto- and paracrine factors, such as growth factors, retinoids, and protein kinase-C activators.

Differentiation and growth of epithelial cells of the tracheobronchial tree is regulated by several growth factors (Vol. 2 – Chap. 3. Growth Factors), such as transforming growth factors-α and -β, insulin-like growth factor-1, and epidermal and fibroblast growth factors (FGF7 and FGF10), that act by an auto- or paracrine mechanisms. Growth factors TGFα and TGFβ synthesized by tracheobronchial epithelial cells operate as autocrine growth factors that influence cell growth. Growth factors TGFα and -β, IGF1, and FGF manufactured by fibroblasts or macrophages control the proliferation of tracheobronchial epithelial cells by a paracrine mechanism.

Epidermal growth factor and EGF-like molecules such as transforming growth factor-α intervene during lung development and repair [125]. Tracheobronchial epithelial cells are able to fabricate and secrete TGFα. TGFα is synthesized as a transmembrane glycoprotein precursor (proTGFα). Upon cleavage of its extracellular domain, this precursor yields TGFα that is secreted in various forms. Factor EGF is also produced as a membrane-bound glycoprotein precursor (preproEGF). Both EGF and TGFα bind to EGFR, a receptor Tyr kinase. Signaling effectors include phospholipase-Cγ1, kinase Raf (MAP3K), and phosphoinositide 3-kinase. Phosphorylated phospholipase-Cγ1 hydrolyzes phosphatidylinositol (4,5)-bisphosphate and generates second messengers inositol (1,4,5)-trisphosphate and diacylglycerol. TGFα not only controls cell proliferation, but also cell differentiation.

Insulin-like growth factors IGF1 and IGF2 promote cell proliferation and differentiation. Isoform IGF1 binds to a specific, glycosylated, tetrameric receptor Tyr kinase (IGF1R). Receptor IGF2R is the mannose-6-phosphate receptor. It does not have protein kinase activity. Both IGF1 and IGF2 are detected in fetal mesenchymal cells of lungs. Expression of IGF1 is predominantly postnatal, whereas that of IGF2 is mainly prenatal [126].

Placental lactogen, or chorionic somatomammotropin, stimulates both IGF1 and IGF2 production in fetuses, whereas after birth, growth hormone is a major stimulator that provokes IGF1 synthesis. Tracheobronchial epithelial cells do not synthesize IGF1. Lung fibroblasts and macrophages produce and release IGF1, which promotes the proliferation of these cells (autocrine activity) as well as growth of tracheobronchial epithelial cells (paracrine function). In addition, IGF1 acts as a chemoattractant for tracheobronchial epithelial cells [125].

Keratinocyte growth factors, i.e., fibroblast growth factors FGF7 and FGF10, are produced by human embryonic lung fibroblasts. They are secreted by stromal cells to stimulate the growth of tracheobronchial epithelial cells, but not fibroblasts. Receptors of FGFs are targeted by fibroblast growth factors FGF1 and FGF2 with lower affinity than FGF7 and FGF10.

Transforming growth factor-β1 and -2 are synthesized by tracheobronchial epithelial cells. Mature, homodimeric TGFβ1 and -2 originate from inactive, glycoproteic precursors owing to proteolytic cleavage by proteases such as plasmin. Lung fibroblasts and macrophages also manufacture and secrete TGFβ. Both TGFβ1 and -2 bind to high affinity TGFβ type-1, -2, and -3 receptors (TβR1–TβR3; Vols. 2 – Chap. 3. Growth Factors and 3 – Chap. 8. Receptor Kinases) that are expressed by lung fibroblasts and tracheobronchial epithelial cells [125]. However, TGFβ2 binds to TβR1 and TβR2 receptors with a lower affinity than TGFβ1. Type-3 TGF receptor is a proteoglycan that contains heparan and chondroitin sulfates and can then serve as a reservoir. At high cell density, TGFβ1 and -2 reversibly preclude cell growth. In addition, TGFβ1 stimulates differentiation into goblet-like cells. TGFβ1 also promotes expression of transglutaminase-2, fibronectin, and collagen-4 by tracheobronchial epithelial cells, as well as proteoglycans, fibronectin and collagen-1 and -2 in fibroblasts [125]. Therefore, TGFβ acts as a regulator of extracellular matrix components, as it also controls the activity of extracellular matrix enzymes. It actually impedes the production of proteinases, such as collagenase and plasminogen activator, and favors that of proteinase inhibitors, such as plasminogen activator inhibitor and metalloproteinase inhibitor.

Retinoids[10] belong to the set of regulators of cell proliferation and differentiation.[11] Retinoids prevents the activity of transglutaminase-1 and cholesterol sulfotransferase to reduce level of cholesterol sulfate in human tracheobronchial epithelial cells. They also hamper synthesis of keratin and crosslinked envelope precursor involucrin [125].[12] On the other hand, retinoids elicit production and secretion of mucins and mucin-like glycoproteins. Retinoic acid targets cytosolic retinoic acid-binding proteins RABP1 and RABP2 and nuclear retinoic acid receptors RARα, RARβ, and RARγ that regulate gene expression in tracheobronchial epithelial cells and lung fibroblasts.

---

[10] Retinoids are related to vitamin-A. They include retinol (vitamin-A), retinal (vitamin-A aldehyde), and retinoic acid (oxidized vitamin-A), as well as pharmaceutical agents.

[11] Vitamin-A-deficient diet leads to the replacement of a mucociliary epithelium by a stratified squamous epithelium [127]. Squamous metaplasia is reversible.

[12] Involucrin is a component of the keratinocyte cytoplasm crosslinked to membrane proteins by transglutaminase. Involucrin yields a structural support to the cell.

Retinoic acid-bound RARs interact with a retinoic acid-specific response element (RARE) in promoters of responsive genes to activate gene transcription (Chap. 5).

Retinoic acid receptors (Vol. 3 – Chap. 6. Receptors) can form heterodimeric complexes among themselves or with other members of the nuclear receptor class that are targeted by steroid and thyroid hormones. These complexes can interact with other nuclear proteins to regulate different sets of genes. Thyroid hormones and glucocorticoid receptors can then participate in the regulation of the tracheobronchial epithelium.

3.3 Airway Stem and Progenitor Cells

Stem cells undergo cell division to generate additional stem cells (self-renewal) or, upon differentiation commitment, to create a given progeny, i.e., various types of progenitor cells for tissue formation, maintenance, and repair.[13] Stem cells most often stay in relative quiescence, and reside in specialized microenvironments, the so-called niches.

The lung is composed of at least 40 differentiated cell types. Tracheal, bronchial, and bronchiolar epithelia contain different populations of stem and progenitor cells that generate different cell lineages. Several cell types in adult tracheobronchial epithelium have proliferative capacity, such as Clara-like, mucous, serous, and basal cells, as well as neuroendocrine cells.

Multipotent progenitor cells of the bronchi include Clara cell secretory protein-expressing cells (CCSP+ cell),[14] and Krt14+ basal cells [128]. Pulmonary, neuroendocrine, CGRP+ (calcitonin gene-related peptide-expressing) cells, bronchiolar progenitor cells, lack the potential for multipotent differentiation. Neuroepithelial body microenvironment is a source of airway progenitor cells [129]. Both Clara cell secretory protein-expressing and calcitonin gene-related peptide-expressing cells proliferate in neuroepithelial bodies.

In the trachea and large bronchi, basal cells (Sect. 3.4.1) act as stem and progenitor cells for mucosal gland development and renewal of the respiratory epithelium. Relatively undifferentiated, basal cells (approximately 30% of the pseudostratified mucociliary respiratory epithelium) express the transcription factor P63 and cytokeratins-5 and -14 (Krt5 and Krt14) as well as 2 cell-surface markers (α_6 integrin and nerve growth factor receptor). During postnatal

---

[13] Stem and progenitor cells contribute in particular to epithelial maintenance in adulthood. During the normal turnover of the tracheobronchial epithelium, cells continuously slough off the respiratory epithelium into the airway lumen and are replaced by the proliferation and differentiation of stem and progenitor cells. Because the respiratory epithelial surface is constantly exposed to potential injury, stem and progenitor cells also quickly respond for epithelial repair.

[14] CCSP+ Cells can be subdivided according to naphthalene susceptibility into naphthalene-sensitive and naphthalene-resistant cytochrome P450– CCSP+ cells. The latter reside within neuroepithelial bodies or the bronchoalveolar duct junction.

growth as well as in adults, in mice as well as in humans, basal cells self-renew and generate differentiated ciliated and secretory cells [130]. Basal cells can generate presecretory and possibly ciliated cells. Presecretory cells that mature into goblet cells also have self-renewal potential and can differentiate into ciliated cells. Mucous cells (pre-goblet cells) that mature into goblet cells have self-renewal capacity and differentiate into ciliated cells.

In bronchioles, pre-Clara cells (Sect. 3.4.4.2) operate as progenitor cells of bronchiolar epithelia, as they also exhibit self-renewal capacity. They mature into fully secretory Clara cells and can differentiate into ciliated cells [125].

A subpopulation of Clara cells at the bronchoalveolar junction in neonatal mouse lungs that express pluripotency markers differentiate into alveolar pneumocyte 1- and 2-like cells [131].[15]

At closed ends of the tracheobronchial tree, pneumocytes 2, or alveolar epithelial type-2 cells, act as precursor cells of alveolar wall. They contribute to the repair of injured alveoli, as they differentiate into pneumocytes 1, or alveolar epithelial type-1 cells. They derive from embryonic stem cells.[16]

Certain types of precursor cells are only present in some animal species.[17] In the early stages of development, the tracheobronchial epithelium is lined by a columnar cell type that likely represents pluripotent precursors of basal, ciliated, mucous, serous, and neuroendocrine cells of the mature epithelium.

A continuous regeneration with cell proliferation, migration, and differentiation as well as intercellular adhesion, maintains the integrity of the respiratory epithelium. The mucosecretory phenotype is controlled by environmental cues. The airway epithelium responds to infectious agents, noxious pollutants, and allergens by increasing production of mucus, cytokines, and reactive oxygen and nitrogen species. Nonetheless, environmental cues can significantly alter epithelial functions, especially the mucociliary clearance, impair epithelial cell polarization, modify the relative contribution of cell types to the respiratory epithelium, as well as cause long-term airway hyperresponsiveness.

---

[15] They express stem cell markers, such as octamer-binding transcription factor-4 (Oct4), stage-specific embryonic antigen-1 (SSEA1), stem cell antigen-1 (SCA1), and Clara cell secretion protein (CCSP), but not stem cell factor receptor (SCFR or Kit), cell-surface glycoprotein cluster of differentiation CD34 that acts as a between-cell adhesion factor, and P63 transcription factor. They also produce angiotensin-converting enzyme.

[16] The precursor alveolar epithelial type-2 cell is endowed with morphological and biological features of the alveolar epithelial type-2 cell, such as lamellar body formation, synthesis of surfactant protein-A, -B, and -C, serpin-A1 (or α1-antitrypsin), cystic fibrosis transmembrane conductance receptor, as well as complement proteins C3 and C5 [132].

[17] For example, Clara-like cells exist in the tracheal epithelium of rabbits and hamsters, but not in that of rats, Rhesus macaque (Macaca mulatta), and sheep [125].

3.4 Cells of the Conduction Airways

The nose is the usual access to lungs at rest or during light exercise. The nasal cavity has a pseudostratified columnar ciliated epithelium with few scattered goblet cells. Its upper region is characterized by the olfactory epithelium. Most anti-inflammatory products are synthesized and secreted by serous cells in the submucous glands [133]. The respiratory epithelium covers the entire larynx, except the vocal folds, which are carpeted by the stratified squamous non-keratinized epithelium.

In the tracheobronchial tree, the height of the respiratory epithelium decreases as the bronchial caliber decays. The number of glands and goblet cells decreases as the bronchus becomes smaller. The bronchioles are lined by a simple columnar epithelium which contains ciliated, goblet, and Clara cells. Respiratory bronchioles have alveoli in their walls. They are lined by either a simple cuboidal epithelium with Clara cells or the squamous epithelium of the alveoli. The epithelial cells have an apical glycocalyx, the thickness of which is difficult to accurately assess.

3.4.1 Basal and Parabasal Cells

The number of basal cells per unit length (mm) of basement membrane of the respiratory epithelium is related to the height of columnar cells, but not to their number. Basal cells that do not extend to the wetted surface have a large surface area in contact with the basement membrane ($51.3 \pm 4.6\,\mu m^2$), i.e., much larger than that of ciliated ($1.1 \pm 0.1\,\mu m^2$), goblet ($7.6 \pm 1.2\,\mu m^2$), and other secretory cells ($12.0 \pm 2.1\,\mu m^2$) [134]. In the first 4 airway generations (G1–G4) distal to the trachea, basal cells account for 30% of the cells in human airway epithelium.[18]

Whereas apical cells are joined by tight junctions (Sect. 7.7.3), basal cells are firmly attached to the basement membrane via hemidesmosomes [135] (Sect. 7.7.1). In addition, they can increase the adhesion of columnar cells via desmosomes between them and columnar cells.

Two roles have been assigned to basal cells: (1) anchoring of the respiratory epithelium to the wall and (2) source of precursor cells. These unspecialized epithelial cells are supposed to be progenitors of ciliated and goblet cells. Basal cells undergo squamous cell differentiation in vivo and in cell cultures that reach confluence. This cellular differentiation is enhanced by protein kinase-C.

Basal cells that are the last major cell type to mature are restricted to large airways. Basal cells are detected in the trachea at week 12 of gestation [111]. Fully mature basal cells are only identifiable in canalicular and saccular developmental stages (Sect. 3.1.1). Basal cells contribute to the growth and

---

[18] The total number of airway epithelial cells was evaluated to be equal to 10.5×10^9 in humans) [134]. In addition, the number of alveolar cells was estimated to be about 18 times greater than that of bronchial epithelial cells.

maintenance of the epithelium and regeneration and remodeling subsequent to damages.

Parabasal cells located just above the basal cells also contribute to cell renewal. In the largest conducting airways (caliber $\geq 4\,mm$), the percentages of basal and parabasal cells are equal to 31% and 7%, respectively [114]. The contribution of basal and parabasal cells to cell proliferation accounts for 51% and 33%, respectively [114]. In the smallest airways (caliber $< 0.5\,mm$), basal cells represent 6% of epithelial cells and 30% of contribution to cell proliferation, whereas parabasal cells are absent.

3.4.2 Ciliated Cells

Ciliated cells line the respiratory tract from the nose down to the terminal bronchioles. They are absent in alveoli and anterior nares. Ciliated columnar cells outnumber the other cell types of the respiratory epithelium. Ciliated cells, in fact, represent approximately 80% of the epithelial surface. In the trachea, 5 ciliated cells are found for every goblet cell.

Ciliated cells are very vulnerable to infection and pollutant exposure. Interleukin-4 and -13 increase the population of goblet cells and decrease that of ciliated cells [136].[19]

Ciliated cells have several types of junctions (Sect. 7.7) between them: (1) zonula occludens (continuous belt-like tight junctions); (2) zonula adherens (intermediate junctions); and (3) macula adherens (discontinuous button-like desmosomes) [137].

Each cell has 200 to 300 cilia at its apical surface (density 6–8 cilia/μm^2) that beat in unison, slowly propelling mucus and entrapped foreign material toward the oropharynx. Some substances such as cigarette smoke are ciliotoxic.

The differentiation of epithelial cells of the airways from basal to columnar epithelial cells, ciliogenesis, maintenance of ciliated cells, and epithelial repair involves Ciliated bronchial epithelium protein-1 (CBE1) synthesized in ciliated cells from the Cbe1 gene, located on chromosome 9 (genetic locus 9p12) [138]. In fact, ciliogenesis involves a set of genes, such as the axonemal dynein and FOXJ1 genes.[20]

Ciliated cells possess receptors of adenosine (A_1), angiotensin-2, nucleotides, and tachykinin (Vol. 3 – Chap. 7. G-Protein-Coupled Receptors). Adenosine impedes ciliary activity by decreasing the intracellular cAMP concentration via A_1 receptors. On the oher hand, angiotensin-2 produced primarily

---

[19] Interleukin-13 induces goblet cell metaplasia, eosinophil infiltration in bronchial mucosa, and bronchial hyperreactivity. It also alters ciliated cell differentiation and reduces ciliary beat frequency in human nasal epithelial cells in culture [136].

[20] Transcription factor FoxJ1 activates the promoters of genes that encode hepatocyte nuclear factor-3β, Clara cell secretory protein, and cytoskeletal protein of cilia tektin.

by endothelial cells of pulmonary capillaries stimulates ciliary activity via Ca^{++}-dependent prostaglandin release without modifying the cAMP level.

3.4.3 Respiratory Cilia

Respiratory cylindrical cilia have a diameter of 200 to 250 nm, a height of 0.5 to 5 μm (5–7 μm in the trachea, 3–4 μm in bronchi [139]), with a mean separation distance between cilia of 2 to 5 μm [140]. The cilium originates in a corpuscle beneath the plasmalemma. Between cilia, cytoplasmic projections (microvilli) can be observed.

The set of cilia of a given ciliated cells and all of these cells of the respiratory epithelium form a coordinated beating cluster. Synchronous beats of cilia generate rythmic waves that propels the mucus layer with its entrapped particles and germs.

Two beating stages can be defined: (1) a preparatory phase with a backward motion and (2) a propulsion phase that yields impulse to the mucus. Cilia are more rigid in the power stroke and more flexible in recovery. The ciliary motion speed differs between the 2 stages, with a quick uprising phase and a slow sagging phase.

The liquid layer at the luminal surface of the respiratory epithelium has 2 layers: (1) a deep stratum that bathes the cilia and (2) a superficial viscous stratum. The cilium tips impinge on the outer mucus layer. The mucus is thus also propelled by momentum transfer from the less viscous liquid film. At the airway branching, ciliary motions impose a particle path lateral from the apex.

The nasal mucus is displaced toward the pharynx at a speed of 0.16 to 0.50 mm/s. The transport speed depends on air flow, on airway bore, on mucous layer thickness, on mucus rheology, and on mucus production rate [141, 142].[21]

---

[21] The air flow rate required for upward transport in vertical tubes of a quasi-uniform liquid layer of rheological properties comparable to human sputum is in the Reynolds number range of 142 to 1132 and 708 to 2830 whether the tube internal diameter is equal to 5 or 10 mm, depending on the liquid types. The mean mucous layer thickness ranges from 0.2 to 0.5 mm and from 0.8 to 1.4 mm whether the internal diameter is equal to 5 mm or 1 cm. The mean mucous layer thickness decays with increasing flow rate and decreasing mucus viscosity. The advancing speed of the leading edge of the mucous layer in a vertical tube ranges from 1.1 to 3.1 cm/min with a mucus feed rate of 0.5 ml/min at airflow rates of 0.33 to 1.17 l/s. The mucus speed increases almost proportionally with increasing mucus feed rate. The mucus speed increases with increasing airflow rates and decreases with increasing mucus viscosity. The transport speed in a horizontal tube is 5 to 60% faster than in a vertical tube.

3.4.4 Secretory Cells

Various types of secretory cells exist in mammalian conducting airways: (1) mucous goblet cells that serve as intra-epithelial glands; (2) serous epithelial cells; (3) Clara cells; and (4) tubuloacinar glands that open to the bronchial surface by an excretory duct. Mucous and serous cells are secretory cells of the trachea and large bronchi. Clara cells are secretory cells of bronchioles. Submucosal glands are present wherever cartilage exists.

Mucous cell can be defined by abundant, large, heterogeneous granules, a developed Golgi body, basally located nucleus, and granular endoplasmic reticulum. Serous cells have homogeneous granules and an extensive granular endoplasmic reticulum. They do not exist in all mammals [143]. Clara cells have homogeneous granules and an abundant agranular- and basal granular endoplasmic reticulum.

In the adult human lung as well as from week 16 of fetal life [144],[22] a low-molecular-weight bronchial antileukoprotease is detected in serous cells of submucosal glands, goblet cells, and bronchiolar Clara cells [145].

In normal bronchi, the relative volume of mucous glands ranges between 7.6 and 16.7% [146]. In asthmatic patients, the mucous gland volume greatly rises. Most patients with chronic bronchitis have an elevated mucous gland volume that can reach up to more than 40%. In approximately half of bronchi of emphysematous lungs, the content of mucous glands is in the normal range.

3.4.4.1 Goblet Cells

The main function of airway goblet cells is to quickly secrete mucin[23] to cover the airway lumen surface with a protective liquid layer, the mucus, exposed to more or less polluted inhaled air. The normal respiratory epithelium contains numerous goblet cells (3–5×10^4/mm^3; $\sim 6,000$/mm^2 in the bronchial epithelium), with a turnover time of ~ 45 days.

The goblet cell takes its name from its characteristic shape in conventionally fixed tissues for histologic examination: a narrow base and a larger apical region, where mucus granules expand. However, the goblet cell is cylindrical. This exocrine cell has a polarized structure, with, at its base, the nucleus, mitochondria, endoplasmic reticulum, and Golgi body.

Their apical secretory vesicles are discharged into the airway lumen. Suitable mucus quality and quantity is necessary for the activity of cilia. Two secretion mechanisms exist. The basal secretion at low level is unregulated. It results from a continuous movement of secretory granules driven by the cytoskeleton. The stimulated secretion corresponds to a regulated exocytosis of

---

[22] Bronchial antileukoprotease is detected in the respiratory epithelium of the trachea as well as in main and lobar bronchi at week 20, in smaller bronchi at 20 to 25 weeks, and in bronchiolar epithelium at weeks 36 to 40 [144].

[23] Exocytosis of a mucin granule is accomplished in less than 100 ms.

granules in response to extracellular stimuli. It heightens the mucus secretion. The mucus secretion is elicited by dusts and smokes.

Owing to their composition, condensed intragranular glycoproteic mucins expand up to several hundred-fold once released in the airway lumen. Mucins then mix with proteins, lipids, and glycoconjugates in order to form a dilute aqueous solution, the mucus gel.

Goblet cells can differentiate into other cell types such as ciliated epithelial cells. The number of goblet cells increases with chronically inhaled aggressive agents, i.e., *hyperplasia* in the large airways, where goblet cells are normally present, and *metaplasia* in the small airways where goblet cells are normally scarce or absent.[24] Goblet cell hyperplasia is a smoking consequence. The number of goblet cells is elevated in several diseases, such as chronic bronchitis and cystic fibrosis (Vol. 6 – Chap. 8. Pathologies of the Respiratory Tract).

3.4.4.2 Clara Cells

Non-ciliated bronchiolar Clara cells are the most abundant secretory cell type in distal airways of the human lung.[25] The average relative number of Clara cells in terminal and respiratory bronchioles has been evaluated to be about 11% and 22%, respectively [115]. In terminal and respiratory bronchioles, 15% and 44% of proliferating airway epithelial cells are Clara cells, respectively [115]. Type-A Clara cells form a major source of progenitor cells for repopulation of airways after injury by inhaled oxidant pollutants.

Mature Clara cells are characterized by a smooth endoplasmic reticulum and secretory granules. Cytochrome-P450 is involved in oxidation of lipophilic compounds in Clara cells [148].

3.4.4.3 Respiratory Mucus

Mucins are secreted not only by goblet cells, but also by submucosal glands of cartilaginous airways. The mucus is composed of *mucins*, glycosylated proteins, suspended in a solution of electrolytes and other molecules. The mucus protects airways against flow stresses and chemical damages, trapping and eliminating particulate matter and microorganisms.

The nasal mucus has a thickness greater than 200 μm. The mucus extends over the tops of cilia in a layer of thickness of about 5 μm. Many mediators trigger mucin secretion, such as cholinergic agonists, lipid mediators, oxidants, cytokines, neuropeptides, neutrophil elastase, ATP and UTP, etc. Interleukins-4, -9, and -13 provoke mucin synthesis.

---

[24] Goblet cell hyperplasia, associated with mucus hypersecretion, is a symptom of chronic obstructive pulmonary disease. The epidermal growth factor pathway and calcium-activated chloride channels are involved in goblet cell hyperplasia. The inhibitor of apoptosis BCL2 is implicated in the maintenance of hyperplasia [147].

[25] Clara cells develop during the second half of gestation from primitive glycogen-containing, non-ciliated cells of distal airways.

The mucus in goblet cell granules is condensed. Negative polyanionic charges within the secretory granules are neutralized by calcium ions. When the granule content is released across a membrane pore into the airway lumen, calcium ions diffuse and polyanionic charges repulse leading to mucus hydration. Once secreted, the mucus then expands almost instantaneously. The mucin gel is indeed able to increase in volume 500-fold in about 20 ms.

Ninety-five per cent of mucus is water. Mucins correspond to less than 2% of the wet weight of the mucus. The respiratory mucus is also composed of carbohydrates, salts, proteins, glycoproteins, proteoglycans, and lipids. Proteoglycans are a major source of mucus hydration. Mucus proteins have a protective effect against H^+ penetration into tissues. The buffer capacity is accompanied by pH-dependency of mucus viscosity.

Mucus rheology influences efficiency of its coupling to sweeping cilia. The respiratory mucus is characterized by a viscoelastic behavior, as well as its adhesiveness and wettability. *Wettability* and *adhesiveness* govern contact properties with the respiratory epithelium. These 2 properties must be high enough in order to avoid sedimentation down to the distal airways, and sufficiently low to be easily mobilized during coughing. An optimal rheology is also required for the mucociliary clearance. The dynamic viscosity of normal mucus is ~ 1 Pa/s. Surface-active phospholipids, such as phosphatidylcholine and phosphatidylglycerol, improve the wettability of mucus. Glycosphyngolipids increase mucus viscosity.

When the respiratory mucus interacts with air flow during coughing, micron-sized droplets form and are expelled to the environment. The droplets can thus participate in possible propagation of respiratory infections.

Exocytosis is done in 3 main steps: (1) motion of mucin granules to the apical plasmalemma; (2) fusion with the plasmalemma; and (3) opening onto the airway surface. Step 1 is regulated by myristoylated alanine-rich C kinase substrate (MARCKS) [149]. Activated protein kinase-C phosphorylates MARCKS causing MARCKS translocation from the plasmalemma to the cytoplasm. MARCKS is then dephosphorylated by protein phosphatase PP2a (Vol. 4 – Chap. 7. Cytosolic Protein Phosphatases) that is activated by cGMP-dependent protein kinase PKG, itself stimulated by the NO–cGMP pathway. A cooperative interaction between protein kinases PKC and PKG is then required. Dephosphorylated cytoplasmic MARCKS associates with the contractile cytoskeleton. Step 2 probably implicates a coordinated interaction between soluble ^{N}ethylmaleimide sensitive factor receptors (SNARE), Unc13 homolog Unc13d[26] and Rab GTPases.

3.4.5 Submucosal Glands

In human adults, trachea and large bronchi possess numerous tubuloacinar glands between cartilage and respiratory epithelium with a duct opening into

---

[26] Protein Unc13d, or Munc13-4, is produced in lungs by goblet cells and type-2 alveolar cells [150].

the airway lumen. Submucosal glands produce most of the mucus in large airways.

In aiways of normal children and adults, glands occupy about 17 and 12% of the wall, respectively. Submucosal gland concentration is higher in proximal than in distal airways. Glands abound in airway bifurcations. In the trachea, gland density is larger in the cartilaginous anteriolateral wall than the posterior region.

Submucosal glands contain 2 cell types: mucous and serous cells that secrete a viscoelastic gel and a fluid, respectively. Epithelial lining fluid is composed of soluble proteins, many of which are produced by serous cells of submucosal glands, such as lactoferrin, lysozyme, secretory immunoglobin-A, neutral endopeptidase, and aminopeptidase. In bronchioles, serous cells tend to outnumber mucous cells. Mucous cells synthesize high-molecular-weight glycoproteic constituents of the mucus and store their dehydrated form before their secretion. Upon release, these substances spontaneously hydrates to produce a gel that floats on epithelial lining fluid.

3.4.6 Pulmonary Neuroendocrine Cells

In human adults, the respiratory epithelium contains a small number of neuroendocrine cells that exist as isolated cells scattered throughout the tracheobronchial epithelium or in clusters, the so-called *neuroepithelial bodies* that are located only in the intrapulmonary airways [121]. The mean quantity of pulmonary neuroendocrine cells is equal to 0.41% ($\pm$ 0.17%) of all epithelial cells, i.e., 12.5 pulmonary neuroendocrine cells per centimeter of basement membrane (3,101 cells per centimeter of basement membrane), in large as well as small conducting airways (but not in alveoli) [151].

Pulmonary neuroendocrine cells can proliferate. However, their mitotic activity is modulated by environmental factors. These cells are characterized by cytoplasmic dense core granules (size 90–150 nm) that are the storage site of amine and peptide hormones. According to the size and morphology of cytoplasmic granules, 3 distinct types of neuroendocrine cells can be defined.

Pulmonary neuroendocrine cells can act as endo- and paracrine regulators of pulmonary vascular or bronchial responses. Neuroendocrine cells secrete local mediators such as serotonin. Pulmonary neuroendocrine cell markers also include peptide hormones, such as bombesin, calcitonin, and Leu-enkephalin (an opioid neurotransmitter) [121]. These signaling molecules act on nerve endings and neighboring cells. They thus regulate the rate of mucus secretion and ciliary beating, as well as the tone of surrounding smooth muscle cells.

Neuroepithelial bodies (but not isolated neuroepithelial cells) comprise non-myelinated nerve endings in contact with granulated cells. A type of Clara cell, the *Clara cell secretory protein-expressing cell*, is located adjacent to and within the neuroepithelial body. The neuroepithelial body thus contains cells (about 10% of the cell population) that express both calcitonin gene-related peptide and Clara cell secretory protein.

Hyperplasia of pulmonary neuroendocrine cells has been described in numerous pulmonary lesions.

3.4.7 Smooth Muscle Cells

In bronchi, smooth muscle cells occupy about 3% of the wall volume in both children and adults, whereas in bronchioles it corresponds to about 10% and 20% of airway walls in children and adults, respectively.

In humans, the development of the cytoskeletal proteins actin, desmin, and vimentin in pulmonary vascular smooth muscle cells of the outer media is larger than that of inner media [152]. In normal humans, the concentration of vimentin decays after birth and then gradually rises, whereas the levels in the cytoskeletal proteins desmin and α-actin increase steadily with age.[27]

Smooth muscle cells account for $4.6 \pm 2.2\%$ of the volume of normal bronchial wall [146]. This percentage increases in asthma ($11.9 \pm 3.4\%$), but not significantly in chronic bronchitis and emphysema.

3.4.8 Cartilage Rings

Cartilage appears in the trachea and main and segmental bronchi in gestational weeks 7, 10, and 12, respectively [153]. Cartilage thus reaches segmental bronchi about 6 weeks after their formation. Cartilage continues to form distally until about month 2 after birth. During infancy ($\leq$ month 18 after birth) and childhood, total cartilage mass progressively increases. Cartilage continues to grow after bronchial branching is complete and new cartilage appears until week 25. The number of cartilaginous bronchi in fetuses after week 26 corresponds to that in adults. However, the ground substance of the cartilage is slow to mature. Even at week 24, some of the most distal cartilage plates are still precartilage.

The number of cartilage-free bronchioles is not correlated with the amount of generations along given bronchial paths from the trachea to bronchioles. In short bronchial networks such as those of upper lobes, 8 to 10 generations of cartilaginous bronchi can be observed. On the other hand, 13 to 21 generations of cartilaginous bronchi can exist in longer bronchial paths such as those in the lingula.

3.4.9 Connective Tissue

The structure and organization of the connective tissue scaffold of the lung parenchyma is determined by various physical agents, such as degree of

---

[27] In the case of pulmonary hypertensive congenital heart disease, levels in α-actin and vimentin strongly heighten between 2 and 8 months in association with intimal proliferation of smooth muscle cells beneath the internal elastic lamina.

lung inflation and associated transmural pressure, perfusion pressure, surface tension, and hydrostatic pressure.

Fibroblasts that correspond to mature, differentiated cells (footnote 44) are the most common resident cells in connective tissue. They maintain the structural integrity of connective tissues by continuously releasing precursors of all of the components of the extracellular matrix, i.e., ground substance (e.g., glycoproteins and proteoglycans) and fibers (collagen and elastin; Chap. 8). Collagen is delivered by fibroblasts as procollagens that are converted extracellularly into tropocollagens which self-assembles into fibers. Collagen type-3, -4, -7, and -8 reinforce the basement membrane.

Fibrocytes also produce constituents of the extracellular matrix, but are much less active than fibroblasts. Fibrocytes express class-2 major histocompatibility complex molecules and costimulatory molecules. They migrate to wound sites for tissue repair.

3.5 Cells of the Gas Exchange Region

The lung parenchyma is composed of: (1) septa between lung acini, lobules, and lobes; (2) sheaths that contain thick-walled airways, blood and lymph vessels, and nerves; as well as (3) the entire collection of alveolar thin walls with pulmonary capillaries. Alveolar thin walls constitute the major part of the lung parenchyma.

Pulmonary acinus is the elementary gas exchange compartment. Pulmonary alveolus is the gas exchange unit that appear on walls of acinal bronchioles, or respiratory bronchioles, distal to terminal bronchioles. Terminal bronchioles actually correspond to the last conducting airway generations. The number of terminal bronchioles is estimated to be about 28,000 [154]. The first generation of respiratory bronchioles marks the beginning of the respiratory compartment, from which gas exchange takes place. Incoming bronchioles at the acinus entrance divide into several generations. The last one yields a set of alveolar ducts, each endowed with 5 to 6 alveolar sacs.

In the acinus, air and blood are conveyed into close proximity, i.e., separated by an extensive, thin barrier. The structure of the alveolar air–blood barrier (harmonic mean thickness 0.62 μm [156])[28] relies on: (1) a pulmonary capillary endothelium with its basement membrane; (2) a thin intertitium

---

[28] The harmonic mean is the reciprocal of the arithmetic mean of the reciprocals, i.e., the reciprocal of the mean of the sum of the reciprocals:

$$\frac{n}{\sum_{i=1}^{n} x_i^{-1}}.$$

For example, it is used to obtain the medium-average hydraulic conductivity for a flow that crosses a set of layers in the perpendicular direction.

Table 3.5. Cell number in alveolar septa and average volume of alveolar wall cells (Source: [154]).

Cell type	Number ($\times 10^9$)	Volume (μm^3)
Pneumocyte 1	19.6	2390
Pneumocyte 2	32.9	820
Endothelial cell	73.2	610
Alveolar macrophage	5.99	1470

(thickness 100–140 nm);[29] and (3) an alveolar epithelium with its basement membrane.

Collagen-4 mainly synthesized by endothelial and epithelial cells is the major constituent of endothelial and epithelial basement membranes.[30] The cell number in alveolar septa and the average volume of alveolar wall cells are given in Table 3.5).

The air–blood barrier withstands changing hemodynamic pressures from the capillary side and surface tension forces from the alveolar side, as well as reacts to physical and chemical agents. It also receives inhaled microorganisms, allergens, carcinogens, toxic particles, and noxious gases.

Although the pulmonary blood flow is pulsatile at the entry of the pulmonary circulation and its exit in the left atrium due to the cardiac activity, in addition to the varying intrathoracic pressure, blood can be assumed to flow quasi-steadily in pulmonary capillaries.

In pulmonary capillaries, the blood flow can be assumed to be steady and modeled by a Stokes flow (Vol. 1 – Chap. 1. Cells and Tissues), as both the Reynolds and Stokes numbers are vey small:

$$- \nabla p + \mu \nabla^2 \mathbf{v} = 0, \tag{3.1}$$

where $\mathbf{v}$ is the fluid velocity; p the pressure; and μ the blood dynamic viscosity.

Air reaches alveolar sacs by gaseous diffusion. An ideal gaseous mixture, composed by N gas species, is fully described by the mole fractions $\chi_i = n_i/n_{\text{tot}}$ $(\sum_{i=1}^{N} n_i = n_{tot})$ of each species i ($i = 1,\ldots,N$) and the total concentration c_{tot} of the mixture. The partial pressure of species i in the mixture (p_i) is given in the case of an ideal gas mixture by the equation

---

[29] In many parts of the air–blood barrier, the thickness of the lamina densa located in the center of the intertitium is smaller than 50 nm [156].

[30] Type-4 collagen serves as an anchoring structure for endothelial and epithelial cells that exhibit $\alpha_1\beta_1$- and $\alpha_2\beta_1$ integrins on their surface. Other intertitium constituents include laminins, nidogen-1, heparan sulfate proteoglycans such as perlecan, and tenascins such as cytotactin [156]. ($\alpha_9\beta_1$ integrin is a receptor for cytotactin, vascular cell adhesion molecule VCAM1, and osteopontin.)

of state $p_i = RTc_{tot}\chi_i$. The force acting on species i in a control volume is given by: $-\nabla p_i$. The force per mole of species i that can be expressed as $-R_gT\nabla\chi_i/\chi_i$ (R_g: ideal gas constant; T: absolute temperature) is balanced by the friction between species i and j acting on species i: $R_gT\chi_i(v_j - v_i)/\mathcal{D}_{ij}$ (v_i: molar average diffusion velocity of species i; $\mathcal{D}_{ij}$: binary diffusion coefficient between the species i and j). Each mole fraction that depends on time t and space variable $\mathbf{x}$ satisfies the continuity equation:

$$\partial_t\chi_i + \nabla \cdot J_i = 0,$$

where $J_i = \chi_i v_i$ is the molar flux of species i. The Stefan-Maxwell diffusion is a model of gas diffusion in multicomponent mixtures [155]:

$$-\nabla\chi_i = \sum_{i\neq j} \frac{\chi_j J_i - \chi_i J_j}{\mathcal{D}_{ij}}. \tag{3.2}$$

The conductance of the air–blood barrier for oxygen is correlated directly with its surface area and inversely with its thickness. The total resistance that oxygen molecules encounters is usually decomposed into resistance of the air–blood barrier, plasma, and erythrocyte. An average partial pressure difference of 57 Pa supports the transfer of oxygen across the air–blood barrier at a rate of $2.3 \times 10^{-5}\,\mathrm{cm^2/s}$ that is then completed in 250–500 ms [156].

3.5.1 Alveolar Epithelial Cells

The alveolar epithelium covers more than 99% of the internal surface area of the lung. It is composed of 2 major specialized epithelial cell types: alveolar type-1 and -2 cells. Type-1 alveolar cells have long, very thin cytoplasmic extensions along the alveolar walls. Type-2 alveolar cells produce the surfactant that lines the alveolar wall.

3.5.1.1 Type-1 Pneumocyte

Type-1 pneumocytes, or type-1 alveolar epithelial cells, are squamous, elongated, and flat cells through which exchanges of O_2 and CO_2 take place. They amount to about one-third of alveolar epithelial cells. They constitute slightly more than 90% of the alveolar surface area. Pneumocyte 1 has a thickness in the perinuclear region of 2 to 3 μm and about 200 nm in the peripheral extensions. These extensions contain very sparse cellular organelles. Pneumocyte 1 contains numerous plasmalemmal invaginations and vesicles.

3.5.1.2 Type-2 Pneumocyte

Type-2 pneumocyte, or type-2 alveolar epithelial cells, are cuboidal, granular cells (thickness ~ 5 μm). They amount to about two-thirds of alveolar epithelial cells and occupy the remaining part of the alveolar surface area (less than 10%). They have a smaller length than that of type-1 pneumocyte. They are characterized by numerous organelles and microvilli at the apical membrane.

Type-2 pneumocytes control the volume and composition of the epithelial lining fluid. They synthezise, secrete, and recycle all components of the surfactant (Vol. 5 – Chap. 13. Surfactant). They also modify hypophase surfactant by regulating pH and calcium concentration of the hypophase.

The main morphological and biological features of alveolar epithelial type-2 cells encompass: (1) the presence of *lamellar bodies* (caliber $\sim$ 600nm; length $\sim$ 900 nm; $\sim$ 13% cytoplasmic volume [157]) and (2) synthesis of surfactant protein-A, -B, and -C,[31] protease inhibitor serpin-A1, or α1-antitrypsin, [32] cystic fibrosis transmembrane conductance receptor,[33] as well as complement proteins C3 and C5 [132].[34]

Type-2 pneumocyte serves as a progenitor cell for type-1 alveolar epithelial cell [158]. It also acts on alveolar fluid balance, coagulation, and fibrinolysis, in addition to immune defense. It participates in epithelial repair by phagocytosis of apoptotic cells. It interacts with resident and mobile cells, either directly by plasmalemmal contact or indirectly via growth factors.

3.5.2 Alveolar Wall Fibroblasts

Alveolar wall fibroblasts link pneumocytes to capillary endothelial cells via apertures in their respective basement membranes [159]. A fibroblast can contact both a pericyte and an endothelial cell through apertures in the endothelial basement membrane. It thus bridges the capillary wall to the alveolar septum. Fibroblasts come into contact with both type-1 and -2 pneumocytes through holes in the epithelial basement membrane. About 54% of type-2 pneumocytes have at least one gap in their basement membrane. Leukocytes can then migrate along fibroblasts without disrupting the extracellular matrix.

---

[31] A major function of type-2 pneumocytes is the synthesis, storage, and secretion of surfactant that reduces surface tension and prevents alveolus collapse during exhalation (Vol. 5 – Chap. 13. Surfactant).

[32] Type-2 pneumocytes secrete the Ser protease inhibitor α1-antitrypsin that impedes the development of pulmonary emphysema.

[33] Type-2 pneumocytes transport ions from the alveolar fluid into the interstitium to minimize alveolar fluid and maximize gas exchange. Channel CFTR participates in the regulation of ion and fluid transfer in the lung alveolus.

[34] Type-2 pneumocyte contributes to pulmonary defense, as it synthesizes and secretes several complement proteins as well as numerous interleukins that modulate the action of many types of leukocytes, especially macrophages, neutrophils, and lymphocytes. Activated C3 and C5 produce the potent complement anaphylatoxins C3a and C5a that intervene in tissue regeneration.

3.5.3 Alveolar Macrophages

The respiratory tract is exposed constantly to inhaled particles and microorganisms. These foreign bodies are usually cleared by the mucociliary transport in the upper airways and tracheobronchial tree on the one hand, and alveolar macrophages (length $\sim 13\,\mu$m) in the alveolar space on the other.

The innate immune system is, in fact, developed in the lung distal regions. The innate immune system is made of: (1) a humoral component, with lactoferrins, lyzozyme, surfactant proteins, collectin-1 (or mannose-binding lectin), and defensins, and (2) a cellular component, mainly alveolar macrophages with numerous receptors for foreign antigens. Toll-like receptors (TLR; Vol. 3 – Chap. 11. Receptors of the Immune System) are microbe-targeting sensors of the innate immunity. Interactions between infectious compounds and TLRs initiate inflammation.

The pulmonary immune response can require both local antigen-presenting cells and regulatory suppressor cells. Dendritic cells from the lung interstitium and alveolar macrophages have a potent antigen-presenting activity. Dendritic cells are separated by a tiny distance (200–500 nm) from alveolar macrophages that adhere to the alveolar epithelium cells. If the non-specific mechanisms fails, dendritic cells trigger the adaptive immune response.

Alveolar macrophages derive from monocytes, both from precursors that originate in the bone marrow and circulates in blood and from local proliferation of precursors. Colony stimulating factor CSF2, or gmCSF, maintains the lung population of alveolar macrophages. Macrophages have defensive functions, digesting bacteria, degrading antigen, synthesizing immunoregulators (interferon, chemoattractants, tumor-inhibiting factors). They also synthesize arachidonic acid metabolites, platelet- and fibroblast-activating factors, and enzyme inhibitors.

However, alveolar macrophages that continuously encounter inhaled materials are not permanently activated in order to avoid collateral damage to type-1 and -2 alveolar cells in response to harmless antigens. They are maintained in a quiescent state. Furthermore, in normal situations, alveolar macrophages prevent the activity of dendritic cells [160], as well as that of T and B lymphocytes [161, 162]. Alveolar macrophages use multiple substances to suppress the intra-alveolar immune response initiated by lung dendritic cells. Alveolar macrophages hence create an immunosuppressive microenvironment within the normal lung via numerous molecules (e.g., transforming growth factor-β, prostaglandin-E2, interleukin-10, and nitric oxide).

Alveolar macrophages are inhibited by transforming growth factor-β bound by $\alpha_V\beta_6$ integrin on the surface of alveolar epithelial cells. Once exposed to microbial products, alveolar macrophages unbind from alveolar epithelial cells [163]. Toll-like receptor ligands reduce $\alpha_V\beta_6$ integrin expression by alveolar epithelial cells and, subsequently, TGFβ-mediated signaling in alveolar macrophages (and thus reduced phosphorylation of effectors SMAD2 and SMAD3; Vol. 3 – Chap. 8. Receptor Kinases). Signaling via TLRs in alveolar macrophages rapidly provokes actin polymerization and disruption of contacts between alveolar macrophages and alveolar epithelial cells.

After clearance of infectious agents by activated alveolar macrophages, matrix metalloproteinase 9 is expressed by alveolar macrophages, probably via interferon-γ. Proteinase MMP9 reactivates latent TGFβ and resets the inhibition on alveolar macrophages. Factor TGFβ may also stimulate collagen synthesis in interstitial fibroblasts in order to restore alveolar wall architecture.

4

Cell Structure and Function

The cell (cellula: small compartment) was described by R. Hooke in 1663. From 1673, A. van Leeuwenhoek wrote his observations on cells. In 1805, L. Oken stated that "all living organisms originate from and consist of cells". In the early 1830s, R. Brown reported on cell nucleus. Slightly later, M. Schleiden and T. Schwann formulated the early cell theory that was improved later by R. Virchow.

The cell theory states that the cell: (1) is the structural and functional unit of living tissues; (2) retains a dual existence as a distinct entity and building block in the construction and organization of living tissues; (3) is a site of metabolism that produces energy; (4) comes from pre-existing cells by division; and (5) contains hereditary information, but access to and processing of the genetic code depends on cell differentiation.

The cell is a self-organizing, self-replicating, basic constituent of organisms. It contains various types of organelles and molecular complexes. It captures and secretes materials from and to the extracellular space. It reacts to environmental stimuli and adapts. Moreover, it is endowed with quality control processes that operate from the translation of genomic information into synthesized proteins to the production of organelles to repair and/or eliminate mistakes and preserve functional molecules and organelles despite experienced challenges. The choice between repair and degradation depends on error type and energetic cost of the process.

4.1 Cell Composition

The cell has a nucleus[1] and several organelle types within its *cytoplasm* (Fig. 4.1). The cytoplasm, or cytoplasma, corresponds to the cell content

---

[1] Anucleated cells include circulating red blood cells, or erythrocytes, that are capsules (mainly a membrane-enclosed hemoglobin solution) and platelets, or thrombocytes, that are cell fragments of megakaryocytes.

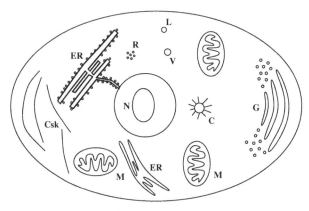

Figure 4.1. Scheme of the cell and its components (N: nucleus; ER: endoplasmic reticulum with its smooth and rough compartments; G: Golgi apparatus; M: mitochondrion; Csk: cytoskeleton; C: centrosome; R: ribosome; V: vacuole, vesicle; L: lysosome). The cytoplasm is surrounded by the plasma membrane, a barrier between the extra- and intracellular fluids. Outside the nucleus, the cytoplasm contains the cytosol, cellular organelles, and large proteic complexes, such as ribosomes and cytoskeleton components (microfilaments, microtubules, and intermediary filaments). In certain cells, especially those that coat duct lumens, the plasma membrane can be carpeted on its external layer by a glycocalyx. The cytosol is composed of soluble molecules (proteins, glucids, nucleotides, etc.). The nucleus contains the hereditary information — the genome — materialized by DNA (cipher molecule), associated with proteins, and embedded in the nucleoplasm. The endoplasmic reticulum is a set of tubules and cisternae that constitute a large surface area. The granular compartment is dedicated to the synthesis of proteins, the smooth compartment to that of lipids as well as to storage. The Golgi stack is composed of saccules and cisternae. (A dictyosome is a small stack of 4 to 5 saccules.) It has a cis and trans face devoted to the reception of immature proteins and secretion of transformed proteins, respectively. Mitochondria are power plants with a double membrane (envelope), like the nucleus. Its internal membrane serves as oxidation sites for glucids and fatty acids. It couples the proton flux to ADP phosphorylation. Its internal matrix contains various enzymes and DNA. The intermembrane space receives protons. Vesicules are used in the intracellular transfer of molecules, mainly along cytoskeletal tracks. The cytoskeleton is also a mechanical signal transducer that adapts the cell configuration to its environment. Lysosomes (1–2 μm) possess enzymes to digest endocytozed substances and senescent organelles.

outside the nucleus. The *cytosol* is the main compartment of the cytoplasm, excluding membrane-bound organelles and very large molecular complexes. The cell that can experience various types of chemical and physical stimuli as well as mechanical stress fields resulting from loadings imposed by the cell environment is wrapped in the plasma membrane, also called plasmalemma (in plants) and sarcolemma in myocytes (Chap. 7).

Table 4.1. Characteristic sizes of the cell structural elements (nm; Source: [164]).

Mitochondrium thickness	500–2000
Lysosome (hydrolases)	200–500
Peroxysome (oxidative enzymes)	200–500
Microtubule caliber	25
Intermediary filament bore	10
Actin filament bore	5–9
Plasma membrane thickness	4–5

4.1.1 Cellular Compartmentation and Organelle Delimitation

Cell organelles are specialized functional components enclosed within their own lipid bilayer. In addition to these membrane-bound cell organelles, the cell possesses other compartments within the cytosol: large proteic complexes, such as cytoskeletal filaments and ribosomes within the cytoplasm, and the nucleolus inside the nucleus.[2] Cell organelles encompass the nucleus, endoplasmic reticulum, Golgi stack, mitochondria, centriole, and various types of vesicles.

The characteristic sizes of cell elements are given in Table 4.1. Shape, location, and motion of organelles within the cell depend on the cell cytoskeleton (Chap. 6) that produces and responds to mechanical forces.

4.1.2 Between-Organelle Communication

The apposition of cell organelles creates exchange paths for signaling molecules and thus facilitates signal transmission. Small between-membrane gaps exist between mitochondria and either the endoplasmic reticulum or the plasma membrane as well as the endoplasmic reticulum and the Golgi stack, endosomes, lysosomes, peroxisomes, and lipid droplets (Table 4.2).

The large tether complex *endoplasmic reticulum–mitochondrion encounter structure* (ERMES) contains numerous Mitochondrial distribution and morphology proteins MDM10, -12, and -34 of the outer mitochondrial membrane and Maintenance of mitochondrial morphology protein MMM1 of the endoplasmic reticulum membrane. The ERMES complex facilitates Ca^{++} and phospholipid exchange between the endoplasmic reticulum and mitochondria.

4.1.3 Molecular Constituents

Water, proteins, and lipids are the main constituents of the cell (Table 4.3). Although the water content is quite high, the deformable cell is often considered as a slightly compressible material. Biopolymers and other solutes make up less than one-third of the mass of a typical cell.

---

[2] Cell organelles, strictly speaking, do not include cytoskeleton, ribosomes, and nucleolus.

Table 4.2. Small between-membrane gaps and main constituents (Source: [165]; ER: endoplasmic reticulum; LE: late endosome; Mit: mitochondrion; NVJ, nucleus–vacuole junction; PM: plasma membrane; 25OCs: 25-hydroxycholesterol; CRAC: Ca^{++}-release-activated Ca^{++} modulator; GRP75: 75-kDa glucose-regulated protein; IP$_3$R: inositol triphosphate receptor; IR: insulin receptor; MDM: mitochondrial distribution and morphology protein; MFn: mitofusin [dynamin-like GTPase]; MMM: mitochondrial morphology protein; ORP: OSBP-related protein; OSH: oxysterol-binding protein homolog; PACS: phosphofurin acidic cluster sorting protein; PITPnm: phosphatidylinositol-transfer protein, N-terminus-interacting, membrane-associated; PMN, piecemeal micro-autophagy of the nucleus; PSS, phosphatidylserine synthase; PI: phosphatidylinositol; PTP: protein Tyr phosphatase; StIM: stromal interaction molecule; TSC: temperature-sensitive CSG2 suppressor; TRP: transient receptor potential; Vac: vacuolar protein; VAP: vesicle-associated membrane protein (VAMP)-associated protein; VDAC: voltage-dependent anion channel).

Type	Components
ER–Mit	MDM10/12/34, MFn1/2, PSS1/2, acylCoA synthase, PACS2, PTP1b, IP$_3$R, VDAC, GRP75
ER–PM	IR, StIM1, CRAC1, IP$_3$R, TRP, Homer ORP3, VAPa
ER–LE	ORP1, VAPa
ER–Golgi	OSBP, CeRT, PITPnm1, 25OCs
NVJ	OSH1, TSC13, NVJ1, Vac8

Table 4.3. Approximate composition of a cell (%; Source: [164]).

Water	70
Proteins	18
Lipids	5
RNA, DNA	1
Polysaccharides	2
Small metabolites	3
Inorganic ions	1

Cell water with its large cohesive energy density (one order of magnitude higher than for most organic liquids) serves as solvent. It modulates intermolecular forces, frames conformational changes of solutes and protein folding, participates in molecular recognition by mediating interactions between binding partners [166] and enzymes, and controls the rates of solute solvation, hydration, and diffusion [167–169].

Water dynamics in the organized intracellular medium can be observed at different time scales using nuclear magnetic resonance technique over a wide magnetic field range (0.2 mT–12 T) [170]. In the heterogeneous intracellular milieu, water diffuses with translational and rotational motions limited by friction, obstruction, and confinement [169].[3] Most water content has bulk dynamics, but a small proportion interacts with molecular surfaces, hence with a retarded motion. A cell water fraction of about 0.1% is related to protein hydration with a microsecond time scale.

4.1.4 Cellular Large Proteic Complexes

The body's cell contains very large proteic complexes in various subcellular compartments. In the cytoplasm, *ribosomes* (Sect. 4.14) involved in protein synthesis contain about 80 proteins. Moreover, additional proteic complexes, such as the *initiation* and *elongation factors*, interact with ribosomes to produce proteins from RNA templates.

Nuclear-pore complex (Sect. 9.1.8); size $\sim 100\,\text{nm}$) is one of the largest proteic conglomerations in the cell, as it consists of hundreds of proteins, although it is made up of repetitive arrangements of about 30 different types of proteins.[4] It serves as both gate and gatekeeper for nuclear import and export of nucleic acids, proteins, and other types of molecules. In addition to the regulation of molecule transfer between the cytoplasm and nucleoplasm, it controls the gene expression, as it coordinates the import of *transcription factors* and export of *translation agents* for cytoplasmic protein synthesis (Chap. 5). Cytoplasmic filaments of nuclear-pore complexes interact with the cytoskeleton and components of protein synthesis.

Spliceosome (Sect. 5.4) chops a newly synthesized messenger RNA, then stitches together selected exons that will be translated into proteins and eliminates introns. It can thus form different proteins from a single DNA sequence. It is composed of about 150 proteins, among which 5 main components are in constant motion, the small nuclear ribonucleoproteins.

4.2 Main Cell Features

Cells are characterized by a *self-organization*[5] and *self-reproduction*. Life is characterized by changes in space and time dictated by *structure–function relationships.*

---

[3] Long-range $(1–10\,\mu\text{m})$ translational apparent diffusion coefficient represents a spatially averaged translational mobility. Measurements by NMR of rotational correlation time corresponds to a spatial averaged inverse rotational mobility. Water rotates at different rates depending on interactions with cell material.

[4] During cell division (Vol. 2 – Chap. 2. Cell Growth and Proliferation), it is dismantled to be later reassembled. It is broken into building blocks that always remain the same.

[5] Self-organization of any system is characterized by system components that interact dynamically with each other to generate a system that acquires emergent properties that cannot be directly predicted from the individual properties of the elements.

4.2.1 Self-Organization

Self-organization concerns dynamical cell shape and coherently associated functions that emerge from regulated molecular interactions within the cell, applied physical and mechanical constraints, and collective behaviors of interacting cells. Formation of internal cell structures also relies on stochastic assembly and self-organization.

The cell nucleus is an organized, compartmentalized structure with nuclear bodies that contains the genetic information that can be read, copied, and repaired. Cell architecture arises from a relatively small number of genes.

Cell morphogenesis is mainly a self-organizing process that in combination with constraints controls the size, shape, position, and number of organelles. Cell shape and organelle organization are polarized. The structure size can be assessed by proteic rulers that have a length equal to that of the structure [171]. Organelles such as the Golgi body with its polarized organization move to suitable positions owing to the cytoskeleton. Organelle position can generate a positive feedback loop, e.g., the Golgi stack contributes further to cell polarity during cell motility. Organelles can also influence the positioning of other organelles. The size may also be determined by the synthesis of only the required quantity of material (*quantal synthesis*).[6] The combination of feedback and regulation of polarizing processes, such as the cytoskeletal dynamics and protein localization, allows fluctuations in cell asymmetry and self-organization.

Organelle shape is defined by the membrane curvature at each point that minimizes the energy of membrane deformation. Membrane-bound organelle morphology results from lipid partitioning combined with protein-induced curvature changes. The local curvature in fact depends on the lipid composition of membrane nanodomains. Integral membrane and membrane-binding proteins can stabilize the local membrane curvature.

The cytoskeleton links to membrane proteins that can then move. The number of organelles depends on the controlled balance between synthesis and fission of organelles on the one hand, and fusion with other organelles and degradation (autophagy) on the other.

4.2.2 Cell Polarity

Cell polarity is determined by the polarized distribution of plasmalemmal proteins. Establishment and maintenance of the cell polarity is needed for correct functioning of numerous cell types. Epithelial and endothelial cells have apicobasal polarity for barrier function. Neurons are polarized to develop dendrites and one long axon for neurotransmission. During migration, cells polarize in the direction of cell movement. Epithelia cell polarity is precisely controlled; its loss contributes to tumor formation and progression.

---

[6] The size of the endoplasmic reticulum can be controlled by feedback associated with the amount of synthezised lipids.

At least for certain cell types, the initial step in polarity commitment is associated with cell migration during embryo- and fetogenesis. In most cases, polarity engagement requires initial sensing of spatial heterogeneities in internal or external signals, their subsequent amplification and stabilization, and transduction of spatial information by cytoskeletal and eventually secretory outputs.

External gradient of chemoattractant elicits intracellular polarization with persistent translocation of selected intracellular molecules and organelles to the cell region close or opposite to the side with the highest receptor occupancy. On the other hand, spatially homogeneous, persistent signals only induce transient responses. Positive feedback loops enable amplification of signals[7]

Some proteins control cell polarization. Polarity proteins, such as Partitioning-defective proteins (Par) and atypical protein kinase-C (Vol. 4 – Chap. 4. Cytosolic Protein Ser/Thr Kinases), operate for apicobasal polarity of epithelial cells and polarized axonal growth, as well as for directional migration (Vol. 2 – Chap. 6. Cell Motility). Clathrin[8] helps to maintain epithelial polarity, as its loss delays the delivery of basolateral membrane proteins and causes their mis-sorting from the Golgi stack into apical carrier vesicles [892].

4.2.3 Potential Difference between Intra- and Extracellular Media

Cell electrical activity allows communication between neurons, contraction of skeletal and cardiac myocytes, sensory perception, and secretion of hormones. Transmembrane currents of ions between the intra- and extracellular spaces via ion carriers, such as ion channels, exchangers, and pumps (Vol. 3 – Chap. 3 Main Classes of Ion Channels and Pumps), and gap junctions (Sect. 7.7.4), which are characterized by their conductance, determine action potential shape and wave propagation velocity of excitable cells, such as neurons, nodal cells, and cardiomyocytes.

4.2.4 Glycocalyx

Glycocalyx is a highly charged layer of extracellular polymeric material attached to the plasma membrane of epithelial and other cells. Membrane-bound macromolecules include polysaccharides and proteins that form glycoproteins and proteoglycans. In particular, the polysaccharide-rich coat of

---

[7] Small GTPases Rho, Rac, and CDC42 (Vol. 4 – Chap. 8. Guanosine Triphosphatases and their Regulators) as well as phosphoinositides (Sect. 7.2.2) are particularly involved in these feedback loops. Temporal and spatial activation of implicated Rho GTPases is achieved by GTPase-activating proteins and guanine nucleotide-exchange factors. Small GTPases can form complexes to amplify signals that generate persistent cell polarization.

[8] Clathrin-coated vesicles are used for intracellular transport (Chap. 9). In particular, clathrin transports hydrolases from the Golgi stack to lysosomes.

the cell outer surface consists of carbohydrate moieties of membrane glycolipids and -proteins. Several constituents are transmembrane proteins that can be linked to the cytoskeleton. The glycocalyx also comprises cell-adhesion molecules (Sect. 7.5) that enable cells to adhere to apposed cells and/or the extracellular matrix, as well as ensure cell migration during embryo- and fetogenesis, as well as tissue repair and remodeling.

This layer functions as a barrier between the cell and its surrounding that must be crossed by nutrients, messengers, and drugs. It cushions the plasma membrane from stresses. Cells have distinct glycocalices that permit their identification. The glycocalyx allows the immune system to recognize the body's cell that then selectively attack transplanted, infected, and cancerous cells. Glycocalyx also serves as a mediator in cell signaling. Its dynamical nature results from thermal fluctuations. In addition, it enables sperm to recognize and bind to eggs (fertilization).

At the surface of endothelial cells, proteoglycans and glycoproteins possess acidic oligosaccharides and terminal sialic acids. The polyanionic nature of these proteoglycan and glycoprotein constituents yields a net negative charge to the glycocalyx. Glycocalyx coating on the wetted endothelial wall in the lumen of blood vessels prevents leukocyte binding in normal conditions.

At the surface of erythrocytes, the glycocalyx determines their transport inside small blood vessels as well as the basis for compatibility of blood transfusions.

4.2.5 Mechanotransduction

The body's cells respond to mechanical stimuli (applied strain and stress fields as well as stiffness of the extracellular matrix). The cell is endowed with mechanosensitivity. Cell contractility adapts immediately to the matrix stiffness, i.e., before any response of signaling cascades.[9] The force-dependent kinetics of myosin binding to actin during cell response to matrix stiffness is similar to that of myocyte adaptation to load [173]. Cells contract in less than 100 ms. The actomyosin cytoskeleton senses and reacts to the rigidity of its environment.

The stiffness of the environment can be modulated using a parallel microplate technique [174]. On soft substrates, cell contractility can generate high strains and low stresses. On stiff substrates, cell contractility can develop high stresses.

Matrix rheology influences cell fate, especially spreading, motility, and differentiation [174]. In a soft matrix that mimics brain consistency, precursor cells differentiate into neurons. In a rigid matrix that ressembles a bony structure, they evolve as osteoblasts.

---

[9] Mechanochemical signaling cascades and subsequent structural reorganization are characterized by a time scale of the minute order.

4.2.6 Self-Destruction

Cells have the capability of self-destruction, as they can undergo a programmed cell death, the so-called apoptosis (Vol. 2 – Chap. 4. Cell Survival and Death). Cell apoptosis is controlled for both tissue maintenance and development. Apoptosis is a cell quality control mechanism that removes unwanted and dysfunctional cells from the organism. On the other hand, apoptosis is prevented during the development and progression of cancers.

The execution of apoptosis relies on caspases. Caspase acticity is controlled by both activators (e.g., cytochrome-C) and inhibitors (e.g., inhibitors of apoptosis) [175]. Caspases are synthesized as zymogens that are cleaved to generate active caspases in response to developmental signals as well as persistent cellular stresses (accumulation of unfolded proteins, cytoskeletal and DNA damages, loss of cellular adhesion, increased concentration of reactive oxygen species, hypoxia, and viral invasion).

4.3 Cellular Membranes

Cellular membranes are organized, flexible, dynamical structures that separate the cell from the extracellular fluid as well as compartmentalize the cell into a nucleus and cell organelles.

Cellular membranes prevent free crossing by ions and molecules. Nevertheless, they act as filters, regulate the active transport across membrane-bound components, and transduce cues, as they contain proteic and polysaccharidic compounds that act in cell transport and signaling.

Cellular membranes communicate with one another. Vesicles bud off from one membrane, travel, and fuse with another. Organelle identity, maintenance, and function require membrane regionalization. Between-organelle cooperation uses many pathways. Moreover, organelles involved in cellular metabolism (endoplasmic reticulum, Golgi apparatus, mitochondria, and peroxisomes) are connected by an endomembrane network that provides functional continuity between organelles.

4.3.1 Membrane Structure

Cellular membranes are phospholipid bilayers[10] (6–10 nm), characterized by fluidity, but relative stability. The lipid composition of a membrane defines the organelle identity, flexibility, and permeability of the bilayer, and its

---

[10] The hydrophobic chains are directed toward the cell and hydrophilic heads outward.

interaction partners.[11] In fact, these structures are mainly mixtures of many types of lipids and proteins.[12]

The lipid bilayer is a composite, dynamical structure built by hydrophobic and between-constituent interactions. Spatial heterogeneity in the lipid bilayer is associated with a lateral pressure field within the membrane thickness. Membrane tension leads to changes in membrane thickness linked to variations in hydration level of the lipid bilayer. Membrane stretching reduces repulsive interactions and elevates interfacial tension because of increased exposure of hydrophobic lipid tails to water, thus modifying the lateral pressure profile [176]. Membrane shearing not only affects the lateral pressure profile, but also the bilayer membrane density profile [177]. Variations in the plasmalemmal lateral pressure pattern are able to activate mechanosensitive membrane proteins, shifting the protein conformation toward the appropriate state.

Both fluidity and lateral organization are needed for membrane functioning. Heterogeneous lipid bilayers are characterized by the existence of dynamical nanodomains with distinct composition. Therefore, patchy membranes ensure many functions according to their structure and lipid–protein matching.

Various techniques have been applied to investigate the lateral heterogeneity of cell membranes. Scanning probe methods, such as x-ray crystallography or atomic force microscopy, give limited information on composition and organization on length scales of tens to hundreds of nanometers (i.e., length scale greater than the characteristic size of protein assembly [$\sim 10\,$nm]). Optical methods (light microscopy, and infrared and coherent Raman imaging techniques) have limited spatial resolution. Fluorescence microscopy is specific to the labeled component and can alter between-compound interactions. Secondary-ion mass spectrometry with isotopic labeling of molecular species allows detection of variations in the gel-phase composition of trapped fluid-phase nanodomains with a lateral resolution of about 100 nm within a phase-separated lipid membrane [178].

4.3.2 Structural Lipids

Lipids are energy storage molecules (e.g., triacylglycerol and steryl esters in lipid droplets) and membrane components that can be used as chemical identifiers of specific membranes.

---

[11] The inner layer of the plasma membrane has a high concentration of cholesterol, whereas the outer leaflet has a high level of sphingomyelins and glycosphingolipids. The membrane of the endoplasmic reticulum has a low cholesterol and large unsaturated lipid content.

[12] Almost all membrane proteins are oligomeric; many are hetero-oligomers.

4.3.2.1 Phospholipids

The main structural lipids in cell membranes are glycerophospholipids: phosphatidylcholine (PtdCho), phosphatidylethanolamine (PtdEtn), phosphatidylserine (PtdSer), phosphatidylinositol (PtdIns), and phosphatidic acid (PA). Phosphatidylethanolamines generate membrane curvature.

The main acidic phospholipids in cell membranes are phosphatidylserine, phosphatidylinositol, and phosphatidic acid. Phosphatidylinositol is phosphorylated at the 3-, 4- and/or 5-positions to generate phosphoinositides, which are much less abundant in cellular membranes despite concentration changes during signaling.[13]

The hydrophobic part of polar lipids (diacylglycerol of glycerophospholipids) self-associates and the hydrophilic region can interact with elements of the aqueous environment. The lipid composition of each leaflet of the membrane bilayer markedly differs. This asymmetry among the bilayer leaflets is maintained by ATP-dependent flippases. Lipid mixtures adopt 2 fluid phases, liquid-disordered and liquid-ordered (membrane raft). Transient rafts, or nanodomains, can coalesce and become stabilized during signaling or formation of vesicles for the intracellular transport. Lipids serve as a barrier to avoid dissemination of products, but allow budding, tubulation, fission, and fusion for membrane trafficking. Besides, lipids can be transported between cellular organelles via zones of apposition between donor and acceptor compartments.

Lipids are quasi-symmetrically distributed between the 2 leaflets of the membrane bilayer of the endoplasmic reticulum. On the other hand, the membrane of the Golgi stack and endosomes as well as the plasma membrane have an asymmetric lipid distribution. The outer layer is enriched with sphingomyelin and glycosphingolipids, and the inner layer enriched with phosphatidylserine and -ethanolamine. Lipids can translocate from the external to the internal leaflet and cross the bilayer according to the local physicochemical state.[14]

Membrane transporters, such as P4-type ATPases and ATP-binding cassette transporters, selectively translocate lipids. ATP-independent scramblase participates in the phospholipid distribution between the 2 leaflets. Lipids either freely flip across the membrane or use flippases.

---

[13] Levels of phosphatidylinositol (3,4,5)-trisphosphate ($PI(3,4,5)P_3$) and phosphatidylinositol (3,4)-bisphosphate ($PI(3,4)P_2$) are strongly and transiently (10–30 mn) increased by phosphoinositide 3-kinases that are activated by receptors [179]. $PI(4,5)P_2$ decays slightly when it is targeted by phospholipase-C or phosphoinositide 3-kinase.

[14] The rate of lipid displacement between bilayer leaflets is fast or slow according to the size, charge, and polarity of the head region of polar lipids, ranging from seconds for ceramide to days for glycosphingolipids [180].

The asymmetric lipid distribution is associated with membrane function, especially in cell–cell or cell–matrix interactions and cell transport.[15]

4.3.2.2 Cholesterol

Cholesterol is a structural component of cellular membranes that generates a semipermeable barrier, regulates membrane fluidity, and modulates the activity of membrane proteins. Cholesterol and sphingolipids form *membrane rafts* with a given affinity for specific proteins. They favor protein assembling for signal transduction, membrane trafficking, and cell adhesion.

Cholesterol originates from 2 sources: diet and de novo synthesis with a ratio of about 30:70. Cells synthesize cholesterol using acetyl CoA and the mevalonate pathway. The rate-limiting enzyme of cholesterol synthesis is hydroxy methyl glutaryl coenzyme-A reductase (HMGCoAR). Cholesterol synthesis and processing are regulated by 2 main nuclear receptors: (1) sterol regulatory element-binding proteins (SREBP) to increase cellular cholesterol level and (2) liver X receptors (LXR) for cholesterol removal (Vol. 3 – Chap. 6. Receptors).

Excess cholesterol in the endoplasmic reticulum is esterified by acylCoA–cholesterol acyltransferase (ACAT) and stored with caveolin and possibly triglycerides in cytoplasmic lipid droplets that bulge from the endoplasmic reticulum [181].

Cholesterol processing occurs in different cell locations, once cholesterol has been delivered into appropriate sites by vesicular (along cytoskeletal routes) or non-vesicular (possibly with lipid transfer proteins) transport.[16]

Cholesterol and its metabolic enzymes are distributed heterogeneously among intracellular membranes (20–25% of plasmalemmal lipids; 1% of total cell cholesterol in the endoplasmic reticulum) and cell types. Cholesterol metabolites (steroids[17] and bile acids) act as signal transducers and solubilizers of other lipids.

---

[15] Phosphatidylserine exposed on the cell surface serves in blood coagulation and phagocytosis. The absence of balance between lipid levels in membrane leaflets contributes to the membrane bending, which is used for vesicle budding (Chap. 9).

[16] Between-cell compartment lipid transfer protein is used by ceramide. The ceramide transfer protein (CerT) belongs to the START protein family (START: steroidogenic acute regulatory protein-related lipid transfer protein). Other START family members bind cholesterol. Between-membrane transport also uses endosomes and low-density lipoprotein receptor.

[17] Synthesis of steroid hormones from cholesterol is catalyzed by cytochrome-P450 CyP11a1 (a.k.a. cytochrome-P450 cholesterol side-chain cleavage [P450SCC]) in the inner mitochondrial membrane. Steroidogenic acute regulatory transfer protein (StART) and translocator protein of the outer mitochondrial membrane (TsPO; a.k.a. peripheral (mitochondrial) benzodiazepine receptor [PBR]) serves in the cholesterol transfer from the outer to inner mitochondrial membrane for steroidogenesis in steroidogenic cells (adrenal glands, brain, and gonads). Other proteins participate in the formation of TsPO–StAR complex, such as PBR-

Sterol sensors and transfer proteins include members of the START family of steroidogenic acute regulatory protein-related lipid transfer proteins (StARTs). Lipoprotein–cholesterol complex serves for secretion and delivery to extracellular acceptors. Circulating cholesterol is largely found in high-density lipoproteins. Several ATP-binding cassette (ABC) transporters cooperate to release cholesterol. Plasma cholesterol level, which is easily measured, reflects the exchange of cholesterol between different cell types.

4.3.3 Membrane Proteins

Membrane proteins either are contained within the bilayer or have structural ectodomains (outside the lipidic layer), which can be larger than the protein endodomain. Certain proteins associate and dissociate with lipids during their activities.

Many membrane proteins are inserted as soon as they emerge from ribosomes via a protein-conducting channel. Protein insertion across the membrane bilayer is catalyzed by a translocon complex. The translocon also acts on the topology of the inserted protein parts, the cytoplasmic side of membrane proteins being positively charged. Afterward, flexible tertiary and quaternary protein structures are built.

Post-translational modifications[18] of proteins, reversible or not, regulate their function and structure. Protein activity, stability, location, and molecular interactions can then be modulated. Glycosylation of plasmalemmal proteins enables recognition and communication between cells.

4.3.3.1 Protein–Lipid Interactions

Membrane lipids interact with adjoining lipids and proteins. Anchoring of the cytoskeleton to the plasma membrane and its cortical dynamics requires the interaction of cytoskeletal proteins with lipid-binding domains. Membrane lipids are needed in membrane trafficking and signal transduction. Stimulation of plasmalemmal receptors induces the transient recruitment of signaling proteins to the cell membrane. Recognition of membrane lipids by proteins allows intracellular transport between cell organelles. Membrane lipids can also act as cofactors for certain membrane proteins. For example, cytochrome-C oxidase requires cardiolipin.

---

associated protein PAP7 and inactive P90 ribosomal S6 kinase RSK1-binding and -regulating PKA regulatory subunit PKAR1α [181].

[18] Post-translational site-specific modifications include conformational changes, phosphorylation either at a specific amino acid or at a small number of amino acids in a given protein, glycosylation, acylation, ubiquitination, redox processes, etc. (Sect. 5.12).

Mutually compatible lipids and proteins of the cell membrane form intimate contacts[19] to create a functioning barrier. The fluidity of the lipid bilayer
membrane can affect the lateral diffusion of molecules and molecular interactions. Higher membrane fluidity can modify the conformation of protein
receptors embedded in the lipid bilayer.

Cell membranes present a mixture of disordered and ordered lipid phases
depending on lipid composition and organization. Cholesterol is one of the
most important regulators of lipid organization. Membrane proteins preferentially associate with a particular type of phase according to the type of
anchors[20] and interactions with membrane components. Protein translocation channel, or *translocon*, mediates the insertion of membrane proteins into
the phospholipid bilayer.

The specificity of protein binding to phospholipids is related to different mechanisms: (1) temporal specificity owing to the presence of a second
messenger; (2) spatial specificity due to membrane curvature magnitude; and
(3) recognition specificity of rare membrane components that are restricted
in their location, time of synthesis, or both.

Phosphatidylethanolamine and cardiolipin incorporate proteins and modulate their activities. Acidic phospholipids, especially phosphatidylserine and
phosphoinositides, are the primary binding targets of proteins for membrane
recognition. Lipids targeted by specific membrane-binding proteins are often
lipid second messengers, such as diacylglycerol and phosphoinositide 3-kinase
products.

4.3.3.2 Membrane Proteic Clusters

Proteins in the plasma membrane are not randomly distributed, but are
mainly partitioned into clusters. Proteins within the cluster can exchange
with freely diffusing molecules. Nanodomains in which membrane proteins are
sorted are stabilized by cholesterol; they support molecular interactions. Signaling molecules can regulate cell functioning by modulating protein translocation, recycling proteins to and from the plasma membrane. Plasmalemmal
proteins are important for communication between the cytosol and the cell
surface on the one hand, the cell and adjoining cells or the extracellular matrix on the other. The membrane is bound to the cytoskeleton proteins (actin,
myosin, fodrin, supervillin,[21] etc.).

---

[19] The lipid bilayer provides the basic barrier into which are plugged membrane
proteins with particular functions (transport and environment sensing). Proteins
inserted in the plasma membrane must operate within the lipid bilayer without
introducing leak and without disturbing the membrane permeability. Lipids can
form a shell around inserted proteins. Phospholipids with their fatty acyl chains
interact with the hydrophobic surface of the protein [182].

[20] Kinases of the SRC family associate with ordered domains, whereas Ras GTPases
prefer disordered domains.

[21] Supervillin is associated with membrane rafts.

Table 4.4. Septins constitute a category of molecules that can be decomposed into several families (Source: [184]).

Family 1A	Sept3, Sept9, Sept12
Family 1B	Sept6, Sept8, Sept10, Sept11, Sept14
Family 2B	Sept1, Sept2, Sept4, Sept5, Sept7, Sept13

4.3.3.3 Septins

Septins segregate membrane areas into discrete domains, intervene in exocytosis and signaling, act as recruitment scaffolds, interact with both actin cytoskeleton and microtubules, maintain cell asymmetry, and coordinate cell growth and division.[22] Septin scaffolds recruit numerous proteins to form proteic complexes involved in cell polarity and growth, membrane trafficking, and organization of the cytoskeleton, in addition to the cell division cycle via the coordination of nuclear division and cytokinesis.

Septins belong to the infraphylum of the guanosine triphosphatase (subphylum of the phosphate-binding (P)-loop nucleoside triphosphate hydrolase (triphosphatase [GTPases and related ATPases]; NTPase)[23] like the RAS hyperfamily of small GTPases, translation factors, the OBG (HflX) class of GTPases,[24] and nanomotor ATPases of the myosin–kinesin superclass [184]. Septins that bind GTP form small complexes that oligomerize to build filaments, rings, and gauzes.

Septins assemble into multimers and form non-polar filaments. In particular, trimeric septins such as the Sept2–Sept6–Sept7 complex assemble into flexible heterohexamers and then apolar filaments. Fourteen types of septin GTPases are known in humans (Table 4.4). Filament assembly is stimulated by septin interaction with membrane components phosphatidylinositol polyphosphates.

Septins that localize to the base, tip, or throughout projections shape and compartmentalize growth [185]. In particular, septins in between-cell partitions compartmentalize pre-existing cellular material (Table 4.5). Septins at the cell cortex or within the cytoplasm are involved in material transfer and organization of the cytoskeleton.

In ciliated epithelial cells, septins contribute to the formation and maintenance of cilia.[25] Septin-2 forms a ring at the base of cilia, where it creates a barrier to compartmentalize the plasma membrane and insulates the ciliary

---

[22] Septins may coordinate chromosome dynamics and cytokinesis.

[23] The P-loop is a common motif in enzymes that hydrolyze nucleoside triphosphate.

[24] In plants and bacteria; OBG: stage-0 sporulation operon-related GTPase; HflX refers to hflX gene of the Escherichia coli genome in the hflA locus.

[25] Cilia form after the recruitment to the cell cortex of a microtubular centriole [186]. The centriole matures into a basal body that assembles an axoneme, the microtubule shaft of the cilium.

Table 4.5. Mammalian septin locations (Source: [185]). Septins are cytoskeletal proteins that assemble into hetero-oligomeric complexes and filaments. Mammalian septins can be detected, in particular, at the neck of dendritic spines and base of sperm flagella and cilia.

Type	Sites
Sept1	Midbody, spindle poles, cleavage furrow
Sept2	Plasma membrane, stress fiber, microtubule, mitotic spindle, cleavage furrow
Sept3	Neuron soma, nerve endings
Sept4	
Sept5	Plasma membrane, microtubule, cytoplasm
Sept6	Microtubule, stress fiber, cytoplasm, mitotic spindle, cleavage furrow
Sept7	Stress fiber, cytoplasm
Sept8	Cytoplasm, vesicles
Sept9	Microtubule, mitotic spindle, cleavage furrow
Sept10	Cytoplasm, nucleus
Sept11	Microtubule
Sept14	Cytoplasm, mitochondrion, stress fiber, cleavage furrow

membrane from the plasma membrane, thereby preventing membrane protein diffusion out of the cilium [187, 188].

In the primary cilium, Hedgehog receptors localize to the ciliary membrane. Septin-2 influences the signaling by secreted Hedgehog. It sets up a lateral diffusion barrier to support Hh receptor confinement as well as other ciliary proteins to the ciliary membrane. The planar cell polarity effector Frizzled-related protein-1 (FRP3 or FrzB)[26] cooperates with Sept2 for the assembling of the septin rings at the base of cilia, i.e., normal ciliogenesis and collective movement of cilia [187].

4.3.4 Organelle Membrane

The cell organelles are membrane-bound compartments with given structures and functions. The composition of a given organelle can vary between different cells. Lipid bilayers form a barrier between organelle lumens and the cytosol. These bilayers contain many transmembrane proteins, which interact with the cytoskeleton and signaling pathways. Organelles are composed of resident and transient molecules for given functions. The composition of organelles thus changes with time. Chemicals are delivered to and transported from

---

[26] A.k.a. Fritz and Frezzled.

the organelles. The transfer is mediated by components of the cytoskeleton and its associated compounds.[27]

Cell organelles synthesize and degrade molecules, and export manufactured substances and import nutrients, using transport vesicles. Accurate identification of target organelles requires activated GTPases[28] (Vol. 4 – Chap. 8. Guanosine Triphosphatases and their Regulators) and label substances such as phosphoinositides[29] (Sect. 7.2.2.2).

---

[27] Mass transfer for absorption, adsorption, and possibly phase change (e.g., evaporation) describes the flux of mass from one site to another. The simplest process, the diffusion of a gas from a high-concentration to a low-concentration compartment is governed by the Fick's law of molecular diffusion:

$$J = -\mathcal{D}\nabla c,$$

where J is the diffusion molar flux (i.e., the amount of substance per unit area per unit time [mol/(m^2.s)]), $\mathcal{D}$ the substance diffusivity, c the substance concentration (mol/m^3), and ∇ a vector differential operator (Sect. 1.9.2). The Fick's second law predicts the rate of concentration change by diffusion:

$$\frac{\partial c}{\partial t} = D\,\nabla^2 c.$$

In the cytosol as well as nucleosol, in which water can dissolve carbohydrates, fatty acids, amino acids, peptides, nucleobases, nucleosides, and nucleotides, diffusion of small molecules, such as ions and gas (e.g., NO and O_2), is the main mechanism of transport. Diffusion of solutes in water depends on the lateral speed of water molecules. Diffusion of small molecules inside the cytosol is slower than in pure water, due mostly to collisions with numerous macromolecules. The intracellular material transfer using cyto- and nucleoskeleton tracks and associated nanomotors is faster than diffusion. The convective mass transfer relies on mass transfer coefficient (h_m; [m/s]):

$$J = h_m \Delta c,$$

where J is the convection molar flux of the substance per unit area. The dimensionless mass transfer coefficient is the Sherwood number Sh $= h_m L/\mathcal{D}$. In a flowing fluid, the concentration boundary layer is a region of the flow characterized by species fluxes and concentration gradients between the surface of the mass source and the free stream, the thickness (δ) of which rises in the streamwise direction.

[28] Rab and ARF GTPases yield organelle identity. Both Rabs and ARFs are located in specific organelles. The GTP-bound (active) form is associated with membranes, the GDP-bound (inactive) form with the cytosol. Proteins (GDP-dissociation inhibitor displacement factors, guanine nucleotide-exchange factors) that regulate the GDP–GTP cycle of Rab GTPases are selection factors rather than linkers between the membrane lipid anchor and Rab GTPase. The restricted distribution of small GTPase ARF is also controlled by their guanine nucleotide-exchange factors.

[29] Phosphatidylinositol 3-phosphate (PI(3)P) localizes to early endosomes, PI(3,5)P$_2$ to late endosomes [183], PI(4)P to the Golgi complex and plasma membrane. The product PI(4,5)P$_2$ serves as a landmark for proteins targeting

Proteins involved in transport include (Chap. 9): (1) coat proteins that generate transport vesicles; (2) motor proteins that move vesicles and organelles along the cytoskeleton tract within the cytosol; and (3) tethering proteins that attach the vesicles to destination organelles before fusion.

4.3.5 Membrane Curvature

Cellular membranes determine the configuration[30] of the cell and its organelles.[31] Six modes of membrane curvature exist at the plasma membrane according to the positive or negative curvature (curvature center outside and inside the cytosol, respectively) and the location of the curvature (tip, neck or shaft of membrane tubulations). Protrusions can be considered as inverted membrane tubulations with a positive curvature at their neck and negative curvatures at their shaft and tip [194]. Invaginations are characterized by a negative curvature at their neck and positive curvatures at their shaft and tip.

Cellular membranes are curved. Membrane shapes can be described by the principal curvatures, κ_1 and κ_2 or the radii of curvature, $1/\kappa_1$ and $1/\kappa_2$. The curvature is small when the radius of curvature is much larger than the membrane thickness. Two indices can be used: (1) total curvature ($\kappa_1 + \kappa_2$) and (2) Gaussian curvature ($\kappa_1 \cdot \kappa_2$).

The curvature of deformable membranes results from interactions (particularly electrostatic ones) of membrane constituents, lipids and proteins, and applied forces.[32] The curvature of membrane nanodomains, or patches, is modulated by: (1) lipid composition,[33] (2) inserted proteins and scaffold proteins at the membrane periphery, and (3) cytoskeletal activity. Membrane patch curvature is required in vesicle formation.

---

the plasma membrane. Diacylglycerol recruits enzymes of the protein kinases-C family (superfamily AGC, subclass of cytoplasmic protein Ser/Thr kinases, class of protein Ser/Thr kinases) at the plasma membrane. Phosphatidylserine is implicated in protein recruitment to the plasma membrane.

[30] The biconcave disc shape of erythrocytes that do not have internal membranes provides the optimal surface-to-volume ratio necessary for fast oxygen exchange between hemoglobin and tissues. Furthermore, erythrocytes change their shape to navigate in blood vessels of various sizes. Cells that generally have internal membranes can also change their shape, especially during migration. Moreover, cell configuration varies when the cell undergoes stresses, particularly sheared endothelial cells at the interface between the flowing blood and the vessel wall.

[31] Organelle shapes maximize the surface area for a low internal volume for efficient transport of molecules. The Golgi stack is rather saccular and fenestrated, the endoplasmic reticulum rather tubular with small cisternae, both configurations having similar transfer efficiency.

[32] Small GTPase SAR1 curve lipid bilayers. Adaptors, coat proteins, amphiphysin, epsin, dynamin, and endophilin are involved in membrane curvature. Moreover, certain proteins might sense membrane curvatures [189].

[33] Lysophosphatidic acid and phosphatidic acid induce opposite curvatures.

Membrane remodeling occurs during molecular transport and organelle genesis. It requires specialized proteins that sense and create membrane curvature. Protein interactions and proteic cluster formation during these processes limit the bending energy [190]. Proteins that possess BAR domains (Bin–amphiphysin–Rvs dimerization domains) generate local membrane curvatures over a tiny length scale (~ 3 nm), especially in vesicle formation and T-tubule organization in myocytes that match the curvature of the BAR domain [191]. BAR domain-containing proteins deform membranes to a geometry that matches their membrane-contact and -binding surface.

Membrane-deforming proteins of the BAR superfamily connect the actin cytoskeleton to cell membranes. They possess [194] (Tables 4.6 and 4.7): (1) BAR; (2) extended Fes-CIP4 homology (EFC; a.k.a. FCH-BAR or simply fBAR); and (3) IRSp53-MIM homology domain (IMD; a.k.a. inverse-BAR or simply iBAR). These domains operate in membrane curvature formation, recognition, and maintenance. BAR domain-containing proteins actually have structurally determined positive and negative curvatures of membrane contact at their BAR, fBAR, and iBAR domains that generate and maintain local membrane curvature by binding to the membrane. In addition, Src homology 3 (SH3) domains of BAR superfamily members bind to actin regulators Wiskott-Aldrich syndrome proteins (WASP) and WASP-family verprolin homologs (WAVe) as well as large GTPase dynamin that pinches membrane into vesicles. The BAR–EFC–IMD–SH3 motif may orient the actin filaments toward the membrane for each subcellular structure. These membrane tubulations are also considered to function in membrane fusion and fission.

Lipids and membrane-anchored proteins sense the curvature of membranes. Membrane folding causes formation of small fissures that can serve for the recruitment and fixation of signaling mediators. Sensing relies predominantly on a higher density of binding sites on curved membranes [192]. Membrane shape may act as a cue for protein localization independent of membrane composition in lipids or proteins. Proteins sense membrane curvature using N-terminal amphiphilic α-helices, but not BAR domains [193].[34] Membrane curvature recognized by proteins can then be utilized for protein sorting, transfer, and signaling.

The plasma membrane contains various types of nano- and microstructures (size $\mathcal{O}[10\,\text{nm}]$–$\mathcal{O}[1\,\mu\text{m}]$) that are supported by the cytoskeleton and membrane-binding proteins. These substructures, such as protrusions (e.g., filopodia and lamellipodia) and invaginations (e.g., caveolae and clathrin-coated pits [width 100–200 nm]), are associated with intracellular transport of molecules (Chap. 9) and cell migration (Vol. 2 – Chap. 6. Cell Motility).

---

[34] Epsins have membrane-deforming capability via the insertion of an α-helix, a common motif of the secondary structure of proteins, into the membrane that causes positive curvature [194].

Table 4.6. Domains that can be observed in members of the superfamily of BAR domain-containing proteins (**Part 1**; Source: [194]). Evolving membrane curvature at membrane contact surface may result from the sequential recruitment of a different domain. In addition, all of the domains within the BAR domain superfamily form homodimers. The BAR domain can form a crescent-shaped dimer with a positively charged concave surface that links to the negatively charged inner surface of the plasma membrane. BAR Domain-containing proteins amphiphysin, endophilin, formin-binding protein FBP17, sorting nexin SNx9, and formin-binding protein-1-like molecule (FBP1L or transducer of CDC42-dependent actin assembly protein TOCA1) have an SH3 domain to bind to nWASP and dynamin, among others. They operate in actin polymerization and membrane scission. Amphiphysin has a N-terminal amphipathic helix that precedes the BAR domain (hence the name nBAR domain). Protein that interacts with C-kinase PICK1 (or protein kinase-Cα-binding protein-1) has BAR (to prevent ARP2–ARP3-induced nucleation of actin filaments) and PDZ domains, but not SH3 domain. The EFC domain of CIP4, FBP17, FCH domain only protein FCHO2, pacsins-1 and -2 (syndapins-1 and -2), and TOCA1 also form crescent-shaped dimers. The EFC domain tethers to phosphatidylserine (PtdSer) and PI(4,5)P$_2$. It creates a much larger curvature than that generated by the BAR domain. It is thus involved in shaft curvature of tubulation. Pacsin EFC domains have a greater concave surface and create narrower membrane tubulations than those induced by the EFC domains of CIP4, FBP17, and TOCA1. The curvature of the membrane contact of FCHO2 is the largest. Members of the Slit-RoboGAP family (SRGAP1–SRGAP4) contain a GAP, EFC, and SH3 domain (SRGAP1 binds to WASP and inactivates CDC42; SRGAP2 to nWASP and inactivates Rac; SRGAP3 to WAVe1 and inactivates Rac). Pro–Ser–Thr phosphatase-interacting proteins PSTPIP1 and -2 have an EFC and SH3 domain. Kinases Fer and Fes of the SRC superfamily have the FX domain. IMD domain binds to the membrane through its convex surface via PtdSer and preferentially PI(4,5)P$_2$ and PI(3,4,5)P$_3$. It can also interacts with small GTPases Rac.

Structural domain	Alias	Function
Actin filament-binding	I/LWEQ	Actin binding
Ankyrin	Ank	Protein interaction
ARF GTPase activating	ArfGAP	Inhibition of ARF GTPases
Bin–Amphiphysin–Rvs	BAR	ARP2–ARP3 complex interaction
CDC42–Rac interactive binding	CRIB	GTPase interaction
Epsin N-terminal homology	ENTH	Binding of AP2, clathrin (endocytosis)
Extended Fes-CIP4 homology (fBAR, FCH-BAR)	EFC	Binding of PtdSer, PIP$_2$ Shaft curvature
F-BAR Extension	FX	Binding of phosphatidic acid
IRSp53-MIM homology (iBAR, inverse BAR)	IMD	Negative membrane curvature (protrusion shaft and tip) Binding of PtdSer, PIP$_2$, PIP$_3$

4.4 Cell Nucleus

The nucleus is bound by an envelope, but remains connected to the cytoplasm by pores. The nucleus is compartmentalized with a nucleoplasm and

Table 4.7. Domains that can be observed in members of the superfamily of BAR domain-containing proteins (**Part 2**; Source: [194]).

Structural domain	Alias	Function
PKC-related kinase homology region 1	HR1	GTPase interaction
Phox homology	PX	Ligand binding
Phospho-tyrosine binding	PTB	Ligand binding
PSD95, DLg, and ZO1	PDZ	Membrane targeting
Rho GTPase activating	RhoGAP	Inactivation of Rho GTPases
Rho guanine nucleotide-exchange	RhoGEF	Activation of Rho GTPases
Src homology 2	SH2	Tyr^P binding
Src homology 3	SH3	Pro-Rich motif binding Binding of WASP, WAVe
Tyrosine kinase	PTK	Ligand binding
WASP (verprolin) homology 2	WH2	Actin binding

molecular complexes, such as the nucleolus (Sect. 4.4.4.1) and nucleolinus. The nucleolinus is an RNA-rich compartment, closely apposed to or embedded within the nucleolus, that operates in mitotic spindle formation [195].

Chromosomes that are mixtures of deoxyribonucleic acid (DNA; the genome) and proteins are generated during the cell division cycle and visible during the mitosis (the last phase of the cell cycle; Vol. 2 – Chap. 2. Cell Growth and Proliferation). On the other hand, chromatin, a set of chromosomic materials, is visible in quiescent cells and during the cell division cycle, only during the interphase (i.e., the 3 first phases of the cell cycle: gap 1, synthesis, and gap 2). Heterochromatin corresponds to dense, inactive regions; euchromatin to light, active zones (transcription areas; Sect. 4.4.2). Nucleolus is the site of the synthesis of ribosomal ribonucleic acids (rRNA) and ribosomal subunit assembling.

4.4.1 Nuclear Envelope and Pores

The nuclear envelope separates the genome from the remaining part of the cell. It is composed of 2 distinct membranes: inner and outer nuclear membranes. These nucleoplasm- and cytoplasm-facing nuclear membranes limit the *perinuclear space*, also called perinuclear cisterna or nuclear envelope lumen (width 20–40 nm). The outer nuclear membrane is contiguous with the rough endoplasmic reticulum.

Molecules between the nucleus and cytosol are carried through cylindrical apertures that join the inner and outer nuclear membranes. These orifices are filled with molecular assemblies, the so-called *nuclear-pore complexes* (caliber ∼ 125 nm). The nuclear-pore complex is composed of multiple copies of about 30 types of *nucleoporins* (Table 4.8).

Table 4.8. Constituents of the nuclear-pore complex (Source: [196]; Aladin, or adracalin: achalasia, adrenocortical insufficiency, alacrimia; GP: glycoprotein; NDC: nuclear division cycle protein; NLP: nucleoporin-like protein; POM: pore membrane protein; SEC13: NuP84 nuclear-pore subcomplex; SeH: SEC13 homolog; TPR, translocated promoter region). The outer (ONM) and inner (INM) nuclear membranes of the nuclear envelope join to form a grommet that insulates the pore. A transmembrane ring anchors the nuclear-pore complex to the nuclear envelope grommet. This ring is connected to inner and outer rings of stable subcomplexes of nucleoporins (NuP). Linker nucleoporins serve as a scaffold that anchors flexible, filamentous Phe-Gly (FG) nucleoporins to fill the central tube. Symmetrical and asymmetrical FG nucleoporins contain multiple Phe-Gly repeats separated by hydrophilic spacers. The nuclear-pore complex is constituted by 8 spokes. The spokes join to form 3 rings that surround a central channel. The inner ring (a.k.a. inner spoke ring and central ring) at the pore equator is sandwiched between 2 outer rings, one cytoplasmic and one nucleoplasmic. This triple-ring pore framework creates a central channel (caliber ~ 35 nm). Cytoplasmic filaments are connected to the cytoskeleton and protein synthesis machinery. The basket is a dynamical platform on the nucleoplasmic side. Nuclear transport factors (NTF) such as karyopherins (Kap) bind cargos that can then cross the channel. The nuclear-pore complex associates with numerous molecules in the cyto- and nucleoplasm via its cytoplasmic filaments and nuclear basket, respectively. Filamentous FG nucleoporins contain Phe-Gly repeat docking sites for most NTF–cargo complexes that move from a binding site to the following across the pore. Nucleoporins can be subdivided into 4 categories according to their structural role: transmembrane, core scaffold (inner and outer rings), linker, and Phe-Gly NuPs.

Structure	Constituents
Transmembrane ring	GP210, NDC1, POM121
Outer ring	Aladin, NuP37, NuP43, NuP75, NuP96, NuP107, NuP133, NuP160, SEC13, SeH1
Inner ring	NuP35, NuP155, NuP188, NuP205
Central filaments	NuP45, NuP54, NuP58, NuP62, NuP98
Linkers	NuP88, NuP93
Cytoplasmic filaments	NLP1, NuP214, NuP358
Basket	NuP153, TPR

The structure of the nuclear-pore complex comprises 2 functional regions: (1) central pore that functions as a sieve to regulate the bidirectional transport of molecules and (2) periphery that extends toward both the nucleoplasm and cytoplasm. The pore central region is constituted by a mesh of relatively small filamentous proteins that select protein types to carry, but are permeable to small molecules, metabolites, and ions. Peripheral extensions consist of asymmetrical filaments that guide cargos transferred between the cyto- and nucleoskeleton. Cytoplasmic filaments of nuclear-pore complexes may mediate the release of shuttling proteins from messenger ribonucleoproteins [196].

The *nuclear basket* is a dynamical, flexible structure that extends from the nuclear-pore complex into the nucleus. It mainly consists of 8 filaments (caliber 8–10 nm, protrusion length 60–80 nm) that converge to a distal ring [196]. It can regulate access to the channel. Fibrils (size similar to basket filaments) emanate from the basket and interconnect to form a lattice that extends along and more or less perpendicularly to the local plane of the nuclear envelope, the so-called *nuclear envelope lattice*. The nuclear basket may intervene in transcriptional control, coupling of small ubiquitin-related modifier to export machinery of ribonucleoproteins, cell cycle progression, chromatin organization, and RNA genesis. It may recruit active genes to the nuclear pore and ensure efficient transport of messenger ribonucleoproteins into the cytoplasm, as multiple components involved in this recruitment of transcribed genes participate in the correct processing, surveillance, and export of messenger ribonucleoproteins [196].[35] On the other hand, it excludes silenced heterochromatin from the transport channel.

4.4.2 Chromatin

The nuclear content is heterogeneous. Transmission electron micrographs of sections of a cell nucleus exhibit dark patches of inactive, highly condensed heterochromatin, especially in the nucleus periphery, at the nuclear edge of the inner membrane of the nuclear envelope.

Nuclear DNA and histones form chromatin. Histones are alkaline proteins that are organized into 2 classes, 6 families, and subfamilies (Table 4.9). Numerous chromatin-modifying enzymes (histone acetyltransferases and methyltransferases, protein kinases, and ubiquitin ligases) and chromatin-remodeling complexes regulate chromatin structure. Methylation of DNA sequences in promoters tends to be repressive. On the other hand, methylation of DNA sequences beyond promoters can promote gene expression.

4.4.2.1 Euchromatin and Heterochromatin

The chromatin state mainly depends on histone modifications. It determines whether genes are active (euchromatin) or silent (heterochromatin).

---

[35] Crosstalk between the nuclear-pore complex-associated Three (3′) repair exonuclease TREx2 complex and the SupTH–Ada–GCN5–AT chromatin remodeling complex (SAGA; SupTH: suppressor of Ty [transposable genetic elements] homolog; Ada: adenosine deaminase; GCN: general control of amino-acid synthesis protein; AT: histone acetyltransferase) might help relocation of transcriptionally active genes to the nuclear periphery. Nascent transcripts recruit shuttling mRNA-coating factors, TREx, and subsequently nuclear RNA export factor NxF1- and nuclear transport factor NTF2-related mRNA export protein NxT1 to construct an export-competent ribonucleoprotein [196]. The association of maturing messenger ribonucleoproteins with components of the basket strengthens in preparation for nuclear export.

Table 4.9. Histone classes, families, and subfamilies in humans. Histones undergo post-translational modifications that influence their interaction with DNA and nuclear proteins. Histones-3 and -4 have long tails that can be covalently modified by acetylation, ADP-ribosylation, citrullination, methylation, phosphorylation, ubiquitination, and sumoylation. The core of histones-2a, -2b, and -3 can also be modified. Histones-3 are activated by acetylation and monomethylation. On the other hand, di- (me^2) and trimethylation (me^3) either activate or repress H3 activity. Linker histones-1 and -5 are involved in the organization and maintenance of high order chromatin structure and in the control of DNA replication during cell division.

Class	Family	Subfamily
Core	H2A	H2A1, H2A2, H2AF
	H2B	H2B1, H2B2, H2BF
	H3	H3A1, H3A2, H3A3
	H4	H41, H44
Linker	H1	H1F, H1H1
	H5	

The genome contains 3 types of gene states [197]: (1) actively transcribed genes with engaged RNA polymerases; (2) potentially active chromatin with promoters ready to respond to activating signals that does not, or rarely, yield stable transcripts; and (3) silent heterochromatin that corresponds to most of DNA in differentiated cells (gene promoters are inaccessible to transcription factors).

Non-coding DNA boundary elements both protect active chromatin regions from heterochromatin repression and restrict interaction of enhancers with promoters. Insulators can mediate long-range DNA interactions.

Chromatin can move in the nuclear space, but with constraints. Mobility restriction of repressed chromatin is caused by nuclear membrane-associated anchor proteins. Silent telomeres can slightly move along the inner nuclear membrane. Transcripts and multiple active polymerases also localize to discrete foci. Active chromosomal loci can move over short distances (0.5–$0.7\,\mu m$) within the nucleoplasm.

Heterochromatin generally marks regions of silenced genes (centromeres, inactive X chromosome in females, telomeres, silenced genes among active genes), whereas euchromatin defines regions of active genes. Heterochromatin corresponding to chromosomal regions, such as centromeres and telomeres, remains condensed whatever the cell life phase.

A competitive balance exists between repression and activation exerted by chromatin regulators of the Polycomb and Trithorax families [198]. Polycomb repressive complex PRC2 catalyzes the trimethylation of histone-3 on Lys27 ($H3K_{27}me^3$). Polycomb repressive complex-1 and -2 (PRC1 and PRC2) can bind to one another and to nucleosomes with $H3K_{27}me^3$ histones. On the other hand, Trithorax complexes are recruited to transcriptionally active promoters, where they also trimethylate histone-3 on Lys4 ($H3K_4me^3$) to promote the gene transcription.

Short non-coding RNAs (Sect. 5.3.2) are frequently produced by RNA polymerase-2 in DNA regions that are rich in cytosine and guanine nucleotides (*CpG islets*) in the immediate proximal vicinity to gene promoters. CpG islets yield sites of histone modifications catalyzed by the proteic complexes of the Polycomb and Trithorax families. An active *transcription initiation apparatus* can recruit proteins of the Trithorax family, whereas transcripts from CpG islets form structures that recruit Polycomb repressive complexes to promoter regions for chromatin-based repression.

Variant histone macroH2a1 that abounds in inactive chromosome X represses transcription of inactivated chromosome X, but can contribute to activation of some autosomal gene expression [199].

Epigenetic reprogramming of the genome involves the removal and replacement of regulatory epigenetic elements such as DNA methylation. Elimination of DNA methylation involves the base excision repair module.

Modifications of histones in nucleosomes can determine gene expression state in specific cells in a given tissue. These marks are maintained during DNA replication that occurs during the cell divison cycle (Vol. 2 – Chap. 2. Cell Growth and Proliferation) owing to the transfer of the parental histone eviction onto nascent daughter DNA strands. The accurate propagation of histone marks during DNA replication relies on the coupling of replication with the recycling of parental histones to the daughter strands. The regulator of DNA translesion synthesis at the replication fork REV1 is required for the maintenance of repressive chromatin marks and gene silencing in the vicinity of DNA G-quadruplex (G4) structures that disrupt DNA replication [200]. The DNA repair protein REV1 is required in the replication of G4-forming sequences, thereby coupling DNA synthesis to histone recycling and preventing the incorporation of newly synthesized permissive histones.

Cell differentiation relies on permissive chromatin structure and the participation of specific transcription factors that govern cellular programs. A given transcription factor can operate in both progenitor and differentiated cells, but temporal and cellular context dictate its mechanism of action. This transcription factor can target different binding sites and function with different chromatin and transcription factors [201].

Centromeres

The centromere is the point of attachment of the mitotic spindle via the kinetochore during the cell cycle (Vol. 2 – Chap. 2. Cell Growth and Proliferation). During chromosome segregation, kinetochores act as couplers between the mitotic spindle, more precisely the plus-ends of microtubules, and chromosomes. Two kinetochores per chromosome assemble, one per centromere of each sister chromatid. The centromere is thus the contact chromosomic region between 2 identical sister chromatids until late metaphase.

The centromere binds to centromere proteins (CenP). A special nucleosome that contains histone-3 variant CenH3 (or CenPa) may serve as the epigenetic marker of the centromere [202]. Heterochromatin formation at specific chromosomal regions such as centromeres involves the *RNA interference pathway*.

Pericentromeric chromatin functions as a spring between sister kinetochores during mitosis, especially in metaphase [203]. The centromere is indeed stretched. The pericentromeric chromatin and centromeric DNA are much more flexible than kinetochores. In addition, transcription of pericentromeric regions leads to recruitment of repressive chromatin regulators to these loci.

Telomeres

Natural ends of chromosomes are protected from degradation, recombination, fusion, and recognition by the checkpoint machinery. Chromosomal ends are protected by nucleoproteic complexes, the so-called telomeres. Telomeres hence are essential for chromosome stability. They can be transcribed into telomeric repeat-containing RNAs (TeRRA) that join telomere heterochromatin [204]. Telomere-associated proteins modulate TeRRA–chromatin association. Effectors of non-sense-mediated mRNA decay at telomeres hinder links between TeRRA and chromatin.

4.4.2.2 Histones and DNA Compaction

Long DNA (length $\sim 2\,$m; $6 \times 10^9\,$base pairs; naked-DNA bore $\sim 2\,$nm; polymer persistence length [length scale over which the polymer is considered stiff] $\sim 50\,$nm) is compacted into the nucleus. Compaction of the double-stranded DNA in structures, such as chromatin and chromosomes, protects the genetic information and allows it to fit into the nucleus. However, gene expression needs access to unwrapped DNA. The opening of DNA structure leads to DNA vulnerability.

Histones participate in DNA packaging into chromatin. Protein complexes that wrap DNA include core and variant histones. Chromatin complexes not only regulate access to specific DNA sequences during gene transcription and DNA replication, but also control chromosome compaction.[36]

DNA Compaction vs. Chromosome Segregation

Cell's packaging strategy needs to retain the organization that delineates gene activity as well as the propensity of chromosomes to reside in specific

---

[36] The condensation state of chromatin varies regionally in chromosomes during interphase. It changes at the onset of mitosis when all chromosomes become highly condensed to facilitate accurate segregation to daughter cells.

nuclear territories on the one hand and chromosome segregation that must be executed with high fidelity during the cell division on the other.

The packaging machinery must distinguish different chromosomes as well as the sister chromatids from homologous chromosomes. The segregation machinery organizes chromosome construction to ensure proper duplication of genomic material and its equal distribution between daughter cells during the cell division.

Condensins

Condensins and cohesin abound in the pericentromeric region during mitosis. Condensins are large proteic complexes that intervene in chromosome assembly and segregation in eukaryotic cells. In vertebrates, 2 known types of condensin complexes exist: condensin-1 and -2. These 2 complexes share the same pair of core subunits, i.e., the structural maintenance of chromosome proteins SMC2[37] and SMC4,[38]) both being members of the group of chromosomal ATPases, but coupled with distinct sets of regulatory subunits.

During the cell cycle, condensin-2 is present within the cell nucleus during interphase. It is involved in an early stage of chromosome condensation within the nucleus in prophase. Condensin-1 is detected in the cytoplasm during interphase. It has access to chromosomes only after the nuclear envelope breaks down at the end of prophase. During pro- and metaphase, both condensin-1 and -2 contribute to the assembly of condensed chromosomes with 2 sister chromatids. These 2 complexes remain associated with chromosomes after separation of sister chromatids in anaphase.

Cohesin

Cohesin complexes bind to DNA to maintain the cohesion of sister chromatids after DNA replication, thereby ensuring their faithful segregation into daughter cells during mitosis. They can also operate in nuclear organization, i.e., chromosome condensation, telomere arrangement, and nucleolus morphology [205]. The mitotic cohesin complex also contains a pair of core ATPase subunits, the Structural maintenance of chromosome proteins SMC1 and SMC3.

Cohesin and condensin rings may yield the elastic properties of the centromere and/or chromosome. They can function as slip rings, or molecular pulleys, that distribute tension from one location to the entire network [203].

4.4.2.3 Histone Post-Translational Modifications

In addition to variant histones that augment the number of histone types, histones experience reversible post-translational modifications, such as acety-

---

[37] A.k.a. chromosome-associated protein CAPe.
[38] A.k.a. CAPc.

Table 4.10. Reversible histone post-translational modifications include acetylation, phosphorylation, methylation (mono-, di-, or trimethylation), ubiquitination, and sumoylation, in addition to variant histones (Source: [206]).

Agent	Recruited molecules
DNA-bound repressor	Histone deacetylases (negative modifiers)
DNA-bound activator	Histone acetylases (positive modifiers)
DNA-bound RNA polymerase	Histone methylases

lation, adpribosylation, citrullination, methylation (mono-, di-, or trimethylation), phosphorylation, sumoylation, and ubiquitination. Stable chromatin modifications during the cell division can be inherited.

Acetylation and sumoylation are activating and repressing process, respectively. Other modifications can have activating and repressing effects depending on targeted residues, combination of proteins, and the context (Source: [206]). Histone deacetylases, Ser/Thr phosphatases, demethylases, and ubiquitin proteases remove acetyl, phosphate, methyl groups, and ubiquitin, respectively. Deiminases alter arginine methylation, converting Arg to citrullin.

Post-translational modifications of histones can directly cause structural changes in chromatin, or indirectly acting via recruitment of effectors, either activators or repressors, both being able to target the same residue.

Post-translational modifications of histones occur during initiation as well as early and mature elongations. They are required for transcription activation. RNA Polymerases cleave nucleosomes, histone–DNA contacts being reformed in the wake of the enzyme. Nucleosome clearance in the promoter and exchange of histone variants in the promoter and the open reading frames occur during transcription.

Dynamical DNA methylation and histone post-translational modifications (Table 4.10) allow the recruitment of proteic complexes that activate or repress transcription (*epigenetic regulation*). Methylation of DNA usually leads to transcription repression by reforming nucleosomes. Both histone modifications at and DNA methylation of the promoters of tumor-suppressor genes and oncogenes are altered in cancers.

Small ubiquitin-related modifier (SUMo) reversibly alters inter- or intramolecular interactions of modified target proteins. SUMo attaches to and detaches from targeted proteins to modify their function. Reversible sumoylation contributes to chromatin structure, DNA repair,[39] transcription (e.g., enhanced sumoylation of corepressor CtBP by Polycomb family member PcG2 for gene silencing; interactions of sumoylated histone acetyltransferase P300

---

[39] For example, non-functional sumoylated DNA-repair thymine DNA glycosylase needs desumoylation to remove mutated bases.

with HDAC6 for gene repression), among other functions (Vol. 3 – Chap. 1. Signal Transduction).

4.4.2.4 Histone Acetyltransferases and Deacetylases

The histone code regulates gene expression via 2 families of enzymes: (1) histone acetyltransferases (HAT) that hyperacetylate histones and relax chromatin, thereby facilitating accessibility of transcription factors to DNA and (2) histone deacetylases (HDAC) that generally antagonize HATs and hence repress gene transcription.

Histone deacetylases are classified into (Table 4.11): (1) widely expressed type-1 and (2) tissue-specific type-2 HDACs. Histone deacetylases are also categorized into 4 subclasses: (1) subclass-1 (HDAC1–HDAC3 and HDAC8); (2) subclass-2 (HDAC4–HDAC7 and HDAC9); (3) subclass-3 HDACs or sirtuins (nicotinamide adenine dinucleotide (NAD^+)-dependent silent information regulator-2); and (4) subclass-4 with HDAC11 (Vol. 2 – Chap. 2. Cell Growth and Proliferation). Histone deacetylases can deacetylate non-histone proteins.[40]

Subclass-2 HDACs possess a limited activity with respect to subclass-1 HDACs. They form a complex with corepressors, such as silencing mediator for retinoid and thyroid receptors and nuclear receptor corepressors, as well as subclass-1 HDACs such as HDAC3. Calcium–calmodulin-dependent protein kinase (CaMK) may phosphorylate subclass-2 HDACs and promote their nuclear exportation, thereby relieving the repression of gene transcription.

HDAC and Vascular Integrity

The activity of HDACs is controlled by transcription factor myocyte enhancer factor-2 (MEF2). Histone deacetylase-7 maintains vascular integrity by repressing matrix metalloprotease-10 expression in endothelial cells. Enzyme HDAC7 controls endothelial cell proliferation, as it prevents the nuclear translocation of β-catenins and the expression of inhibitor of DNA binding ID2 [207].[41] The unspliced HDAC7 isoform precludes vascular smooth muscle cell proliferation, but not the spliced HDAC7 isoform [207].

The inactivation of HDAC7 causes defaults in blood vessel patterning with loss of tight junctions between endothelial cells.[42]

---

[40] Enzyme HDAC3 deacetylates the RelA subunit of NFκB, thus promoting its association with IκBα and enhancing NFκB nuclear exportation. Moreover, transcription factors, such as P53, E2F1, and STAT1, may also be directly deacetylated by HDACs [207]).

[41] Inhibitor of DNA binding ID2 regulates cell proliferation and differentiation. The Id2 promoter is activated by the T-cell factor TCF1 that associates with β-catenin. The TCF–β-catenin pathway also activates the transcription of cyclin-D1. In addition, HDAC7 binds to and may retain β-catenin in the cytoplasm.

[42] Histone deacetylase 7 represses MEF2 transcriptional activity and decreases matrix metalloprotease MMP10 activity (Chap. 8), thus inhibiting degradation of

Table 4.11. Classes of histone deacetylases (HDAC; Source: [207]). In humans, 18 known HDACs are categorized into 4 subclasses according to the similarity in function and targeted DNA sequences. Histone deacetylases operate in cell activation, differentiation, proliferation, migration, and apoptosis, especially in vascular endothelial and smooth muscle cells during atherogenesis. They can deacetylate non-histone proteins.

Subclass	HDAC	Location
1	HDAC1	Nucleus
	HDAC2	Nucleus
	HDAC3	Nucleus, cytoplasm
	HDAC8	Nucleus
2A	HDAC4	Nucleus, cytoplasm
	HDAC5	Nucleus, cytoplasm
	HDAC7	Nucleus, cytoplasm
	HDAC9	Nucleus, cytoplasm
2B	HDAC6	Mainly cytoplasm
	HDAC10	Nucleus, cytoplasm
3	Sirt1–Sirt7	Nucleus, cytoplasm
4	HDAC11	Nucleus, cytoplasm

Enzyme HDAC3 contributes to the maintenance of endothelial cell survival via protein kinase B. Sirtuin-1 prevents apoptosis of endothelial cells via PKB1 and FoxO3a, and may even foster their proliferation via LKB1 [207]. Histone deacetylases are also involved in apoptosis of smooth muscle cells.

Enzyme HDAC5, a repressor of the synthesis of fibroblast growth factor FGF2, may impede the migration of endothelial cells [207]. On the other hand, HDAC7, an inhibitor of platelet-derived growth factor PDGFb and its receptor PDGFRβ, and HDAC9 may support the migration of endothelial cells.

HDAC and Cardiac Hypertrophy

Histone deacetylase-5 and -9 operate as stress-responsive inhibitors of cardiac growth. Inactivation of HDAC5 and HDAC9 predisposes to cardiac hypertrophy. Subclass-2 HDAC members repress cardiac hypertrophy via gene expression that depends on myocyte enhancer factor-2. On the other hand, subclass-1 HDACs repress antihypertrophic pathways. Histone deacetylase 2 allows reactivation of a fetal gene program that leads to cardiac hypertrophy. In contrast, upregulation of the gene encoding inositol polyphosphate

proteins involved in vessel integrity [208]. Moreover, HDAC7 increases expression of tissue inhibitor of metalloproteinase TIMP1.

Table 4.12. Aliases of transcription factors of Runt-related (RUNX) and core-binding factor (CBF) family.

Type	Aliases
Runx1	CBFα2, PEBP2αa, AML1
Runx2	CBFα1, PEBP2αb, AML3
Runx3	CBFα3, PEBP2αc, AML2
CBFβ	PEBP2β

5-phosphatase-f that degrades PIP_3 yields activation of glycogen synthase kinase GSK3β and reduces cardiac hypertrophy [209].

HDAC in Hematopoiesis and Angiogenesis

Heterodimeric transcription factor constituted of Runt-related transcription factors Runx1[43] (Table 4.12)[44] and its non-DNA binding partner CBFβ intervenes in hematopoiesis and formation of blood vessels.

Nuclear Ser–Thr homeodomain-interacting protein kinase-2 (HIPK2) forms a complex with Runx1–CBFβ and P300 histone acetyltransferase. It phosphorylates (activates) Runx1 and, subsequently, P300,[45] to stimulate transcription and activities of histone acetyltransferases [211]. Deficits in P300 and Runx1 induce defects in hematopoiesis and vasculo- and angiogenesis.

4.4.2.5 Nucleosome

The core structure of chromatin is the nucleosome. The nucleosome is composed of histones H2a, H2b, H3, and H4, the so-called *core histones*, that wrap

---

[43] A.k.a. acute myeloid leukemia protein AML1, core-binding factor CBFα2, and polyomavirus enhancer-binding protein PEBP2αa

[44] The transcription factor Runx1 is associated with a type of leukemia. It belongs to the RUNX–CBF family. The non-DNA-binding partner CBFβ increases DNA-binding ability and stability of the Runx–CBFβ complex. The transcription factor Runx1 interacts with many transcription factors, such as Activator protein AP1, erythroblastosis virus E26 proto-oncogene product homolog ETS1, CCAAT/enhancer-binding protein C/EBPα, GATA1, and spleen focus forming virus (SFFV) proviral integration proto-oncogene product SPI1 [210]. It also links to coactivators, such as histone acetyltransferases P300, cAMP-responsive element-binding protein (CREB)-binding protein (CBP), and MYST3 (MOZ, YBF2 [SAS3], SAS2 and TIP60-related protein; a.k.a. Runx-binding protein RunxBP2), and Yes-associated protein YAP and WW domain-containing transcription regulator WWTR1 of the Hippo pathway, as well as corepressors transducin-like enhancer of split TLE1, histone methyltransferase suppressor of variegation 3-9 homolog SuV39H1, and histone deacetylase subunit Sin3a.

[45] Histone acetylation is regulated by histone acetyltransferases and deacetylases. Histone acetyltransferases P300 and related CREB-binding protein are coactivators of transcription factors.

DNA around them for DNA compaction. Histone octamers are composed of 2 H3–H4 and 2 H2a–H2b dimers. Assembly occurs via the $(H3–H4)_2$–DNA *tetrasome*, i.e., using $(H3–H4)_2$ tetramer and then by addition of 2 H2a–H2b dimers.

Histone-3 and -4, in particular, bear a series of covalent modifications for progression into the cell-division cycle. Trimethylation of Lys4 (K4) of histones-3 $(H3K_4me^3)$ is found in euchromatins that are open to transcription. On the other hand, trimethylation of Lys9 (K9) of histones-3 $(H3K_9me^3)$ happens in heterochromatins. Euchromatic and heterochromatic structures are early and late replicating during S phase, respectively.

On the other hand, *linker histone* H1 binds to DNA in the nucleosome and linker DNA (DNA between each nucleosome), thereby favoring the compaction of nucleosomes into chromatin structures. Therefore, linker histone H1 is an important modulator of chromatin structure.

In humans, 5 major variants of histone H1 exist (H1.0 and H1.2–H1.5). Monomethylation of Lys residues and phosphorylation of Ser and Thr residues are major post-translational modifications of histone H1 that regulate its binding to DNA [212]. Most prominent methylation sites reside in the N-terminus and globular domain of histone H1.

DNA Methylation generally silences gene expression. Like other epigenetic marks, DNA methylation may be self-propagating, thereby causing inappropriate spreading and silencing of nearby active genes. Suppressors, such as specific DNA sequences and DNA methylation modulators (DMM), antagonize this process. The DMM proteins might remove ectopic marks directed by $H3K_9me^3$ that recruits heterochromatin protein HP1 [213].

Exons (sequences that are selected from precursor mRNAs and ligated in mature mRNAs) have elevated nucleosome occupancy w.r.t. introns (removed sequences of precursor mRNAs). Exons are enriched in some histone modifications. According to the operation mode, several alternatively spliced transcripts arise.

Histones regulate not only the gene expression, but also DNA replication and repair.[46] As nucleosomes package DNA, their assembly and disassembly regulate access to the genome. Nucleosome turnover can be observed at sites of epigenetic regulation.

4.4.3 Genome

Synthesized proteins (Chap. 5) enable numerous cellular functions according to type of mediators and intensity and duration of signaling that targets gene regulators. These regulators include multiple interacting molecules.

---

[46] Histones are subjected to post-translational modifications that influence their interaction with DNA and nuclear proteins, and thus, their activity in gene regulation.

4.4.3.1 Genomic Information

The human body's cells use 2 different genetic codes to translate their hereditary information: (1) the standard code for nuclear-encoded proteins and (2) mitochondrial code. The mitochondrial genetic code has been designed, at least partially, by oxidative stress.[47]

The genetic code encodes information in primary genetic materials — DNA — and secondary elements (after DNA translation) — messenger RNA (mRNA) — using triplets of nucleotides — *codons* — that are read by complementary triplets — *anticodons* — in amino acid-carrying transfer RNAs (tRNA). Trinucleotide sequences are then translated into amino acid chains, i.e., peptides and proteins.

4.4.3.2 Genome Self-Organization

Chromosomal content resides in a particular location in the nucleus w.r.t. the periphery of the nucleus, so that it can easily condense during the cell cycle. Gene activity of a given chromosome contributes to its position. Chromosomes with low overall transcriptional activity preferentially localize close to the nuclear envelope, whereas chromosomes with highly transcribed genes are situated in the center of the nucleus. Moreover, at least some genes can move as they become activated or repressed [215].

Self-organization of chromosomes means that gene arrangement depends on functional interactions between genomic regions, such as clustering of coregulated genes to form transcription hot spots and association of regulatory elements on one chromosome with genes of other chromosomes for their activation.[48]

Multipotent hematopoietic precursor cells differentiate either into erythrocytes or neutrophils depending on growth conditions. Homologous chromosomes tend to associate during hematopoiesis according to the distribution of coregulated genes. The chromosome distribution patterns and expression profiles are tightly linked during the differentiation of a multipotent hematopoietic progenitor [216]. Computational results are consistent with experimental observations.

---

[47] Oxidant-labile methionine is a sulfur-containing proteinogenic amino acid that enriches many mitochondrially encoded proteins to ensure anti-oxidant and cytoprotective properties. Methionine can be oxidized by reactive oxygen species to form (R-form oxidizing [R])-sulfoxide (S)-methionine, which can be rereduced by methionine sulfoxide reductase using $NADPH/H^+$. Methionine serves as an anti-oxidant of respiratory chain complexes [214]. This increase in methionine content in mitochondrially encoded proteins is due to the use of a second codon for methionine by non-standard genetic code.

[48] For example, activation of the IFNβ gene on human chromosome 9 requires its connection to regulatory enhancers on chromosomes 4 and 18 [215].

Table 4.13. DNA Compartments in humans (approximated fraction of genome [%]; Source: [217]). Genes are translated into both exons and introns. Unedited mRNA transcript, or pre-mRNAs, consist of $5'$ cap, $5'$ untranslated region ($5'$ UTR, a part of $5'$ exon), coding sequence (CdS) of $5'$ exon, introns and exons, $3'$ exon with its CdS and $3'$ untranslated region, and polyadenylated tail:

$$\text{cap--}5' \text{ UTR--CdS--}(I\text{--}E\text{--}I)_{i=1}^{N}\text{--CdS--}3' \text{ UTR--PolyA.}$$

Because untranslated regions are non-coding regions of first and last exons that enables efficient translation of the transcript and control of the rate of translation and life duration of the transcript, the exon fraction is slightly overestimated.

Genic	Exonic	Non-coding	
		Intronic	Intergenic
45	3	42	55

4.4.3.3 Genome Content

The nucleus contains the genome that encodes thousands of gene products aimed at synthesizing proteins or controlling this synthesis. Distant regions of the genome of a given or different chromosomes are able to interact (*long-range communication*). Most chromosomes contain clusters of highly transcribed genes that are interspersed with regions of weakly expressed genes [218, 219].

Notations $5'$ (upstream end) and $3'$ (downstream end) of an RNA refer to the direction of the DNA template in the chromosome.[49] However, most of the human genome consists of non-protein-coding DNA (Table 4.13).

A large number of transcripts arise from intronic and intergenic regions, where RNA polymerase-2 can be detected. Another important pool of noncoding RNAs (ncRNA) correspond to known open reading frames, either in the same (sense) or opposite orientation (antisense) to the coding transcript.

4.4.3.4 Mobile Genetic Elements

Mobile genetic elements (MGE) are types of DNA sequences that can move within the genome. They include transposons (transposons, retrotransposons, and insertion sequences), plasmids (in bacteria and lower eukaryotes), bacteriophage elements, and group-2 introns (self-catalytic, self-splicing ribozymes in lower organisms).

Transposons are DNA sequences that can travel to different positions within the genome (transposition). Transposons are grouped according to the mechanism of transposition. Class-1 mobile genetic elements (retrotransposons) move in the genome once they have been transcribed to RNA and

---

[49] The $3'$–$5'$ non-coding template DNA strand guides RNA synthesis. As transcription proceeds, RNA polymerase uses base pairing complementarity with the $5'$–$3'$ coding strand to create the RNA. Transcription is thus described as occurring from $5'$ to $3'$, hence $5'$ end corresponds to upstream RNA extremity.

back to DNA by reverse transcriptase. Class-2 transposable elements move directly from one position to another within the genome using a transposase. Class-3 transposons are also called miniature inverted-repeats transposable elements (MITE).

During transposition, transposons can cause mutations and change the DNA amount. Transposons correspond to a large genome fraction. Repetitive DNA derives from transposable elements that are able to replicate within the genome. Non-coding elements originate from transposable elements. MicroRNAs, small interfering RNAs, repeat-associated small interfering RNAs, and piwi-interacting RNAs very likely derive from transposons [220]. DNA-binding proteins and transcription factors can also originate from transposases.

Transposons influence the transcriptional and post-transcriptional regulation of nearby gene expression [220]. Transposons located upstream from a gene can insert promoter sequences and introduce an alternative transcription start site, disrupt existing cis-regulatory element, or generate a new cis element. Transposons located within an intron can conduct antisense transcription, interfere with sense transcription and with pre-mRNA splicing pattern, and be incorporated as an alternative exon. Transposons can serve as a nucleation center for heterochromatin, hence silencing the transcription of adjacent genes. At the post-transcriptional level, transposons can provide an alternative polyadenylation site and a binding site for microRNAs or RNA-binding proteins.

4.4.3.5 Non-Coding Elements

Non-coding RNAs

Long-range RNA regulators are able to move inside the nucleus, hence to modulate the activity of distant loci among the different nuclear compartments. Small double-stranded RNAs participate directly in the regulation of gene expression and/or heterochromatin formation. MicroRNAs cause post-transcriptional silencing, as they stimulate the degradation of target mRNAs or inhibit their translation.

Approximately 1% of the human genome is covered with small unannotated RNAs (< 200 nucleotides) that cluster at the 5'- and 3' ends of protein-coding genes [221]. In mice, short RNAs (20–90 nucleotides) localize near the transcription start sites of more than 50% of protein-coding genes. In humans, transcription initiation RNAs (tiRNA; < 22 nucleotides) use transcription start sites of several protein-coding genes.

Non-coding elements (NCE) are conserved at varying degrees among mammalian genomes. Promoter-associated, non-coding RNAs (PAR) are unannotated transcripts that comprise both cryptic unstable transcripts (CUT) and stable unannotated transcripts (SUT). All spurious transcripts are not unstable; PARs are stable in humans. Promoter upstream transcripts (Prompt) are short, polyadenylated, highly unstable transcripts.

Both PARs and Prompts can be transcribed from the gene promoter, 5' nucleosome-depleted (5' NDR; a.k.a. nucleosome-free region), and intergenic regions in either sense or antisense orientation [221]. Bidirectional PARs and mRNAs might originate from different pre-initiation complexes (PIC) and compete for the same pool of transcription factors to initiate transcription.

Most of the promoters can lead to a transcription in either direction owing to their bidirectional character. A minority of promoter-associated, noncoding RNAs are transcribed in the same orientation as that of the coding transcript (mRNA). Yet, most of the transcripts are in antisense orientation w.r.t. the downstream gene and generate divergent transcripts [221]. Noncoding cryptic unstable transcripts (200–800 nucleotides are transcribed by RNA polymerase-2, capped, and polyadenylated. Promoter upstream transcripts are also bidirectionally transcribed from a region upstream from the transcription start sites of protein-coding genes and depend on the same promoter as that of the downstream mRNA. Antisense ncRNAs can control gene expression post-transcriptionally by inhibiting the translation of protein-coding RNAs.[50]

In humans, transcription of numerous ncRNAs may start at the 3' end of genes [221]. A 3' nucleosome-depleted region (3' NDR) resides near the mRNA cleavage and polyadenylation sites. This region may elicit non-coding, antisense transcription. It is repressed by chromatin-remodeling factor and nucleosomes. Exon-originated cryptic transcripts have also been identified. Histone chaperones and histone modification factors repress this type of transcription [221].

Telomeric repeat-containing RNAs (TeRRA) are involved in telomere maintenance and genomic stability. In humans, RNA polymerase-2-dependent telomeric transcripts (> 100 nucleotides) have been observed [221]. Production of long non-coding RNAs (lncRNAs) is regulated during growth or development by transcription factors. PIWI-associated RNAs probably mature from a transposon-derived long RNA precursor.

Regulatory ncRNAs can act in cis, at the site of transcription, or in trans. They regulate gene expression at both the transcription site and other genomic loci [221].[51]

Among small, regulatory, non-coding RNAs, microRNAs (miR; 22 nucleotides) bind predominantly to the 3' untranslated regions (3' UTR) of mRNA to silence the gene expression [217]. Small interfering RNAs (siRNA; 21 nucleotides) operate in the degradation of mRNAs. PIWI-interacting RNAs

---

[50] In humans, a ncRNA transcribed from a transcription start site upstream from the Dhfr promoter inhibits the synthesis of dehydrofolate reductase (DHFR), as it binds to GTF2b, destabilizing the pre-initiation complex, and prevents transcription [221].

[51] The long ncRNA XIST (X inactive-specific non-protein coding transcript) operates in cis by coating the chromosome X and recruiting Polycomb family proteins that form $H3K_{27}me^3$ and subsequent epigenetic gene silencing.

(piRNA; 21–28 nucleotides) repress transcription of transposons in the germ line.

RNAs with Short Open Reading Frames

Some RNAs contain only short open reading frames (< 100 codons). They can produce small peptides with unknown activity. Small ORF-derived peptides can control transcriptional programs during embryogenesis. At least, some long non-coding RNAs are involved in gene silencing and imprinting.

Small functional peptides could evolve rapidly. Random mutations that introduce start codons in existing long non-coding RNAs, or within untranslated regions of coding mRNAs, could generate small peptides that are easily selected to perform a specific function. Small peptides are able to quickly modifiy activities of target proteins without elaborate, time-consuming translation of large proteins. They can serve as temporal switches.

Non-Coding and Pseudo-Non-Coding DNA Elements

Non-coding DNA elements[52] include: (1) small components, such as short repeats, regulatory factor-binding regions and small RNA-coding sequences [217]; (2) medium-sized components, such as broad histone marks, transcripts, transposable elements, and pseudogenes; and (3) large components, such as regulatory forests and deserts, segmental duplications, and structural variants.

Simple repeats are probably generated by polymerase slippage errors. Tandem repeats are often detected at centromeres and telomeres of chromosomes, where they can support the structure [217].

Transposable elements that are active or not are grouped into DNA–based transposons and RNA-based retrotransposons. Long interspersed elements (LINE) are retrotransposons that encode reverse transcriptase. Short interspersed repeats (SINE) are fragments of RNA polymerase-3-transcribed genes that rely on LINEs for propagation. Long terminal repeat (LTR) retrotransposons are flanked on both ends by direct LTRs [217]. Pseudogenes constitute several categories, such as duplicated, processed, and unitary pseudogenes.

A large part of segmental duplications occur in tandem runs spaced less than 1-Mb apart on the same chromosome [217]. Structural variants can be generated by insertion, deletion, reciprocal translocation, or inversion. Duplications and deletions cause copy-number variation across the population.

Regulatory elements are binding sites of transcription factors. The human genome produces 17 to 19×10^2 transcription factors. The class of regulatory elements include promoters, enhancers, silencers, insulators, and locus-control regions (LCR). Promoters are regulatory sites that influence the expression of the nearest gene. The other elements act on more distant genomic locations.

---

[52] Misnomers due to limited knowledge.

Large intergenic non-coding RNAs are spliced like pre-mRNAs [217]. They function in several cellular processes. Small nucleolar RNAs ($\sim$ 90% of cell RNA content) that are generated by RNA polymerases-1 and -3 help to synthesize the translational apparatus.

4.4.3.6 Primary Transcripts

Genes that encode various RNAs are mainly transcribed in the nucleus, as some are transcribed in mitochondria. Gene transcription is carried out by RNA polymerase-1, -2, or -3. Gene expression is strongly regulated. In eukaryotes, thousands of genes that encode messenger RNAs are transcribed by RNA polymerase-2. Core promoter recognition is the first step in transcription initiation.

RNA Polymerase-2 requires 5 cofactors for promoter recognition and initiation of transcription. Initiation of transcription by RNA polymerase-2 begins by the binding of specific activators to their cognate binding sites. This initial phase triggers the recruitment of coactivator complexes and general transcription factors at promoters that serves as a platform for the assembly of the *transcription pre-initiation complex*. This complex includes general transcription factors with subunits GTF2a, GTF2b, GTF2d to GTF2f, and GTF2h that recruit RNA polymerase-2 [222].[53] This complex specifies the transcription start site. General transcription factor GTF2b is involved in start site selection and stabilization of the initial transcript.

The primary prototypical core promoter recognition factor for genes is the general transcription factor GTF2d that binds multiple core promoter elements to begin the formation of pre-initiation complexes that contain RNA polymerase-2. In higher eukaryotes, core promoters are highly diverse. The most recognizable core promoter element is the TATA box, but TATA-containing promoters correspond to a minority w.r.t. many families of TATA-less promoters.

The formation of the transcription pre-initiation complex commences with GTF2d binding to the TATA box (a DNA sequence 5′–TATA–3′ or a variant in the core promoter of approximately 24% of human genes), initiator, and/or downstream promoter element of most core promoters [222]. General transcription factor GTF2d comprises TATA box-binding protein (TBP) and 13 to 14 TBP-associated factors. Multiple subunits of GTF2d bind core promoter elements. TATA Box-binding protein binds TATA boxes. TBP-Associated factors TAF1 and TAF2 bind the initiator element. TBP-Associated factors TAF6 and TAF9 link to the downstream promoter element [224]. Some TBP-associated factors are targets of transcriptional activators that allow GTF2d to integrate signals from activators to the core promoter.

---

[53] General transcription factors operate in the transcription of class-2 genes into mRNAs. A class-2 gene codes for a protein upon transcription by RNA polymerase-2. Class-2 gene possess a promoter that often contains a TATA box. General transcription requires the formation of a pre-initiation complex.

Table 4.14. Components of core promoter recognition complexes and their functions (Source: [224]; TAF: TBP-associated factor; TBP: TATA box-binding protein; TRF: TBP-related factor). Oogenesis and spermatogenesis are the female and male type of gametogenesis.

Factor	Functions
TBP	Mitotic bookmarking
TAF3	Anchorage of GTF2d to H3K4me$_3$, hematopoiesis, myogenesis
TAF4b	Oogenesis, spermatogenesis
TAF7L	Spermatogenesis
TAF8	Adipogenesis
TAF10	Hepatogenesis
TRF2	Early embryogenesis
TRF3	Gastrulation, hematopoiesis, myogenesis, oogenesis

The general transcription factor GTF2a helps GTF2d in binding core promoters. Afterward, the remaining components of the general transcription factors (GTF2b, GTF2e, GTF2f, and GTF2h) connect to the precluster. The next step is not limited to the connection of other general transcription factors, but also the mediator coactivator complex as well as RNA polymerase-2 via either a sequential assembly or a pre-assembled RNA polymerase-2 holoenzyme.

As DNA is associated with nucleosomes, transcription activators, coactivators, chromatin-modifying factors and transcription elongation factors are involved (Table 4.14). In particular, any regulated transcription requires general cofactors that transmit regulatory signals between gene-specific activators and the general transcription machinery. During the transcription initiation, coactivators assist in both catalysis and recruitment. They transform chromatin for DNA access as well as enhance the formation of initiation complexes.

Many types of general cofactors, such as TATA box-binding protein (TBP)-associated factors (TAF), Mediator, upstream stimulatory activity (USA)-derived positive cofactors (PC1–PC5)[54] and negative cofactor NC1,[55] operate independently or cooperatively to fine-tune the promoter activity in a gene- or cell type-specific manner [222].

The transcription coactivator complex Mediator (Med) of the transcription mediated by RNA polymerase-2 bridges transcriptional activators, coactivators, general transcription factors, and RNA polymerase-2 to enhance or

---

[54] Upstream stimulatory factor (USF) homo- and heterodimers that pertain to the basic helix–loop–helix (bHLH)–leucine zipper class (bHLHzip) of transcription factors bind to the E-box element. Positive cofactor PC1 corresponds to polyADP-ribose polymerase PARP1, PC2 with Med26 to Mediator, and PC3 to DNA topoisomerase Tpo1.

[55] A.k.a. high-mobility group box protein HMGB1.

repress the formation of the transcription pre-initiation complex. Constituents of the Mediator complex participate in the formation of several functional complexes, such as the activator-recruited cofactor (ARC) cluster recruited to nuclear receptors,[56] and stimulator of RNA-binding (SRB)–Med cofactor complexes (SMCC), that, in particular, works for transcriptional activation by hepatocyte nuclear factor HNF4, another SRB–Med complex (with different components such as Med6), the NAT complex, that represses activation of transcription by RNA polymerase-2, cofactor required for SP1 transcriptional activation (CRSP) that is required together with TAF2s to activate the transcription factor Specificity protein SP1, and positive cofactor complex PC2. Mediator complex subunit Med1[57] is a component of Mediator that associates with nuclear hormone receptors (Vol. 3 – Chap. 6. Receptors). With the Mediator kinase module, Mediator forms the TRAP complex that represses transcription. With the subunit Med26, the Mediator complex known as PC2 activates transcription.

Other cofactors, such as TAF1, bTF2d-associated factor BTAF1[58] and negative cofactor NC2, can also modulate the binding to the core promoter of the transcription factors TATA-binding protein and GTF2d [222]. In general, these cofactors repress basal transcription when activators are absent and stimulate transcription in the presence of activators.

However, the primary transcripts are most often immature (inactive). These precursor RNAs undergo post-transcriptional processing. Pre-messenger

---

[56] A.k.a. thyroid hormone receptor-associated proteic (TRAP) and vitamin-D receptor-interacting proteic (DRIP) complex.

[57] A.k.a. 220-kDa thyroid hormone receptor-associated TRAP220 or -interacting protein TRIP2, peroxisome proliferator-activated receptor (PPAR)-binding protein [PPARBP], 205-kDa vitamin-D receptor-interacting protein complex component DRIP205, activator-recruited cofactor component ARC205, and P53 regulatory protein RB18a.

[58] A.k.a. TATA box-binding protein [TBP]-associated factor TAF172 and ATP-dependent helicase.

RNAs bear 5' *capping*,[59] 3' *polyadenylation*,[60] *splicing*,[61] and sometimes *RNA editing*.[62]

The concentration of mature mRNA in the cytoplasm that determines the rate of translation depends on mRNA quality-control, transport, storage, and degradation. In eukaryotes, mRNA degradation results from 3'- and 5' terminal events, such as deadenylation, decapping, and exonucleolytic cleavage [223] (Table 4.15).

4.4.3.7 Cell-Specific Programs of Transcription

The transcription pre-initiation complex, also called the RNA polymerase-2-based core promoter recognition complex, mediates interactions with chromatin to prime transcription. It can also maintain the gene expression during the cell division. Furthermore, the pre-initiation complex contributes to the regulation of cell-specific programs of transcription during development.

General transcriptor factor GTF2d is a complex of TATA box-binding protein (TBP) and TBP-associated factors (TAF) that intervenes in transcription initiation. In addition to prototypical components of pre-initiation complex, non-prototypical factors of the core promoter recognition complex are required to initiate transcription of specific sets of genes, such as TBP-associated and TBP-related factors (TRF). In particular, these factors direct cell type-specific transcription initiation. Both TRF2 and TRF3 that are highly expressed in germ cells operate in germ cell differentiation [224].

4.4.3.8 Regulation of Gene Expression

Various short non-coding RNAs — microRNAs, small interfering RNAs, and PIWI-interacting RNAs — operate as regulators. These non-coding RNAs

---

[59] After transcription of the 5' end of an RNA transcript, the cap-synthesizing complex on the C-terminus removes the γ-phosphate and attaches a methylated guanosine monophosphate (GMP) connected by a (5',5')-triphosphate linkage.

[60] In eukaryotes, mRNAs typically end with a long, 3' terminal polyadenylated tail.

[61] The sets of genes (arguments) and unedited mRNA transcripts, or pre-mRNAs (images), is related by a bijective mapping. A bijective mapping can be used to illustrate the relationship between genes or pre-mRNAs and protein types, excluding isoforms of each protein type. This mapping is lost between mRNAs and proteins, as the one-to-one function (injection or injective relationship) between a given type of pre-mRNA and produced protein is destroyed by the intermediary stage of mRNA maturation for most protein types due to alternative transcript (pre-mRNA) splicing that causes synthesis of many protein isoforms.

[62] Post-transcriptional modification of the information content in an RNA (tRNA, rRNA, or mRNA) that use diverse mechanisms. For example, modification of mRNA by adenosine deaminases that converts adenosine into inosine rewrites the informational output.

Table 4.15. Mechanisms of cytoplasmic mRNA decay (Source: [223]; CNOT: CCR4-associated negative regulator of transcription [NOT] transcription complex; DCP2: decapping enzyme subunit-2; DCPS: decapping enzyme, scavenger; miR: microRNA; PABP: polyadenylate-binding protein; PANi: polyadenylate-specific ribonuclease subunit i; PARN, polyadenylate-specific RNase; siRNA: small interfering RNA; SMG6: SMG6 homolog, nonsense mediated mRNA decay factor; XRN1: $(5',3')$ exoribonuclease-1 ZCCHC: zinc finger CCHC domain-containing protein [a.k.a. polyadenylate RNA polymerase (PAP)-associated domain-containing protein PAPD3]). Rapid degradation of mRNAs that contain a premature termination codon (PTC) results from the assembly of the PTC surveillance proteins Up-frameshift suppressors UpF1, UPF2, and UPF3 at the site of translation termination under guidance of eukaryotic release factors eRF1 and eRF3.

Enzyme	Target and function
	Endonucleases
Argonaute	Cleavage of mRNA–siRNA or mRNA–miR duplexes
SMG6	Degradation of premature termination codon-containing mRNAs
	$3'$ exonucleases
Exosome	$3'$ End not protected by PABP
	$5'$ exonucleases
XRN1	Monophosphorylated $5'$ end
	$5'$ end modification
DCP2	Decapping of RNA polynucleotides
DCPS	Decapping of RNA oligonucleotides
	$3'$ end modification
CNOT	Deadenylation
PAN2–PAN3	Deadenylation
PARN	Deadenylation
ZCCHC11	Oligouridylation

regulate gene expression upon base pairing between them and a segment of mRNA. They then cause translational repression and accelerate mRNA decay.

In eukaryotes, ribosomes are usually recruited to mRNA owing to the affinity of small ribosomal subunits for the complex composed of eIF3 and eIF4F that assembles on the mRNA cap to prime translation initiation.

Genes associated with the vasculature include 2 main sets: (1) genes involved in cell signaling and (2) genes regulating the vascular structure via connected cell and extracellular components (cell cytoskeleton, cell membrane, and extracellular matrix) [225]. Transcriptional regulation is achieved not only by import into and export from the nucleus of transcription factors, but also by interactions with DNA-binding factors. Transcriptional regulation is responsible for gene expression patterns that are specific to cell type in combination with post-transcriptional control (mRNA processing, export, and translation).

Cell differentiation is associated with stable silencing of certain gene domains and expression of relevant gene clusters that bear coordinated transcriptional regulation. Tissue-specific genes that are coexpressed during cell differentiation can be grouped in distinct chromosomal regions [226].[63]

Non-coding regions of the genome, the so-called *enhancers*, control the tissue-specific expression of a close and remote gene. Enhancers recruit regulators to decondense chromatin and promote the assembly of transcriptional complexes at gene promoters.

4.4.3.9 Replisomes

Replisomes are proteic complexes that replicate DNA. DNA Polymerases can only add deoxynucleotides to 3′ ends, i.e., DNA strands are synthesized only in the 5′-to-3′ direction. Replisomes contain several replicative polymerases and clamp loaders that oligomerize the core polymerase and activate helicase activity.

During DNA replication, a complementary new strand of DNA is synthesized along each parental strand to form a double-stranded molecule that contains old (parental) and new DNA. In parallel, chromatin duplicates to ensure memory of silencing of particular genomic regions, i.e., epigenetic information. New nucleosomes result from de novo assembly of newly synthesized histones H3–H4 dimers, in addition to recycled H3–H4 dimer from disrupted parental nucleosomes. The latter undergo splitting such as H3.3–H4 tetramers or remain intact such as H3.1–H4 tetramers [227].

4.4.3.10 Maintenance of Genome – DNA-Damage Response and DNA Repair

Genome maintenance is based on chromatin structure and dynamics as well as genome organization. DNA mismatches can be occasionally introduced during DNA replication as can DNA strand breaks caused by abortive activity of topoisomerases-1 and -2. Yet, numerous DNA damages occur per day. Genome is continuously assaulted by endogenous (e.g., reactive oxygen species) as well as exogenous (e.g., ultraviolet light and ionizing radiation) stresses that cause base damage as well as single- and double-stranded DNA breaks that are hazardous for genome integrity. Chromosome breakage can also result from viral infection and cellular transformation. In addition, genome damage and transcriptional noise increase during aging.[64]

---

[63] Six liver- and 5 colon-related chromosomal domains are highly transcribed in corresponding organs, but no brain-related chromosomal domains have yet been identified.

[64] Although gene expression levels vary among cardiomyocytes of young heart, variation in transcript levels are much more pronounced in cardiomyocytes of old heart [228].

Nevertheless, natural ends of linear chromosomes that resemble double-stranded DNA breaks are packaged into protective telomeres that suppress DNA repair. Furthermore, DNA breaks are normal intermediates in meiosis, DNA replication, as well as recombination for adaptive immunity.

In any case, single- and double-stranded DNA breaks can prevent the genome replication and transcription. When these breaks are not repaired or repaired incorrectly, mutations and chromosomal translocations that cause cancer are created. Therefore, any genetic damage is very dangerous to cells.

Immediate repair is thus needed. DNA and chromatin remodeling aims at restoring genomic stability and function. Cells are endowed with different mechanisms to repair double-stranded DNA breaks according to the nature of the breaks and cell-cycle phase during which the damage is detected.[65]

Among all the possible repair processes, both accidental and meiosis-specific double-stranded DNA breaks can be repaired by homologous recombination that involves the interaction between homologous DNA sequences, i.e., using the genetic information stored in the sister chromatid or homologous chromosome to accurately restore lost genetic information at the break site. On the other hand, non-homologous end joining directly rejoins 2 chromosomal ends with no or minimal base pairing at the junction and can generate mutations at the end joining sites. Double-stranded DNA breaks also trigger activation of DNA damage and recombination checkpoints, respectively, that regulate DNA repair and recombination and coordinate progression through mitosis and meiosis [229].

Both telomeres and double-stranded DNA breaks undergo a resection of DNA 5'-ends, to enable repair by homologous recombination or action of telomerase at telomeres.[66] The initial stage of homologous recombination and telomerase-mediated elongation are related [229]. Failure to protect the natural chromosome ends leads to chromosomal rearrangements, cell death or cancers.

PIKK Family

The family of phosphatidylinositol 3-kinase-related protein kinases (PIKK) includes catalytic subunits of DNA-dependent protein kinases (DNAPK), ataxia telangiectasia mutated (ATMK), ataxia telangiectasia and Rad3-related (ATRK) kinases, target of rapamycin (TOR), suppressor with morphogenetic

---

[65] The choice of the repair mechanism of double-stranded DNA breaks depends on the cell cycle phase, more precisely on cyclin-dependent protein kinases. Non-homologous end joining is used during G1 phase. Homologous recombination in haploid cells occurs during S and G2 phases, when DNA has replicated and the sister chromatid is available as a repair template.

[66] During the first phase of homologous recombination, mitotic and meiotic double-stranded DNA breaks undergo degradation of their 5'-ending strands to create 3'-ended single-strand DNA.

effect on genitalia SMG1,[67] as well as the catalytically inactive member transformation–transcription domain-associated protein (TrrAP). The latter is involved in transcription and DNA repair as a core component of a histone acetyltransferase complex.

The PIKK family members participate in DNA- and RNA-based processes, such as DNA-damage response, messenger RNA quality control, transcription, and translation. They thus contribute to the maintenance of genome integrity and accurate gene expression. Furthermore, they control protein production in response to nutrient supply.

Proteins of the RuvB-like class of the superclass of adenosine triphosphatases associated with diverse cellular activities (AAA) RuvBL1 and RuvBL2[68] are involved in transcription, RNA modification, DNA repair, and telomere maintenance. Together, RuvBL1 and RuvBL2 form a double hexamer. They operate in various chromatin-remodeling clusters, a histone acetyltransferase complex, small nucleolar ribonucleoprotein huddles, and telomerase reverse transcriptase complex. Moreover, they interact with various transcription factors. Proteins RuvBL1 and RuvBL2 associate with each PIKK family member [230]. Both RuvBL1 and RuvBL2 control the abundance of PIKK family members and stimulate the formation of PIKK-containing complexes.

Fanconi Anemia-Related Family of Proteins

The DNA repair response is defective in the genetic illness Fanconi anemia. Fanconi anemia patients have an augmented sensitivity to DNA interstrand crosslinking chemicals, in addition to congenital malformations, bone-marrow deficiency, and cancer. These agents crosslink DNA strands of the double helix together. The Fanconi anemia results from mutations in 13 genes involved in a DNA repair pathway that processes the damage caused by erroneous chemical crosslinks between the 2 strands of the DNA double helix. Repair of DNA double-stranded break uses the error-free pathway or error-prone repair pathway. Loss of function of these genes fosters a state of cancer predisposition. Chromosome defects in Fanconi anemia cells result from altered DNA repair as well as chromosome segregation defects during mitosis and unequal chromosome partitioning [232].

---

[67] The protein SMG1 pertains to the SMG1C complex with SMG8 and SMG9. The SMG1C complex intervenes in non-sense-mediated mRNA decay, an mRNA quality control mechanism that occurs in the cytoplasm, detects, and degrades mRNAs with premature termination codons.

[68] Proteins RuvBL1 and RuvBL2 are also called 49-kDa (TIP49 or TIP49a) and 48-kDa (TIP48 or TIP49b) TATA box-binding protein-interacting protein, respectively, as well as 54-kDa (ECP54) and 51-kDa (ECP51) erythrocyte cytosolic protein and INO80 complex subunit-H (INO80h) and -J (INO80j), respectively. In addition, RuvBL1 is termed nuclear matrix protein NMP238 and pontin, whereas RuvBL2 is named repressing pontin or reptin (or reptin52).

Table 4.16. Fanconi anemia proteins: subfamilies and function. Partners of the FANC pathway include breast cancer susceptibility protein BrCa1 and Bloom syndrome, RecQ DNA helicase-like protein (Blm; a.k.a. RecQ helicase-like protein RecQL2 or RecQL3).

Subfamily	Members	Effect
1	FancA/B/C/E/F/G/L/M	Ubiquitin ligase complex
		Target group-2 FANC proteins
2	FancD2/I	DNA repair
3	FancD1/J/N	DNA repair

The family of Fanconi anemia proteins include 13 members (FancA to -C, -D1, -D2, -E to -G, -I, -J, and -L to -N) that can be grouped into 3 subfamilies according to their function (Table 4.16). The FANC subfamily 1 comprises 8 Fanc proteins (FancA to -C, -E to -G, and -L to -M) that form the nuclear Fanc complex that acts as a ubiquitin ligase. A major event in the Fanconi anemia pathway is monoubiquitination of the FANC subfamily 2 members FancD2 and FancI that dimerize to form the so-called ID complex. Fanconi anemia subfamily-1 proteins assemble into the nuclear Fanconi anemia core complex that monoubiquitinates FancD2 and FancI. The ID complex then forms DNA damage-induced nuclear foci together with other DNA damage-response proteins. The FANC subfamily 3 encompasses FancD1 (a.k.a. BrCa2), FancN[69] and FancJ[70] that may promote DNA repair.

In addition, Fanconi anemia-associated nuclease FAN1, a monoubiquitin-binding protein, connects to the FancD2–FancI complex [231]. The monoubiquitinated ID complex localizes to sites of DNA damage, in which it recruits FAN1 protein. FAN1 Agent is a DNA branch-specific nuclease that acts in the removal and subsequent repair of the DNA interstrand crosslinks.

Upon DNA damage, the Fanc complex causes a monoubiquitination of both FancD2 and FancI in synergy with agents of cell cycle checkpoints and DNA repair, such as replication protein-A, ataxia telangiectasia and Rad3-related kinase, checkpoint kinase-1, and breast cancer-associated gene product BrCa1 [232]. FancN as well as partners of the FANC pathway BrCa1 and Bloom syndrome, RecQ DNA helicase-like protein (Blm)[71] contribute to chromosome segregation, spindle assembly, cytokinesis, and sister chromatid decatenation at anaphase [232].

Fanconi anemia proteins are components of a DNA repair pathway that facilitates homologous recombination as well as intervene in non-homologous

---

[69] A.k.a. partner and localizer of BrCa2 (PALB2).

[70] A.k.a. BrCa1-interacting protein C-terminal helicase-1 (BrIP1).

[71] A.k.a. RecQ helicase-like protein RecQL2 or RecQL3.

end joining.[72] They are particularly involved in interstrand crosslink repair during replication. In the absence of functional Fanconi anemia proteins, DNA double-strand breaks accumulate after exposure to crosslinking agents.

The Fanconi anemia pathway promotes homologous recombination repair of DNA double-stranded breaks, as it counteracts x-ray repair cross-complementing protein XRCC6,[73] a non-homologous end-joining factor. Protein FancD2 may antagonize XRCC6 activity. The Fanconi anemia pathway is thus aimed at diverting double-strand break repair away from abortive NHEJ for homologous recombination [233].

The FANC pathway is involved in rescuing abnormal ana- and telophase and preventing aneuploidy. Protein FancD2 targets spots on mitotic chromosomes [232]. Chromosomal localizations of FancD2 comprise mainly certain chromatid gaps and breaks.

In addition to the family of 13 Fanc caretaker genes that are mutated in Fanconi anemia, the Bloom syndrome characterized by mutations of the gene that encodes Bloom syndrome, RecQ DNA helicase-like protein causes cancer predisposition following genome fragility. The protein Blm collaborates with Fanconi anemia proteins during the S phase of the cell cycle to prevent chromosome instability. Moreover, crosstalk between the FANC and BLM pathways beyond the S phase rescues damaged DNA, as it leads to BLM recruitment via Fanc proteins to non-centromeric abnormal structures caused by replicative stresses [232].

DNA-Damage Response

The maintenance of the genome relies on damage sensing, and, once any damage has been detected, on cell cycle regulation, and DNA repair. The *DNA-damage response* aims at detecting DNA lesions, signaling their presence, and eliciting their repair. DNA-damage signaling and DNA repair operate synergistically and share many components. *DNA-damage sensors* recruit mediators that amplify the response via transducers and other effectors [234]. In particular, DNA-damage sensors recruit specific kinases to the sites of damage.

The major involved kinases include ATMK and ATRK enzymes. Both ATMK and ATRK phosphorylate components of the DNA damage response network, leading to the formation of proteic complexes.[74] The major targets of

---

[72] Breaks of the DNA double-strand are preferentially repaired by non-homologous end joining (NHEJ) in the G1 phase of the cell cycle and homologous recombination during replication.

[73] A.k.a. ATP-dependent 70-kDa subunit of 5′-deoxyribose 5-phosphate lyase Ku70.

[74] Complexes that contain tumor suppressor breast cancer BrCa1 ubiquitin ligase and ubiquitin-binding protein RAP80 target sites of DNA damage. These complexes bind either Abraxas, DNA repair protein BACH1, or tumor suppressor protein CtIP [235].

ATMK and ATRK are the checkpoint kinases CHK1 and CHK2 that reduce the activity of cyclin-dependent kinases. The diversity of types of DNA lesion necessitates multiple, distinct DNA-repair mechanisms (Table 4.17).

Upon recognition of DNA lesion, a DNA-damage response leads to histone modification and exchange as well as chromatin remodeling (rapid local and global decondensation to facilitate genome surveillance). Chromatin environments influence DNA repair efficiency. Double-stranded DNA break repair in heterochromatin has slower kinetics than that in euchromatin. Activation of ATMK provokes chromatin relaxation at sites of double-stranded DNA break.

Furthermore, DNA repair depends on appropriate spatiotemporal organization of repair complex assembly on damaged chromatin. *DNA-repair focus* formed by recruitment and accumulation of repair factors at sites of DNA damage amplifies and transmits the damage signal [236]. In particular it leads to activation of cell cycle checkpoint kinases.

Proteins that detect DNA alterations amplify damage, as they fold and separate 2 DNA strands. This process facilitates subsequent intervention of other proteins that recognize and cut DNA altered parts. Repair of DNA can use the normal strand as a template.

Double-stranded DNA break repair is carried out by 2 main pathways: (1) non-homologous end joining (NHEJ) for repair of non-replication-associated breaks[75] occurs predominantly in G1 phase of the cell cycle and (2) homologous recombination repair (HRR)[76] that uses a homologous template and happens mainly during late S and G2 phases of the cell cycle [236].

DNA damage triggers the DNA-damage response (DDR) that activates protein kinases, such as DNA-dependent protein kinase (DNAPK)[77] as well as ATMK and ATRK kinases, that all phosphorylate histone variant H2ax[78] on chromatin that flanks double-stranded DNA break sites (Ser139) [234,237]. Subsequently, phosphorylated H2ax recruits mediator of DNA damage checkpoint MDC1 that, in turn, promotes the accumulation of ubiquitin ligase complex constituted by RING finger protein RNF8 and ubiquitin-conjugase

---

[75] A double-stranded DNA break is sensed by the Ku80–Ku70 heterodimer that recruits DNA-dependent protein kinase catalytic subunit for assembly and activation of DNAPK complex. The DNAPK complex regulates DNA end processing and facilitates the recruitment of agents that carry out the final rejoining reaction [236]. The direct ligation of DNA ends is sometimes accompanied by a loss of genetic information.

[76] Primary damage sensor MRE11–Rad50–NBS1 complex (MRN; MRE11: meiotic recombination-11 homolog; Rad50: radiation sensitivity DNA-repair Rad50 homolog; NBS1: Nijmegen breakage syndrome-1 homolog or nibrin) is recruited to double-stranded DNA breaks and generates single-stranded DNA by resection. Single-stranded ends are bound by Replication protein-A and Recombination factors Rad51 and Rad52 and can subsequently use homologous templates to prime DNA synthesis and to restore the genetic information [236].

[77] A.k.a. PrKDC.

[78] A.k.a. H2afx.

Table 4.17. Mechanisms of the DNA-damage response and their components (Source: [234]; DSB: double-stranded DNA break; NER: nucleotide excision repair; NHEJ: non-homologous end-joining; HRR: homologous recombination repair (restricted to S- and G2 phases); SSB: single-stranded DNA break; APE1: apurinic-apyrimidinic endonuclease; ATRIP: ATRK-related-interacting protein; BrCa: breast-cancer susceptibility protein; ChK: checkpoint kinase; DDB: damage-specific DNA binding protein; DNAPK: DNA-dependent protein kinase; ERCC: DNA-excision repair protein, cross-complementing repair deficiency, complementation group; Exo1: exo–endonuclease-1; FEN1: flap structure-specific endonuclease-1; MLH1: DNA-mismatch repair MutL homolog-1; MRN: MRE11–Rad50–NBS1 complex [MRE11: meiotic recombination-11 homolog; Rad50: radiation sensitivity DNA-repair Rad50 homolog; NBS1: Nijmegen breakage syndrome-1 homolog or nibrin]; MSH2: DNA-mismatch repair MutS homolog-2; PARP: poly(ADP-ribose) polymerase; PCNA: proliferating cell nuclear antigen; PMS: postmeiotic segregation increased homolog; Pol: polymerase; Rad: radiation sensitivity protein; Rev: reversionless, damaged DNA-binding homolog; RPA: replication protein-A; XP: xeroderma pigmentosum DNA-repair complementing protein [XPe: DDB1; XPg: ERCC5]; XRCC: X-ray repair complementing defective repair in Chinese hamster cells [double-strand DNA break rejoining]).

Process	Main mediators
Direct lesion reversal	Methylguanine methyltransferase
Mismatch repair	Sensors MSH2–MSH3/–MSH6, MLH1–PMS1/–PMS2, DNA Polδ/ϵ, ligase-1, Exo1, PCNA, RPA
Base excision and SSB repair	Sensors DNA glycosylases, endonucleases APE1, FEN1; DNA Polsδ/ϵ, ligases-1/3, PARP1/2, Polynucleotide kinase, aprataxin
NER	Sensors elongating RNA polymerase, DNA Pols, GTF2H, Ligases-1/3, DDB1/2, XPa/c, ERCC4/5/6/8, PCNA, RPA
Translesion bypass	DNA Pols-η/ι/κ, REV1/3
NHEJ	Sensor Ku, DNA Pols, DNA ligase-4, DNAPK, XRCC4,
HRR	Rad51, Rad51b/c/d, Rad52/54, XRCC2/3, FEN1, BrCa2, DNA Polymerases, nucleases, helicases, RPA
Fanc pathway	FancA/C/D1/D2/E/F/G/I/J/L/M/N, PALB2, HR
ATMK-mediated signaling	ATMK, MRN, ChK2
ATRK-mediated signaling	Sensors ATRK, ATRIP, RPA, Rad1–Rad9–HUS1, Rad17, ChK1

UbC13 for histone ubiquitination at or near sites of DNA breaks [238]. Ubiquitin ligase RNF168 (or Riddlin) recognizes ubiquitinated histones and, with UbC13, amplifies local ubiquitination. Receptor-associated protein RAP80 (a.k.a UIMC1) recruits the BrCa1a complex to DNA double-stranded breaks.

In summary, the DNA-damage response primes the accumulation at DNA damage loci of diverse proteins, such as BrCa1, transcriptional corepressor C-terminus-binding protein (CTBP)-interacting protein (CTIP),[79] ATMK recruiter mediator of DNA damage checkpoint protein MDC1, P53BP1, RNF8, and RNF168, to amplify DNA double-strand break signaling and repair. In addition, small ubiquitin-related modifier SUMo1 to SUMo3[80] accumulate at DNA double-strand break sites with SUMo conjugase UbC9 and Ub ligases PIAS1 and PIAS4 to elicit the linkage of BrCa1, P53BP1, and RNF168 proteins. SUMo Ligases PIASs are components of the double-strand break response required for both homologous recombination repair and non-homologous end-joining [239].

Breast and ovarian cancer susceptibility protein BrCa1 intervenes in cell cycle checkpoint control[81] and DNA damage repair. Ubiquitin ligase BrCa1 is regulated by SUMo (SRUbL: SUMo-regulated ubiquitin ligase) [239].[82] SUMo Ligases PIASs colocalize with and cause sumoylation of BrCa1 that is required for BrCa1 ubiquitin ligase activity in cells. Moreover, sumoylation by PIAS1 and PIAS4 of BrCa1 as well as various other DNA-damage response proteins allows the stable association of BrCa1 with DNA-damage sites. Ubiquitin ligase BrCa1 has 2 motifs: (1) the RING sequence that confers the ubiquitin ligase activity and mediates the interaction with ubiquitin conjugating enzymes for mono- and polyubiquitination and (2) BRCT domain that binds to phosphorylated proteins involved in DNA damage repair. Distinct BrCa1-based complexes govern specific responses to DNA damage. Ubiquitin ligase BrCa1 forms a heterodimer — the core complex — with BrCa1-associated RING domain protein BARD1 that promotes ubiquitin ligase activity and is implicated in the maintenance of genomic stability. This core complex is inhibited by BrCa1-associated protein BAP1 [238]. Sumoylation of the BrCa1–BARD1 heterodimer greatly increases the ligase activity.

---

[79] A.k.a. retinoblastoma-binding protein RBBP8.

[80] Agents SUMo1 to SUMo3 are covalently attached to target proteins by a SUMo-conjugation system that consists of SUMo activase (SAE1 or SAE2), SUMo conjugase UbC9 (a.k.a. UbE2i), and various SUMo ligases for different target proteins.

[81] Cell cycle checkpoints are control mechanisms that result in cell cycle arrest at the G–S transition, during S phase, and at the G2–M transition to ensure faithful inheritance of the genome as well as mitosis checkpoints such as the spindle assembly checkpoint to ensure proper segregation of sister chromatids.

[82] Known post-translational modifications of BrCa1 comprise phosphorylation, ubiquitination, and sumoylation.

Ubiquitin ligase BrCa1 constitutes 3 known complexes [238]: (1) *BrCa1a complex* with abraxas,[83] RAP80, BrCa1 coiled-coil domain-containing proteins BRCC36 and BRCC45, and 40-kDa Mediator of RAP80 interactions and targeting subunit (MeRIT40, or NBA1) that control of the G2–M checkpoint and BrCa1 accumulation at damage-induced loci; (2) *BrCa1b complex* with FancJ[84] and topoisomerase-2-binding protein TopBP1 that is involved in DNA replication and S-phase progression; and (3) *BrCa1c complex* with CTBP-interacting protein (CTIP or RBBP8) and the meiotic recombination MRe11–RAD50–nibrin (MRN) complex that participates in DNA resection and G2–M checkpoint control [238].

In addition, BrCa1 links to BrCa2, or FancD1, via FancN[85] for homologous recombination repair. Ubiquitin ligase BrCa1 participates in DNA crosslink repair in concert with FancD1, FancJ, and FancN. Ubiquitination that involves RING finger proteins RNF8 and RNF168 and Ub conjugase UbC13 and ubiquitin recognition by RAP80 governs BrCa1 localization in the vicinity of double-stranded DNA breaks [238]. BrCa1 might also act as a scaffold protein at damage-induced foci to facilitate ATMK- and ATRK-mediated phosphorylation of many proteins, such as P53, nibrin,[86] and ChK1 and ChK2 checkpoint kinases.

During meiosis and DNA repair, 4-stranded DNA intermediates, the so-called *Holliday junctions*, covalently link 2 DNAs. Holliday junction resolvase GEN1 of the Rad2–XPg family of nucleases promotes Holliday junction resolution by symmetrical cleavage [240].

The group of RecA–Rad51 homolog ATPases mediates homologous recombination of DNA during double-strand break repair that maintains gene integrity and generates genetic diversity.[87] Agent Rad51 (BRCC5), ATP, and single-stranded DNA form a helical nucleoproteic filament that binds to double-stranded DNA, searches for homology, and then catalyzes the exchange of the complementary strand, producing a new complementarity-dependent heteroduplex [241]. Homologous pairing of a undamaged DNA with a damaged DNA mediated by strand-exchange proteins is one mechanism to repair DNA damage.

---

[83] A.k.a. CCDC98 and FAM175a.

[84] A.k.a. BrCa1-interacting protein C-terminal helicase BaCH1 and BRIP1.

[85] A.k.a. Partner and localizer of BrCa2 (PALB2).

[86] A.k.a. Nijmegen breakage syndrome protein NBS1.

[87] Recombination involves exchange of DNA strands between 2 homologous segments of chromosomes catalyzed by the RecA class of ATPases. RecA–ATP and RecA–DNA interactions are coupled and cooperate to induce a new RecA–RecA interface and a conformational change that activates the nucleoprotein filament for strand exchange.

Other Functions of the DNA-Damage Response

Programmed genome alterations, such as V(D)J recombination, class-switch recombination, and somatic hyper-mutation, occur in developing B and T lymphocytes to generate new immunoglobulins and T-cell receptors. Exons that encode the antigen-binding portions of these molecules are composed of V, D, and J segments that are combined in various ways to generate mature immunoglobulin and TCR genes. The DNA-damage response controls the quality of these processes [234]. In addition, tDNA-damage response is also involved in the generation of genetic diversity during meiosis, i.e., species reproduction.

Ends of linear chromosomes cannot be replicated to their termini, hence terminal sequences can be lost. Moreover, these ends can be mistakenly sensed as DNA double-strand breaks, thereby activating DNA repair pathways that cause serious genome derangement. Therefore, telomeres, repeat sequences at the ends of chromosomes that are shielded by the proteic complex shelterin, are added. The shelterin complex sequesters telomeres to prevent their engagment in NHEJ-mediated fusions or activation of ATMK–ATRK signaling [234]. Nevertheless, DDR components intervene in normal functioning of telomeres.

4.4.4 Nuclear Subdomains

The nucleus is compartmentalized. Many nuclear factors involved in DNA replication and transcription, RNA processing, and ribosome subunit synthesis are organized in distinct nuclear domains: chromosome territories, interchromatin granule clusters, nucleoli, and nuclear bodies.

The nuclear envelope harbors both silent chromatin and active genes, but at distinct sites. The nuclear cortex enables the access to pools of repressors [197]. The *nuclear lamina*, a layer of intermediate filaments at the interface between chromatin and inner nuclear membrane between pores, can tether chromatin at the nuclear cortex, thus contributing to heterochromatin distribution.

Active genes are located near nuclear pores, where they could be crosslinked and modified by proteins of nuclear pores. Certain transcription activators are able to, at least in some species, preferentially stimulate genes bound to nuclear-pore proteins, such as nucleoporins, myosin-like proteins, chromosome-segregation proteins, and karyopherins [870] (Sect. 9.1.8). Transcription regulators with their coactivators are capable of binding nucleoporins for full transcriptional induction (at least in some species, but it can be a widespread eukaryotic process) [871]. However, transcription does not necessarily occur close to nuclear pores.

The nucleus contains nuclear bodies that assemble at sites of high gene activity, where freely diffusing proteins accumulate. Interchromatin granule clusters assemble various proteins involved in gene expression, pre-mRNA splicing, and RNA metabolism. They can operate as assembly, modification, and/or storage sites for proteins involved in pre-mRNA processing [244]. The nucleolus is a nuclear body that contains clusters of ribosomal DNA that produces ribosomal RNA (rRNA).

4.4.4.1 Nucleolus

In the nucleoplasm, with its active euchromatin and dense, inactive heterochromatin composed of DNA, histones, and acidic nuclear proteins, nucleoli form around sites of rRNA transcription and pre-mRNA splicing. Its primary function is rRNA transcription and processing, as well as ribosome subunit assembly.

The nucleolus is composed of different regions: fibrillar center (FC), and dense fibrillar (DFC) and granular (GC) components [245]. This structure is maintained during ribosome biogenesis.

Ribosome genesis is organized into nucleolar organizer regions. Ribosome genesis and processing are coupled, as transcriptional U three protein (tUTP), a component of the small subunit processing complex, i.e., the *SSU processome*, is required for efficient transcription of preribosomal RNAs [246].

Ribosomal gene transcription is modulated by the transcriptional apparatus and epigenetic silencing that involves the nucleolar remodeling complex (noRC). Processing of 18S ribosomal RNA involves the SSU processome that comprises U3 small nucleolar ribonucleoprotein (snoRNP), tUTP, B subunit of UTP (bUTP),[88] and M-phase phosphoprotein MPP10, among other factors [246].

The nucleolus is a dynamical nuclear compartment that is implicated not only in genesis of ribosome subunits, but also in the cell cycle, stress response, cell growth and death, and ribonucleoprotein production (RNP; Table 4.18).

The nucleolus disassembles during the cell division and reassembles after cell division. During mitosis, the nucleolus disassembles and components of the rRNA transcription complex migrate with ribosomal genes, whereas the processing machinery is distributed at the chromosome periphery [246]. When cell division ends, processing proteins complex in prenucleolar bodies. Resulting complexes are then recruited to sites of rRNA transcription and new nucleoli are formed.[89]

Various proteins associate with the nucleolus at different stages of the cell cycle, hence, regulating the cell-cycle evolution. Post-translational modifications (sumoylation, desumoylation, phosphorylation, and dephosphorylation) that occur throughout the cell cycle are regulated via the nucleolus. The nucleolus also sequesters proteins (e.g., telomerase reverse transcriptase by

---

[88] Proteic subcomplex bUTP of the SSU processome contains several proteins that interact. For example, UTP6 interacts with UTP18 and UTP21.

[89] Pre-rRNA transcripts exist in prenucleolar bodies. Late processing protein B23 rapidly relocates to prenucleolar bodies, but not early processing protein fibrillarin.

Table 4.18. Role of the nucleolus (Source: [246]; DM2: human double minute-2 Ub ligase; TIF: transcription-initiation factor). Jun N-terminal kinase JNK2 and factor P53 are cell-survival proteins that participate in the nucleolar response to toxic stresses. Nucleophosmin (Npm, also called nucleolar phosphoprotein B23) is a nucleolar protein that binds to P53 and alternate reading frame product of the cyclin-dependent kinase inhibitor-2A (CDKN2A) locus (ARF) and serves in ribogenesis, cell survival after DNA damage, and cell proliferation. Nucleophosmin travels between nucleolus, nucleoplasm, and cytosol. Nucleolar tumor-suppressor protein ARF operates in protein localization. Subunit RelA of nuclear factor-κB translocates from nucleoplasm to nucleolus for cell apoptosis.

Function	Effect
Ribosome genesis	Transcription from ribosomal DNA
	and processing of preribosomal RNA
RNA processing	Processing of some transfer RNAs
	Maturation of small nucleolar and small nuclear RNAs
	Transport of messenger RNAs
Production of ribonucleoproteins	Processing of nucleic acids
Stress sensing	JNK2–TIF1A-mediated inhibition of rDNA transcription
	Stabilization of P53
Cell proliferation and apoptosis	Sequestration of regulators:
	P53 (the ARF–DM2–P53 complex),
	subunit RelA of nuclear factor-κB

nucleolin) to regulate the cell cycle [245]. In cells subjected to stress (DNA damage, heat shock, or hypoxia), P53 (Vol. 4 – Chap. 9. Other Major Signaling Mediators) accumulates owing to the action of nucleolar cyclin-dependent kinase inhibitor-2A (CKI2a) and inhibits the CcnD1–CDK4 and CcnE1–CDK2 complexes via CKI1a, causing cell cycle arrest.[90]

The nucleolus responds to changes in metabolic activity by modifying the rate of ribosome production. The nucleolus contains ribosomal gene clusters. The 47S ribosomal RNA precursor is transcribed by RNA polymerase-1. This single precursor of mature 28S, 18S, and 5.8S ribosomal RNAs is cleaved and post-transcriptionally modified via interaction with small nucleolar ribonucleoproteins and with other processing factors. Finally, it is assembled with many ribosomal proteins into ribosome subunits. The 5.8S and 28S ribosomal RNAs[91] assemble with 5S rRNA to form the large 60S ribosome subunit, whereas 18S rRNA constitutes the small 40S ribosome subunit. The ribosome subunits are exported into the cytoplasm.

---

[90] Ubiquitin ligase DM2 specific to P53 reduces P53 concentration due to degradation of ubiquitinated P53 in the cytoplasm. Nucleolar CKI2a links DM2 and inhibits P53 ubiquitination and subsequent degradation.

[91] Both 5.8S and 28S ribosomal RNAs are parts with 18S rRNA of the 45S precursor.

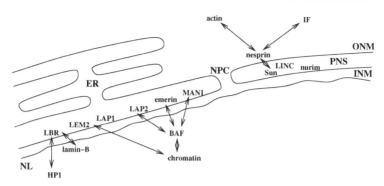

Figure 4.2. Nuclear envelope made of inner (INM) and outer (ONM) nuclear membranes separated by the perinuclear space (PNS) (Source: [248]). It is connected to the endoplasmic reticulum (ER) and nuclear lamina (NL). The nuclear envelope is perforated by pores at sites of fusion between inner and outer nuclear membranes that are filled with nuclear-pore complexes (NPC) for molecular transport. Nuclear-pore complexes are large molecular complexes made of nucleoporins and other elements. Lamin-B receptor (LBR) interacts with lamin-B and chromatin-associated heterochromatin protein HP1. Lamina-associated protein LAP2, emerin, and inner nuclear-membrane protein MAN1 bind to lamins and interact with chromatin through DNA-binding barrier-to-autointegration factor (BAF). The Sad1 and Unc84 homology proteins (SUn) in INM interact with nesprins in ONM to form the linker of nucleoskeleton and cytoskeleton complexes (LINC) that connect to actin and intermediate filaments (IF).

4.4.4.2 Nuclear Envelope

The nuclear envelope plays an active role in mechanotransduction. The nuclear envelope is composed of inner and outer nuclear membranes that are separated by the perinuclear space (thickness 30–50 nm [247]; Fig. 4.2). At least 60 types of nuclear envelope proteins have been identified; most of them reside in the inner nuclear membrane.

The inner nuclear membrane lies on the nuclear lamina, which is mainly composed of lamin-A and -B of the intermediate filament class (Table 4.19). The inner nuclear membrane comprises an array of integral membrane proteins that include lamina-associated polypeptides LAP1 (LAP1a–LAP1c) and LAP2, emerin, lamin-B receptor (LBR), among others. These proteins interact with heterochromatin protein HP1 involved in chromatin silencing and barrier-to-autointegration factor (BAF) and provide links to chromatin.

Several integral proteins of the inner nuclear membrane interact with transcription factors and/or chromatin modifiers.[92] In the inner nuclear

---

[92] Emerin links to gene-regulatory proteins such as BAF, germ cell-less (GCL), β-catenin, death-promoting transcriptional repressor Btf, mRNA splicing factor YT521B, and LIM domain only protein Lmo7 [247]. Lmo7 is an emerin-binding protein and also an afadin- and alpha-actinin-binding protein that connects the

Table 4.19. Proteins of the nuclear envelope (Source: [252]).

Protein	Interacting molecules
	Lamins
Lamins-A, -B	LAP1/2, histones, DNA
Lamin-A	Emerin, RB1
Lamin-B	LRB, LAP2β
	Integral inner nuclear membrane component
LBR	Lamin-B, DNA, HP1
LAP1α, β, γ	Lamins
LAP2β	Lamin-B, DNA, BAF
LAP2γ, δ, ε	Lamins
Emerin	Lamin-A, -B
MAN1	
Nurim	
LUMA	
RFBP	
Nesprins	Actin
	Nuclear-pore complex
NuP153	Lamin-B
gp210	
POM121	
	Chromatin
BAF	Emerin, LEMD3, LAP2β
HP1α, β, γ	LBR
LAP2α	Lamin-A
H2a, H2b	Lamin
Methyl H3	

membrane, LEM domain-containing protein LEMD3[93] binds to rSMADs to repress transforming growth factorβ and bone morphogenetic protein signaling [251] (Vols. 2 – Chap. 3. Growth Factors and 3 – Chap. 8. Receptor Kinases).

---

nectin–afadin and E-cadherin–catenin complexes in adherens junctions [249]. Shuttling transcription activator Lmo7 activates emerin gene [250]. Lmo7 is inhibited by binding to emerin.

[93] A.k.a. human antigen MAN1. The LEM domain is a binding motif for the chromatin protein BAF. LEM-domain proteins constitute a family of non-related inner nuclear membrane and intranuclear proteins. This family of nuclear proteins comprises lamina-associated polypeptide LAP2 that has several alternatively spliced isoforms, emerin, and LEMD2 (or NET25) to LEMD5. Whereas LAP2β lodges in inner nuclear membrane, LAP2α resides in the nucleoplasm.

The outer nuclear membrane is continuous with the membrane of the endoplasmic reticulum. It contains emerin and cytoskeleton-associated nesprins. Nesprins are tethered by Sad1 and Unc84 homology proteins SUn1 and SUn2[94] of the inner nuclear membrane that also interact with lamin-A. Nesprins are encoded by 3 identified genes. The largest isoforms nesprin-1 and -2 contain actin-binding domains, whereas nesprin-3 has a binding site for intermediate filament-associated plectin.

Transmembrane proteins interact across the perinuclear space and link to both the lamina and cytoskeleton to favor transluminal interactions. Outer nuclear membrane actin-binding protein Anc1 joins with proteins of the SUN family at the inner nuclear membrane. SUn Proteins and nesprins form the linker of nucleoskeleton and cytoskeleton complexes (LINC). Emerin connects with nesprins and lamins, as well as nuclear actin, myosin-1, and $\alpha 2$-spectrin [247].

Annular junctions between the inner and outer nuclear membranes form pores that traverse the nuclear envelope. These pores are occupied by *nuclear-pore complexes* (NPC) that contain *nucleoporins* (NuP; Chap. 9). Nucleoporins act as soluble carriers and accessory proteins, such as transport cofactors and assembly factors. Nucleoporins have been classified into 3 groups: (1) membrane nucleoporins; (2) non-membrane nucleoporins that contain multiple repeats of a Phe-Gly motif; and (3) non-membrane proteins without these multiple repeats. The nucleoporin NuP107–NuP160 complex links to the coat protein complex CoP1 that is associated with the Golgi apparatus.

Nuclear-pore complexes are characterized by compartment-specific assembly and disassembly. Nuclear-pore complexes are symmetrically located on both nucleoplasmic and cytoplasmic faces of the nuclear envelope, although peripheral elements have distinct compositions. Mammalian cells typically contain 3,000 to 5,000 nuclear-pore complexes. The number of nuclear-pore complexes per nucleus varies with the cell type and growth conditions. Nuclear-pore complexes form a channel (caliber 9 nm; length 45 nm).

During mitosis, the nuclear envelope undergoes structural reorganization. Phosphorylation of its components mediates nuclear envelope disassembling. Nuclear envelope disassembly allows spindle microtubules to have access to the chromosomes. Disruption of the nuclear membranes and dispersal of their constituents in the endoplasmic reticulum is favored by microtubules and their nanomotor dynein that resides in the cytoplasm, but connected to the outer nuclear membrane [253]. Proteins NuP358 and NuP153 coordinate the nuclear envelope breakdown [254]. Following nuclear envelope demolition, NuP358 of the nuclear-pore complexes relocates to kinetochores and participates in chromatid segregation [255]. NuP53 is involved in the regulation of cytokinesis via sumoylation of septins. During the late stage of cell division that commences in anaphase, disassembled components are reused to assemble nuclear envelopes in each daughter cell.

---

[94] SUn1 and SUn2 are also termed Unc84a and Unc84b.

Chromatin mobility in the nucleus allows rapid localization of genes in response to stimuli and nuclear positioning of genes controls their expression. On the inner nuclear membrane, active genes localize to nuclear pores, whereas silent chromatin is located at non-pore sites [197]. Nuclear-pore components not only recruit RNA-processing and RNA-export elements, but also contribute to the transcription regulation.

4.4.4.3 Cajal Bodies

Cajal bodies are spherical suborganelles in the nucleus of proliferative cells or metabolically active cells such as neurons discovered by S. Ramón y Cajal in 1903 (0.1–2.0 μm). Cajal body number varies over the cell cycle and in different types of cells. Like all other nuclear compartments, Cajal bodies do not have a membrane. These dynamical suborganelles rapidly and continuously exchange proteins with the surrounding nuclear space.

Cajal bodies either can assemble according a deterministic way, as Cajal-body proteins can bind to scaffold proteins, such as coilins and survival of motor neuron gene product (SMN), to build up structures or aggregate in a stochastic fashion (self-organization). Because any Cajal-body protein can initiate the structure formation and coilin and SMN that are required for efficient formation can self-assemble, Cajal body formation relies on self-organization [256]. Cajal bodies are bound to nucleolus by coilins. Cajal bodies are preferentially located near clusters of genes encoding histones or small nuclear RNAs.

Nuclear ribonucleoproteins that form complexes with small nucleolar and small nuclear RNAs initially concentrate in nuclear domains distinct from their function sites. Maturation and displacement of these small nucleolar and small nuclear ribonucleoproteins for splicing and ribosomal RNA modification involve a transient localization in Cajal bodies [257]. Cajal bodies could then be involved in nuclear RNA modifications and RNA-processing factor recycling. A subset of Cajal bodies located near the histone genes contains many of the components required for histone gene expression.[95]

4.4.5 Nuclear Filaments

The nuclear envelope is contiguous with the endoplasmic reticulum. It possesses nuclear-pore complexes (Sect. 9.1.8). It is lined by the nuclear lamina (thickness 10–20 nm), which is composed of *lamins*. The inner nucleus membrane is related to a network of intermediate filaments that include not only lamins, but also lamin-associated proteins (emerin, lamina-associated

---

[95] Histone genes that produce histone pre-mRNAs are clustered in the genome. U7 small nuclear ribonucleoprotein, stem-loop binding protein, and the cleavage complex process histone pre-mRNA adjacent to a Cajal body to form mature histone mRNA [258].

polypeptides LAP1 and LAP2, and lamin receptor). Lamin-B is expressed during all embryonic stages, whereas lamin-A is only produced during cell differentiation.

Lamin-A and -B form stable filaments in the nucleus. They anchor many nuclear-membrane and soluble proteins, as well as proteic complexes [259]. Lamins are implicated in mechanical stability and shape of the nucleus, DNA replication and transcription, chromatin organization, cell cycle regulation, cell development and differentiation, nuclear anchoring and migration, centrosome positioning, and apoptosis.

Lamins and matrin-3 in the nuclear matrix orientate many elements of the nuclear regulatory machinery. Lamins connect to gene regulators, such as retinoblastoma protein (RB), barrier-to-autointegration factor, histones, sterol response element-binding protein SREBP1, and Fos [247].

Lamins bind LEM domain-containing proteins (emerin, LEMD3, and LAP2) that function as scaffolds for various proteins, including transcription regulators (e.g., germ cell-less and BCL2-associated transcription factor), chromatin proteins (e.g., barrier-to-autointegration factor), and proteins responsive to transforming growth factor-β such as SMADs.

Mutations in lamin-A[96] or lamin-binding membrane proteins, such as emerin and lamin-B receptor, cause laminopathies (muscular dystrophy, cardiomyopathy, and accelerated aging). Types of defective assembly or function of lamin-dependent proteic complexes depend on altered tissue type.

During cell division, many nuclear components assemble and may form a spindle matrix, i.e., a molecular mesh that provides a platform upon which the spindle can assemble, along which proteic nanomotors can move. These nanomotors operate in combination with microtubule growth and contraction during chromosome displacement. Lamin-B, nuclear isoform of titin, tubulin, skeletor, megator, and chromator appear at the same site within a dividing cell.[97] Both lamin-A and -B are dispersed during mitosis. Lamin-A and -C behave as soluble proteins, but lamin-B, which is constitutively farnesylated, remains largely associated with the membrane [261]. Lamin-B operates in mitotic progression by contributing to the formation of a spindle matrix [262]

Nesprins are nuclear membrane-associated proteins that bind emerin and lamin [263]. Nesprins can be involved in cellular compartmentation of organelles, particularly in smooth muscle cells and cardiomyocytes.

---

[96] Progerin that is a lamin-A mutant causes Hutchinson-Gilford progeria syndrome. Progerin generates nuclear defects and increases DNA damage. In addition, progerin promotes premature differentiation of mesenchymal stem cells by upregulating the Notch pathway [260]. Progerin enhances osteogenesis, but impedes differentiation of human mesenchymal stem cells into adipocytes.

[97] Megator belongs to complexes that regulate transport across the nuclear membrane. In quiescent cells, chromator and megator are dispersed within the nucleus or reside on chromosomes. When a cell starts to divide, it rearranges into an oval spindle-like structure. Lamin-B that is dispersed owing to disassembly of the nuclear lamina is used to form a matrix that harbors spindle assembly factors.

Nuclear actin and actin-related proteins (ARP) participate in chromatin-remodeling complexes. Due to the presence of specific binding domains in the structure of chromatin regulators, these chromatin-remodeling complexes recruit actin and ARPs, bind to actin–ARP or ARP–ARP complexes, and regulate chromatin activity [264].

Nuclear actin participates in the control of gene expression, especially in the regulation of the coactivator myocardin-related transcription factor MRTFa[98] of serum response factor. Cytoplasmic MRTFa runs in the nucleus and binds monomeric, globular (^{G}actin) [265]. Binding of MRTFa to actin in the nucleus leads to its nuclear export. Dissociation of MRTFa from actin (as well as ^{G}actin depletion) fosters SRF activation by nuclear MRTFa coactivator. Emerin also caps minus-ends of filamentous actin.

4.5 Endoplasmic Reticulum

Both the endoplasmic reticulum and Golgi body participate in the first 2 stages of protein exocytosis. The endoplasmic reticulum is made of interconnected convoluted tubules and flattened sacs, which extends throughout the cytoplasm. The lumen (cisternal space) of central elements is connected to the nucleus. It serves as a transport channel and reservoir for certain chemical species.

The endoplasmic reticulum with its oxidizing luminal tool manufactures, stores, carries, and releases materials, such as matured proteins and lipids. Messenger RNAs can be translated at the endoplasmic reticulum. Approximately one-third of cellular proteins, such as proteins destined for the plasma membrane or intracellular organelles, Golgi compartment, and lysosomes, travel into the endoplasmic reticulum lumen to undergo post-translational modifications, folding, and oligomerization. In addition to proteins, lipids also enter the secretory pathway via the endoplasmic reticulum. Under cellular stress that lead to protein misfolding, the endoplasmic reticulum triggers the unfolded protein response.

The endoplasmic reticulum cannot be generated de novo by cells. During the cell division, each daughter cell must be provided with its own endoplasmic reticulum membranes that serve as a template for future growth.

4.5.1 Types and Functions of Endoplasmic Reticulum

Whether ribosomes cover its surface or not, the endoplasmic reticulum is rough or smooth. The tubular, *rough endoplasmic reticulum* is the site of protein synthesis.[99] These proteins become resident or are released by the

---

[98] A.k.a. megakaryoblastic leukemia-1 fusion protein MKL1, MAL, and RNA-binding motif protein-15.

[99] Cytoplasmic proteins are synthesized in free ribosomes. Ribosomes bind to the endoplasmic reticulum to produce glycoproteins.

smooth endoplasmic reticulum and transferred to vesicles for intracellular use or export from the cell. On the other hand, the sheet-like, *smooth endoplasmic reticulum* synthesizes lipids.

The endoplasmic reticulum produces most of the lipids required for the generation of new segments of the cell membrane. These lipids include cholesterol, phosphatidylcholine, and phosphatidic acid. The latter is produced by the cleavage of phosphatidylcholine by phospholipase-D1.

The endoplasmic reticulum sequesters calcium from the cytosol, especially in myocytes that have an abundant smooth endoplasmic reticulum, the *sarcoplasmic reticulum* (SR; Vol. 5 – Chap. 6. Heart Wall). The endoplasmic reticulum thus is a major signal-transducing organelle that responds to environmental cues to release calcium.

Among proteins of rough and smooth endoplasmic reticulum with a known function, most are implicated in protein synthesis, folding, and modification, whereas others interact with the actin cytoskeleton or operate in metabolism [266].

The exocytosis from the endoplasmic reticulum to the Golgi body is mediated by the tubulovesicular *endoplasmic reticulum–Golgi intermediate compartments*, smooth components of the rough endoplasmic reticulum.

Reticulon localizes to the rough endoplasmic reticulum, but not the smooth endoplasmic reticulum. It influences the local membrane curvature [171].

4.5.2 Protein Folding

The endoplasmic reticulum is the site for correct folding and maturation of proteins and glycoproteins that are destined for secretion toward the plasma membrane as well as intracellular organelles. Numerous proteins reside in the lumen of the endoplasmic reticulum, such as chaperones and enzymes of protein folding (e.g., protein disulfide isomerase [PDI] and cis–trans peptidyl-prolyl isomerase [PPIase]). The latter interconverts the cis and trans isomers of peptide bonds with the amino acid proline. FK506-binding protein (FKBP) is a protein-folding chaperone for proteins that contain proline residues.

Among chaperones, endoplasmic reticulum luminal Ca^{++}-binding protein heat shock protein HSPa5 is produced during glucose deprivation.[100] It prevents the secretion of incompletely assembled immunoglobulins, as it precludes the release of immunoglobulin heavy chains in the absence of light chains in pre-B lymphocytes.[101] Chaperones 90-kDa heat shock proteins HSP90β1[102] and HSP90β2[103] are paralogs of HSP90 in the endoplasmic reticulum. HSP90β2 intervenes in the transfer of ligands to class-1 major histocompatability complex molecules that assemble in the endoplasmic reticulum. Two additional glucose-regulated proteins that have chaperone functions

---

[100] Hence its name 78-kDa glucose-regulated protein GRP78.
[101] Hence its name immunoglobulin heavy chain-binding protein (BiP).
[102] A.k.a. glucose-regulated protein GRP96 and endoplasmin.
[103] A.k.a. GRP94.

and participate in protein quality control are calreticulin[104] and the uridine diphosphate (UDP)–glucose glucuronosyltransferase (UGT).

Incorrectly folded proteins are bound by HSPa5 to be removed from the lumen by transmembrane stress sensors, such as Ser/Thr kinase and RNase inositol-requiring kinase IRe1, endoplasmic reticulum-bound RNA-dependent eukaryotic translation initiation factor eIF2α kinase-3 (PERK), and Activating transcription factor ATF6.[105] This mechanism activates the unfolded protein response (UPR).[106]

Once malfolded polypeptides have been discovered in the lumen of the endoplasmic reticulum, a proteolysis occurs in the cytosol, which ensures the secretion of correctly folded proteins. This degradation requires the ubiquitin–proteasome system.

4.6 Golgi Body

The Golgi body[107] processes, i.e., post-translationally modifies (in particular via glycosylation and phosphorylation), sorts for destination, and packages, newly synthesized macromolecules, such as proteins, oligosaccharides, and lipids, destined for exocytosis from the cell.[108] The Golgi body is also

---

[104] A.k.a. calregulin and endoplasmic reticulum-resident protein ERP60. Calreticulin and calnexin are homologous lectin chaperones that transiently and selectively bind to newly synthesized glycoprotein-folding intermediates to prevent their transit via the secretory pathway. Calnexin is an integral transmembrane protein that is not regulated by endoplasmic reticulum stress.

[105] Activated RNase IRe1 cleaves mRNA of X-box binding protein XBP1 to produce an active transcription factor XBP1 that binds to the UPR-responsive element of gene promoters to synthesize unfolded protein response chaperones. Transcription factor ATF6 travels to the Golgi network, where it is cleaved into an active cytosolic fragment that migrates to the nucleus and binds to endoplasmic reticulum stress-responsive element-containing gene promoters. Splitting of PERK induces PERK dimerization that then phosphorylates translational initiation factor eIF2α to prevent formation of the translational initiation complex and lower the protein folding function. Phosphorylated eIF2α precludes translation of most mRNAs, but promotes translation of a subset of stress response genes such as Activating transcription factor ATF4. Misfolded proteins can be eliminated by endoplasmic reticulum-associated proteasomal degradation and autophagy (Vol. 2 – Chap. 4. Cell Survival and Death).

[106] Endoplasmic reticulum stress and activation of unfolded protein response contribute to the development of obesity-induced insulin and leptin resistance and type-2 diabetes.

[107] A.k.a. Golgi apparatus and Golgi complex. The terms apparatus and complex are here avoided, as they preferentially refer to a physiological system (macroscopic scale) and molecular aggregates (nanoscopic scale). The Golgi body was discovered by C. Golgi in 1898.

[108] In endo- and exocrine cells, the elaboration of a secretory protein is constituted of 6 steps: (1) synthesis on ribosomes associated with the endoplasmic reticulum;

involved in the creation of lysosomes. The Golgi body can also operate as a calcium store [267].

4.6.1 Structure of the Golgi Body

The Golgi body is made of a stack of membrane-bound, elongated, flattened, pancake-like, cisternae associated with lateral, membrane-bound saccules (Latin sacculus) and membrane-bound vesicles. A cisterna is, indeed, connected at both ends to a dilated saccule. Type-1 Golgi saccules are spherical and continuous with inner Golgi cisternae. Type-2 Golgi saccules are ellipsoidal. Type-3 Golgi saccules are more tubular. As type-3 saccules mature, interconnecting cisternae progressively shorten. A cell can contain 40 to 100 layered, tubular cisternae. Cisternae can split to generate presecretory vesicles, which then mature into secretory vesicles.

Vesicles from the endoplasmic reticulum target the Golgi body via *vesiculotubular clusters* and fuse with the Golgi membrane. Vesicles can progress through the stack up to the trans-Golgi network, where they are packaged and sent to the required destination. Therefore, proteins are delivered to the cis-face of the Golgi body, i.e., the closest edge of the Golgi body to the endoplasmic reticulum and exit from the trans-face, i.e., the closest edge to the plasma membrane. Protein transport from the endoplasmic reticulum to the Golgi body is the first step of intercompartmental transfer.

Exocytotic vesicles carry proteins destined for extracellular release. After packaging, vesicles bud off and immediately move towards the plasma membrane, where they fuse and release their content into the extracellular space (*constitutive secretion*). On the other hand, *secretory vesicles* (Vesiculae secretoriae) store their cargos inside the cell. When a proper signal is received, these vesicle move toward the plasma membrane and expel their content (e.g., hormones and neurotransmitters) in the extracellular space (*regulated secretion*). In particular, the Golgi body is the site of synthesis of glycosaminoglycans destined to the extracellular matrix. Calcium ions provoke the detachment of cortical vesicles from the cytoskeleton to which they are bound. In addition, some vesicles contains proteases destined for lysosomes.

The Golgi body is composed of 3 major subcompartments: (1) an entry region, the *cis-Golgi reticulum* facing the endoplasmic reticulum (the closest to the nucleus); (2) the *endoplasmic reticulum–Golgi intermediate compartment* (ERGIC); and (3) the *trans-Golgi network* (TGN), which is oriented toward the cell periphery. The tubular trans-Golgi network is connected to the stacked, flat Golgi cisternae.

---

(2) segregation within the cisternal space of the rough endoplasmic reticulum; (3) transport to the vicinity of the Golgi body; (4) concentration and additional processing in Golgi saccules; (5) storage in exocytotic and secretory vesicles that are created from the trans-Golgi saccule; and (6) transfer to the plasma membrane and discharge from vesicles to the extracellular medium.

Table 4.20. Functional subcompartments of the Golgi body. The tubulovesicular trans-Golgi network is a sorting station for newly synthesised proteins destined to the plasma membrane or endosomes (Man: mannose, NANA: ^{N}acetylneuraminic acid, the predominant sialic acid). O-linked glycosylation, or O-glycosylation, is the covalent attachment in the Golgi body of ^{N}acetyl-galactosamine (GalNAc) to the hydroxyl group of serines or threonines of side chains by UDP-^{N}acetyl ^{D}galactosamine:polypeptide ^{N}acetylgalactosaminyltransferase (as well as ^{N}acetyl-glucosamine (GlcNAc) on serines and threonines, fucose, glucose, and mannose).

Compartment	Activity
cis-Golgi network	Sorting
	Phosphorylation of oligosaccharides on
	lysosomal proteins
	Golgi stack
cis-Golgi stack	Removal of Man by mannosidase
medial-Golgi stack	Removal of Man,
	attachment of GlcNAc, fucose
trans-Golgi stack	Glycosylation (Gal and NANA)
trans-Golgi network	Sorting
	Sialylation and fucosylation of glycans
	Sulfation of tyrosines and carbohydrates
	(sulfation of glycosaminoglycans)
	Protein maturation by endoproteases
	Sphingolipid synthesis

In fact, the stack of cisternae can be decomposed into 5 functional regions (Table 4.20): (1) the cis-Golgi network, (2) cis-Golgi stack, (3) medial-Golgi stack, (4) trans-Golgi stack, and (5) trans-Golgi network. The *Golgi stack* includes the cis-, medial-, and trans-Golgi stack. The *Golgi body* encompasses the Golgi stack and the cis- and trans-Golgi network. Each functional region of the Golgi body contains different enzymes. Carried proteins visit each type of Golgi cisternae, but not all cisternae of a given type.

4.6.2 Function of the Golgi Body

Ribosome clusters can link to the Golgi body. Therefore, some proteins can be synthesized at the Golgi body. However, its main function is to temporarily store proteins and process various newly synthesized substances (proteins, oligosaccharides, and lipids) arising out of the rough endoplasmic reticulum. It then packages processed molecules in sorted vesicles that are carried toward suitable organelles or the plasma membrane, where the content can be exported outside the cell. Therefore, the Golgi body selects the destination of the modified molecules (sorting according to the functioning site of the substance). Post-translational modifications such as phosphorylation (Fig. 4.3)

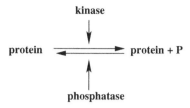

Figure 4.3. Reversible protein phosphorylation is aimed at controlling protein activities. Phosphates are added by kinases either on serine (Ser), threonine (Thr), or tyrosine (Tyr) of the protein from ATP to switch protein activity on or off (phosphorylation either activates or inactivates the substrate). Phosphates are removed by phosphatases. Protein can undergo single or multiple phosphorylation.

are aimed at regulating cell processes. Final reactions and sorting take place in the trans-Golgi network.

Cisternae of the Golgi stack are interconnected by transfer vesicles. Vesicles bud and fuse within minutes. During membrane trafficking, transport vesicles are first tethered to the target membrane prior to fusion. Membrane tethering between both types of membranes precludes vesicle escape from the Golgi body.[109]

Long, string-like, coiled-coil *golgins* (Table 4.21) tether cisternae and transport vesicles at the Golgi stack via the coat protein complex-1 (CoP1). Golgins not only interact with proteic coats, but also with Rab GTPases and the cytoskeleton. Vesicles of the Golgi body differ according to the tether type (e.g., golgin-tethering factor Uso1, or general vesicular transport factor P115, vs. golgin-A5–CASP tether) [270].[110]

Another member of the golgin family, Golgi-associated microtubule-binding protein GMAP210, bridges highly curved vesicular membrane to flat cisterna

---

[109] The member of the golgin family of coiled-coil proteins GolgA2 (or GM130) is involved in membrane tethering at the endoplasmic reticulum to Golgi intermediate compartment and cis-Golgi network. It interacts with syntaxin-5, a target soluble ^{N}ethylmaleimide-sensitive factor attachment protein receptor (tSNARE; Chap. 9) of the early secretory pathway [268]. Binding of GolgA2 to syntaxin-5 is impeded by the binding of the vesicle-docking protein Uso1 (a.k.a. general vesicular transport factor P115, transcytosis-associated protein [TAP], and vesicle-docking protein [VDP]) to GolgA2. Protein Uso1 is a cytoskeleton-related molecule that contributes to the protein transport from the endoplasmic reticulum to the Golgi body. The interaction between GolgA2 and the small GTPase Rab1 is also prevented by Uso1 binding. Enzyme Rab1 recruits Uso1 to coat protein complex-2 (CoP2)+ vesicles during budding from the endoplasmic reticulum, where it interacts with a select set of CoP2+ vesicle-associated SNAREs to form a cis-SNARE complex that promotes targeting to the Golgi body [269].

[110] Golgin-A5, or golgin-84, interacts with an alternatively spliced product of the CutL1 gene CASP (CDP/Cut alternatively spliced product), a.k.a. CCAAT-displacement protein CDP1, and Cut-like homeodomain-containing protein Cux1.

Table 4.21. Golgins and their partners. Members of the golgin family of cytoplasmically orientated coiled-coil proteins belong to the category of numerous tethering factors at the Golgi body (ACBD3: acyl-coenzyme-A-binding domain containing protein-3 [or Golgi resident protein GCP60]; GCP170: 170-kDa Golgi complex-associated protein; GMAP: Golgi-associated microtubule-binding protein; GoRASP1: 65-kDa Golgi reassembly-stacking protein 1 [or GRASP65]; RB: retinoblastoma protein; TRIP11: thyroid hormone receptor [TR] interactor-11 or TR and retinoblastoma-interacting protein TRIP230). Golgins are targeted to the Golgi membrane via their GRIP (Golgin-97, RanBP2a, IMH1 [integrin–myosin homolog], and P230/golgin-245) domains. The vesicle-docking protein Uso1 is also called general vesicular transport factor P115, transcytosis-associated protein (TAP), and vesicle-docking protein (VDP). Protein Uso1, which is predominatly cytosolic, is essential in the docking of endoplasmic reticulum-derived vesicles to the cis-Golgi network.

Golgin	Alias	Location	Binding partner
GolgA1	Golgin-97	Golgi cisternæ	ARL1
GolgA2	GM130	Cis-Golgi ERGIC	Myristoylated GoRASP1, GoRASP2, Uso1, syntaxin-5, Rab1a/1b/2a
GolgA3	Golgin-160 GCP170	Golgi body	Caspase-2
GolgA4	Golgin-240/245 Trans-Golgi P230	Golgi body	Rab6, ARL1
GolgA5	Golgin-84		Rab1a
GolgB1	Macrogolgin, giantin	Vesicles	GolgA2, Uso1, ACBD3
GMAP210	TRIP11/230	Golgi body (cis-Golgi)	RB1, TR, microtubule ends

envelope [271].[111] The Golgi body thus keeps a constant configuration despite permanent remodeling due to membrane transport.

In the trans-Golgi network, many proteins and polysaccharides bear final post-translational modifications (e.g., sialylation and fucosylation of glycans and sulfation of glycosaminoglycans) and lipids undergo synthesis completion before export.

The trans-Golgi network synthesizes sphingolipids [272]. Ceramide synthesized in the endoplasmic reticulum is converted into sphingomyelin by sphingomyelin synthase-1.[112]

---

[111] Asymmetric tethering relies on motifs that sense membrane curvature. These motifs localize to the N-terminus of GMAP210 and to ArfGAP1 that controls the interaction of the C-terminus of GMAP210 with ARF1. The tethering mode maintains the Golgi structure without disturbing material transport.

[112] Ceramide is also glycosylated into glucosylceramide in the cis-Golgi network by glucosylceramide synthase. Ceramide and glucosylceramide are delivered from

The trans-Golgi network is a major sorting center for cell lipids and proteins at the junctions between endocytic and exocytic pathways. Its tubular membranes possess carriers that target molecules to different destinations. Cargo segregation into different TGN domains requires specific molecular machineries that decipher sorting motifs and create lipid and protein nanoenvironments that have very high affinity for selected targets to be transported. The destination of each cargo type is selected by sorting signal of the cargo, which is decoded by docking and fusion factors, coat adaptors (AP1a, AP1b, AP3, and AP4), small GTPases of the ARF and RAB superfamilies (Table 4.22), golgins, as well as lipids (phosphoinositides, sphingolipids, and cholesterol).

The main destinations of TGN-derived carriers are [272]: (1) plasma membrane, (2) early and late endosomes, and (3) secretory granules and other specialized compartments. Each destination is mapped by at least one carrier type. Different carriers can indeed carry given cargoes to the same acceptor organelle.

Tubule size and number depends on transport magnitude. In the trans-Golgi network, a cargo directed to a given destination is sorted into a specific carrier. After cargo sorting into a forming tubule, tubular carrier precursors are extruded along microtubules using nanomotors down to fission of the elongated tubule into a free carrier [272].

Microtubule-binding, cytoplasmic linker-associated proteins (CLAsP) recruited to TGN by golgins elicit formation of microtubules at the Golgi body. Nanomotors include kinesins KIF5b and KIFc3 for distinct apical cargo, and KIF13a that interacts with AP1 for endosomal cargo. Dynamin (Chaps. 6 and 9), Gβγ dimeric subunit (Vol. 4 – Chap. 8. Guanosine Triphosphatases and their Regulators), phospholipase-Cβ (Vol. 4 – Chap. 1. Signaling Lipids), protein kinase-D (Vol. 4 – Chap. 4. Cytosolic Protein Ser/Thr Kinases), and brefeldin-A-ribosylated substrate are involved in fission that occurs at the thinnest section of elongated tubule. Actin, spectrin, ankyrin, and actin-related proteins are recruited onto Golgi membranes in an ARF-dependent process. Small GTPase CDC42a that binds to the Golgi body regulates actin polymerization.

Sorting of transmembrane proteins to endosomes and lysosomes is mediated by motifs[113] within the cytosolic domains of the proteins [273].[114] These sequences are recognized by components of protein coats tethered to the cytosolic face of membranes. Phosphorylation regulates signal recognition. Ubiquitination of cytosolic lysine residues also serves as a signal for sorting directed to the endosomal–lysosomal pathway.

---

synthesis sites to the late Golgi compartment by cytosolic lipid-transfer proteins, such as ceramide-transfer protein CERT and FAPP2.

[113] These short, linear sequences of amino acid residues are also termed signals.

[114] Proteic sequences for transport sorting include: (1) tyrosine-based sorting signals that are recognized by adaptor complexes AP1 to AP4 and (2) dileucine-based signals that are detected by adaptors GGAs.

Table 4.22. Small GTPases of the trans-Golgi network (Source: [272]; Vol. 4 – Chap. 8. Guanosine Triphosphatases and their Regulators).

GTPase	Regulators and effectors (function)
GTPases of the ADP-ribosylation factor superfamily	
ARF	Brefeldin A-sensitive exchange factors BIG1 and BIG2
	GBF1
	GTPase-activating proteins ARAP1
	Coat adaptors AP1 and AP3
	(endosomal sorting)
	AP1B and AP4 complexes
	(basolateral sorting
	Golgi-localized, γ-ear-containing, ARF-binding proteins
	Lipid metabolism enzymes PI4K3β and PLD
	(production of PI(4)P)
	Lipid-transfer protein CERT and FAPP2
	(membrane tubulation;
	transfer of ceramide and glucosylceramide)
	Lipid-transfer protein OSBP1
	(cholesterol level sensor)
	Arfaptins (membrane tubulation)
	ARF-like protein ARL1 and ARF-related protein ARPR1
	GRIP domain-containing golgin-97 and -245
	(possible membrane tubulation)
GTPases of the RAB superfamily	
Rab6	Kinesin
Rab8	(plasmalemmal destination)
	Coat adaptor AP1B (endosome recycling)
	Optineurin (that binds myosin-6 and FIP2)
	Hungtintin (that binds actin)
Rab10	(plasmalemmal route regulation)
Rab11a	(endosome recycling; basolateral trafficking)
Rab14	
Rab22b	(endosome trafficking)
GTPases of the RHO superfamily	
CDC42	ARF1-regulated CDC42 GTPase-activating protein
	Fgd1

Post-translational modifications of cargos influence their sorting. Ubiquitination is used to target endosome in cooperation with Golgi-localized, γ-ear-containing, ADP ribosylation factor-binding proteins (GGA) [272]. The trans-Golgi network also operates as an acceptor for endosomal vesicles that have a high cholesterol and sphingolipid content.

4.7 Mitochondria

The mitochondrion (20–1000 per cell) has an envelope made of outer and inner membranes. The inner membrane has infoldings to increase the reaction surface area. Mitochondrial activity depends on the number and working efficiency of mitochondria. This activity varies according to the cell type.

The number of mitochondria increases during both cardiogenesis and cardiac hypertrophy. In cardiomyocytes, the lack of the mitochondrial chaperone HSP40[115] causes progressive respiratory chain deficiency and decreased level of mitochondrial DNA, which lead to cardiomyopathy [274].

4.7.1 Mitochondrial Remodeling

In addition, mitochondria are dynamical organelles that can change in shape and size. Mitochondrial shape indeed ranges from small organelle to large interconnected tubular network. They move toward suitable locations within the cell, as they are attached to the microtubule cytoskeleton. Mitochondrial morphology and size within cells are maintained through a balance between 2 opposing processes, fission and fusion. Fusion and fission dynamics permanently remodel the mitochondrial network.

Numerous genes are involved in mitochondrial morphogenesis. Fission protein dynamin-related protein DRP1 is a cytosolic dynamin GTPase. Protein DRP1 is phosphorylated by protein kinase-A. This phosphorylation is regulated via the cAMP–Ca^{++} pathway[116] to induce mitochondrion elongation and apoptosis resistance, although DRP1-induced mitochondrial fission occurs early during cell death due to BAX–BAK activation, possibly due to loss in DRP1 availability for interaction with other regulators [276]. Dephosphorylation of DRP1 by protein phosphatase-3 (or calcineurin) promotes mitochondrial fragmentation and increases cell vulnerability to apoptosis (Vol. 2 – Chap. 4. Cell Survival and Death). Regulation of DRP1 is thus linked to apoptosis command.

---

[115] A.k.a. Dnaja3 or Tid1.

[116] Recruitment of DRP1 at scission sites occurs simultaneously with calcium uptake by mitochondrion [275].

4.7.2 Mitochondrial Function

Mitochondria produce adenosine triphosphate.[117] They also synthesize metabolites and reactive oxygen species. Mitochondria also act as buffers of cytosolic Ca^{++} concentration.

Mitochondria regulate cell apoptosis (programmed death of damaged cells) and protect against cancer. Involved proteins are located in the mitochondrial outer membrane. Various anti-apoptotic molecules are counteracted by pro-apoptotic factors when cell apoptosis becomes imperative.

Mitochondria contain their own genome (mtDNA). However, mitochondrial proteins are mostly encoded in the nucleus.[118] Mitochondrial genes can be damaged by free radicals generated during energy production. Ribonucleotide reductase catalyzes the synthesis of deoxyribonucleotides from corresponding ribonucleotides.[119]

Maternally inherited mtDNA mutations cause cardiomyopathy and neurodegenerative disorders. Female germline filters out the most deleterious mtDNA mutations prior to conception, but moderately deleterious mtDNA mutations are transmitted [277].

Most mitochondrial proteins are synthesized as preproteins in the cytosol and must be imported across outer and inner mitochondrial membranes. Proteins are selectively transported across the mitochondrial inner membrane through the translocase of the inner membrane TIM23, such that the electrochemical proton gradient, which drives adenosine triphosphate synthesis, is conserved.[120] Protein TIM50 maintains the membrane potential. It induces a rapid closing of TIM23 translocation channel [278]. Mitochondrial preproteins relieve the TIM50-mediated inhibition to be translocated.

Reversible phosphorylation of mitochondrial proteins regulates respiratory function, energy metabolism, and mitochondrion-mediated cell death. Various

---

[117] Mitochondria store the energy in the form of ATP, which is used to power cell functioning, from protein synthesis to muscle contraction. ATP is generated when charged protons are transferred. In mitochondria, flavin mononucleotide is a primary electron acceptor within the respiratory chain of the membrane. A set of iron–sulfur clusters transport electrons from dihydronicotinamide adenine dinucleotide (NADH) into the hydrophobic proton-pumping domain of the complex, avoiding generation of deleterious reactive oxygen species.

[118] Heme, part of the oxygen-carrying hemoglobin, is made by mitochondria in bone marrow cells.

[119] Ribonucleotide reductase is made of 2 subunits R1 and R2. Subunit R2, a homolog of ribonucleotide reductase P53R2, is generated by P53 factor. In the S phase of the cell cycle, most deoxyribonucleotides are synthesized by R1/R2 ribonucleotide reductase. However, in a non-proliferating cell, deoxyribonucleotides are produced by R1/P53R2 ribonucleotide reductase. Agent P53R2 is involved in mitochondrial DNA synthesis and repair.

[120] Tim23 complex contains the pore-forming protein Tim23 and 3 membrane proteins Tim17, Tim21, and Tim50. Tim23 has a transmembrane domain and a domain in the intermembrane space.

kinases (protein kinase-A, -B, -C, and -G, mitogen-activated protein kinases, glycogen synthase kinase-3β, hexokinase, creatine kinase-2, Src kinases) resides in or are in close association with mitochondria.[121]

Mitochondrial outer membrane permeability regulates not only metabolite and energy exchange between mitochondria and cytosol, thereby controlling cell respiration, but also apoptosis by releasing apoptogenic factors from the mitochondrial intermembrane space into the cytosol at the early stage of apoptosis. Mitochondrial outer membrane permeability depends on substance carriers.

The outer membranes of mitochondria contain channel proteins with β-barrel proteins that are required for communication between mitochondria with the rest of the cell. β-Barrel precursors are synthesized in the cytosol and imported into mitochondria by translocase of mitochondrial outer membrane (TOM) and the sorting and assembly machinery complex (SAM) [279]. Import pore TOM and SAM insertase enable insertion of nearly all newly synthesized proteins destined for mitochondria. The SAM complex contains 2 proteins essential for cell viability, channel-forming SAM50 and SAM35. Insertion of β-barrel precursors is initiated by binding to SAM35, which is linked to SAM50. SAM35 recognizes the β-signal and generates a conductance increase in SAM50 channel.

The outer membrane of mitochondria contains 3 families of integral membrane proteins that act for full communication between mitochondria and the rest of the cell. In addition to TOM and SAM, voltage-dependent anion channels (VDAC), or mitochondrial porins, constitute the most abundant proteins in the mitochondrial outer membrane. Channels VDACs are responsible for the energy maintenance of the cytosol. Oxidative phosphorylation requires transport of metabolites (cytosolic ADP, ATP, and inorganic phosphate) across both mitochondrial membranes to generate ATP in mitochondrial matrix. Voltage-dependent anion channel can adopt a unique fully open state and multiple states with smaller conductance impermeable to ATP, but still permeable to small ions. Heterodimeric tubulin interacts with voltage-dependent anion channel to increase its voltage sensitivity and induce voltage-sensitive, reversible closure at low transmembrane potentials [280].

Heating that results from the activity of the brown adipose tissue is due to mitochondrial proton gradients that are uncoupled from ATP production,

---

[121] Phosphorylation of mitochondrial proteins can intervene in various mitochondrial functions, such as proteins of oxidative phosphorylation (e.g., NADH dehydrogenase Fe-S protein-7, NADH dehydrogenase-1β-4, ubiquinol–cytochrome-C reductase complex core protein-2, cytochrome-C oxidase subunit-4 isoform-1, and succinate dehydrogenase complex subunit-B), lipid metabolism (e.g., propionyl-CoA carboxylase β-chain, trifunctional enzyme subunits α and β, (2,4)-dienoyl-CoA reductase, (3,2)-trans-enoyl-CoA isomerase, short chain 3-hydroxyacyl-CoA dehydrogenase, and carnitine palmitoyltransferase-2), molecular transport, and chaperone activity.

the chemical energy being converted directly into heat under certain environmental circumstances. Uncoupling protein-1 is required for thermogenesis.

A cytosolic longevity-determinant protein that translocates into the mitochondria can cooperate in cell death signaling. Mitochondrial accumulation of adaptor isoform Src homology-2 domain-containing (SHC) transforming protein-1 (SHC1, or P66SHC) results from its phosphorylation by protein kinase-Cβ activated by oxidative stress and recognition by protein peptidyl prolyl isomerase NIMA-interacting PIN1 [281]. Protein SHC1 acts as a reactive oxygen species producer within mitochondria, leading to apoptosis.

Mitochondria have diverse tissue-dependent metabolic functions. In the liver, mitochondria process a wide variety of molecules. Because enzymes reside predominantly in the matrix, substrates are transported across the mitochondrial inner membrane. The mitochondrial metabolite transport system requires manifold specific carriers.[122] The mitochondrial membrane is impermeable to small ions and permeable to small uncharged molecules such as oxygen. Calcium uptake is catalyzed by a Ca^{++} channel in the inner membrane that exchanges Ca^{++} for H^+. The mitochondrial membrane is impermeable to nicotinamide adenine dinucleotides (NAD^+, NADH, $NADP^+$, and NADPH). Two main routes are then employed, the glutamate–aspartate and the dihydroxyacetone phosphate paths. The former uses 2-oxoglutarate–malate and glutamate–aspartate exchangers, as well as NADH and a transaminase enzyme. The latter relies on 2 different glycerol 3-phosphate dehydrogenase enzymes, the first one using an NAD–NADH couple and the second using a FAD-linked membrane-bound enzyme.

4.7.3 Oxidative Phosphorylation

Mitochondria are responsible for cell respiration, using delivered oxygen, as mitochondria function in oxidative phosphorylation (Sect. 4.7.6). Oxidative phosphorylation is the process of ATP synthesis due to an electron transfer from nicotinamide adenine dinucleotide (NADH) and flavin adenine dinucleotide (FADH2) to oxygen by a set of electron carriers.[123] Relevant proteic complexes are located at the inner mitochondrial membrane. Oxidation and phosphorylation are coupled by a proton level difference across the inner mitochondrial membrane. Respiratory chain complexes assemble into respiratory

---

[122] These carriers include the PO_4^{3-}–OH^- exchanger, the adenine nucleotide transporter, the tricarboxylate transporter, the transporters for citrate–isocitrate, malate–succinate, pyruvate, and 2-oxoglutarate carrier, etc.

[123] Energy that is produced in oxidative phosphorylation results from a set of oxidoreduction reactions during which electron transfer is mainly carried out by 4 large membrane protein assemblies, the so-called *respiratory chain complexes*, and some small electron carriers, such as quinones and cytochrome-C. The tortuous electron path through the respiratory chain triggers the transport of protons from the inside to outside of the mitochondrial inner membrane that leads to an electrochemical gradient used by the ATP synthase complex.

supercomplexes or *respirasomes* such as complexes that involve homodimeric cytochrome-BC1 and cytochrome oxidase (i.e., respiratory chain complex-3 and -4).

Mitochondrial oxidative phosphorylation is controlled by both mitochondrial and nuclear genes. Oxidative phosphorylation is composed of multiple proteins ($\sim$90), among which 13 proteins are encoded by the mitochondrial genome [282]. During cell growth, transcription and translation of nuclear and mitochondrial genes of oxidative phosphorylation are orchestrated to achieve metabolic adaptation. On the other hand, degenerative diseases such as type-2 diabetes mellitus are characterized by a decline in oxidative phosphorylation and a rise in ROS levels. Moreover, microtubule stabilizers and destabilizers increase oxidative phosphorylation and decrease ROS via PGC1α–ERRα-mediated expression of oxidative phosphorylation genes.

4.7.4 Coordination between Mitochondria and Nucleus

Mitochondria contain their own genomes, but most of their proteins are encoded in the nucleus. The coordination of separate genomes is achieved by both anterograde (nucleus-to-organelle) and retrograde (organelle-to-nucleus) mechanisms [283]. The nucleus genome encodes mitochondrial proteins, hence controls mitochondria. In return, mitochondria send signals to the nucleus to coordinate nuclear and mitochondrial activities.

The anterograde control of mitochondrial gene expression in response to endogenous and environmental stimuli (stress, nutrient availability, developmental cues, and physical factors) received by the nucleus involves regulators of mitochondrial gene expression proteins (RMGE) that target specific RNA transcripts. Mitochondria use a nuclear-encoded RNA polymerase to produce mitochondrial proteins.

The retrograde control from mitochondria uses numerous mechanisms to regulate nuclear gene expression and adapt the anterograde control. The activity of basic leucine zipper transcription factor BACH1 involved in heme metabolism and oxidative-stress response is regulated by mitochondrial heme, a sensor for O_2 availability.

4.7.5 ATP Synthesis

Mitochondria supply cells with energy, as they produce ATP, the main energy supplier for multiple biochemical reactions within the cell (mitochondrial respiration, as O_2 is required). Adenosine triphosphate is synthesized from adenosine diphosphate and phosphate by ATP synthase. During cellular respiration, a transmembrane proton flux through the electron transport chains yields the driving force of ATP synthesis. Back flow of protons caused by the so-called *protonmotive force* is confined to proton-translocating ATP synthase to be coupled to ATP formation. These protons are supplied by few freely diffusing protons and numerous proton-buffering groups at the membrane surface. ATP synthase is composed of 2 rotary nanomotors coupled by a central rotor and an eccentric stator [284]. It combines the electrical, mechanical, and chemical features of enzyme function. The electrical rotary nanomotor drives the chemical rotary nanomotor by elastic power transmission to produce ATP with high kinetic efficiency. Proton flow through electrical generator yields the torque needed to power chemical generator.

Mitochondria convert substrate to energy using 2 different complexes. *Complex-I* reflects the global cellular NADH production from fatty acid oxidation, the tricarboxylic acid (TCA) cycle, and glycolysis (Fig. 4.4). The TCA cycle-dependent *complex-II* receives FADH2 directly from succinate dehydrogenase. The ATP production-to-oxygen consumption ratio varies in mitochondria of different cell populations. Cardiac mitochondria use more oxygen, but produce ATP at a faster rate than liver ones [285]. In the heart, the TCA energy-converting cycle maximizes the oxidative phosphorylation production of ATP in mitochondria (Fig. 4.5). Type-F ATPases in mitochondria are membrane-bound complexes that couple the transmembrane proton motor to ATP synthesis. The latter depends on a conformational change driven by the passage of protons from the intermembrane space into the matrix via channels.

Electron transfer in cell respiration is coupled to proton translocation across mitochondrial membranes.[124] The resulting electrochemical proton gradient is used to power energy-requiring reactions such as those of ATP synthesis. Cytochrome-C oxidase is a major component of the respiratory chain. It attaches oxygen as a sink for electrons and links O_2 reduction to proton pumping. Electrons from cytochrome-C are transferred sequentially from donors (CuA and heme-A sites) to O_2 reduction site of cytochrome-C oxidase. This electron transfer initiates the proton pump [286]. Cytochrome-C in the mitochondrial intermembrane space thus serves as an electron shuttle between complex III and complex IV of the respiratory chain and links to cardiolipin. Cytochrome-C–mediated electron transfer is due to the iron atom within the heme prosthetic group.

The energy production by mitochondria generates by-products, the *reactive oxygen species* (ROS), electrons reacting with oxygen to form superoxide. Reactive oxygen species[125] are highly reactive molecules that include free radicals, anions composed of oxygen, such as superoxide and hydroxyl, or compounds containing oxygen such as hydrogen peroxide, which can produce free radicals or be activated by them.

---

[124] Proton transfer to the O_2 reduction site is associated with electron transfer from heme-A to the O_2 reduction site. This initial driving step of proton pump is followed by proton release on and uptake from the 2 aqueous sides of the membrane.

[125] Sources of ROSs are mainly aerobic respiration, peroxisomal oxidation of fatty acids, microsomal cytochrome-P450 metabolism of xenobiotics, stimulation of phagocytosis by pathogens or lipopolysaccharides, arginine metabolism, and specific enzymes.

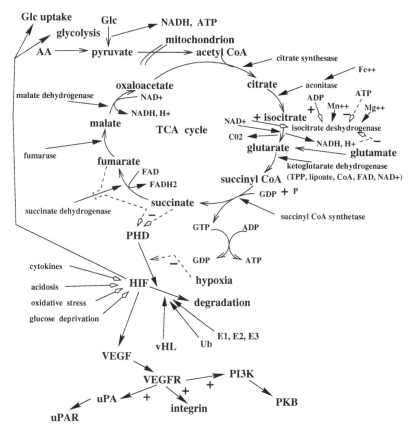

Figure 4.4. The tricarboxylic acid cycle (TCA or Krebs cycle) serves as a metabolic platform. During each TCA, 3 nicotine adenine dinucleotide (NAD^+) are converted into NADH and 1 flavine adenine dinucleotide (FAD) in FADH2. TCA leads to electron transport and manufactures ATP. The hypoxia-inducible factor (HIF) is degraded by the von Hippel-Lindau protein (vHL), after hydroxylation by the prolyl hydroxylase (PHD). The latter is inhibited by increased concentrations in fumarate and/or succinate as well as local hypoxia. Factor HIF is promoted not only by hypoxia, but also by glucose deprivation, acidosis, and cytokines. It upregulates vascular endothelial growth factor (VEGF) and promotes glucose (Glc) uptake and glycolysis, hence ATP and pyruvate formation. Once bound to its receptor, VEGF upregulates integrin expression and stimulates several pathways, particularly via urokinase-type plasminogen activator (uPA), phosphatidylinositol 3-kinase (PI3K), and nitric oxide. Degradation of HIF involves the ubiquitin–proteasome system with Ub-activating (E1), Ub-conjugating (E2), and Ub-ligating (E3) enzymes.

The NADH–ubiquinone oxidoreductase complex, which produces superoxide rather than hydrogen peroxide, is a major source of reactive oxygen species in the mitochondrion. The production rate of superoxide is determined by electron transfer between fully reduced flavin to oxygen, the reacting flavin level

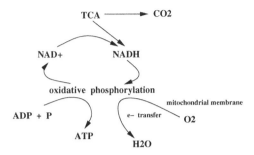

Figure 4.5. Last stage of catabolism of nutrients into wastes. This stage produces a great quantity of ATP from phosphate and ADP. NADH produced by the tricarboxylic acid cycle is used for ATP synthesis during oxidative phosphorylation in the mitochondrion.

depending on concentrations of NADH and NAD^+ [287]. Superoxide production is enhanced when the NAD^+ pool is reduced.

Cardiolipin (diphosphatidyl glycerol) is a component of the mitochondrial membrane, mainly on the inner membrane, especially in the cardiomyocyte. Several types of cardiolipin exist, a major cardiac cardiolipin, tetralinoleoyl–cardiolipin, and minor cardiolipin species. This phospholipid stabilizes the activity of protein complexes implicated in the mitochondrial electron transport system or the respiratory chain. It generates an electrochemical potential for substrate transport (cardiolipin is required for translocation, but not for binding) and ATP synthesis. It interacts with multiple mitochondrial proteins, such as NADH, ubiquinone oxidoreductase, cytochrome-C oxidase, and cytochrome-C. It thus acts on enzymes involved in oxidative phosphorylation. The early oxidation during cell apoptosis is catalyzed by a cardiolipin-specific peroxidase activity of cardiolipin-bound cytochrome-C. Furthermore, cardiolipin is involved in the osmotic stability of the mitochondrial membrane. Affazin is an enzyme involved in the synthesis of cardiolipin.[126]

Mitochondrial oxidative phosphorylation depends on the intramitochondrial reducing power, i.e., on (1) the concentrations of NADH ([NADH]) and of NAD+ ([NAD+]), indexed by the [NADH]/[NAD+] ratio; (2) the cytoplasmic energy state (the relative concentrations in ATP [ATP] and ADP [ADP]) given by [ATP]/([ADP][Pi]) ratio; and (3) intracellular oxygen pressure [288].

The mitochondrial respiratory rate is determined by the rate of cell ATP use. The cellular energy state at a given rate of ATP utilization is determined by the activity of mitochondrial dehydrogenases expressed as [NADH]/[NAD+] and the local oxygen pressure. Concentration of NADH in the mitochondrium is influenced by the activity of mitochondrial dehydroge-

---

[126] The Barth syndrome (cardiomyopathy) results from mutations in the affazin gene.

nases, cytosolic phosphorylations related to the energy state, and cytochrome-C oxidation in the respiratory chain.[127]

The synthesis or hydrolysis of ATP by the mitochondrion occurs in its matrix space. Nevertheless, ADP and ATP can cross the mitochondrial membrane, via the adenine nucleotide transporter[128] (ANT) and phosphate transporter[129] (Fig. 4.6).

Creatine kinase (CK), located at sites of energy demand and production, catalyzes the reversible transfer of phosphate from phosphocreatine (PCr or Cr^P) to ADP, which generates ATP and creatine (Cr). It then controls the energy homeostasis of cells with high, fluctuating energy requirements such as cardiomyocytes. Creatine kinases have a subcellular compartmentation to be coupled to sites of either energy production (glycolysis and mitochondrial metabolism) or energy consumption (actomyosin ATPase, sarcoplasmic reticulum Ca^{++} ATPase). They form a regulated energy distribution network, the so-called phosphocreatine circuit (or shuttle).

The main pools of creatine kinases include mitochondrial (miCK), cytosolic[130] (cyCK), and myofibrillar (mmCK) creatine kinases (Fig. 4.6). Mitochondrial creatine kinase binds to mitochondrial membranes and forms a complex with porin and adenine nucleotide translocase[131] (ANT) for export of phosphocreatine into the cytosol. Adenosine monophosphate-activated protein kinase (AMPK), an energy sensor, phosphorylates CK and inhibits its activity. Kinase AMPK is regulated by both ATP/AMP and Cr^P/Cr ratios.

Under some circumstances, substantial amounts of acetate are produced that must be processed by acetyl-CoA synthase. Moreover, acetate must be converted for energy production. Acetyl-CoA synthases are abundant in the heart and skeletal muscles. The mitochondrial matrix acetyl-CoA synthase-2 is activated by reversible deacetylation induced by sirtuin [289].

The mitochondrial nitric oxide synthase produces nitric oxide (NO). Nitric oxide inhibits cytochrome-C oxidase, and may act as an oxygen sensor. Low oxygen supply can inhibit nitric oxide production, and hence increase oxygen affinity of cytochrome-C oxidase to maintain NADH oxidation and electron transfer. The mitochondrial nitric oxide synthase also leads to the production of superoxide anion (O_2^-), especially after ischemia–reperfusion episodes.

---

[127] Oxidation of NADH is related to cytochrome-C oxidation.

[128] The adenine nucleotide transporter catalyzes the exchange of ATP for ADP across the inner mitochondrial membrane. Entry of ADP and exit of ATP are favored.

[129] The transport involves either exchange of $H_2PO_4^-$ for OH^-, or cotransport of $H_2PO_4^-$ with H^+.

[130] Several cytosolic creatine kinase isoforms exist: MM in skeletal and cardiac myocytes, MB in cardiomyocytes, and BB in smooth muscle and most non-muscle cells.

[131] The adenine nucleotide translocator is the most abundant protein of the inner mitochondrial membrane. This transmembrane channel is responsible for the export of ATP in exchange with ADP (antiport) across the inner membrane.

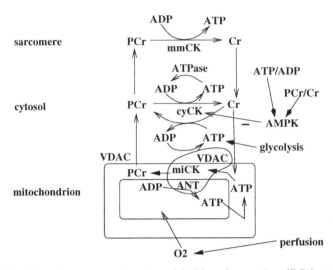

Figure 4.6. Phosphocreatine circuit model. Phosphocreatine (PCr) and creatine (Cr) are involved in cell energy production, distribution, and consumption in diverse subcellular compartment: mitochondrion, cytosol, and sarcomere (ATP-consuming site). Adenosine triphosphate (ATP) losing a phosphate group is transformed in adenosine diphosphate (ADP). Conversely, phosphocreatine can give its phosphate group, thereby regenerating ATP from ADP at high rates. The PCr–Cr circuit connects localized creatine kinase (CK), especially in cells with high energy demands. Creatine kinase catalyzes the reversible transfer of phosphate group from PCr to ADP. It specifically localizes to regions of energy demand (cytosol and sarcomere, where ATPases act, such as actomyosin ATPase and sarcoplasmic reticulum Ca^{++} ATPase particularly in the cardiomyocyte) and energy production (mitochondrion and glycolysis sites). Cytosolic creatine kinase (cyCK), in conjunction with Ca^{++} pumps, is required for the energetics of Ca^{++} homeostasis. Cytosolic CKs (e.g., muscle [MM], cardiac [MB], and smooth muscle and most non-myocyte type [BB]) are coexpressed in a tissue-specific fashion with mitochondrial isoforms. Mitochondrial creatine kinase (miCK) in the mitochondrial intermembrane space crosslinks mitochondrial inner and outer membranes. It forms a complex with porin (VDAC) and adenine nucleotide translocase (ANT) for export of PCr into the cytosol. It phosphorylates ATP produced by the mitochondrion into PCr, subsequently expelling to the cytosol via VDAC. The main myofibrillar CK (mmCK) converts PCr into ATP. The PCr/Cr reaction is regulated by AMP-activated protein kinase (AMPK) via ATP/AMP and PCr/Cr ratios. Enzyme AMPK is activated by a low PCr/Cr ratio. In myocytes, the CK system corresponds to an energy buffer, allowing nearly constant ATP levels during muscle activity.

Reactive oxygen species such as peroxynitrites ($ONOO^-$) are released by the reaction between NO and O_2^-. Peroxynitrites inactivate mitochondrial creatine kinases.

Reactive oxygen species that react with nitric oxide generate *reactive nitrogen species*, which include not only peroxynitrite, but also nitrogen dioxide

(NO_2) and dinitrogen trioxide (N_2O_3). These substances induce a nitrative stress.

Mitochondrial respiration is linked to glycolysis and cell growth, especially for adaption to prolonged food deprivation by reducing glucose utilization. Impaired mitochondrial respiration causes NADH accumulation via the *pentose phosphate pathway*. Accumulation of NADH inactivates phosphatidylinositol 3-phosphatase PTen as it prevents thioredoxin NADP(H)-dependent PTen activation, and thus activates protein kinase-B (PKB) [290]. Insulin and growth factor receptor signaling pathways that cooperatively stimulate effector PKB using the PIP_3 pool are amplified by PTen inactivation. Thioredoxin-interacting protein (TRxIP) maintain sufficient thioredoxin–NADPH activity[132] to reactivate oxidized (inactive) PTen to oppose PKB activity, thereby switching fuel away from glucose toward 3-hydroxybutyrate and fatty acids and adapting to nutrient deprivation. Thioredoxin-interacting protein thus counteracts metabolic and growth signaling. In the absence of TRxIP, oxidative tissues (skeletal muscles and heart) but not lipogenic tissues (liver and adipose tissue) exhibit increased PKB signaling associated with impaired mitochondrial oxidation and oxidative inactivation of PTen, as well as elevated insulin sensitivity and glycolysis.

4.7.6 Main Regulators of Cell Energy

4.7.6.1 AMP-Activated Protein Kinase and Cell Energy

Adenosine 5′-monophosphate-activated protein kinase (AMPK) is a cellular energy sensor that controls the balance between ATP production and consumption in cells according to fluctuations in intracellular ATP/AMP ratio. Upon attenuation of this ratio, this fuel gauge stops ATP-consuming pathways and primes ATP-producing catabolic pathways. Once switched on by stresses that disturb the energy balance, it triggers both acute and longer-term adaptations. Enzyme AMPK mediates transcriptional responses to metabolic perturbations to regulate cellular energy homeostasis.[133]

---

[132] Anti-oxidant thioredoxin is an oxidoreductase kept in the reduced state by thioredoxin reductase, in a NADPH-dependent reaction. Thioredoxin substrates include ribonuclease, coagulation factors, glucocorticoid receptor, insulin, and choriogonadotropins.

[133] Members of the AMPK family are heterotrimers with catalytic α-, β-, and γ subunits. Isoforms α1 and 2, both with Ser/Thr kinase domains, auto-inhibitory sequence, and binding site for β- and γ-subunits, are encoded by the PRKAA1 and PRKAA2 genes in mammals [291]. Regulatory β- and γ-subunits bind to glycogen and nucleotides AMP or ATP, respectively. Isoforms β1 and β2 are encoded by the PRKAB1 and PRKAB2 genes in mammals. Isoforms γ1 to γ3 are encoded by the genes PRKAG1 to PRKAG3 in mammals. Most of the AMPK activity in most cell types relies on γ1 isoform. Mutations in γ2-subunit cause Wolff-Parkinson-White syndrome. γ-Subunit has 2 Bateman domains that bind either AMP, ATP, or ATP^{Mg}, whereas a third site contains a tightly bound AMP [292].

Free ATP is present in cells at concentrations that are almost 2 orders of magnitude lower than ATP^{Mg}. AMP in the micromolar range competes with ATP even though total cell concentration of ATP is in the millimolar range, because ATP is much less abundant than ATP^{Mg} [291].

Phosphorylated AMPK has a substantial basal activity that can be moderately enhanced by AMP binding. Furthermore, AMP binding markedly decreases the rate of dephosphorylation of activated AMPK. Increase in intracellular AMP level inhibits dephosphorylation (Thr172) in the activation loop of the kinase domain. Binding of AMP triggers subunit interactions that are not possible when ATP is bound [292].

When energy is needed, catabolism that converts ADP into ATP is activated and metabolism (protid, glucid, and lipid synthesis) that hydrolyzes ATP into ADP is stopped. Binding of AMP inhibits dephosphorylation of the activating site that is phosphorylated by upstream kinases, such as liver kinase-B LKB1 and calmodulin-dependent kinase kinases CaMKKα and particularly CamKKβ.[134]

Kinase AMPK is activated in circumstances that inhibit ATP production (e.g., hypoxia and hypoglycemia) or stimulate ATP consumption (e.g., synthesis and activation of motor proteins and ion carriers). Enzyme AMPK is targeted by cytokines, especially adipokines (Vol. 2 – Chap. 3. Growth Factors). Leptin secreted by adipocytes stimulates oxidation of fatty acids and glucose uptake as it activates AMPK (subunit α2 in skeletal muscle) [293]. Adiponectin also secreted by adipocytes stimulates AMPK phosphorylation required for glucose utilization and fatty acid oxidation in skeletal muscle and liver [294].[135] Interleukin-6 [295] and ciliary neurotrophic factor [296][136] activate AMPK.

Kinase AMPK phosphorylates many substrates, thereby: (1) activating catabolic enzymes, transcription factors, and coactivators involved in ATP production and (2) inactivating anabolic enzymes to impede ATP consumption (Table 4.23). Consequently, AMPK prevents cell growth. AMPK indeed phosphorylates (inhibits) effectors of the target of rapamycin pathway (Vol. 2 – Chap. 2. Cell Growth and Proliferation) and favors cell-cycle inhibitors, such as P53 and CKI1b cyclin-dependent kinase inhibitor. When the cell is subjected to a metabolic stress, AMPK mediates the stress response partly

---

[134] Kinase LKB1 is ubiquitous, but CaMKKs have a more restricted distribution. Both CamKK isoforms are expressed predominantly in nervous tissue. Isoform CaMKKβ is also found in endothelial and hematopoietic cells.

[135] In addition to AMPK activation, adiponectin stimulates the phosphorylation of biotin-dependent acetyl coenzyme-A carboxylase (ACC) and provokes (1) fatty-acid oxidation, glucose uptake, and lactate production in myocytes, and (2) reduction of gluconeogenesis in the liver. Kinase AMPK is the main inhibitor of ACC1 and ACC2.

[136] Ciliary neurotrophic factor transmits signal via the CNtFRα–IL6R–GP130β receptor complex. It increases fatty acid oxidation and reduce insulin resistance in skeletal muscle by activating AMPK kinase.

Table 4.23. Functions of AMP-activated protein kinase (AMPK; Source: [291]; ACC: acetyl coenzyme-A carboxylase; CHREBP: carbohydrate-responsive element-binding protein; CRTC: CREB-regulated transcription coactivator; eEF2K: calcium–calmodulin-dependent eukaryotic elongation factor-2 kinase; GluT: glucose transporter; HMGCoAR: hydroxy methylglutaryl coenzyme-A reductase; NOS: nitric oxide synthase; PGC: peroxisome proliferator-activated receptor-γ coactivator; SREBP: sterol regulatory element-binding protein). Kinase AMPK switches on ATP-generating catabolism (uptake and metabolism of glucose and fatty acids) and switches off ATP-consuming anabolism (synthesis of fatty acids, cholesterol, glycogen, and proteins).

	Catabolism activation
Glucose uptake	GluT4 translocation to the plasma membrane
Glycolysis	6-Phosphofructo-2-kinase upregulation
Lipolysis	Hormone-sensitive lipase downregulation
Fatty acid oxidation	ACC1 and ACC2 inhibition
Mitochondrial genesis	PGC1α upregulation
	Anabolism inhibition
Gluconeogenesis	Cytoplasmic translocation of CRTC (reduced transcription of gluconeogenic genes)
Lipogenesis	SREBP1c downregulation
	Decreased transcription of lipogenic genes by CHREBP (liver)
Sterol synthesis	HMGCoAR downregulation
Glycogen synthesis	Glycogen synthase downregulation
Protein synthesis	eEF2K activation
	TSC1–TSC2 complex phosphorylation (TOR inactivation)
	Cell signaling
Nitric oxide production	NOS3 activation
Insulin signaling	Opposition
	Ion transport
Cystic fibrosis transmembrane conductance regulator	Channel opening inhibition (airway and gut)

via P53 phosphorylation and activation of P53-dependent transcription. Kinase AMPK can also cause cell-cycle arrest by preventing the nuclear export of RNA-binding protein human antigen HuR. Besides, it also maintains cell polarity in epithelial cells [291].

In response to a reduced ATP/AMP ratio, AMPK phosphorylates (activates) proliferator-activated receptor-γ coactivator PGC1α that promotes transcription of genes involved in cellular metabolism (Vol. 2 – Chap. 2. Cell Growth and Proliferation). However, PGC1α-dependent transcription of genes of mitochondrial function and ATP-producing lipid catabolism requires NAD$^+$-dependent deacetylase sirtuin-1 that is able to deacetylate (activate)

PGC1α. Kinase AMPK responds to energy deprivation by raising cellular concentration of another metabolic sensor, NAD^+, that activates sirtuin-1 to stimulate expression of genes implicated in mitochondrial genesis and fatty-acid oxidation in skeletal muscle cells [297]. The higher the $NAD^+/NADH$ ratio, the greater the PGC1α deacetylation. Kinase AMPK phosphorylates PGC1α that subsequently can undergo Sirt1-mediated deacetylation.[137]

Under hypoxia or low nutrient input, the transcriptional activity of AMPK is mediated not only by phosphorylation of transcription factors and coregulators (P53, FoxO3a, PGC1α, and CREB-regulated transcription coactivator CRTC2[138] that recruit RNA polymerase-2 to promoters of AMPK-dependent genes, but also phosphorylation of the chromatin protein histone H2b [298]. Phosphorylation of histone H2b (Ser36) by AMPK at P53-binding sites activates the gene transcription. Both AMPK and its upstream activator liver kinase-B LKB1 colocalize at metabolic stress-responsive genes for transcription. The LKB1–AMPK energy homeostasis pathway thus operates directly in the nucleus to promote energy conservation and cell survival. In particular, AMPK and LKB1 link to the promoter of the genes that encode cyclin-dependent kinase inhibitor CKI1a and carnitine palmitoyltransferase CPT1c during a metabolic stress [298].

4.7.6.2 Target of Rapamycin Complex, Cell Energy, and Protein Synthesis

In addition to AMPK, target of rapamycin complex TORC1 that contains the regulatory associated protein of TOR (Raptor)[139] couples cellular energy status, growth factor abundance, and nutrient availability to protein synthesis. It supports cell survival decision, i.e., mRNA translation initiation (Chap. 5), ribosome genesis, cell cycle progression (Vol. 2 – Chap. 2. Cell Growth and Proliferation), and inhibition of apoptosis (Vol. 2 – Chap. 4. Cell Survival and Death). Signaling by TOR causes phosphorylation of the TORC1 substrate P70 ribomal S6 kinase S6K1.

Quality of newly synthesized polypeptides for proper cellular functioning relies on: (1) molecular chaperones such as HSP90 that ensure the proper folding of newly synthesized proteins (Sect. 5.10) and (2) ubiquitin-induced proteasomal degradation. Chaperone HSP90 is an activation-induced binding

---

[137] Transcription factors FOXO1 and -3a are also phosphorylated by AMPK and then deacetylated by Sirt1.

[138] A.k.a. transducer of regulated CREB activity TORC2.

[139] The immunosuppressive and anticancerous agents rapamycin and rapalogs everolimus and temsirolimus inhibits TORC1, as they disrupt the TOR–Raptor interaction. In addition, rapamycin pertains to the set of HSP90 inhibitors. Heat shock protein HSP90 is a chaperone that intervenes in the proper folding of proteins, hence supporting cell survival. Inhibitors of HSP90 promote T-cell anergy [299]. In particular, they preclude full stimulation of T_H cells.

partner of Raptor at least in T lymphocytes [299]. On the other hand, cellular stresses can lead to accumulation of misfolded proteins.

Agent TORC1 rapidly detects, integrates, and responds to environmental signals. Metabolic stresses, such as nutrient deprivation and hypoxia, decrease TORC1 activity and protein production. Furthermore, TORC1 senses protein misfolding. Therefore, TORC1 links protein quantity to quality, i.e., protein generation to quality control, as it both recognizes environmental cues and chaperone availability [300].

However, stress effects on TORC1 depend on their magnitude and duration. Cells indeed distinguish moderate reductions in protein quality from severe protein misfolding using molecular chaperones to differentially regulate TORC1 signaling [300]. When the availability of molecular chaperones is moderately reduced, TORC1 signaling heightens and fosters S6K1 phosphorylation. On the other hand, high-intensity stresses and loss of chaperones suppress TORC1 activity and prevent S6K1 phosphorylation.

4.8 Cytoskeleton

The cytoskeleton is composed of networks of various types of filaments (microfilaments, intermediate filaments, and microtubules), associated motor proteins, and regulatory molecules (Chap. 6). It is particularly involved in cargo transport within the cytosol (Chap. 9), cell adaptation to its environment (Chaps. 7 and 8), cell adhesion and motility (Vol. 2 – Chap. 6. Cell Motility).

4.9 Centriole

A centriole is a barrel-shaped cell structure. It is composed of an outer set of 9 triplets of microtubules. Centrioles are components of centrosomes, cilia, and flagella.

The *centrosome* is a pair of centrioles arranged perpendicularly and surrounded by a dense pericentriolar material. Centrosomes are involved in the organization of microtubules in the cytoplasm. During the cell division, these microtubule-organizing centers function as spindle poles, organizing the mitotic spindle and cytokinesis.

The centriole assembly and duplication is supported by the scaffold 152-kDa centriolar (centromeric) protein[140] (CeP152) [301]. This scaffold recruits another centriolar (centromere) protein CenPj[141] and Polo-like kinase-4 to regulate centriole self-assembly and duplication of the centriole and organization of pericentriolar material.

---

[140] Human ortholog of Asterless (Drosophila).
[141] A.k.a. centrosomal P4.1-associated protein (CPAP).

4.10 Primary Cilium

Primary cilium is an ubiquitous, single, immotile, microtubule-based, cortical organelle that protrudes from the plasma membrane of almost every eukaryotic cell and can detect optical, mechanical, and chemical cues.[142] In fact, the primary cilium is an evolutionarily conserved cellular organelle that organizes diverse signaling pathways. It transduces mechanical or chemical stimuli into intracellular signals that, in particular, control developmental pathways. It actually reacts to developmental signals and serves in the patterning of tissues during the body's development.

Primary cilium is constituted by a huge number of cilium- or basal body-associated proteins, the so-called *ciliome*. Primary cilium has an arrangement of 9 outer doublet microtubules. Proteins and lipids also travel to and from the primary cilium. This organelle enables transport of substances along the axoneme (central cilium structure) using the intraflagellar transport apparatus. The primary cilium can thus act as assembler of pathway components.

4.10.1 Primary Cilium and Signaling

Primary cilium contains components of the Hedgehog[143] and wingless-type (Wnt) pathways (Vol. 3 – Chap. 10. Morphogen Receptors).[144] Transcription factors Gli abound in cilia. Kinase Fused binds to kinesin KIF7, hence linking intraflagellar transport to the Hedgehog pathway. Protein KIF7 localizes to the base of the primary cilium and moves to the tip of the cilium in response to activation of the Hh pathway. After Hedgehog activation, KIF7 carries Hedgehog signaling members to the cilium tip, such as Suppressor of Fused and Gli factors, to prevent Gli activation [302]. Kinase Fused not only phosphorylates KIF7, but also Suppressor of Fused.

---

[142] Photoreceptors are modified primary cilia that sense light. Primary cilia of renal epithelial cells are mechanosensors that respond to fluid flow. Olfactory cilia detect odors and initiate the activity of olfactory neurons.

[143] Almost all Hedgehog signaling components localize to the primary cilium. Sonic Hedgehog induces delocalization from the primary cilium of its receptor Patched-1 that can then act as a transmembrane receptor. Binding of Sonic Hedgehog to Patched-1 abolishes inhibition by Patched-1 of the G-protein-coupled receptor Smoothened that accumulates in the primary cilium. Activated Smoothened then transduces cues to the nucleus via Glioma (Gli) transcription factors.

[144] In kidneys, inversin may control the balance between canonical and non-canonical signaling during tubule morphogenesis. The canonical β-catenin signaling regulates cellular proliferation, whereas non-canonical signaling modulates the correct orientation and elongation of cells during tubule formation. Several Bardet-Biedl syndrome (BBS) proteins that localize to the primary cilium and basal body cooperate with non-canonical Wnt signaling in planar cell polarity and tissue morphogenesis. In addition, BBS1 and BBS6 proteins interact with Van Gogh-like protein VangL2 that operates in β-catenin-independent, Frizzled-mediated signaling.

The primary cilium might also switch the canonical to non-conventional Wnt signaling pathways. Primary cilia also operate in signaling from platelet-derived growth factor receptor-α in cultured fibroblasts [302]. This receptor localizes to the primary cilium in neural stem cells of the adult rat subventricular zone. The primary cilium not only acts during embryogenesis, but also in adults. However, the formation of cilia is regulated by other signaling pathways.

Mechanotransduction that converts mechanical stimuli into biochemical signals relies in particular on the primary cilium. The bending of the primary cilium operates on target of rapamycin (TOR) to limit the cell size, independently of calcium transients and protein kinase B [303]. The primary cilium indeed act as a filiform mechanosensor of fluid flow.[145] Primary cilium bending activates [304]: (1) G-protein-coupled receptor-like polycystin-1 and mechanosensitive cation channel polycystin-2, thereby igniting calcium transients at the cilium; and (2) liver kinase-B1 (LKB1) located at the axoneme and basal body, thus phosphorylating AMP-activated protein kinase and reducing cell size. Enzyme AMPK may then phosphorylate inhibitor tuberin (tuberous sclerosis complex TSC2) to activate the RHEB GTPase-activating TSC1–TSC2 complex. The TSC1–TSC2 dimer inactivates RHEB GTPase, preventing it from activating the TORC1 complex. Alternatively, AMPK may also directly antagonize TORC1. Polycystin-1 colocalizes with target of rapamycin. It interacts with TSC2 and may thus repress the TORC1 pathway.

4.10.2 Formation of the Primary Cilium

The construction and maintenance of cilia requires intraflagellar transport along axonemal microtubules by the heterotrimeric complex KIF3a–KIF3b–KAP (KAP: kinesin [KIF]-associated protein) and KIF17 homodimer. In addition to kinesin-2 family members, the KIF3a–KIF3b nanomotor complex and KIF17, as well as KIF-associated protein-3 (KIFAP3 or KAP3), proteins that participate in the formation of the primary cilium include Fused, components of the intraflagellar transport complexes IFTa and IFTb, and a subset of basal body-associated, centrosomal, or intraflagellar transport proteins. These proteins include Bardet-Biedl syndrome complex (Bbsome), Meckel syndrome MKS1, fantom (Ftm),[146] oral-facial-digital syndrome OFD1, and Talpid³ [302].[147]

---

[145] In the nephron tubule, fluid flow provokes the bending of primary cilia of epithelial cells.

[146] A.k.a. retinitis pigmentosa GTPase regulator (RPGR)-interacting protein-1-like protein (RPGRIP1L).

[147] Talpid³ intervenes in Hedgehog signaling that is involved in the development of the vasculature. Patterning of venous and arterial endothelial identity is regulated by both hemodynamic stresses and molecular cues. The Hedgehog pathway controls the endothelial arterial identity of blood vessels. Sonic Hedgehog

Many (36 positive and 13 negative) modulators of primary ciliogenesis have been identified [306]. Ciliogenesis modulators encompass actin partners and transport vesicle agents. Inhibition of actin assembly (e.g., silencing of actin-related protein ARP3 involved in actin polymerization at filament branches) increases cilium length, as the *pericentrosomal preciliary compartment* is stabilized. This vesiculotubular structure stores transmembrane proteins destined for the cilium during the early phase of ciliogenesis. Among the ciliogenesis modulators, 2 gelsolin family proteins — gelsolin (Gsn) and advillin (Avil) — sever actin filaments (Chap. 6).

Cilia are nucleated by the basal body that is made up of the mother centriole and associated pericentriolar proteins. In addition, ADP-ribosylation factor-like protein ARL13b that localizes to the cilium is required for axoneme structure.

The formation of the primary cilium at the cell periphery depends on polarized transport vesicles of ciliary components (Chap. 9). Small GTPase Rab11 binds and activates Rab8-specific guanine nucleotide-exchange factor Rab8-interacting protein (RabIn8) [307]. Both Rab8 and Rab11 regulates anterograde transfer of materials from the trans-Golgi network and recycling endosomes to the plasma membrane. In addition, Rab8, RabIn8, and Rab11 colocalize to the basal body. The activator RabIn8 binds to the Bardet-Biedl syndrome proteic complex BBS1.

Fibroblast growth factor participates in the regulation of the ciliary length [302]. Inositol polyphosphate 5-phosphatase InPP5e that abounds in the ciliary axoneme may also contribute.

Like nuclear import, ciliary entry of KIF17 nanomotor is regulated by importin-β2 (Ipo2)[148] and a ciliary–cytoplasmic gradient of small GTPase Ran [308].[149] The C-terminus of KIF17 contains ciliary (CLS) and nuclear (NLS) localization signals. Hence, transportin-1 can recruit KIF17 to the primary cilium. The KIF17–transportin-1 complex crosses the ciliary transition zone (through the ciliary pore complex) and is dissociated by Ran^{GTP} in the proximal cilium.

4.10.3 Intraflagellar Transport

Intraflagellar transport, i.e., the bidirectional transport along axonemal microtubules maintains and extends the primary cilium, in particular, its membrane that corresponds to a specialized region of the plasma membrane, which protrudes from the cell. Materials are carried between the ciliary base and

---

provokes artery-specific neuropilin-1 expression, but has no effect on vein-specific neuropilin-2A [305]. Sonic Hedgehog also stimulates angiogenic growth factors VEGF and angiopoietin-2A in limb buds for vascular remodeling.

[148] A.k.a. karyopherin-β2 (KPnβ2) and transportin-1 (Trn1 or Tnpo1).

[149] Nuclear import requires recognition of nuclear localization signals by importins, translocation through the nuclear-pore complex (Sect. 9.1.8), and dissociation of the imported protein–importin complex in the nucleus by active Ran^{GTP}.

tip along the axoneme by the intraflagellar transport complexes IFTa and IFTb.

The intraflagellar transport complex IFTb acts in anterograde transport. On the other hand, IFTa and the minus-end-directed dynein nanomotor[150] operate in retrograde transfer, i.e., to the base of the cilium [302].

Intraflagellar transport is driven by the KIF3a–KIF3b–KAP trimer and KIF17 homodimer. Kinesin KIF17 interacts with transportin-1 under the control of Ran GTPase. In addition to KIF17, importin-β also targets the atypical transmembrane protein CRB3-CLPI, an alternate splice form with a novel C-terminal sequence ending in CLPI of the Crumbs homolog Crb3 that localizes to the cilium membrane [309].[151]

Ciliary membrane protein fibrocystin[152] possesses a ciliary targeting signal (CTS) that serves for Rab8 interaction and proper ciliary targeting.[153] Fibrocystin then colocalizes with polycystin-2 at the basal body of primary cilium [310].

4.11 Motile Cilia of Ciliated Cells

Motile cilia are components of certain epithelial cells, such as those that cover the brain ventricles, fallopian tubes, and respiratory tract. All cilia orient in the same direction to generate flow of fluids, such as cerebrospinal and airway lining fluids. The latter ensure clearance of the mucus layer in the respiratory tract that entraps small inhaled foreign bodies.

Cell polarization is programmed by the *planar cell polarity* pathway. Cilia orientation depends on ciliary motility. It can be further refined via a cilia-generated, positive, flow-polarity feedback loop. Therefore, both planar cell polarity signaling and fluid flow influence ciliary orientation [311].

Asymmetrical location of planar cell polarity proteins modulates the position and orientation of centrioles and cilia according to the direction of flow. Direction of rotational beating of cilia is indeed dictated by orientation of centrioles that act as basal bodies to form cilia. Centrioles first dock apically with random orientations and subsequently reorient in the flow direction [311]. Centriole orientation is directed by polarity signals before ciliogenesis. The planar cell polarity pathway operates not only upstream of ciliary orientation, but also during flow generation by cilia.

Hydrodynamic forces are coupled to the planar cell polarity pathway via Van Gogh-like protein VanGL2 [311]. The latter localizes to the entire cilium,

---

[150] I.e., dynein-2 heavy (DynC2H1) and light intermediate chain-1 (DynC2LI1).

[151] Members of the Crumbs family operate in cell polarity, as they link to tight junctions.

[152] A.k.a. polycystic kidney and hepatic disease protein PKHD1.

[153] Fibrocystin may be involved in tubulogenesis and/or maintenance of duct epithelia, such as those in neural tubules, gut, and bronchi.

from tip to base, i.e., between mechanosensors such as polycystins and basal-body prepatterning agents such as Disheveled. It colocalizes with intraflagellar transport proteins. Moreover, VanGL2 interacts with Bardet-Biedl syndrome proteins. The protein VanGL2 is stimulated by hydrodynamic forces to reorient basal bodies in the streamwise direction.

Airway epithelial cilia ensure mucociliary clearance. Ciliary beating is characterized by effective and recovery strokes separated by a rest period (Vol. 5 – Chap. 12. Airway Surface Liquid and Respiratory Mucus). Beating mode of respiratory cilia relies on the axonemal curvature. The latter requires tubulin glutamylation by tubulin Tyr ligase-like protein family member TTLL1 [312].[154]

4.12 Vesicles

Vesicles and vacuoles transport molecules. Certain vesicles carry proteins and lipids between the endoplasmic reticulum and the Golgi body. *Phagosomes* internalize large macromolecular complexes, especially pathogens, which are bound to plasmalemmal receptors for degradation.

4.12.1 Exosomes

Exosomes (size 30–90 nm) are vesicles secreted into the extracellular medium. The exosomal composition of proteins varies with the producing cell type; however, most exosomes contain soluble 70-kDa heat shock protein HSPa8 and cytosolic 90-kDa heat-shock protein HSP90αA1, among others. Exosomes selectively remove many plasmalemmal proteins after endocytosis and fusion of late *endosomes* with the plasma membrane (Chap. 9). They carry proteins and functional RNAs, such as mRNA and miRNA. They act in intercellular signaling.

4.12.2 Macropinosomes

Macropinosomes lead to massive membrane internalization from the cell surface. Clathrin-coated vesicles and calveolae are used in the endocytosis of various substances, such as signaling receptors, membrane pumps, and nutrients. Their formation at the Golgi body or the plasma membrane involves Adaptor proteins AP1 and AP2.

4.12.3 Peroxisomes

Peroxisomes are specialized for oxidative reactions that produce hydrogen peroxide. They are implicated in the lipid metabolism; catabolism of very-long

---

[154] A.k.a. tubulin polyglutamylase complex subunit PGS3.

and branched-chain fatty acids as well as plasmalogen synthesis. Peroxisomes contain oxidative enzymes, such as catalase and amino acid oxidase. Catalase uses H_2O_2 to oxidize substrates.

Peroxisomes can derive from the endoplasmic reticulum. Peroxisomal matrix proteins possess a peroxisomal targeting signal (PTS) to be imported. Protein receptors, such as peroxins Pex5 and Pex7, release their cargos and recycle. Peroxisomes are linked to the actin cytoskeleton.

4.12.4 Endosomes

Endosomes (Sect. 9.1.11) carry internalized molecules within the cytoplasm. These carrier vesicles recruit coat proteins. Early endosomes mature into late endosomes. In particuler, they lose small GTPase Rab5 and recruit small GTPase Rab7 (Sect. 9.1.13.10 and Vol. 4 – Chap. 8. Guanosine Triphosphatases and their Regulators). Late endosomes that contain luminal vesicles are called multivesicular bodies.

4.12.5 Lysosomes

Lysosomes (<5% of intracellular volume) are degradative elements of endocytosis. They contain many different hydrolytic enzymes and signaling proteins. Proteins of the endoplasmic reticulum and membrane rafts are observed in the lysosomal membrane. Their lumen is maintained at a pH of 4.6 to 5.0 by proton-pumping vacuolar ATPases.

Lysosomes contain acid hydrolases to degrade received molecules from various origins: exo- and endocytosis, autophagy, and phagocytosis [314] (Chap. 9). They differ from endosomes by the absence of mannose-6-phosphate receptors.

Many newly synthesized hydrolases and membrane proteins are delivered from the trans-Golgi network to endosomes after tagging with mannose-6-phosphate and subsequent binding to mannose-6-phosphate receptors. This process uses clathrin, adaptor protein AP1, and adaptors Golgi-localized, γ-ear-containing, ADP ribosylation factor-binding proteins. Afterward, hydrolases dissociate from the receptors and receptors recycle back to the trans-Golgi network. Newly synthesized lysosomal membrane proteins incorporate lysosomes from the trans-Golgi network using adaptor protein AP3 or tubular endosomes.

Some cell types contain specialized secretory lysosomes (melanosomes, class-2 major histocompatibility complex compartments, basophil granules, neutrophil azurophil granules, platelet-dense granules, mast-cell secretory granules, eosinophil-specific granules, and cytotoxic T lymphocyte lytic granules) that store newly synthesized secretory proteins. Many pathogens use endocytosis to enter into cells, but avoid degradation by lysosomes.

Endosomes and lysosomes can undergo a continuous cycle of transient contacts followed by a dissociation (fusion–fission cycle). Fusion delivers endocytosed molecules to lysosomes that form hybrid organelles resulting from mixing of contents of endosomes and lysosomes. After tethering, trans-SNARE complexes (SNARE assembly into tight, 4-helix bundles) bridges the 2 organelles using ^{N}ethylmaleimide sensitive factor (NSF), soluble NSF attachment proteins (SNAP), and small GTPase Rab. Trans-SNARE receptor complex forms using syntaxins. Lysosomes also fuse with autophagosomes and phagosomes after fission. Lysosome reformation from hybrid organelles requires recycling of some membrane proteins by transport vesicles.

4.13 Other Granules and Droplets

4.13.1 Lipid Droplets

Lipid droplets form cell dynamical stores of fatty acids and sterols. They consist of a core of neutral lipids, predominantly triacylglycerols or cholesteryl esters, surrounded by a monolayer of phospholipids and associated proteins (Rab and ADP-ribosylation factor GTPases, caveolins, phospholipase-D, and perilipin) [313]. The neutral lipids stored in lipid droplets are used for metabolism and for membrane and steroid synthesis, especially in adipocytes and steroidogenic cells.

4.13.2 Proteasome Storage Granules

When a cell exits from its division cycle and enters into quiescence, the proteolytic activity of 26S proteasome that degrades proteins involved in cell cycle progression decays. Its regulatory element is released from the core. Subunits from 20S core and 19S regulatory elements are then located into cytosolic proteasome storage granules [315].

When cells again enter into proliferation, proteasomes quickly reassemble and relocalize to the nucleus without requiring protein synthesis. Other molecular stores are created upon cell quiescence such as P-bodies, which contain RNA and RNA-binding proteins. Cellular components thus reorganize according to cell activity.

4.13.3 Stress Granules

Cytoplasmic RNA granules contain various ribosomal subunits, translation factors, decay enzymes, helicases, scaffold proteins, and RNA-binding proteins, to control the location, stability, and translation of their RNA cargoes [316].

Assembling of cytoplasmic stress granules represents an adaptive defense mechanism. In response to certain environmental stresses,[155] cells produce cytoplasmic stress granules that store mRNAs to halt translation. Stress granule assembly is initiated by RNA-binding silencers T-cell internal antigen-1 (TIA1) and TIA1-related protein (TIAR) that bind to 48S pre-initiation complex and promote polysome disassembly and mRNA recruitment into stress granules. Upon cell stress, phosphorylation of eukaryotic initiation factor eIF2α, a translation initiation regulator, by eIF2α kinase promotes the formation of stress granules [317]. Stress granules also recruit proteins that elicit mRNA stability (e.g., protein HuR) or destabilize mRNA (tristetraprolin). Furthermore, these aggregates that contain stalled translation pre-initiation complexes prevent the accumulation of misfolded proteins.

Growth factor receptor-bound protein GRB7 is an RNA-binding translational regulator that is a component of stress granules [318]. The adaptor GRB7 interacts with Hu antigen-R to stabilize TIA1 aggregates and form stress granules. When stress ends, GRB7 is phosphorylated by focal adhesion kinase and thus loses its ability to interact with antigen HuR and is dissociated from stress granule components. Several growth factors, such as axon guidance factor, netrin-1, epidermal growth factor, and insulin, regulate GRB7–FAK interactions.

Stress granule formation prevents misfolded protein aggregation and apoptosis by suppressing stress-responsive P38MAPK and Jun N-terminal protein kinase pathways and sequestering scaffold RACK1 (receptor for activated C-kinase) that maintains mitogen-activated protein kinase kinase kinase MAP3K4[156] in an inactive state (preformed MTK1 dimer ready for activation) in unstressed cells [319]. Pro-apoptotic kinase MAP3K4 activates P38MAPK and JNK and interacts with RACK1 that promotes MAP3K4 dimerization, but not its activation. On the other hand, growth arrest and DNA damage-inducible protein GADD45 binds to and activates MAP3K4 by relieving its autoinhibition, as it not only enables further dimerization, but also autophosphorylation. Prosurvival stress granule formation impedes GADD45-mediated MAP3K4 activation via RACK1 sequestration in stress granules.

4.14 Ribosomes

Ribosomes (15–20 nm in caliber) are molecular complexes, or *ribozymes*, but not organelles (in the absence of membrane). These large ribonucleoprotein

---

[155] Type-1 stresses, such as hypoxia and heat shock, trigger the formation of cytoplasmic stress granules. Type-2 stresses, such as x-rays and genotoxic drugs, prime apoptosis.

[156] A.k.a. MAP three kinase MTK1 (Vol. 4 – Chap. 5. Mitogen-Activated Protein Kinase Modules).

Table 4.24. Ribosomal constituents. Ribosomal proteins are encoded by the nuclear genome (e: eukaryotic; eTF: eukaryotic translation termination factor). Eukaryotic cells contain at least 2 distinct types of ribosomes: (1) cytoplasmic ribosomes (cytoribosomes with cytoplasmic ribosomal proteins [cRP]), mainly on the endoplasmic reticulum and in the aqueous cytoplasm; and (2) mitochondrial ribosomes (mitoribosomes with mitochondrial ribosomal proteins [mRP]), which synthesize proteins involved in oxidative phosphorylation encoded by genes of the mitochondrial genome. Ribosome, site of protein synthesis, consists of a small 40S subunit and large 60S subunit, which join together during the translation initiation. Ribosomal subunits are composed of up to 80 structurally distinct cytoplasmic proteins. Ribosomal proteins are named according to the ribosome subunit to which they belong (RPS and RPL: ribosomal protein of the small [S] and large [L] subunit). Ribosomal subunits also comprise ribosomal RNA species. Mammalian cells possess 2 types of mitochondrial rRNAs (12S and 16S rRNAs) and 4 types of cytoplasmic rRNA (5S, 5.8S, 18S, and 28S rRNAs). Ribosomal RNAs (large [LSu rRNA] and small [SSu rRNA]) form 2 subunits, between which mRNA is sandwiched. In addition to their role in ribosome assembly and function, several ribosomal proteins carry out extra-ribosomal activities, such as DNA repair, transcriptional regulation, and apoptosis.

Type	Members
Ribosomal proteins	RPS1–RPS31; RPL1–RPL44
Initiation factors	eIF1–eIF6 (eIF1, eEF1a1–eEF1a2, eEF2a–eEF2b, eIF3, eEF4a–eEF4b, eEF4e–eIF4f, eEF1g1–eEF1g2, eIF4h, eIF5a–eIF5b, eIF6)
Elongation factors	eEF1–eEF2 (eEF1a1–eEF1a3, eEF1b1—-eEF1b4, eEF1d, eEF1e1, eEF1g)
Release factors	eRF1 (or eTF1)
Ribosomal RNAs	Cytoribosome: 5S, 6S, 30S LSu rRNAs, and 18S SSu rRNAs Mitoribosome: 12S and 16S

particles contain ribosomal ribonucleic acid ($\sim 65\%$ RNA and $\sim 35\%$ ribonucleoproteins; Table 4.24).

Ribosomes are assembled in the nucleolus. They disassemble into subunits when they do not synthesize proteins. These subunits are recycled from one synthesis to the next. 80S Ribosomes are constituted of 40S and 60S subunits.[157] The large subunit is composed of a 5S rRNA (120 nucleotides), 6S rRNA (160 nucleotides), 30S rRNA (4700 nucleotides), and about 49 proteins. The small subunit possesses a 18S rRNA (1900 nucleotides) and approximately 33 proteins.

---

[157] The S unit corresponds to sedimentation rate.

Ribosomes are sites of messenger RNA translation to manufacture proteins using amino acids delivered by transfer RNA (Chap. 5). Messenger RNAs serve as templates; aminoacyl-transfer RNAs as substrates. Ribosomes are either free in the cytoplasm that produce intracellular proteins, or attached to the membrane of organelles, especially the rough endoplasmic reticulum, synthesizing export proteins.

During the translocation step of protein synthesis, 2 tRNAs move at a distance of at least 2 nm to adjacent sites in the ribosome. This motion is coupled to one-codon displacement of mRNA. Translocation is catalyzed by elongation factor-G that hydrolyzes GTP.

5

Protein Synthesis

Proteins, or polypeptides, are generated using a more or less numerous set of elementary components selected among 20 amino acid types, post-translationally modified,[1] folded (into their three-dimensional, functional configuration), transported (through the Golgi body and, then, cytosol up to destination), cleaved (from a cytosolic or membrane-embedded precursor), assembled (to create a functional proteic complex), disassembled, and degraded for a proper cell functioning.

Proteins can be classified into many categories according to their function in: (1) cell structure; (2) cell adhesion and motility; (3) cellular metabolism, in which they intervene at least as enzymes; (4) transcription and translation of genetic informations; (5) intra- and intercellular transport; (6) local and remote cell communications, especially signaling that triggers major events of the cell life, such as growth, division (or proliferation; i.e., the active part of the cell cycle), differentiation, aging, and death; and (7) immune defense.

A single gene can produce several substances due to alternative splicing. Moreover, post-translational modifications influence stability, location, and activity of proteins. A given protein can act inside and/or outside of the cell. Multifunctional proteins operate in a context-dependent manner. Proteins that have dual, related activity in both intra- and extracellular spaces can coordinate the organization and maintenance of a global tissue function [320] (Table 5.1).

---

[1] A post-translational modification, a step of the protein synthesis, is a chemical change of a protein after the translation from a mature messenger RNA, mainly in the endoplasmic reticulum, via: (1) attachment of a chemical group (e.g., acetyl, methyl, or phosphate) or substance (e.g., lipid or carbohydrate); (2) variation of the chemical nature of an amino acid; or (3) structural change (e.g., formation of disulfide bridges; Sect. 5.12).

Table 5.1. Proteins with dual, related activity in both intra- and extracellular spaces (Source: [320]; AMF: autocrine motility factor [a.k.a. glucose 6-phosphate isomerase, neuroleukin (Nlk), phosphoglucose isomerase (PGI), and phosphohexose isomerase; not AMFR: autocrine motility factor receptor]; ERK: extracellular signal-regulated kinase; HMMR: hyaluronan-mediated motility receptor [a.k.a. receptor for hyaluronan-mediated motility (RHaMM)]). These molecules include syntaxin-2 (or epimorphin) that intervenes in the morphogenesis of secretory organs; amphoterin, or high-mobility group box-1 protein (HMGB1), that coordinates tissue inflammation, as it regulates gene expression; and tissue transglutaminase that influences delivery of as well as response to apoptotic signals on both sides of the plasma membrane. Distinct motifs can mediate different functions of a given protein.

Protein	Intracellular function	Extracellular activity
Amphoterin	DNA-binding protein	Cytokine
Annexin-A2	Vesicular transport	Cell-surface receptor
AMF	Glycolysis	Cytokine, morphogen
HMMR	ERK1/2 signaling	Hyaluronan receptor
Syntaxin-2	Vesicular transport	Morphogen
Thioredoxin	Redox reactions	Cytokine
Tissue transglutaminase	Cell signaling	Matrix modification

5.1 Epigenetic Printing

Epigenetics refers to the inheritance of changes in gene expression without variation in the nucleotide sequence. The epigenetic control enables the establishment, maintenance, and reversal of transcriptional states that can be remembered by the cell and transmitted to daughter cells. Epigenetics deals with the self-sustaining, self-propagating information reversibly carried by the genome, but not encoded in the DNA. Epigenetic traits are inherited as regulatory signals in addition to genetic information. Transgenerational inheritance and cellular memory refer to meiotic and mitotic epigenetics [321]. In the early embryo, cells decide between somatic fate or germ line. During maturation, germ cells undergo a transition from cell division to meiosis to differentiate as sperm or egg.

Epigenetic states are organized by the molecular memory of interactions of genes with their environment, such as transcription factors, non-coding

RNAs,[2] and enzymes of DNA methylation[3] and histone modifications [321].[4] Epigenetic program determines the types of genes that are expressed in any specialized cell.

Most epigenetic states are established by factors that are transiently synthesized or activated to respond to environmental stimuli, developmental cues, or internal events (e.g., the reactivation of a transposon). These initiating signals target chromatin[5] that can then remodel the transcriptional landscape.

Transcription factors orchestrate cell lineage-specification programs. They recruit factors that modulate transcription transiently as well as influence epigenetic states (DNA methylation and histone modifications). Transcription factors can transiently control the cell fate, and, once they disappear, the cell identity is maintained and transmitted by transcriptional activator such as homologs of the trithorax group (mixed-lineage [myeloid–lymphoid] leukemia factors [MLL]) and repressor such as homologous members of the Polycomb group that remodel the chromatin.

Trans-epigenetic signals are transmitted via the synthesis of proteins that distribute in the partitioning cytosol during cell division and maintain feedback loops and networks of transcription factors that operated in mother cells. A transcription factor that activates its own transcription and/or represses antagonistic agents can yield a self-sustaining epigenetic state after the removal of the initiating stimulus. After each cell division, inherited transcription factors resume their trans function on regulatory DNA sequences. In particular, a simple regulatory loop refers to an epigenetic signal that provokes its own expression (positive feedback) [321]. Some small RNAs can also act as trans-epigenetic signals.

Cis-epigenetic signals are molecular markers associated with the DNA that are inherited with the chromosome on which they act via DNA replication and proper chromosome segregation during cell division, as covalent modifications of the DNA (e.g., DNA methylation) or as changes in histones [321]. Histones,

---

[2] Non-coding regions of the genome are transcribed and generate non-coding RNAs that often have regulatory functions (Sect. 5.3.2). Long non-coding RNAs can recruit chromatin modifiers such as the *chromatin-modifying complex* to target certain DNA loci and achieve signals for epigenetic states [321].

[3] DNA methylation satisfies the 3 independent criteria that define epigenetic molecular signal [321]: (1) transmissibility, as symmetrical methylation marks on both DNA strands generate 2 hemi-methylated double strands; (2) faithful reproducibility after DNA replication and chromosome segregation; and (3) impingement on gene transcription.

[4] Histone modifications result from a balance between antagonistic pairs of histone-modifying enzymes, such as histone acetyltransferases and deacetylases as well as methyltransferases and demethylases. These antagonisms are sources of reversible chromatin changes.

[5] Resulting histone post-translational modifications comprise trimethylation of histone-H3 on lysine (K) in position 4 ($H3K_4me^3$) and 36 ($H3K_{36}me^3$), among others.

which constitute the protein backbone of chromatin, can carry information by their primary sequence (histone variants), post-translational modifications (often on their N- and C-termini), or remodeling (position) relative to the DNA sequence.

5.2 DNA and Gene Expression

Deoxyribonucleic acid contains sequences that have protein synthesis, structural, or regulatory roles. The genetic code serves for cell development and functions. Deoxyribonucleic acid is composed of repeating units — the nucleotides — based on 4 nucleobases (purines or pyrimidines): adenine, cytosine, guanosine, and thymine.[6] Deoxyribonucleic acid is arranged in 2 long chains that twist around each other to form a double helix.

The genes are DNA segments with genetic instructions. The genetic code is a set of codons (3-nucleotide sets). Codons are mapped to amino acids, building blocks) of proteins. Some amino acids are mapped by more than one codon (redundancy). The base sequence of a gene determines the composition of messenger RNA (mRNA), hence the sequence of amino acids of the encoded protein. Proteins are actually assembled from amino acids using information encoded in genes. The linear sequence of amino acids contains sufficient information to determine the three-dimensional structure of proteins.

Ribonucleic acids (RNA) are linear polymers that are constituted of tens to thousands of nucleotide residues. Similarly to DNA, they contain 4 nucleotide bases: adenine, cytosine, guanine, and uracil (hence corresponding nucleosides adenosine, guanosine, cytidine, and uridine). In addition to messenger RNAs, transfer RNAs contribute to decipher the genetic information to synthesize proteins.

The main stages of the protein synthesis include transcription, post-transcription, translation, post-translation, and modifications (Fig. 5.1). Proteins are characterized by their atomic interactions formed by the backbone and side chains of amino acids in a polypeptide that allow proteins to carry out specific functions in living cells and tissues, such as sensing of chemical, physical, or mechanical cues (sensors and receptors), recognizing foreign substances (immune mediators), generating mechanical forces (nanomotors), transmitting information (messengers), and regulating cellular processes (signaling effectors) and metabolism of nutrients (enzymes and substrates).

---

[6] Nucleobases adenine and guanine are purines; cytosine, thymine, and uracil are pyrimidines. A base attaches to a deoxyribose sugar to form a nucleoside. A phosphorylated nucleoside is a nucleotide. DNA Nucleoside include deoxyadenosine, deoxyguanosine, deoxycytidine, and deoxythymidine. Nucleosides are enzymatically modified to form derivatives with distinct chemical structures.

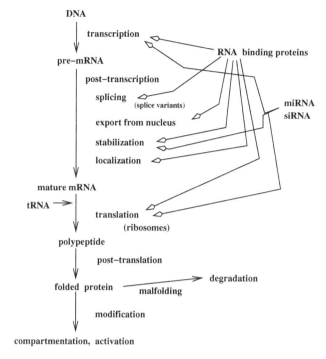

Figure 5.1. Main stages of the protein synthesis: transcription, post-transcription, translation, post-translation, and post-translational modifications. The genetic instructions are transcribed in the nucleus from DNA to single-stranded, pre-messenger ribonucleic acid (mRNA) and conveyed into the cytoplasm by mRNA complementary coding sequences. Long pre-mRNA are cut into much shorter mature mRNAs. The genetic information is translated from RNA into protein using transfer RNA (tRNA) as an adaptor between mRNA and the amino acid. The exons, mRNA sequences, can be distributed discontinuously in the genome. Exons can be trimmed from primary transcripts and spliced, thereby generating different mRNAs and leading to various proteins that emanate from a given primary transcript. Introns are degraded.

5.2.1 Gene Transcription

The genetic code in DNA is transcribed into pre-messenger RNA by RNA polymerase-2 once the latter has bound to the promoter (initiation), run along the DNA strand (elongation), and terminated to avoid degradation.[7]

---

[7] Transcription by RNA polymerase-2 into a permissive chromatin environment involves the recruitment of transcription factors, coactivators, and RNA polymerase-2 to the promoter to form a *pre-initiation complex*. Upon transcription initiation, RNA polymerase-2 is phosphorylated by CDK7–cyclin-dependent kinase. The *initiation complex* formed by promoter-bound RNA polymerase opens the DNA duplex and synthesizes short transcripts (abortive synthesis). After

The chromatin state that controls gene expression is regulated by transcription factors. Transcription factors bound to specific gene sites recruit chromatin-remodeling factors and enzymes that modify histones, thus allowing binding of other regulators that operate with chromatin to create a permissive or non-permissive environment for gene expression.

5.2.2 mRNA Processing

Pre-messenger RNA is processed to form mature messenger RNA (post-transcription), translocated into the cytosol to become a template for protein synthesis in the ribosome. Translation also comprises 3 steps: initiation starting with fixation of transfer RNA (tRNA) and mRNA on specific ribosomal sites, elongation, and termination when stopping codon is detected. The messenger RNA is linked to the ribosome and each codon is read and matched by anticodon of transfer RNA, which carries the corresponding amino acid. Post-translational modification and protein folding follow the synthesis.

Nascency refers to the period between initiation of the nascent chain and completion of its assembly into a functioning product. A nascent protein is in a delicate situation, as it can bear inappropriate interactions, misfolding, and aggregation. Molecular chaperones, appropriate enzymes, and accessory factors guide the protein maturation. In addition, during elongation, ribosomes may work with a varying translation speed and even transient translational pauses to facilitate mRNA localization [322].

5.2.3 Regulation from Gene Transcription to Protein Folding

Gene regulation occurs at multiple stages (Table 5.2). Pre-transcriptional regulation depends on cell signaling. Transcriptional regulation is associated with particular DNA sites. Post-transcriptional regulation of gene expression that is achieved, in particular, by microRNAs, operates via: (1) mRNA polyadenylation, localization, splicing (Table 5.3), and degradation;[8] (2) chromatin modifications; as well as (3) protein localization, modification, and degradation.

5.2.3.1 Regulation of Gene Transcription

Promoters regulate DNA transcription. Augmenters, or enhancers, and silencers elicit and impede gene transcription, respectively. Augmenter and

---

the synthesis of approximately 20 to 50 bases of RNA, RNA polymerase-2 is often repressed by negative elongation factor (NElF) and dichloro $1\beta^D$ribofuranosylbenzimidazole sensitivity-inducing factor (DSIF) to cause a pause. Transcription continues when cyclin-dependent kinase CDK9 phosphorylates RNA polymerase-2, NElF, and DSIF. After the transition to the elongation phase and release of the promoter, RNA polymerase-2 proceeds progressively down the DNA template to produce a complete RNA transcript.

[8] Messenger RNA is cleaved into fragments by endoribonuclease.

Table 5.2. Gene regulation and protein modifications. Regulatory sites are often located upstream from protein-coding genes. Transcription factors control transcriptional regulation by binding to cis-regulatory sites. Post-transcriptional regulation is mediated by microRNAs, also by binding to cis-regulatory sites. Other stages of gene regulation include cell signaling, mRNA splicing, polyadenylation, and localization, chromatin modifications, and protein localization, modification, and degradation. Protein modifications, such as acetylation, farnesylation, myristoylation, palmitoylation, and phosphorylation, assist in localizing the enzyme to various cell compartments. Except for phosphorylation, these changes do not significantly affect the protein activity. Ubiquitination is a reversible post-translational modification of proteins in which ubiquitin is attached to lysine residues of target proteins. Monoubiquitination corresponds to the binding of a single ubiquitin to a protein, and multiubiquitination to the binding of a single ubiquitin to several Lys residues of a given protein. Mono- or multiubiquitination is required for endocytosis. Deubiquitination can follow monoubiquitination and enable protein activity. The protein–ubiquitin complex can attach further ubiquitins. Polyubiquitination directs the ubiquitinated protein to the proteasome. Subsequently, ubiquitin is recycled and ubiquitinated protein is degraded.

Transcriptional control	Post-transcriptional control
Transcription factors	mRNA
Chromatin state	miR
Promoters	Polyadenylation
	RNA-binding proteins
	Cell signaling
	Protein location
	Ubiquitination
Cofactors	mRNA splicing

Table 5.3. Example of alternative splicing of mRNA exons. Alternative splicing generates protein diversity using various combinations of gene exons of a pre-messenger RNA transcript. It occurs during the majority of gene translation. It is regulated by standard and tissue-specific splicing factors, reaction kinetics, and histone modifications. Chromatin structure and histone modifications actually influence the splicing outcome via the recruitment of chromatin-binding proteins and splicing regulators. Histone modifications correspond to adaptors for the pre-mRNA splicing machinery.

Original exons	1	2	3	4	5	6	7
Isoform 1	1	2	3	4	5	6	
Isoform 2		2	3	4	5	6	
Isoform 3		2	3	4	5	6	7
Isoform 4	1	2	3	4	5	6	7

silencer sorting is specific to each gene. Distinct combinations of activators and repressors control the transcription of different genes. Coactivator–activator complexes bind to augmenters; corepressors and repressors to silencers. Basis factors must aggregate on the minimal promoter in order to start the transcription. RNA polymerase-2 and -3 synthesize mRNA and tRNA associated with class 2 and -3 genes, respectively. Diseases occur by dysregulation of gene transcription and protein under- or overproduction.

During gene transcription, some activators bind periodically to their promoters. Binding cycles are either fast (30–60 s) or slow (15–90 mn). The same transcription activator can use both cycling types on the same promoter for transcription initiation (fast cycling) and to modulate promoter accessibility, and thus the number of mRNA produced (slow cycling) [323].

Cooperative binding of interacting and non-interacting transcription factors to promoters and other regulatory regions enables precise gene expression [324]. Cis-regulatory regions (DNA elements that regulates the expression of genes) contain 200 to 3,000 base pairs (bps) and clusters of 3 to 50 transcription factor-binding sites. Transcriptional synergy, i.e., cooperative recognition or simultaneous contact of transcription factors and other elements of the transcription machinery rather than cooperative binding of transcription factors to DNA allows the gene expression in the presence of rearrangements of transcription factor-binding sites. In addition, synergistic binding of non-interacting transcription factors on nucleosomal DNA yields an alternative mechanism of cooperativity that involves neither chromatin modification nor direct between-protein interactions [324]. Moreover, competition between histones and transcription factors for a DNA region that contains an array of transcription factor-binding sites provokes a cooperative binding of transcription factors and nucleosome eviction. This nucleosome-mediated mechanism is compatible with remodeling of transcription factor-binding sites such as turnover of sites in DNA enhancers.

Three possible targets in gene expression changes due to environmental factors include promoters, transposable elements adjacent to genes with metastable epialleles (primitively identical genes that differ in the extent of methylation), and regulators of imprinted genes [325].[9]

The role of genotype in human diseases has been demonstrated. In addition, epigenetic modifications can associate alterations in gene expression leading to chronic diseases in association with the environment, particularly

---

[9] Metastable epialleles are loci that bear variable, reversible, epigenetical modifications leading to a phenotype distribution in genetically identical cells. The expression of imprinted genes results from only one of the 2 parental alleles. In mice, the paternal allele of the gene coding for insulin-like growth factor-2 is expressed, whereas the maternal allele remains silent. On the other hand, insulin-like growth factor-2 receptor is expressed only by the maternal chromosome. However, both alleles yield IGF2R in humans. Imprinted genes are implicated in fetomaternal physiology. The alleles of imprinted genes differ mainly by DNA methylation, and also by chromatin conformation, histone modification, replication timing, and recombination rate [325]. Aberrant imprinting disturbs development and causes various syndromes. Regulation alterations of imprinted genes, as well as their mutations, lead more easily to pathologies than when both alleles are expressed.

prenatal and early postnatal environmental (nutritional, chemical, and physical) factors [326]. During gestation, several coexisting generations (pregnant female, embryo, and its germ line) are exposed to environmental factors. Transmission of modified phenotypes is then possible.

5.2.3.2 Microprotein Interference

Many proteins achieve their function as functional di- and oligomers, or multimers. Microproteins (miP) contain protein-binding domains, but lack functional domains. They behave as post-translational regulators, as they dimerize with their targets, mainly transcriptional regulators, that then cannot complex with their usual partners to form functional mediator.

Inhibitors of DNA binding that contain a HLH domain interact with the basic helix–loop–helix (bHLH)-containing transcriptional regulators, such as myogenic differentiation factor (MyoD) and transcription factors TcFE2a (or E12) and TcFE3 (or E47) and prevent their binding to DNA [327].[10] Protein ID2 may sequester the bHLH domain-containing transcription factors BMAL and CLOCK of the circadian clock.

Microproteins can be generated by alternative splicing (Sect. 5.4). A splice variant of (E26) factor ETS1 encodes a 27-kDa protein ($ETS1_{P27}$) that represses the ETS1-induced gene transcription. Moreover, the $ETS1_{P27}$–ETS1 dimer translocates to the cytoplasm [327].[11]

5.2.3.3 Regulation of mRNA Translation

Heterogeneous ribonucleoprotein particles (hRNP), which are complexes of RNA and nuclear proteins, bind to pre-mRNA for export to the cytoplasm. After splicing, hRNPs remain bound to spliced introns for degradation.

Riboswitches are mRNA elements that bind small molecules to regulate its own gene expression.[12] Riboswitches are made of 2 domains, a binding site

---

[10] Factor MyoD forms heterodimers with TcFE2a or TcFE3 proteins.

[11] The short alternatively splicing variant $SIAH1_S$ of the Ub ligase Seven in absentia homolog (SIAH) dimerizes with SIAH1 and impedes SIAH1 action.

[12] A signal-mediated, protein-dependent riboswitch that corresponds to 3′ untranslated region (UTR) of the human vascular endothelial growth factor-A mRNA causes conformational change to integrate signals from interferon-γ and hypoxia and modulate VEGFa translation in myeloid cells. Conformational changes following VEGFa 3′ UTR switch are independent of metabolites, but depend on mutually exclusive linkage to Ifnγ-activated inhibitor of translation complex and heterogeneous nuclear ribonucleoprotein-L (hnRNPl) [328]. In atherosclerotic lesions, VEGFa expression is induced by hypoxia, but repressed by Ifnγ produced by inflammatory cells as well as Ifnγ-activated inhibitor of translation complex (GAIT). Macrophages are thus simultaneously exposed to opposing inflammatory and hypoxic signals. However, hypoxia overrides GAIT-mediated repression of VEGFa translation observed in normoxia. Signal integration thus regulates protein synthesis in response to diverse environmental stimuli.

for ligands and a regulation site that controls the gene expression. Regulation is achieved via ligand-dependent conformation changes in the binding site.[13]

5.2.3.4 Protein Folding and Translocation

Newly-made proteins require assistance to convert from a linear chain of amino acids to a functional three-dimensional protein. Protein folding (Sect. 5.10) is assisted by *chaperonins*. These proteic complexes are made of components that belong to the class of molecular chaperones. Chaperonins are ubiquitous.[14] These oligomers are classified into 2 families. Family-1 and -2 chaperonins reside in mitochondria and cytosol, respectively.

Many unfolded proteins are recognized by and bind to these chaperonins. Binding of ATP and subsequent hydrolysis induce a series of structural changes that release the unfolded protein inside the cavity for its folding. However, some chaperonins exert another procedure such as Chaperonin containing T-complex protein TCP1 (CCT), an oligomer that regulates the folding of cytoskeletal proteins, such as actin and α- and β-tubulins, as well as β-transducin and von Hippel-Lindau protein. The interaction between CCT in its open, substrate-receptive conformation and unfolded tubulin takes place at both the apical region devoted to substrate recognition and equatorial, sensor domain close to the ATP-binding site [329].

The main chaperone classes that prevent the accumulation of misfolded proteins include heat shock proteins HSP60 and HSP70. Misfolded proteins that are not refolded are generally processed by either cytosolic ATP-dependent AAA+ proteases such as the 26S proteasome or acidic hydrolases after their transport into the lysosomal compartment [330].

The chaperone 90-kDa heat shock protein is aimed at preventing unspecific aggregation of proteins. Each monomer of the HSP90 dimer comprises 3 domains that serve for cofactor interaction: (1) the N-terminal domain that possesses an ATP-binding motif;[15] (2) the M-domain;[16] and (3) the C-terminal dimerization domain. It is more selective than the other chaperones. Substrates of HSP90 are mainly signal transduction components, such as

---

[13] Ligand binding to non-coding regions of mRNA modulates gene expression. Free metabolites and other small molecules bind and induce conformational changes in regulatory RNAs. RNA-ligated metabolites such as aminoacyl-charged tRNA in the T-box system and protein-bound metabolites in the glucose- or amino acid-stimulated terminator–anti-terminator systems also act as riboswitches.

[14] Their structure comprises 2 rings that enclose a cavity in which substrate folding takes place.

[15] Hydrolysis of ATP by HSP90 is rather slow. In humans, HSP90 hydrolyzes one molecule of ATP every 20 mn [331]. Upon ATP binding, the ATPase domain undergoes a conformational change leading to a closure. Upon ATP hydrolysis, the closed conformation returns to the open state.

[16] Activator of HSP90 ATPase homolog AHSA1 (or AHA1), the only known activator of the HSP90 ATPase activity, binds to the M-domain.

kinases[17] and transcription factors. Chaperone HSP90 causes conformational changes in folded substrates that lead to their proper folding as well as activation or stabilization. It interacts with numerous cochaperones that support HSP90 activity [331]. Protein HSP90 requires the cooperation of the chaperone HSP70 and its cofactor HSP40. Several cofactors interact with HSP90 in a sequential manner to assemble a functional chaperone.[18] The major subclass of HSP90 cochaperones is the tetratricopeptide repeat (TPR) domain-containing cofactors,[19] such as HSP70–HSP90 complex-organizing protein (HOP), the catalytic subunit of protein phosphatase-5 (PP5), and the large prolyl isomerase,[20] among others (Table 5.4)).[21] Several of the HSP90 cofactors modulate its ATPase activity. For example, HSP90 cochaperone P23 binds to the ATPase domain and stabilizes the dimerized conformation at the late stage of the ATPase cycle.[22] The activity of HSP90 is also regulated by post-translational modifications (acetylation, S-nitrosylation, and phosphorylation). Kinases that phosphorylate HSP90 include casein kinase-2, protein kinase-B, and DNA-dependent protein kinase. Some HSP90 cofactors are phosphatases such as PP5.

In the endoplasmic reticulum, oxidoreductases operate in both degradation of misfolded proteins and folding of newly synthesized proteins [322]. Cochaperones contribute to the endoplasmic reticulum quality control (Sect. 5.7).[23] Chaperones also participate in the clearance of damaged proteins, as dedicated chaperones permit the proteasome assembling.

Cytoplasmic *aggresomes* correspond to a cell response to misfolded proteins. Aggresome formation is associated with the redistribution of cytoskeleton proteins and the recruitment of proteasomes. Aggresome formation is characterized by the following features: (1) aggresomes occur either in response

---

[17] The kinase-specific HSP90 cofactor CDC37 associates with substrate kinases in their inactive forms. This complex may then be loaded onto HSP90 for substrate maturation [331]. Cofactor CDC37 slowers the ATPase activity of HSP90.

[18] This ordered linkage succession relies on the strong binding preference of HSP90 cofactors for a specific HSP90 conformation.

[19] The tetratricopeptide repeat (TPR) motif is a protein interaction domain.

[20] A.k.a. peptidylprolyl isomerase (PPIase).

[21] The progressive construction of the HSP90 complex requires different cochaperones. The HSP70–HSP90 complex-organizing protein (HOP), an HSP90 ATPase inhibitor, is later replaced by a prolyl isomerase (PPIase). Therefore, the bound HOP is expelled from the ternary HSP90–HOP–PPIase intermediate owing to ATP [332].

[22] Cofactor P23 is a HSP90 cochaperone that localizes both to the cytoplasm and nucleus. It also stabilizes unliganded steroid receptors as well as controls the catalytic activity of certain kinases and protein–DNA dynamics [333]. Proteins HSP90 and P23 form a 2:2 complex ($HSP90_2P23_2$) [334]. The molecular chaperones P23 and HSP90 bind to the catalytic subunit of telomerase that maintains the chromosome length [335].

[23] In particular, cochaperones of the DNAJ-like membrane chaperone family DnaJb6 and DnaJb8 are suppressors of aggregation and toxicity of polyglutamine proteins [322].

Table 5.4. Numerous cochaperones support heat shock protein-90 (HSP90) in the folding and activation of substrates (Source: [331]). The progressive construction of the chaperone HSP90 complex involves different cochaperones (AHR: aryl hydrocarbon receptor; AHSA1: activator of HSP90 ATPase homolog [or AHA1], AIP: aryl hydrocarbon receptor-interacting protein [a.k.a. FKBP16, FKBP37, immunophilin homolog ARA9, and HBV X-associated protein XAP2]; CDC37: cell-division cycle-37 homolog; CHIP: C-terminus heat shock cognate-70-interacting protein [a.k.a. STIP1 homology and U-box-containing protein-1 (STUB1), ubiquitin ligase]; HBx: porcine homeobox; HOP: HSP70–HSP90 complex-organizing protein; FKBP4 (or FKBP52): 52-kDa (59-kDa) FK506-binding protein [a.k.a. peptidyl-prolyl cis-trans isomerase FKBP4, immunophilin FKBP52 or FKBP59, and HSP-binding immunophilin]; FKBP5 [or FKBP51]: 51-kDa FK506-binding protein [a.k.a. peptidyl-prolyl cis-trans isomerase, HSP90-binding immunophilin, and 54-kDa progesterone receptor-associated immunophilin]; PP5: protein phosphatase-5; PPId: peptidylprolyl isomerase-D [a.k.a. 40-kDa cyclophilin-D (Cyp40 or CypD)]; SCF: SKP1–Cul1 (CDC53)–F-box ubiquitin ligase complex; SuGT1: suppressor of G2 (two) allele of S-phase kinase-associated protein SKP1; TTC4: tetratricopeptide repeat domain-containing protein-4; TOM70: 70-kDa translocase of outer membrane; TPR: tetratricopeptide repeat motif; Unc45: Uncoordinated-45 homolog).

Cochaperone	Function
	Organizing protein
HOP	Adaptor for HSP90 and HSP70, substrate maturation
	Large peptidylprolyl isomerase
AIP	Formation of complex with AHR and HBx
FKBP4	Substrate maturation
FKBP5	Substrate maturation
PPId	Substrate maturation
	Ubiquitin ligase
CHIP	Protein degradation
	Phosphatase
PP5	Substrate maturation
	Other TPR domain-containing cofactors
SuGT1	Formation of complex
TOM70	Mitochondrail protein import
TTC4	
Unc45	Assembly of myosin fiber
	Non-TPR domain-containing cofactors
AHSA1	ATPase activator
CDC37	Kinase-specific HSP90 cofactor, ATPase inhibitor
P23	Substrate maturation, ATPase inhibitor

Table 5.5. Molecular chaperones and proteases of the aggresome disaggregation (Source: [330]; AAA: ATPase associated with diverse cellular activities; ER: endoplasmic reticulum; HSP: heat shock protein; VCP: valosin-containing protein [ATP- and ubiquitin-dependent AAA+ chaperone]).

Molecule	Effect
HSPa (HSP70)	Folding of newly synthesized and misfolded proteins
AAA+ proteases	Degradation of misfolded and aggregated proteins
26S proteasome	Degradation of polyubiquitinated proteins
VCP	Degradation of misfolded ER proteins and membrane fusion

to unfolded or misfolded proteins or to regulate protein functioning by quick sequestration and elimination, as aggresomes contain proteasomes, ubiquitins, and heat-shock proteins; (2) aggresomes form in the neighborhood of the microtubule-organizing center, near the centrosome, using retrograde transport on microtubule tracks using the dynein–dynactin microtubule nanomotor; (3) aggresome genesis is associated with the collapse of intermediate filament around the aggresome; and (4) aggresome production is accelerated by proteasomal inhibition.

Therefore, protein aggregation allows for efficient solubilization and refolding or degradation of defective proteins by components of the protein quality control [330]. The protein quality control prevents prolonged protein aggregation. Proper refolding of aggregated proteins leads to disaggregation. Aggresome disaggregation relies on molecular chaperones and proteases (Table 5.5). On the other hand, untreated aggregation of misfolded proteins is associated with disturbed cell functions, aging, and diverse human disorders.

The aggresome is a transient, vimentin-coated structure at an indentation of the nuclear envelope at the microtubule-organizing center often surrounding the centriole. Aggresome formation requires the adaptor histone deacetylase HDAC6 that binds to ubiquitinated proteins on the one hand and the microtubule minus-end nanomotor dynein on the other to carry aggregated proteins to aggresomes [330]. In mammals, protein quality-control compartments comprise the *juxtanuclear quality-control compartment* (JUNQ), the *perivacuolar insoluble protein deposit* (IPOD), as well as *perinuclear aggresomes*, where misfolded, ubiquitinated proteins accumulate.

Newly-made, well conditioned proteins then translocate toward their destination, possibly across membranes of organelles. For example, mitochondria import about 1,000 proteins from the cytosol. Post-translational modifications of transport components influence the genesis and function of preprotein translocases [322]. Besides, the local membrane compositions of proteins and phospholipids are coupled.

5.3 RNA Classes

Ribonucleic acids can be categorized into large and small as well as coding and non-coding (ncRNA) classes. Three large RNA categories include messenger (mRNA), transfer (tRNA), and ribosomal (rRNA) RNA that are necessary for protein synthesis. Specific genomic sequences in DNA are transcribed into messenger RNAs that are then translated into proteins, whereas ribosomal and transfer RNAs are non-coding RNAs. Small RNAs regulate gene transcription by interacting with the promoter region and modifying the histone code.

Gene transcription is more complicated than the simple conversion of genetic code into mRNAs. Genes that encode RNAs are transcribed in the nucleus and mitochondria by RNA polymerase-1, -2, or -3. Normally, RNA polymerase-2 binds to the promoter sequence of a given gene to generate mRNA. Cap markers, or cap-analysis of gene expression (CAGE) tags, are supposed to be transcription start sites, as they mark the 5′ end of capped, long RNA transcripts [336]. In a few cases, the 5′ ends of genes on opposite DNA strands share the same promoter. Binding of RNA polymerase-2 can then trigger bidirectional transcription and generally generate long mRNAs. Therefore, transcription can be carried out in opposite orientation to genes.[24] Moreover, mature mRNAs processing generates long and short RNAs.

Primary transcripts are immature, inactive elements. RNA precursors undergo post-transcriptional modifications. In particular, the processing of pre-mRNAs involves 5′ capping,[25] 3′ polyadenylation, splicing, and sometimes RNA editing. Mature mRNA translation rate then depends on mRNA concentration, i.e., cytoplasmic transport, storage, and degradation, as well as RNA quality control.

5.3.1 Modifications of Non-Coding RNAs

Modifications of non-coding RNAs, such as transfer, ribosomal, and spliceosomal small nuclear RNAs (Sect. 5.4.1), as well as microRNAs (Sect. 5.3.3.5) modulate their activities. At least, 119 different modifications have been found [337]. Enzymes (deaminases, dihydrouridine synthases, hydroxylases, methyltransferases, methylthiotransferases, pseudo-uridine synthases, thiolases, transglycosylases, etc.) can be categorized according to the type of

---

[24] Long and short non-coding RNAs can be formed by bidirectional transcription and unusual processing. Genomic sequences upstream from gene promoters can be transcribed into 2 main types of long non-coding RNAs that serve as transcription regulators: cryptic unstable transcripts and stable unannotated transcripts. Cryptic unstable transcripts can be degraded soon after synthesis by an RNA surveillance system or control gene expression.

[25] Immediately after the transcription begins, a cap is added to the 5′ end of mRNA transcribed by RNA polymerase-2. This structure is required for RNA stability and protein translation.

reaction they trigger. These modifications are either constitutive or conditionally induced. Methylation and pseudo-uridylation outnumber the other modification types.

Pseudo-uridine, an isomer of uridine, participate in the control of ncRNA functions. Ribosomal RNA contains about 100 pseudo-uridines at specific sites for proper genesis and functioning of ribonucleoproteins [338].

Pseudo-uridylation relies on 2 mechanisms [338]: (1) RNA-independent pseudo-uridylation that is catalyzed by pseudo-uridine synthases that convert uridine to pseudo-uridine and (2) RNA-dependent pseudo-uridylation that uses small nucleolar or Cajal body RNAs (collectively named H/ACA RNAs), which specify the target uridine, whereas associated pseudo-uridine synthases ensure the conversion in cooperation with 3 proteins that form an H/ACA ribonucleoprotein. Spliceosomal small nuclear RNA U2 (Sect. 5.4.4) possesses 13 pseudo-uridines. Nutrient deprivation causes additional pseudo-uridylations in small nuclear RNA U2 that impinge on pre-mRNA splicing [339].

5.3.2 Non-Coding RNAs

Non-coding RNAs can be divided into: (1) *structural* (ribosomal, transfer, small nuclear, and small nucleolar RNAs) and (2) *regulatory* (micro, Piwi-interacting, small interfering, and long non-coding RNAs) non-coding RNAs, including promoter-associated (PAR) and enhancer (eRNA) RNAs (Table 5.6).

Regulatory ncRNAs intervene during embryo- and fetogenesis, as well as in response to environmental stimuli. Non-coding RNAs can serve as sensors of environmental cues as they act as transcriptional coactivators and corepressors. Moreover, some ncRNAs regulate the transcription via the control of the nucleocytoplasmic transfer of transcription factors. They also contribute to the regulation of post-transcriptional processing (splicing, transport, translation, and degradation).

The genome is transcribed into long intergenic non-coding RNAs (lincRNA; ~ 200 nucleotides), such as *promoter-associated long RNAs* (PALR) and less stable *promoter upstream transcripts* (Prompt), and short non-coding RNAs (< 50 nucleotides), such as *promoter-associated short RNAs* (PASR) and *terminus-associated short RNAs* (TASR) [336]. The transcription factor P53 primes the production of numerous lincRNAs.

Long non-coding RNAs may regulate gene transcription by recruiting chromatin-remodeling complexes. Long intergenic non-coding RNAs can recruit histone-modifying enzymes to specific genomic locations, thereby directing epigenetic modifications in chromatin states.

Non-coding RNAs are transcribed in the down- (like mRNAs) or upstream direction from a promoter. Some Prompts[26] as well as transcription start

---

[26] Short, polyadenylated, unstable RNA promoter upstream transcripts are produced approximately 0.5 to 2.5 kilobases upstream of active transcription start sites.

Table 5.6. Features and functions of regulatory non-coding RNAs (Source: [346]); P[I]M: perfect [imperfect] match; ds: double stranded; ss: single stranded; eRNA (E-RNA): enhancer RNA; lncRNA: long non-coding RNA; miR: microRNA; pri-miR: primary microRNA; PAR: promoter-associated RNA [i.e., pasR: promoter-associated small RNA; Prompt: promoter upstream transcript; tssaRNA: transcription start site-associated RNA; and tiRNA: transcription initiation RNA] piRNA: P-element-induced wimpy testis-interacting (PIWI) RNA; siRNA: small interfering RNA).

Type	Number of nucleotides	Features	Function
MiR	20–24	Capped and polyadenylated pri-miR; mature ssRNA	PM: Ago2-mediated cleavage of mRNA; IM: suppression of translation or mRNA degradation
PiRNA	20–31	Precursor ssRNA; complexes with Piwi proteins (Argonaute family)	Silencing of transposable elements in the germline
SiRNA	20–24	Perfectly base-paired dsRNA; complexes with RISC	PM: endonucleocytic cleavage; IM: translational repression or exonucleocytic degradation
PAR	16–200	Weakly expressed ssRNAs; short half-life	
E-RNA	100–900	ssRNA, short half-life	
LncRNA	>200	Precursor ssRNA; post-transcriptionally modified; mostly nuclear	Chromatin remodeling; transcriptional and post-transcriptional regulation; precursors for siRNAs

site-associated RNAs (tssaRNAs)[27] are transcribed in the opposite direction, i.e., from an upstream sequence toward a promoter, without starting from a promoter. The interplay between polymerases and regulators over broad promoter regions dictates the orientation of transcription [342].

Most of short half-life *promoter-associated RNAs* are associated with highly expressed genes. They act as transcriptional activators and repressors.

Prompt transcription occurs in both sense and antisense directions w.r.t. the downstream gene. It requires the presence of the gene promoter and is positively correlated with gene activity [340].

[27] Transcription start site-associated RNAs (20–90 nucleotides) non-randomly flank active promoters, with peaks of antisense and sense short RNAs at 250 nucleotides upstream and 50 nucleotides downstream of TSSs, respectively [341].

Non-coding RNAs share with mRNA several features, such as promoter-primed transcription, 5' capping, and possible post-transcriptional modification. Whereas capped coding RNAs acquire a cap during transcription, secondary capping of short non-coding RNAs can follow cleavage of longer capped RNAs.

Enhancer RNAs (eRNA) are produced from regions enriched in monomethylated Lys4 of histone-3 ($H3K_4me^1$) and poor in trimethylated histone-3 at Lys4 ($H3K_4me^3$), in which abound RNA polymerase-2 and transcriptional coregulators such as P300. These short half-life transcripts are produced on signaling with neighbor mRNAs and may serve as transcriptional activators.

5.3.3 Regulatory Non-Coding RNAs

Various types of small non-coding RNAs regulate various biological processes, especially gene expression. They can be grouped in several subclasses: (1) microRNAs (miR); (2) P-element-induced wimpy testis (PIWI)-associated RNAs (piRNA; originally called repeat-associated small interfering RNAs [rasiRNA]);[28] (3) small interfering RNAs (siRNA); and (4) small nuclear (snRNA) and nucleolar (snoRNA) RNAs (Table 5.7).

Although small RNA pathways are interconnected, these RNA classes differ according to synthesis, mode of substrate regulation, and regulated pathways. Small RNA pathways collaborate, as they share proteins, or compete for substrate loading to regulate gene expression and protect the genome from external and internal threats.

Non-coding RNAs participate in the regulation of the chromatin state and epigenetic inheritance, as they specify the pattern of histone modifications on target genes. They control developmental gene expression, as they establish chromatin domains in an allele- and cell-type specific manner, once associated with chromatin-remodeling complexes. In embryonic stem cells, many genes such as the homeobox-containing genes (HOX) that encode homeodomain-containing morphogens and developmental regulators, are transcriptionally silent but possess bivalent histone-H3 lysine$_4$ ($H3K_4$) and lysine$_{27}$ ($H3K_{27}$) methylation, which are resolved into univalent $H3K_4$ or $H3K_{27}$ methylation domains upon differentiation [343].[29]

---

[28] The P-element is a single transposable DNA-element, the uncontrolled activity of which causes defects in gametogenesis and sterility. Small RNA-silencing is a regulatory mechanism that controls transposon activity, particularly in germ-cell lineage, which that transmits the genetic information onto the next generation.

[29] The large intergenic non-coding (lincRNA) HotAIR transcribed from the type-C HOX locus binds to the histone methyltransferase Polycomb repressive complex-2 (PRC2) to repress transcription of the HOX-D locus [344]. In fact, HotAIR serves as a scaffold that is able to tether 2 different histone-modification complexes [343]. A 5' domain binds to trimethylated Lys27 of histone-H3 ($H3K_{27}me^3$). The Polycomb repressive complex PRC2 comprises enhancer of zeste homolog EZH2, suppressor of zeste-12 homolog SUZ12, and

Table 5.7. Small regulatory RNAs. MicroRNAs repress the expression of target mRNAs, but they can activate protein translation. Small nucleolar RNAs regulate alternative splicing. Telomerase that maintains chromosome integrity over numerous cycles of DNA replication during cell division uses a small RNA primer associated with a ribonucleoprotein complex. Spliceosomal small nuclear RNAs form small nuclear ribonucleoproteins that remodel during spliceosome assembly and functioning. Alternative pre-mRNA splicing generates mRNA isoforms. Signaling pathways influence alternative splicing by regulating the activity of enzymes on splicing factors. Some small inhibitory RNAs, such as small interfering RNAs and microRNAs, derive from double-stranded RNA, but not PIWI-interacting RNAs.

Type	Function
MicroRNAs (miRs)	Translation repression
	Argonaute protein association (miRISC)
	Regulation of transcription factors
	Transcription reduction
	Regulation of mRNA stability
	Alternative splicing
PIWI-associated RNAs (piRNAs)	Transposon silencing
	Gametogenesis
Repeat-associated small interfering RNAs (rasiRNAs)	Transposon silencing
	Histone methylation
Small interfering RNAs (siRNA)	Gene expression silencing (transcriptionally and post-transcriptionally)
	Argonaute protein association (RISC) (functioning modulation)
	RNA-induced silencing complex activity
	Chromatin remodeling
	Degradation of mRNAs
Small nuclear RNAs (snRNA)	Regulation of transcription factors
	Telomere function
	Regulation of mRNA splicing
Small nucleolar RNAs (snoRNA)	Nucleotide modification

---

embryonic ectoderm development protein (EED). A $3'$ domain tethers to the KDM1a–CoREST–REST repressor complex (KDM1a: lysine-specific histone demethylase-1A [or LSD1]; REST: RE1-silencing transcription factor; RCoR1: REST corepressor-1 [or CoREST]). The lincRNA HotAIR recruits PRC2 to HOX-D and other genes, where its histone methylase activity prevents the gene transcription. Demethylase KDM1a that demethylates bivalent Lys4 of histone-H3 ($H3K_4me^2$) acts to remove transcription-activating histone marks, thereby reinforcing the repressive activity of the PRC2 complex. LincRNA-mediated

5.3.3.1 Types of Small-Silencing RNAs

Small regulatory RNAs function both in the cytoplasm, where they inhibit expression from messenger RNAs, and in the nucleus, where they silence heterochromatin and prevent genome rearrangement. In the nucleus, short interfering RNAs can associate with nuclear RNA interference (RNAi)-defective proteins NRDe2 and NRDe3 of the Argonaute family on nascent transcripts [345]. This complex prevents the action of RNA polymerase-2 during the elongation phase of transcription.

Small Interfering RNAs and MicroRNAs

Small-silencing RNAs guides specialized RNA-binding Argonaute proteins. Among small-silencing RNAs, small interfering RNAs are produced in many animal species to defend against viruses and transposons. Mammals synthesize siRNAs to regulate genes. The second subclass of small-silencing RNAs — the microRNAs — selectively reduce the stability and rate of translation of mRNAs.

PIWI-Interacting RNAs

PIWI-interacting RNAs are the longest small RNAs. They derive from very long, single-stranded RNAs, transcribed from discrete genomic loci — the piRNA clusters — that contain transposon and repeat fragments [347]. PiRNA precursors are processed into primary piRNAs. Primary piRNAs initiate a feed-forward amplification loop ("ping-pong" cycle) that generates secondary piRNAs.[30] The cell then distinguishes transcripts from piRNA clusters, transposons, and actual genes. Resulting amplified piRNAs silence the set of parasitic genetic elements of the genome, especially mobile genetic elements such as transposons. The mechanism of piRNA production in the somatic cells can differ from that in germ cells.[31]

The PIWI–piRNA complexes monitor and silence their targets such as mRNA transcripts of transposons, as they guide a specialized subclass of

---

assembly of PRC2 and KDM1a coordinates connections of PRC2 and KDM1a to chromatin for coupled histone-H3 Lys27 methylation and Lys4 demethylation.

[30] In the so-called ping-pong cycle, 2 PIWI family members (Aubergine and Argonaute-3 in flies) cleave reciprocally transcripts generated from sense active DNA segments and antisense piRNA clusters, respectively. Each cleavage produces a novel piRNA from the cleaved RNA.

[31] In Drosophila melanogaster, PIWI proteins and bound PIWI-interacting RNAs form the core of the small RNA-mediated defense against parasitic genetic elements such as transposons. In the fly germ line, piRNAs are generated via an Aubergine- and Argonaute-3-dependent piRNA amplication cycle. In somatic cells, piRNA are created via a PIWI-dependent, Ago3-independent pathway [348].

Argonaute proteins — the PIWI proteins — especially in the germ line [349].[32]
PIWI-associated RNAs protect germ cells from genetic parasites and contribute to their development. PIWI-like proteins PIWIL1[33] and PIWIL2[34]
repress transposon activity, as they prevent transposons from producing proteins required for their transposition [350].

5.3.3.2 RNA Interference by Double-Stranded RNAs

Precursors of miRs and siRNAs are double-stranded RNAs (dsRNAs) that are
processed to small RNAs by dedicated enzymes assisted by other proteins.
Double-stranded RNAs trigger suppression of gene activity in a homology-
dependent manner, the so-called RNA interference (RNAi) effect. They specifically target mature mRNA sequence homologous to dsRNA for degradation,
other mRNAs being unaffected. In contrast, antisense suppression of gene expression does not degrade mRNA. Single-stranded RNA fragments bind to
mRNA, then block translation. RNA interference is a process can be used as
a protective mechanism to keep transposons (mobile elements of the genome)
silent and to prevent the production of defective proteins, as well as possibly
acting in resistance to viral infection.[35]

RNA Interference also represents a technique in which exogenous, double-
stranded RNAs complementary to targeted mRNAs are introduced into a
cell to specifically destroy these mRNAs to reduce or abolish a given gene
expression.

Double-stranded RNAs can spread between cells. They exert their effects
at the post-transcriptional level prior to translation. Therefore, RNA interference corresponds to a post-transcriptional gene silencing. Furthermore, RNAi-
like mechanisms keep chromatin condensed and suppress transcription in the
vicinity of the condensed blocks of chromatin.

5.3.3.3 MicroRNAs and Endogenous Small Interfering RNAs

Both single-stranded antisense and sense RNAs either cannot cause or weakly
induce gene silencing. Double-stranded RNA is processed into double-stranded
fragments, the *small inhibitory RNAs*, or *small silencing RNAs* (20–30 nucleotides), that cleave mRNA. Small inhibitory RNAs are either called *small
interfering RNAs* (siRNA) when they are exogenous or *microRNAs* (20–25
nucleotides) when they are produced by RNA-coding genes within the cell.

---

[32] Although some transposons have lost the ability to jump to new locations, others remain active. PiRNAs ensure the genomic integrity of eggs and sperm, as
they protect the germ cell DNA from double-stranded breaks and insertional
mutagenesis caused by active transposons.

[33] A.k.a. MIWI.

[34] A.k.a. MILI and HILI.

[35] In plants, RNA interference produces an antiviral response.

Yet, *endogenous small interfering RNAs* (endo-siRNA) have been identified in mammals.[36] They are distinct from miRs and piRNAs. Small silencing RNAs are characterized not only by their short length, but also by their association with members of the Argonaute family of proteins for target guidance.

Mature endogenous miRs are nearly similar to siRNAs produced from exogenous dsRNAs.[37] Pre-microRNAs are processed into single-strand, antisense RNA and then target mRNAs for degradation. Dicer Ribonucleases produce microRNAs from their precursors and cut long double-stranded RNAs into small double-stranded interfering RNAs.

5.3.3.4 RNA-Induced Silencing Complexes

Small RNAs contain both sense and antisense RNA sequences. Small noncoding RNA molecules form with proteic partners ribonucleoprotein complexes, which affect the transcription, translation, replication, chromosome structure, and regulation of RNA processing.

The *RNA-induced transcriptional silencing complex* (RITS) maintains the genome structure and regulates histone methylation. The *RNA-induced silencing complex* (RISC) is the protein-containing complex responsible for mRNA destruction.

The RISC complex contains endonucleases Argonautes (Sect. 5.3.4) that cleave mRNA strands complementary to bound small inhibitory RNAs. The small antisense RNA (guide strand) — the mature miR — links the RISC complex to the targeted mRNA strand by base-pairing. The passenger strand of small double-stranded RNA (duplex; Sect. 5.3.3.5) is degraded during RISC activation. In the presence of highly imperfect base pairing, antisense RNA blocks translation rather than triggers degradation of sense mRNA.

5.3.3.5 Production and Nuclear Processing of MicroRNAs

A huge number of different miR genes exist (500–1000). Most of miR genes are found in intergenic regions or in antisense orientation to genes. The majority of the remaining miR genes is detected in introns; corresponding miRs are generated by intron processing of protein-coding genes. MicroRNAs regulate deadenylation, translation, and decay of their target mRNAs. MicroRNA expression is often tissue specific and developmentally regulated.

---

[36] In cultured human cells, full-length Line-1 retrotransposon possesses both sense and antisense promoters in its 5′ UTR that produces overlapping complementary transcripts that result from bidirectional transcription of Line-1 retrotransposon. These transcripts can be processed into endogenous small interfering RNAs by Dicer [351]. These endogenous small interfering RNAs suppress retrotransposition.

[37] Typical siRNAs have a perfect base pairing for mRNA, thus cleaving a single, specific target. MiRNAs instead have incomplete base pairing to targeted RNAs, thereby inhibiting the translation of many different mRNAs with similar sequences.

Although RNA-dependent RNA polymerases, which transcribe single-stranded RNA to make double-stranded RNA, produce endogenous small interfering RNAs that suppress the expression of mobile genetic elements (transposons), the latter can derive from double-stranded RNAs [352].

Conventional MiR Synthesis

Long primary transcripts of microRNAs (*pri-miRs*) are transcribed by RNA polymerase-2 within the nucleus. Subsequent cleavage generates RNA duplexes and then functional RNA fragments (miRs) [353] (Table 5.8). After transcription, capping, and polyadenylation, pri-miRs are cleaved by the ribonuclease Drosha. Primary microRNAs are processed by Drosha complex with double-stranded RNA-binding proteins into hairpin precursor microRNAs (*pre-miR*; ~ 70 nucleotides). These pre-miRs are transported to the cytoplasm, where they are also cleaved by the Dicer complex [354]. Multiple miR-containing transcripts are co-transcriptionally cleaved by Drosha [355], similarly to other RNA processing, such as pre-mRNA capping, splicing ,and 3′ end cleavage and polyadenylation that occur in synergy with RNA polymerase-2 transcription.

Control of MicroRNA Synthesis

Various growth and transcription factors participate in the regulation of each step of the miR genesis. DNA-binding factors that regulate miR transcription comprise ubiquitous Myc and P53 transcription factors as well as cell-type specific transcription factors, such as MEF2, SPI1 (or PU1) and RE1-silencing transcription factor[38] (REST) [356]. Furthermore, transcription of pri-miRs can be regulated by growth factors, such as platelet-derived (PDGF) and transforming (TGFβ) growth factor as well as bone-derived neurotrophic factor.

Epigenetic control with DNA methylation and histone modifications such as acetylation also regulates the expression of miR genes. In addition, heterogeneous nuclear ribonucleoprotein hnRNPa1 is required for the processing of miR18a, a member of the miR17-92 cluster [356].

KH-Type splicing regulatory protein (KSRP) binds to miRs and promotes both the Drosha and Dicer processing [356]. Cell lineage proteic homolog (Caenorhabditis elegans) Lin28 prevents the association of Drosha and Dicer with pri- and pre-miRs, respectively. It also acts as a scaffold for the noncanonical poly(A)[39] polymerase Terminal uridine transferase-4 (TUT4) that promotes the 3′-uridinylation of pre-miR that is then rapidly degraded. Lethal

---

[38] The transcriptional repressor REST is also termed neuron-restrictive silencer factor (NRSF) and X2 box repressor (XBR).

[39] The poly(A) tail consists of multiple adenosine monophosphates. Polyadenylation is the addition of a poly(A) tail to an RNA molecule.

Table 5.8. Stages of microRNA genesis and fate (Source: e.g., [356]; dsRNA: double-stranded RNA [mature miR guide strand–passenger strand duplex]; pre-miR: precursor microRNA [60–100 nucleotide, hairpin structure]; pri-miR: primary microRNA [long (several thousand nucleotides), capped, polyadenylated structure]). Drosha needs cofactors to enhance its catalytic activity. It associates with at least 20 distinct polypeptides to form the Drosha microprocessor complex. Like Drosha, Dicer forms the RISC-loading complex with several partners. Following Dicer-mediated cleavage (Dicing), the resulting duplex is dissociated and the mature miR is incorporated into the RNA-induced silencing complex (RISC) to cause a translational inhibition or promote the degradation of target mRNAs. Messenger RNAs are selected from imperfect base pairing between the miR and miR-recognition site in the 3'-UTR of mRNA. In RISC complexes, mature miRs serve as guides orienting RISCs to target mRNAs.

Stage	Product	Factor
	Nucleus	
Transcription	Pri-miR	RNA polymerase-2
Cropping	Pre-miR	Drosha ribonuclease-3
		Drosha microprocessor complex
Export	Pre-miR	Exportin-5, Ran^{GTP}
	Cytoplasm	
Dicing	DsRNA	Dicer ribonuclease-3
Strand selection	MiR	
Aggregation	RISC	
Targeting		
Repression		
Deadenylation		
Degradation		

homologs (Caenorhabditis elegans) of the Let7 family of miRs repress Drosha processing via nuclear factors NF45 and NF90. These nuclear factors tether to pri-miRs and preclude their binding with the Drosha processing complex. The proteic NF45–NF90 complex also associates with small dsRNAs.

The *Drosha microprocessor complex* is strongly controlled.[40] In particular, Drosha is able to regulate the production of its own cofactors such as DiGeorge syndrome critical region gene product DGCR8 by cleavage. Drosha processing is also controlled by the DEAD-box RNA helicase-interacting proteins DDx5 and DDx17 that are components of the Drosha microprocessor complex [356]. They may enhance the recruitment and linkage of pri-miR to Drosha or serve

---

[40] The Drosha microprocessor complex contains DiGeorge syndrome critical region gene product DGCR8, a double-stranded RNA-binding domain (dsRBD)-containing protein and several auxiliary factors, such as hnRNPp2 (a.k.a. RNA-binding proteins Ewing sarcoma breakpoint region protein EWSR1 and Fused in sarcoma [FuS]), numerous heterogeneous nuclear ribonucleoproteins, and DEAD-box helicases DDx5 and DDx17 [357].

as scaffolds for the recruitment of other factors. They can also participate in the regulation of pri-miR processing mediated by multiple agents, such as SMADs, P53, and estrogen receptor-α. Besides, SMAD nuclear-interacting protein SNIP1 interacts with Drosha. It serves as Drosha cofactor in pri-miR processing of certain miRs, such as Let7i and miR21 to miR23 [357].

Cleavage by Dicer produces an unstable dsRNA composed of an active guide strand (miR) and a passenger strand. The active guide strand is incorporated into the *RNA-induced silencing complex* (RISC) that contains Argonaute proteins. Dicer builds the *RISC-loading complex* (RLC) with several partners. Several Dicer-associated proteins, such as trans-activation response (TAR) RNA-binding protein (TRBP) and protein kinase-R-activating protein (PAct), stabilize Dicer and enhance its activity [356]. Extracellular signal-regulated protein kinase phosphorylates TRBP to improve miR production. Lin28 Homolog impedes Dicer processing of Let7 family members [356].

RNA-binding KH-type splicing regulatory protein (KHSRP or KSRP)[41] interacts with single-strand AU-rich-element-containing mRNAs and mediates mRNA decay. It is also a component of both Drosha and Dicer complexes, hence participating in the synthesis of miRs [358].

In addition to pre-miR, Dicer cleaves small nucleolar RNA, short and long hairpin RNA, as well as viral RNA to form miR-like molecules. Type-2 tRNA-derived small RNAs (tsRNAs or tRNA-derived RNA fragments [tRF]) are cleaved from tRNA precursors by t-ribonuclease-Z; type-1 tsRNAs are then produced by Dicer cleavage [357].

In addition to their role in miR-mediated mRNA silencing, Argonaute proteins contribute to the maturation of several (but not all) miRs. Passenger strands of pre-miRs can be nicked by Argonaute proteins. Additional intermediates that arise from Ago-cleaved pre-miR before Dicer processing stabilize mature some types of miRs [357].

Non-Canonical MiR Synthesis

Several miR synthesis pathways deviate from the conventional miR genesis pathway [357]. Alternative pathways of miR maturation bypass Drosha or Dicer. Small nucleolar RNAs also escape ribonuclease-3 Drosha-mediated processing.

Several miR classes bypass ribonuclease-3 Drosha-mediated processing. Introns are a common source of canonical miRs. They can also encode a separate subclass of miR precursors, the so-called *mirtrons*, that are processed by a distinct, Drosha-independent pathway. In the mirtron pathway, spliced lariat introns (Sect. 5.4.4) become pre-miRs after linearization by the lariat debranching enzyme and RNA refolding [357]. These pre-miRs translocate to the cytosol owing to exportin-5, where they can be cleaved by Dicer.

---

[41] A.k.a. Far upstream element-binding protein FUBP2 and FUSE-binding protein FBP2.

MicroRNA-451 involved in erythropoiesis is processed independently of Dicer. After Drosha cleavage, pre-miR451, the stem of which is shorter than those of canonical miRs, is cleaved by Argonaute and then uridylated and further trimmed (independently of Dicer) to generate a functional miR [357].

5.3.3.6 Nuclear Export of MicroRNAs

Large mRNAs and numerous small RNAs, such as tRNA and pre-miRs, are transported to the cytoplasm through nuclear pores by exportins that are specific for each class of RNAs. Both pre-miRs and tRNAs are generated from larger precursor RNAs. Nuclear processing of both tRNAs and miRs needs to be completed before their nuclear export. They are then exported using exportins of the karyopherin superfamily in conjunction with Ran guanosine triphosphatase (GTPase). Pre-microRNA is exported primarily by exportin-5 (Exp5) and tRNA by exportin-T (XpoT) [359]. The pre-microRNAs complex links to Exp5 and small nuclear GTPase Ran^{GTP} to be protected from degradation by nucleases and to undergo facilitated transport to the cytoplasm.

5.3.3.7 MicroRNA Functions

Mature miRs constitute ribonucleoproteic complexes (*miRISC*) with Argonaute proteins, among others. MicroRNA functions include inhibition of individual target transcripts as well as coordinated regulation of target set and fine-tuning of target activity. MicroRNA activities can be classified according to detectable effects on morphology, function, or behavior.

MicroRNAs regulate almost every cellular process and mediate translational repression or degradation of mRNAs selected from base pairing to mRNAs. MicroRNAs modulate particularly cell proliferation, differentiation, and apoptosis[42] They are required in developing timing control.[43] However, miR absence only causes minor effects in most circumstances. MicroRNAs might modulate rather than regulate target gene expression or could be incorporated into regulatory networks that can lose one element without detectable consequences, except when the network is impaired [360]. RNA-binding proteins are able to counteract miR-mediated silencing, possibly under certain conditions and in specific tissues.

MicroRNAs interact with 3′ untranslated regions of mRNAs. Inactive mRNAs assemble into repressive messenger ribonucleoproteins (mRNP) that accumulate in P-bodies or stress granules. However, miRs can reversibly switch on and off mRNA translation. Microribonucleoproteins (miRNP) can even act as translational activators. In addition, *mirtrons* that do not bear Drosha-mediated cleavage mainly operate as translational repressors.

---

[42] Apoptosis is a programmed cell death mode (απoπτωσιζ: falling off with fragments).

[43] For example, miR expression changes during T-lymphocyte maturation.

A single miR can directly downregulate production of hundreds of proteins [361,362]. MicroRNAs repress protein synthesis both translationally and post-translationally. The seed sequence in the 3' UTR is a primary motif of miR-mediated regulation of protein production [361].

MicroRNAs can modulate target activity, as they set a limit on target mRNA concentration. In addition, they can determine a threshold activity level for target gene transcription to ensure that corresponding pathway activity is not deployed inappropriately by transcription noise [363]. In addition, miRs tune targets that remain functional. Futhermore, their repression can be reversible for a quicker mRNA reactivation than transcription activation of repressed genes. MicroRNA-mediated repression can indeed be antagonized by RNA-binding proteins.[44] MicroRNAs can thereby control timing of cellular events.

MicroRNA–protein complexes can activate the translation of reporters and specific mRNAs in quiescent cells. In particular, they maintain oocytes (at least in Xenopus laevis) in an immature state owing to the cAMP–PKA2 axis [364].

The equilibrium level of miRs is determined by opposing activities of miR synthesis and degradation. Exoribonucleases of the family of small RNA-degrading nucleases are specific to short single-stranded RNAs to avoid miR accumulation [365].

MicroRNA as Endocrine Regulators

MicroRNAs such as miR150 can be detected in microvesicles shed from cells in blood. These microvesicles may then enable miR transport and delivery to target cells after they have been internalized by endothelial cells [366].

5.3.3.8 MicroRNA Expression Pattern

The expression pattern of microRNAs varies according to cell type. MiR1, miR16, miR27b, miR30d, miR126, miR133, miR143, and the Let7 family are abundantly (but not exclusively) expressed in cardiomyocytes [367]. Coronary arterial smooth muscle cells synthesized large amounts of miR23, miR125a, miR125b, miR143, miR145, and the Let7 class. Endothelial cells produce high levels of the Let7 family, miR126, miR221, and miR222. Fibroblasts mainly express miR16, miR21, miR22, miR23a, miR24, miR27a. In addition, miR expression changes according to cell activity.

---

[44] Liver-specific miR122 targets CAT1 amino acid transporter. When CAT1 activity must increase, CAT1 inhibition by miR122 is opposed by RNA-binding protein HuR that allows translation from pre-existing mRNAs for CAT1 synthesis [363].

5.3.4 Argonaute Proteins

Small non-coding RNAs of exo- or endogenous origins, such as small interfering RNAs and microRNAs, bind to Argonaute proteins (Ago), thereby contributing to transcriptional and post-transcriptional gene regulation. In mammals, endogenous small interfering RNAs abound only in oocytes and embryonic stem cells. Double-stranded RNAs constitute a major source of small non-coding RNAs. RNA Duplexes (small interfering RNA and microRNA) bind to Argonautes that separate RNA strands.

Argonautes participate in siRNA and miR maturation as well as the production of P-element-induced wimpy testis-interacting (PIWI) RNAs that are primarily expressed in the germ-line. Argonaute proteins also impede protein synthesis and mRNA stability (RNA silencing). In addition, Argonautes contribute to the maintenance of chromosome integrity and formation of heterochromatin. As they control the activity of mobile genetic elements, Argonautes yield an innate immune defense against viruses and transposons.

Argonaute proteins contain 4 conserved domains: N-terminus, PAZ- and Mid binding sites for small non-coding RNA, such as siRNAs and miRs, and PIWI catalytic sequence for endonuclease "slicing" activity as well as binding site for Dicer. The PAZ and Mid domains anchor the 3'- and 5' end of small RNAs.

Argonaute proteins are classified into 2 families: Argonaute-like and PIWI-like proteins.[45] In humans, 8 Argonaute genes exist: 4 Argonaute-like and 4 PIWI-like genes [368]. Argonaute-like proteins (Ago1–Ago4) are also called type-2C eukaryotic translation initiation factors (eIF2c1–eIF2c4). In humans, PIWI-like proteins encompass PIWIL1 to PIWIL4 [349].[46]

Argonaute-like proteins participate in the maturation of siRNAs, as they eliminate the non-active siRNA strand and initiate cleavage of target RNAs. In animals, microRNAs regulate their targets without Ago-mediated cleavage. Mammalian Ago families possess catalytic and non-catalytic members. In mammals, only Ago2 has retained the endonuclease activity. Proteins encoded by the Ago1/3/4 locus have lost catalytic competence, but can support miR-mediated silencing. MicroRNAs are the exclusive partners of Ago proteins. PIWI-like proteins are involved in the maturation of repeat-associated small interfering RNAs and PIWI-interacting RNAs.

Argonaute associated with miR binds to target mRNA for degradation (*microRNA pathway*). Argonautes actually are the primary component of the RISC complex. The Argonaute–miR complex represses target mRNA and can also affect the formation of ribosomes. Furthermore, it can recruit enzymes

---

[45] PIWI species comprises PIWI, Aubergine, and Ago3 in flies as well as MILI, MIWI, and MIWI2 (a.k.a. PIWIL1, PIWIL2, and PIWIL4, respectively) in mice [349]. Argonaute proteins can be decomposed into Ago found in both plants and animals, PIWI, and worm-specific Wago classes.

[46] PIWI-like proteins PIWIL1 to PIWIL4 are also called HIWI1, HILI, HIWI2 and HIWI3, respectively.

that degrade growing polypeptides. Argonaute associated with siRNA forms RNA-induced silencing complex (RISC) that can cleave target mRNA (*RNA interference or siRNA pathway*).

Argonautes Ago1 and Ago2 interact with various proteins to form distinct complexes that are involved in RNA processing, maturation, transport, and regulation of RNA stability and translation. Some interactors are implicated in the processing of small nuclear and small nucleolar RNAs.

Enzyme Ago2 can cleave siRNAs and pre-miRs.[47] Precursors cleaved by Ago2 can serve as Dicer substrates. In addition, Ago2 can intervene in a non-canonical miR processing pathway that is independent of ribonuclease-3 Dicer. Argonaute-2 indeed cleaves pre-miRs to generate functional miRs [369]. Argonaute-2 is ubiquitinated and degraded upon interaction with tripartite motif-containing TriM71[48] that is expressed primarily in stem and undifferentiated cells [356]. Reduction of Ago2 level by TriM71 may prevent expression of miRs and cellular differentiation.

5.3.5 Other Regulators of Non-Coding RNAs

Ada-Two-A-containing histone acetyltransferase (ATAC)[49] connects with the coactivator complex Mediator (Sect. 4.4.3.6 and Vol. 3 – Chap. 6. Receptors) and regulates independent and distinct steps during transcription initiation and elongation. This coactivator supercomplex can also contain leucine zipper motif-containing protein-1 to regulate a subset of RNA polymerase-2-transcribed non-coding RNA genes [371].

5.4 Pre-Messenger RNA Splicing

Gene expression requires the excision of introns from pre-messenger RNAs during splicing, i.e., elimination of sequences that do not encode proteins and interrupt the translation open-reading frame in mRNAs. Many gene transcripts, i.e., 95 to 100% of human pre-mRNAs, contain several introns that separate exons. In the human genome, typical pre-mRNAs consist on average of 8 introns and 9 exons. Alternative splicing of human pre-mRNAs that contain more than one exon select exons in pre-mRNAs. The number of isoforms of mRNAs and corresponding proteins encoded by a single gene can vary from a few to a huge number ($\mathcal{O}[10^2]$–$\mathcal{O}[10^3]$) [372].

---

[47] MicroRNA-451 is involved in erythropoiesis. Its primary transcript is processed by Drosha. Its maturation does not rely on Dicer, but requires Ago2 catalysis [370].

[48] A.k.a. abnormal cell lineage homolog Lin41.

[49] Ada-Two-A-containing coactivator complex is related to, but functionally distinct, the histone acetyltransferase SAGA complex. Both are constituted by lysine (K) acetyltransferases KAT2a (or General control of amino acid synthesis protein GCN5) and paralog KAT2b (or P300–CBP-associated factor [PCAF]).

Excision at intron–exon boundaries of pre-mRNA is catalyzed by ribonucleoproteic complexes, the so-called *spliceosomes*. The spliceosome, indeed, recognizes introns and exons, hence splicing sites, and removes introns that are deleted, thereby separating exons of pre-mRNAs that are then reconnected to yield a mature exon sequence, i.e., a mature mRNA. However, RNA splicing variation produces different ribonucleotide arrangements of exons. In other words, recombination of different exons generates distinct mature mRNA isoforms.

Alternatively spliced transcripts thus expands the structural diversity of proteins synthesized from a single gene and, hence, the coding capacity of the genome. Less than 25×10^3 protein-coding multi-exon genes and pre-mRNAs of the human genome are estimated to create approximately 10^5 protein isoforms. This diversity is further heightened by alternative transcription start sites, alternative polyadenylation, and RNA editing, as well as post-translational modifications.

Alternative splicing of pre-mRNAs with its differential inclusion and exclusion of exons depends on a cell type-specific regulation. Alternative splicing is also often regulated according to the developmental stage. Extracellular signals, nutritional status, and membrane depolarization influence the splicing location. In particular, alternative splicing of pre-mRNAs can be controlled by growth factors and hormones.[50]

Protein isoforms that result from alternative splicing are different versions of a given protein that arise from transcription of a single gene. In particular, alternative splicing explains the existence of membrane-bound and secreted forms of certain molecules as well as that of tissue-specific peptide hormones such as those encoded by the calcitonin–calcitonin-related polypeptide gene.

Introns are excised from pre-mRNAs before export from the nucleus. The splicing reaction consists of 2 consecutive trans-esterifications. Cells synthezise splicing factors, use phosphorylation, and regulate the intranuclear location of splicing factors[51] to achieve alternative splicing patterns [373].

5.4.1 Spliceosome

Spliceosome is a large complex composed of 5 small nuclear ribonucleoproteins (snRNP), i.e., 5 small nuclear RNAs (snRNA; U1–U2 and U4–U6) that assemble with specific proteins (to form snRNPs) and more than 100 non-snRNP splicing factors [375]. The spliceosome thus involves a set of RNAs with a pre-mRNA and small nuclear RNAs that interact with introns. Any small

---

[50] Insulin-activated PI3K is able to regulate alternative splicing. Fibronectin has 3 alternatively spliced regions: ED1, ED2, and 3CS. Hepatocyte- (HGF; a.k.a. scatter factor), keratinocyte- (KGF1 and KGF2 correspond to FGF7 and FGF10, respectively), and fibroblast growth factor FGF1 cause ED1 and 3CS, but not ED2 inclusion into fibronectin mRNA [374].

[51] Cells select alternative splicing sites by changing the concentration and activity of splicing regulatory proteins.

nuclear ribonucleoprotein-Ui participates in several RNA–RNA and RNA–protein interactions. RNA-binding proteins can operate as splicing repressors or activators.

During spliceosome assembly, U4 interacts with the spliceosome as a preformed U4–U5–U6 triple small nuclear ribonucleoproteic complex. Subsequently, U4 becomes loosely associated with the spliceosome, whereas U5 and U6 remain tightly associated [376].

Recognition and Binding Sites

Four motifs enable recognition of the pre-mRNA by the spliceosome: (1) exon–intron junctions at the 5' and 3' ends of introns, i.e., 5' (5' SS) and 3' (3' SS) splice sites (SS), branch site sequence (BSS), and the polypyrimidine tract (PPT) located between the 3' SS and the branch site [377]:

$$-5'\ SS–BSS–PPT–3'\ SS–.$$

Exons and introns contain short, degenerate binding sites for splicing auxiliary proteins, the so-called *exonic* (ESE) and *intronic* (ISE) *splicing enhancers*, as well as exonic (ESS) and intronic (ISS) *silencing silencers*.

$$(I\text{-}5'\ end)–ISE–ISS–(I\text{-}3'\ end)–(E\text{-}5'\ end)–ESE–ESS–(E\text{-}3'\ end).$$

Splicing was considered sufficiently accurate owing to the 3 splicing signals (5' SS, 3' SS, and BSS) recognized by the spliceosome. However, the splicing is not optimal, as unspliced precursors can be observed.[52] Quality control mechanisms (Sect. 5.7) are thus aimed at removing RNAs that have escaped a proper splicing process. *Non-sense-mediated mRNA decay* (NMD) eliminates transcripts that contain premature termination codons and thus most unspliced RNAs when they reach the cytoplasm (Sect. 5.5.4).

Types of Spliceosomes

Two types of spliceosomes have been identified — major and minor — that contain different snRNPs. The major spliceosome composed of the U1, U2, and U4 to U6 snRNPs as well as U2AF and SF1 proteins splices introns containing GU at the 5' splice site and AG at the 3' splice site (*the lariat pathway* with GU–AG intronic flanking sequences).

The minor spliceosome splices rare introns with different splice site sequences (*non-canonical splicing*). Both the minor and major spliceosomes contain the same U5 snRNP. Yet, the minor spliceosome has different, but functionally analogous snRNPs for U1, U2, U4, and U6: U11, U12, U4$_{atac}$, and U6$_{atac}$.

---

[52] The open-reading frame is not preserved in unspliced RNAs. Consequently, a premature stop codon may occur in the intron and the translation of unspliced RNAs may generate truncated polypeptides with potential deleterious effects [378].

Table 5.9. Ribonucleoproteic regulators of pre-mRNA splicing (Sources: [379], IHOP; AUF: AU-rich element RNA-binding protein; hnRNP: heterogeneous nuclear ribonucleoprotein; hnRNPlL: hnRNPl-like; PTBP: polypyrimidine tract-binding protein; RBMX: RNA-binding motif protein, X-linked; APP: amyloid-βprecursor; BCL2L: BCL2-like; CALCA: calcitonin-related polypeptide-α; Fgfr: fibroblast growth factor receptor; GRIN3B: glutamate receptor, ionotropic, NMDA 3B; IK-BKAP: inhibitor of κ-light polypeptide gene enhancer in B cells, kinase complex-associated protein; Nos: nitric oxide synthase; Ptpr: protein Tyr phosphatase receptor; SMN: survival motor neuron; TNFSF: tumor-necrosis factor superfamily ligand; TNTT2: troponin T type-2; TPM1: α-tropomyosin).

Name	Other aliases	Target genes
hnRNPa1		SMN2
hnRNPa2	hnRNPb1	IKBKAP
hnRNPc1	hnRNPc2	APP
hnRNPd	AUF1	
hnRNPf		BCL2L2
hnRNPg	RBMX, RBMXP1, RBMXRT	SMN2, TPM1
hnRNPh1	HNRPh	
hnRNPh2	hnRNPh', FTP3	BCL2L1
hnRNPH3		
hnRNPi	PTBP1, PTB1, PTB2, PTB3, PTB4, PTBt	CALCA, GRIN3B, Ptbp TNFSF6, TNTT2
hnRNPl		Nos, PtprC
hnRNPlL		PtprC
hnRNPm	hnRPM4, HTGR1, NAGR1	Fgfr2
hnRNPq		SMN2

5.4.2 Splicing Regulators

Splice-site recognition of alternative exons is mediated by RNA-binding proteins that link to the splice-recognition regulatory sequences, such as the Ser–Arg proteins (SR), heterogeneous nuclear ribonucleoproteins (hn-RNP), polypyrimidine tract-binding proteins (PTB), T-cell-restricted intra-cellular antigen TIA1,[53] hexaribonucleotide-binding proteins HRNBP1[54] and HRNBP2,[55] and neurooncological ventral antigens NOVA1 and -2 [377] (Table 5.9). These factors frequently interact.

Ser–Arg-rich RNA-binding proteins form a superfamily of molecules in-volved in both constitutive and alternative pre-mRNA splicing. They activate splicing by binding to exonic splicing enhancers and recruiting components of the spliceosome [372]. In particular, they stimulate the binding of U1 to the

---

[53] T-cell-restricted intracellular antigen TIA1 (cytotoxic granule-associated RNA-binding protein) tethers to intronic splicing enhancers (ISE) immediately down-stream from 5' SS to facilitate U1 binding.

[54] A.k.a. ataxin-2-binding protein A2BP1 and Fox1 homolog A or FOX1.

[55] A.k.a. RNA-binding motif protein RBM9 and Fox1 homolog B or FOX2.

5′ SS and that of U2AF to the 3′ SS. Proteins SR also cooperate with other activators to form larger splicing enhancing complexes, such as transformer Tra2 and SR-related nuclear matrix proteins SRRM1 (a.k.a. SRM160) and SRRM2 (a.k.a. SRM300) [379].

Proteins SR recruit not only splicing factors for the spliceosome assembly, but also bridging factors targeting 3′ and 5′ splice sites that define alternative exons. They contain RS-binding domains for proteins and RNA recognition motifs for exonic or intronic splicing enhancer sequences of pre-mRNAs. The activity of SR proteins is more efficient on co-transcriptional than on post-transcriptional splicing, as they associate with RNA polymerase-2.

The 3 pre-mRNA processing stages (capping, splicing, and cleavage and polyadenylation) can indeed be coupled to RNA polymerase-2 transcription, because pre-mRNA processing factors can interact with RNA polymerase-2 or other transcription components. In addition, certain proteins have dual activities (recruitment transcription–RNA processing coupling).[56] Furthermore, changes in rate of transcriptional elongation affect timing of splice site presentation to the spliceosome (kinetic transcription–RNA processing coupling). Finally, processing factors influence RNA polymerase-2 transcription.[57]

On the other hand, hnRNPs repress splicing by binding to sequences corresponding to exons or introns and interfering with the ability of the spliceosome to engage splice sites. Polypyrimidine-tract-binding protein (PTB) is another inhibitory splicing factor.

Repressors FOX1 and FOX2 hinder the binding of splicing factor SF1 to the branch point and those of TRA2 and SRP55 to ESEs. They thus hamper the spliceosomal E′ and E complex assembly (Sect. 5.4.4).

However, NOVA1, NOVA2, FOX1, FOX2, hnRNPf, hnRNPh, hnRNPl, and hnRNPlL can act as repressors or activators according to the location of their binding site. Agents NOVA1 and NOVA2 can bind to ESS, thereby inhibiting the formation of the pre-spliceosomal E complex [379]. On the other hand, NOVA1 and NOVA2 can link to ISE downstream from the alternative exon, hence promoting the formation of spliceosomal complexes A, B, and C.

Therefore, pre-mRNA splicing is regulated by binding of repressors such as hnRNPs and activators such as SRs to silencers (ESS and ISS) and enhancers (ESE and ISE) on the RNA, respectively.

Many factors such as chromatin structure can influence splicing patterns, hence transcription rate, as well as spliceosome concentration, intron size, and competition between splice sites [372].

---

[56] Coactivator PGC1 is also an alternative splicing regulator. Its splicing activity is exerted only when it is tethered to promoters by binding to transcription factor.

[57] SR Splicing factor SC35 binds to CDK9, the kinase component of the transcriptional elongation factor Positive transcription elongation factor-B (pTEFb), i.e., the complex made of CDK9 and cyclin-T1, -T2, or -K, and promotes RNA polymerase-2 phosphorylation that stimulates transcriptional elongation by the RNA polymerase-2 complex in many genes [380].

Post-translational modifications of splicing factors enable cells to switch rapidly between alternative splicing isoforms after environmental stimuli. Phosphorylation can change the intracellular localization of splicing factor, as well as between-protein and RNA–protein interactions, in addition to intrinsic splicing factor activity [379].

5.4.3 Alternative Splicing Modes

Alternative splicing modes can be classified into 4 main types [377]: (1) exon skipping, in which a cassette exon (E) is spliced out of the transcript together with its flanking introns (I–E–I); (2) alternative 3' (3' SS; I–EI) and (3) 5' splice site (5' SS; IE–I) selection, when at least 2 splice sites are recognized at one end of an exon; and (4) intron retention (IEI), in which an intron remains in the mature mRNA transcript. Nearly all alternative splicings result from the use of at least one of these 4 basic modalities.[58] Other much less frequent splicing patterns that give rise to alternative transcript variants include mutually exclusive exons (I–E–E–I; one of 2 exons is retained in mRNAs after splicing, but not both), alternative promoter usage, and alternative polyadenylation. Distinct regulatory programs can be used in different cell types.

5.4.4 Spliceosome Assembly

Initially, U1 recognizes the 5' splice site (base pairing) and the splicing factor SF1[59] binds to the branch site sequence near to the 3' end of intron and the U2 auxiliary factor (U2AF) to the polypyrimidine tract and 3' intronic end (Table 5.10). This assembly forms the *E complex*. Afterward, U2 replaces SF1 at the intronic branch site. The E complex is the converted into the pre-spliceosomal *A complex* [379]. The U4–U5–U6 complex is then recruited to form the *B complex*. The latter is converted into catalyzer *C complex* after extensive conformational changes and remodeling. Element U1 is then displaced by U6 that couples with the 5' splice site and U2, and U5 contacts the 5' exon.

During each splicing stage, the spliceosome recognizes splice-site sequences of the pre-mRNA and assembles on the substrate via coordinated rearrangements of its components via specific ATPases that are modulated by other factors such as processing proteins that can be components of small nuclear

---

[58] The human gene KCNMA1 (or SLO1; Vol. 3 – Chap. 3. Main Classes of Ion Channels and Pumps) contains numerous modules, such as alternative 3' SS and 5' SS and cassette exons, and more than 500 mRNA isoforms of KCNMA1 can be generated [372].

[59] A.k.a. mammalian branch point-binding protein (mBBP).

Table 5.10. Stages of spliceosome assembly (Source: [379]; SS: splice site; BSS: branch site sequence; PPT; polypyrimidine tract; SF: splicing factor [SF1 a.k.a. branch site-binding protein (BBP); SF2 a.k.a. alternative splicing factor (ASF)]; snRNP: small nuclear ribonucleoprotein; U2AF: small nuclear RNA U2 auxiliary factor). (**Stage 1**) small nuclear ribonucleoprotein U1 binds to the GU sequence at the 5′ SS with accessory protein SF2, SF1 to BSS, and U2AF to the PPT site and AG sequence at the 3′ SS. (**Stage 2**) U2 binds to BBS and ATP is hydrolyzed. (**Stage 3**) The U4–U5–U6 trimer connects to the intron; U5 binds to exon at the 5′site and U6 to U2. (**Stage 4**) U1 is released; U5 shifts from the exon to intron, and U6 tethers to the 5′ SS. (**Stage 5**) U4 is released; U5 binds to the exon at 3′ SS; the 5′ intron end is cleaved to form the lariat (lasso-like shape). (**Stage 6**) The U4–U5–U6 trimer remain bound to the lariat; the 3′ intron end is cleaved, and exons are ligated. The spliced RNA is then released and the lariat debranches.

Step	Event
1	Fixation of snRNA U1 on 5′ SS (GU sequence)
	Commitment of pre-mRNA to splicing
2	Binding of SF1 to BBS
	Connection of U2AF heterodimer to PPT and 3′ SS (AG sequence)
	E Complex; (U1–SF2–GU)–(SF1–BBS)–(PPT–U2AF–AG)
3	Replacement of SF1 by U2
	(pre-spliceosome)
	A Complex; (U1–SF2–GU)–(U2–BBS)–(PPT–U2AF–AG)
4	Recruitment of the U4–U5–U6 complex
	B1 Complex; U1–U4–U5–U6–U2
5	Release of U5
	B2 Complex; U1–U4–U6–U2 and U5–exon
6	Release of U4 and reintegration of U5
	Cleavage of intron 5′end
	C1 Complex; U5–U6–U2 and lariat-like intron
7	Cleavage of intron 3′end
	C2 Complex; U5–U2–U6–5′ SS

ribonucleoproteins.[60] Between-protein interactions involve modifications, such as phosphorylation or ubiquitination.[61]

---

[60] mRNA Processing protein PrP8 (U5 component) contacts the 5′ splice site, branch point sequence, and 3′ splice site of pre-mRNA. It interacts with many spliceosomal proteins, small nuclear ribonucleoproteins, and splice signals in the intron, as it contains a motif that binds one or more RNA components of the spliceosome and an RNase-H domain that binds to the intron at both 5′-

Small nuclear ribonucleoproteins U1 and U2 form the *A complex* before the U4–U6–U5 complex that generate the pre-catalytic *B complex*. Ribonucleoprotein rearrangements lead to the displacement of U1 and U4. Both U4–U6 and U1–5′ SS complexes are disrupted for alternative interactions to generate the active spliceosome. Activated B spliceosome catalyzes the first stage of splicing (leading to cleaved 5′ exon and intron–3′ exon intermediate). Afterward, the spliceosomal C complex is built to catalyze the second step of splicing (intron excision and ligation of 5′ and 3′ exons to form mRNA). The SF3a–SF3b complexes are absent from the ribonucleoproteic core of the *C complex* [375]. The PrP19–CDC5 complex and related proteins are more stably associated in C complexes. The transition from B to C complex thus occurs with exchange of proteins, stabilization of the Prp19–CDC5 complex, and destabilization of the SF3a–SF3b complex when the active site of the spliceosome is formed (splicing stage 1).

Distinct types of spliceosomes exist: low-abundance, minor, U12-dependent and major, U2-dependent spliceosomes. U12-dependent spliceosome catalyzes the removal of a minor class of introns ($<1\%$). U12-Dependent spliceosome contains U11, U12, U4$_{atac}$, and U6$_{atac}$ snRNAs, which are paralogs of U1, U2, U4, and U6 snRNAs of the U2-dependent spliceosome, respectively. The 2 spliceosome types share U5. Components of both U12-dependent and U2-dependent spliceosomes are nearly exclusively detected in the nucleus [383]. Minor U11 and U12 colocalize with U4 and U2. Most of snRNP and non-snRNP spliceosomal proteins are common to both spliceosomes, except those associated with U11 and/or U11–U12 snRNPs.

5.4.5 Tissue-Specific Alternative Splicing

Tissue-specific alternative splicing can result from, at least partly, tissue-specific expression of splicing factors (Table 5.11).[62]

---

and 3′splice sites [381]. (RNase-H is a metalloenzyme). Processing protein PrP8 controls rearrangements of small nuclear ribonucleoproteins to form the active spliceosome. It orients the pre-mRNA substrate in the active site and participates in catalysis. Then, PrP19 and the PrP19–CDC5 complex associate with pre-mRNA. Finally, proteins of the SF3a and SF3b heteromers that are essential in early splicing link to the pre-mRNA and stabilize U2–branch point sequence interaction.

[61] Splicing factor Prp8 is ubiquitinated within triple U4–U6–U5 snRNP complex [382].

[62] Several brain-specific splicing factors have been identified [379]. Polypyrimidine-tract binding protein is expressed in neural progenitor cells, but its expression is downregulated in differentiated neurons. Conversely, the brain paralog neurally-enriched homolog of PTB (nPTB or polypyrimidine tract-binding protein PTBP2) is upregulated in differentiated cells.

Table 5.11. Tissue-specific alternative splicing factors (Source: [379]; A2BP: ataxin-2-binding protein; CELF: CUGBP- and ETR3-like factor; ELAV: embryonic lethal, abnormal vision, Drosophila-like KHDRBS: KH domain-containing, RNA-binding, signal transduction-associated protein; MBnL: muscleblind-like; NOVA: neurooncological ventral antigen; PTB: polypyrimidine tract-binding protein; RBM: RNA-binding protein; SLM: Sam68-like mammalian protein; TIA1: T-cell-restricted intracellular antigen; TIAR: TIA1-related protein).

Name	Other names	Tissue expression
CELF1	BrunoL2	Brain
CELF2	BrunoL3, ETR3	Heart, skeletal muscle, brain
CELF4	BrunoL4	Muscle
CELF5	BrunoL5, NAPOR	Heart, skeletal muscle, brain
CELF6	BrunoL6	Kidney, brain, testis
ELAV2	HUb	Neurons
HRNBP1	A2BP1	Muscle, heart, neurons
HRNBP2	RBM9	Muscle, heart, neurons
KHDRBS3	SLM2, TSTAR	Brain, heart, testis
MBnL		Muscle, uterus, ovary
NOVA1		Neurons (hindbrain, spinal cord)
NOVA2		Neurons (cortex, hippocampus, dorsal spinal cord)
nPTB	brPTB, PTBP2	Neurons, myoblasts, testis
Quaking	QK1/3, QKL	Brain
RBM35a	ESRP1	Epithelial cells
RBM35b	ESRP2	Epithelial cells
TIA1		Brain, spleen, testis
TIAR	TIAL1	Brain, spleen, lung, liver, testis

5.5 Translation Initiation

Maturation of mRNAs involves nuclear processing before cytoplasmic translation. Nuclear export of mRNAs follows a successful pre-mRNA splicing. Surveillance of mRNA corresponds to the final quality-control stage. Different proteic complexes actually form on newly spliced mRNA to ensure accuracy and efficiency of gene expression. The *post-splicing exon-junction complex* that is associated with mRNA surveillance illustrates involved proteic complexes.

5.5.1 Initiation Complex

After pre-mRNA splicing and nuclear export, an *initiation complex* is assembled for the first stage of translation on newly spliced, mature mRNAs. Protein synthesis is principally regulated at the initiation stage rather than during elongation or termination to ensure a rapid, reversible, spatially controlled procedure.

Translation initiation is a multistep process that starts the assembly of elongation-competent 80S ribosomes, in which the initiation codon is base-paired with the anticodon of initiator tRNA ($tRNA_i^{Met}$) in the ribosomal P site. Translation initiation requires: (1) the full set of eukaryotic initiation factor (eIF; Table 5.12); (2) initiator $tRNA_i^{Met}$ bound to the corresponding amino acid Met–$tRNA_i^{Met}$ in higher eukaryotes;[63] and (3) 5′ capped mRNA to assemble the 80S ribosome at the AUG start codon.

An initial complex is composed of $tRNA_i^{Met}$, guanosine triphosphate (GTP), and the guanosine triphosphatase (GTPase) initiation factor eIF2 (Table 5.13). Heterotrimer eIF2 is composed of an α (subunit 1), β (subunit 2), and γ (subunit 3) subunit. Factor eIF2 mediates the binding of $tRNA_i^{Met}$ to the ribosomal complex using GTP. Activity of initiation factors is modulated by sequence-specific RNA-binding proteins and miRs. Translation initiation comprises 2 main steps: (1) the formation of 48S initiation complexes with established codon–anticodon base-pairing in the P-site of the 40S ribosomal subunit and (2) connection of 48S complexes with 60S ribosomal subunit (Table 5.13).

The canonical initiation mode commences with the recruitment of the 43S complex (i.e., 40S ribosomal subunit, eIF1, eIF1a, eIF3, probably eIF5, and the $tRNA_i^{Met}$–eIF2^{GTP} ternary complex) to the mRNA 5′ cap via the 5′ cap-binding eIF4f complex. In parallel, mRNA 5′ terminal secondary structure is unwound by eIF4a, eIF4b, and eIF4f factors.

From 43S Pre-Initiation Complex to 80S Ribosomal Complex

The eIF2^{GTP}–$tRNA_i^{Met}$ ternary complex binds to the 40S ribosomal subunit to construct the 43S *pre-initiation complex* (PIC) with initiation factors eIF1, eIF1a, and eIF3.

5′ untranslated region (5′ UTR) of mRNA under reading procedure possesses a structure that must be processed to allow reading commencement. Factors eIF4f and eIF4b or eIF4h cooperatively unwind the 5′ capped proximal region of mRNA for ribosomal attachment. Factor eIF4f is composed of the DEAD-box RNA helicase eIF4a, cap-binding protein eIF4e, and scaffold eIF4g that binds to eIF4e, eIF4a, PABP, and eIF3.

After attachment is achieved by the cap–eIF4e–eIF4g–eIF3–40S chain of interactions, the 43S complex scans mRNA downstream of the cap to the

---

[63] In bacteria and mitochondria, initiator tRNA is $tRNA^{fMet}$. ^{N}Formylmethionine (fMet), a derivative of methionine to which a formyl group is attached, initiates protein synthesis. It is removed upon synthesis completion. The start (or initiation) codon AUG (nucleotides that consists of a phosphate, ribose, and nitrogenous nucleobases adenine [A], uracil [U], and guanine [G]) on mRNA encodes both Met and fMet. The translation initiation codon encodes fMet, the first amino acid (N-terminus) of the nascent peptide chain, delivered to the ribosome–mRNA complex by a specialized fMet-$tRNA^{fMet}$. Later, AUG codon in the mRNA encodes for Met.

Table 5.12. Eukaryotic initiation factors (Source: [384]; $5'$ m^{7}GpppG cap: post-transcriptionally modified mRNA $5'$ end [$5'$ cap] by added 7-methylguanosine [m^{7}G], i.e., methylated and phosphorylated guanosines [m^{7}GpppG: P^1-7-methylguanosine-$5'$ P^3-guanosine-$5'$triphosphate] eIF2 TC: eIF2^{GTP}–tRNA$_i^{Met}$ ternary complex; MIF4GD: MIF4G domain-containing protein [a.k.a. SLBP-interacting protein SLIP1]; PABP: mRNA polyadenylated sequence-binding protein). Two paralogs of eIF4a (eIF4a1 and eIF4a2) and eIF4g (eIF4g1 and eIF4g2) are encoded by different genes and functionally similar (eIF4g1 and eIF4g2 exhibit selectivity toward mRNA types.

Type	Function
eIF1	Selection of initiation codon
	Promotion of ribosomal scanning
	Stimulation of binding of eIF2 TC to 40S ribosomal subunit
	Prevention of premature hydrolysis of eIF2^{GTP} by eIF5 GAP
eIF1a	Stimulation of binding of eIF2 TC to 40S ribosomal subunit
	Cooperation with eIF1 to promote ribosomal scanning
	and initiation codon selection
eIF2	Formation of eIF2 TC
eIF2b	Guanosine nucleotide-exchange factor for eIF2
eIF3	Binding to 40S ribosomal subunit, eIF1, eIF4g, and eIF5
	Stimulation of binding of eIF2 TC to 40S ribosomal subunit
	Promotion of attachment of 43S complex to mRNA
	Preclusion of 40S–60S subunit connection
eIF4a	DEAD-box ATPase and ATP-dependent RNA helicase
eIF4b	RNA-binding protein
	Enhancement of eIF4a helicase activity
eIF4e	Binding to the mRNA $5'$ m7GpppG cap
eIF4f	Unwinding of mRNA $5'$ proximal region
	Attachment of 43S complexes to mRNA
eIF4g	Binding of eIF3, eIF4a, eIF4e, MIF4GD, PABP, and mRNA
eIF4h	RNA-binding protein
	Enhancement of eIF4a helicase activity
eIF5	GTPase-activating protein for eIF2
eIF5b	Ribosome-dependent GTPase
	Connection of ribosomal subunits
eIF6	Preclusion of 40S–60S ribosomal subunit connection

initiation codon. Scanning is supported by eIF4a and its cofactors eIF4b and eIF4g [384]. It may involve additional DEAD box- or DExH box-containing proteins such as DEAH box polypeptide-29 (Dhx29).[64] Scanning consists of

---

[64] The DExH/D family proteins comprise the DEAD, DEAH, and DExH sub-families. All members contain at least 8 characteristic motifs, in particular the

Table 5.13. The canonical mode of translation initiation comprises several stages. Eukaryotic initiation factor eIF2 is required in the initiation of translation. It forms a ternary complex with GTP and tRNA$_i^{Met}$ and mediates the binding of initiator tRNA$_i^{Met}$ to the ribosome. This complex binds the 40S ribosomal subunit to form the 43S pre-initiation complex (PIC) with initiation factors eIF1, eIF1a, and eIF3. The PIC complex then binds mRNA that has previously been unwound by eIF4 to build the 48S complex on the mRNA. Upon base pairing of the start codon with tRNA$_i^{Met}$, GTPase-activating protein eIF5 is recruited to the complex. Following eIF2^{GTP} hydrolysis into eIF2^{GDP}, eIF2 is released from the 48S complex and translation begins after recruitment of the 60S ribosomal subunit and formation of the 80S initiation complex. Translation is a cyclical process that reuses ribosomal subunits. Post-termination ribosomal complexes are indeed recycled. They contain mRNA-bound 80S ribosomal subunit, P-site deacylated tRNA, and at least one release factor such as eukaryotic release factor eRF1.

Complex	Contributors
Formation of 43S complex	Ribosome 40S subunit
	tRNA$_i^{Met}$–eIF2^{GTP}
	eIF1, eIF1a, eIF3
Formation of 48S complex	mRNA (start codon recognition)
	mRNA-binding regulator eIF4
Formation of 80S complex	GTPase-activating eIF5
	Release of eIF2
	Ribosome 60S subunit
	Release of eIF1, eIF1a, eIF3, eIF5

2 coupled processes: (1) unwinding of the 5′ untranslated region (5′ UTR) and (2) ribosome movement along it with assistance of eIF1 and eIF1a. The PIC complex scans the 5′ UTR from the 5′ cap (proximal point that serves as mRNA initial attachment) to the initiation codon, i.e., in the 5′-to-3′ direction, once the 40S ribosomal subunit has undergone a suitable conformation induced by eIF1 and eIF1a. Efficient scanning needs Dxh29 protein.

To ensure the recognition of the correct initiation codon, the scanning complex must discriminate the first AUG triplet among codons and prevent partial base pairing of triplets in the 5′ UTR with tRNA$_i^{Met}$ anticodon. Factor eIF1 is a major player in this process. Ribosome then commences initiation at the AUG initiation codon owing to eIF2-specific GTPase-activating protein eIF5.

Once the codon–anticodon base pairing is established, eIF1 is displaced from the P-site, the 40S ribosomal subunit switches from a closed to an open conformation [384]. In addition, after initiation codon recognition and

---

ATP-hydrolysis motif 2 from which their names are derived: DEAD, DEAH and DExH, in single-letter amino acid code (DEAD: Asp–Glu–Ala–Asp; DEAH: Asp–Glu–Ala–His). DExH/D proteins constitute a subset of the SF2 helicase superfamily (SF), which is related to the SF1 helicase superfamily.

48S complex formation, the repression of eIF5-mediated hydrolysis of eIF2^{GTP} is relieved by eIF5b. Factors eIF5 and eIF5b also promote the displacement of eIFs and the joining of the 60S ribosomal subunit. Factors eIF1 and eIF1a are removed from the interface of the 40S ribosomal subunit and link to the 60S ribosomal subunit to form the 80S ribosomal complex with initiator tRNA in the P-site.

Presence of Short Upstream ORF

About 50% of mammalian genes encode mRNAs have at least one short upstream open reading frame (uORF; typically < 30 codons), i.e., upstream from the main protein coding open reading frame. In these cases, ribosomes that have translated the uORF resume scanning and reinitiate at downstream sites [384].

eIF2^{GTP}–eIF2^{GDP} Cycle

Initiation factor eIF2, a GTPase, associates with mRNA in ribosomes. This phenomenon is accompanied by GTP hydrolysis that is required with eIF1 and eIF1a during the recognition of the start codon. The GTP cycle involves eIF2, GDP–GTP-exchange factor eIF2b, and GTPase-activating protein eIF5. Once the initiation is completed, eIF2^{GDP} is released from the ribosome and must be reactivated into eIF2^{GTP} (i.e., GDP must be exchanged for GTP) to participate in a new translation initiation.

Ribosome Recycling

Ribosomes that participate in initiation are recycled from post-termination ribosomes, i.e., ribosomes that have completed the previous translation. Post-termination ribosomes first dissociate into free 60S and 40S ribosomal subunits owing to eIF3 in cooperation with its associated eIF3j subunit as well as eIF1 and eIF1a, at low free Mg^{++} concentration (1 mmol) [384].[65]

ATP-binding cassette protein ABCe1 splits post-termination ribosomes into free tRNA- and mRNA-bound 40S and 60S subunits [384]. Then, tRNA and mRNA are released from 40S ribosomal subunits. This release requires eIF1, eIF1a, and eIF3. Factor eIF3, and probably eIF1 and eIF1a, remains associated with recycled 40S subunits to prevent their reassociation with 60S subunits. Removed deacylated tRNA is replaced by tRNA$_i^{Met}$ that re-forms with eIF2^{GTP} a ternary complex.

---

[65] Nucleotide-unbound Mg^{++} stabilizes association of ribosomal subunits.

5.5.2 Initiation by Internal Ribosome Entry Sites

Although most mRNAs use the scanning mechanism, initiation on a few mR-NAs is mediated by an internal ribosome entry site (IRES) [384].[66] Initiation on type-1 and type-2 IRESs involves their specific binding to eIF4g that is enhanced by eIF4a. The eIF4a–eIF4g complex recruits 43S complexes to type-1 and type-2 IRESs without eIF4e [384]. Initiation on type-3 IRESs implicates their interaction with eIF3 and 40S ribosomal subunit components of 43S complexes. Type-3 IRESs directly attach 43S complexes to the initiation codon independently of eIF1, eIF1a, eIF4b, and eIF4f. Initiation on type-4 IRESs involves their connection to 40S ribosomal subunit. Type-4 IRESs initiate the process without eIFs or $tRNA_i^{Met}$. The P-site of the 40S ribosomal subunit is occupied by an IRES domain that mimics codon–anticodon base pairing.

5.5.3 Regulation of Protein Synthesis Initiation

Eukaryotic initiation factors correspond to the rate-limiting step of initiation of translation. Two eIFs prime translation upon extracellular stimuli: (1) eIF4e intervenes in the eIF4f complex to regulate binding of capped mRNA to 40S ribosomal subunits upon growth factor stimulation and (2) eIF2 controls loading of the ternary complex on 40S ribosomal subunits and is inhibited by stress.

5.5.3.1 Regulation of eIF Activity by Reversible Phosphorylation

Regulation of eIF activity by reversible phosphorylation influences the translation of most mRNAs by the canonical scanning-dependent mode of initiation. Factors eIF2s can be phosphorylated by any of 4 stress-activated kinases that hence modify the concentration of the $eIF2^{GTP}$–$tRNA_i^{Met}$ complexes and 4E-binding proteins.

Protein kinases that phosphorylate eIF2 include: (1) heme-regulated kinase eIF2aK1 that mainly operates in erythroid cell lines; (2) protein kinase-R (PKR or eIF2aK2) that is activated by double-stranded RNAs; (3) PKR-like endoplasmic reticulum kinase (PERK or eIF2aK3) that is activated by endoplasmic reticulum stress due to misfolded proteins in the organelle lumen; and (4) GCN2 homolog eIF2aK4 that is activated by starvation of certain amino acids.

Phosphorylated eIF2 ($eIF2^P$) sequesters the guanine nucleotide-exchange factor eIF2b, thereby abrogating its activity and lowering mRNA translation.

---

[66] Internal ribosome entry sites are RNA elements that mediate end-independent ribosomal recruitment to internal locations in mRNA. Initiation on some IRESs also requires RNA-binding IRES transacting factors (ITAF) that may stabilize the optimal IRES conformation.

Yet, protein synthesis from certain mRNAs with at least 2 uORFs of appropriate type and position can be stimulated, such is the case of transcription factors ATF4 and ATF5 once PERK as been activated [384].

In addition, mitogen-activated protein kinase-interacting kinases MNK1 or MNK2 (or MKNK1 and MKNK2; Vol. 4 – Chap. 5. Mitogen-Activated Protein Kinase Modules) phosphorylates eIF4e bound to eIF4g to which they can link [384].

The eIF4f cap-binding complex that mediates the initiation of mRNA translation is composed of eIF4e, which binds to mRNA cap, eIF4g, which indirectly links mRNA cap with the 43S pre-initiation complex, and eIF4a, which is a helicase necessary for initiation. In response to mitogens and growth factors, target of rapamycin releases eIF4e for linkage to eIF4f and promotes translation.

eIF4e-Binding proteins (4eBP1–4eBP3) can experience a hypophosphorylation that allows their binding to eIF4e to prevent eIF4e tethering to eIF4g [384]. On the other hand, hyperphosphorylation of 4eBPs on multiple sites, mainly by TOR, liberates eIF4e that can then connects to eIF4f.

Additionally, eIF6 in response to extracellular stimuli initiates the translational activity of 60S ribosomal subunit [385]. Translational control is implicated in malignancy. Both eIF4e downstream from the TOR pathway and eIF6 are upregulated in various cancer types.

5.5.3.2 Regulation by Sequence-Specific RNA-Binding Proteins

Sequence-specific RNA-binding proteins selectively regulate specific classes of mRNA. 3′ UTR-binding proteins commonly repress translation, as they form an inhibitory closed loop with 5′ cap-binding proteins such as eIF4e and an intermediate bridging protein [384].

On the other hand, polyadenylated sequence-binding protein (PABP) bound to the 3′ polyadenylated tail enhances initiation, possibly by tethering eIF4f. Hence, despite its dissociation from the 5′ cap, eIF4f remains available to promote further initiation on the same mRNA without recruitment from outside.

5.5.3.3 Regulation by MicroRNAs

MicroRNAs bind to the 3′ UTR and, via trinucleotide repeat-containing protein TNRC6a[67] repress mRNA translation and accelerate the rate of deadenylation-dependent mRNA degradation [384].

Argonaute proteins (Ago1–Ago4), or eukaryotic translation initiation factors (eIF2c1–eIF2c4) link to the miR–mRNA couple. Many other proteins

---

[67] A.k.a. CAG repeat-containing protein CAGH26, 182-kDa glycine–tryptophan protein GW182, GW1, and EMSY interactor protein.

participate in this association, such as ATP-dependent DEAD (Asp-Glu-Ala-Asp) box-containing helicase Ddx6 (or RCK) and TNRC6 isoforms (TNRC6a–TNRC6c) [384]. MicroRNAs act as adaptors that confer binding ability to mRNA of Argonaute. Moreover, tethering of any TNRC6 isoform can by-pass the requirement for miR and Argonaute in transcript silencing.

5.5.4 mRNA Surveillance Pathway

The mRNA surveillance pathway, or mRNA quality control, aims at checking the translation efficiency of spliced mRNAs with respect to non-spliced mRNAs. The mRNA surveillance pathway is constituted by *non-sense codon recognition* and then *non-sense-mediated mRNA decay*. Mechanisms of mRNA quality control control whether the mRNA stop codon and $3'$ end are appropriate and search for potential errors in splicing as well as mutations. Non-sense-mediated decay primes translation or degradation. It involves translation initiation factor eIF3 and mRNA decay factors. The mRNA surveillance pathway requires the exon–junction complex that is composed of eIF4a3.

The non-sense-mediated mRNA decay is triggered when translation terminates sufficiently upstream from the exon–junction complex, i.e., when ribosomes encounter a premature translation termination- or non-sense codon. Subsequent binding of human non-sense-mediated mRNA decay factor Upf1 to the exon–junction complex primes Upf1 phosphorylation. Phosphorylated Upf1 interacts with eIF3 and impedes eIF3-dependent conversion of the $40S^{Met}$–tRNAi^{Met}–mRNA complex to the translationally competent $80S^{Met}$–tRNAi^{Met}–mRNA initiation complex [386]. Translational repression that precedes mRNA degradation avoids production of defective proteins by ending the process in the case of incompletely processed mRNA or mRNA that lacks complete open reading frames.

5.5.5 Exon–Junction Complex

The exon–junction complex influences the first step of protein synthesis. Protein translation starts with the first passage of ribosomes on newly spliced mRNAs that are associated with the cap-binding proteic CBP20–CBP80 heterodimer. The target of rapamycin, which operates as a nutrient-, stress-, and energy-sensing checkpoint kinase, and its effector S6 protein kinase-1 intervene in assembling the translation initiation complex.

S6K1-Specific interactor SKAR connects to CBP80-bound mRNPs in a splicing-dependent manner and binds to the exon–junction complex [387]. Agent SKAR interacts with the exon–junction complex to recruit S6K1 enzyme. Then, SKAR and S6K1 increase the translation efficiency of spliced mRNA. The target of rapamycin thus stimulates protein synthesis, but is able to repress splicing-dependent translation efficiency. It indeed promotes or prevents phosphorylation of proteic substrates of CBP-bound mRNPs.

5.6 Elongation Cycle and Translation Termination

Translation, i.e., protein synthesis, proceeds in 4 phases: activation, initiation, elongation, and termination. Once the strongly controlled initiation phase terminates, the elongation stage starts. The elongation cycle comprises the sequential steps that add amino acids to the polypeptide chain.

The mRNA sequence serves as a template to guide the synthesis of a chain of amino acids that will form a protein. The interface between the 2 ribosomal subunits lodges RNAs. The template mRNA binds in a cleft of the 30S subunit, where its codons interact with the anticodons of tRNA.

The ribosome contains 3 RNA-binding sites: (1) A (acceptor) site that binds the incoming amino-acylated tRNA, i.e., tRNA charged with its amino acid; (2) P (peptidyl-tRNA) site that holds a peptidyl-tRNA (tRNA connected to elongating peptidic chain) and attaches it to the nascent polypeptide chain (peptidyl-transferase reaction); and (3) E (exit) site that receives a free (deacylated) tRNA after peptide-bond formation before its ejection from the ribosome.

The ribosome uses amino acyl-tRNA that matches the read codon on the mRNA over which the ribosome is running to add an amino acid to the polypeptide chain. The next loaded tRNA in line binds to the ribosome along with GTP and an elongation factor. During decoding, the next amino acid is indeed delivered in a ternary complex that contains eukaryotic translation elongation factor eEF1α1,[68] GTP, and amino acyl-tRNA. Decoding is followed by peptide-bond formation that elongates the polypeptide chain by one amino acid.

Elongation factors facilitate the translational elongation. Translation elongation factors are responsible for 2 main processes during protein synthesis on the ribosome: (1) selection and binding of the cognate amino acyl-tRNA to the A-site and (2) translocation of the peptidyl-tRNA from the A site to P site of the ribosome.

In addition to the initiation factor eIF2 (or eIF5b), elongation factors eEF1 and eEF2 are guanosine triphosphatases involved in translation. Factor eEF1α1 brings the cognate amino-acylated tRNA and promotes its binding to the A site of the ribosome. Elongation factor-2 catalyzes the shift of the

---

[68] Eukaryotic translation elongation factor eEF1α1 corresponds to prokaryotic translation elongation factor-Tu (EFTu). Alias TuFM designates the mitochondrial translation elongation factor. The eukaryotic elongation factor-1 consists of several subunits (eEF1α–eEF1δ). Active eEF1α^{GTP} acts as the carrier of amino acyl-tRNA into the ribosome. Factor eEF1α1 is a GTPase. The binding of amino acyl-tRNA stimulates the GTP hydrolysis by eEF1α1 that provokes a conformational change in EF1a, hence detaching eEF1α1^{GDP} from the ribosome, but leaving the amino acyl-tRNA attached at the A site. Inactivated eEF1α1 (eEF1α1^{GDP}) returns to its active state (eEF1α1^{GTP}) owing to eEF1βγδ (or translation elongation factor EFT) that functions as a guanine nucleotide-exchange factor (GEF) for eEF1α.

amino acyl-tRNA from the A to the P site of the ribosome, thereby freeing the A site for the next amino acyl-tRNA.

Successively, tRNAs bring amino acids identified by anticodons that correspond to codons read as the ribosome moves along the mRNA strand. During each cycle iteration of amino acid-chain elongation, codon–anticodon pairing actually allows entry of the correct tRNA into A site. The growing polypeptide chain bound to tRNA at P site is then transferred to tRNA with the new amino acid at A site. Afterward, tRNAs at site P and site A move to site E and site P, respectively. The next tRNA can occupy free site A and mRNA moves by one codon toward E site.

This cycle continues for each mRNA triplet that is read until the ribosome reaches the mRNA 3′ end. In other words, termination of translation happens when the A site of the ribosome reads the final stop codon (UAA, UAG, or UGA). No tRNA can match the stop codon that is recognized by a releasing factor. In eukaryotes, a single releasing factor eRF1 recognizes all 3 stop codon types. Therefore, when any stop codon reaches site A, protein synthesis stops and the release factor promotes hydrolysis of mature polypeptide chain from the P site.

5.7 Protein Synthesis Error Rectification

Mistakes that occur during protein synthesis are usually prevented, as ribosomes carefully monitor amino acids added into a growing polypeptide chain. After peptide-bond formation, a ribosome can detect codon–anticodon mismatches and terminate protein synthesis. Despite this check, errors (e.g., erroneous translation of AAU codon into amino acid lysine rather than asparagine) arise. Resulting misfolded, dysfunctional proteins must then be refolded or destroyed after translation.

Ubiquitin ligases support quality control-directed degradation. This process happens in several subcellular locations. In the endoplasmic reticulum, Ub ligases synovial apoptosis inhibitor Syvn1,[69] a transmembrane protein of the endoplasmic reticulum, and nuclear envelope[70] carry out the job.[71]

Messenger RNAs that lack stop codons (*non-stop mRNA*) can accidentally arise from errors in gene expression and encode aberrant *non-stop proteins*.

---

[69] A.k.a. synoviolin and HMGCoA reductase degradation protein HRD1 and membrane-associated ring finger-CH-containing protein MARCH6.

[70] A.k.a. membrane-associated ring finger (C_3HC_4) MARCH6 RING finger protein RNF176, TEB4, and Doa10 homolog.

[71] In yeast, San1 Ub ligase mediates destruction of misfolded nuclear proteins. In the cytoplasm, ubiquitination of numerous misfolded cytoplasmic proteins employs 2 strategies: chaperone-assisted ubiquitination by Ub ligase component N-recognin-1 (UbR1) and chaperone-dependent delivery to nuclear San1 ligase [388].

These non-stop proteins must be eliminated to avoid their deleterious accumulation.[72]

5.7.1 Quality Control by the Ribosome

Ribosomes contain 3 binding sites for transfer RNAs: amino acyl (A), peptidyl (P), and exit (E) sites (Sect. 5.6). For each amino acid, a specific codon on mRNA is paired with a complementary anticodon on a tRNA that, at its other end, carries the corresponding amino acid. Upon codon–anticodon pairing, the corresponding amino acid is added to the growing polypeptide chain by a peptide bond.

Protein synthesis relies on combined accuracy of 2 basic processes for acceptance of appropriate transfer RNA into site A before peptide-bond formation: (1) amino-acylation of tRNA with cognate amino acid by amino acyl-tRNA synthase and (2) selection of specific amino acyl-tRNA by ribosome facilitated by GTPase elongation factor eEF1a (or eEF1α).

However, even after peptide-bond formation, a quality-control mechanism ensures premature termination of erroneous growing polypeptide chains [390]. Upon translation error, ribosomes become much less efficient at further adding amino acids and nascent peptide chain with encoding errors can be prematurely released owing to a set of releasing factors. Furthermore, when site A loses its specificity, a mismatched codon–anticodon pair in site P produces an additional error by incorporating a new amino acid possibly due to ribosome conformational changes that regulate mRNA decoding and termination. Additional miscoding favors peptide release.

5.7.2 Quality Control of Messenger RNAs

Cytoplasmic mRNA that transfers genetic information from the nucleus to ribosomes is either translated, stored for later translation, or degraded. Messenger RNAs initially translated can be temporarily translationally repressed. All mRNAs are ultimately degraded.

Messenger ribonucleoproteins mediate the subcellular localization, translation, and decay of mRNAs. RNA-Processing bodies, the so-called P or GW bodies, correspond to cytoplasmic foci that sequester translationally silenced mRNPs and possibly degrade mRNAs. P-Bodies contain the GW182 protein[73] that is required for miR-mediated translational repression, activity of deadenylases, exonucleases, DCp1–DCp2 decapping complex, decay factors, and Argonautes, and premature termination codon-containing mRNAs to increase the efficiency of mRNA degradation [391].

---

[72] In Saccharomyces cerevisiae, RING-domain-type ubiquitin ligase (Ltn1, a yeast Listerin homolog) that is predominantly associated with ribosomes acts in the quality control of non-stop proteins [389]. They indeed marks nascent non-stop proteins with ubiquitin to prime their proteasomal degradation.

[73] A.k.a. trinucleotide repeat-containing protein TNRC6.

The non-sense-mediated mRNA decay process detects and eliminates abnormal mRNAs that harbor premature termination codons and can generate truncated proteins, which are often deleterious.[74] A large exon–junction protein complex upstream of exon–exon junctions during RNA splicing triggers non-sense-mediated mRNA decay.

Abnormal 3′ -untranslated regions immediately downstream from the stop codon impede normal remodeling required for normal translation termination. Core factors required for non-sense-mediated mRNA decay include homologs of regulator of non-sense-transcript Up-frameshift suppressors (UpF1–UpF3a/b) that are ATP-binding helicases.[75] Remodeling of mRNPs initiated by various elements downstream from the stop codon controls the gene expression quality [391].

MicroRNAs, after incorporation into RNA-induced silencing complex, guide the RNA interference machinery to target mRNAs by forming RNA duplexes, thereby causing repression of productive translation and mRNA decay. MicroRNA-mediated mRNA silencing involves base pairing of miRs with 3′ -untranslated regions of target mRNAs to ensure a protein expression that fits normal cellular function.

5.7.3 Peripheral Protein Quality Control

Chaperones also promotes the clearance of defective proteins from the plasma membrane that either escape previous components of protein quality control and reach their destination or are generated by environmental stresses. At the endoplasmic reticulum, the quality control facilitates folding and eliminates misfolded polypeptides using the ubiquitin–proteasome degradation axis.

The peripheral quality control of proteins uses a cytosolic ubiquitin conjugase UbCh5c and Ub ligase C-terminus heat shock cognate HSC70-interacting protein (CHIP) to synthesize Ub chains that confer a recognition signal for degradation, which primes the retrotranslocation of abnormal proteins from the plasma membrane to lysosome for destruction [392]. On the other hand, the cytosolic chaperones heat shock proteins HSP70 and HSP90, the HSP40 family of folding chaperones (e.g., heat shock cognate HSC70 cochaperones DnaJa1, DnaJb2, and DnaJc7), the coupling factor HSP70–HSP90 complex-organizing protein (HOP), and BCL2-associated anthanogene BAG1 contribute to protein stabilization at the plasma membrane.

The balance between protein folding and degradation is ensured owing to the adenosine triphosphatase cycles of HSP70 and HSP90 [393]. Both HSP70

---

[74] Most normal mRNAs contain a stop codon at the end of the coding region, but abnormal mRNAs can have a stop codon upstream from the normal stop codon, and thus are recognized by the non-sense-mediated mRNA decay machinery.

[75] A.k.a. regulator of non-sense transcripts (RENT1–RENT3a/b) and nonsense mRNA reducing factor (NORF1–NORF3a/b).

and HSP90 cycle between active and inactive forms under the control of regulatory ATP-loading and ATPase-activating cochaperones. Agent BAG1 operates as a nucleotide-exchange factor (ATPase activator) for HSP70 and HSP40.

A partial overlap exists between constituents of the quality control at the plasma membrane and endoplasmic reticulum. The peripheral (plasma membrane) and endoplasmic reticulum quality control work with similar surveillance agents, but different degradative axes [392]. The cytosolic components of the peripheral quality control, UbcH5, CHIP, HSP70, and HSP90 behave like their homologs in the lumen of the endoplasmic reticulum that also foster protein folding, export, and degradation at the endoplasmic reticulum membrane during protein synthesis. On the other hand, protein degradation results from chaperone-mediated translocation to the proteasome from the endoplasmic reticulum, whereas it is associated with the vesicular delivery to lysosomes from the plasma membrane.

5.8 Protein Synthesis Control by External Cues

Extracellular signals can regulate protein translation within specific cellular subregions. The transmembrane receptor Deleted in colorectal cancer (DCC) forms a binding complex that contains multiple translation components, such as eukaryotic initiation factors, ribosomal large and small subunits, and monosomes (single ribosomes combined with mRNAs that can relocalize from polysomes to monosomes upon transient disaggregation of polysomes) [394].[76]

5.9 Protein Interactions

Protein dynamics are characterized by the magnitude, direction, and time scale of structural fluctuations in a given chemical and physical context (concentration, temperature, pressure, energy, etc.). The cell metabolism and homeostasis are based on interacting proteins. Colocalized proteins interact with other proteins by binding directly to and/or modifying another protein. Proteins can form transient functional complexes, each component having a given relative position and orientation. The fraction f of protein B bound to protein A is assessed by the following equation:

$$f = \kappa_a[A]/(1 + \kappa_a[A]) = ([A]/\kappa_d)/(1 + [A]/\kappa_d), \qquad (5.1)$$

where $[A]$ is the concentration of A (the activity of A), κ_a and κ_d the association and dissociation constants for the binding of A to B (κ_a equals $1/\kappa_d$).

---

[76] In neuronal axons and dendrites, the extracellular ligand netrin promote the translation initiated by Deleted in colorectal cancer protein.

Signals are transduced owing to interactions between signaling mediators. Proteins interact either briefly to modify proteins or for a long time to carry cargos. Proteic complexes are characterized by conformational change with respect to free proteins.

An enzyme is a protein with a flexible structure that can only interact with its specific substrate. Most often, the substrate concentration is low and the environment is characterized by high protein concentration. Enzyme structure favors atomic vibrations that produce or break bonds. Enzymes accelerate the rate of a chemical reaction, at they bridge reacting species, generating enzyme–substrate and enzyme–intermediate complexes. Target residues usually adjust to permit binding of the specific substrate. Enzymes can distort the substrate that adopts a conformation with increased reactivity.

Various quality control mechanisms, such as unfolded protein and heat shock responses, enable proteins to adequately fold and to assemble into functional complexes, but avoid misfolding and formation of uncontrolled, nonspecific, harmful proteic aggregates. However, many interactions that stabilize interproteic interfaces and promote the formation of functional complexes, such as hydrophobic and electrostatic interactions, can cause abnormal intermolecular association. Nonetheless, disulfide bonds and salt bridges in proximity to the interface of homodimers allow stabilizing of aggregation-prone interfaces into their functional conformations to prevent uncontrolled assembly [395]. In addition, charged residues can disrupt hydrophobic patterns at interface and protect aggregation-prone segments.

5.10 Protein Folding

Protein homeostasis relies on the maintenance of functional proteins as well as removal of destabilized and misfolded proteins from synthesis sites such as the endoplasmic reticulum (Sect. 5.11) to the plasma membrane by the so-called *peripheral protein quality control*. Aberrant polypeptides result from errors in transcription, translation, and/or folding, mainly due to genetic mutations and environmental stresses.

The sequence of amino acids in a protein defines its primary structure. The initial linear chain of amino acids bears an intermediate secondary structure (α-helix and β-sheet) built by hydrogen bonds before folding. Protein folds into a three-dimensional (tertiary) structure with functional domains.

Amino acids with given features (water attracting or repelling, i.e., hydrophylic or hydrophobic, acidic or basic, electrically charged or polar uncharged, etc.) interact with each other and their surroundings to generate a 3D structure that depends on environmental factors. In less than 10 ms, the motion of the water network is changed due to influence of folding protein and the protein is simultaneously restructured [396]. After this initial phase, a slower stage ($\sim 900\,\text{ms}$) occurs during which the protein folds to its final structure. Folding that is assisted by chaperones is a co-translational process.

Most folded proteins have a hydrophobic core and charged side chains exposed to the neighboring water molecules.

Both single chain polypeptides and complexes undergo covalent and non-covalent assembly into a correct 3D structure. Regions of a single-domain protein approach each other for protein folding. During formation of proteic complexes, binding occurs by docking of rigidly folded monomers or in many cases concomitantly with folding from unfolded monomers to a folded complex.

Small, single-domain proteins represent less than 30% of proteins. Proteins actually often possess several domains that interact during protein folding. Between-domain interactions in the native conformation determine kinetics of cooperative folding [397]. Chain connectivity and distribution of between-domain interactions govern the structural features of folding intermediates.

Disordered protein that does not adopt any definite 3D structure can be characterized by rapid turnover, fast or specific binding kinetics,[77] and functional promiscuity. A disordered protein can function, especially in between-protein interactions, when only a high-affinity, small motif of the protein is specifically recognized. On the other hand, a defined 3D structure can be involved in protein binding with low energy when protein folding and binding energy patterns are coupled [398].

In cells, protein homeostasis relies on maintenance of protein conformation, refolding of misfolded proteins, and degradation of damaged proteins using regulators of protein synthesis and degradation. Molecular chaperones participate in protein folding and prevention of protein aggregation of misfolded proteins to avoid disease development. In particular, prolonged binding of proteins to chaperone HSP70 primes the recruitment of Ub conjugases and Ub ligases that launch degradation.

The endoplasmic reticulum contains mRNAs. These mRNAs encode secretory or membrane proteins. Only correctly folded proteins are allowed to travel via secretory vesicles for eventual release at the cell surface. The endoplasmic reticulum not only serves as a synthesis site for secretory and membrane proteins, but also as a binding site for many other mRNAs that are translated on ribosomes.

Some inherited diseases are characterized by the spontaneous polymerization and toxic accumulation of a misfolded, genetic variant. Instead of a soluble itinerant protein that can be released from the endoplasmic reticulum to travel into the secretory route to the cell surface, insoluble polymers can undergo spatial separation into domains of the endoplasmic reticulum that can be selectively eliminated by autophagy (Vol. 2 – Chap. 4. Cell Survival and Death). Autophagy of subcellular compartments with aberrant molecules can then eliminate the toxic effects of misfolded proteins. Endoplasmic reticulum-associated degradation (ERAD) is able to destroy molecules that cannot get their correct structure.

---

[77] Time scale of mobility of enzyme active sites ranges from microsecond to second.

On the other hand, deficiency in protein-folding capacity of the endoplasmic reticulum and degradation of misfolded proteins lead to endoplasmic reticulum stress. Accumulation of misfolded proteins triggers the *unfolded protein response.*

5.11 Unfolded Protein Response

The unfolded protein response results from 3 main signaling mediators: (1) eukaryotic translation initiation factor-2α kinase-3 (eIF2αK3);[78] (2) 2 inositol-requiring enzyme-1 homologs (IRE1α and IRE1β) that phosphorylate eIF2; and (3) activating transcription factor-6 (ATF6).

Activated eIF2αK3 during endoplasmic reticulum stress enables phosphorylation of eIF2α (Ser51) and subsequently causes global translational attenuation. Transmembrane IRE1 is an endoplasmic reticulum bifunctional kinase and endoribonuclease that senses misfolded protein accumulation in the endoplasmic reticulum lumen and subsequently activates the transcription factor X-box-binding protein XBP1 in the cytosol [399]. Enzyme IRE1 excises an intron from XBP1 mRNA to give rise to highly active transcription factor XBP1$_S$ [400]. It upregulates gene expression of endoplasmic reticulum chaperones as well as components of the endoplasmic reticulum-associated degradation pathway. This unconventional mRNA splicing event initiates the unfolded protein response to relieve endoplasmic reticulum stress. In response to endoplasmic reticulum stress, IRE1 forms clusters. Oligomers IRE1 recruit unspliced XBP1 mRNA and serve as stress signaling sites.

Numerous endoplasmic reticulum stresses activate the *endoplasmic reticulum overload response* (EROR), i.e., transcription factor nuclear factor-κB to provoke transcription of pro-inflammatory and immune response genes, in addition to signaling via glucose-regulated proteins [401] (Sect. 4.5). Moreover, the endoplasmic reticulum generates 2 second messengers in response to stress: calcium and reactive oxygen intermediates. The latter are produced by several oxidases of the endoplasmic reticulum membrane, such as cyclooxygenases, lipoxygenases, and cytochrome-P450.

Responses to endoplasmic reticulum stress comprise not only unfolded protein (UPR) and endoplasmic reticulum overload (EOR) responses, but also eIF2α phosphorylation.[79] Endoplasmic reticulum stress indeed causes a translational control, especially at the initiation stage, but also via phosphorylation

---

[78] A.k.a. protein kinase-R (PKR)-related endoplasmic reticulum kinase (PERK), PKR being also called eIF2αK2.

[79] Four specific eIF2α kinases have been identified in the mammalian genome: (1) hemin-regulated inhibitor of protein synthesis in reticulocytes; (2) eukaryotic translation initiation factor-2αkinase-4 (eIF2αK4), also called general control of amino acid synthesis kinase GCN2, that is activated upon amino acid deprivation to launch the production of GCN4; (3) ubiquitous, interferon-inducible, dsRNA-activated eIF2αK2, or protein kinase-R (PKR), that localizes to rough

of eIF2 heterotrimer. Increased eIF2α phosphorylation impedes protein synthesis. Activation of eIF2αK3, and possibly eIF2αK2, results from stress in the endoplasmic reticulum [401].

The regulatory subunits of phosphatidylinositol 3-kinase P85α (PI3KR1) and P85β (PI3KR2) form heterodimers. Insulin targets its receptor and prevents PI3KR1–PI3KR2 heterodimerization. In the absence of heterodimerization, P85 monomer interacts with the spliced form of X-box binding protein XBP1$_S$ and increases its nuclear translocation for the resolution of endoplasmic reticulum stress, as XBP1 is one of the main regulators of the unfolded protein response [402].

5.12 Post-Translational Protein Modifications

Protein synthesis include various steps — transcription, nuclear export, and translation — that involve many regulators and cofactors, as well as RNA binding proteins. Genes sometimes have many promoters and, thus, different regulations of their expression. Due to alternative splicing and RNA editing, a single gene can give rise to a set of alternative proteins. Post-translational protein modifications yield additional sources of protein variation and govern protein behavior.

Post-translational modifications of proteins enable the modulation of protein function, location, and turnover. Conjugation can occur at many sites of a given protein. In many cases, it forms chains of varying lengths. Conjugations often modulate each other.

Post-translational modifications relies on attachement of small chemical groups, such as acetyl, methyl, nitric oxide, and phosphate groups, as well as entire proteins (e.g., ubiquitins and ubiquitin-like proteins), glucids, and lipids (Tables 5.14 and 5.15).

5.12.1 Attachement of Small Chemical Groups

Post-translational modifications include, in addition to the widely used, studied, and modeled phosphorylation, acetylation, hydroxylation, methylation, nitrosylation, and polyglutamylation, among others.

5.12.1.1 Phosphorylation

Reversible phosphorylation was the first discovered post-translational protein modification. Kinases and phosphatases attach and remove a phosphate group to target proteins, respectively, A protein can have many phosphorylation sites. Each phosphorylation site can be targeted by several kinases that either activate or inactivate the protein function.

---

endoplasmic reticulum membrane and protects cells against viral infections; and (4) eIF2αK3, or PKR-related endoplasmic reticulum kinase (PERK).

Table 5.14. Post-translational modifications of proteins (**Part 1**).

Type	Added agent
Acylation	Amide (carbonyl group [R-C=O]; N-acylation)
	Ester (R-CO$_2$-R$'$; O-acylation)
	Thioester (C-S-CO-C; S-acylation)
Acetylation	Acetyl group (COCH$_3$)
Amidation	Amide (R$_1$(CO)NR$_2$R$_3$)
Arginylation	tRNA-mediated addition of Arg
Biotinylation	Biotin (vitamin-H or -B7)
Carbamylation	Conversion of lysine to homocitrulline
γ-carboxylation	Carboxyglutamic acid
Citrullination	Conversion of arginine into citrulline
Deamidation	Conversion of glutamine to glutamic acid
	Conversion of asparagine to aspartic acid
Formylation	Formyl group (CHO)
Glutamylation	Glutamic acid
Glycosylation	Glycosyl
Glycylation	Glycine
Glypiation	Glycosylphosphatidylinositol anchor
Heme attachment	Heme
Hydroxylation	Hydroxyl group (OH)
Iodination	Iodine
Isgylation	Interferon-stimulated gene product ISG15
Isoprenylation	Farnesol, geranylgeraniol
Lipoylation	Lipoate

Before phosphorylating proteins, protein Tyr kinases must first be activated by autophosphorylation of their own specific tyrosine residues. Activation involves the displacement of an amino acid loop, which initially blocks access to the active site, but moves away upon autophosphorylation.

Reversible phosphorylation yields not only a on–off switch, but also precise tuning of transactivating potential of transcription factors, such as P53 (Vol. 4 – Chap. 9. Other Major Signaling Mediators) and nuclear factor of activated T cells. Multisite phosphorylation actually enables signal integration, as several agents can operate with a single factor [403].

5.12.1.2 Acetylation

Acetylation is the post-translational addition of acetyl groups to given residues of proteins. In particular, lysine acetylation is the transfer of acetyl groups from acetyl coenzyme-A as acetyl group donor onto lysine residues, which is associated with charge neutralization, and influences the properties of proteins. Acetylation also influences between-protein interactions.

In the nucleus, acetylation of lysine residues of histones and nuclear transcription regulators contributes to the control of gene transcription. Histone

Table 5.15. Post-translational modifications of proteins (**Part 2**).

Type	Added agent
Methylation	Methyl group (CH_3)
Myristoylation	Myristoyl group ($CO(CH_2)_{12}CH_3$)
Neddylation	NEDD8
Nitrosylation	Nitrosyl group
Oxidation	Loss of electrons
Palmitoylation	Fatty acid ($CO(CH_2)_{14}CH_3$)
Prenylation	Farnesyl, geranylgeranyl
Pegylation	Polyethylene-glycol chain
Inositidation	Phosphatidylinositol
Phosphopantetheinylation	Phosphopantetheinyl
Phosphorylation	Phosphate group (P)
Polyadpribosylation	PolyADPribose (^{ADP}ribose)
Polysialylation	Polysialic acid
Racemization	Proline (targeted by prolyl isomerase)
Sulfation	Sulfate group
Selenoylation	Selenium
Sumoylation	Small ubiquitin-related modifier (SUMo)
Thiolation	Sulphur
S-thiolation	Protein SH groups form mixed disulphides with low-molecular-weight thiols (e.g., glutathione)
Ubiquitination	Ubiquitin
Urmylation	Urm1

acetylation weakens histone–DNA interaction and, hence, contributes to the activation of gene transcription. In addition, acetylation of transcription factors, such as P53 and E2F1, increases binding to DNA. Acetylation of SMAD7 by the histone acetyltransferase P300 competes with ubiquitination on the same lysine residue, hence avoiding degradation.

Acetylation enables the control of the activity of metabolic enzymes, therefore the response to changes in metabolic demands. Target metabolic enzymes include those involved in glycolysis, gluconeogenesis, fatty acid and glycogen metabolism, and the tricarboxylic acid and urea cycles.

The retained N-terminal methionine residue of a nascent protein is often acetylated. The removal of N-terminal Met by Met-aminopeptidases frequently leads to N-terminal acetylation of resulting N-terminal alanine, cysteine, serine, threonine, and valine residues. More than 80% of human proteins are N-terminally acetylated. N-terminally acetylated residue may act as a degron.

5.12.1.3 Hydroxylation

Hydroxylation, i.e., addition of hydroxyl groups, of proline residues, happens in secreted proteins and components the cellular oxygen sensor. Hydroxylation of proline residues is catalyzed by prolyl hydroxylases. Hydroxylated hypoxia-inducible factor-α is recognized by von Hippel-Lindau ubiquitin ligase.

5.12.1.4 Methylation

In addition to reversible methylation of nucleic bases of DNA,[80] various amino acids of proteins can undergo methylation. Many classes of protein carboxyl methyltransferases exist according to the methyl acceptor.

Methyltransferases act in transcriptional repression, as they promote the formation of heterochromatin by methylating tails of histone proteins.[81] Histone lysine methyltransferases can cause gene silencing.

Methionine methyltransferase has 2 substrates: ^{S}adenosyl methionine, a common cosubstrate involved in methyl group transfers, and ^{L}methionine, to produce ^{S}adenosylhomocysteine and ^{S}methyl ^{L}methionine. 5-Methyltetrahydrofolate-homocysteine methyltransferase is also called methionine synthase, as it produces methionine from homocysteine. Glycine methyltransferase catalyzes the chemical reaction from ^{S}methyl ^{L}methionine and glycine to ^{S}adenosyl ^{L}homocysteine and sarcosine.

^{L}Isoaspartate (^{D}aspartate) methyltransferase (or protein carboxyl methyltransferase PCMT1) catalyzes the transfer of a methyl group from ^{S}adenosyl ^{L}methionine to the free carboxyl groups of ^{D}aspartyl and ^{L}isoaspartyl residues. It then converts abnormal ^{L}isoaspartyl and ^{D}aspartyl residues to normal ^{L}aspartyl residues to maintain proper protein conformation.

Leucine carboxymethyltransferases methylate the carboxyl group of leucine residues. ^{S}adenosyl ^{L}methionine-dependent arginine ^{N}methyltransferases modulate intra- or intermolecular interactions of target proteins and regulate their function. Leucine carboxyl methyltransferase LCMT1 causes reversible carboxy methylation of C-terminal leucine of catalytic subunit of protein Ser/Thr phosphatase PP2a (Vol. 4 – Chap. 7. Cytosolic Protein Phosphatases). Methyltransferase LCMT2 is involved in post-translational modification of phenylalanine tRNA.

Arginine methylation is a post-translational modifications caused by arginine methylase used in signaling pathways. Three main forms of methylarginine have been identified: N^G-monomethylarginine (MMA) and $N^G N^G$ asymmetric (aDMA) and symmetric (sDMA) dimethylarginines. Arginine methylation modifies transcription factors and histones (transcriptional phase), as well as splicing and ribosomal proteins that interact with RNAs (translational phase) [404]. The majority of nuclear asymmetric dimethylarginine residues is found in heterogeneous nuclear ribonucleoproteins that operate

---

[80] DNA methylation on cytosine residues regulates the gene expression. Methylation can also serve to protect DNA from enzymatic cleavage.

[81] Distinct histone acetyl- and methyl-transferases modify histone tails at different positions and yield a combinatorial interplay between these modification types.

in pre-mRNA processing and nucleocytoplasmic RNA transport. RNA binding proteins that are methylated on arginine residues are monomethylated or asymmetrically dimethylated rather than symmetrically dimethylated. However, certain spliceosomal small nuclear ribonucleoproteins are symmetrically dimethylated. Myelin basic protein that belongs to the first group of detected arginine-methylated proteins contains monomethylarginine and symmetrically dimethylated arginine residues.

The family of protein arginine methyltransferases includes multiple types (PRMT1–PRMT5) [404]. These methyltransferases have been classified as type-1 or -2 PRMT enzymes whether, after monomethylating arginine, further dimethylation is asymmetrical or symmetrical, respectively. Most PRMTs are type-1 PRMTs, whereas PRMT5[82] is a type-2 Arg methyltransferase. Arginine methyltransferases are involved in transcription, nucleocytoplasmic transport, protein sorting, and cell signaling. They interact with interferon receptors, Janus kinases, transcription factors STAT, and transcriptional coactivators of nuclear hormone receptors that are members of the P160 steroid receptor coactivator (SRC) family (SRC1–SRC3).

The activity of the transcription factor P53 that triggers either cell-cycle arrest or apoptosis is regulated by numerous post-translational modifications[83] in addition to accessory proteins and cofactor binding. Transcription factor P53 bears lysine methylation that can be reversed by demethylation. Methyl lysine provides docking sites for effector proteins. Methylation can also serve to prevent alternate post-translational modifications on the same lysine residue. Lysine can be mono-, di-, or trimethylated. Methylation level is correlated with distinct genomic locations and functions [405]. Protein arginine methyltransferase PRMT5 targets P53 that then forms a defense complex with stress-responsive activator of P300 (STRAP)[84] and junction-mediating and regulatory protein (JMY) in response to DNA damage to promote cell cycle arrest rather than death [406]. Transcription cofactor JMY of P53 binds to the transcription coactivator and histone acetyltransferases cAMP responsive element-binding (CREB) protein-binding protein (CBP) and P300 to activate P53-dependent transcription. Upon DNA damage, P53 arginine methylation stimulates the CDKN1A gene to prime G1 arrest. Furthermore, Arg methylation of P53 regulates its activity by modulating its subcellular location (predominately nuclear or cytoplasmic) and oligomerization. (PRMT5 decreases formation of P53 dimers and tetramers.)

---

[82] PRMT5 is also called Janus kinase-binding protein-1.

[83] Phosphorylation of tumor suppressor P53 by related kinases ATM (ataxia-telangiectasia mutated) and ATR (ataxia-telangiectasia and Rad3-related) is a response to DNA damage.

[84] A.k.a. Ser/Thr kinase receptor-associated protein.

5.12.1.5 Nitrosylation

S-Nitrosylation is the covalent addition of a nitric oxide group to a cysteine thiol.[85] Hence, this post-translational modification transmits nitric oxide-based signaling. Nitric oxide signaling is often achieved by the binding of nitric oxide synthase to target proteins, either directly or via scaffold proteins, such as C-terminal PDZ domain-containing ligand of NOS1 (CaPON) and Disc large homolog DLg4. In general, S-nitrosylation operates as an allosteric effector.[86]

S-Nitrosylation is used in signal transduction cascades. Hemoglobin auto-S-nitrosylates, i.e., transfers a nitric oxide group from heme iron (Fe–NO) to a Cys residue in its oxygenated R-state. The transition from R-state to T-state coupled to deoxygenation promotes Hb^{SNO} binding to and transnitrosylation of the Cl^-–HCO_3^- exchanger SLC4a1.[87] The allosteric regulation of hemoglobin permits to propagate a vasodilatory signal via transfer of nitric oxide groups [407].

Whereas other types of post-translational modifications, such as phosphorylation and ubiquitination, are precisely regulated by many hundreds of kinases and ubiquitine ligases, respectively, only 3 isoforms of nitric oxide synthases exist (Vol. 4 – Chap. 9. Other Major Signaling Mediators). Binding of nitric oxide synthases to its substrates provokes nitrosylation. Yet, the amount of nitrosylated molecules outnumbers that of NOS binding partners. S-Nitrosylation is achieved not only by nitric oxide synthases, but also numerous types of nitrosylases such as nuclear nitrosylase S-nitrosylated glyceraldehyde 3-phosphate dehydrogenase (GAPDH) [408].[88] Therefore, nuclear GAPDH not only activates the acetyltransferases P300 and CBP, but also regulates the activity of deacetylases.

In fact S-nitrosylated proteins can function as nitrosylases after being transnitrosylated. During cysteine-to-cysteine transnitrosylation, a nitric oxide group is transferred between S-nitrosylated proteins ($protein^{SNO}$).[89] Between-protein transnitrosylation is a mode of allosteric regulation. Consequently, the huge number of nitrosylases and substrates reach the same order of magnitude than those of enzymes and targets involved in other major types of post-translational modifications.

---

[85] The nitric oxide group interacts with protein thiols with high reactivity.

[86] The allosteric regulation relies on the binding of an effector at an allosteric site that differs from the active site to modify the function of the substrate.

[87] A.k.a. plasmalemmal anion exchanger AE1.

[88] After receptor-mediated activation of NOS, GAPDH is transnitrosylated and binds to Seven in absentia homolog SIAH1 ubiquitin ligase and then cotranslocates with SIAH1 to the nucleus. In the nucleus, $GAPDH^{SNO}$ transnitrosylates various substrates, such as histone deacetylases HDAC2 and SIRT1 as well as DNA-dependent protein kinase. Ligase SIAH1 primes the degradation of nuclear proteins.

[89] $Caspase-3^{SNO}$ binds and transnitrosylates X-linked inhibitor of apoptosis (XIAP) Ub ligase [407]. Mitochondrial $caspase-3^{SNO}$ transnitrosylates also thioredoxin.

5.12.1.6 Polyglutamylation

Reversible polyglutamylation adds glutamate side chains of variable lengths (from 1 to about 20 glutamyl units) at specific sites on proteins. Polyglutamylation of α- and β-tubulins enable the discrimination among microtubule species. Tubulin polyglutamylases are members of the tubulin Tyr ligase-like protein family [409]. Other substrates of polyglutamylation include nucleosome-assembly proteins NAP1 and NAP2 and chromatin-associated proteins. Both tubulins and NAPs also undergo *polyglycylation*.

5.12.1.7 Sulfation

Tyrosine sulfation is a post-translational modification in which a sulfate group is added to a Tyr residue of a protein. A sulfate group is added either via oxygen (O-sulfation) or nitrogen (N-sulfation).

Sulfation contributes to the strengthening of between-protein interactions. Tyrosylprotein sulfotransferase (TPST) catalyzes the transfer of sulfate from 3'-phosphoadenosine 5'-phosphosulfate (PAPS) to the hydroxyl group of Tyr residues to form a Tyr sulfate ester [410].

Two types of tyrosylprotein sulfotransferases have been identified (TPST1–TPST2). In the trans-Golgi network, protein Tyr sulfation is catalyzed by TPST1 enzyme. This post-translational modification is restricted to proteins of the secretory pathway. In humans, proteins that experience Tyr sulfation include adhesion molecules, G-protein-coupled receptors, coagulation factors, serine protease inhibitors, extracellular matrix proteins, and hormones. Sulfation can also refer to the generation of sulfated glycosaminoglycans.

5.12.2 Attachment of Ubiquitins and Ubiquitin-like Proteins

Attachment of ubiquitins and ubiquitin-like proteins are called ubiquitination, isgylation (conjugation of interferon-stimulated gene product ISG15), neddylation (conjugation of neural precursor cell expressed, developmentally downregulated protein NEDD8), sumoylation (conjugation of small ubiquitin-related modifiers [SUMo]), and urmylation (conjugation of ubiquitin-related modifier-1 homolog [Urm1]), among others.

5.12.2.1 Ubiquitination

Reversible ubiquitination (or ubiquitylation) tags proteins with ubiquitin. Ubiquitin is a small protein modifier that regulates, in addition to protein stability, many biological processes, such as gene transcription, cell cycle progression, DNA repair, apoptosis, receptor endocytosis, and immune response. Ubiquitin has proteolytic as well as numerous non-proteolytic functions. Ubiquitination is used in protein transport and signal transduction (Table 5.16).

Table 5.16. Ubiquitination functions. Protein ubiquitination is carried out by enzyme E1 (activase), -E2 (conjugase), and -E3 for degradation or other cellular fate. In humans, about 650 ubiquitin ligases have been detected.

Protein degradation
Protein transport
Protein–protein interaction
Protein activation
DNA repair
Gene replication and transcription

Ubiquitination usually requires the successive action of 3 enzymes: Ubiquitin activase (E1) that needs ATP to attach and activate ubiquitin that is subsequently recruited by ubiquitin conjugase (E2) that transfers the activated ubiquitin to a substrate bound to an ubiquitin–protein ligase (E3). Ubiquitin ligase binds to both an ubiquitin conjugase and substrate to catalyze ubiquitin transfer from the Ub conjugase to substrate.

Two major categories of ubiquitin ligases exist: RING and RING-like ligases[90] The HECT subclass of Ub ligases includes members of the neuronal precursor cell expressed developmentally downregulated NEDD4 family (9 identified members), HERC family (6 known members), and others (13 members) [413].[91] Ubiquitin ligases of the RING subclass act as scaffolds to bring ubiquitin conjugases near the substrates and facilitate the transfer of ubiquitin to targets. Ubiquitin ligases of the HECT subclass can directly transfer ubiquitin to the substrate. Enzymes of the HECT subclass determine ubiquitination specificity for degradation by the 26S proteasome or transfer in the case of receptors, channels, and transporters [413]. Ligases of the HECT subclass can be activated by phosphorylation.[92] Activity of HECT can be

---

[90] Ligases of the RING subclass have a RING domain that uses Cys and His residues and a pair of zinc ions. Ligases of the RING-like subclass possess an U box domain that achieves the same goal without any metal ions.

[91] Mammalian NEDD4 family members comprise NEDD4 (a.k.a NEDD4-1), NEDD4-like protein NEDD4L (or NEDD4-2), Itch, WW domain-containing ubiquitin ligases WWP1 and WWP2, SMAD ubiquitination regulatory factors SMURF1 and SMURF2, and NEDD4-like ubiquitin ligases NEDL1 (or HECW1) and NEDL2 (or HECW2). Members of the NEDD4 family participate in the regulation of cell growth and proliferation. Other HECT ligases include E6-associated protein, EDD (also called UBR5), HUWE1, among others.

[92] Ligase ITCH is stimulated by Jun N-terminal kinase JNK1 and, once bound to Notch1 receptor, inhibits Notch1. It also degrades JunB that regulates transcription of the interleukin-4 gene as well as FLIP. Conversely, several kinases, such as SGK1 and PKB1, phosphorylate NEDD4-like protein, thereby promoting the recruitment of 14-3-3 proteins. Once bound to and sequestered by 14-3-3, NEDD4L cannot bind to its target epithelial Na^+ channel [413].

regulated by auxiliary factor.[93] Enzymes of the HECT subclass can bind to phospholipids as well as arrestin-related adaptors,[94] annexins, proteins 14-3-3, and growth factor receptor-bound proteins GRB7, GRB10, and GRB14. Adaptors can facilitate substrate binding.[95]

Ubiquitin can be conjugated to target proteins either as monomers or chains. *Monoubiquitination*, when protein is tagged with a single ubiquitin, regulates DNA repair and receptor endocytosis. *Multimonoubiquitination* is defined by the association of a single ubiquitin at multiple sites. *Polyubiquitination*, i.e., protein tagging by a chain of ubiquitins, serves for protein degradation. Limited ubiquitination modifies the activity of some proteins and can be reversible. Whereas ubiquitin chains linked by their lysine residues at position 48 (Lys_{48}) tag proteins for proteasomal degradation, ubiquitination can use other lysine linkers (Lys_6, Lys_{11}, Lys_{27}, Lys_{29}, Lys_{33}, and Lys_{63}) of ubiquitin chains.[96] These atypical ubiquitin chains are homotypic (using the same ubiquitin Lys for conjugation), of mixed-linkage type (using several distinct Lys to connect ubiquitins), or heterologous (connecting ubiquitin with ubiquitin-like modifiers such as SUMo chains) [412].

In cells, protein homeostasis relies on protein synthesis and destruction. The ubiquitin–proteasome axis is responsible for most of protein degradation. It destroys proteins that are coupled to a polyubiquitin chain and contain degradation signals (*degrons*) into small peptides and recycles ubiquitin tags.[97] In particular, the ubiquitin–proteasome axis removes signaling regulators destined for elimination for a proper process (cell-cycle progression, transcriptional regulation, temporospatial control of signal transduction, and circadian rhythms), in addition to aberrant proteins with synthetic errors

---

[93] Ligase SMURF1 is regulated by casein kinase-2-interacting protein CKIP1 [413].

[94] Arrestin-related proteins regulate ubiquitination and endocytosis of plasmalemmal proteins.

[95] NEDD4 family-interacting proteins NDFIP1 and NDFIP2 (also called N4WBP5 and N4WBP5a, respectively) that are located in intracellular membranes can interact with NEDD4 ligases [413]. Adaptor NDFIP1 facilitates ITCH interaction with JunB. Adaptors NDFIP1 and NDFIP2 also interact with DMT1 that is responsible for iron uptake in enterocytes and release from hepatocytes for ubiquitination by WWP2. Conversely, binding protein N4BP1 competes for linking ITCH that cannot then interact with its substrate.

[96] Lys_{63}-linked ubiquitin chains intervene in the recruitment of repair machinery to sites of DNA damage. Lys_{29} and Lys_{33}-linked mixed chains operate in the regulation of AMP-activated protein kinase-related kinases, such as AMPK family member-5 and microtubule-affinity-regulating kinase kinases.

[97] Any degron, i.e., unstable protein motif for recognition and degradation) of signaling termination-tagged, misfolded, or misassembled proteins involves a specific E3-binding determinant, ubiquitin acceptor motif, and proteasomal degradation initiation site. Post-translational modifications (phosphorylation, hydroxylation, and proteolytic cleavage), as well as specific conformation or assembly state activate degrons [411].

or misfolding (protein quality control) [411]. In addition to tagging proteins for degradation by 26S proteasome, ubiquitination actually mediates various non-degradative functions. Once attached to a polyubiquitin chain of at least 4 ubiquitins, the substrate protein can bind to: (1) ubiquitin receptors of 19S regulatory complex of 26S proteasome or (2) adaptors that contain both polyubiquitin- and proteasome-binding domains and recruit substrates to the proteasome for cleavage into short peptides [411]. In addition, ubiquitination can serve to direct certain proteins (often membrane proteins) to lysosomes for proteolysis independently of proteasome.

Ubiquitination is a reversible process that is regulated by both ubiquitinating and deubiquitinating enzymes (DUb). Deubiquitinating enzymes cleave ubiquitin chains from ubiquitinated proteins. According to their structures, deubiquitinating enzymes can be classified into 5 families: (1) ubiquitin C-terminal hydrolases (UCH); (2) ubiquitin-specific proteases (USP); (3) ovarian tumor-related proteases (OTU), Machado-Joseph disease protein domain-containing proteases (MJD), and JAB1/PAB1/MPN-domain-containing metalloenzymes (JAMM). Deubiquitination involves specific adaptor proteins.

Deubiquitinase CYLD of the USP family controls microtubule assembly, calcium channel activity, cell migration, and cell cycle, among other functions [414]. Deubiquitinase A20 of the OTU family operates as both a DUb and a ligase. The OTU Deubiquitinase cellular zinc finger anti-NFκB (Cezanne) represses NFκB in the tumor-necrosis factor pathway.

Ubiquitination and deubiquitination regulate immune responses, such as innate immune responses to viruses and bacteria, proliferation and activation of lymphocytes, and maintenance of immunological tolerance. Deubiquitinase DUbA targets tumor-necrosis factor receptor-associated factor TRAF3, an adaptor that connects the antiviral effector TRAF family member-associated NFκB activator (TANK)-binding kinase TBK1 and inhibitor of nuclear factor-κB kinase IKKε to upstream signaling molecules [414]. Deubiquitinase A20 represses innate immune-receptor signaling mediated by Toll-like receptor TLR3 and retinoic acid-inducible gene RIG1 and prevents inflammation. Deubiquitinase CYLD regulates the proliferation and activation of B and T lymphocytes. It impedes activation of NFκB that regulates both innate and adaptive immune responses, but promotes thymocyte T-cell receptor signaling by deubiquitinating LCK kinase. Expression of DUbs can be induced in lymphocytes by cytokines, such as interleukins IL2 to IL5 and gmCSF.

Ubiquitin-like proteins (UbL) regulate interactions of proteins with other macromolecules, such as proteasome and chromatin. Ubiquitin-like proteins are involved in the regulation of gene transcription, signal transduction, autophagy, and cell-cycle control. They use related enzymes to, most often transiently, attach specific UbLs to proteins. Ubiquitin-like proteins include small ubiquitin-like modifier (SUMo), interferon-stimulated gene product ISG15, autophagy-related gene product AtG8, and neural precursor cell expressed, developmentally downregulated NEDD8. Ubiquitin-like modifiers SUMo2, SUMo3, and NEDD8 participate in chain formation.

5.12.2.2 Neddylation

Neddylation is a reversible post-translational protein modification carried out by the covalent attachment of *neural precursor cell expressed, developmentally downregulated protein* NEDD8 to target proteins similarly to ubiquitination [415]. Neddylation: (1) induces conformational change of substrates; (2) competes with other post-translational modifications;[98] and (3) provides a novel binding surface to recruit new partners.[99] Deregulated neddylation may be involved in neurodegenerative disorders and cancers.

Neural precursor cell expressed developmentally downregulated protein NEDD8 is attached to a lysine of proteic subtrates. The NEDD8 genes encode precursors that are cleaved into conjugatable proteins. After processing, NEDD8 is activated by an ATP-dependent catalysis based on activase E1^{NEDD8}, a heterodimer of amyloid β-precursor protein binding protein APPBP1 and ubiquitin-like modifier activase UbA3. Activated NEDD8 is then transferred by conjugase E2^{NEDD8} to ligase E3^{NEDD8} for specific conjugation of NEDD8 to its substrate. Unlike ubiquitination, which uses multiple E2 enzymes, a single Ub conjugase UbC12 is exclusively associated with NEDD8 [415].

Several neddylated proteins are either substrates or components of Ub ligases. Proteins of the cullin family are regulated by neddylation. Cullins act as scaffolds for the assembly of multisubunit ubiquitin ligases [415]. Cullin neddylation stimulates their ubiquitin activity. Tumor suppressors P53 and P73 are both neddylated and ubiquitinated by double minute DM2 that also self-neddylates. Upon stimulation, casitas B-lineage lymphoma protein (CBL) can neddylate and ubiquitinate epidermal growth factor receptor. Several ribosomal proteins can be modified by NEDD8. Protein neddylation and deneddylation cycles are required for correct cullin functioning. Deneddylation is mainly carried out by NEDD8 isopeptidase that is a subunit of COP9 signalosome CSn5. NEDD8 Protease NedP1 not only processes NEDD8 precursors, but also operates as a specific NEDD8 isopeptidase. Some proteases, such as ataxin-3, ubiquitin C-terminal hydrolases UCHL1 and -L3, USP21, 54-kDa Plasmodium falciparum ubiquitin C-terminal hydrolase (PFUCH54) have dual specificity for ubiquitin and NEDD8 [415].

5.12.2.3 Sumoylation

Small ubiquitin-related modifiers (SUMo1–SUMo4) couple to target proteins and reversibly changes inter- or intramolecular interactions of modified proteins. Sumoylation regulates the location, activity, and stability of substrates.

---

[98] Cullin neddylation prevents interaction with cullin-associated and neddylation-dissociated CAND1.

[99] Neddylated epidermal growth factor receptor recruits endocytosis proteins like monoubiquitination to downregulate EGFR signaling.

Attachment of SUMo to a protein involves SUMo (E1) activase (the AOS1–UbA2 heterodimer forms a SUMo–adenylate conjugate), SUMo (E2) conjugase UbC9, and usually one among several SUMo (E3) ligases that facilitates the transfer of SUMo to the substrate. SUMo Ligases bind UbC9 and SUMo, as well as their targets.

Many proteins undergo rapid cycles of SUMo attachment and deconjugation. SUMo-specific proteases (SenP1–SenP3 and SenP5–SenP7) remove SUMo from substrates, with distinct activity, targets, and locations. In particular, SenP3 and SenP5 preferentially deconjugate SUMo2 and SUMo3 from substrates. Sumoylation and desumoylation cycles sometimes occur only at certain times (e.g., at a certain stage of the cell cycle) and places (such as the nucleus).

5.12.2.4 Urmylation

Ubiquitin-related modifier-1 homolog (Urm1) is used to modify both proteins and RNAs [416]. It can be covalently conjugated to other proteins owing to E1-like Ub activase Uba4. Humans have a single Uba4-like E1 enzyme encoded by MOCS3 gene. The human genome contains at least 2 genes URM1 and MOCS2 that encode 2 ubiquitin-like proteins. Protein Urm1 also acts as a sulfur carrier in the thiolation of cytoplasmic transfer RNAs.

5.12.3 Attachment of Carbohydrates and Carbohydrates–Lipids

Carbohydrate-based modifications encompass glycosylation, acetylglucosamination, and polyADP-ribosylation, among others.

5.12.3.1 Glycosylation

Protein glycosylation regulates the final destination of proteins within or outside the cell. Glycosylation also determines half-life of proteins that travel through the body. Glycosylation tags proteins with linear and/or branched moieties.

Glycosylation is the attachment of glycans to proteins (as well as lipids and other organic substances), especially membrane and secreted proteins. Glycosylation is a co- and post-translational modification. The majority of proteins synthesized in the rough endoplasmic reticulum is glycosylated. Yet, glycosylation can also occur in the cytoplasm and nucleus.

Glycosylation is a site-specific enzymatic process (unlike glycation). Five types of glycosylations are produced: (1) *N-linked* glycosylation by attachment of glycans to a nitrogen (N) of asparagine or arginine of protein side-chains; (2) *O-linked* glycosylation in the Golgi body by binding of glycans to the oxygen (O) of serine, threonine, tyrosine, hydroxylysine, or hydroxyproline of protein side-chains;[100] (3) *P-linked* glycosylation that targets the phosphate of serine^P; (4) *C-linked* glycosylation via tie to a carbon of tryptophan of protein side-chain; and (5) *glypiation*, i.e., the addition of a glycosylphosphatidylinositol (GPI) anchor linking proteins to lipids via glycans.

Multiple proteins are post-translationally modified by a GPI anchor at their C-termini that enables the connection to the plasma membrane via lipid segments. Glycosylphosphatidylinositol-anchored proteins (gpiAP)[101] are mainly associated with plasmalemmal nanodomains (10–200 nm), the so-called membrane rafts,[102] that are enriched in sphingolipids and cholesterol. Proteins gpiAPs act as enzymes, receptors, regulators, adhesion molecules, and immunological mediators [417].

The GPI-anchored heparan sulfate proteoglycan glypicans regulate morphogens, such as Wnts and Hedgehogs (Vol. 3 – Chap. 10. Morphogen Receptors). Notum pectinacetylesterase homolog cleaves and releases glypicans and other gpiAPs from the plasma membrane. Released glypicans impede interaction between Wnt and its receptor. In addition, angiotensin converting enzyme not only cleaves angiotensin-1 and bradykinin, but also gpiAPs.

5.12.3.2 Acetylglucosamination

Protein acetylglucosamination, or O-Glc-N-acylation (not the classical N- and O-glycosylation of the secretory pathway), i.e., glycosylation with β^Nacetylglucosamine, is an ubiquitous protein modification that serves as a nutrient and stress sensor to modulate the activity of many nuclear and cytoplasmic proteins. Protein acetylglucosamination by attachment of O-linked monosaccharide β^Nacetylglucosamine (^{O}Glc^{N}Ac; β-O-linked 2-acetamido 2-deoxy ^{D}glycopyranose) is involved in the regulation of many cell processes, such as nuclear transport, translation and transcription, signal transduction, cytoskeletal reorganization, proteasomal degradation, and apoptosis.

The addition of a single O-linked ^{N}acetylglucosamine to Ser and Thr residues is catalyzed by O-linked ^{N}acetylglucosaminyltransferase[103] (OGT),

---

[100] Glycans can be also tethered to the oxygen of lipids such as ceramide.

[101] A glycosylphosphatidylinositol has a phosphatidylinositol backbone with ethanolamine phosphate, mannose, and glucosamine and at least a glycan branch. Precursors of gpiAPs synthesized in the endoplasmic reticulum have a GPI-attachment signal peptide. They are transported by vesicles from the endoplasmic reticulum to the Golgi body, where they are remodeled.

[102] Lipid-modified proteins of membrane rafts are either located in the outer leaflet such as gpiAPs or in the inner leaflet such as Src Tyr kinase.

[103] O-linked ^{N}acetylglucosaminyltransferase is ubiquitously expressed. Nonetheless, it is particularly abundant in the pancreas, adipose tissue, brain, and skeletal and cardiac muscles. This enzyme acts as a dimer. It is able to bind phosphatidylinositol (3,4,5)-trisphosphate for recruitment at the plasmalemma. It is phosphorylated (activated) by calcium–calmodulin-dependent protein kinase-4 in response to depolarization.

uridine diphospho (UDP) ^{N}acetylglucosamine:polypeptide β^Nacetylglucos-aminyl transferase (UDPNAc transferase). The cycling of ^{O}Glc^{N}Ac is enabled by its removal by β^Nacetylglucosaminidase (^{O}Glc^{N}Acase [OGA]). Different nucleoplasmic, cytosolic, and mitochondrial isoforms of O-linked ^{N}acetylglucosaminyltransferases exist. β^NAcetylglucosaminidase is encoded by a single gene, but has 2 splice variants. In addition, O-linked ^{N}acetyl-glucosaminyltransferases can form complexes with catalytic subunits of protein phosphatase-1.

Acetylglucosamination–deacetylglucosamination cycles most often target the serine and threonine residues at the same or proximal sites as that used by phosphorylation, although acetylglucosamination sites can be distant from phosphorylation sites. Consequently, these post-translational protein modifications are additional sources of crosstalks between pathways to regulate protein function and signal transduction. Attachment of ^{O}Glc^{N}Ac competes with that of phosphate and disturbs signaling, especially in cells of the immune system [419].

In hepatocytes, phosphorylation of cAMP-response-element-binding coactivator CRTC2 promotes its interaction with 14-3-3 proteins and subsequent cytosolic retention, whereas O-glycosylation at the same residues primes CRTC2 translocation into the nucleus to regulate the expression of genes of gluconeogenesis. ^{N}acetylglucosaminyltransferase particularly targets phosphatidylinositol 3-kinase, protein kinase-B, insulin receptor substrate-1, glycogen synthase, and endothelial nitric oxide synthase, as well as transcription factor P53, nuclear factor-κB, nuclear factor of activated T cells, etc. (Vol. 4 – Chap. 9. Other Major Signaling Mediators).

Unlike extracellular glycans, acetylglucosamination is done rapidly (at a rate comparable to that of phosphorylation) on regulatory proteins in response to metabolic and environmental signals. In some cases that impede phosphorylation such as glycogen synthase kinse-3 inhibition, acetylglucosamination can also decay. However, when acetylglucosamination increases, phosphorylation loci can change. Elevated acetylglucosamination causes lower phosphorylation at some sites and higher phosphorylation at others [420]. Interplay between acetylglucosamination and phosphorylation can result from diverse mechanisms: (1) competition for occupancy on same residues; (2) competition via hindrance by residence on proximal sites; (3) phosphorylation of ^{O}Glc^{N}Ac transferase or ^{O}Glc^{N}Acase; and (4) acetylglucosamination of kinases or phosphatases.

Acute protein acetylglucosamination corresponds to a stress response to numerous stress stimuli (e.g., glucose deprivation, heat shock, and oxidative stress) that can aim at enhancing cell survival. In the cardiovascular system, protein acetylglucosamination promotes cell protection, but also contributes to insulin resistance [418] (Table 5.17). Insulin stimulates the formation of the OGT–insulin receptor complex that phosphorylates, acetylglucosaminates, and increases the OGT activity. Hyperglycemia is associated with an increase in protein acetylglucosamination in vascular smooth muscle

Table 5.17. Knock-on and -off effects of acetylglucosamination in cardiomyocytes and vascular smooth muscle and endothelial cells (Source: [421]). O-linked attachment of the ^{N}acetylglucosamine to nuclear and cytosolic proteins is catalyzed by O-linked ^{N}acetylglucosaminyltransferase (OGT) and its removal by β^Nacetylglucosaminidase (^{O}Glc^{N}Acase [OGA]). This modification is similar to and can interact with phosphorylation.

Positive effects	Negative effects
Reduction of oxidative stress	Elevation of angiotensin-2 synthesis
Attenuation of Ca^{++} overload	Impairment of Ca^{++} signaling
Diminution of endoplasmic reticulum stress-induced cell death	Alteration of phosphorylation of mitochondrial proteins
	Potentiation of insulin resistance

cells as well as acetylglucosamination (inactivation) of nitric oxide synthase NOS3 in endothelial cells [421]. On the other hand, protein acetylglucosamination attenuates intimal hyperplasia after balloon-caused injury of rat carotid arteries. In cardiomyocytes, protein acetylglucosamination impairs their contractility, but acute acetylglucosamination is cardioprotective [421]. Therefore, protein acetylglucosamination can have both positive and negative repercussions on the function of cardiovascular cells according to the context (more than 600 oglcnacetylated proteins have been identified, such as transcription factors, signaling effectors, such as sarcoplasmic reticulum Ca^{++} ATPase, phospholipase-C, and protein kinase-C, as well as cytoskeletal components, including constituents of the contractile apparatus).

In vascular smooth muscle cells, endothelin-1 binds to ET_A receptor (predominant type in vascular smooth muscle cells; Vol. 3 – Chap. 7. G-Protein-Coupled Receptors) and heightens protein acetylglucosamination, thereby causing a vasoconstriction[104] via an augmented expression of guanine nucleotide-exchange factors RhoGEF1, -11, and -12 and activation of the RhoA–RoCK pathway [422] (Vol. 4 – Chap. 8. Guanosine Triphosphatases and their Regulators).[105] Proteins of the RhoA–RoCK pathway that undergo an acetylglucosamination include myosin light chains MLC1 and MLC2, Rho GDP-dissociation inhibitor, and RhoGEF11.

---

[104] Contraction of smooth muscle cells results mainly from the phosphorylation of myosin light chains.

[105] Endothelin-1 stimulates several signaling mediators that support Ca^{++} sensitization of smooth muscle cells, such as protein kinase-C, mitogen-activated protein kinases, and RhoA GTPase. It also activates transcriptional factors that increase the synthesis of enzymes and cytokines involved in the regulation of the vasomotor tone, inflammation, oxidative stress response, and wall repair and remodeling.

5.12.3.3 Polyadpribosylation

Polyadpribosylation occurs in response to DNA damage. PolyADPribose polymerases PARP1 and PARP2 actually catalyze ADP ribosylation of various factors for DNA repair after DNA-strand breakage.[106] In addition, PARP1 controls chromatin architecture, nucleosome removal, and transcription activation in response to corresponding stimuli [423]. Enzymes PARPs recruit base excision repair factors that remove damaged bases and fix single-strand breaks. Single-strand breaks that persist into S phase of the cell cycle produce replication fork collapse, requiring BRCA1- and BRCA2-mediated homology-dependent repair for resolution. PolyADPriboses interact not only with many repair factors, but also with checkpoint proteins to control cell division. Furthermore, PARP1, PARP2, PARP9, and PARP14 act as transcription cofactors, as they influence the formation and/or activity of various transcription-factor complexes, without necessarily synthesizing polyADPriboses. PolyADPriboses regulate gene expression by targeting insulators that limit heterochromatin–euchromatin borders and prevent inappropriate cross-activation of neighboring genes. PolyADPriboses inhibit DNA methyl transferase DNMT1 for chromatin insulation.

Polyadpribosylation also represents a post-translational modification of proteins. The polymerization of ^{ADP}ribose units from NAD$^+$ is catalyzed by polyadpribose polymerases[107] to create branched polyADPriboses [423]. PolyADPriboses can covalently link to acceptor proteins (mainly via glutamic residues). PolyADPriboses can be degraded either by polyADPribose glycohydrolase[108] or ADPribosyl hydrolase ARH3.

Polyadpribosylation also regulates protein function via non-covalent binding as in interactions of proteins with proteins or nucleic acids. Non-covalent PAR-binding may attract proteins to specific subcellular sites. In addition, a given protein can both non-covalently and covalently bind to polyADPribose for its recruitment to a specific site where the recruited protein can be

---

[106] PolyADPribose polymerases PARP1 and PARP2 are stimulated by DNA breaks. However, PARP1 could also be activated by some types of DNA structures, such as hairpins, cruciforms, and supercoiled DNAs. Activation of PARP1 by tolerable amounts of DNA breaks favors DNA repair. Conversely, a severe DNA injury that overactivates PARP1 triggers cell apoptosis via apoptosis-inducing factor.

[107] PolyADPribose polymerases catalyze successive addition of ^{ADP}ribose units to either ^{ADP}ribose or proteins. The polyADPribose polymerase family contains 17 identified members. Certain members could act as monoADPribose transferases. Activity of PARP1 can be modulated by post-translational modifications. Phosphorylation of PARP1 by extracellular signal-regulated kinases ERK1 and ERK2 yields maximal activation after DNA damage [423]. Acetylation of PARP1 by histone acetyl transferases P300 and cAMP-responsive element-binding protein-binding protein (CBP) enhances the synergistic activity of PARP1 and P300–CBP in inflammation.

[108] PolyADPribose glycohydrolase is encoded by a single gene, but different isoforms exist with various subcellular locations.

modified [423]. Diverse functional outcomes can arise from different sizes and branching types of polyADP-riboses.

PolyADPribose can serve as a local ATP source [423]. AMP-activated protein kinase, which is activated when ATP concentration decays, phosphorylates PARP1. PolyADPribose is involved in the sequential activation of calpain and the pro-apoptotic factor B-cell lymphoma-2-associated X protein that causes to translocation of apoptosis-inducing factor from mitochondria to the cell nucleus and cell apoptosis (Vol. 2 – Chap. 4. Cell Survival and Death). [ADP]Ribose operates as a second messenger that activates transient receptor potential melastatin channel TRPM2 (Vol. 3 – Chap. 2. Membrane Ion Carriers) for Ca^{++} influx in response to oxidative stress.

5.12.3.4 Sialic Acid Attachment and Detachment

Sialidases are glycosidases that catalyze the removal of sialic acids[109] from glycoproteins and glycolipids. Sialidases that are involved in lysosomal catabolism also modulate many biological processes. Four types of mammalian sialidases have been identified (Neu1–Neu4) [424]. They are encoded by different genes. They also differ in their subcellular localization and enzymatic properties. They are expressed in a tissue-specific manner. Sialidase Neu1 targets oligosaccharides and glycopeptides. Sialidases Neu2 and Neu4 hydrolyze glycoproteins and gangliosides at near-neutral pH and at pH 4.6, respectively. Sialidase Neu3 preferentially degrades gangliosides, but has other substrates, such as oligosaccharides and glycoproteins.

5.12.4 Attachment of Lipids

Lipid modifications comprise reversible palmitoylation as well as irreversible myristoylation and prenylation.

5.12.4.1 Palmitoylation

Palmitoylation (or palmitylation) consists of reversible addition of palmitate to a cysteine residue of membrane proteins. Palmitoylation–depalmitoylation cycles allow the cell to regulate the location of specific proteins. Palmitoylation enhances the hydrophobicity of proteins and contributes to membrane linkage. Palmitoylation also intervenes in cellular transport of proteins between membrane compartments and modulates between-protein interactions.

Palmitoylation is undergone by many signaling mediators and affectors, such as hRas, $G\alpha_s$ subunit of G proteins, β2-adrenergic receptor, and endothelial nitric oxide synthase.

---

[109] Sialic acids are negatively charged acidic monosaccharides generally attached to ends of glycoproteins and glycolipids.

5.12.4.2 Myristoylation

Myristoylation is an irreversible protein modification that happens either post-translationally or during translation (co-translational modification). A myristoyl group derived from myristic acid is attached to the N-terminal amino acid (often glycine) of a nascent polypeptide. Myristoylation is catalyzed by ^{N}myristoyltransferase.

5.12.4.3 Prenylation

Post-translational modification with isoprenoids modulates protein–protein and –lipid interactions. Protein prenylation is attachment of either a *farnesyl* or a *geranylgeranyl* moiety from a soluble isoprenoid pyrophosphate to target proteins. Most of the prenylated proteins are subunits of G proteins and Ras GTPases that act in signaling pathways (Vol. 4 – Chap. 8. Guanosine Triphosphatases).

Protein prenylation is carried out by 3 heterodimeric enzymes: farnesyl transferase, geranylgeranyl transferase-1, and Rab geranylgeranyl transferase [425]. Unlike other 2 protein prenyltransferases, Rab geranylgeranyl transferase does not recognize its protein substrate directly, but operates in coordination with Rab escort protein.

5.12.5 Reduction–Oxidation

Reduction–oxidation cycles regulate enzyme actvity. Oxidation of specific residue can cause homodimerization by a disulfide bridge as well as regulate the catalytic activity of kinases and phosphatases.

5.13 Proteolysis

Regulated proteolysis is required in many cellular processes, especially when the cell must react quickly. Gene expression and signaling particularly undergo quality control. Irreversible proteolysis is initiated after a specific post-translational cleavage using specific adaptors. Protease substrates localize in the cytoplasm, plasma membrane, and extracellular matrix. Substrate cleavage release reactants, convert structural matrix proteins to signaling molecules, regulate growth factors, and modify signaling.

Proteolysis at the cell surface includes 2 kinds of enzymes: (1) membrane-tethered proteases, which are soluble enzymes and (2) integral membrane proteins responsible for the regulated intramembrane proteolysis. The transmembrane serine proteases are involved in cell signaling and interaction with its environment. Proteases of the ADAM and BACE families cleave plasmalemmal proteins.

Intramembrane-cleaving proteases (ICliP) form a family of enzymes that cleave hydrophobic substrates within the lipid bilayer. This family includes: (1) the set of site-2 proteases — zinc metalloproteases and presenilin and presenilin-like aspartyl proteases and (2) the rhomboid set of serine proteases. Presenilin is the catalytic component of γ-secretase complex.

6

Cell Cytoskeleton

Cell cytoskeleton enables cell deformation, in particular when the cell is immersed in a mechanical stress field, as well as transport of intracellular cargos (Chap. 9), cell division (Vol. 2 – Chap. 2. Cell Growth and Proliferation), and motility (Vol. 2 – Chap. 6. Cell Motility).

The cytoskeleton actually possesses 3 main functions: (1) it spatially organizes the cell content; (2) it connects the cell to its environment; and (3) it generates coordinated forces that enable the cell to adapt and react. Cells are connected to their environment by their cytoskeleton, either in direct contact with apposed cells (Chap. 7) or with the polymer mesh of the extracellular matrix (Chap. 8).

Cytoskeleton includes 3 interconnected filament networks — microfilaments, microtubules, and intermediary filaments — as well as their nanomotors, and regulatory, crosslinking, and partner proteins. These networks are characterized by their architecture and function, as well as mechanical stiffness, assembling and disassembling dynamics, polarity, and types of associated nanomotors.

6.1 Cytoskeleton and Mechanical Stresses

6.1.1 Cytoskeleton Crosslinks for Optimal Response

Different types of cytoskeletal polymers are intricately linked together, despite their distinct properties and types of structural network. The organization of these links and resultant architecture of the cytoskeleton permit the detection of the mechanical environment and the transmission of compressive and tensile stresses. Structural meshes of microtubules and micro- and intermediate filaments interact with each other and other cellular structures either non-specifically via proper interactions or specifically via crosslinkers [426].

6.1.2 Polymerization–Depolymerization Cycle

The cytoskeleton is a dynamical, adaptive structure, as its polymeric constituents and regulatory proteins are in constant flux. The regulated self-assembly of cytoskeleton components is guided by spatial chemical cues and physical constraints. Permanent remodeling, i.e., cycles of polymerization and depolymerization, or fluidization and solidification, determines cell shape, resists mechanical stress and exerts mechanical forces on cell surroundings, optimizes location and structure–function relationship of cellular organelles and other constituents, supports exchanges of substances between cell and its environment, and transmits signals.

6.1.3 Stress Adaptation and Minimization

The cytoskeleton experiences mechanical stresses and responds to minimize local stresses. Moreover, the cytoskeleton contracts and forms stiffer bundles, the so-called *stress fibers*, to stiffen the cell. Cytoskeletal networks generate, transmit, and respond to mechanical signals over both short and long time scales. In general, the cytoskeleton is an array of self-organizing constituents that carries out cellular functions over diverse length and time scales.

The cytoskeleton is malleable in order to accommodate large strains and sufficiently stiff to transmit stresses (Table 6.1). Hence cells supported by their cytoskeleton adapt to imposed loads. The cytoskeleton is in a constant state of rearrangement driven by the applied loads, thereby accommodating relatively important changes in loading. In addition, cytoskeleton anchorage on adjacent tissue elements allows the deformation of cellular ensemble.

6.1.4 Filament Mesh and Poroelasticity

Dense cytoskeletal meshes slow fluid flow. Poroelasticity deals with resistance to flow through deformable porous media. The extracellular matrix, cytoplasm, and nucleoplasm are aqueous media that contains bodies and meshes of various types of filaments. The extracellular matrix can be considered as a medium with a ground substance, cells, and filaments of elastin and collagen; the cell as a medium with a cytosol, organelles, large proteic complexes, and a cytoskeleton; the cell nucleus as a medium with a nucleosol, nucleoli, proteic aggregates, and a nucleoskeleton. These media can be modeled by a poroviscoelastic material. Short- and long-time relaxation behaviors are determined by the viscoelastic properties of the medium and interstitial fluid, respectively.

The poroelastic constitutive response expresses the dependence of strain and fluid content on stress and pore pressure. M.A. Biot (1905-1985) developed a theory of poroelasticity, i.e., of the coupling of the fluid transport into a porous medium and the medium deformation, using the following assumptions: (1) isotropy of the material; (2) reversibility of the deformation

Table 6.1. Rheology of fiber networks, such as collagens in the extracellular matrix and actin inside the intracellular space. Polymer lattices are connected by crosslinkers. Actin cytoskeleton consists of fibers, i.e., serial arrangements of monomers that are more ore less branched and crosslinked to produce bundles (parallel fibers) or meshs (grids). The persistence length is a mechanical property that quantifies the bending stiffness of a long polymer with respect to the thermal energy. The transition between a bent and a stretched configuration is associated with a water loss. Fiber stretching evolves in 3 phases: non-linear, small; linear, large; and non-linear, very large deformations. For a given fiber, tension develops at a finite curvature and then at a zero curvature (infinite curvature radius; rod-like shape). For a fiber network, a stretch reorientates the fibers according to its application direction and then imposes a buckling.

Parameter	Symbol
Crosslinker distance	d_χ
Persistence length	d_p

Gel (depolymerization state): $d_\chi \gg d_p$
Fiber network (polymerization state): $d_\chi \ll d_p$

(absence of energy dissipation), the medium undergoing quasi-static deformations; (3) linearity of the stress–strain relation (linearly elastic solid); (4) small strains; (5) incompressible fluid (possibly biphasic, i.e., water may contain air bubbles; (6) fluid flows through pores according to the Darcy law.

Any physical quantity of the process is the average over a certain length scale that is large with respect to the length scale of the microstructure of the fluid-filled porous material of porosity that is the fluid volume fraction ($\alpha_f = V_f/V$).

The process is mainly described by: a solid (small) average displacement vector $\mathbf{u}^s$, which tracks the motion of the porous solid with respect to a reference configuration; the fluid flow average velocity $\mathbf{v}$; the relative average displacement ($\mathbf{w} = \alpha_f(\mathbf{u}^f - \mathbf{u}^s)$); the relative average velocity between the fluid and solid ($\dot{\mathbf{w}} = \alpha_f(\mathbf{v} - \dot{\mathbf{u}}^s)$); the stress tensor ($\mathbf{C}$; i.e., the sum of the effective stress in the purely eleastic isotropic structure given by the Hooke law and effective pressure stress of the fluid on the solid); the small (linearized) strain tensor ($\mathbf{S} = 1/2(\nabla\mathbf{u} + (\nabla\mathbf{u})^T)$), which is a measure of the local deformation of the object; the pore hydraulic pressure (scalar p); and the dimensionless fluid content change (per unit volume of porous material [ζ]; i.e., the spatial variation of the rate of fluid volume that crosses a unit area of porous solid in a given direction).

The force equilibrium (momentum conservation) in the transport–deformation process in the transient regime states that:

$$\rho_s(\ddot{\mathbf{u}}^s + \ddot{\mathbf{w}}) = \mathbf{f} + \nabla \cdot \mathbf{C},$$
$$\rho_s\ddot{\mathbf{u}}^s + \frac{\rho_f}{\alpha_f}\ddot{\mathbf{w}} = \mathbf{f} - \nabla p + \mu\,\nabla^2\mathbf{v} - \frac{\mu}{\mathcal{P}}\mathbf{v}, \tag{6.1}$$

where **f** are the body force per unit volume, $\rho^{s(f)}$ is the mass density of the solid (s) and fluid (f) parts, and $\mathcal{P}$ the permeability of the porous material. The additional dissipation (the last term of the right hand side of the second equation of the set 6.1) quantifies the friction effect due to the fibrous matrix that partially occupies the pore space. The Navier-Stokes equation that takes into account both convection and diffusion (the Stokes equation when inertia forces can be neglected, i.e., in a stationary state at a low Reynolds number) then leads to a modified Brinkman equation.

The three volumetric material constants that characterize the elastic response with fluid infiltration include: (1) the bulk modulus of elasticity in 2 limit conditions, i.e., *drained bulk modulus* (K_d; related to a deformation at fixed p, with the fluid flowing in or out of the deforming element without change in the pore pressure)[1] and *undrained bulk modulus* (K_u, the fluid being trapped in the porous solid with associated changes in p, i.e., ζ remain constant); and (2) the Biot coefficient (the ratio of K to a poroelestic constant $[H]$).[2]

The *Biot coefficient* (B) is the ratio of the fluid volume variation (gain or loss) in a material element to the volume change of that element, when loaded under the drained condition, i.e., when the pore pressure can return to its initial state, or the pore pressure remains constant. The variation in fluid volume equals the change in pore volume. In particular, the Biot coefficient measures the ratio of the fluid volume squeezed out to the volume change of the compressed porous medium.

The *Biot modulus* (M_B) can be defined as the change of the amount of fluid per unit volume of medium that results from a unit variation of pore pressure under constant volumetric strain [427]:

$$M_B = (K_u - K_d)/B^2 = \frac{2G(\nu_{\mathrm{P}u} - \nu_{\mathrm{P}d})}{B^2(1 - 2\nu_{\mathrm{P}d})(1 - 2\nu_{\mathrm{P}u})}. \tag{6.2}$$

The relationship between the dimensionless fluid content ζ and the stress is given by [428]:

---

[1] The pore fluid drains entirely to the environment, so that the pore fluid pressure equilibrates with the environmental (reference) pressure ($p = 0$; $\zeta = B\,\mathrm{tr}(\mathbf{S})$; $\mathrm{tr}(\mathbf{S}) = \sum_{k=1}^{3} S_{kk} \equiv S_{kk}$: trace of tensor $\mathbf{S}$). Drained conditions correspond to the deformation at a fixed pressure p, with the fluid flowing in or out of the deforming control volume. The compressibility of the material (K_d^{-1}) is obtained by the measure of the change in volumetric strain due to changes in applied stress at constant pressure ($\delta S/\delta C|_p$.

[2] The parameter H^{-1} represents the poroelastic expansion coefficient that describes the amount of bulk volume changes due to a pore pressure change at constant applied stress ($\delta S/\delta p|_S$). The relationship between stress and strain is given by:

$$\mathrm{tr}(\mathbf{S}) = \frac{1}{3K_d}\,\mathrm{tr}(\mathbf{C}) + \frac{1}{H}p.$$

$$2G\zeta = B\frac{1 - 2\nu_{Pd}}{1 + \nu_{Pd}}\left(\operatorname{tr}\mathbf{C} + \frac{3p}{C_{Sk}}\right). \tag{6.3}$$

The linear theory can be introduced with the fundamental set of material constants: drained and undrained Poisson ratios, the proper shear modulus (G) for both drained and undrained conditions, and the Biot coefficient [427]. The drained (ν_{Pd}) and undrained (ν_{Pu}) Poisson ratios are related to G, K_d, and K_u:

$$\nu_{Pd} = \frac{3K_d - 2G}{2(3K_d + G)}, \qquad \nu_{Pu} = \frac{3K_u - 2G}{2(3K_u + G)}. \tag{6.4}$$

The compressibility coefficient, or Skempton pore pressure coefficient (C_{Sk}) is a measure of the relative compressibilities of the fluid and solid phases. It is related to the drained and undrained solid and fluid properties by [428]:

$$C_{Sk} = \frac{3(\nu_{Pu} - \nu_{Pd})}{B(1 - 2\nu_{Pd})(1 + \nu_{Pu})}. \tag{6.5}$$

In addition to the Biot modulus and Skempton pore pressure coefficient, a poroelastic stress coefficient C_C has been introduced [427]:

$$C_C = \frac{B(1 - 2\nu_{Pd})}{2(1 - \nu_{Pd})}. \tag{6.6}$$

Whereas the Biot modulus is defined under constant volumetric strain, a storage coefficient (C_S) has been defined for uniaxial strain and constant normal stress in the direction of the strain $(\delta\zeta/\delta p|_C)$:

$$C_S = \frac{(1 - \nu_{Pu})(1 - 2\nu_{Pd})}{M_B(1 - \nu_{Pd})(1 - 2\nu_{Pu})}. \tag{6.7}$$

The pressure–deformation coupling can be given by the following constitutive equation:

$$\zeta = \frac{B}{3K}\operatorname{tr}(\mathbf{C}) + C_S p, \tag{6.8}$$

6.1.5 Cytoskeleton and Adhesion Sites

Transmembrane adhesion molecules (Sect. 7.5) and linked cytoskeletal filaments transmit localized mechanical stresses applied at the cell surface to distant cytosolic sites, cell organelles, and nucleus, as well as opposite plasma membrane. Mitochondria associated with microtubules can sense mechanical stimuli and release signaling molecules. Conversely, increased cytoskeletal tension is transmitted to the extracellular matrix, where it can provoke conformational change of constituents that can promote fibril assembly.

6.1.6 Nanofracture and Resealing

Strong, acute mechanical perturbations can generate nanofractures in the cell membrane followed by membrane resealing that is controlled by depolymerization of cortical filamentous actin (^{F}actin) [429]. Mechanical stress can thus release plasmalemmal receptors and cytosolic agents. Intervesicular and vesicle–plasma membrane fusion that occurs at the cell cortex surrounding the plasmalemmal defect initiates a rapid resealing in the presence of calcium.

6.1.7 Cytoskeleton and Cell Structure–Function Relationship

Cells, especially smooth muscle cells, can deform while maintaining their optimal functions. Vascular and airway smooth muscle cells can undergo large strains due to vessel deformation occurring during the cardiac and respiratory cycles. Manifold molecules and fibers form the dynamical cell framework that determines cytosol organization and intracellular displacements [430]. The energy supply and between-protein interactions that reorganize the cytoskeletal are highly regulated.

6.1.8 Prestress

Cytoskeleton is generally prestressed to immediately respond to any mechanical loading and increase its stiffness in direct proportion to the applied mechanical stress [431]. Prestress that allows adaptation to mechanical stimuli arises from force and reaction exerted between a cell and its neighborhood (cell, matrix, and other environmental elements such as blood vessels, which exert hemodynamic stress), as well as between tension induced by contractile microfilaments and compression generated by microtubules. Cellular response to mechanical stimuli depends on prestress in the cytoskeleton.

Tension applied by magnetic microbeads bound to integrins generates 4 types of mechanical responses: (1) immediate viscoelastic response; (2) early adaptive, attenuated response to oscillatory forces; (3) later adaptive stiffening with sustained (>15 s) static stimuli; and (4) large-scale repositioning to prolonged (>1 mn) loading [432]. Immediate and early responses, but not later adaptive reactions, depend on dissipating cytoskeletal prestress.[3]

---

[3] Cells use multiple mechanisms to sense and transduce static and dynamical loading. Generalized membrane distortion, particularly integrin distortion, in the absence of mechanoreceptor remains unproductive. Immediate and early responses as well as repositioning depend on small GTPase Rho (Vol. 4 – Chap. 8. Guanosine Triphosphatases and their Regulators). Assembly of focal adhesions induced by local mechanical stimuli indeed requires activated small GTPase Rho and its effectors RoCK kinase that phosphorylates myosin light chain and thus fosters cell contractility and Diaphanous that causes actin polymerization. Cell strengthening and repositioning are influenced by mechanosensitive ion channels (Vol. 3 – Chap. 2. Membrane Ion Carriers) and protein Tyr phosphatase PTPn11 that

Small, externally applied, localized oscillatory force induces strain and stress with phase lags and stress-concentration sites within the cytoplasm that depend on prestress in the cytoskeleton and cytoskeletal anisotropy [433]. Nonetheless, mechanical activation of cAMP signaling (Vol. 4 – Chap. 10. Signaling Pathways) via plasmalemmal G proteins (Vol. 4 – Chap. 8. Guanosine Triphosphatases and their Regulators) occurs at sites of stress concentration within the plasma membrane regardless of magnitude of cytoskeletal prestress [434].

6.1.9 Mechanotransduction

Cells are able to sense and transduce mechanical signals. Mechanical stimuli modulate cellular functioning by priming adequate signaling pathways. Mechanotransduction converts experienced mechanical stresses and strains into a chemical response.

The cell response is initiated at the plasma membrane from embedded mechanosensitive molecules (mainly proteins) that undergo proper conformational changes to become activated. In particular, a fast activation (<300 ms) of Src kinase at remote cytoplasmic sites results from small, local, external stress imposed on integrins [435]. Sites of Src activation caused by mechanical stress correspond to large deformation sites of microtubules.

Microtubules constitute structures for stress transmission and cytoplasmic protein activation. Prestressed structures enable force transfer and focus stress on cytosolic load-bearing molecules. Mechanotransduction can happen at concentrated stress sites. Moreover, mechanical loading increases microtubule polymerization [431]. Also, linkage between myosin and actin filaments depends on applied mechanical force.

Like plasmalemmal mechanosensitive molecules, certain cytoskeletal proteins can undergo sufficient conformational change under stretch applied with adequate magnitude and direction to prime their activity. In addition, cytoskeletal tension can exert traction on transmembrane mechanosensitive ion channels enough to open their pores, as well as on other mechanotransducers to trigger their activity by exposing their binding sites.

Connections between the extracellular matrix and/or adjacent cells, plasma membrane with its adhesomes and other mechanosensors and transducers, cytoskeleton, nuclear envelope with its ion channels and linkers, and nucleoskeleton participate in cellular response to mechanical cues. The nucleoskeleton indeed is strongly related to nuclear structures to which it can transmit sensed mechanical signals. The activity of enzymes such as RNA polymerase and that of nanomotors such as myosin partly depends on the experienced mechanical loading field.

---

activates protein Tyr kinase Src (Vol. 4 – Chaps. 3. Cytosolic Protein Tyrosine Kinases and 7. Cytosolic Protein Phosphatases).

6.2 Cytoskeletal Filaments

Three classes of cytoskeleton filaments exist: microfilaments, microtubules, and intermediary filaments (Tables 6.2 and 6.3). Microtubules are the stiffest of these 3 polymers. In addition, they have the most complex assembly and disassembly dynamics. The *spanning network* is a fourth element that has a tiny weft and fills the whole cytosol. It can determine the sites of protein synthesis and the assembling locations of filaments and microtubules. It acts on cell organelle motility.

Several classes of regulators, especially of actin filaments, include [426] (Tables 6.2 and 6.3): (1) *nucleation-promoting factors* that initiate filament formation; (2) *capping proteins* that terminate filament growth by binding the barbed ends of actin filaments; (3) *polymerizing agents* that promote faster or sustained filament growth; (4) *depolymerizing and severing factors* that disassemble filaments; and (5) *stabilizing proteins* and *crosslinkers* that organize and reinforce higher-order cytoskeletal structures. The cytoskeleton is actually built as networks of fibrillar polymers with articulation nodes from which the cytoskeleton can reorganize itself. Actin-based structures include stress fibers, endocytic pits, peri-organellar actin, cortical spectrin–actin mesh, nuclear actin network, and cytokinetic ring during mitosis.

In particular, among actin-binding scaffolds, myotilin and myopalladin abound in striated myocytes (skeletal and cardiac muscle) and palladin is ubiquitously produced in cells of mesenchymal origin, especially during morphogenesis.[4] They all bind to α-actinin. In addition, palladin links to vasodilator-stimulated phosphoprotein and ezrin, myotilin to filamin, and myopalladin to nebulin and cardiac ankyrin repeat-containing protein (CARP or AnkRD1) [436].

Ezrin–radixin–moesin proteins (ERM; Sect. 6.3.3.12) act as both effectors and affectors of Rho GTPases [437].

Ubiquitous tetrameric adducin (made of α–β or α–γ heterodimers) localizes at spectrin–actin junctions. It binds calmodulin and is a substrate for protein kinase-C (PKC) and Rho-associated, coiled-coil-containing protein kinase (RoCK). Adducin caps the fast-growing ends of actin filaments and recruits spectrin to filament ends [438]. Calmodulin, gelsolin, and phosphorylation impede its activity.

Dynein, kinesin, and myosin constitute the 3 major classes of cytoskeleton-based motor proteins. These nanomotors share several features (Table 6.4). Dynein is much bigger than myosin-2 and kinesin. Motor domains use the energy derived from ATP hydrolysis to undergo conformational changes at each progression step along the filament track. Conformational changes are similar in myosins and kinesins, but different in dyneins. The step size ranges from

---

[4] Myotilin localizes to the Z disc of sarcomeres. It supports the formation of actin bundles. Myopalladin tethers nebulin in skeletal myocytes and nebulette in cardiomyocyts to α-actinin at the Z disc.

Table 6.2. Types of cytoskeleton filaments (**Part 1**; NA: not applicable; ERM: ezrin, radixin, and moesin; GFAP: glial fibrillary acidic protein; TRIOBP: TRIO and [F]actin-binding protein; WHAMM: WASP homolog associated with actin, membranes, and microtubules). The plakin family includes desmoplakin (Dsp), envoplakin (Evpl), periplakin (Ppl), plectin (Plec), dystonin (Dst), corneodesmosin (Cdsn), and microtubule–actin crosslinking factor (MACF). Both α- and β-catenins anchors the actin cytoskeleton to cadherins of adherens junctions. Plakophilin that interacts directly with several desmosomal components (desmoplakin, plakoglobin, desmoglein, and desmocollin) compete with cadherin for signaling mediator and connector β-catenin. The subsarcolemmal costamere (made of dystrophin, dystrobrevin, syncoilin, synemin, sarcoglycan, dystroglycan, and sarcospan) couple the sarcomere to the sarcolemma in striated myocytes. On the other hand, fibrillin constitutes extracellular microfibrils and -filaments (caliber 10 nm).

Protein type (function)	Microfilaments	Intermediate Filaments	Microtubules
Constituent	Actin	Keratin, desmin, GFAP, peripherin, vimentin, neurofilaments, synemin, syncoilin, nestin, lamin	Tubulin, tektin
Nanomotor Motor linker	Myosin	NA	Dynein, kinesin Dynactin
Stabilizer	TRIOBP		Tau
Remodeler	Profilin, supervillin, myopalladin, myotilin, palladin		Stathmin
Cytoskeleton linker	Plectin, WHAMM, MACF, coronin	Plectin	Plectin, MACF, WHAMM
Adhesion site linker and anchor	Plakin, talin, vinculin, plectin, catenin	Desmoplakin, periplakin envoplakin, dystonin, plectin	Plectin
Membrane linker Costamere	Spectrin, ERM, Adducin, ankyrin Dystrophin		

a few nanometers for myosin-2, to 8 nm for kinesins and dyneins and 36 nm for myosin-5 and -6 [439]. Non-motor domains participate in motor function. Coiled-coil segments serve for dimerization and regulate cargo-binding domains. Nanomotors can generate a force of several pN. Dynein is responsible

Table 6.3. Types of cytoskeleton filaments (**Part 2**) Other types of acting-binding partners (ADF: actin-depolymerizing factor; ARP: actin-related protein; ENA–VASP: Enabled homolog and vasoactive stimulatory phosphoprotein; WASP: Wiskott-Aldrich syndrome protein). Actin-binding proteins regulate the rate of assembling and disassembling of actin filaments. Cofilin and profilin promote actin depolymerization behind and polymerization close to the leading edge, respectively. Cofilin can bind both ^{G}- and ^{F}actin. It controls actin polymerization and depolymerization in a pH-sensitive manner. Profilin sequesters actin, but also fosters the conversion of poorly polymerizing, ADP-bound ^{G}actin (^{G}actin^{ADP}) into readily polymerizing, ATP-bound actin monomers (actin^{ATP}). In addition, the ARP2–ARP3 complexes give birth to actin filament branches. Filaments then elongate and can push the plasma membrane forward. Thymosin-β4 complexes with and sequesters globular (^{G})actin. Capping proteins block assembly and disassembly at filament ends, tropomodulin and CapZ at the minus- and plus-end, respectively. Myristoylated alanine-rich C-kinase substrate (MARCKS) is an actin filament crosslinker. Myosins form regular rotating spots in the cytosol over which run actin filaments that are then displaced to cause a local cell contraction. Calponin is a calcium-binding inhibitor of the ATPase activity of myosin. Phosphorylation of calponin upon calcium binding to calmodulin relieves inhibition of calponin.

Partner (function)	Protein types
Polymerizers	ENA–VASP
Nucleators	WASP, WAVe, WASH, formin cortactin, dynamin
Bundling crosslinkers	Espin, actinin, plastin, fascin, filamin, scruin
Branching crosslinkers	ARP2–ARP3 complex
Connectors	Titin Actinin
Cappers	Tropomodulin, CapZ, adducin
Depolymerizers	ADF (cofilin, destrin), gelsolin, villin
Sequestrators (^{G}actin)	Cofilin, profilin, thymosin-β4
Contraction	Myosin, troponin, tropomyosin Caldesmon

for maintaining endosome localization near the cell center, kinesins probably transport vesicles from the perinuclear compartment to the cell cortex. Myosin

Table 6.4. Properties of molecular motors (Source: [439]). Processivity is the property of nanomotors to move over long distances without dissociating from the filament track; processive motors are dimers that have, at any given time, one head firmly bound to the track while the other is stepping toward the next binding site. Cargo association involves non-motor domains and most often accessory proteins.

Energy source	ATP hydrolysis
Processive motor	Dynein, kinesin-1, myosin-5/6
Material transport	Vesicular cargos
Membrane lipid binding	Subclass-1 myosins,
	some members of the kinesin-3 family
Membrane protein binding	Kinesin-1, myosin-1a/5a/7a
Adaptor binding	Kinesin-1 to -3,
	myosin-5 and -6, dynein

and kinesin have similar structure, whereas dynein is an ATPase associated with diverse cellular activities.

6.3 Microfilaments

Microfilaments contain several proteins, especially actin. *Actin filaments* (length $0.1–1\,\mu m$, thickness $7–10\,nm$) are involved in cell configuration, adhesion, and motility.[5] Permanent exchanges exist between actin filaments and cytosolic non-polymerized actin [440], associated with a sol–gel transition.[6]

Actin filaments form bundles. Actin filaments are peripherally concentrated to build a felting in the cell cortex, which is connected to the plasma membrane. Cortical actin filaments are directly anchored on the cell membrane, using mooring proteins, *talin* and *vinculin* [441].

Microfilaments can serve as mooring and transmission lines in a stress field, and as towlines during motion. Moreover, myosin-5 transports various cellular cargos along actin filaments (Chap. 9).

Myosin and actin not only function as molecular motors for muscle contraction, cell motility, and cell division, forming actomyosin complexes, but are also involved in transcription of DNA into RNA [442].[7] Motions of transcriptional complexes and/or DNA are necessary to transcribe the genetic

---

[5] The cell motility requires contractile proteins (actin and myosin) with adequate ATP supply under the control of regulatory proteins.

[6] The sol state corresponds to a suspension of large particles (concentrated solution of actin monomer, i.e., depolymerized actin). The gel is stiffer. Gelation is associated with the conversion of actin into actin filaments (actin polymerization).

[7] Transcription requires chromatin remodeling, formation of preinitiation complexes, binding of transcription factors to regulatory regions of DNA, recruitment of RNA polymerase, etc. (Vol. 1 – Chap. 5. Protein Synthesis). β-Actin and actin-related protein BAF53 associate with chromatin-remodeling complex barrier-to-auto-integration factor (BAF) [443]. Myosin-1 is observed in the nucleus [444].

Table 6.5. Actin isoforms and genes (with eventual alternative symbol).

Gene	Protein	Cell Type (predominant expression)
	Subfamily-2, α group, muscle actins	
ACTA1	Actin α1	Skeletal muscle cell
ACTA2	Actin α2	Smooth muscle cell
ACTC1	Cardiac actin α1	Cardiomyocyte
ACTG2 (ACTA3)	Actin α3 (γ2)	Smooth muscle cell
	Subfamily-1, β and γ groups, non-muscle actins Cytoplasmic actins	
ACTB	Actin β(β_{cyto})	All non-muscle cells
ACTG1	Actin γ1 (γ_{cyto})	All non-muscle cells

code. Nuclear actin and myosin-1 may act as auxiliary motors for RNA polymerase-2 that can move transcriptional complexes relative to DNA. They also might have a structural role. Actin and actin-related proteins remodel chromatin. Actin interacts with heterogeneous nuclear ribonucleoprotein complexes.[8] Conversely, the eukaryotic translation elongation factor 1A (eEF1A or EF1), a nucleotide-binding protein that delivers aminoacyl-tRNA to the ribosome, is implicated in the actin cytoskeleton [445].

6.3.1 Actin Types

Actin is a monomeric, globular protein (^{G}actin) observed in all eukaryotic cells. Actin polymerizes to form filamentous actin (^{F}actin), a helical filament, following a conformational transition, as well as filament assembly into higher-order structures. Actin activity is modulated by more than 150 different actin-binding proteins.

Six genes encode distinct vertebrate actin isoforms (Table 6.5). Three main groups of actin isoforms (α–γ) exist. The actin family can then be divided into 2 broad subfamilies: subfamily-1, cytoplasmic, non-muscle (β and γ groups) and subfamily-2, muscle (α group) actins. Actin types are also classified according to the cell type in which they are predominantely expressed. More than one actin species exist in the same cell.

α1 Skeletal muscle, α1 cardiac muscle, and α2 (aortic type) and α3 (γ2; enteric type) smooth muscle actins are primarily found in myocytes. They are encoded by the ACTA1, ACTC1, and ACTA2 and ACTG2 (or ACTA3) gene, respectively (Table 6.5).[9]

---

[8] Heterogeneous nuclear ribonucleoproteins are involved in newly transcribed RNA processing and transport.

[9] In humans, the genes ACTBL2 and ACTBL3 generate β-actin-like proteins 2 (κ-actin) and 3 (κ-actin or POTE ankyrin domain family member K). The genes

Cytoplasmic β- and γ-actins (also denoted β_{cyto}- and γ_{cyto}-actins) are building blocks of the cytoskeleton in all non-muscle cells. They are encoded by the ACTB and ACTG1 genes, respectively. Cytoplasmic β_{cyto}- and γ_{cyto}-actin are among the most abundant proteins in every mammalian cell.

Cytoplasmic β- and γ-actins participate in all actin-based cytoskeletal structures.[10] γ-Actin is required for reinforcement and long-term stability of ^{F}actin-based structures, but is not an essential component of the developing cytoskeleton.

β-Actin preferentially localizes at the protruding leading edge of a migrating cell, where it forms a loose, branched mesh of relatively short actin filaments [447]. Nearly identical γ-actin resides in the cell body in a dense, non-branched network and long contractile stress fibers that support cell adhesion.

Both nascent β- and γ-actins can be arginylated. Unlike arginylated β-actin, arginylated γ-actin is rapidly degraded by the ubiquitin–proteasome axis [448]. γ-Actin is translated more slowly than β-actin, hence more easily exposing a lysine residue to ubiquitination before folding.

6.3.2 Actin Filaments

Actin filaments are polarized polymers with dynamical *barbed ends* and less active *pointed ends*. Actin filaments are components of many cell structures. They are much less rigid than microtubules. Actin monomers are able to self-assemble into helical filaments. Helical actin filaments polymerize from a pool of actin monomers in response to different stimuli in the presence of assembly factors. Subsequent monomer addition is much easier than di- and trimerization.

High concentrations of actin monomers (200–400 µmol) is contained in resting cells, i.e., an amount 600 to 1200 times greater than the critical concentration for assembly of pure actin.

Actin-based structures include nuclear actin mesh, peri-organellar actin grid, cortical spectrin-actin network, and stress fibers, as well as endocytic pits, and cytokinetic ring during the cell division (mitosis phase).

6.3.2.1 Central Actin Structures

Assembly (nucleation) factors are required to initiate assembly of any actin-based structure. Once nucleated, the type of actin structure depends

---

ACTBP1 to ACTBP11 correspond to actin-β pseudogene 1 to 11 and the genes ACTGP1 to ACTGP10 to actin-γ pseudogene 1 to 10.

[10] In particular, β- and γ-actins localize to mechanosensory stereocilia of hair cells in the inner ear sensory epithelium for hearing. γ-Actin maintains the integrity of the stereocilium actin core. It localizes along the length of stereocilia, but redistributes to sites of ^{F}actin core disruptions after exposure to damaging noise [446].

on regulators. Mature structures require additional factors for crosslinking, bundling, and interacting with cellular membranes. High concentrations of crosslinkers promotes the assembly of organized, stiff structures, such as bundled and branched filaments. In the presence of capping factors, filaments are capped to prevent further elongation.

A 3D fiber *geodesic dome* (geodome) surrounds the cell nucleus. From its vertices, filament bundles project toward the cell cortex. Geodome foci contain α-actinin and edges tropomyosin, whereas fibers attached to the vertices that extend peripherally contain actin, α-actinin, and tropomyosin.

6.3.2.2 Peripheral Actin Structures

At the cell periphery, actin filaments are able to form sheet-like (lamella, lamellipodia, ruffles, and phagocytic cups) and finger-like protrusions (filopodia and microvilli; caliber <200 nm), as well as podosomes and invadopodia.[11]

Cytonemes or Membrane Nanotubes

Some cytoskeletal structures can span distances much larger than that of the cell. Cytonemes, or membrane nanotubes (a.k.a. nanotubules), that contain actin filaments are specialized filopodia that can grow over long distances (to lengths of millimeters) to mediate intercellular signaling. Two types of membrane nanotubes have been observed, at least between human macrophages [449]: (1) thin membrane nanotubes (bore <0.7 μm) that contain only ^{F}actin and transfer portions of the plasma membrane between cells in both directions and (2) thick membrane nanotubes (caliber >0.7 μm), in which both actin and microtubules carry components of the cytoplasm between cells, such as vesicles and organelles.

Microvilli

Microvilli contain parallel bundles of 10 to 30 actin filaments with barbed ends toward the plasma membrane. Microvilli of epithelial cells and lymphocytes contain parallel bundles of actin filaments crosslinked by villin, fimbrin (or plastin), and espin.

Cell Protrusions during Migration

Bundles of aligned filaments support filopodial protrusions during chemotaxis and intercellular communication. Branched filaments assist the leading edge of motile cells. They also generate forces involved in changes in cell shape such as those during phagocytosis.

---

[11] Podosomes are dynamical actin-containing adhesion structures in osteoclasts, macrophages, and fibroblasts. Invadopodia are invasive protrusions in tumoral cells.

Cells crawl in response to lipid and ionic signals elicited by external sensory stimuli by reorganizing their peripheral actin lamellae in the direction of locomotion. The cytoskeleton remodeling includes vectorial assembly of actin subunits into linear polymers at the leading edge and filament crosslinking. Simultaneously, the disassembly of crosslinked filaments into short fragments and actin monomers at the trailing region supplies components for the actin assembly that directs cell protrusion. A subclass of actin-associated exocytic myosins guides sensory receptors to the plasma membrane. Another myosin subclass imposes contraction on actin networks [450].

Ruffles are transient (half-life $\mathcal{O}[mn]$) cellular protrusions that do not attach to the matrix. Peripheral ruffles assemble at the leading edge of moving cells and move rearward. Dorsal ruffles constrict into a circular structure before disappearing. The ARP2–ARP3 complex and formins intervene in ruffle assembly and dynamics.

Filopodia, like microvilli, contain parallel bundles of 10 to 30 long actin filaments with barbed ends toward the plasma membrane. Filopodia adhere to the matrix or another cell, whereas microvilli do not. Filopodia are enriched with actin regulators of the ENA–VASP (Enabled homolog and vasodilator-stimulated phosphoprotein) family that function as anticapping proteins, myosin-10, and formins emanate from lamellipodia. Fascins act as bundling proteins.

Lamellum and *lamellipodium* are formed during cell spreading and migration (Vol. 2 – Chap. 6. Cell Motility). At the leading edge of moving cells, lamellipodia occupy several micrometers back from the cell front and lamella extend from the lamellipodium to the cell body. Lamellipodium is thinner (thickness 100–160 nm) than lamellum (>200 nm) and weakly adherent, as strong adhesion begins at the lamellipodia–lamella border [451]. Two distinct actin-based networks with different kinetics constitute the lamellipodium and lamellum. Lamellipodial actin filaments are branched, whereas lamellar actin filaments form a less-defined network that persists many micrometers behind the lamellipodium. Lamellipodial filaments can lie above lamellar filaments. Diaphanous or other formins are involved in lamellar filament assembly as well as in focal adhesion turnover. Myosin-2 activity at adhesion site causes retrograde flow of the lamellipodial actin network.

Podosomes (caliber and height $\sim$0.5 μm) on the basal cell surface contain integrins, matrix metalloproteases, ARP2–ARP3 complexes, and an actin-rich core surrounded by a ring of several actin-associated proteins and signaling proteins.

Invadopodia (caliber $\sim$2 μm) that have a similar architecture and protein composition to podosomes are formed by migrating cancerous cells for matrix degradation.

Cell Protrusions during Phagocytosis

Phagocytosis (particle size $> 0.5\,\mu$m) requires actin polymerization. Macro-phages use 2 processes: (1) Fc receptor-mediated phagocytosis during which the macrophage membrane protrudes around the target particle (*phagocytic cups* that involve WASP and the ARP2–ARP3 complex) and (2) complement receptor-mediated phagocytosis similar to endocytosis (*phagocytic pits* that implicate formins Dia1 and Dia2).

6.3.2.3 Interactions with Other Cytoskeletal Meshes

Various molecules interconnect different cytoskeletal networks to create a con-tinuous mechanical coupling between cytoskeleton components. These con-nections ensure the transmission of internal or external forces as well as their distribution throughout the cell.

Wiskott-Aldrich syndrome protein homolog-associated with actin, Golgi membranes, and microtubules (Sect. 6.3.3.26), an actin nucleation-promoting factor, binds not only to actin, but also microtubules and cell membranes [426]. Monomeric GTPase Rac1 that is activated by the growth of microtubules, in turn, stimulates the polymerization of actin in lamellipodia.

6.3.2.4 Dynamics of Actin Filament Assembly and Disassembly

Actin monomers are added to or removed from either filament end, but both processes are more than 10-fold faster at the barbed end. Profilin that adds actin monomers only to the barbed end effectively limits elongation to barbed ends. Actin filaments further assemble into structures with specific functions.

The growth velocity of a branched actin network does not depend on load over a wide range of applied forces [452]. Furthermore, when the force exerted on a growing network decreases, the velocity increases to a value greater than the previous velocity at the same load (hysteresis). Growth velocity depends on loading history (not only on the instantaneous load).

The actin cytoskeleton continually assembles and disassembles in response to signaling mediators. The assembly of contractile actin-filament bundles — the stress fibers — is triggered locally when plasmalemmal integrins are bound to their ligands [426].

During endocytosis of extracellular molecules, signals from the invaginat-ing plasma membrane trigger local assembling of actin filament to support membrane internalization as an endocytic vesicle.

Most actin-based structures undergo polymerization–depolymerization cy-cles (i.e., transformation of monomeric, globular actin [^{G}actin] to poly-meric, filamentous actin [^{F}actin] and vice versa) with monomers adding at barbed ends and removal at pointed ends (Fig. 6.1). Actin polymerization–depolymerization cycle is required to fit the needs of processing cells.

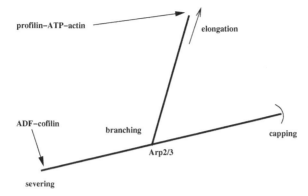

Figure 6.1. Actin filament fate. The initial nucleation phase that requires assembly (or nucleation) factors correponds to the assembly of actin dimers and trimers from a pool of actin monomers. Profilin adds actin monomers to the barbed end of actin filament. Actin filaments further assemble into actin structures. Capping factors impede filament elongation. Depolymerization particularly caused by complexes made of actin-depolymerizing factor (ADF) and cofilin targets the pointed end of actin filament. Actin-related protein complex ARP2–ARP3 allows actin filament branching.

Filament assembling necessitates adenosine triphosphate (ATP). Monomeric, ATP-bound, globular actin can undergo cycles of *self-assembly* into filamentous actin, ATP hydrolysis, and depolymerization.

The rate-limiting step in actin polymerization is the self-association of actin monomers into short-lived dimers and trimers. When kinetic barriers are overcome, an actin tetramer is formed and yields a stable seed for polymerization that progresses until actin monomers reach the critical concentration [453].

Spontaneous actin assembly is inefficient. To initiate actin assembly, free barbed ends are created and act as templates for polymerization by nucleating new monomers, whereas existing filaments are uncapped and severed [454].

Polymerization of ^{G}actin into ^{F}actin activates the ATPase function of actin. The nucleotide-binding site is actually closed in ^{G}actin. The actin nucleotide-binding site for ATP or ADP is located between the 2 domains that undergo rotation when actin is polymerized.

A simple relative rotation of the 2 major domains that are adjacent to the nucleotide-binding site by about 20 degrees with resulting flattening appears in ^{F}actin subunits [455]. Domain rotation that leads to flat actin is essential for the formation of stable, helical ^{F}actin. ^{F}Actin enables cell retraction or expansion (i.e., contraction and relaxation).

Stabilization of actin filaments arises from rearrangement of the filament structure from a relatively disordered state immediately after polymerization to a stable helix. The unstable conformation of newly polymerized filaments is useful in dynamical actin assemblies, whereas stabilizing structural rear-rangements with aging confer enhanced strength to oppose mechanical stress.

Actin filament reorganization can be primed by activated plasmalemmal receptors. These receptors target Rho GTPases via Rho guanine nucleotide-exchange factors (GEF) directly or kinases, such as Src and focal adhesion kinases, as well as integrin-linked (pseudo)kinase [456]. Subsequently, Rho activate their effectors, such as Rho-associated kinases (RoCK), formins, Wiskott-Aldrich syndrome protein and their homologs, the actin-related protein ARP2–ARP3 complex, and other actin-binding proteins, to trigger actin polymerization. Moroever, LIM domain kinases stimulated by RoCKs hinder actin-depolymerizing factors.[12]

Stimulatory plasmalemmal receptors pertain to 5 classes [456]: (1) G-protein-coupled receptors (GPCR; Vol. 3 – Chap. 7. G-Protein-Coupled Receptors) that are mainly coupled to α subunits of the G12/13, Gq/11, and Gi/o families;[13] (2) receptor Tyr kinases (RTK; Vol. 3 – Chap. 8. Receptor Kinases);[14] (3) receptor Ser/Thr kinases, i.e., transforming growth factor-β receptors (TβR; Vol. 3 – Chap. 8. Receptor Kinases);[15] (4) integrins at focal adhesions (Sect. 7.6);[16] and (5) cadherins at adherens junctions (Sect. 7.7.2).[17]

---

[12] Rho GTPases can also activate Jun N-terminal kinases (Vol. 4 – Chap. 5. Mitogen-Activated Protein Kinase Modules) via MAP3K11 kinase or CRK adaptor [456]. In addition, the RhoA–RoCK pathway primed by regulators of actin dynamics can stimulate nuclear factor-κB.

[13] G-Protein-coupled receptors are activated once bound to their specific agonists, such as hormones, chemokines, bioactive lipids (e.g., lysophosphatidic acid and sphinghosine 1-phosphate), platelet activators, some morphogens, etc. In the non-canonical, β-catenin-independent planar cell polarity Wnt signaling in polarized cell movements during gastrulation and organogenesis, Disheveled activates regulators of actin dynamics, such as Rho, Rac, and CDC42 GTPases, and their effectors RoCK kinase and formin-related protein Disheveled-associated activator of morphogenesis DAAM1 [456] (Sect. 6.3.3.31).

[14] Numerous receptor Tyr kinases link to Rho GTPases, such as those for insulin, epidermal (EGF), fibroblast (FGF), hepatocyte (HGF), platelet-derived (PDGF), and vascular endothelial (VEGF) growth factors, as well as ephrin-A and neurotrophic Tyr receptor kinases.

[15] Some growth factors, such as transforming growth factor-β, bone morphogenetic proteins, and activins, signal once bound to heterotetrameric Ser/Thr kinase receptors.

[16] Integrins participate in actin dynamics, as they activate the small GTPases RhoA and Rac1, once RhoGEFs have been activated by integrin-linked (ILK), focal adhesion (FAK), and Src kinases.

[17] Cadherins are linked to the actomyosin mesh via catenins. Assembly and disassembly of adherens junctions activate the MRTF–SRF pathway via Rac1 (Sect. 6.3.2.6).

6.3.2.5 Actin in the Nucleus

Three types of actin-binding proteins (ABP) contain nuclear localization and export signals to shuttle between the cytoplasm and nucleus and possess a transcriptional activation capacity [456]:[18] (1) *globular actin-binding proteins* (GABP), such as myocardin-related transcription factors, actin-binding Rho-activating protein (ABRA),[19] junction-mediating and regulatory protein (JMy; Sect. 6.3.3.26), β4-thymosin,[20] profilin (Sect. 6.3.3.13), neural Wiskott-Aldrich syndrome protein (Sect. 6.3.3.23), the ARP2–ARP3 complex (Sect. 6.3.3.21), and Spire (Sect. 6.3.3.38); (2) *filamentous actin-binding proteins* (FABP), such as actin-binding LIM proteins (ABLIM1–ABLIM3),[21] α-actinin (Sect. 6.3.3.5), filamin (Sect. 6.3.3.8), cofilin (Sect. 6.3.3.18), gelsolin (Sect. 6.3.3.17), supervillin, and some other LIM domain proteins (paxilin, zyxin, lipoma-preferred partner [LPP],[22] cysteine and glycine-rich proteins [CSRP1–CSRP3],[23] TGFβ1-induced transcript product TGFβ1I1,[24]

---

[18] Among these proteins, LIM domain proteins are committed in proteic complexes that link focal adhesion to actin fibers. They also can act as transcription cofactors or regulators of mRNA export from the nucleus [456]. Cysteine and glycine-rich proteins and Four and a half LIM domain-containing proteins (FHL) operate as cofactors of transcription factor serum response factor (SRF). Protein FHL1 is strongly expressed in skeletal and cardiac myocytes. Zyxin and FHL2 are encoded by SRF-regulated genes.

[19] A.k.a. striated muscle activator of Rho-dependent signaling (StARS). Protein ABRA activates serum response factor by priming the nuclear translocation of myocardin-related transcription factors that are coactivators for the serum response factor (Sect. 6.3.2.6). Nuclear import of these ABRA-dependent coactivators requires RhoA GPTase and actin polymerization [457]. Protein ABRA localizes to sarcomeres of striated myocytes. It interacts with monomeric actin and promotes nuclear translocation of MRTFa coactivator [456].

[20] Thymosin-β4 is an actin-sequestering protein (Sect. 6.3.3.2).

[21] Proteins ABLIMs act as linkers between actin cytoskeleton and cell signaling pathways. They serve as scaffolds for signaling modules of the actin cytoskeleton and modulate transcription. Protein ABLIM1 is expressed in retina, brain, and muscles. Isotypes ABLIM2 and -3 have distinct tissue-specific expression patterns with the highest expression levels in muscular and nervous tissues [458].

[22] LIM domain-containing lipoma-preferred partner, a member of the zyxin family, localizes to cell adhesion sites and stress fibers as well as transiently to the nucleus. It can interact with α-actinin and vasodilator-stimulated phosphoprotein [459]. It is particularly expressed in vascular and visceral smooth muscle cells (to a much lesser extent, in cardio- and skeletal myocytes and other cells) [460].

[23] Cysteine and glycine-rich proteins are also named Cys-rich proteins (CRP1–CRP3). However, alias CRP is used not only for LIM domain-containing Cys-rich proteins (CRP), but also C-reactive protein. These adaptors are involved in cytoskeletal remodeling and transcriptional regulation.

[24] Transforming growth factorβ-1-induced transcript product-1 interacts with androgen receptor, among other molecules.

antileukoproteinase [ALP],[25] thyroid receptor-interacting protein TRIP6,[26] and LIM and SH3 domain-containing protein LASP1);[27] and (3) ^{F}*actin complex-associated proteins* (FACAP) that assemble at cytoplasmic sides of focal adhesions (Sect. 7.6) as well as adherens (Sect. 7.7.2) and tight (Sect. 7.7.3) junctions, such as Abl1 kinase, integrin cytoplasmic domain-associated protein ICAP1, some LIM domain proteins (LIM kinases, Particularly interesting new Cys-His protein PINCH1 [or LIMS1], and Four and a half LIM domains protein FHL2), and catenin-δ1.

6.3.2.6 Stimulation of Gene Transcription by Actin

Actin cooperates with all of the 3 types of RNA polymerases [456]. It contributes to chromatin remodeling and binds different ribonucleoprotein complexes (small nuclear RNAs and heterogeneous nuclear RNPs).

The activity of intracellular (nuclear) hormone receptors (Vol. 3 – Chap. 6. Receptors), especially androgen receptors, is modulated by actin and actin-binding proteins (α-actinin-2, filamin, gelsolin, supervillin, and transgelin), in addition to their cognate hormones.

Information on actin remodeling can be transmitted to the nuclear genome via 3 modes [456]: (1) changes in the concentration of cytoplasmic ^{G}actin, and, subsequently, its nuclear level; (2) release of ^{G}actin-binding proteins upon actin polymerization followed by their nuclear translocation and modulation of nuclear transcription factors; and (3) release of ^{F}actin-binding and ^{F}actin complex-associated proteins and their nuclear translocation to control transcriptional activity.

Globular actin polymerization liberates myocardin-related transcription factors MRTFa and MRTFb[28] that are coactivators of the nuclear transcription factor serum response factor (SRF).[29]

---

[25] Antileukoproteinase, or secretory leukocyte protease inhibitor SLP1, is a secreted agent that protects cells against serine proteases, such as trypsin, leukocyte elastase, and cathepsin-G.

[26] Thyroid receptor-interacting protein-6 is a member of the zyxin family that localizes to focal adhesions and along actin stress fibers.

[27] Protein LASP1 accumulates in cell membrane protrusions [461].

[28] A.k.a. myocardin-like proteins MKL1 and MKL2.

[29] Members of the myocardin family engage various partners in addition to or conjunction with serum response factor, such as histone acetyltransferase P300, class-2 histone deacetylases, small mothers against decapentaplegic homolog SMAD4, forkhead box protein FoxO4, and GATA factors [456]. These combinatorial associations of SRF with coactivators and coinhibitors yields cell-type specificity and temporal control of the response of SRF target gene. The regulation of different classes of SRF target genes according to the signal type relies on selection of cofactor recruited to SRF target genes, either the ternary complex factor (TCF) of ETS DNA-binding domain-containing cofactors that is involved in the MAPK pathway or myocardin and MRTFs. Class-1 SRF target genes that primarily encode proteins with immediate–early functions are aided

Serum response factor modulates the expression of genes that encode structural and regulatory effectors of actin dynamics [456].[30] However, nuclear ^{G}actin promotes nuclear MRTF export and prevents nuclear MRTF activation of SRF target genes.

Globular actin can enter into and exit out of the nucleus. ^{G}Actin then controls both nucleus–cytoplasm shuttling and nuclear activity of transcriptional cofactors of the myocardin protein family (i.e., myocardin itself and myocardin-related transcription factors). Inside the nucleus, actin acts jointly with both ^{G}- (e.g., MRTFs) and ^{F}actin-binding proteins [456]. ^{G}Actin nuclear export is carried out owing to its nuclear export sequence, partly in conjunction with members of the exportin family or profilin.

Myocardin-related transcription factors can be exported from the nucleus bound to ^{G}actin [456]. On the other hand, the concentration of MRTF retained in the cytoplasm is correlated to the cytoplasmic concentration of ^{G}actin. Incorporation of ^{G}actin into actin filament provokes MRTF release and translocation into the nucleus, where MRTF cofactors can interact with SRF factor [456].

Class-2 SRF target genes that are transcribed upon MRTF–SRF activation include those that encode actins as well as many genes that modulate actin dynamics, such as the GSN (gelsolin) and VCL (vinculin) genes, as well as microRNAs that yields a feedback control [456].[31]

---

by TCF cofactors. Class-2 SRF target genes are regulated by SRF in synergy with myocardin or MRTFs. They encode at least 3 types of proteins that operate in: (1) cell contraction functions; (2) microfilament dynamics and cell motility; and (3) microRNA activities.

[30] Synthesis of cytoskeletal components associated with actin filament rearrangements happens in various processes, such as intracellular transport, cell adhesion to adjacent cells and extracellular matrix, protrusions (ruffles, filopodia, and lamellipodia), cell invagination (endocytic pits and phagocytic cups), formation of stress fibers and maintenance of the cellular tone, cell shape adaptation, cell spreading, neurite and vascular tip cell extension, axon guidance, and epithelial–mesenchymal transitions. Serum response factor regulates genes involved in cytoskeletal dynamics, cell proliferation, migration, survival and apoptosis, as well as myogenesis. Homodimer SRF builds a regulatory platform with recruited cofactors and target gene promoters.

[31] Serum response factor activates miR genes that encode miR1-1, miR1-2, miR133a1, and miR133a2 [456]. These miRs influence the SRF–MRTF signaling (negative and positive feedback loops). Two bicistronic miR genes encode pairs of homologous miRs: miR1-1 and miR133a2 as well as miR1-2 and miR133a1. In mammals, miR1 and miR133a genes are expressed specifically in cardio- and skeletal myocytes, miR133b in skeletal myocytes. MiR133a inhibits SRF expression. Serum response factor also activates transcription of miR143 and miR145 in cardiomyocytes and smooth muscle cells that regulate the expression of numerous mRNAs of regulators of actin remodeling and MRTFs (SRF inhibitors Krüppel-like factor KLF4 and KLF5, MRTFb, phosphatase slingshot-2 that dephosphorylates (activates) cofilin, capping protein adducin-3, and RHO (CDC42

Coactivators of serum response factor include DNA-binding, homeodomain-containing protein[32] of the Nkx2-5 family,[33] transcription factors of the GATA family (GATA1–GATA6), and Cys-rich LIM-only proteins of the CSRP family, in addition to MRTF coactivators [456].

Inhibitors of SRF comprise the LIM-only protein FHL2,[34] histone deacetylase HDAC4, muscle segment homeodomain-containing (MSH) transcriptional repressor MSx1,[35] and Krüppel-like factor KLF4, as well as homeodomain-only protein (HOP or HOP homeobox [HOPx]) in cardiomyocytes [456].

6.3.2.7 Actin Polymerization at the Cell Cortex

When growing filament barbed ends encounter a wall, they are stalled and must reorient away from it [462]. Once activated by a nucleation-promoting factors at the cell cortex, the ARP2–ARP3 complex (Sect. 6.3.3.21) captures a free polar seed (with its pointed and barbed ends) that is attached to its side to generate a branch point to nucleate a daughter filament. This branch can take any direction.

Inwardly developing filaments as well as filaments tangential to the surface elongate until stopped by capping proteins. Outwardly developing filaments (that grow toward the plasma membrane, i.e., free barbed end toward the cell surface) are blocked until they escape, reorient, and restart growth tangential to the obstacle [462].

Proteins that bind to actin barbed ends at or near the protrusion leading edge of migrating cells such as formins can contribute to the forward orientation of the barbed ends. Tangentially oriented filaments act as primers for further branching events at the cell cortex.

A cell protrusion is produced at the cell cortex by an actively polymerizing zone that provides pressure. This zone is constrained to grow between the

---

and Rac) inhibitors RhoGAP13 (or Slit–Robo Rho GTPase-activating protein SRGAP1) and RhoGAP14 (SRGAP2). MicroRNA-486 is regulated by SRF together with MRTF. It resides particulalrly in cardiac and skeletal myocytes. MiR486 promotes the PI3K–PKB axis, as it inhibits PTen and FoxO1a.

[32] The homeodomain is a protein domain encoded by a homeobox, a DNA sequence, of the Hox gene clusters (HoxA–HoxD) and other homeobox family genes, in particuler Short stature homeobox gene (SHOX), as well as other genes. A homeodomain protein can bind DNA, i.e., transcription factors that generally act in the promoter region of target genes complexed with other transcription factors.

[33] NK2 Transcription factor related, locus-5 homolog is also called cardiac-specific homeobox (CSx).

[34] LIM-only protein FHL2 is a p53-inducible protein also called Downregulated in rhabdomyosarcoma LIM domain protein (DRAL). It is a coactivator of β-catenin. It is also a corepressor of the transcriptional repressor Krüppel-like factor expressed in promyelocytic leukemia, the zinc finger and BTB domain-containing-factor ZBTB16 (a.k.a. zinc finger protein ZnF145 and promyelocytic leukemia zinc finger protein (PLZF).

[35] A.k.a. homeobox protein Hox7.

surface and a inner dead zone of capped filaments due to the high density of capping proteins that cover the entire surface at a certain distance from the wall [462]. Rupture of the dead zone yields space for the formation of a newly polymerizing zone that pushes the corresponding cell region forward. This polymerization zone, in turn, bears capping, leading to a new stress–rupture cycle.

6.3.2.8 Actin Filament Nucleator in Sarcomere Assembly

Sarcomeres are the smallest contractile units of striated muscle cells (Vol. 5 – Chap. 5. Cardiomyocytes). Sarcomeric filaments are in permanent renewal. Sarcomere formation requires the precise alignment and coordinated integration of actin and myosin filaments owing to the action of at least 2 giant scaffold proteins, titin and nebulin. Initiation of muscle actin assembly in sarcomeres is mediated by filament nucleation factors.[36]

Leiomodin is an actin-filament nucleator that shares 2 actin-binding sites with tropomodulin and caps actin filament pointed ends [463] (Sect. 6.3.3.40). Leiomodin activity is enhanced by tropomyosin. Tropomyosin binds to leiomodin[37] to enable its localization to the middle of muscle sarcomeres.

6.3.3 Regulators of Actin Dynamics

Rho-associated, coiled-coil-containing protein kinase (RoCK) phosphorylates myosin light chain and inhibits MLC phosphatase, thereby promoting myosin–actin binding and bundling of actin filaments into stress fibers. Kinase RoCK also phosphorylates LIMK that phosphorylates (inactivates) actin-depolymerizing cofilin. Furthermore, RoCK phosphorylates formin homology domain-containing protein FHoD1 (Sect. 6.3.3.32), an auto-inhibited endothelial formin, to induce the formation of stress fibers [464]. Thrombin activates via its G-protein-coupled receptor Rho and RoCK and fosters the synthesis of stress fibers (Fig. 6.2).

6.3.3.1 Actin-Binding Proteins

More than 150 proteins are binding partners of actin (Tables 6.6 and 6.7). Actin-binding proteins operate as positive and negative actin partners. Actin-binding proteins, indeed, include: (1) actin partners that participate in monomer sequestration; (2) those that are involved in monomer delivery; (3) those that promote filament polymerization, i.e., elongation and branching of actin filaments; (4) those that cause filament capping; and (5) those that support actin filament depolymerization. Actin-binding proteins can thus transform

---

[36] In non-muscle cells, filament nucleators comprise the ARP2–ARP3 complex and formins.

[37] λεῖος: smooth; λειότης: smoothness; λειόω: to smooth.

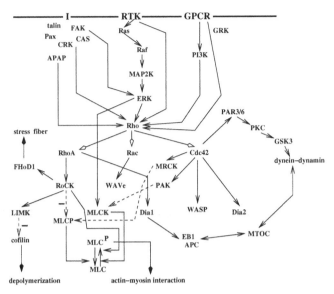

Figure 6.2. Regulation of actin polymerization (Sources: cmckb.cellmigration.org, [451, 464]; APAP: ArfGAP with PIx- and paxillin-binding domains APC: adenomatous polyposis coli protein; CAS: CRK-associated substrate; CRK: CT10 regulator of kinase; Dia: Diaphanous; EB: end-binding protein; ENA–VASP: Enabled homolog and vasoactive stimulatory phosphoprotein family member; ERK: extracellular signal-regulated protein kinase; FAK: focal adhesion kinase; MLC^P: phosphorylated myosin light chain (MLC); MRCK: myotonic dystrophy kinase-related CDC42 binding kinase; MTOC: microtubule organizing center; PAK: P21-activated kinase; PK: protein kinase; WASP: Wiskott-Aldrich syndrome protein; WAVe: WASP verprolin homolog [a.k.a. suppressor of cAMP receptor (SCAR)]). Actin filaments assemble and grow from a pool of actin monomers in response to different stimuli via integrins (I), G-protein-coupled receptors (GPCR), and receptor Tyr kinases (RTK). All activated plasmalemmal receptors stimulate members of the RHO family (RhoA, Rac, and CDC42). RTKs can operate via the mitogen-activated protein kinase (MAPK) cascade and GPCRs via phosphatidylinositol 3-kinase (PI3K). Kinase RoCK, an effector of small GTPase Rho, inactivates myosin light-chain phosphatase (MLCP), phosphorylates myosin light chain (MLC^P), LIM kinase (LIMK) that inactivates actin-severing cofilin, and FHoD1. Formin FHoD1 enhances the formation of stress fibers in vascular endothelial cells.

actin sol into gel by lengthening and branching of actin filaments and vice versa. They can recruit both actin-polymerizing and -depolymerizing proteins for suitable actin filament turnover.

The actin cytoskeleton dynamics are governed by the balance between actin-polymerizing and actin-depolymerizing proteins that are also known as actin-binding proteins (ABP) and actin-severing proteins (ASP; Table 6.8).

Table 6.6. Examples of actin filament constituents, regulators, and partners (**Part 1**; Sources: [456, 465]). Transgelin-1 (Tagln) is also called smooth muscle protein-22α (SM22A gene). Actin-related protein ARP3 is encoded by the ACTR3 gene (or Arp3). Cofilin-1 is a widely distributed, 18-kDa, non-muscle, actin-modulating protein that depolymerizes filamentous actin (^{F}actin) and inhibits the polymerization of monomeric globular actin (^{G}actin). Gelsolin as well as cofilin bind to polyphosphoinositides. Gelsolin activity is stimulated by calcium ions. It is one of the most potent members of the actin-severing gelsolin–villin family. Villin resides in microvilli of the brush border of epithelia of gut and nephron that possesses gelsolin-like domains and phosphatidylinositol (4,5)-biphosphate binding sites. It operates in the bundling, nucleation, capping, and severing of actin filaments. Supervillin associates with both actin filaments and the plasma membrane.

Molecule	Gene	Function
Structural components of actomyosin microfilaments		
Actin		Actin filament
Caldesmon-1	CALD1	Cell contraction
Dystrophin	DMD	Membrane anchorage
$β_1$ Integrin	ITGB1	Focal adhesion linkage
Myosin heavy chains-i	MYHi	Nanomotor
Myosin light chains-i	MYLi	Nanomotor
Transgelin	TAGLN	Cell contraction
Smoothelin	SMTN	Cell contraction
Tropomyosin-1	TPM1	Cell contraction
Vinculin	VCL	Integrin–actin linking
Actin polymerization and depolymerization		
Actin-related protein-3	ACTR3	Actin filament branching
Cofilin-1	CFL1	Depolymerization (turnover)
Gelsolin	GSN	Depolymerization (filament fluidization)
Villin-1	VIL1	Filament mesh organization
Supervillin	SVIL	Filament rearrangement

6.3.3.2 Actin-Sequestering Proteins

Actin-sequestering proteins prevent spontaneous assembly of monomeric actin. Cells other than myocytes possess 2 sequestering proteins: profilin and β4-thymosin.

Profilin binds actin monomers and can also weakly interact with filament barbed ends. Profilin catalyzes the conversion from actin^{ADP} and free ATP to actin^{ATP} and free ADP. Profilin supports the transfer of actin monomers to the polymerization zone.

Thymosin-β4 binds actin monomers and creates a sequestered pool of monomers to supply actin^{ATP} during periods of filament growth. It fails to accumulate in actin polymerization regions.

Table 6.7. Examples of actin filament constituents, regulators, and partners (**Part 2**; Sources: [456,465]). Filamin-A crosslinks actin filaments into orthogonal networks in the cell cortex and participates in the anchoring of membrane proteins to the actin cytoskeleton. It interacts with integrins. Myosin light-chain kinase (MLCK) is encoded by the MYLK gene (or Mlck). Four different MLCK isoforms exist: smooth muscle (MLCK1, smMLCK, or telokin), skeletal–cardiac (MLCK2 or skMLCK), cardiac (MLCK3), and novel (MLCK4). Myosin light-chain phosphatase (MLCP) is a heterotrimer with a catalytic protomer PP1cδ and a regulatory (inhibitory) subunit PPP1R12a (or myosin phosphatase targeting subunit MyPT1). It dephosphorylates phosphorylated myosin light chain (MLC^P). Phosphorylation of PPP1R12a by Rho-activated kinase (RoCK), Death-associated protein kinase DAPK3 (or ZIPK), and protein kinase-C (PKC) mediate calcium sensitization for cell contraction. Phosphorylation of protein phosphatase-1 regulatory subunit PPP1R14a (a.k.a. 17-kDa C-kinase potentiated PP1c inhibitor CPI17) is also promoted by calcium sensitization. Calcium desensitization depends on activated cGMP-dependent protein kinase PKG that causes PPP1R12a phosphorylation and PPP1R14a dephosphorylation by PP2a phosphatase. Globular actin polymerization liberates cofactors of the myocardin-related transcription factor (MRTF), thereby priming the activity of the serum response factor (SRF) that targets genes encoding structural and regulatory effectors of actin dynamics.

Molecule	Gene	Function
Regulators of actin dynamics		
Myosin light-chain kinase	MYLK	Cell contraction
MLC Phosphatase	PPP1R	Cell relaxation, Calcium sensitization
Filamin-A	FLNA	Crosslinking and integrin signaling
Talin-1	TLN1	Integrin signaling
Four-and-half LIM domain-	Fhl1	Adherens junction signaling
containing proteins	Fhl2	and transcriptional regulation via SRF
Connective tissue GF	Ctgf	Integrin binding, Cell migration
Serum response factor	Srf	Transcriptional regulation

6.3.3.3 Actin Nucleators

Nucleators (assembly factors) and/or their cofactors bind to and recruit multiple actin monomers to support polymerization. The FH2 domain does not serve to bind actin monomers, but to stabilize spontaneously formed actin dimers and trimers that are used in actin polymerization [453].

Classification of Actin Nucleators

Mammalian actin nucleators constitute a class of nucleation-promoting factors that can be decomposed into 3 subclasses [451, 454] (Table 6.9):

Table 6.8. Actin polymerization–depolymerization cycle.

Process	Factors and pathways
Polymerization	CDC42–WASP–ARP2/3
	CDC42–Dia2
	Rho–WAVe–ARP2/3
	RhoA–Dia1
	PAK–LIMK⊖ ⟶ cofilin
	RoCK–LIMK⊖ ⟶ cofilin
	Ena–VASP
	PAK–MLCK
	RoCK–MLCP
Depolymerization	Gelsolin
	Cofilin

subclass-1 with the actin-related protein ARP2–ARP3 complex that links a new filament from the side of an existing filament with a filament branching at a 70-degree angle and its 9 nucleation-promoting factors (NPF), i.e., 7 NPF infraclass-1 Wiskott-Aldrich syndrome proteins and 2 NPF infraclass-2 members, cortactin and hematopoietic cell-specific LYN substrate HCLS1 (or HS1);[38] subclass-2 with 15 formins that share 2 formin homology domains (FH1 and FH2) at their C-termini, move processively with the barbed end as it elongates, and include Diaphanous-related formins (DRF);[39] and subclass-3 with 4 WASP homology 2 (WH2) domain-containing nucleators, i.e., Spire,[40] Cordon-bleu homolog (CoBl), and leiomodin (LMod1–LMod3; in myocytes).[41]

Actin Nucleation-Promoting Factors

Actin polymerization is carried out in synergy with nucleation-promoting factors that stimulate ARP2–ARP3 activity (Table 6.10). Actin nucleation-

---

[38] Cortactin is a substrate of Src kinase.

[39] Diaphanous-related formins constitute a family of the subclass of formins that serves as effectors of Rho GTPases. DRF Proteins are auto-inhibited by intramolecular interactions between the C-terminal Diaphanous-autoregulatory domain (DAD) and N-terminal Diaphanous inhibitory domain (DID) in the regulatory region that also encompasses a GTPase-binding and dimerization domain.

[40] Formin- and Spire-nucleated filaments are not branched.

[41] Leiomodin-1 is also called thyroid and eye muscle auto-antigen D1, 64kD, and SMLMod. Leiomodin-2 and -3 correspond to the cardiac and fetal isoforms, respectively. Leiomodin localizes to the middle of sarcomeres and contributes to sarcomere assembly. Translocated actin-recruiting phosphoprotein (TARP) is a bacterial nucleator of actin (not the human phosphoprotein messenger thymocyte cAMP-regulated phosphoprotein [TARPP; a.k.a. 21-kDa cAMP-regulated phosphoprotein (ARPP21)] produced in the brain).

Table 6.9. Class of actin nucleators (Source: [454]; DRF: Diaphanous-related formins; ER: endoplasmic reticulum; HCLS: hematopoietic cell-specific LYN substrate; JMy: junction-mediating regulatory protein; WASH: WASP and SCAR homolog; WASP: Wiskott-Aldrich syndrome protein; WAVe: WASP-family verprolin homolog [a.k.a. suppressor of cAMP receptor (SCAR)]; WHAMM: WASP homolog associated with actin, membranes, and microtubules).

Category	Members	Function
Subclass 1	ARP2–ARP3	Branched actin filament meshes
	Nucleation-promoting factors (NPFs)	
	WASP, WAVe	NPFs at plasmalemmal protrusions
	WASH	NPF at endosomes
	WHAMM	NPF at ER-to-Golgi body vesicles
	JMy	NPF in the nucleus and ruffles
	Cortactin	Cell protrusion, vesicular transfer,
	cell migration	
	HCLS1	T-cell activation
Subclass 2	Formins	Regulation of actin structures
	(stress fibers, long actin fibers,	
	and filopodial actin network)	
	DRFs	Effectors of Rho GTPases
Subclass 3	Spire	Membrane trafficking
	Cordon-bleu	Neuron morphology
	Leiomodins	Sarcomere assembly

promoting factors of the Wiskott-Aldrich syndrome protein (WASP) and WASP-family verprolin homolog (WAVe)[42] subfamilies support branched filament networks at plasmalemmal protrusions. In other cellular loci (endosomes, secretory route from the endoplasmic reticulum to the Golgi body, and nucleus), actin is assembled by actin nucleators in cooperation with WASP and SCAR homolog (WASH), WASP homolog associated with actin, membranes, and microtubules (WHAMM), and junction-mediating regulatory protein (JMy), respectively [454].

6.3.3.4 Actin-Crosslinking Proteins

The mechanical properties of the actin cytoskeleton are defined by local activation of different actin crosslinking proteins [466]. These crosslinkers, which belong to a subset of actin-binding proteins, determine the structure of actin networks.

Actin-binding proteins that crosslink actin filaments differ according to the actin-assembly types. Crosslinkers vary: (1) according to the actin-binding

---

[42] A.k.a. suppressor of cAMP receptor (SCAR).

Table 6.10. Localization of actin nucleation-promoting factors (Source: [454]; ARP: actin-related protein; CoBl: Cordon-bleu homolog; DAAM: Disheveled-associated activator of morphogenesis; Dia: Diaphanous; FHoD: formin homology-2 domain-containing protein; FRL: formin-related in leukocytes; InF: inverted formin; JMy: junction-mediating regulatory protein; nWASP: neuronal WASP; WASH: WASP and SCAR homolog; WASP: Wiskott-Aldrich syndrome protein; WAVe: WASP-family verprolin homolog; WHAMM: WASP homolog associated with actin, membranes, and microtubules).

Location	Actin partners
Cell junction	ARP2–ARP3, Dia1–Dia2, formin-1, WAVe2
Golgi body	ARP2–ARP3, nWASP, WHAMM
Stress fiber	DAAM1–DAAM2, Dia1–Dia3, FHoD1, InF1
Endocytic structure	ARP2–ARP3, Dia1–Dia2, nWASP, Spire-1, WASH
Phagocytic structure	ARP2–ARP3, Dia1, FRL1, nWASP, WAVe2
Membrane ruffle, lamellipodium	ARP2–ARP3, CoBl, Dia1–Dia2, formin-1, JMy, nWASP, WAVe1–WAVe2,
Filopodium, cell spike	ARP2–ARP3, Dia1–Dia2, nWASP

affinity of their binding domains and (2) structure, number, and organization of their spacing rod domains that separate the binding domain.

Bundling crosslinkers are classified as short crosslinkers, such as fascin and plastin, and long crosslinkers, such as α-actinin and filamin. Short crosslinkers form filaments from isolated bundles. In addition, crosslinkers can be grouped as rigid (e.g., scruin) and compliant types.

The more flexible crosslinker filamin-A, together with the nanomotor myosin, augments the rigidity of the cytoskeletal network w.r.t. the entangled filament network [426].[43]

The architecture of the actin network is determined by the kinetics of the interaction between actin crosslinkers and actin [426]. If the dissociation rate of the crosslinker from the actin filaments is high, then filaments are aligned into bundles. Otherwise, filaments are stabilized in a more randomly ordered state.

Actin assemblies have very different turnover rates, ranging from tens of seconds (e.g., in lamellipodia)[44] to days (e.g., in cochlear hair cell microvilli). Actin dynamics also depend on bound nucleotide state. Actin filaments that

---

[43] Filament entanglement impedes displacement of one filament due to the presence of another filament.

[44] Lamellipodia are three-dimensional cytoskeletal meshes that resist deformation and generate forces to displace the cell membrane. Filopodia are parallel bundles

are first polymerized from actin^{ATP} monomer have slower shrinkage rate than that of filaments polymerized directly from actin^{ADP} monomer. In a population of young filaments, actin^{ATP} dissociates rapidly from an actin^{ATP} lattice. However, structure and stability of an actin filament is not determined uniquely by the state of its bound nucleotide.)[45]

6.3.3.5 α-Actinin

Long actin crosslinker α-actinin supports the formation of actin stress fibers. α-Actinin is found in adhesion plaques, in filament–membrane binding sites, and within actin filaments. Unlike fascin, α-actinin possesses filament-binding sites that rotate much more freely. It can stabilize orthogonal actin-filament networks that reinforce the plasma membrane or parallel bundles.

6.3.3.6 Espin

Espin (ectoplasmic specialization protein)[46] is a multifunctional actin-bundling protein. It abounds in microvillus-like projections of stereocilia and microvilli, where it is linked to parallel actin bundles. It intervenes in the regulation of the organization, dynamics, and signaling of ^{F}actin-rich, microvillus-type, transducers in various mechano- and chemosensory cells. Espin is encoded by a single gene; several espin isoforms exist.

6.3.3.7 Fascin

Short actin crosslinker fascin preferentially stabilizes parallel bundles of filaments such as those in filopodia, owing to the rigid coupling between filament-binding sites on fascin. Fascin localizes to the cytoplasm in an inactive phosphorylated and active dephosphorylated form. Fascin is a substrate of protein kinase-Cα (Ser39) and possibly PKCβ and PKCγ.

Fascin interacts with many partners, such as β- and γ-catenins, as well as low-affinity nerve growth factor receptor and small GTPase Rab35 [468]. Fascin colocalizes on actin bundles with α-actinin and LIM and SH3 domain-containing proteins LASP1 and LASP2, as well as glial fibrillary acidic protein.

In mammals, fascin lodges in dendritic cells rather than monocytes, macrophages, granulocytes, and B and T lymphocytes [468]. Fascin participates in the regulation of dendritic cell maturation. Fascin is also expressed at high

---

of actin filaments that sense chemical gradients and interact with appropriate cellular targets.

[45] In vitro, actin filaments constitute 2 populations according to shrinkage rates. Newly polymerized filaments shrink rapidly, primarily from barbed ends (shrinkage rate ∼1.8/s), whereas old, stable filaments shrink slowly, mainly from pointed ends (shrinkage rate ∼0.1/s) [467].

[46] A.k.a. autosomal recessive deafness protein Dfnb36.

levels in the adult mammalian nervous system (neurons of the cortical, hippocampal, and cerebellar regions) [468].

Three mammalian fascin isoforms are encoded by different genes: ubiquitous 55-kDa fascin-1, retinal 56-kDa fascin-2, and testis 56-kDa fascin-3 [468].

Fascin-1 localizes to microspikes and stress fibers. Thrombospondins, laminin, and fibronectin can trigger the formation of fascin-containing $^{\rm F}$actin microspikes. These microspikes form when cells initiate spreading and disappear when cells are fully expanded. Microspike dynamics involve protein kinase-C that phosphorylates fascin as well as small GTPases Rac and CDC42 that recruit fascin to lamellipodia, filopodia, and cell edge [468]. Fascin-1 abounds in stromal fibroblasts and vascular endothelial cells.

Fascin-2 is a retina and ear specific isoform.[47] It stabilizes actin bundles in these structures. It cooperates with other actin crosslinkers, such as espin, and plastin-1 and -3.

In migrating cells, fascin is associated with actin dynamics in leading-edge protrusions [468]. When filaments in the periphery extend, active dephosphorylated fascins are recruited to the clustered barbed ends of growing actin filaments, link the filaments into bundles, and form the nascent filopodia. Actin crosslinks by fascin are dynamical during filopodal extension. The association and dissociation of fascin from the shaft of filopodia allow the direct interaction between myosins and actin filaments. Consequently, filaments in filopodia can cycle back toward the cell body through myosin-dependent retrograde flow. The disassembly of bundles at the basal region of filopodia is rapid. Besides, actin–fascin bundles slide on myosin-2 and -5 as fast as single actin filaments.

6.3.3.8 Filamin

Actin aggregation to produce a filamentous gel is induced by actin filament crosslinker filamin. Filamin not only binds actin, but also other molecules, such as small RHO GTPases CDC42, Rac, Ral1, and Rho, as well as effector RoCK kinase. Its other partners include enzyme Trio, caveolin-1, and tumor-necrosis factor receptor-associated factor TRAF2 (Ub ligase).

Filamin stabilizes the membrane-cytoskeletal interface in platelets. In endothelial cells, filamin regulates the distribution of $^{\rm F}$actin between the cortical actin band and actin stress fibers. Filamin can thus regulate the endothelial barrier, as it stabilizes cortical actin as well as controls the organization and distribution of actin filaments at intercellular junctions. Integrin β_1 and β_2 (Sect. 7.5.4) recruit filamin to the plasma membrane and associate with it. Filamin also links to E-selectin (Sect. 7.5.3) upon leukocyte binding to the luminal surface of endothelial cells. In unstimulated endothelial cells, membrane-associated filamin is constitutively phosphorylated by protein kinase-A to protect filamin from proteolysis by calpain and hence avoid

---

[47] Fascin-2 abounds in stereocilia on specialized hair cells of the inner ear that detect and transduce mechanical signals.

disruption of cortical actin band. On the other hand, calmodulin-dependent kinase CamK2 phosphorylates filamin in the cortical actin band to destabilize its membrane conjugation.[48]

6.3.3.9 Plastins

Plastins, or *fimbrins*,[49] constitute a family of actin-binding proteins (Pls1–Pls3) that crosslink actin filaments. They are encoded by 3 genes on chromosome-3 (PLS1), -13 (PLS2), and -X (PLS3).

Plastin-1, or I-plastin (intestine-specific plastin), is expressed at high levels in the small intestine, colon, and kidneys, as well as at relatively lower levels in lungs and stomach [469]. Plastin-1 colocalizes with parallel actin filaments. It serves to bundle actin filaments in parallel orientation.

Plastin-2, or L-plastin,[50] is produced predominantly in hematopoietic cells. Plastin-2 participates in leukocyte function. In macrophages, it colocalizes with actin particularly in podosomes and filopodia. It is used as a marker of cancers. (It was identified in human neoplastic fibroblasts.)

Plastin-3, or T-plastin, is synthesized in cells from solid tissue. It may be involved in DNA repair [470]. In epithelial cells, overexpression of plastin-3 increases the length and density of microvilli. Plastin-3 and -1, but not plastin-2, can be detected in stereocilia of cochlear auditory hair cells.

6.3.3.10 Scruin

Scruin heightens markedly the magnitude of the elastic modulus (a measure of the resistance of the cytoskeletal network to deformation; stress-stiffening behavior) [426].[51] The actin elastic modulus depends on the concentration of crosslinkers.

6.3.3.11 Spectrin

Spectrin is an actin crosslinker that can bind zonula occludens protein ZO1 to interact with adherens junction protein α-catenin and gap junction protein connexin-43 (Chap. 7), as well as vasoactive stimulatory phosphoprotein.

---

[48] Bradykinin and thrombin activate CamK2 and cause filamin translocation from the cell cortex to the cytosol, thereby increasing endothelial permeability.

[49] The name "fimbrin" refers to the association of this protein with surface structures and protrusions, such as focal adhesions, membrane ruffles, microvilli, and microspikes.

[50] A.k.a. lymphocyte cytosolic protein LCP1.

[51] In cytoskeletal meshes that have a highly organized architecture, non-linear stress stiffening can be followed by stress softening at high stresses that is reversible. Actin filaments that bear compression and stretch can bend like a tube. Actin filament bending can be the main strain.

Table 6.11. Ezrin–radixin–moesin (ERM) proteins and interacting kinases and GTPases (Source: [437]; $\oplus \longrightarrow$: stimulation; $\ominus \longrightarrow$: inhibition; MRCK: myotonic dystrophy kinase-related CDC42-binding kinase; PI(4)P5K: phosphatidylinositol 4-phosphate 5-kinase; PIP$_2$: phosphatidylinositol bisphosphate; PP: protein phosphatase; RoCK: Rho-associated, coiled-coil-containing protein kinase). A reciprocal regulation exists between ERMs and RHO family GTPases (CDC42, Rac, and Rho).

RHO Effectors	CDC42–MRCK
	Rho–RoCK, Rho–PI(4)P5K–PIP$_2$
	Rac–PP
RHO Target	ERM–RhoGDI$\ominus \longrightarrow$ Rho
	ERM–RhoGEF21$\oplus \longrightarrow$ Rho
Other kinases	PKC, LCK, CDK5

Table 6.12. Ezrin–radixin–moesin (ERM)-binding partners in leucocytes (Source: [437]; ABCb1: ATP-binding cassette sub-family-B [multidrug resistance protein] member-1 [a.k.a. P-glycoprotein-1]; HCAM: homing cell adhesion molecule, heparan sulfate proteoglycan [a.k.a. epican and CD44]; ICAM: intercellular adhesion molecule; LCK: lymphocyte-specific protein Tyr kinase; NCF1/4: 47- and 40-kDa neutrophil cytosolic factor-1 and 4; SelPLg: selectin-P ligand [a.k.a. P-selectin glycoprotein ligand PSGL1]; SLC9a3R1: solute carrier family-9 [sodium–hydrogen exchanger] member-3 regulator-1 [a.k.a. NHERF1]; SYK: spleen Tyr kinase). Sialophorin is also called CD43, galactoglycoprotein (GalGP), leukocyte sialoglycoprotein and leukosialin.

Type	Species
Integral membrane proteins	L-selectin, ICAM1/2/3, HCAM, sialophorin, SelPLg, Syndecan-2, TNFRSF6a, ABCb1
Cytosolic proteins	Calpain, SLC9a3R1, NCF1/4, LCK, SYK, PI3K

6.3.3.12 Ezrin–Radixin–Moesin

Actin-binding proteins ezrin, radixin, and moesin (ERM; Sect. 7.2.4.12)) crosslink actin with the plasma membrane upon phosphorylation (activation). The phosphorylation status of ERMs is regulated by the RhoA–RoCK axis or phosphatidylinositol 4-phosphate 5-kinase (Table 6.11). Ezrin–radixin–moesin proteins have many binding partners (Table 6.12).

Ezrin–radixin–moesin proteins act as both effectors and affectors of Rho GTPases. Phosphorylation of a threonine residue (Thr558 in moesin, Thr567 in ezrin, Thr564 in radixin) of the actin-binding domain of ERM proteins primes their activation [437]. In addition, ezrin is phosphorylated (Tyr353) on its middle sequence and then interacts with phosphatidylinositide 3-kinase that can activate Rac. All 3 members of the ERM family are phosphorylated on other residues (N-terminal Tyr145 and Thr235 and C-terminal Thr558). Protein kinases PKCα and PKCθ target ezrin and moesin, respectively. Kinases RoCK and MRCK (myotonic dystrophy kinase-related CDC42-binding kinase) can also phosphorylate ERM proteins.

6.3.3.13 Profilin

Profilin (Pfn) precludes actin polymerization. However, this actin-sequestering protein has a subtle activity, as it is able to support polymerization of actin onto the barbed ends of actin filaments. Profilin can stimulate assembly of actin filaments; once complexed with ^{G}actin, it attracts monomers to the actin end. Polyphosphoinositides impede the fixation of actin-severing proteins on actin, thereby enabling actin filament lengthening.

Profilin that binds actin monomers maintains a pool of unpolymerized ATP-bound actins. Incorporation of actin^{ATP} into a filament promotes ATP hydrolysis (reaction half-life of few seconds) [471]. Profilin elicits the exchange of ADP for ATP, then binds tightly to actin^{ATP} monomers, refilling the actin monomer pool.

6.3.3.14 Cortactin and HCLS1

Translocated cortactin as well as the related Hematopoetic cell-specific Lyn kinase (SrcB subfamily) substrate HCLS1 (or HS1) pertain to the weak infraclass-2 activators (subclass-1 [ARP2–ARP3 and WASP] actin nucleators) that increases actin polymerization.[52] Cortactin is a prominent Src substrate. It is implicated in cellular protrusion, particularly during cell migration, and endocytosis. Protein HCLS1 is involved in T-cell activation.

These proteins bind ARP2–ARP3, but not ^{G}actin. Instead, they interact with ^{F}actin. They also link to nWASP and WASP-interacting protein (WIP). Albeit a weak ARP2–ARP3 activator, cortactin can enhance nWASP-mediated activation of the ARP2–ARP3 complex. Moreover, cortactin stabilizes ARP2–ARP3-bound actin branches, as it prevents spontaneous dissociation of filament–branch junctions. Activated GTPase ARF1 stimulates the recruitment of actin, cortactin, and dynamin-2 to Golgi membranes [472].

---

[52] Cortactin and HS1 promote ARP2–ARP3-dependent actin assembly by a different mechanism with respect to infraclass-1 nucleation-promoting factors, such as the ARP2–ARP3 complex and 7 members of the Wiskott-Aldrich syndrome protein family, including WASH and WHAMM activators. Mammalian subclass-1 actin nucleation-promoting factors share a C-terminal catalytic motif: the verprolin or WH2 domain, central connector and acidic domain (VCA) that enables to form a tripartite module with ^{G}actin and the ARP2–ARP3 complex.

6.3.3.15 Coronin

Most coronins are ^{F}actin-binding proteins that interacts with microtubules. Coronin homologs constitute a subfamily among WD repeat-containing proteins.

Seven coronin isoforms (Crn1–Crn7) have been detected in mammals. Coronins are classified into: (1) short conventional coronins (e.g., Crn1 and Crn2) with a C-terminus that mediates homophilic dimerization or olimerization and (2) long coronins. Coronin-1A and -1B preclude the formation of actin filaments via the ARP2–ARP3 complex. On the other hand, coronin-7 does not interact with actin, but participates in the cargo transfer in the Golgi body.

Ubiquitous coronins are direct inhibitors of the ARP2–ARP3 complex. They bind to ^{F}actin and protect newly formed actin^{ATP} filaments from severing cofilin. Nevertheless, they can synergize with cofilin and actin-interacting protein AIP1 to promote severing or bursts of disassembly of old actin^{ADP} filaments [454]. Inhibition of ARP2/3-mediated actin nucleation promotes the recycling of actin and the ARP2–ARP3 complex for efficient turnover of materials during cell migration.

6.3.3.16 Capping Protein

Heterodimeric actin filament-capping protein is composed of α and β subunits. ^{F}Actin-capping protein subunits $\alpha 1$ and $\alpha 2$ that are encoded by the CAPZA1 and CAPZA2 genes are detected in myocyte Z-lines (hence the alias CapZ). ^{F}Actin-capping protein subunit β is encoded by the CAPZB gene.

Capping protein (CP) participates in various actin-containing structures. It bind to the end of the actin-related protein ARP1 minifilament in the dynactin complex. In migrating cells, lamellipodia contain a mixture of uncapped and capped actin filaments close to the leading edge.

Actin-filament capping by capping proteins is controlled by proteic and lipidic inhibitors. Many proteins contain the capping protein interaction (CPI) motif, such as complex ARP2/3 myosin-1 linker (CARMIL), CK2-interacting protein CKIP1, CD2-associated protein (CD2AP), SH3 domain-containing kinase-binding protein SH3KBP1,[53] CapZ-interacting protein (CapZIP), and Fam21-containing WASH complex and capping protein[54] (WashCap) [473].

Protein CKIP1 links CK2 kinase that is involved in the regulation of cell polarity, morphology, and differentiation to actin remodeling. Protein CKIP1 inhibits interaction between capping protein and actin filaments.

The homologous protein CD2AP and related SH3KBP1 that have uncapping activity connect to endocytic components and cortactin. Protein

---

[53] A.k.a. 85-kDa CBL-interacting protein CIN85.
[54] A.k.a. family with sequence similarity-21 (Fam21).

SH3KBP1 also binds WIP, WAS–WASL-interacting protein (WIP) family protein WIPF2[55] nWASP, and CARMIL [473]. On the other hand, CD2AP is indirectly linked to nWASP via Pro–Ser–Thr phosphatase-interacting protein PSTPIP1. Both CD2AP and SH3KBP1 associate endocytosis and cadherin-based junctions with actin remodeling.

CapZ-interacting protein and myotrophin impede actin-filament capping by capping proteins. Family with sequence similarity-21 members (FAM21a–FAM21d) are components of the WASH complex that locates to endosomes and activates the ARP2–ARP3 complex.[56]

Capping protein interaction motif may act as an allosteric modulator that restricts capping protein to a low-affinity, filament-binding conformation [473]. Peptides that possess these CPI motifs are able to inhibit capping protein and to uncap capping protein-bound actin filaments.

6.3.3.17 Gelsolin

Actin-depolymerizing protein gelsolin caps barbed ends of actin filaments in a Ca^{++}-dependent and pH-sensitive fashion. Once capped, growing actin filaments spontaneously depolymerize from their pointed ends.

6.3.3.18 Cofilins and Destrin

Actin-depolymerizing protein cofilin not only caps barbed ends of actin filaments, but also breaks down fully polymerized actin filaments and prevents their lengthening. Severing proteins of the actin depolymerization factor–cofilin family modulates the structure, dynamics, and mechanical properties of actin filaments such as filament bending stiffness.

Cofilin constitutes with actin-depolymerizing factor (ADF) a family of 3 members: non-muscle cofilin-1 (or n-cofilin); muscle cofilin-2 (or m-cofilin); and destrin, or ADF, that are encoded by the CFL1, CFL2, and DSTN genes, respectively.

Cofilin regulates actin filament dynamics that are required for membrane dynamics as well as cell polarity, motility, division, and apoptosis. Cofilin binds to both ^{G}- and ^{F}actins. The activity of cofilin is driven by a phosphorylation-dephosphorylation cycle. Cofilin severs actin filaments and thus generates free actin barbed ends.

Cofilin binds to actin^{ATP} and accelerates phosphate dissociation. Dissociation of phosphate favors dissociation of branches from the ARP2–ARP3

---

[55] A.k.a. WASP-interacting protein (WIP)-related protein (WIRE) and WIP-and CR16-homologous protein (WICH).

[56] The larger WASH complex comprises WASH, capping protein, tubulins, coiled-coil domain containing-53 proteins (CCDC53), strumpellin, and HSP70. As the WASH complex represses capping by capping proteins, WASH (Fam21) can be renamed WashCap [473].

complex and binding of cofilin to actin^{ADP} subunits. Cofilin bound to filaments severs and depolymerizes actin^{ADP} filaments.

Cofilin-decorated actin filament is about 5-fold more flexible than an actin filament. The persistence length of actin filaments (9.8 μm) that corresponds to a flexural rigidity of 0.040 pN/μm^2 is reduced to a value of 2.2 μm and the filament flexural rigidity to 0.0091 pN/μm^2 [474]. Molecular dynamics simulations yield similar results [475].

Small GTPases Rho activate P21-activated protein kinase (PAK; Vol. 4 – Chap. 4. Cytosolic Protein Ser/Thr Kinases) that stimulates LIM kinase to phosphorylate (inactivate) cofilin and slows down the rate of filament disassembly.

Chronophin, or pyridoxal phosphate phosphatase (PLP),[57] that is encoded by the PDXP gene, is a phosphatase that dephosphorylates cofilin [476]. Both chronophin and LIMK are required to control cofilin activity.

The depolymerizing activity of cofilin is actually inhibited by phosphorylation by LIM kinases (LIMK).[58] Reversible phosphorylation of cofilin by LIMK is supported by the 14-3-3ζ scaffold. Cofilin^P is dephosphorylated (re-activated) by slingshot phosphatases. In addition, cofilin^P stimulates phospholipase-D1 [477]. Muscarinic receptors (Vol. 3 – Chap. 7. G-Protein-Coupled Receptors) and Rho–RoCK-dependent activation actually control phospholipase-D1 activity via LIMK, slingshot phosphatase, 14-3-3ζ, and cofilin. Activated phospholipase-D1 hydrolyzes phosphatidylcholine of cell membranes into phosphatidic acid that elicits stress fiber formation in certain cell types.

Slingshot phosphatases[59] (Ssh) and LIMK regulate actin dynamics via a reversible inactivation of the actin-depolymerizing factor–cofilin component. Its activity is regulated by ^{F}actin and 14-3-3ζ protein. The activity of slingshot phosphatases is controlled by multiple factors (calcium ions, cAMP, and phosphatidylinositol 3-kinase). A proteic complex that consists of Ssh1L, LIMK1, actin, and a scaffold protein regulates, with kinase PAK4, ADF–cofilin activity [478].

6.3.3.19 ENA–VASP Family Members

Members of the ENA–VASP family comprise Enabled homolog and vasoactive stimulatory phosphoprotein. These actin-binding proteins participate in the regulation of cell adhesion and motility, especially axon guidance and platelet aggregation.[60] The C-terminus elicits the self-assembly of ENA–VASP

---

[57] Pyridoxal is a natural form of vitamin-B6.

[58] Enzyme LIMK is a Ser/Thr kinase that contains a Lin1, Isl1, and Mec3 (LIM) domain. Two LIM-kinases — LIMK1 and LIMK2 — are expressed in most tissues. They are activated by Rho GTPases.

[59] Protein phosphatase slingshot dephosphorylates both LIM kinase and cofilin.

[60] Protein VASP was initially identified as a target of cAMP- and cGMP-dependent kinases during platelet activation.

proteins into stable tetramers. Actin remodelers of the ENA–VASP family connect to both monomeric and filamentous actin.

The ENA–VASP family proteins are encoded by 3 genes: (1) the ENAH (Enabled homolog) gene,[61] (2) the Vasp (vasoactive stimulatory phosphoprotein) gene, and (3) the Evl (ENA–VASP-like) gene.[62] Alternatively spliced isoforms have been detected, particularly during various phases of embryo- and fetogenesis.

Actin remodelers of the ENA–VASP family are able to move with and protect actin-growing barbed end upon clustering [453]. Oligoproline sequence of actin regulator VASP represents the profilin-binding site. Profilin potentiates actin filament growth. High concentration of profilin in the polymerization zone boosts actin assembly. Oligoproline sequence repeats may contribute to the force-producing and position-securing steps of the formation of lamellipodia and filopodia at the leading edge of migrating cells.

Protein VASP acts as a processive actin polymerase that binds to actin monomers and targets them to the growing barbed end of actin filaments. In the absence of ^{G}actin, VASP tetramer uses its ^{F} (FAB) and ^{G}actin-binding (GAB) domains to bind along the side of ^{F}actin [479]. Monomeric actin binds to the GAB domain, impedes lateral VASP–^{F}actin interactions, and targets VASP to the barbed end of actin filaments. Actin partner VASP associates with ^{F}actin processively, as it delivers actin monomers to the barbed end. The rate of filament elongation is enhanced by ^{G}actin-binding profilin, but precluded by capping proteins [479].

6.3.3.20 Caldesmon

Actin-binding caldesmon inhibits actin-activated myosin ATPase in a Ca^{++}–calmodulin-dependent manner. Caldesmon phosphorylation relieves its inhibition on myosin ATPase and yields actomyosin filament contraction. As caldesmon binds both actin and myosin, it can stabilize contraction when myosin light-chain phosphorylation level is low. In addition, caldesmon regulates capping and severing activity of gelsolin.

6.3.3.21 Actin-Related Proteins

The actin-related protein complex ARP2–ARP3 is composed of 7 polypeptides (ARPC1–ARPC5, ARP2, and ARP3). Like actin, ARP2 and ARP3 nucleators bind ATP for conformational changes. The ARP2–ARP3 complex binds to actin filaments. It undergoes phosphorylation.

The ARP2–ARP3 complex binds to the side of actin filaments to generate an actin branch on a pre-existing filament. Repeated branching leads to a

---

[61] Enabled homolog (EnaH; a.k.a. mammalian Ena [MEna]) is a mammalian ortholog of the Drosophila protein Ena.

[62] Proteins EnaH and EVL are highly expressed in neurons and the spleen and thymus, respectively.

dendritic network. The ARP2–ARP3 complex remains at the branch segment between the new filament and the side of the existing filament.

Nucleation-promoting factors elevate the activity of the ARP2–ARP3 complex. It can actually be activated by one of several nucleation-promoting factors, such as Wiskott-Aldrich syndrome protein (WASP) and WASP verprolin homolog (nWASP and WAVe1–WAVe3).

Extracellular stimuli activate plasmalemmal receptors to trigger signaling down to Rho effectors. Small GTPases Rho bind to and activate auto-inhibited WASP proteins that bring together actin monomers and ARP2–ARP3 complexes [471]. The ARP2–ARP3 complex then caps actin pointed ends and primes the growth of a new actin filament as a branch on the side of an old actin filament. The branch grows rapidly at its barbed end by addition of actin-profilin complexes. The new filament elongates until lengthening is impeded by capping proteins. Two types of inhibitors designated as CK548 and CK636 bind to the ARP2–ARP3 complex and prevent its actin nucleation activity [480].

6.3.3.22 Wiskott-Aldrich Syndrome Protein

Wiskott-Aldrich syndrome protein is restricted to hematopoietic cell lineages, whereas neural WASP (nWASP) and WASP-family verprolin homologs (WAVe1–WAVe3)[63] are more widely expressed. Binding of WASP-interacting protein (WIP) to WASP protects WASP from degradation by calpains.

In cells of the innate and adaptive immunity, Wiskott-Aldrich syndrome protein operates as a signaling scaffold and actin regulator. WASP Proteins bind the ARP2–ARP3 complex and recruits monomeric actin to stimulate nucleation of branched actin filaments. Activity of WASP is attenuated by multiple signaling pathways. Protein WASP undergoes conformational change, phosphorylation, and degradation. Protein WASP interacts with many partners (Table 6.13).

In mammals, 7 members of the Wiskott-Aldrich syndrome protein family (nucleators of the class-1 of cytoskeleton regulators; infraclass-1 of the subclass of actin nucleation-promoting factors) have been identified: WASP, nWASP, WAVe1 to WAVe3, JMy,[64] WHAMM,[65] and WASH.[66] Actin nucleators WASP and nWASP are involved in receptor endocytosis [482]. Proteins WAVes regulate lamellipodium formation and cell migration and adhesion. Agent JMy influences cell migration as well as transcriptional regulation. Factor WHAMM intervenes in molecular transfer from the endoplasmic reticulum to the Golgi body.

---

[63] A.k.a. products of the Wiskott-Aldrich syndrome protein family genes (WASF1–WASF3).

[64] JMy: junction-mediating and regulatory.

[65] WHAMM: WASP homolog associated with actin, membranes, and microtubules.

[66] WASH: WASP and SCAR homolog.

Table 6.13. WASP Partners (Source: [481]; ARP: actin-related protein; CDC42: small GTPase cell division cycle; CK: casein kinase; PI(4,5)P$_2$: phosphatidylinositol (4,5)-bisphosphate; PTPn: protein Tyr phosphatase non-receptor; FnBP1L: formin-binding protein 1-like protein (a.k.a. transducer of CDC42-dependent actin assembly TOCA1; VASP: vasodilator-stimulated phosphoprotein; WIP: WASP-interacting protein) Src family kinases (SFKs) include Fyn, viral feline Gardner-Rasheed sarcoma oncogene homolog (FGR), hematopoietic cell kinase HCK, leukocyte-specific kinase (LCK), and viral yes-1 Yamaguchi sarcoma-related oncogene homolog (Lyn). TEC family kinases (TFKs) comprises: Bruton Tyr kinase BTK, IL2-inducible T-cell kinase (ITK), and Tyr kinase expressed in hepatocellular carcinoma (TEC). Adaptors encompass non-catalytic region of Tyr kinase (NCK), growth factor receptor-bound protein (GRB), CRK avian sarcoma virus CT10 homolog-like (CRKL), syndapin, intersectin-2, and Pro–Ser–Thr phosphatase-interacting protein PSTPIP1.

WASP Partner	Effect
WIP	WASP stabilization
PI(4,5)P$_2$	WASP activation with CDC42 and NCK1
ARP2, ARP3	Branching actin polymerization
CDC42	WASP activation
FnBP1L	CDC42 docking for WASP activation
Rac1, RhoQ	WASP activation
SFKs	WASP activation
TFKs	WASP activation
Adaptors	WASP activation
PTPn12	WASP dephosphorylation (inactivation)
VASP	WASP activation
CK2	WASP activation
Monomeric actin	Activation of ARP2–ARP3 complex

Three subfamilies constitute the WASP family: (1) subfamily 1 with Wiskott-Aldrich syndrome proteins, including nWASP; (2) subfamily 2 with WASP-family verprolin homologs (WAVe), also termed suppressors of cAMP receptor (SCAR); and (3) subfamily 3 with WASP and SCAR homolog (WASH).

Actin nucleation-promoting members of the WASP family couple the ARP2–ARP3 complex to actin on the one hand and to small GTPases CDC42 and Rac to support the growth of branched actin filaments on the other. They also bind to PIP$_2$. The latter agent located in membranes promotes WASP colocalization and dimerization.

6.3.3.23 Neural Wiskott-Aldrich Syndrome Protein

Ubiquitous neural Wiskott-Aldrich syndrome protein (nWASP) pertains to the WASP subfamily of activators of the ARP2–ARP3 complex. Activators WASP and nWASP are encoded by the gene WAS and WASL (WASP-like), respectively.

Protein nWASP mediates actin assembly at membrane interfaces to promote or stabilize membrane tubulation in vesicular transport. It accumulates at clathrin-coated pits, where it recruits the ARP2–ARP3 complex for actin polymerization at the pits. It also interacts with several endocytic proteins. It is involved in fission of endocytic vesicles.

Moroever, nWASP supports with CDC42 GTPase the actin assembly during formation of membrane ruffles, filopodia, lamellipodia, and invadopodia. In particular, it is required for membrane ruffles, especially during internalization of epidermal growth factor receptors [483].

Spontaneously, nWASP has a slight activity because of its auto-inhibition. This inactive conformation is stabilized by WASP homology-1 domain (WH1)-interacting proteins such as WASP-interacting protein (WIP), a ^{G}- and ^{F}actin-binding molecule, which complexes with nWASP [454]. Furthermore, nWASP oligomerization increases nWASP activity and its affinity for the ARP2–ARP3 complex [454].

Binding of small GTPase CDC42 changes the conformation that allows an allosteric activation, which is enhanced by connection to phosphatidylinositol (4,5)-bisphosphate. Adaptors Non-catalytic kinases NCK1 and NCK2, formin-binding protein-1-like FnBP1L,[67] and Abelson kinase interactor AbI1 (or AblBP4), also activate nWASP. Protein Tyr kinases phosphorylate nWASP to stabilize its active form.

Scaffold nWASP connects to many molecules, such as the ARP2–ARP3 complex, ^{G}- and ^{F}actins, negatively charged phospholipids (e.g., phosphatidylinositol (4,5)-bisphosphate), heat shock protein HSP90, CDC42, many kinases, the WIP family proteins, members of the Bin–Amphiphysin–Rvs (BAR) domain-containing protein superfamily,[68] heterogeneous nuclear ribonucleoprotein hnRNPk, WISH,[69] Disabled homolog Dab1, IQGAP1, ABP1, NCK1, NCK2, GRB2, AbI1, intersectin-1, profilin, and cortactin. Some of these partners synergetically activate nWASP, such as CDC42 with GRB2 or NCK1 with PI(4,5)P$_2$.

Protein nWASP is regulated by auto-inhibition that is relieved by multiple ligands, especially CDC42^{GTP}. Two other small GTPases — RhoJ and RhoT — as well as IQGAP1 activate nWASP [483]. Binding of phosphoinositides enhances nWASP activation of the ARP2–ARP3 complex in synergy with

---

[67] A.k.a. transducer of CDC42-dependent actin assembly TOCA1.

[68] The BAR domain-containing protein superfamily includes several families: classical BAR, extended Fes–CIP4 homology BAR (fBAR), N-terminal amphipathic helix BAR (nBAR), inverse-BAR (iBAR), and PhoX BAR (pxBAR). BAR Domain-containing proteins can also bind to dynamin that is involved in fission of membrane tubules for vesicle formation. Interactors of nWASP that belong to the BAR domain-containing proteins comprise: (1) fBAR members syndapins (pascins), CIP4, FBP17, TOCA1, PSTPIP1 and -2, and nostrin; (2) nBAR proteins amphiphysin, Tuba and endophilin; (3) iBAR protein IRSP53; and (4) pxBAR protein SNX9.

[69] A.k.a. SPIN90 and DIP.

CDC42. On the other hand, formin-like protein-3 (FmnL3)[70] interacts with nWASP in the nucleus, hence causing its nuclear accumulation and precluding its cytoplamic effects.

Scaffold nWASP is phosphorylated by numerous kinases, such as SRC family kinases Fyn and Lck (Tyr256), as well as focal adhesion (FAK; Tyr256), Abelson (Abl), and casein kinase CK2 (Ser484 and Ser485) [483]. nWASP Phosphorylation (Tyr291) by Fyn kinase leads to its ubiquitination and degradation. Conversely, proteasomal degradation of nWASP is impeded by binding to HSP90 [483]. Lipid PIP_2 facilitates nWASP phosphorylation by Fyn kinase. Protein HSP90 increases the level of phosphorylation by SRC family kinases. nWASP Phosphorylation regulates binding with the ARP2–ARP3 complex.

Members of the WIP family (WIP, CR16, and WICH, or WIRE) interact with nWASP to prevent its activation by CDC42 [483]. Whereas the WIP–nWASP complex is inhibitory, the ternary complex with Transducer of CDC42-dependent actin assembly TOCA1 relieves the WIP-induced inhibition.

nWASP Dimerization explains the behavior of multivalent nWASP partners [483]. Phosphorylation and interaction with HSP90 and FBP11 promotes nWASP localization inside the nucleus. In the nucleus, nWASP also interacts with the PSF-NonO complex that is associated with RNA polymerase-2 [483].

Agent nWASP is involved in actin assembly in sarcomeres. It interacts with nebulin [484]. It is rapidly recruited to the Z-disk, where actin filaments are inserted, when the cell is subjected to insulin-like growth factor IGF1 that signals via the PI3K–PKB pathway.

6.3.3.24 WASP Family Verprolin Homologs

The subfamily of Wiskott-Aldrich syndrome protein (WASP) family verprolin homologs encompasses 3 mammalian members (WAVe1–WAVe3) encoded by the WASF1 to WASF3 genes. They are expressed in numerous cell types, but WAVe1 and WAVe2 are the most broadly distributed isoforms. All 3 isoforms abound in the central nervous system.

The WAVe complex promotes the assembly of actin-filament networks at the leading edge of motile cells. Movement of the WAVe complex correlates with cell protrusion.

Unlike WASPs, WAVes are intrinsically active. However, the WAVe complex is intrinsically inhibited due to an auto-inhibition by a component of the WAVE complex. The activation of the WAVe complex relies on binding of the WAVe complex to Rac^{GTP} and acidic phospholipids in cell membranes, particularly PIP_3 on vesicles, and phosphorylation of the WAVe complex.

---

[70] A.k.a. formin-binding protein FBP11.

They link to a complex that contains chromosome-3 open reading frame C3ORF10,[71] AbI1, NCK-associated protein NAP1, and Specifically Rac-associated protein SRA1 [454].[72]

Protein WAVe2 can bind to 53-kDa insulin receptor substrate protein (IRSP53)[73] and PI(3,4,5)P$_3$. Its Tyr phosphorylation modulates its activity [454].

6.3.3.25 WASP and SCAR Homolog

Seven WASH copies exist on different human chromosomal ends, but only the 9p copy encodes a full-length functional WASH protein. Yet, multiple WASH might be splice variants [483].

Ubiquitous WASP and SCAR homolog (WASH) interacts with multiple proteins, such as capping protein CapZ, which caps filament barbed ends, and Family with sequence similarity FAM21, which links WASH to endosomes [454]. Mammalian WASH, indeed, localizes to early and recycling endosomes. The ARP2–ARP3 activity mediated by WASH influences retromer-dependent transfer to the trans-Golgi network, recycling to the plasma membrane, and transport to late endosomes.

Agent WASH may also collaborate with dynamin-2 to regulate endosomal scission. Moreover, WASH associates with retromer components, sorting nexins SNx1 and SNx2 that are involved in retrograde transport [482]. It connects also to strumpellin and coiled-coil domain-containing protein CCDC53.

6.3.3.26 ARP2–ARP3 Activators WHAMM and JMy

Both WASP homolog-associated with actin, membranes, and microtubules (WHAMM) and junction-mediating regulatory protein (JMy) are expressed in various cell types. Activator WHAMM is less potent than nWASP. On the other hand, JMy can nucleate actin even in the absence of the ARP2–ARP3 complex.

Protein WHAMM localizes to the cis-Golgi network and tubulovesicular membranes. It influences the morphology of the Golgi body and anterograde transport from the endoplasmic reticulum. In addition, WHAMM permits crosstalks between the actin and microtubule cytoskeletons [454].

Protein JMy is abundant in the nucleus. It may be recruited in response to DNA damage [454]. In migrating cells, JMy relocalizes to lamellipodia.

---

[71] A.k.a. BRICK1 [BRK1] homolog and hematopoietic stem–progenitor cell protein HSPC300.

[72] A.k.a. cytoplasmic fragile X mental retardation FMR1-interacting protein Cy-FIP1.

[73] A.k.a. brain-specific angiogenesis inhibitor-1-associated protein BAIAP2.

6.3.3.27 The Formin Subclass

Formins are involved in the remodeling of actin and microtubule cytoskeletons to coordinate actin and microtubule dynamics, direct cell shape, and control cell motility and division as well as tissue morphogenesis.

Formins facilitate filament elongation and, hence, polymerize unbranched actin structures.[74] Formins associate with the fast-growing barbed end of actin filaments. Unbranched actin filaments constructed by active formins are used to form stress fibers, actin cables, microspikes, and contractile rings. In addition, formins can bundle actin during plasma membrane protrusion and cell division.

Formins constitute a large subclass of proteins characterized by formin homology domains (FH1 and FH2). The large formin subclass can be subdivided into 2 infraclasses according to the presence or absence of a N-terminal transmembrane domain. The subclass of 15 mammalian formins is composed of 7 families [454] (Table 6.14): family-1 with Diaphanous (Dia); 2 with formin-related proteins in leukocytes (FRL); 3 with Disheveled-associated activators of morphogenesis (DAAM); 4 with formin homology domain-containing proteins (FHoDs); 5 with formins (Fmn); 6 with delphilin; and 7 with Inverted formins (INF).

Formins govern 2 distinct phases of actin polymerization: nucleation and elongation. Formins nucleate actin and cooperate with profilin. Formin homodimers bind to barbed ends and act as processive caps on elongating filaments. They then prevent other capping proteins from terminating filament lengthening. Two formin monomers are connected by flexible tethers to form a ring [454].

Once actin filament is nucleated, the dimeric FH2 domain moves processively with the growing barbed end and shields it from capping proteins, but allows the rapid addition of new subunits [453]. The rate of FH2 movement matches the rate of actin subunit addition that can exceed 100 subunits per second. The rate of elongation at FH2-capped barbed ends is influenced by interactions of adjacent FH1 domains with actin-linked profilin.

6.3.3.28 Formins

Mammalian formins encompass Fmn1 and Fmn2. Multiple formin isoforms exist. Formin-1 localizes to cadherin–catenin-based adherens junctions (or zonula adherens). Formin-2 contributes to spindle and chromosome movement to subcortical regions [454].

Formins are major actin nucleators for intracellular transport as well as cell migration and division. These single polypeptides dimerize actin. Some formins can also bundle or sever actin filaments.

---

[74] Unlike the ARP2–ARP3 complex, but like all other known actin nucleators, formins produce unbranched filaments.

Table 6.14. Mammalian formins and their effects (Source: [453]; DAAM: Disheveled-associated activator of morphogenesis; Dia: Diaphanous; FHoD, formin homology (FH1–FH2) domain-containing protein; Fmn: formin; FmnL: formin-like protein; InF: inverted formin).

Type	Functions
DAAM1–2	Gastrulation, morphogenesis, neuronal growth cone dynamics
Delphilin	Binding to glutamate receptors at the postsynaptic side of neuromuscular junctions
Dia1–3	Lammellipodium, filopodium, stress fiber formation, cell adhesion, migration, cytokinesis, endosomal transport, phagocytosis, synaptic growth and stability, microtubule stabilization, kinetochore attachment,
FHoD1/3	Membrane protrusion, stress fiber formation, microtubule alignment with stress fibers, cell adhesion, transcriptional activation
Formin-1–2	Membrane protrusion, cell adhesion, mitotic spindle positioning, cytokinesis
FmnL1–3	Phagocytosis, cell adhesion and migration
InF1–2	Microtubule stabilization, organization of the endoplasmic reticulum

Formin dimers associate with actin filament tips and promote the elongation of these cytoskeletal filaments (growth rate $\sim$100 actin molecules per second; elongation speed $\sim$0.25 μm/s [485]). Formins remain bound at the fast-growing, barbed end and moves processively as the filament elongates. Formin-mediated elongation is further enhanced by linking to profilin.[75]

6.3.3.29 Diaphanous-Related Formins

Diaphanous-related formins (Dia1–Dia3, or Diaph1, Diaph3, and Diaph2, respectively) are dimers. They are regulated by autoinhibition. Their actin nucleation function is stimulated by binding to GTPases of the RHO family, such as RhoA, RhoB, and RhoC, as well as CDC42, which disrupts the inhibitory interaction between the Diaphanous autoregulatory and inhibitory domains [454]. In fact, different GTPases (RhoA to RhoD, RhoF, CDC42, and Rac1) signal to multiple Diaphanous-related formins. Moreover, SRC family kinases interact with Diaphanous-related formins.

---

[75] Formin homology-1 domain (FH1) of formin interacts with profilin. FH1 Domain serves to recruit ^{G}actin monomers via profilin binding, but does not support actin nucleation. FH2 Sequence serves for actin polymerization. The flexible FH1 domain is positioned next to the FH2 domain and, like a tentacle, binds to the actin–profilin complex.

Diaphanous Dia1 rotate spirally on the end of actin filaments, both during growth and depolymerization of the filament [486]. In addition to their role in actin assembly, Dia1 to Dia3 can bind to microtubules as well as proteins that interact with microtubule plus-ends to assist in microtubule stability [454].

6.3.3.30 Formin-Related Proteins in Leukocytes

Formin-related proteins in leukocytes (FRL1–FRL3) bundle ^{F}actin. Actin bundling is especially important in the construction of actin arrays with parallel crosslinked filaments, such as stress fibers, cables, and filopodia. Formins FRL1 and FRL2 have an auto-inhibited conformation, but FRL3 seems to be constitutively active [454]. Protein FRL1 is involved in phagocytosis mediated by Fc immune receptors (Vol. 3 – Chap. 11. Receptors of the Immune System).

Formin-like proteins FmnL1 (or FRL1), FmnL2 (or FRL2), and FmnL3 (or FRL3), as well as Dia2 and inverted formin-2, are able to sever and depolymerize actin filaments, in addition to their nucleation and elongation effects [453]. Severing supports disassembly and turnover of actin structures.

6.3.3.31 Disheveled-Associated Activators of Morphogenesis

Disheveled-associated activators of morphogenesis include 2 gene products DAAM1 and DAAM2. Formin DAAM1 is also regulated by an auto-inhibitory interaction. Its inhibition can be relieved by the binding to small GTPase RhoA, RhoB, or RhoC as well as phosphoprotein Dishevelled [454]. Formin DAAM1 is required in non-canonical Wnt signaling (Vol. 3 – Chap. 10. Morphogen Receptors), during which it translocates to stress fibers to influence cell shape. It also intervenes in RhoA activation during cell polarization.

6.3.3.32 Formin Homology Domain-Containing Proteins

Formin homology domains (FH1 and FH2)-containing proteins FHoD1[76] FHoD2 (or FmnL2, but not FRL2), FHoD3 (or FmnL3),[77] and FHoD4 (or FmnL1) do not share primary sequence similarity in their N-termini with Dia, FRL, and DAAM proteins. Protein FHoD1 that is auto-inhibited by interactions between its C- and N-termini associates with Rac1 GTPase [454].

Activated FHoD1 promotes the formation of stress fibers. It operates via small GTPase RhoA, kinase RoCK1, and possibly Rac1 GTPase. Kinase RoCK1 is implicated in Rho-mediated actin organization. It phosphorylates FHoD1, thereby relieving its autoinhibition.

---

[76] Formin FHoD1 is also called FH domain-containing homolog overexpressed in spleen FHoS1.

[77] A.k.a. formin-related actin-organizing protein formactin-2, FHOS2, and WW domain-binding protein WBP3.

6.3.3.33 Delphilin

In mammals, delphilin is highly expressed in the central nervous system. Different splice variants contain a palmitoylation site and PDZ domains that enable linkage to cellular membranes [454]. Delphilin can contribute to synaptic adaptivity (plasticity).

6.3.3.34 Inverted Formins

Inverted formins include InF1[78] and InF2.[79] Protein InF1 that localizes to microtubules can induce the formation of stress fiber [454]. On the other hand, InF2 that localizes to the endoplasmic reticulum[80] participates in actin polymerization as well as severing and disassembly [454].

Diaphanous (Dia1–Dia3) interact via their diaphanous autoregulatory domains (DAD) with InF2 via its diaphanous inhibitory domain (DID). Subsequently, this binding, impede Dia-mediated, Rho-activated actin polymerization, as well as SRF-responsive gene transcription, especially in glomerular podocytes [487].

6.3.3.35 Formin Regulatory Processes

Numerous factors locally recruit, activate, or inactivate formins to prime or unleash formin effects on actin polymerization. Formin activities are regulated from their initial recruitment and activation at membranes to actin polymerization as well as their displacement, inactivation, recycling, and turnover. Cells thus use various spatial and temporal regulatory mechanisms to control formin effects on actin and microtubule dynamics.

Phase 1: Formin Inhibition Abrogation

Many formins are maintained in an inactive state in the cytosol, primarily by auto-inhibition, before being recruited to and activated at membranes [453]. Auto-inhibition has been observed in 5 among the 7 formin families. Formin-like protein FmnL2, inverted formins, and delphilin are actually not regulated by classic auto-inhibition. Formin inhibitory ligands include: (1) Dia-interacting protein DIP1[81] and (2) Spire.

---

[78] A.k.a. FH domain-containing protein FHDC1.

[79] Mutations in the INF2 gene cause kidney disease characterized by focal and segmental glomerulosclerosis.

[80] Formin InF2 is farnesylated to be recruited to endoplasmic reticulum membranes.

[81] A.k.a. NCK-interacting protein with SH3 domain NCKIPSD, 54-kDa VacA-interacting protein VIP54, 90-kDa SH3 protein-interacting with NCK SPIN90, transcriptional activator of acute leukemias and fusion partner of the mixed lineage leukemia gene at chromosome 3p21 (AF3p21), oncogenic determinant for Ad9 ORF1, Wiskott-Aldrich syndrome protein-like (human homolog)-binding protein (WASLBP) and WASP-interacting SH3 protein (WISH).

Phase 2: Formin Activation and Recruitment

Small GTPases Rho participates in formin activation. Yet, some formins are relieved from auto-inhibition by post-translational modification, even in the absence of Rho binding [453]. Kinase Src binds to delphilin, DAAM1, Dia2, and formin-1. P21-Activated kinase PAK1, a P53 regulator, can target some formins. Enzyme RoCK supports FHoD1 full activation. Dimerization of Dia1 may help to maintain Dia1 localized at the cell cortex. The effector of Rac1 and CDC42 IQ motif-containing GTPase-activating protein IQGAP1 binds to Dia1 for Dia1 recruitment to phagocytic cups. Furthermore, IQGAP1 may cooperate with Rho GTPase to activate Dia1. Microtubule plus-end tracking cytoplasmic linker protein CLiP1 binds to Dia1 during phagocytosis and may recruit Dia1 to the cortex. Abelson kinase interactor AbI1, a component of the WAVe complex, recruits Dia1 to adherens junctions.

Phase 3: Actin Nucleation and Elongation

The ARP2–ARP3 complex has a poor nucleation activity and requires a co-factor, i.e., a nucleation-promoting factor such as its actin monomer-binding cofactor WAVe., to efficiently polymerize actin. The rate of elongation is improved by Dia1 in the presence of profilin. Formin-binding proteins may compete with profilin [453].

Phase 4: Formin Action Attenuation and Termination

Many of the actin structure assembled by formins consist of short filaments (0.3–2.3 µm), such as cables, stress fibers, and cytokinetic rings, on a time scale of less than one second [453].

6.3.3.36 Formin Partners

Formin-binding partners regulate formin functions (Table 6.15). Many signaling pathways converge toward formins to trigger cytoskeletal remodeling.

Protein DIP1 localizes in the cell nucleus and acts in signal transduction, stress fiber formation, myofibril assembly into sarcomeres, and maintenance of sarcomeres. It interacts with Dia1 and Dia2. It enhances nWASP-induced ARP2–ARP3 complex activation independently of CDC42 GTPase. It is required for proper formation of the cortical actin network [453].

Formin Dia1 interacts with 53-kDa SH3 domain-containing insulin receptor substrate protein (IRSP53)[82] that coordinates membrane and actin cytoskeleton remodeling.

---

[82] A.k.a. brain-specific angiogenesis inhibitor-1-associated protein BAIAP2.

Table 6.15. Formin-binding proteins that regulate their localization and activity (Source: [453]; AbI: Abelson kinase interactor; CLiP: cytoplasmic linker protein; Dia: diaphanous; FHoD: formin homology (FH1–FH2) domain-containing protein; Fmn: formin; FmnL: formin-like protein; IqGAP: IQ motif-containing GTPase-activating protein; NCKIPSD: NCK-interacting protein with SH3 domain (DIP1); RoCK: Rho-activated kinase).

Partner	Formin	Cell type	Effects
Profilin	All	All	Recruitment of actin monomers
Spire			Actin assembling
AbI1	Dia1	Epithelial cells	Dia1 recruitment to cell adhesions, lamellipodia, and filopodia
CLiP1	Dia1	Macrophages	Dia1 recruitment to phagocytosis sites
G12/13	Dia1	Fibroblasts	Dia1 recruitment to leading edge of migrating cells
IqGAP1	Dia1	Fibroblasts, macrophages	Dia1 recruitment to leading edge and phagocytic cups
CDC42	FmnL1	Macrophages	FmnL1 recruitment to cell cortex
RhoA	Dia1	Epithelial cells	Dia1 recruitment to adherens junctions
RhoB	Dia1	Melanoma cells,	Dia1 recruitment to endosomes
	Dia2	fibroblasts	Dia2 recruitment to endosomes
RhoF	Dia2	Fibroblasts	Dia2 recruitment to filopodial tips
NCKIPSD	Dia2	HEK and HeLa	Dia2 inhibition
RoCK1	FHoD1	HeLa	FHoD1 phosphorylation (activation)
α-catenin	Fmn1	Epithelial cells	Fmn1 recruitment to adherens junctions

6.3.3.37 WH2 Domain-Containing Nucleators of the Spire, Cordon-bleu, and Leiomodin Families

Tandem WASP homology (WH2) domain-containing nucleators of the Spire, Cordon-bleu homolog, and leiomodin families cause actin nucleation by tethering at least 3 actin monomers in either a single-stranded, long-pitch multimer or a short-pitch trimer [454]. Nucleators of the Cordon-bleu homolog and leiomodin families use multiple monomer-binding sequences to assemble trimeric actin nuclei [454]. Nucleators of the Spire, Cordon-bleu, and leiomodin families nucleate actin to control membrane transport, neuron morphology, and myocyte organization, respectively [454].

6.3.3.38 Spire Family Members

In mammals, regulators of actin dynamics of the Spire family comprise Spire-1 and Spire-2 . Spire is able to bind cooperatively to 4 ^{G}actins. At low Spire/actin ratios, it can polymerize actin filaments. At a high Spire/actin ratio, Spire can cap and sever actin filaments as well as sequesters actin monomers [454]. In humans, Spire-1 interacts with barbed ends, where it inhibits profilin–actin assembly, and cuts filaments at pointed end. Actin monomer-binding protein Spire that nucleates actin assembly cooperates with formins in actin polymerization. Mammalian Spire-1 and Spire-2 actually bind to Fmn1 and Fmn2 formins.

Protein Spire is a single polypeptide that stabilizes longitudinal actin tetramers.[83] It induces rapid formation of filamentous actin structures [488]. Furthermore, Spire can crosslink actin filaments and microtubules. Spire can also inhibit actin nucleation by binding to formin-1.

6.3.3.39 Cordon-bleu Homolog

Cordon-bleu homolog binds ^{G}actin and promotes actin filament growth from trimeric seeds. Cordon-bleu also binds to ^{F}actin and contributes to the protection of barbed ends from disassembly [454]. It interacts with actin-binding protein ABP1 and syndapin that both support actin polymerization.

6.3.3.40 Leiomodin Family Members

Leiomodins are specific ^{F}actin nucleators of myocytes. Whereas Lmod1 and Lmod3 reside in smooth muscle cells, Lmod2 localizes to skeletal and cardiomyocytes [454]. Leiomodin N-terminus contains tropomyosin- and actin-binding motifs. ^{F}Actin-binding tropomyosin enhances actin polymerization by Lmod2. In cardiomyocytes, Lmod2 localizes with tropomyosin in sarcomeres, where ^{F}actin pointed ends are positioned, to participate in sarcomere organization.

6.3.3.41 Kinase and Phosphatase Regulators

Protein Ser/Thr calcium–calmodulin-dependent kinase CamK2β (but not CamK2α; Vol. 4 – Chap. 4. Cytosolic Protein Ser/Thr Kinases) binds to ^{F}actin filament to regulate ^{F}actin stability and support the generation of ^{F}actin-rich structures [489]. Binding of CamK2β to stable ^{F}actin is disrupted by calcium–calmodulin binding and phosphorylation.

6.3.3.42 Small GTPases

Small GTPases Rho control cytoskeletal dynamics via cytoplasmic effectors and myocardin-related transcription factors (MRTF) that are coactivators for serum response factor (SRF) to regulate assembly, activity, and turnover of actin-based structures. Rho Signaling controls MRTF activity, as it regulates the availability of ^{G}actin that binds to MRTFs to: (1) inhibit MRTF nuclear import; (2) promote MRTF nuclear export; and (3) hamper activation by the

---

[83] Spire contains a cluster of 4 WH2 domains, each of which binds an actin monomer.

MRTF–SRF complex. Target genes of SRF encode cytoskeletal components such as actin.

Small GTPases CDC42, Rac1, and RhoA cooperate to control cytoskeletal dynamics, especially during cell protrusion. In particular, all these GTPases are activated at the front of migrating cells (Vol. 2 – Chap. 6. Cell Motility). These monomeric GTPases regulate one another. Small GTPase CDC42 can activate Rac1. Small GTPase Rac1 and RhoA are mutually inhibitory.

In migrating cells, RhoA is activated at the cell edge synchronously with edge advancement, whereas CDC42 and Rac1 are activated 2 μm behind the edge with a delay of 40 s [490]. Small GTPases Rac1 and RhoA operate antagonistically with a spatial separation and precise activation timing during the protrusion–retraction cycle. The activation of RhoA during dynamical events at the leading edge is confined to a 2-μm layer from the leading edge. The activation of RhoA increases and decreases synchronously with cell protrusion and retraction. The activation of CDC42 and Rac1 is initiated at a distance from the plasma membrane at the leading edge. The maximal activation of CDC42 and Rac1 is reached with a 40-s delay relative to protrusion. Therefore, RhoA intervenes in the initial events of protrusion, whereas Rac1 and CDC42 activate signaling for reinforcement and stabilization of newly expanded protrusions.

6.3.3.43 14-3-3 Proteins

Proteins 14-3-3 interact with the actin-depolymerizing factor–cofilin complex as well as its regulators, LIM kinase and Slingshot phosphatase. LIM kinase phosphorylates (inactivates) ADF–cofilin complex.

6.3.3.44 Non-Catalytic Region of Tyr Kinase Adaptors

The NCK adaptor family consists of Src homology SH2–SH3 (3 SH3 and 1 SH2 motifs) domain-containing non-catalytic region of Tyr kinase adaptor. This adaptor family includes not only NCK family members that regulate Tyr kinase signaling, but also closely related SH2–SH3 domain-containing GRB2 and CRK adaptors. These adaptors connect protein Tyr phosphorylation to various pathways. They recruit proline-rich effectors to Tyr^P kinases or their substrates. Adaptor GRB2 links activated receptor Tyr kinases to the mediators SOS and Ras for cell proliferation [491].

Members of the NCK family are implicated in the organisation of the actin cytoskeleton. Members NCK1 and NCK2 contribute to integrin signaling from the extracellular space into the cell and interact with receptor Tyr kinases to regulate cell shape, survival, and proliferation. Adaptor GRB2 links plasmalemmal receptors to the Ras pathway that triggers protein synthesis, whereas NCK connects plasmalemmal receptors to the actin cytoskeleton [492]. In the GEF–Rac–NCK–PAK signaling axis, NCK permits the interaction of P21-activated kinase with Rac. The Nck genes have broad and overlapping expression pattern.

In particular, NCK family members recruit WASP and nWASP to sites of Tyr phosphorylation for actin assembly. The SH2 domain of NCK associates with Tyr^P proteins on the cell surface. The SH3 domain binds to signaling proteins, such as WASP, WIP, and P21-activated kinase, which regulate actin dynamics-dependent processes, such as membrane ruffling and vesicle motion. All 3 NCK SH3 domains are required for maximal activation. The Wiskott-Aldrich syndrome protein and nWASP that are coordinately activated by both CDC42 and phosphatidylinositol (4,5)-bisphosphate can stimulate the nucleation of actin filaments via the ARP2–ARP3 complex. On the other hand, NCK can fully activate nWASP in cooperation with PIP_2, but not CDC42. Therefore, the NCK-dependent, CDC42-independent activation of the ARP2–ARP3 complex that causes actin polymerization at Tyr phosphorylated, NCK-binding sites complements the $CDC42^P$- and PIP_2-dependent stimulation axis of the ARP2–ARP3 complex [493].

NCK-Interacting kinase (NIK or MAP4K4)[84] regulates the lamellipodium formation by growth factors and phosphorylation of ezrin, radixin, and moesin [494]. ERM Proteins regulate cell morphology and plasma membrane dynamics, as they reversibly anchor actin filaments to integral plasma membrane proteins. Kinase NIK binds to and phosphorylates ERM proteins. It localizes in particular with phosphorylated ERM proteins at the distal margins of lamellipodia. It is necessary for increased phosphorylation of ERM proteins in response to growth factors, such as EGF and PDGF, hence controlling growth factor-induced membrane protrusion and cell morphology.

6.3.3.45 Dynamins

Large GTPase dynamin links to the actin cytoskeleton either directly, as it possesses a ^{F}actin-binding site [495], or via actin-binding proteins such as cortactin. In addition to its role in the regulation of cell motility and endocytosis, particularly in clathrin-coated vesicle formation, dynamin colocalizes with actin filaments, often in locations where membranes undergo remodeling and actin rapid cycles of assembly and disassembly, such as cortical ruffles and podosomes. Moreover, dynamin binds numerous actin-binding proteins, such as cortactin, NCK adaptors (NCK1–NCK2), and profilin.

Short actin filaments created by capping proteins, such as gelsolin and actin-severing proteins, promote dynamin oligomerization and accelerate dynamin's GTPase activity,[85] whereas both ^{G}actin and long actin filaments have no effect [495, 496]. Oligomerized dynamin removes the high-affinity capping

---

[84] Not NFκB-inducing kinase (also alias NIK) or MAP3K14. NCK-Interacting kinase belongs to the subfamily-1 Sterile-20-related kinases–germinal center kinases (SLK–GCK), i.e., MAP4K1 to MAP4K5 and subfamily 4 of cytosolic Ser/Thr kinases with NRK, NIK, and TNIK (Vol. 4 – Chap. 4. Cytosolic Protein Ser/Thr Kinases).

[85] Dynamin's GTPase activity is involved in particular in podosome structure and function.

protein gelsolin at the barbed ends of filaments [495] (whereas ENA–VASP proteins delay gelsolin binding to ^{F}actin [496]), thereby contributing to filament elongation and actin dynamics. However, the uncapper oligomerized dynamin neither impinges on ^{F}actin formation nor reverses the inhibition of high-affinity, heterodimeric capping protein on actin polymerization. Actin-binding dynamin links actin remodeling to membrane dynamics, especially during molecular transfer. Actin is indeed required for dynamin-mediated membrane fission.

The dynamin family includes classical dynamins, dynamin-like proteins, Mx proteins such as GTP-binding protein Mx1 (Mx1: Myxovirus [influenza virus] resistance-1,[86] Optic atrophy protein,[87] mitofusins, and Atlastin-related proteins, or guanylate-binding proteins (GBP).

Classical dynamins include dynamin-1 to dynamin-3. Dynamin-1 is synthesized in neurons and neuroendocrine cells; dynamin-2 in most cell types; dynamin-3 highly in the testis, but is also in the brain, heart, and lung.

Dynamin-related proteins that are involved in organelle fission comprise dynamin-like protein and OpA1, a major organizer of the mitochondrial inner membrane required for the maintenance of cristae. Agent OpA1 in the mitochondrial inter membrane space and Mitofusin on the outside of the outer membrane are involved in mitochonrial fusion. They antagonize dynamin-related proteins.

Proteins Mx stimulated by type-1 interferons are involved in resistance against viral infections. Guanylate-binding proteins induced by type-2 interferons and anti-inflammatory cytokines have also an antiviral effect, in addition to an antiproliferatve function.

Dynamin has a similar affinity for actin than that of other actin-binding proteins, but exhibits a lower actin-bundling activity than α-actinin [496]. However, dynamin's bundling activity is enhanced in the presence of lipids.

6.3.4 Myosin Filaments

Myosin filaments (thickness 6 nm) localize along actin filaments. These actin-based nanomotors form spots and filaments that have about the same bore as actin filaments; myosin filaments in non-myocytes are thinner than those in muscle cells. Myosins are mechanoenzymes that convert the chemical energy derived from ATP hydrolysis into mechanical work.

Numerous different myosin types constitute the myosin class ($\sim$139 members [498]). The myosin class consists of at least 35 subclasses. In humans, 40 myosin genes exist that can be classified into 12 subclasses based on analysis of the structure of their head and tail [499] (Table 6.16). Myosin-1 (basic-tail monomers) and -2 (conventional) are the most abundant.

---

[86] A.k.a. interferon-inducible P78 protein (IFI78).

[87] Optic atrophy type-1 is a dominantly inherited optic neuropathy that results from mutations of the OPA1 gene, which encodes a dynamin-related protein localized to mitochondria [497].

Table 6.16. Human myosin class (Source: [499]; FERM domain: four point one–ezrin–radixin–moesin). The conventional class-2 myosins comprises the cluster of 6 skeletal muscle and 2 non-muscle myosin-2 heavy chains (MyH9 and MyH10). Cardiac myosin heavy chains myosin-2α and -2β corresponds to MyH6 and MyH7, respectively. Smooth muscle myosin-2 heavy chain tallies with MyH11. Embryonic and perinatal myosin-2 dovetail with MyH8 and MyH3, respectively.

Subclass	Members	Features
1	8 (Myo1a–Myo1h)	Monomeric, basic tail that binds phospholipids
2	14 (MyH1–MyH14)	Conventional; tails form bipolar filaments
3	2 (Myo3a–Myo3b)	NinaC-like; N-terminal kinase domain;
5	3 (Myo5a–Myo5c)	Vesicular transport; "dilute" domain, IQ motifs
6	1 (Myo6)	Reverse direction
7	2 (Myo7a–Myo7b)	Associated with deafness and blindness; with domains myosin tail homology MyTH4 and band FERM)
9	2 (Myo9a–Myo9b)	GAP domain for Rho, IQ motifs
10	1 (Myo10)	Pleckstrin homology (PH), MyTH4, and FERM domains
15	2 (Myo15a–Myo15b)	Deafness syndrome; MyTH4 and FERM domains
16	1 (Myo16)	N-terminal ankyrin repeats
18	2 (Myo18a–Myo18b)	N-terminal PDZ-myosin domain
Orphan	1	Short tail

Each myosin carries out its own functions. Muscle subclass-2 myosins yield appropriate forces and velocities during muscle contraction. Non-muscle myosin-2 intervene in cytokinesis. Myosin-1, -6, and -7 are involved in the inner ear function. Myosin-5 and -6 contribute to membrane trafficking.

Myosin-1, -2, -5, and -9 to -11 move toward the barbed plus-end of actin filaments. On the other hand, myosin-6 moves toward the pointed minus-end of actin filaments. Myosin-5 and -6 operate in exo- and endocytosis, respectively.

Most myosins are composed of a head, neck, and tail. The myosin head binds filamentous actin. It uses ATP hydrolysis to generate force and to travel along the filament toward the barbed plus-end (except myosin-6). The neck acts as a linker and lever arm that transduces force generated by the catalytic motor domain. The neck can also serve as a binding site for regulatory myosin light chains. The tail generally mediates interaction with cargos and/or other myosin subunits.

Myosin catalytic head binds to both actin and nucleotides. It contains a light chain-binding domain that extends out from the myosin catalytic domain like a lever arm to amplify small movements in the catalytic head.

The kinetic cycle of myosin-driven movement consists of 4 basic steps: (1) ATP binding to myosin catalytic head that releases the head from actin; (2) ATP hydrolysis that causes a conformational change of the catalytic head into a pre-stroke state; (3) upon phosphate release, the head rebinds strongly to actin and undergoes a transition from the pre-stroke state to a post-stroke state; (4) finally, ADP is released from the catalytic head, hence allowing ATP to rebind to complete the cycle.

Myosin heavy chains are major component of the contractile apparatus in myocytes. The MHC isozymes expressed in different muscle types are correlated with the specific activity of the Ca^{++} ATPase that reflects the maximal velocity of contraction of these muscle types. Sarcomeric myosin heavy chains are encoded by embryonic, perinatal, slow, and fast skeletal MHC genes. The muscle-specific expression can be modulated by neurological, hormonal, and mechanical stimuli. Muscle-specific, sarcomeric MHC isozymes differ from cytoplasmic isotypes. Some isoforms of these ubiquitous proteins also reside in non-muscle cells.

In mammals, at least 9 distinct sarcomeric myosin heavy chain genes (MyHC or MYH) exist (MyHC1, MyHC2A–MyHC2B, MyHC2D, MyHC3, MyHC6–MyHC8, and MyHC13). The MYH1 gene encodes the myosin heavy chain-1 found in the skeletal muscle of human adults. Cardiomyocytes possess both the MYH6 (a.k.a. MyHC2α and α-MyHC) and MYH7 (a.k.a. MyHC2β and β-MyHC) genes that produces the fast and slow isoforms of cardiac myosin heavy chain, respectively. Fast, type-2 skeletal muscle fibers express one or more of the 6 type-2 skeletal muscle myosin genes. Type-2 skeletal muscle myosin genes include: (1) the developmental MyHC genes, i.e., MYH3 (embryonic MyHC [MYHC-EMB]) and MYH8 (perinatal MyHC [MyHC-PN]); (2) the adult-fast MyHC genes, i.e., MyHC2A, MyHC2B, and MyHC2D; and (3) the specialized MYH13 gene (extraocular MyHC [MyHC-EO]).

Non-muscle myosin heavy chain-2 isoforms are encoded by 3 known genes: MYH9, MYH10, and MYH14 (or NMMHC2A– NMMHC2C). The functions of the N-terminal motor and C-terminal rod domains of non-muscle myosin-2A and -2B explains the different roles played in cell–cell and cell–matrix adhesion [500].

Numerous genes (MYL or MyLC) encode myosin light chains (MyLC1–MyLC5, MyLC6a–MyLC6b, MyLC7, and MyLC9) as well as myosin regulatory light chain-interacting protein (MyLIP) and cardiac muscle myosin light polypeptide-like MyLL1 that are considered as components of functional myosin ATPase complexes, rather than myosins, in addition to the MYLK1 (or simply MYLK) and MYLK2 genes that encode smooth (smMLCK or MLCK1) and skeletal (skMLCK or MLCK2) muscle myosin light-chain kinase.

Subclass-5, -6, and -10 myosins transport cargos along cytoskeletal tracks. Myosin-5 contributes to vesicle transfer; myosin-6 to organelle anchoring and endocytosis; myosin-10 to filopodial trafficking. Myosin-5 and -10 move toward the barbed end of actin filaments; myosin-6 toward the pointed end. In some types of mammalian cells, zones of actin either enhance or inhibit myosin of specific subclasses (cellular region selectivity of myosins) [501]. In some cells, myosin-6 moves along stress fibers, but motility of myosin-5 and -10 is suppressed (cytoskeletal structure selectivity of myosins). Furthermore, the processing velocity of myosins can differ according to subcellular localization due to a possible regional control of nanomotor motility by the actin cytoskeleton (cellular region selectivity of myosin motility). Certain cell types display a uniform motility of all myosin-5, -6, and -10 throughout the entire cell surface.

6.3.4.1 Myosin-1

Myosin-1 in association with actin displaces cell organelles. Myosin-1 nanomotor is a single-headed myosin molecule that acts in the regulation of cell membrane structure and dynamics. Myosin-1 is a linker of cellular membranes to the actin cytoskeleton for mechanical stress sensing, transduction, and generation, as well as membrane trafficking. Myosin-1 can sense tension to control cell shape and transport of materials. It also yields tension to sensitize mechanosensitive ion channels, especially for hearing. Small loads ($<2\,pN$) cause 75 times lower rates of myosin-1 detachment from actin [502].

Widely expressed, tension-sensitive, myosin-1 isoform Myo1b is alternatively spliced (Myo1ba–Myo1bc; Myo1bc: shorter isoform) during development as well as in various cell types in its light chain-binding domain (myosin region that acts as the lever arm), thereby yielding proteins with lever arms of different lengths. Actin-detachment kinetics of Myo1b splice variants depends on tension [503]. Magnitude of tension sensitivity is related to splicing.[88] Therefore, the light chain-binding domain regulates step size, motility rate, myosin activation, and force sensing. In addition, Myo1b is significantly more tension-sensitive than other myosins with similar length lever arms.

6.3.4.2 Myosin-2

Myosin filaments can assemble and disassemble in non-muscle cells and some types of smooth muscle cells. Like actin that has 2 states (monomeric, globular and polymeric, filamentous), smooth muscle myosin-2 forms 2 distinct pools: (1) an assembled myosin-2 pool in equilibrium with actin filaments to build stress fibers and (2) a disassembled myosin-2 pool in which the 10S self-inhibited, monomeric conformation of myosin-2 prevails. The monomeric myosin-2 can be tranferred to the assembled myosin-2 pool. In human airway smooth muscle cells, the 10S monomeric smooth muscle myosin-2 also exists in the assembled myosin-2 pool, in which it can assemble into filaments according to the context [504].

Myosin-2 is also detected in striated myocytes. Human skeletal myocytes express 5 type-2 myosin heavy chains in distinct spatial and temporal patterns. In adults, skeletal myocytes muscles are characterized by a predominant expression of the MyHC2A, -2B, and -2D genes, as well as MYH7 (β-MyHC). Yet, expression patterns differ among species and are influenced by exercise

---

[88] Myosin-2 and -5 have light chain-binding domains that differ in size and hence have different force sensitivity.

and age. Moreover, the 3 adult subclass-2 fast myosin isoforms have ^{F}actin-activated ATPase rates that are close, but with the following increasing order [505]:

$$MHC2a < MHC2d < MHC2b.$$

These diverse isoforms are characterized not only by distinct biochemical activities, but also actin-filament velocities. In addition, their ATPase activity and motility are generally correlated. The MyHC2B gene is not expressed under most circumstances. The corresponding myosin MHC2b has a high ATPase activity, but slow motility [505].

Non-muscle myosin heavy chain-2A is a subunit of non-muscle myosin-2A that controls cell adhesion and migration. It abounds in membrane rafts.

Myosin-2 and actin bind weakly in the presence of ATP, and strongly with ADP^{Mg}. Actin binding leads to conformational changes in myosin and subsequent sequential release of hydrolysis products of ATP, first phosphate, and then ADP. Myosins move toward the barbed plus-end of actin filaments (except myosin-6 implicated in cellular transport).

An index of the mechanical performance of actin–myosin-2 interaction is the *duty ratio*, i.e., the fraction of the cycle time during which each myosin head is attached to actin. The slower the kinetics of the ATPase cycle, the longer the duration of actin–myosin attachment, and the larger the duty ratio. Muscle myosin-2 duty ratio is low to suit rapid contraction. Long, bipolar filaments of muscle myosin-2 can sustain loads even when a small fraction of heads are attached. Short filaments (with a restricted number of attachment modules) of non-muscle myosin-2 that operate in cytokinesis, tension maintenance, and cell contractility, have high duty ratio. Non-muscle myosin-2A and -2B exhibit load-dependent changes in kinetics of ADP release, but ATP uptake is insensitive to load [502].[89] Mechanosensitive ADP release rate (that rises and decays upon assisting and resisting load, respectively) that is faster for myosin-2A (lower duty ratio) than for myosin-2B, allows maintenance of tonic contraction without disturbing cell contractility driven by other myosin types. Under assisting loads, acceleration of ADP release prevents myosin-2 from impeding contraction imposed by muscle myosin. Under resisting loads, longer duration of actin–myosin-2 binding enables tension maintenance.

The activity of the actin–myosin cytoskeleton of the cell cortex participates in cell morphogenesis, cell locomotion, and cytokinesis during the cell cycle. Cortical tension is exerted by ^{F}actin and myosin-2 anchored in the plasma membrane.

Myosin-2 is activated by phosphorylation of the regulatory light chain by RoCK kinase most often activated by Rho GTPases and Rho guanine nucleotide-exchange factor RhoGEF2. G-Protein subunits of the Gα_{12}–Gα_{13} family, actin-binding protein Shroom,[90] and Abelson kinase Abl1 regulate

---

[89] Myosin-2 and -5 are characterized by load-sensitive nucleotide release, whereas myosin-6 and kinesin display load-sensitive nucleotide binding [506].

[90] In humans, the SHROOM family comprises 4 members (Shroom-1–Shroom-4). Protein Shroom is an actin-associated determinant of cell morphology that inter-

cortical remodeling. The actin–myosin cytoskeleton is also involved in cell adhesion reorganization.

The force–velocity relation of myosin-2 implies that myosin generates greater force when adhesion complexes slide slowly. When external forces are applied to the cytoskeleton, actin filaments polymerize and orient parallel to the force direction. The greater the external force, the faster the actin polymerization [508]. Large actin bundles correspond to stress fibers. Due to time scale of actin bundle formation and degradation, more bundles are formed when the cytoskeleton time scale is small, i.e., the matrix is stiffer. Mechanosensation can rely on adhesion complexes, non-muscle myosin-2, and stress fibers. Adhesion complexes yield a viscous drag for expansion and retraction of actin-based cellular protrusions. This viscous drag for both linear and rotational movements arises from integrins and actin crosslinking proteins. The drag coefficient depends on attachment and detachment rates as well as matrix stiffness. On stiff surfaces, adhesion complexes mature into focal adhesions. Conversely, on soft surfaces, cell–matrix adhesions remain small, dynamical adhesion complexes.

Myosin-2 nanomotors on actin filaments generate forces that direct cell and tissue shape. In epithelia, contractile forces are balanced by adhesive forces at E-cadherin-based apical adherens junctions. Myosin-2 can concentrate in junctions, where it transmits the cortical tension [509]. Crosslinkers between filaments transmit tension to different parts of the cytoskeleton and couple the cortex actomyosin mesh to E-cadherin–β-catenin–α-catenin complexes. Epithelial morphogenesis relies on polarized, contractile activity of actomyosin filaments that results from interactions between E-cadherins and the actomyosin mesh.

Cell locomotion relies on a spatial regulation of actin network turnover with assembly of a network of actin filaments in protrusion of the leading edge and its disassembly at the cell rear. Actin-binding proteins of the ADF–cofilin family (actin-depolymerizing factor, non-muscle cofilin-1 [Cfl1], muscle cofilin-2 [Cfl2], cofilin-like-2 [CflL2], cofilin pseudogene proteins [CflP1, CflP3–CflP5]) contribute to actin disassembly. In keratocytes, myosin-2 concentrates in regions at the cell rear with high rates of network disassembly [510]. The contribution of myosin-2-driven disassembly of the actin meshwork to bulk cell motility may vary according to the cell type and spatial arrangement of the cytoskeleton. In fibroblasts and larger epithelial cells, myosin-2 accumulates

---

acts with myosin-2. Shroom-Protein 1 is also called apical-like protein-2 (ApxL2); Shroom-2 apical-like protein (ApxL); Shroom-3, or ApxL3, Shroom-related protein (ShrmL); and Shroom-4 Second homolog of apical protein (SHAP). Shroom-1, -2, and -4 have distinct abilities to interact with the actin cytoskeleton. In fibroblasts, Shroom-1 associates with stress fibers and induces bundling; Shroom-2 colocalizes with cortical actin; and Shroom-4 resides with cytoplasmic ^{F}actin [507]. In epithelial cells, Shroom-1 provokes apical constriction, but not Shroom-2 and -4. Nevertheless, Shroom-2 and -4 have the capacity to do so, when they localize to the apical junctional complex.

in the lamella behind the lamellipodium, where it may contribute to disassembly of the actin network in the transition zone between the lamellipodal and lamellar networks.

6.3.4.3 Myosin-3

Myosin-3 (Myosin heavy chain-3) detected in embryonic skeletal muscle cells is encoded by the MYH3 gene. This hexameric protein is composed of a pair of myosin heavy chains and 2 pairs of different light chains.

6.3.4.4 Myosin-5

Myosin-5 is a 2-headed processive nanomotor that acts as a cargo transporter inside cells. Myosin-5 takes diverse steps along its track without dissociating. The motor domains alternate binding to releasing from track. This dimer coordinates the enzymatic state of its 2 catalytic domains to prevent premature detachment from its track. Gating is achieved by slowing ADP release from the leading head that remains strongly bound, until the trailing head detaches from actin. Intramolecular strain produced when both myosin heads are attached to actin prevents completion of the lever arm swing of the leading head and blocks ADP release.

Dimeric myosin-5 walks in a hand-over-hand fashion (step ~ 36 nm per each hydrolyzed ATP) along actin filaments. A step takes place through tethered diffusion of a detached head and detachment of the 2 heads is coordinated through intramolecular strain to facilitate hand-over-hand forward stepping and avoiding motor detachment [511]. The rate of tethered diffusion that depends on nanomotor size and stiffness must match the binding and unbinding rates of the nanomotor head.

Myosin-5 targets the plus-end of growing microtubules by associating with plus-end-tracking protein End-binding protein-1 (EB1) and melanophilin. Myosin-5 is able to couple long-range vesicle movement on microtubules with short-range actin-based displacement in the cell periphery in the presence of calcium and ATP owing to the myosin-5–melanophilin–EB1 complex and/or possibly by binding to kinesin. Myosin-5 handles junctions of actin filaments as well as actin–microtubule intersections (Chap. 9).

6.3.4.5 Myosin-6

Myosin-6 participates in the transfer of endocytic vesicles from clathrin-coated pits to endosomes, along the cortical actin network. Full-length myosin-6 exists as a monomer and dimer for optimal effect. Cargo binding promotes myosin-6 dimerization [512]. The myosin-6-binding cargo adaptors monomeric Dab2 and dimeric optineurin prime myosin-6 dimerization. In the monomer, intramolecular interactions that involve cargo-binding domain may inhibit dimerization.

These inhibitory interactions are destabilized by the binding of mono- and dimeric adaptors.

Myosin-6 that moves toward the minus-end of the actin filament operates as a processive transporter as well as an actin-based anchor. Myosin-6 uses the energy derived from ATP hydrolysis to transport cargo toward the minus-end of actin filaments, i.e., the opposite direction w.r.t. that used by other myosins. It also intervenes in cell migration and maintenance of the Golgi body and microvilli, as it can shape and anchor cellular substructures within the actin cytoskeleton.

It has a unique gating mechanism. Myosin-6 is gated by blocking ATP binding to leading head once ADP has been released due to an insert near the nucleotide-binding domain. Strong binding of the leading head is maintained by intramolecular rearward strain and blocks ATP binding [513]. Forward strain on the trailing head accelerates ATP binding.

Numerous cargo adaptors recruit myosin-6. Optineurin enables myosin-6 localization to the Golgi body. Disabled homolog Dab2 and Disc large homolog DLg1[91] mediate the recruitment of myosin-6 to clathrin-coated pits and vesicles. Regulator of G-protein signaling-19-interacting protein GIPc (synectin) allows the recruitment of myosin-6 to uncoated vesicles [512].

In addition to endocytosis, myosin-6 at the cell periphery operates in cell migration, during tissue development as well as cancer metastasis. The converter (between the catalytic head and light chains) of myosin-6 rotates about 180 degrees (about 30 nm stroke), i.e., much more than the 70 degrees (10–20 nm stroke) in other myosin types [514].[92] The kinetic mechanism of processive stepping by dimeric myosin-6 relies on the medial tail [518]. The medial tail contributes to the stride of dimeric myosin-6. In addition, intramolecular tension blocks ADP release in the front head of processively stepping myosin-6 dimers. Similarly to myosin-5, ADP gating may serve to increase effective affinity of myosin-6 for actin in the presence of applied load, hence improving its performance as both a transporter and cytoskeletal anchor.

6.3.4.6 Myosin-7

Myosin-7 is an unconventional myosin with FERM domains. Myosin tail homology-4–band 4.1, ezrin, radixin, and moesin domain-containing myosins (MyTHFERM) is an unconventional myosin involved in cell adhesion, polarization, and chemotaxis [519]. It is required for extension of actin-filled

---

[91] A.k.a synapse-associated protein SAP97.

[92] The exact process remains controversial [515–517]. The semi-rigid, stable, single α-helix, the so-called medial tail, may extend the lever arm by about 10 nm in single-headed constructs to cause a lever arm swing of 30 nm. Other observations suggest that the proximal part of the medial tail may act as a dimerization motif, and that the small, globular, proximal tail instead unfolds to yield the framework for the observed 30-nm steps of dimeric myosin-6.

projections such as filopodia. A simple complement of MyTHFERM myosins MyoG participates in cell polarization and chemotaxis in response to cAMP.

Myosin-7A can move processively when it dimerizes. This dimeric nanomotor carries its cargo to the tip of filopodia. However, the association of monomeric myosin-7A with membrane via the MyRIP–Rab27a complex enables cargo transport by myosin-7A at the membrane, where it possibly forms a dimer [520].[93]

6.3.4.7 Myosin-9

In humans, 2 genes — MYH9 (NMMHCA or NMMHC2A) and MYH10 (NMMHCB or NMMHC2B) — encode non-muscle myosin heavy chain-9 and -10 (myosin-9 and -10). These 2 myosin heavy chain isoforms are differentially expressed in myocytes as well as numerous non-muscle cell lines. On the other hand, the MYL9 gene encodes 20-kDa myosin regulatory light chain-9.[94]

Mammalian unconventional, single-headed, subclass-9 myosins comprise myosin-9A (Myo9a)[95] and -9B (Myo9b).[96] Isoform Myo9a is mainly expressed in testes and brain; Myo9b in the immune system.

Directional migration of immune cells to sites of pathogen invasion or tissue damage is primed by chemoattractants. During motion, cells successively generate membrane protrusions at their leading edges and retract their trailing ends under the control of monomeric GTPases of the RHO family. In immune cells, such as macrophages and neutrophils, which rapidly migrate toward sites of inflammation or tissue damage, myosin-9B acts as a Rho GTPase-activating protein, hence Rho inhibitor, that is required for coordination of membrane protrusions and retraction [521].

6.3.4.8 Myosin-10

Myosin-10 is an unconventional myosin. It localizes primarily to actin restructuring zones such as actin-rich protrusions (filopodium tip, lamellipodium edge, and membrane ruffles). Myosin-10 operates in filopodium formation, extension, and sensing, as it transports actin-binding proteins to the tips and links membranes to actin [522]. Myosin-10 also interacts with microtubules.

Myosin-10 step size (average step size 27 nm) and velocity are smaller on actin bundles than individual filaments (average step size 34 nm) [522]. A single domain extends the lever arm. In addition, the neck–tail junction or tail is flexible. This flexibility together with the extended lever, in conjunction

---

[93] Cargo myosin-7A- and Rab-interacting protein (MyRIP) may act as an activator of myosin-7A transporter.

[94] A.k.a. LC20 and myosin regulatory light chain-2, smooth muscle isoform (MLC2c).

[95] A.k.a. MyR7 (MyR: myosin, rat, homolog).

[96] A.k.a. MyR5.

with the membrane- and microtubule-binding domains, enables myosin-10 to perform multiple functions on various cytoskeletal structures.

When crossed actin filament density is high, myosins can switch actin filament tracks in the middle of a processive run and travel in a new direction. Myosin-10 displays track switching at least in some cells [501]. The cellular actin architecture neither enhances nor inhibits myosin processivity. Complexes of actin and actin-binding proteins can promote myosin motility such as monomeric actin crosslinker (actin filament bundling protein) fascin for myosin-10.

6.3.4.9 Myosin-15

Unconventional myosin-15 is encoded by the MYO15A (MYO15) gene. On the other hand, myosin heavy chain-15 that is encoded by the MYH15 gene and corresponds to a slow-twitch myosin that serves in contraction for long periods, but with little force, pertains to subclass-2 myosins.

6.3.4.10 Myosin-16

Unconventional myosin-16 is encoded by the MYO16 gene. Subclass-16 myosins Myo16a and Myo16b are characterized by an N-terminal ankyrin repeat domain that mediates the association with protein phosphatase-1 catalytic subunits PP1Cα and PP1Cγ. Isoform Myo16b is the main isoform that is predominantly expressed in developing neural tissue.

6.3.4.11 Myosin-18

The subclass of myosin-18 possesses 2 members: Myo18a and Myo18b. Myo18b Isoform may regulate muscle-specific genes when it resides in the nucleus. This homodimer may influence intracellular transport when it localizes to the cytoplasm.

6.3.4.12 Cardiomyocyte Myosin Filaments

Cycles of myocardium contraction–relaxation results from interaction between thick myosin and thin actin filaments that builds with other proteins sarcomeres that fill cardiomyocytes (Vol. 5 – Chap. 5. Cardiomyocytes). Cardiac myosin filaments consist of nanomotor myosin-2, sarcomeric template titin, and cardiac modulatory myosin-binding protein-C. Mutations in myosin and myosin-binding protein-C are the most common causes of inherited hypertrophic cardiomyopathy.

Myosins assemble into bipolar filaments, with α-helical coiled-coil tails in the filament axis and paired heads on the surface. Myosin filaments are characterized by conformations of myosin heads in the relaxed state into a set of periodic crowns (~43 nm) and axial rise between levels of heads of about 14.3 nm [523]. Antiparallel overlap between myosin tails at the filament core creates regions free of myosin heads. Myosin heads have a near-helical distribution on the surface of myosin filament in the relaxed state. Myosin heads interact with each other. Phosphorylation of myosin-binding protein-C modulates myosin activity. Phosphorylated regulatory light chains increase head mobility and activate myosin, probably by breaking between-head interactions that maintain the relaxed conformation.

Cardiac myosin-binding protein-C has a periodic structure with 11 domains (43 nm-long repeat of 7 fibronectin-like and 4 immunoglobulin-like motifs). At least 3 of its 11 domains run along the filament surface parallel to the axis interacting with titin and probably with myosin tails. It modulates cardiac contraction.

Titin acts as a template for sarcomere assembly. It is a giant protein that extends from the M line to the Z line along the myosin filament. Elongated titin strands lie on the surface of myosin filament backbone, approximately parallel to the filament axis. Titin interacts longitudinally with myosin tails and myosin-binding protein-C. Titin forms a simple linear scaffold for myosin filament formation; myosins assemble on the interior of this scaffold during sarcomere development and turnover [523].

6.4 Microtubules

Microtubules operate in intracellular transport, organelle positioning, cell polarization, and formation of the mitotic spindle required for chromosome segregation during the mitosis phase of the cell division.[97] A single microtubule can form a track that span the cell length. A microtubule can switch between 2 states: progressively growing in the presence of nucleotide-bound monomers and rapidly shrinking [426].

Microtubules (internal and external diameters ~15 and 24–25 nm, length 10–50 μm) are long, hollow, cylindrical polymers of α- and β-*tubulins* with polarity. Tubulins in microtubules are arranged in many protofilaments, parallel to the microtubule axis.[98]

Microtubules are characterized by a dynamical instability, evolving between growth and shrinking. The microtubule length adjusts to the cell size. The filament, indeed, collapses when it comes into contact with the cell edge. Microtubule dynamics can be regulated by applied stresses. Microtubules are

---

[97] During mitosis, the microtubule cytoskeleton rearranges itself into a DNA-segregating network, the so-called mitotic spindle that aligns chromosomes.

[98] Stable αβ-tubulin heterodimers are aligned to form protofilaments. The cylindrical and helical microtubule wall (helical pitch 12 nm, longitudinal repeat between αβ-tubulin subunits 8 nm) typically comprises 13 parallel protofilaments that generates a lattice.

thicker and less stable than microfilaments. They are the stiffer element of the cytoskeleton.

The microtubule architecture is often determined by central organizing centers, the centrosomes. The *centrosome* is a cell structure from which radiate the microtubules [524]. Because the microtubules control actin and intermediary filament distribution, the centrosome organizes the cytoskeleton. It is also involved in intracellular transport. The centrosome contains 2 *centrioles*, each composed of 9 cylindrical elements like a paddle wheel and 3 microtubules. The microtubules are thus involved in cell division.

During the interphase of the cell cycle, the Golgi body also intervenes in the nucleation of perinuclear microtubules [525].[99] Yet, microtubules nucleated at the Golgi body differ from those polymerized at centrosomes. They are asymmetrically organized, with preferential growth toward the leading edge of a migrating cell. They become rapidly acetylated, hence more stable. Moreover, they are coated with the cytoplasmic linker-associated protein CLASP2, a microtubule plus-end-binding protein that is specifically recruited to the trans-Golgi network to stabilize microtubule seeds by golgin GRIP and coiled-coil domain-containing protein GCC2[100] of the membrane of the trans-Golgi network [526]. Microtubule nucleation at the Golgi body requires AKAP9, a centrosomal γtubulin ring complex (TuRC)-interacting protein [527]. The cis-Golgi protein golgin GolgA2 enables recruitment of AKAP9 to the Golgi membranes and adequate organization of AKAP9 network.

Microtubules form a scaffold for the positioning of myosin filaments during sarcomere formation, when myoblasts give birth to striated muscle cells [528]. Myosin indeed moves toward microtubule plus-ends. Furthermore, microtubules are required for the stabilization of myosin-containing elements prior to their incorporation into mature sarcomeres.

6.4.1 Assembling and Disassembling

Microtubules are characterized by a dynamical instability due to alternating phases of tubulin polymerization and depolymerization. A protective guanosine triphosphate-bound cap at the polymerizing microtubule plus-extremity prevents microtubule depolymerization [529]. Tubulin^{GTP} remnants inside the polymer owing to incomplete GTP hydrolysis during tubulin polymerization are able to launch tubulin repolymerization.

Tubulin subunits assemble into microtubules that, in turn, aggregate to form more complex structures. Tubulin polymerizes in the presence of guanosine triphosphate and calcium. According to the allosteric and lattice (better)

---

[99] Golgi-dependent microtubule nucleation requires γ-tubulin and the γ-tubulin ring complex (γTuRC) that may be recruited to the Golgi body via interactions with Golgi proteins such as Golgi microtubule-associated protein GMAP210 (a.k.a. thyroid receptor-interacting protein TRIP11) and A kinase anchor protein AKAP9 (a.k.a. AKAP350 and AKAP450) [525].

[100] A.k.a. Ran-binding protein-2-like-4 (RanBP2L4).

model, unpolymerized $\alpha\beta$-tubulin adopts a conformation upon GTP binding suitable for polymerization; conformational changes occur only upon recruitment into the growing lattice. Main structural rearrangements, thus, do not occur in tubulins in response to GTP binding [530]. Heterodimers α–β-tubulins associate laterally[101] and longitudinally to form polar, cylindrical microtubules. Polymerization is initiated from a pool of GTP-loaded tubulins, as tubulin^{GTP} favors longitudinal contacts within the microtubule lattice. Growing microtubule ends fluctuate between slightly bent and straight protofilament sheets.

Tubulin^{GTP} is added to the microtubule end. Microtubule growth results from head-to-tail assembly of dimers. When tubulin^{GTP} is incorporated into the microtubule lattice, it is hydrolyzed shortly after incorporation because microtubules are mainly built from tubulin^{GDP} at interdimer contacts.[102] Microtubules continuously grow and shrink during the polymerization and depolymerization of tubulin.

Hydrolysis of GTP associated with the microtubule depolymerization promotes protofilament outward curvature [531]. In the presence of microtubule-associated proteins or certain divalent cations, tubulins^{GDP} bend back on themselves to self-assemble into ring-like structures. (Straightening is the consequence of assembling.) GDP-bound subunits of the main microtubule are straight (preferred conformation) because of intra- and inter-dimer interactions in the tubulin lattice, except at the ends, which can capture GTP and store strain energy. Binding of GTP induces a conformational change that straightens the dimer and allows formation of lateral contacts.[103] Depolymerization releases the strain energy, even in the presence of high concentrations of tubulin^{GTP}.

Tubulin-α undergoes a post-translational detyrosination–tyrosination cycle. This cycle generates 2 tubulin pools: tyrosinated (tubulin^{Tyr}) and detyrosinated (tubulin^{Glu}).[104] Old and new microtubules have tubulin^{Glu} and tubulin^{Tyr}, respectively [532]. Furthermore, α-tubulin tyrosination allows the microtubule end to recruit plus-end-tracking proteins, such as dynactin-1 and cytoplasmic CAP-Gly domain-containing linkers CLiP1 and CLiP2.[105]

Members of 3 formin families DIA, FMN, and INF (Diaphanous, formins, and inverted formins) can be involved in the regulation of polarized microtubule dynamics. Microtubules grow and shrink primarily at their plus-ends.

---

[101] Long, wide tubulin sheets close to form cylindrical microtubules.

[102] Guanosine triphosphate bound to β-tubulin is hydrolyzed to guanosine diphosphate during microtubule assembly. In the presence of tubulin and constant GTP levels, microtubule ends grow and then switch to shortening (structural instability).

[103] Tubulin^{GDP} curved dimers hamper the formation of lateral contacts. Furthermore, tubulin^{GDP} cannot be incorporated into microtubules.

[104] Tyrosine residues are cleaved and added by tubulin Tyr carboxypeptidase that preferentially acts on microtubules and ATP-dependent tubulin Tyr ligase that only recognizes free tubulin subunits, respectively.

[105] Proteins CLiP1 and CLiP2 are also called CLip170 and CLiP115, respectively.

Activation of Dia1 and Dia2 markedly increases the number of stabilized microtubules [453]. Inverted formin InF1 localizes predominantly to microtubules rather than actin structures. It contributes to the stabilization of the microtubule. Stable microtubules are typically oriented with their plus-ends toward the plasma membrane. They yield tracks on which nanomotors transport various cargos.

6.4.2 Microtubules as Transport Tracks

The microtubules are organized as a scaffold within the cytoplasm. They control the distribution of cellular organelles. Mitochondria and the endoplasmic reticulum are located along the microtubule network. The microtubules are required for vesicle formation and traveling across the cytosol.

6.4.2.1 Bidirectional Transfer

Microtubules serve as information transmission lines. Moving microtubules are used for simultaneous displacement in the 2 directions of vesicles and other organelles, in particular mitochondria with their ATP stores, using ATP as an energy source owing to associated ATPase [533]. The heavier the moving body, the slower the motion.

Material transport along microtubules is bidirectional despite dynamic asymmetry of plus- and minus-end-directed microtubule motors. Although microtubule displacements may be limited by microtubule attachments, microtubules undergo sliding, bending, and buckling that lead to fluctuations, which modify the cargo motion. Microtubule sliding velocity measured from buckling is in the range of typical motor speeds (0.3–$1.0\,\mu m/s$). These nanomotor-dependent longitudinal microtubule oscillations contribute to cargo displacement along microtubules.

6.4.2.2 Coupled Microtubule Motion and Nanomotor-Mediated Transfer

Organelle motion results not only from attached nanomotor action, but also microtubule motion. In addition, the majority of microtubules are crosslinked with each other and other cellular structures. Therefore, a vesicle may be linked to several motors that simultaneously couple it to several microtubules in a bundle [534]. Motors bound on the surface of organelles generate forces between organelles and microtubules. Organelle motion causes various strains to microtubules.

Peroxisome motility in a cell is characterized by [534]: (1) sharp changes in speed that is initially high and quickly decaying; (2) motion of peroxisomes associated with back-and-forth movement of the microtubule; and (3) synchronous nanomotor-driven displacement of peroxisome pairs over long duration.

6.4.2.3 Coupled Microtubule and Microfilament Transfer

Long-range transport takes place on microtubules, whereas final delivery involves shorter actin-based movements. Therefore, organelle transport to the cell periphery involves coordinated transport between the processive motors kinesin and myosin-5. Transfer of myosin-5 on microtubules improves the activity of kinesin when both motors are present on the same cargo, whatever the type of track (actin filament or microtubule) [535]. One nanomotor could tether the other and prevents its diffusion away from the track, hence increasing the processive run length.

6.4.2.4 Transfer of Mitochondria

Appropriate movement and distribution of mitochondria contributes to their coordinated functioning. Like other organelles, mitochondria link to specific nanomotors via specific adaptors. Mitochondrial motion that depends on microtubules and eventually actin microfilaments controls organelle localization, which maintains adequate local ATP concentration.

Actin can contribute to retrograde motion. Actin-dependent anchoring of mitochondria is especially observed in neuronal growth cones.[106]

Kinesin-associated adaptors Milton, syntabulin, and APP-like protein interacting protein APLIP1[107] are specific for coordinated microtubule nanomotors plus-end-directed kinesins-1 and -3 and minus-end-directed dynein. In addition, nitric oxide, local ADP concentration, and zinc and calcium levels influence the mitochondrial distribution.

RhoT GTPase (or Miro) of the RHO family operates with Milton to regulate mitochondrial transport, whereas Rab GTPases are involved in myosin-5-based mitochondrial displacement [536]. Enzyme RhoT interacts with kinesin-binding proteins $GABA_A$ receptor-interacting factor GRIF1 and O-linked ^{N}acetylglucosamine transferase-interacting protein OIP106 [537]. Both RhoT-1 and -2 influence mitochondrial aggregation and transport.

6.4.3 Microtubule-Associated Nanomotors

Two classes of motor proteins move along microtubules — kinesins and dyneins — using their ATPase sites. Most kinesins taxi toward the microtubule plus-end, whereas some kinesins and all dyneins move toward the minus-end.

Dynein and kinesin move along the microtubule at rates of about 0.5 to $3\,\mu m/s$ and generate forces of about 1 to 5 pN.[108] Members of the myosin

---

[106] In neurons, microtubule minus-ends are generally oriented toward the microtubule-organizing center in the soma. Plus-end-directed motion hence represents anterograde movement down to neuron peripheral regions, whereas minus-end-directed movement corresponds to retrograde movement toward the soma.

[107] A.k.a. JNK-interacting protein JIP1 in Drosophila melanogaster.

[108] In vivo velocity for kinesin-2 homo- and heterodimers is equal to 0.4 μm/s [538].

and kinesin class undergo conformational changes coupled with ATPase cycle, rotating their heads around the cargo-binding tail site.

Dynein can change its transport mode (forward or backward motion) and motion rate (slow or fast). Dynein undergoes ATP-dependent rotation around the head–tail junction; the tail motion is responsible for quick sliding along the microtubule, whereas the head drives slow motions and the ATP hydrolysis rate [539].

Intracellular transport of organelles involves both dynein and kinesin. The family of ATPases associated with various cellular activities (AAA)[109] are involved particularly in the motion of the microtubule nanomotor dynein [540].

6.4.3.1 Dynein

Dynein is the main nanomotor that transports cargos to the minus-end of microtubules. It serves in the retrograde transport and motility of cilia. This cytoplasmic, minus-end-directed microtubule nanomotor operates also at the nuclear envelope, where it is recruited before entry into mitosis [541].

Dynein has its motor domain in the heavy chain that also contains a ring of 6 mechanoenzymes, the ATPases associated with diverse cellular activities. Dynein uses its mechanoenzymes for ATP hydrolysis to amplify conformational changes and generate motion along microtubules toward the minus-end.

Dynein heavy chain folds and forms 3 structural domains: (1) a tail with binding sites; (2) the AAA1–AAA6 ring; and (3) long ($\sim$15 nm), antiparallel, coiled-coil stalk with its globular tip that contains the microtubule-binding site.

Dynein functions as a holoenzyme that assembles with several smaller, non-catalytic subunits that help its connection to cargos as well as regulatory factors.

Dynein has 2 modes of microtubule sliding: (1) one driven by the power stroke of the tail and (2) the other for slower sliding associated with Brownian motion or active tilting of the stalk against the AAA ring [539].

Minus-end-directed dynein transport different types of cargos owing to several multifunctional adaptors, such as dynactin, platelet-activating factor acetylhydrolase isoform PAFAH1b1, nuclear distribution gene-E homolog (NuDe), and NuDe-like protein (NDEL1) [542],[110] as well as Bicaudal-D homolog (BicD), kinetochore-associated homolog ZW10,[111] kinetochore-associated protein Kntc1,[112] and Zwilch kinetochore-associated homolog that regulate dynein function and localization [543]. Two dynein adaptors, dynactin and the PAFAH1b1a–NuDe or PAFAH1b1a–NDEL1 complex are required for

---

[109] Proteins AAAs form a set of enzymes that have an ATPase domain, which induce conformational changes (protein remodeling).

[110] A.k.a. mitosin-associated protein MitAP1.

[111] A.k.a. mitotic-15 protein.

[112] A.k.a. Rod.

all dynein functions. The PAFAH1b1a–NuDe and PAFAH1b1a–NDEL1 complexes may act as switches for dynein activity. Bicaudal-D homolog serves as a modular link between dynein and cargo. It interacts with dynein and cargo-specific factors such as Rab6 GTPase. Small GTPases regulate dynein association with some of its cargos.

6.4.3.2 Dynein Activator Dynactin

Dynein activator dynactin is composed of 11 different subunits, including actin-related protein ARP1. Dynactin helps to link dynein to its cargos. Moreover, ARP1 component of dynactin links to β3-spectrin on the cytosolic surface of cellular membranes [543].

Dynactin also connects to secretion-associated and Ras-related protein SAR1 GTPase-activating protein Sec23 homolog [543]. Protein transport Sec23-related protein is a component of the CoP2 coat on transport vesicles that originate from the endoplasmic reticulum. Dynactin interacts with Rab-interacting lysosomal protein (RILP), a Rab7 effector on late endosomes. Protein RILP also promotes the interaction of Rab7 with another of its effectors, oxysterol-binding protein-related protein OSBPL11 [543].[113]

6.4.3.3 Kinesins

Members of the kinesin class[114] transport various cargos, such as organelles, proteic complexes, and mRNAs, to specific destination along microtubule tracks in an ATP-dependent manner (Tables 6.17 to 6.20). They also participate in chromosomal and spindle movements during mitosis and meiosis.

Different kinesins recognize and bind to specific cargos, determine the direction of transport, and unload their cargos. The spatiotemporal delivery of cargos by KIF-based transport can be regulated by kinases, G proteins, Rab GTPases, and Ca^{++} ions [544]. In general, kinesins use scaffold and adaptor proteins to bind to transport vesicles, but they can sometimes bind to their cargo directly.

The human genome encodes at least 45 kinesin types (Tables 6.21 and 6.22). that have specialized cellular functions, among which 38 are detected in the brain. The human kinesin class includes: (1) kinesin heavy chain members KIF1a, KIF2a and -2c, KIF3b and -3c, KIF4a and -4b, KIF5a to -5c, KIF6, KIF7, KIF9, KIF11, KIF12, KIF13a and -13b, KIF14, KIF15, KIF16b, KIF17, KIF18a and -18b, KIF19, KIF20a and -20b, KIF21a and -21b, KIF22 to KIF25, KIF26a and -26b, KIF27, and KIFc1 to -c3 as well as (2) kinesin light chains KLC1 to KLC4.

---

[113] A.k.a. ORP1L.

[114] Usually defined either as superfamily (KISF) or family (KIF). Aliases entirely written with upper case letters refer to a molecule set. The alias KIF is conserved in the present text.

Table 6.17. Cargo complexes transported by proteins of the kinesin class (usually family [KIF]; Source: [544]; ER: endoplasmic reticulum; PM: plasma membrane; TGN, trans-Golgi network; Anx: annexin; AP: adaptor protein complex; EGF: epidermal growth factor; EGFR: epidermal growth factor receptor; ERGIC58: 58-kDa ER–Golgi intermediate compartment protein; KAP: KIF-associated protein; KDELR1: KDEL [Lys–Asp–Glu–Leu] ER protein retention receptor; L1 [L1CAM]: neuronal cell adhesion molecule [neither L1/Fam49b (family with sequence similarity 49, member B), L1/GKV1-16 (immunoglobulin-κ variable 1-16), 60S ribosomal protein L1 (RPL1 or ribosomal protein RPL4)]; M6PR: mannose 6-phosphate receptor; PARP: polyADP-ribose polymerase; PI: phosphatidylinositol; TNFRSF16: tumor-necrosis factor receptor superfamily member 16 [a.k.a. low-affinity nerve growth factor receptor (LNGFR) or P75 vesicular membrane protein neurotrophin receptor]; P115: general vesicular transport factor or vesicle docking protein [VDP; a.k.a. USO1 homolog]; Rab11FIP5: RAB11 family-interacting protein-5 [family 1; a.k.a. Rab11-binding protein RIP11]).

Type	Cell component and Cargo	Adaptor or scaffold proteins, and eventual KLC
	Conventional transport	
KIF3a/b	Golgi–ER Vesicles (KDELR1)	KAP3
	Early/late ndosomes (Rab4/7)	
	Recycling endosomes (Rab11)	Rab11FIP5
KIF4	PARP1	
	Vesicles (L1)	
KIF5	ER (kinectin)	
	Golgi–ER Vesicles	
	(ERGIC58 and P115)	
	Endosomes (Rab4)	
	TGN–PM vesicles (TNFRSF16)	
KIF13a	TGN–PM vesicles (M6PR)	AP1 (β1/γ-adaptin)
KIF16b	Early endosomes (Rab5,	
	PI(3,4,5)P$_3$, EGF, EGFR)	
KIF20a	Golgi body (Rab6)	
KIFC2	Early endosomes (Rab4)	
KIFC3	Golgi body	
	TGN–PM vesicles (Anx13b)	

In addition, multiple isoforms can be generated by alternative mRNA splicing. The kinesin class can be decomposed into 15 families [544] (Table 6.23).

Chromokinesins constitute a category of DNA- and/or chromatin-binding kinesin motors that act in various stages of mitosis (chromosome condensation, metaphase alignment, chromosome segregation, and cytokinesis). Some microtubules interact with chromosomes at kinetochores, whereas others connect to chromosome arms. All chromokinesins belong to 2 kinesin families: kinesin-4 and -10 (Table 6.24). Kinesin-4 and -10 family members contain nuclear localization signals and DNA-binding ability [547]. Kinesin nanomotors

Table 6.18. Cargo complexes transported in neurons by proteins of the kinesin class (usually family [KIF]; **Part 1** Dendritic transport; Source: [544]; FMRP: fragile X mental retardation protein [FMR1]; GluR: AMPA [α-amino 3-hydroxy 5-methyl 4-isoxazole propionic acid]-type glutamate receptor; GRIP: glutamate receptor-interacting protein; hnRNP: heterogeneous nuclear ribonucleoprotein; KLC: kinesin light chain; K_V: voltage-gated K^+ channel; mRNP: messenger ribonucleoprotein; NR2b: NMDA [^{N}methyl ^{D}aspartate]-type glutamate receptor-2b; Pur: purine-rich element binding protein). The CASK–Lin7a–Lin10 complex (LIN2–LIN7–LIN10) comprises calcium–calmodulin-dependent protein serine kinase (CASK), Lin7a homolog, and members of the family A of β-amyloid precursor-binding proteins (ABPa). Kinase CASK (Lin2 homolog) is related to the MAGUK (membrane-associated guanylate kinase) family. It is expressed in epithelia and neurons and binds to neurexin, syndecan, and protein 4.1.

Type	Cell component and cargo	Adaptor, scaffold, and eventual KLC
	Dendritic transport	
KIF5	Vesicles (GluR2)	GRIP1
	mRNP Complex (hnRNP-U, Purα/β)	
	mRNP Complex	KLC, FMRP
KIF17	Vesicles (NR2b)	CASK–Lin7a–Lin10 Complex
	Vesicles (GluR5)	
	Vesicles (K_V4.2)	
KIFc2	Multivesicular body	

of the KIF4 chromokinesin family generally associate with chromosome arms and function as nanomotors for chromosome transport.[115] Moreover, kinesin-8 family member KIF18a intervenes in chromosome alignment during mitosis. In addition to its chromokinesin role, KIF4a localizes to the nucleus during the interphase of the cell cycle and can operate in the DNA damage response [546].

The kinesin class with its 15 families can be grouped into 3 subclasses according to the position of the motor domain in the molecule [544]: (1) N-kinesins that have a motor domain in the N-terminus; (2) M-kinesins (or I type) that possess a motor domain in the middle region (internal region); and (3) C-kinesins that contain a motor domain in the C-terminus. In general, N- and C-kinesins serve for microtubule plus- and minus-end-directed transport, respectively. M-kinesins depolymerize microtubules. All kinesin families consist of N-kinesins, except KIF2 members of the kinesin-13 family (M-kinesins) and KIFC1 to KIFC3 (C-kinesins). For example, KIF14 is a N-type kinesin that pertains to the kinesin-3 family.

---

[115] Kinesins KIF4a and KIF4b are a.k.a. chromokinesin-A and -B and KIF22 as kinesin-like DNA-binding protein (KiD) and kinesin-like protein-4.

Table 6.19. Cargo complexes transported in neurons by proteins of the kinesin class (usually family [KIF]; **Part 2** Axonal transport; Source: [544]; ADAP: adhesion and degranulation-promoting adaptor protein; APC, adenomatous polyposis coli protein; BIG: brefeldin-A-inhibited guanine nucleotide-exchange protein [Arf1GEF]; Cdh: cadherin; Ctn: catenin; DLg: Disc large homolog; KAP: KIF-associated protein; KBP: KIF1-binding protein; MADD: mitogen-activated protein kinase-activating death domain [a.k.a. Differentially expressed in normal and neoplastic cells DENN]; PI: phosphatidylinositol; Syp: synaptophysin; Syt: synaptotagmin; SVP: synaptic vesicle precursor).

Type	Cell component and Cargo	Adaptor or scaffold proteins, and eventual KLC
	Axonal transport	
KIF1a or	SVP (Rab3a, Syp, Syt)	MADD
KIF1bβ	Vesicles ($PI(4,5)P_2$)	
KIF1bα	Mitochondria	KBP
KIF3a/b	Vesicles (APC, Par3)	KAP3
	Vesicles (NCdh, βCtn)	
	Vesicles	KAP3, Fodrin
KIF13b	Vesicles ($PI(3,4,5)P_3$)	Centaurin-α1 (ADAP1)
	DLg	
KIF21a		BIG1

The 3 subclasses (C, M, and N) of the kinesin class can be subdivided into 14 infraclasses (Table 6.25). N-Kinesins are indeed classified into 11 infraclasses (N1–N11 kinesins) and C-kinesins into 2 infraclasses (C1–C2 kinesins). Among the 39 N-kinesins, 2 are monomeric and 37 multimeric. Most infraclasses consist of members of a single family. Orphan KIF members encompass KIF6, KIF7, and KIF9.

N1 Kinesin infraclass is constituted by the KIF5 (heterotetramers) subfamily, i.e., the kinesin-1 family; N2 kinesin infraclass by the KIF11 (homotetramers) subfamily, i.e., the kinesin-5 family; N3 infraclass by the KIF1 (mostly monomers), KIF13, and KIF16 (multimers) subfamilies of the kinesin-3 family; N4 infraclass by the KIF3 (heterotrimers) and KIF17 (homodimers) subfamilies, i.e., the kinesin-2 family; N5 infraclass by the KIF4 and KIF21 subfamilies, i.e., members of the kinesin-4 family; N6 infraclass by the KIF20 and KIF23 subfamilies, i.e., members, i.e., the kinesin-6 family; N7 infraclass by the KIF10 subfamily, i.e., the kinesin family-7; N8 infraclass by the KIF18, KIF19, and KIF22 subfamilies, i.e., the kinesin family-8 and -10; N9 infraclass by the KIF12 subfamily of the kinesin family-12; N10 infraclass by the KIF15 subfamily also of the kinesin family-12; N11 infraclass by the KIF24, KIF25, and KIF26 subfamilies of the kinesin family-11, -13, and -14B.

Table 6.20. Cargo complexes transported in neurons by proteins of the kinesin class (usually family [KIF]; **Part 3** Axonal transport [Cont.]; Source: [544]; APP: β-amyloid precursor protein; AtG: autophagy-related gene product; CRMP: collapsin response mediator protein; Ctn: catenin; DISC: disrupted in schizophrenia; DLg: Disc large homolog; GolSyn: Golgi body-localized protein [a.k.a. syntabulin and syntaxin-1-binding protein]; KLC: kinesin light chain; LAMP: lysosome-associated membrane protein; LRP8: low-density lipoprotein receptor-related protein 8 [a.k.a. apolipoprotein-E receptor ApoER2]; MAPK8IP: mitogen-activated protein kinase-8-interacting protein 1 [a.k.a. Jun N-terminal kinase (JNK)-interacting protein (JIP)]; NDEL1: nuclear distribution gene-E product-like-1 [a.k.a. NuDeL and mitosin-associated protein MITAP1]; NF(H/M/L): neurofilament protein; PAFAH1b1: platelet-activating factor acetylhydrolase isoform-1b subunit-1 [a.k.a. lisencephaly Lis1]; RanBP: Ran-binding protein; SNAP: soluble ^{N}ethylmaleimide-sensitive factor attachment protein; Stx: syntaxin; Syt: synaptotagmin; SVP: synaptic vesicle precursor; Unc: uncoordinated receptor). The MAGUK family includes disc large homologs DLg1 (SAP97) and DLg4 (PSD95 or SAP90). In mammals, DlgA binds to K_V1 channel and DLg4 to NMDA glutamate receptor and K_V1 channel. Adaptors APBa1 to APBa3 (a.k.a. X11α to X11γ and neuronal Munc18-1-interacting protein MInt1 to MInt3) bind to the syntaxin complex involved in synaptic vesicle fusion.

Type	Cell component and Cargo	Adaptor or scaffold proteins, and eventual KLC
	Axonal transport	
KIF5a/c	Mitochondria	Milton, Miro, GolSyn, RanBP2
	Lysosomes (LAMP2)	
	SVP (Syt, AtG1)	Unc76
	SVP (Stx1)	GolSyn
	SVP (SNAP25)	
	NDEL1, PAFAH1b1	DISC1
	Vesicles (LRP8)	MAPK8IP1/2
	Vesicles (APP^P)	MAPK8IP1
	Tubulin	KLC, CRMP2
KIF5a	NFH/M/L	

The N-terminal motor kinesin class has also been divided into the KIF1A, KIF3, KIF5, KIF23, and chromokinesin infraclasses (Table 6.26).[116] The *KIF1A (UNC104) infraclass* comprises KIF1a and ubiquitous KIF1b monomer. Members KIF1a and KIF1b carry synaptic vesicle precursors and mitochondria toward the microtubule plus-end at about 1 and 0.5 µm/s, respectively [548]. The *KIF3 infraclass* consists of two-headed heterodimers KIF3a and KIF3b that heterotrimerize with kinesin-3-associated protein KIFAP3 (or KAP3) [549]. The *KIF5 (KHC) infraclass* includes 3 members *Homo sapiens* ubiquitous (HsuKHC or KIF5b) and neuronal (HsnKHC, i.e., KIF5a and

---

[116] Among these infraclasses, the BimC infraclass has not been identified in mammals.

Table 6.21. Kinesin types in humans (**Part 1**; Source: [545]; ATSV: axonal transporter of synaptic vesicles; CenPe: centromere protein-E; GaKin: Guanylate kinase-associated kinesin; HKLP: human kinesin-like protein; HKSP: human kinesin-like spindle protein; KnsL: kinesin-like protein; mCAK: mitotic centromere-associated kinesin; nKHC: neuronal kinesin heavy chain; uKHC: ubiquitous kinesin heavy chain; TRIP: thyroid receptor-interacting protein).

Type	Old alias (Homo sapiens)
KIF1a	ATSV
KIF1b	
KIF1c	
KIF2a	KIF2
KIF2b	
KIF2c	mCAK, KnsL6
KIF3a	
KIF3b	
KIF3c	
KIF4a	KIF4, chromokinesin-A
KIF4b	chromokinesin-B
KIF5a	nKHC
KIF5b	uKHC
KIF5c	xKHC, nKHC2
KIF6	
KIF7	
KIF8	
KIF9	
KIF10	CenPe
KIF11	Eg5, HKSP, TRIP5, KnsL1
KIF12	
KIF13a	
KIF13b	GaKin
KIF14	HUMORFW
KIF15	HKLP2

KIF5c) forms of conventional kinesin heavy chain.[117] The *KIF23 (MKLP) infraclass* (MKLP: mitotic kinesin-like protein) contains a single member in humans (MKLP1). Nanomotor KIF23 is involved in mitotic spindle elongation as well as the formation of contractile ring and completion of cytoplasmic division.

Conventional kinesin (kinesin-1) and members of at least the KIF1 to KIF3 subfamily are involved exclusively in between-membrane transport. Some members of 2 other subfamilies — KIF4 and KIFC — previously assigned to solely mitotic or meiotic spindles, are in fact implicated in organelle transport.

---

[117] Isotype HsnKHC localizes to the neuronal body and HsuKHC to the axon.

Table 6.22. Kinesin types in humans (**Part 2**; Source: [545]; KiD: kinesin-like DNA-binding protein; KLPMPP: kinesin-like protein M-phase phosphoprotein; KnsL: kinesin-like protein; MKLP: mitotic kinesin-like protein; Rab6Kin or Rab6KIFL: Rab6-interacting kinesin-like protein).

Type	Old alias (Homo sapiens)
KIF16a	
KIF16b	
KIF17	
KIF18a	
KIF18b	
KIF19a	
KIF19b	
KIF20a	Rab6Kin (Rab6KIFL), MKLP2
KIF20b	KLPMPP1
KIF21a	
KIF21b	
KIF22	KiD, KnsL4
KIF23	MKLP1, KnsL5
KIF24	
KIF25	KnsL3
KIF26a	
KIF26b	
KIFC1	HSET, KnsL2
KIFC2	
KIFC3	

All kinesins contain a head that acts as a motor and a variable tail. For example, kinesin-1 is organized as a heterotetramer with 2 heavy chains (KHC) and 2 light chains (KLC) [551]. Kinesins participate in cell displacement, providing coordination between cell regions. Kinesin nanomotors are major microtubule-dependent transporters in neurons and non-neuronal cells. Family-1 displace many different types of cargos in neuronal axons. Kinesin-1 motor–cargos linkers include Sunday driver and amyloid precursor protein [551]. Kinesin-2 are required for transport in motile and non-motile cilia.

Using optical tweezers, the interaction force of a single kinesin with the microtubule has been found to be about 1 pN [552]. Several motor and non-motile microtubule-associated proteins compete for binding to microtubules. Total internal reflection fluorescence microscopy shows which kinesin-1 waits in a strongly bound state on the microtubule with high bond densities until obstacle molecules unbind and frees the binding site for further kinesin processing [553].

Table 6.23. Kinesin families, their subfamilies, and members (Source: [544]). In particular, the kinesin-2 family includes the KIF3 and KIF17 subfamilies. These microtubule-based mechanochemical enzymes (ATPases) move cellular cargos toward the fast- (plus-; most often) or slow-growing (minus-) ends of microtubules.

Family	Name	Member(s) and eventual subfamily
Family 1	Kinesin-1	KIF5a–KIF5c
Family 2	Kinesin-2	KIF3a–KIF3c, KIF17
Family 3	Kinesin-3	KIF1a–KIF1c, KIF13a–KIF13b, KIF14, KIF16a–KIF16b
Family 4	Kinesin-4	KIF4a–KIF4b, KIF7, KIF21a–KIF21b, KIF27
Family 5	Kinesin-5	KIF11
Family 6	Kinesin-6	KIF20a–KIF20b, KIF23
Family 7	Kinesin-7	KIF10
Family 8	Kinesin-8	KIF18a–KIF18b, KIF19a–KIF19b
Family 9	Kinesin-9	KIF6, KIF9
Family 10	Kinesin-10	KIF22
Family 11	Kinesin-11	KIF26a–KIF26b
Family 12	Kinesin-12	KIF12 and KIF15
Family 13	Kinesin-13	KIF2a–KIF2c, KIF24
Family 14A	Kinesin-14A	KIFC1
Family 14B	Kinesin-14B	KIF25, KIFC2–KIFC3

Table 6.24. Chromokinesins (Source: [547]).

Type	Localization
KIF4a	Nucleus (interphase),
	chromosome arms, mitotic spindle, mid-body (mitosis)
KIF4b	
KIF22	Nucleus (interphase),
	chromosome arms, spindle (mitosis)

Regulation of Cargo–Nanomotor Complexes

Cargos need to be delivered to adequate destinations by specific kinesins and unloaded from these kinesins at an appropriate instant for efficient cell behavior. Phosphorylation state of kinesins by kinases regulate their association with and dissociation from cargos and microtubules.

Members of the RAB superfamily of small GTPases that are regulated by inhibitory GTPase-activating proteins (GAPs) and stimulatory guanine nucleotide-exchange factors (GEFs) control organelle location. Small GTPases Rab participate in the regulation of the association and dissociation between kinesins and organelles.

Mitochondria are bidirectionally transported along microtubules by kinesins of the KIF5 subfamily (kinesin-1 family) as well as KIF1bα [544].

Table 6.25. Kinesin sub- and infraclasses (Source: [545]). These microtubule-associated nanomotors possess a motor domain in the middle region, or C- or N-terminus.

Infraclass	Member(s)
	C-Subclass
C1	KIFC1
C2	KIFC2, KIFC3
	M-Subclass
M	KIF2a–KIF2c
	N-Subclass
N1	KIF5a–KIF5c
N2	KIF8, KIF11
N3	KIF1a–KIF1c, KIF13a–KIF13B, KIF14, KIF16a–KIF16b
N4	KIF3a–KIF3c, KIF17
N5	KIF4a–KIF4b, KIF21a–KIF21b
N6	KIF20a–KIF20b, KIF23
N7	KIF10
N8	KIF18a–KIF18b, KIF19a–KIF19b, KIF22
N9	KIF12
N10	KIF15
N11	KIF24, KIF25, KIF26a–KIF26b

Calcium ions prevent mitochondrial motility. Adaptors Milton and Ca^{++}-dependent RhoT bridge KIF5 to the mitochondrion.

6.4.3.4 Kinesin-1

Kinesin-1 dimer that carries cellular cargos along microtubules consists of 2 identical polypeptide chains. Kinesin head has binding sites for ATP and microtubule. Kinesin processive movement requires hydrolysis of one ATP molecule per step (size 8 nm) [554].[118] Dimeric kinesin steps by alternating head attachments to the microtubule. One head remains firmly bound to the track, whereas the other head moves forward, as the ATPase cycles in the 2 kinesin heads are coordinated during processive motion. The detached head behind the microtubule-bound head does not release its bound ADP when it interacts with the rear tubulin-binding site. Release of ADP can occur after the detached head takes a forward position, moving to a forward tubulin-binding site by ATP-triggered, neck-linker docking in the bound partner head [554]. At high ATP levels, kinesin spends most of its time bound to the microtubule with both heads. However, at low ATP concentrations, kinesin waits for ATP linked with a single head and makes brief transitions to a 2-head-bound intermediate as it walks along the microtubule.

---

[118] Motion step size corresponds to tubulin subunit spacing.

Table 6.26. Standardized kinesin families (Source: [550]; CENPe: centromere protein-E [KIF10]; MCAK: mitotic centromere-associated kinesin [KIF2c] MKLP: mitotic kinesin-like protein).

Standardized name	Other name	Infraclass		Example
Kinesin-1	Conventional	KHC	N1	KIF5a
Kinesin-2	Heterotrimeric	KRP85/95	N4	KIF3a/b
Kinesin-3	Monomeric	UNC104	N3	KIF1a
Kinesin-4	Chromokinesin		N5	KIF4
Kinesin-5	Bipolar, tetrameric	BIMC	N2	KIF11
Kinesin-6		MKLP	N6	KIF20/23
Kinesin-7		CENPe	N7	KIF10
Kinesin-8		KIP3	N8	KIF19
Kinesin-9		MKLP		KIF9
Kinesin-10	Chromokinesin	KID	N8	KIF22
Kinesin-11	Divergent kinesin-1		N11	KIF26
Kinesin-12		MKLP	N10	KIF15
Kinesin-13		MCAK	M	KIF2
			N11	KIF24
Kinesin-14	C-terminal motor		C1	KIFC1

During the cell cycle, microtubules buckle and loop owing to microtubule sliding against one another. Between-microtubule sliding that occurs during anaphase is caused by the activity of kinesin-1 heavy chain, but not kinesin-1 light chain or dynein [555].

Small GTPase Rab3 resides in synaptic vesicles and controls their exocytosis. Protein Rab3 is transported to the axon terminal from the cell body, using vesicles carried by KIF1a and KIF1bβ [544]. However, KIF1a and KIF1bβ are necessary, but not sufficient, to transport cargos toward axon terminals. Mitogen-activated protein kinase-activating death domain (MADD) binds to both the Rab3^{GTP} and KIF1a and KIF1bβ.

6.4.3.5 Kinesin-4

Calmodulin-dependent kinase CamK2 phosphorylates polyADPribose polymerase PARP1 that binds to KIF4, thereby causing dissociation of KIF4 from PARP1 [544].

6.4.3.6 Kinesin-5

The function of KIF5 can be regulated by at least 2 kinases. Protein kinase A phosphorylates the KIF5–KLC dimer to preclude the association of this nanomotor with synaptic vesicles [544]. Glycogen synthase kinase GSK3 phosphorylates KLC and inhibits the association of KIF5–KLC complexes with organelles. Jun N-terminal kinase may also phosphorylate KIF5 to weaken its binding to microtubules [544].

In addition, phosphorylation of adaptors regulates nanomotor–cargo binding. Adaptor Unc76 intervenes in axonal transport of synaptotagmin-carrying vesicles. Protein Ser/Thr kinase Unc51-like kinase ULK1 enables axonal elongation, as it binds and phosphorylates Unc76 [544]. Phosphorylated Unc76 then associates with synaptotagmin-1, but not the unphosphorylated form.

6.4.3.7 Kinesin-8

Members of the kinesin-8 family possess dual motile and depolymerizing function, as they move toward the microtubule plus-end and depolymerize this ends on arrival, hence releasing tubulin from microtubules to control the microtubule length [556].

6.4.3.8 Kinesin-13

Only the kinesin-13 family contains M-kinesins. Isotype KIF13b is also called guanylate kinase-associated kinesin (GAKin). ATP-Dependent depolymerization of microtubules by motor proteins can use the kinesin-13 family. Kinesin motors reach microtubule ends by ATP-dependent translocation along microtubules. Kinesin-13 transiently interacts with microtubules at rates exceeding those of standard enzyme–substrate kinetics. Kinesin-13 follows a random path on the microtubule lattice during its transient interaction with the microtubule. Kinesin-13 moves switching from a three-dimensional mechanism to a one-dimensional one to find the microtubule ends [557].[119]

6.4.3.9 Kinesin-16

Kinesin KIF16b that contains a phosphatidylinositol 3-phosphate-binding Phox (PX)-homology domain, transports early endosomes to the plus-end of microtubules, the process being regulated by the small GTPase Rab5 and its effector, the phosphatidylinositol 3-kinase PI3KC3 (VPS34) [558]. Small GTPase Rab5 controls endocytosis. Active Rab5 locally recruits PI3KC3 to endosomes. The latter synthesizes PI3P in endosomes and recruits KIF16b. In addition, $Rab5^{GTP}$ recruits dynein to transport endosomes in both directions along microtubules [544].

---

[119] The average MCAK time diffusion is equal to 0.83 s with a diffusion coefficient of $0.38\,\mu m^2/s$.

6.4.3.10 Kinesin-17

Ca^{++}–calmodulin-dependent protein kinase CamK2 phosphorylates KIF17 that transports vesicles that contain ionotropic glutamate receptor GluN2b (NMDA type) [544].

6.4.3.11 Kinesin-20

Kinesin KIF20a can specifically bind to $Rab6^{GTP}$ and control the motility and localization of the Golgi body [544]. When Rab6 is inactivated, KIF20a dissociates from the Golgi body.

6.4.4 Microtubule-Associated Proteins

Microtubules in cells are coated with microtubule-associated proteins (MAP) that regulate microtubule dynamics, i.e., the balance between microtubule polymerization and depolymerization. Certain can form a scaffold for the recruitment of other proteins that exert growth and/or stabilizing functions. Microtubule-associated proteins hence facilitate microtubule assembling. Both MAP1 and MAP2 enhance the microtubule growth rate. Other microtubule-associated proteins only stabilize the microtubule skeleton. Among the microtubule-associated proteins, plus-end-tracking proteins such as protein end-binding-1 accumulates at the growing microtubule plus-end.

Microtubule-associated protein Tau influences the motility of the 2 microtubule-based nanomotors dynein and kinesin that transport cellular cargos toward opposite ends of microtubule tracks. In neurons, microtubules are densely coated by microtubule-associated proteins such as Tau, so that nanomotors frequently encounter MAPs along their path. Dynein tends to reverse direction and kinesin to detach at patches of bound Tau [559].

6.4.5 Microtubule Plus-End-Tracking Proteins

Microtubule plus-end-tracking proteins ($^{+}$TP) are specialized microtubule-associated proteins that accumulate at growing plus-ends of microtubules to form dynamic networks. These multisubunit proteins comprise motor and non-motor proteins, the latter including transmembrane proteins. They often colocalize and share common activities. Many $^{+}$TPs interact with each other to exert different and sometimes opposite effects, such as stabilization and destabilization.

Microtubule plus-end-tracking proteins can affect the shape and positioning of microtubule networks by: (1) regulating microtubule dynamics; (2) linking microtubules to different cell structures; (3) interacting with signaling factors; and (4) exerting forces at the microtubule ends [532]. Therefore, $^{+}$TPs are involved in cell division, polarity, and differentiation, as well as morphogenesis.

The plus-end-tracking protein family includes: end-binding proteins (EB); 115 and 170 kDa cytoplasmic linker proteins (CLiP115 and CLiP170); structurally related large subunit of the dynactin complex; 150-kDa dynein-associated polypeptide DAP150,[120] CLiP-associated proteins CLAsP1 and CLAsP2; adenomatous polyposis coli protein (APC); microtubule-actin cross-linking factor (MACF),[121] stromal interaction molecule STIM1; lissencephaly protein LIS1; microtubule-associated protein XMAP; mitotic centromere-associated kinesin (MCAK); and dynein heavy chain (HC). Other representatives are melanophilin, neuron navigator-1, RhoGEF2, among others. Several $^+$TPs (CLiP170, EB1, etc.) bear intramolecular interactions that lead to auto-inhibition [532]. This auto-inhibition can be relieved by $^+$TP clustering.

Both plus- and minus-end-directed nanomotors that track growing microtubule ends comprise microtubule-depolymerizing kinesin-13 family members such as MCAK and representatives of the kinesin-14 family. Dyneins and kinesins accumulate at microtubule plus-ends in association with $^+$TPs. Myosin-5 indirectly interacts with microtubule tips.

Microtubule plus-end-tracking proteins bind to the plus-ends of depolymerizing microtubules, to the tips of stable microtubules, or (most of them) growing microtubule plus-ends. Some $^+$TPs such as CLiPs may copolymerize with tubulin dimers or oligomers. Several $^+$TPs can be transported to growing microtubule ends by plus-end-directed nanomotors. However, most $^+$TPs have a weak affinity for the microtubule lattice. Nevertheless, moderate, but specific affinity allows suitable interactions between $^+$TPs that are implicated in regulated cellular processes, during which transient $^+$TPs complexes are broken and reformed in association with motions of target structures.

The regulation of $^+$TP clustering and tethering to microtubule ends can give to microtubules a specific identity. This regulation results from specific motor-based loading, regional control of post-translational modifications, and/or intramolecular interactions. Phosphorylation of $^+$TPs alters their affinity for microtubules.[122]

Microtubule plus-end-tracking proteins control the microtubule dynamics [532]. End-binding proteins enhance the microtubule dynamics. Members of the CLIP family help to convert shrinking microtubules into growing structures. Partners CLAsPs, spectraplakins, and APC usually are microtubule-stabilizing factors. On the other hand, kinesin-13 family members (e.g.,

---

[120] A.k.a. dynactin subunit-1 and P150^{glued}.

[121] MACF is also called actin crosslinking family ACF7

[122] Microtubule binding by APC and CLAsPs is inhibited by phosphorylation by glycogen synthase kinase-3β. In addition, cyclin-dependent kinase CDK1 impedes binding of APC to EB1. Target of rapamycin and protein kinase-A regulate microtubule-binding activities of CLiP170 and DAP150, respectively. Aurora-B controls the localization and activity of MCAK during mitosis [532].

MCAK) destabilize microtubules. Most $^+$TPs (e.g., XMAP) have at least 2 tubulin-binding domains to promote tubulin multimerization.

Microtubule plus-end-tracking proteins attach and stabilize microtubule ends at the cell cortex, either by linking microtubule ends directly to actin (e.g., spectraplakins, APC, and CLAsPs) or to cortically bound factors (e.g., APC and CLAsPs) [532]. They stabilize microtubules independently of each other or cooperatively. Moreover, they can link microtubule ends to intracellular membranes. Consequently, they can favor delivery of molecules and vesicles to cellular subdomains.

Microtubule plus-end-tracking proteins participate in the recruitment of signaling factors, as they concentrate transiently at the growing microtubule ends and are subsequently released for a new cycle of binding and dissociation. Furthermore, thay can link microtubule ends to the actin cytoskeleton and adhesion sites.

6.5 Intermediate Filaments

Intermediate filaments (mature bore of 10–12 nm) cross the cytoplasm either as bundles or isolated elements, often in parallel to the microtubules. They can be crosslinked to each other as well as actin filaments and microtubules by *plectins* [426].

Intermediate filaments provide resistance to mechanical forces, although they are the least stiff of the 3 types of cytoskeletal polymers. Intermediate filaments cope with mechanical stresses in conjunction with cell junctions to ensure both rigidity and flexibility. They resist tensile forces much more effectively than compressive forces [426].

Intermediate filaments are assembled from fibrous proteins. Unlike microtubules and microfilaments, intermediate filaments do not exhibit structural polarity and do not form seeds to which subunits add for filament growth. Cytoplasmic intermediate filament proteins laterally tether to form filaments that can secondarily elongate and reduce their bore. Moreover, intermediate filaments form distinct molecular species in different cell types. Nuclear lamins that construct nuclear intermediate filament, assemble by simultaneous lateral and longitudinal association of dimers. Intermediate filaments are specific to the developmental stage and cell type (mesenchymal vimentin, muscular desmin, epithelial keratin, etc.).

Unlike assembling of actin and tubulin polymers that is coupled to ATP and GTP hydrolysis, intermediate filaments do not hydrolyze nucleotides. In addition, intermediate filaments do not possess molecular motors. They have slower dynamics that those of ^{F}actin and microtubules.

Intermediate filaments build 2 networks. In the nucleus, intermediate filaments attach to the inner nuclear membrane. In the cytoplasm, they connect cell junctions at the plasma membrane via integrins to the outer nuclear membrane via plectin and nesprin-3 [560]. The properties of cytoplasmic and nuclear intermediate filaments are different. These 2 networks withstand tensile and bending stress. Intermediate filaments thus adapt and stabilize the cell shape to environmental conditions. In the nucleus, intermediate filaments provide a skeleton for the assembly of nuclear proteic complexes with emerin, lamina-associated proteins, lamin-B receptor, heterochromatin protein-1, SMADs, and nesprins. Intermediate filaments can then contribute to the regulation of gene expression.

6.5.1 Intermediate Filament Components

In opposition to microtubules and microfilaments, intermediate filaments are made of many different proteins. The class of intermediate filament proteins is encoded of more than 70 genes. Cytoplasmic intermediate filament proteins include *desmin, keratins, lamins, nestin, syncoilin, synemin*, or desmuslin, and *vimentin* (Vim), among others.

Desmogleins and *desmocollins* anchor different intermediate filaments according to the cell type (keratins in epithelial cells, desmins in cardiomyocytes, and vimentin in endothelial cells).

Six families of intermediate filament proteins have been defined according to their amino acid composition: families 1 and 2 correspond to cytoplasmic acidic and basic keratins; 3 to cytoplasmic desmin, glial fibrillary acidic protein, peripherin, and vimentin; 4 to cytoplasmic neurofilament proteins, α-internexin, synemins-α and -β, and syncoilin; 5 to nuclear lamins; and 6 to cytoplasmic nestin mostly synthesized transiently in neuro- and myogenic cells in mammals.[123]

Another functional classification leads to 3 sets according to the assembly mode: heteropolymeric keratins, vimentin-like proteins, and lamins. These 3 kinds can coexist in the same cell, contributing to the cell rheology. Desmin- and vimentin-based intermediate filaments can form homopolymers. However, family-3 intermediate filament proteins can assemble in combination with proteins of the family-4 and -6 intermediate filaments. Family-6 intermediate filament proteins cannot self-assemble into filaments and need other components to build intermediate filaments. Nestin can form homodimers and -tetramers (but not intermediate filaments). It preferentially coassembles with vimentin and internexin. In the nucleus, intermediate filaments are assembled from lamins.

6.5.1.1 Family-1 and -2 Intermediate Filament Proteins – Keratins

Keratins constitute a family of structural proteins that dimerize, form filamentous polymers and then intermediate filaments. Keratins thus provide mechanical support in epithelial cells, in addition to other functions. They

---

[123] The family 6 of intermediate filament proteins also includes tanabin in amphibians and transitin and its splice variant paranemin in birds.

have a cell differentiation-specific expression pattern. Keratins-α and -β correspond to hard hair (H) and very hard nail keratins.

Keratins have been grouped in family-1 acidic (Krt9–Krt19) and family-2 basic-to-neutral keratins (Krt1–Krt8). However this classification is incomplete. In humans, 54 functional keratin genes exist (28 class-1 and 26 class-2 keratins), in addition to pseudogenes (KRTiPj; 5 class-1 and 8 class-2 genes). Keratins are then categorized into 3 other types of families: *epithelial* (17 family-1 and 20 family-2 genes) and *hair* (11 family-1 and 6 family-2 genes) keratins and associated genes, and keratin pseudogenes (class-1 and -2 pseudogenes) [561].[124]

6.5.1.2 Family-3 Intermediate Filament Proteins

Desmins, abundant near Z-discs, connect sarcomeres to desmosomes in cardiomyocytes. The coordination of microfilaments and intermediate filaments in intercalated discs between cardiomyocytes involves γ-catenin (plakoglobin) and plakophilin. Muscle intermediate filament proteins, synemin and muscle-specific syncoilin, integrate into intermediate filaments via dimerization with vimentin, desmin, α-internexin, or light neurofilament NeFL (or NFL). Synemins can bind to α-actinin and vinculin, thus connecting with focal adhesions and microfilaments.

Glial fibrillary acidic protein (GFAP) forms intermediate filaments not only in astrocytes and other cell types of the central nervous system, but also cells of kidney and bronchus, among other organs.

Peripherin is another type of family-3 intermediate filament protein found in the developing peripheral neurons. This neuron-specific protein abounds in the neural crest.

Binding of intermediate filaments to desmosomes and hemidesmosomes is a common feature of intermediate filament organization in cells (Sect. 7.7.1). Desmosome is a typical feature of epithelial cells. Vimentin intermediate filaments interact with desmosomal proteins such as desmoplakin, the major constituent of desmosome.

In the perinuclear region of endothelial cells, intermediate filaments are composed of vimentin. Vimentin associates with integrins, particularly at adhesion sites. The intermediate filament network thus transmits mechanical forces sensed by plasmalemmal integrins. Vimentin participates in vascular adaptation to blood flow behavior.

---

[124] Class-1 human epithelial keratins encompass Krt9, Krt10, Krt12 to Krt20, and Krt23 to Krt28. Class-1 human hair keratins include Krt31, Krt32, Krt33a and Krt33b, as well as Krt34 to Krt40. Class-2 human epithelial keratins comprise Krt1 to Krt5, Krt6a to Krt6c, Krt7, Krt8, and Krt71 to Krt80. Class-2 human hair keratins incorporate Krt81 to Krt86. Class-1 keratin gene family is composed of 3 epithelial keratin pseudogenes (KRT221P–KRT223P). Class-2 hair and epithelial keratin pseudogenes contain KRT121P to KRT124P and KRT125P to KRT128P, respectively.

6.5.1.3 Family-4 Intermediate Filament Proteins

Neurofilaments (size 10 nm) are intermediate filaments in neurons. The family of neurofilament proteins includes 3 member types (subfamily): light (NFL; ~70 kDa), medium (NFM; ~150 kDa), and heavy (NFH; ~200 kDa) polypeptides. In adult mammals, neurofilament proteins coassemble and form heteropolymers that contain NFL and NFM or NFH component.

Internexin[125] can form homopolymers. It is expressed in neuroblasts as well as adult cells of the central nervous system. In addition to self-assembly, it coassembles with neurofilament proteins.

In myocytes, syncoilin may serve as a linker between the sarcomere Z-disc, where it binds to desmin, and the dystrophin-associated complex, where it tethers to α-dystrobrevin.

Synemin-α and -β are synthesized from 2 alternatively spliced transcripts of the SYNM gene in myocytes. Synemin-β is also called desmuslin (Dmn). Their binding partners include α-dystrobrevin, α-actinin, and desmin. Synemins can transmit force laterally, especially between the sarcomere and extracellular matrix.

6.5.1.4 Family-5 Nuclear Intermediate Filament Proteins – Lamins

Nuclear lamins contribute to the structure of the cell nucleus and transcriptional regulation (Sect. 4.4.5). They interact with membrane-associated proteins to form the nuclear lamina on the cortex of the nucleoplasm. They are involved in the breakdown and reformation of the nuclear envelope during mitosis.[126] They also participate in the positioning of nuclear pores.

In humans, 3 subfamilies of lamin genes exist. Subfamily-A lamins are only expressed after gastrulation. Subfamily-B lamins (B1 and B2) reside in every cell. Subfamily-C lamins have a tissue-specific expression. Lamin-A and -C form homodimers that link head to tail.

6.5.1.5 Family-6 Intermediate Filament Protein – Nestin

Nestin is produced mostly in neurons and myocytes with other intermediate filament proteins, such as neurofilament proteins (NFH, NFL, and NFM), synemin-α and -β, and syncoilin. Nestin is expressed by many cell types during embryo- and fetogenesis. Its expression is usually transient. In adults, it can be found in stem and precursor cells. It can be re-expressed during healing in the central nervous system and muscle regeneration.

---

[125] A.k.a. α-internexin, neurofilament-5 (NeF5), and 66-kDa neurofilament protein (NF66).

[126] The nuclear envelope acts as a selective barrier around the genome and scaffold that contributes to the DNA organization in the nucleoplasm. During cell division, the nuclear envelope breaks down to liberate chromosomes that can travel along microtubules, which form the mitotic spindle. After chromosome segregation, a new nuclear envelope reassembles in each daughter cell.

6.5.2 Intermediate Filament Interactions

Intermediate filaments are linked to intercellular junctions such as desmosomes by desmoplakin and cadherins. Intermediate filaments are also connected to cell–matrix adhesions, such as hemidesmosomes and focal adhesions by plectin, dystonin, and integrins.[127]

Many cell types assemble intermediate filaments in response to mechanical stresses. In airway epithelial cells, keratin intermediate filaments form a network that assists in withstanding mechanical stress [426].

Polymerized nuclear lamins fortify the nucleus. Phosphorylation of nuclear lamins by cyclin-dependent kinases bolsters triggering of breakdown of the nuclear envelope at the beginning of mitosis.

Intermediate filament remodeling and functioning as stress absorbers depend on the combined action of kinases, phosphatases, and chaperones. Intermediate filament extensibility is much higher than that of microfilaments and microtubules. They can bear 250% tensile strain and up to 400% shear strain [560].

In vascular endothelia, intermediate filaments are implicated in structural responses of blood vessels to changes in hemodynamic quantities. Due to lack in vimentin, the endothelial production of either nitric oxide or endothelin (Vol. 5 – Chap. 9. Endothelium) is defective. Vimentin also intervenes in molecule localization within the cell. Intermediate filaments can interact with molecular complexes involved in signaling initiated by mechanotransduction.

6.6 Septins

During migration, cells form protrusions only at the leading edge and suppress inappropriate protrusions at the trailing edge [562]. Septin GTPases (Sect. 4.3.3.3) constitute a family of 13 detected cytoskeletal proteins distinct from microfilaments, microtubules, and intermediate filaments.

Septins modulate cell shape and provide cortical rigidity in T lymphocytes during their migration [563]. They form ring-like filaments arranged perpendicular to the axis of migration. In addition, ^{F}actin crosslinked by cortical filamin is also responsible for cortical rigidity in many cells. However, septins stabilize stress fibers.

6.7 Cytoskeleton–Integrin Link

Integrins serve as transmembrane receptors. They mediate cell adhesion through intracellular linker proteins that connect to the actin cytoskeleton.

---

[127] Integrins are connected to membrane-associated collagens, hence collagen fibers of the extracellular matrix on the one hand and microfilaments and intermediate filaments on the other.

Integrin-mediated adhesions undergo changes in composition and configuration from small focal complexes to stress fiber-associated focal contacts and mature fibronectin-bound fibrillar adhesions according to experienced forces generated from both actomyosin cytoskeleton and extracellular matrix components.

More than 50 proteins are involved in focal contacts. Several membrane molecules localize to focal contacts, such as proteoglycans and signaling molecules. Most focal contact proteins interact with different partners, as they contain multiple binding sites. The resulting complexes form regulated submembrane plaques. Focal contacts include [747]: (1) cytoskeletal proteins, such as α-actinin, α-parvin,[128] paxillin, talin, tensin, and vinculin; (2) protein Tyr kinases, such as Abl, CSK, FAK1, FAK2, and Src kinases; (3) protein Ser/Thr kinases, such as ILK, PAK, and PKC kinases; (4) modulators of small GTPases, such as ArfGAP ASAP1 and RhoGAP10 and -26; (5) protein Tyr phosphatases, such as plasmalemmal PTPRf and cytoplasmic PTPn11; and (6) other enzymes, such as PI3K and calpain-2. Some of these proteins can bind, cap, bundle, or nucleate actin filaments (α-actinin, α-parvin, tensin, VASP, vinculin, and ERM-proteins) and/or bind to integrins (α-actinin, FAK, ILK, talin, and tensin).

Among numerous linkers, few are essential. Talin, tensin, integrin-linked kinase, and PINCH (particularly interesting new Cys–His protein) coordinate signaling with cytoskeletal changes. Talin is required for initial formation and strengthening of integrin adhesion complexes, possibly by recruiting other proteins (ILK and tensin). Focal adhesion kinases have tyrosines that are phosphorylated to enable the binding to Src kinases, phosphatase PTen, adaptors GRB2 and GRB7, and PI3K, as well as Pro-rich domains that bind to CRK-associated substrate (CAS), RhoGAP10 and- 26, and PLCγ. Vinculin contains: (1) a globular head that binds α-actinin and talin; (2) a neck that interacts with VASP, ponsin, and vinexin; and (3) a long flexible tail that links to vinculin head, ^{F}actin, paxillin, and phosphatidylinositol (4,5)-bisphosphate. Head–tail interaction masks binding sites of α-actinin, ^{F}actin, talin, and VASP. Binding of PI(4,5)P$_2$ primes vinculin open conformation for the assembly of focal contact complexes.

Focal complexes that are induced by small GTPase Rac develop into focal contacts owing to small GTPase Rho or applied forces. Combined action of 2 Rho GTPase targets, RoCK kinase and Diaphanous Dia1, stimulates the development of focal complexes into focal contacts.

Adhesion of endothelial cells to basement membrane ensures vascular integrity. Integrins intervene in endothelial cell adhesion, as they form transmembrane links between the basement membrane and the actin cytoskeleton.

---

[128] A.k.a. actopaxin, matrix remodeling-associated protein-2, and calponin homology (calponin-like) integrin-linked kinase-binding protein (CHILKBP). It is encoded by the PARVA gene.

Moreover, $\alpha_6\beta_4$ integrin binds to laminin and vimentin of intermediate filaments owing to plectin [565].

6.8 Cytoskeleton and Nucleoskeleton

The nuclear envelope that serves as a barrier between the nucleus and cytoplasm is composed of inner and outer membranes separated by a perinuclear space. Yet, the nuclear envelope links the nucleus content to the cytoskeleton, therefore to cell membrane and external medium. Both inner and outer nuclear membrane actually contain specific integral proteins (e.g., Sad1–Unc84 homology proteins SUn1 and SUn2 and nesprin-2 giant, respectively) that connect contents of cyto- and nucleoplasm across the nuclear envelope. The nucleoskeleton is actually connected to the cytoskeleton via *linkers of nucleoskeleton and cytoskeleton* (LiNC). These proteic complexes contain lamins, nesprins, and SUn proteins.

The nucleus is approximately 10 times stiffer than the cytoplasm. However, localized loading (surface deformation $\sim 0.4\,\mu$m) on integrins (Sect. 7.5.4) of the plasma membrane causes displacement of nuclear and nucleolar structures [566]. The larger the pre-existing tension (prestress) in the cytoskeleton, the higher the stress propagation efficiency between the loaded plasmalemmal site and the nucleoskeleton owing to connecting nodes at the cell cortex (e.g., spectrin of the cortical actin cytoskeleton), perinuclear region, and nucleoplasmic peripheral layer (e.g., lamin of the nucleoskeleton).

The nuclear matrix contains multiple components, such as DNA, DNA regulatory proteins, chromatin structure-modifying proteins, scaffolds that link different regions of chromosomes to facilitate combinatorial control of gene transcription, RNA, and ribonucleoproteins. The genome is organized into functional territories via *nuclear scaffold–matrix attachment regions* that bind the nuclear skeleton to DNA [567].

The plasma membrane in strain and stress fields not only transduces these signals into chemical stimuli, but also transmits mechanical cues within the cell using the cytoskeleton linked to the plasma membrane on the one hand and nuclear envelope and nucleoskeleton on the other. External mechanical signals transmitted to the nucleus by cytoskeletal networks can hence influence nuclear events.

Forces applied to the cell can also generate distortion of mechanosensors of the nuclear envelope that are able to convert mechanical signals into chemical cues. Mechanotransduction mediators bear conformational changes between stretched and rest form that rearrange internal domains, especially those responsible for their activity. Mechanical loads transmitted by the cytoskeleton to the nucleoskeleton can influence nucleolar state and chromatin folding, as well as deform DNA and excite regulatory factors of gene transcription.

Furthermore, mechanical forces transferred via the cytoskeleton from the external medium to the nucleus can activate mechanosensitive ion channels on the nuclear envelope, thereby influencing gene transcription. Mechanosensitive Ca^{++} channels of the nuclear envelope gated by mechanical stimuli actually initiate calcium entry to induce gene transcription [568]. Increased concentration of nuclear calcium stimulates calmodulin-dependent kinase-4 that activates myocyte enhancer factor-2.

6.8.1 Lamins

Lamin-A, -B and -C form the nuclear lamina on the nucleoplasmic surface of the inner nuclear membrane as well as the internal nuclear skeleton. Lamins can be divided into 2 categories: A-group (i.e., 2 subfamilies of lamin-A and -C) and B-group (lamin-B) lamins. Lamins have different turnover rates. A-group lamins have a major role in mechanotransduction. Lamins interact with most integral inner nuclear membrane proteins, such as lamin-B receptor, lamina-associated polypeptide LAP1 and LAP2, emerin, MAN1, and nurim.

Lamins link to the genetic machinery directly and indirectly via nuclear proteins such as inner nuclear membrane protein *emerin* (Sect. 6.8.4). Lamin-A and -C bind scaffold emerin that links on DNA-bridging protein *barrier-to-autointegration factor* (BAF) as well as other possible partners for chromatin attachment to the nuclear envelope [569].

Lamins-A and -C interact with the nuclear matrix and the transcription factor Krüppel–TF3A-related zinc finger protein ZnF239 (or MOK2) that has dual affinity for DNA and RNA to repress transcription [570].[129] Lamin-A and emerin participate in the regulation of heart-specific gene expression. Lamin-A regulates the expression and localization of gap junctions and intercalated disc components [571]. Lamin-A and emerin are also required to maintain nuclear envelope integrity. Lamin-B receptor binds to DNA, histone, and various chromatin-associated proteins. Lamins and lamin-associated proteins such as emerin thus contribute to the control of nuclear organization, mRNA splicing, and gene functioning.

Lamin-B and hetero-oligomer lamin-A–lamin-B bind vimentin of intermediate filaments to anchor their network to the nuclear lamina, hence providing a connection between the nucleoskeleton and cytoskeleton from the nucleus surrounding to the cell cortex [572]. On the other hand, lamin-A, but not lamin-B, interacts with integral membrane protein SUn1 of the inner nuclear membrane that associates with nesprin-1 and- 2 of the outer nuclear membrane for nucleo-cytoskeletal linkage [573].

Loss of lamin-A and -C reduces nuclear stiffness and can lead to cell death under mechanical stress, whereas lamin-B1 contributes to nuclear integrity, but not stiffness [574].

---

[129] Zinc finger protein-239 is a DNA- and RNA-binding protein.

6.8.2 Nesprins

The nuclear envelope spectrin repeat proteins nesprins are able to bind SUn proteins as well as actin filaments. Large isoforms of nesprins-1 and -2[130] on the outer nuclear membrane connect actin filaments to SUns on the inner nuclear membrane. On the other hand, nesprin-3 binds to plectin of the plakin family member [575]. Nesprin-3 actually connects to intermediate filaments by binding to plectin-1. Nesprin-3 and plectin can lodge in the perinuclear region of the cytoplasm with keratin-6 and -14. In addition, plectin also links on $\alpha_6\beta_4$ integrin at the cell surface.

Short isoforms of nesprins that lack actin-binding domain are located in the inner and outer nuclear membranes. These isoforms could interact with cytoskeletal components via spectrin. Like nesprin-1 and -2, nesprin-3α links to the outer nuclear membrane by interacting with SUn proteins at the inner nuclear membrane. Deletion of SUn-binding domain yields a redistribution of nesprin-3α into the endoplasmic reticulum, where it colocalizes with cytoskeletal crosslinker plectin [576]. In addition, nesprin-3α and plectin form dimers in the nuclear envelope. Actin cytoskeleton influences the binding of plectin dimers to nesprin-3α dimers, as their stabilization prevents plectin recruitment to the nuclear envelope, so that nesprin-3α is retained in the nuclear envelope via interaction with proteins SUn1 and -2.

Nesprin-4 is an outer nuclear membrane and kinesin-1-binding protein that can cause kinesin-mediated cell polarization with relocation of the centrosome and Golgi body relative to the nucleus [577]. Nesprin-1 also binds kinesin-2 for transport of membrane to the midbody of a cell that undergoes cytokinesis.

6.8.3 SUn Proteins

Sad1 and Unc domain-containing proteins (SUn) bind to lamins of the nuclear lamina and skeleton. SUn Proteins include 2 known isoforms SUn1 and SUn2 on the inner nuclear membrane.

Protein SUn1 links to nucleoskeletal lamin-A as well as nuclear-pore complexes [578]. It is an important determinant of the distribution of nuclear pore complexes along the nuclear surface, as its depletion causes clustering of nuclear-pore complexes.

Isoform SUn2 is also located close to nuclear pores, but not intimately associated with nuclear-pore complexes. It has its SUN domain in the perinuclear space between the inner and outer nuclear membranes. It resides in small clusters [579].

Both inner nuclear membrane proteins SUn1 and SUn2 contribute to the localization of nesprin-2 giant in the outer nuclear membrane, as proteins SUns tether nesprins via interactions that span the perinuclear space [580].

---

[130] Nesprin-1 and -2 are also called SYNE1 and -2.

6.8.4 Emerin

Emerin is a component of the LiNC complex that connects the nucleoskeleton to the actin cytoskeleton. Emerin that is located at the outer nuclear membrane and endoplasmic reticulum and interacts with tubulin associates the centrosome with the outer nuclear membrane [581].

Emerin binds not only complex LiNC via nesprins and lamins, but also many different regulators involved in chromatin modification, transcriptional regulation, and mRNA processing. Lamin-binding emerin actually interacts directly or indirectly with β-actin, nuclear myosin-1, nuclear α2-spectrin, non-muscle myosin heavy chain-α, transcription regulator LIM domain-only-7 protein LMO7,[131] transcription repressor germ cell-less (GCL),[132] death-promoting BCL2-associated transcription repressor BTF1 [584],[133] splicing factor YT521b [586], calponin-3, and SIKE. Emerin by linking on barrier to autointegration factor is connected to DNA. Emerin forms proteic complexes with lamin-A and repressor germ cell-less or lamin-A and barrier to autointegration factor [583]. Emerin-associated proteic complexes that have distinct components are involved in nuclear architecture (e.g., nuclear myosin-1, α2-spectrin, and lamins) or gene or chromatin regulation (barrier to autointegration factor, transcription regulators, histone deacetylases, nuclear corepressor complex) [587].

6.8.5 Matrin

Matrin-3 is a nuclear matrix protein that has RNA-binding domains. This protein interacts with other nuclear proteins to anchor edited RNAs to the nuclear matrix and to modulate the activity of promoters. Matrin-3 binds to calcium–calmodulin and serves as a substrate of caspases. Matrin-3 and lamins extend throughout the nucleus and orientate many elements of the nuclear regulatory machinery.

6.8.5.1 Actin and Myosin

Actin and myosin contribute to nuclear structure and functions, such as chromosome displacement and transcription. Emerin and lamin-A and -C can bind to both nuclear α- and β-actins and cytoplasmic actin. In addition, emerin and BAF (barrier-to-auto-integration factor) connect LiNC (linker of

---

[131] Emerin-binding protein LMO7 shuttles between the cell surface and nucleus, where it regulates emerin transcription. Protein LMO7 targets other genes, especially myocyte genes (e.g., CREBBP, NAP1L1, LAP2, RBL2) [582].

[132] Transcriptional repressor GCL competes with barrier to autointegration factor for binding to emerin [583].

[133] Protein kinase-Cδ activates and interacts with death-promoting transcription factor BTF to transactivate the p53 gene and trigger P53-dependent apoptosis in response to DNA damage [585].

nucleoskeleton and cytoskeleton) and DNA to nuclear actin. Emerin preferentially binds polymerized actin and stimulates its polymerization. Protein Ser/Thr phosphatase-1 that links to lamin-A and -C inhibits emerin–actin interaction, whereas another Ser/Thr protein phosphatase increases emerin–actin binding [588].

Actin associates with all of the 3 types of RNA polymerases and cooperates with nuclear myosin-1 to promote gene transcription, in particular that mediated by RNA polymerase-1 [589]. Whereas monomeric nuclear actin operates in signal transduction, polymeric actin intervenes in RNA polymerase-1-induced gene transcription [590]. Moreover, nuclear myosin-6, the single myosin that moves toward the minus-end of actin filaments, enhances RNA polymerase-2-dependent transcription [591]. Myosin-6 is recruited to the promoter and intragenic regions of genes that encode urokinase plasminogen activator, eukaryotic initiation factor-6, and low-density lipoprotein receptor.

Proteic complexes that comprise actin, α2-spectrin, and other partners interact with emerin. Emerin binds and stabilizes the pointed end of actin filaments and increases the actin polymerization rate [592]. Emerin contributes to the formation of an actin-based cortical network at the nuclear inner membrane.

6.9 Connections between Cytoskeletal Components

Interaction between microtubules and filamentous actin regulates intracellular transport (Chap. 9) as well as cell shape and migration (Volume I, Part B – Chap. 6. Cell Motility). Microtubule linkage to focal adhesions allows focal adhesion turnover and, hence, cell migration.

Spectraplakins constitute a family of proteins that are able to crosslink microtubule and ^{F}actin networks. Suitable focal adhesion dynamics (formation, maturation, and disassembly) is required during cell migration. Assembly and disassembly rates as well as focal adhesion features (size and strength) depend on spectraplakins.

Actomyosin filaments coordinate the cell response to applied stress and strain fields throughout the cell. Filamentous actin is connected to microtubules via actin crosslinking factor ACF7 and to intermediate filaments via plectin-1. Plectin-1 also links intermediate filaments to microtubules as well as to outer nuclear membrane via nesprin-3.

Cadherins of adherens junctions also link the extracellular matrix to the cytoskeleton, as associated β- and γ-catenins tether to actin- and intermediate filaments, respectively. Intermediate filaments connect transmembrane integrins to the nucleus, as their component vimentin binds nuclear lamin-B.

6.10 Cytoskeleton and Cell Polarity

Most cells exhibit shape and functional asymmetry such as apicobasal (or front–back) polarity. Oriented lattices of actin filaments and microtubules are formed by dynamical, polar polymers that yield structural support for cell polarization.[134] Usually, the actin cytoskeleton initiates the polarized distribution of regulatory molecules, whereas microtubules maintain the stability of the polarized organization [593]. Cell polarity partially results from interactions between microfilaments and microtubules.

Cell cytoskeleton not only mediates the activity of polarization signals, but also localizes regulators to specific cortical sites via nanomotors and dynamical assembly of mutually exclusive structures. Formin-nucleated actin cables and cortical actin patches are involved in cell polarization. Capture of microtubules by cortical factors that increase the stability of plus-ends and/or generate pulling forces on microtubules is frequently used during cell polarization [593]. Other mechanisms include changes in microtubule assembly properties and microtubule bundling. Cortical capture involves 2 categories of proteins: (1) plus-end-associated proteins ([+]TP) and (2) cortical factors that are controlled by Rho GTPases and signaling agents.

The microtubule network organizes the intracellular distribution of organelles, the positioning of which participates in cell polarity. However, the microtubule organization and anchorage depends on the cell type [594].[135]

Small GTPases of the RHO family that regulate actin and microtubule cytoskeleton interact with polarity proteins to induce spatially restricted cytoskeletal remodeling required for cell polarization in different cellular contexts [595].

Three sets of polarity proteic complexes include: Crumbs-3, Partitioning-defective (Par), and Scribble complexes.[136] Polarity complexes control

---

[134] Myosins move unidirectionally along actin filaments that are composed of the ATP catalyzers actins oriented in the same direction with structurally different barbed and pointed ends. Kinesins and dyneins also run unidirectionally along microtubules that are constituted of the GTP catalyzers α- and β-tubulins oriented in the same direction with plus- and minus-ends, the latter being often anchored at the centrosome or other microtubule-organizing centers. Polarity results from head-to-tail association of protein subunits. Actin filaments and microtubules are also dynamical polymers, as they undergo polymerization–depolymerization cycle according to available concentration of free subunits. Growth and collapse rates differ at opposite ends. Nevertheless, actin and microtubules reorganize rapidly and locally in response to polarity cues.

[135] In lymphoid cells, microtubules are anchored at the tip of a set of long appendages around the axis of the centriole. In columnar epithelial cells, microtubules form bundles parallel to the apicobasal axis, with all plus-ends oriented toward the basal cell side. In somatic cells, the centrosome can localize close to the plasma membrane, thereby inducing organelle repositioning.

[136] Polarity proteic complexes either cooperate or antagonize depending on target process. The Par complex is constituted by isoforms Par3 and Par6 and

polarization processes in many different cell types, especially epithelial cells, neurons, and T lymphocytes.[137]

Local increase in concentration of phosphatidylinositol (3,4,5)-trisphosphate stimulates Rap1 GTPase that activates CDC42 GTPase. Active CDC42 binds to partitioning-defective protein Par6 and enables Par6 to activate atypical protein kinase-C [595]. In addition, Rac-specific guanine nucleotide-exchange factor T-cell-lymphoma invasion and metastasis-1 TIAM1 (Vol. 4 – Chap. 8. Guanosine Triphosphatases and their Regulators) interacts with Par3 and leads to Rac1 activation. Active $Rac1^{GTP}$ also hastens atypical protein kinase-C. Activated atypical protein kinase-C phosphorylates its substrates for cell polarization and epithelium maturation.

Proteins Par6 and aPKC activate RhoGAP35 that enables Rac1 to impede RhoA activity (Rac–Rho antagonism). Atypical protein kinase-C may also indirectly increase RhoGAP35 activity by phosphorylating (inactivating) GSK3β, an inhibitor of RhoGAP35. On the other hand, small GTPase RhoA suppresses Rac1 activity via the Par complex and Rho-associated coiled-coil-containing protein kinase (RoCK) that phosphorylates Par3 and dissociates the Par3 complex. Conversely, the Par3 complex controls regulatory proteins of RhoA and thus Rho–Rac antagonism.

---

atypical protein kinase-C, as well as small GTPase CDC42 that are ubiquitously expressed [595]. The Crumbs complex comprises the transmembrane protein Crumbs-3 and cytoplasmic scaffold Protein associated with Lin-7 PALS1 (a.k.a. membrane protein, palmitoylated MPP5 and MAGUK P55 subfamily member-5) and PALS1 associated with tight junction protein (PATJ). Crumbs-3 that is mainly expressed in epithelial cells serves as an apical membrane determinant. The Crumbs (Crb–PALS–PATJ) as well as Par (Par3–Par6–aPKC–CDC42) complexes are predominantly located in tight junctions. The Scribble complex include Scribble, Disc large homolog (DLg), and Lethal giant larvae (LGL) proteins. The Scribble complex participates in the identity of basolateral regions of the plasma membrane. Mutual exclusion of the Scribble complex and apical polarity complex controls apicobasal polarity as well as phosphorylation of LGL2 and Par1 by atypical protein kinase-C.

[137] Epithelial cells establish a barrier to protect tissues from their environment and regulate molecule transport. Neuron that is characterized by multiple dendrites and a single axon transmits signals received at dendrites and processed at the soma via its long axon ($>$1 m in motoneurons). During adhesion, migration, or intercellular interactions, T lymphocytes that have a round morphology in blood and lymph vessels polarize with a lamellipodium (leading edge enriched in actin, integrins, chemokine receptors, small GTPases Rap1A, CDC42, and Rac1, RacGEF TIAM1, and Par complexes), a midbody (with small GTPase RhoA, its effector RoCK, myosin, and Crumbs-3 complexes), and a large uropod (tail with T-cell receptors, intercellular adhesion molecules, CD3, CD43, CD44, ezrin, and Scribble complexes) [595].

6.11 Modeling of the Cytoskeleton Mechanics

At rest, short actin filaments are surrounded by a pool of actin monomers bound to profilin. Myosin-2 exists in the bent state. Integrins (Sect. 7.5.4) are dispersed over the cell surface. The formation of stress fibers is triggered by an activation signal. Several parallel intracellular pathways exist. Adhesion to the extracellular matrix (Chap. 8) triggers a signaling pathway with profilin, cofilin, and gelsolin that activates phospholipase-C, leading to calcium release from stores. Calcium influx activates gelsolin, which cleaves the capped actin filaments into tiny fragments for formation of long filaments. Calcium-triggered phosphorylation favors myosin-2 extended state for assembly of myosin filaments and formation of stress fibers. These fibers generate tension by cross-bridge cycling between the actin and myosin filaments. When the tension disappears, the stress fibers disassemble.

A model for the cell contractility hence takes into account 3 processes that govern the dynamic remodeling of the cytoskeleton: (1) the activation of actin polymerization and myosin phosphorylation; (2) the tension-dependent assembly of the actin and myosin into stress fibers; and (3) the cross-bridge cycling between actin and myosin that generates tension [596]. The fiber formation rate that depends on the activation signal is coupled to a dissociation rate dependent on the tension. The stress fiber contraction rate depends on the tension via the cross-bridge dynamics. The model predicts the main features: (1) the dependence on substrate compliance; (2) the influence of cell shape and boundary conditions, and (3) the high concentration of the stress fibers at the focal adhesions (Sect. 7.6).

Rounded cells commonly have a microfilament mesh in the cortex, just below the membrane. A bi-dimensional model of cytoskeleton dynamics has been developed to describe stress-induced interactions between actin filaments and anchoring proteins [597]. A small shear induces rearrangment of the 4 filament populations[138] toward an orientation parallel to the streamwise direction.

Reactive flow model of contractile networks of dissociated cytoplasm under an effective stress (magnitude c_{eff}) in a square domain is associated with a system of non-linear partial derivative equations with boundary conditions [598]. This system requires the following variables: space- and time-dependent network ϕ_n and solution ϕ_s volume fractions, an effective pressure and velocity field, with the given rheological properties, the network and solution shear $(G_{n/s})$ and dilation $(D_{n/s})$ viscosities, and the network-solution drag coefficient $(C_D$, which are involved in the mass and momentum conservation equations. Two main dynamical modes are observed during contraction of a dissociated wholly polymerized contractile network.[139] (1) Rending-type contraction

---

[138] The population of actin filaments include 2 subpopulations, whether filaments are moored or not on the cell membrane, each family having 2 subsets according to presence or absence of connections to other actin filaments.

[139] Experimentally, various components of the cytosolic network are put in a solution in a depolymerized form and polymerize secondarily.

is related to violent contractions that tear the network. (2) Squeezing-type contraction describes progressive global contractions with gradual expulsion of fluid. Three dimensionless parameters govern the dynamical behavior of the network. According to the value of the rending number, which depends on C_D, G_n, and D_n, the network contracts into a single mass (small values) or breaks up into clumps (large values). Crucial dynamic factors of cytoskeleton mechanics are: (1) the viscosity of the contractile network associated with an automatic gelation as the network density enlarges, without undergoing large deformation; (2) a cycle of polymerization–depolymerization; and (3) a control of network contractibility and plasmalemmal adhesion.

The continuum hypothesis is supposed to be valid because the problem length scale, although small with respect to the cell size, remains greater than the cell organelle size. It is then assumed that the cell microstructural elements can be neglected. However, cytoskeleton elements are in general smaller than a required size of 50 times greater than the typical cell component. Actin filament behavior is then investigated in a domain that contains a solution of cytoskeleton components rather than the cell itself. The stripped cytoskeleton can also be considered as a discrete structure of stress-bearing components.

Filamentous proteins arranged in a homogeneous, isotropic, crosslinked mesh, uniformly loaded, stiffen from low to intermediate strains, without requiring a specific architecture or multiple elements with different stiffnesses, assuming affine deformations [599]. Stiffer filaments, like ^{F}actin or collagen, stiffen at a few percent strain, whereas more flexible filaments like vimentin stiffen only at larger strains, approaching 100%. Biological tissues then adapt to loading, using not only the non-linear passive behavior, i.e., the strain stiffening, but also stiffness changes associated with contraction of proteic nanomotors of the cytoskeleton.

Tensegrity models[140] of cytoskeleton mechanics are aimed at depicting the essential features of stress-subjected cytoskeleton distortion. Tensegrity models consider deformable cells as a set of beams and cables that sustain tension and compression [600]. For instance, the cytoskeleton model, subjected to 1D traction, can contain 6 rigid struts (compression-resistant elements that mimic the microtubules), to which pulling forces are applied. The struts are connected to 24 elastic prestressed cables (tension-resistant elements that represent actin filaments), either via frictionless looped junctions

---

[140] Tensegrity stands for tensile integrity. Tensegrity structures made of bars linked at nodes by cables have several advantages, resistance and high deformability, for a minimal material weight. Tensegrity structures can easily remodel. Tensegrity models are used in civil engineering (in addition to serve as toys). This building principle was first described in architecture and constructed by a Fuller's student, the sculptor K. Snelson [601].

(free slip) [602].[141] or pin joints (crosslinks)[142] The stiffness depends on the prestress level, and for a given prestress state, to the applied stretch, in agreement with experimental findings [603]. Non-linear elastic elements can be used to take into account the cell stiffening (strain-hardening) response due to the non-linear stress–strain relationship. Using a theoretical model of a 30-element tensegrity structure, normalized element length and elastic tension are found to govern the mechanical response of the structure for 3 types of loadings (extension, compression, and shear) [604]. The tensegrity model can also be composed of viscoelastic prestretched cables (Voigt bodies), arranged in a network of 24 cables, associated with 6 rigid bars [605]. The normalized viscosity and elasticity moduli are found to be almost independent or dependent of changes in normalized initial internal tension, respectively. Both quantities depends on the normalized length of the structure.

---

[141] The tension in the cables depends on the total cable length, and only tension and compression are transmitted to the model components.

[142] The tension in each cable depends on its length and forces acting at element ends reduced to a single force.

7

Plasma Membrane

The cell membrane is also called plasma membrane and plasmalemma (mainly in plant cells), or sarcolemma in striated myocytes (i.e., myocardial [heart] and skeletal muscle cells). Cellular membranes also refer to membranes and envelopes of cell organelles.

The plasma membrane constitutes a barrier between the extra- (ECF) and intracellular (ICF) fluids. Nevertheless, the plasma membrane connects the cell to its environment, i.e., (1) apposed cells, especially in epi- and endothelia; (2) the extracellular matrix (Chap. 8); as well as, (3) in the case of vascular endothelia, flowing blood (Vol. 5 – Chap. 9. Endothelium) and respiratory epithelia, epithelial lining fluid that separates epithelial cells from the mucus layer and breathed air. Cells sense the matrix or cells via specialized adhesion molecules on the cell surface. Among these adhesion molecules, some are mechanosensors. In addition, membrane bending generates lateral forces that modulate membrane function.

7.1 Recruitment to and Molecular Motions within the Plasma Membrane

Effective diffusion coefficients of lipids in cell membranes range from 0.1 to 1 $\mu m^2/s$, i.e., about 2 orders of magnitude less than in the cytoplasm for three-dimensional diffusion [606].

Random molecular movements contribute to cell membrane heterogeneity.[1] Molecular transfer in cell membrane differs from Brownian diffusion.

---

[1] Under thermal agitation, molecules permanently and randomly move and collide. Molecular diffusion that corresponds to the macroscopic effect of the Brownian motion, thus favors homogeneity. Homogeneity results from the random continuous spreading of molecules from regions with high molecular concentrations to those with lower concentrations. The spreading rate is the diffusion coefficient. The smaller the molecular size, the lower the molecule density, the smaller the friction, or the higher the temperature, the greater the diffusion coefficient.

In a multiphase medium, molecules are trapped in domains in which positive interactions occur with the nearest neighbors, and negative interactions are minimized. Membrane protein transport is affected by interactions with molecular complexes.

The *picket–fence model* states that membrane proteins and lipids are confined in their movement by actin cytoskeleton fences and actin-anchored protein pickets [607]. Monomeric proteins can move between plasmalemmal nanodomains, whereas oligomers are trapped in nanodomains, where they began to form. Cells can thus impede the diffusion of receptor complexes. Immobilization of receptor oligomers may facilitate the assembly of signaling complexes.

Molecule colocalization at the cell membrane can increase the local concentration of substances by a factor of about 1,000 with respect to freely moving forms in the cytoplasm [606]. This increase in apparent affinity appears only when both molecules of the interacting pair are connected to the membrane. If only one of the 2 proteins is anchored to the membrane, then the complex dissociates and one partner rapidly diffuses away from the membrane before being captured by other membrane-bound proteins.

7.2 Main Constituents of the Plasma Membrane

The cell membrane has multiple functions. (1) Its continuous sheet separates the cytosol from the extracellular space, but certain membrane elements connect these 2 interacting compartments. (2) It controls molecule transport, providing a selectively permeable barrier. It contains carriers for substance transport across it, from both compartments. (3) It responds to external stimuli, via signal transduction. (4) It has components for cell recognition and adhesion. (5) It organizes cellular biochemical activities via vesicle transport associated with a submembrane and cytoplasmic scaffold. (6) It maintains cell polarity. (7) It is involved in chemical energy transfer from glucids and lipids to adenosine triphosphate (ATP).

The phospholipid bilayer of the plasma membrane embeds proteins and glucids. Proteins partly immersed in the phospholipid bilayer and carbohydrates attached to the membrane surface are required for cell communication and transport across the membrane. The membrane has specialized sites such as sphingolipid–cholesterol–protein clusters for exchange of information, energy, and nutrients; these are essentially made from transmembrane proteins. Membrane proteins not only serve as transporters and transducers, but also enzymes.

The architecture of the plasma membrane is suitable for signal transduction, as it participates in the space and time organization of signaling mediators. The spatiotemporal dynamics of the plasma membrane can regulate signaling complexes that are tethered over a long period or transiently recruited. Temporary plasmalemmal nanoclusters can contain kinases, phosphatases, and other signaling proteins that are anchored directly to the membrane or via lipid-anchored proteins.

7.2.1 Plasmalemmal Nanodomains

The 2 layers of the cell membrane that contain different types of lipids with numerous proteins, are asymmetrical. Moreover, the asymmetric lipid bilayer contains numerous lipid and protein species with regions of structural and compositional heterogeneity. Cell membranes are able to laterally segregate their constituents to initiate and coordinate vital and adaptive functions. Structural and compositional heterogeneities of the lipid bilayer determines various types of specialized functional domains at the nanoscopic scale (nanodomains; size <200 nm).

Cell membrane organization into nanodomains impedes the diffusion of transmembrane proteins. Moreover, these membrane compartments are connected to the submembrane cortical actin cytoskeleton. The cytoskeleton hinders lateral transport of transmembrane proteins, thereby confining materials into nanodomains. Nevertheless, enzyme activities, membrane recycling, and signaling events permanently disturb the equilibrium of cell membranes, but support nanodomain formation [608]. In particular, the cortical actin cytoskeleton contributes to cluster assembly aimed at regulating signaling.

Among nanodomains that compartmentalize the plasma membrane, fluctuating membrane rafts[2] (length scale <20 nm) result from specific assemblies of sphingolipids, cholesterol, and proteins. These platforms operate in cell signaling and material transfer between the extra- and intracellular media. They exist transiently (time scale ~20 ms). Membrane rafts can be stabilized by lipid-anchored proteins and crosslinking that can adapt length and time scales.

In addition to the lipid-enriched membrane rafts, many receptors and signaling molecules form clusters in the plasma membrane. Most or all plasmalemmal proteins are incorporated in cholesterol-enriched domains (size 30–700 nm; raft or non-raft regions) connected to the cortical cytoskeleton and separated by protein-free membrane segments poor in cholesterol [610]. However, lipid compartmentalization in nanodomains is independent of the cytoskeleton. Most plasmalemmal proteins self-organize in nanoclusters. Certain cell junction zones and signaling pathways use these nanodomains. Lipid-based nanodomains in the plasma membrane allow proteins and lipids to assemble or separate with a time scale of tens to hundreds of milliseconds.

---

[2] Membrane rafts are also called "lipid rafts". However, this term is not appropriate as proteins also constitute these nanodomains. Rafts of the plasma membrane evolve according to 3 main stages: small, short-lived assembling, platform stabilization by oligomerization to form larger rafts with variable size and lifetime, and separation into ordered and disordered phases by the clustering of raft components such as the ganglioside GM1 [609].

The prototypal cluster is composed of receptor Tyr kinase (Vol. 3 – Chap. 8. Receptor Kinases), phospholipids, and small GTPase Ras (Vol. 4 – Chap. 8. Guanosine Triphosphatases and their Regulators).

Membrane nanodomains represent connecting platforms of transport and signaling via clusters of transfer and signaling molecules [961]. In particular, membrane rafts and caveolae are membrane nanoregions used for cell transport. Membrane rafts allow protein attachment. Membrane rafts depend on sphingolipid and cholesterol that repress lateral diffusion of raft-associated proteins. These transient nanodomains are generated and stabilized by protein crosslinking. Caveolae, flask-shaped invaginations of the plasma membrane, differ from membrane rafts. Caveolae contain caveolin that distinguishes caveolae from other membrane rafts.

7.2.2 Membrane Lipidic Composition

The deformable, dynamical plasma membrane has constituents that regulate the flux of materials, react to ligand binding, and interact with the cell environment. The membrane is essentially a fluid bilayer with homogeneous lipid distribution. It contains thousands of lipid species that can be grouped into 3 major superclasses: *glycerophospholipids*, *sphingolipids*, and *sterols*.[3] Lipid transfer is achieved by both vesicular and non-vesicular transport.

Some lipids are exclusive, whereas others are binding partners, hence supporting distinct lipid structures. Phospholipids with relatively long saturated acyl chains and cholesterol pack together to create a liquid-ordered phase that can give rise to membrane rafts [612]. Bulk phospholipids composed of phosphoglycerolipids with polyunsaturated fatty acids build a liquid-disordered phase. The approximate composition of the plasma membrane is given for the erythrocyte (or red blood cell) in Table 7.1. Phospholipids possess a hydrophilic and hydrophobic end (Table 7.2).

Cholesterol is a major structural and functional component of cellular membranes. Minor disturbances in membrane cholesterol content cause strong changes in membrane physical properties, affecting intracellular signaling and transport. The phospholipid bilayer is impermeable to water-soluble molecules and ions.

7.2.2.1 Sphingolipids

Sphingolipids include sphingomyelin (SM), ceramide (Cer), ceramide 1-phosphate (C1P), glucosylceramide, dihydroceramide, sphingosine (Sph),

---

[3] Glycerophospholipids are also called phospholipids. Sphingolipids are synthesized using acyl-coenzyme-A (long-chain fatty acid-attached coenzyme-A), then converted into ceramides, phosphosphingolipids, glycosphingolipids, and other compounds. Ceramide phosphocholines are sphingomyelins. Glycosphingolipids include cerebrosides and gangliosides.

Table 7.1. Approximate composition of an erythrocyte plasma membrane (Source: [164]). Three main kinds of lipids reside in cell membranes: phospholipids, with hydrophilic and hydrophobic ends, cholesterol, and glycolipids. The erythrocyte plasma membrane has an asymmetrical lipid distribution between the bilayer leaflets. The outer plasmalemmal leaflet is enriched in sphingomyelin and phosphatidylcholine, whereas the inner leaflet is enriched in phosphatidylethanolamine, phosphatidylserine, and phosphatidylinositol.

Cholesterol	23
Phosphatidylethanolamine	18
Phosphatidylcholine	17
Phosphatidylserine	7
Sphingomyelin	18
Glycolipids	3
Others	14

Table 7.2. Phospholipid distribution in the 2 layers of the erythrocyte plasma membrane (Source: [613]).

	External layer	Internal layer
Sphingomyelin	∼20	∼5
Phosphatidylcholine	∼20	∼10
Phosphatidylethanolamine	∼10	∼25
Phosphatidylserine		∼10
Phosphatidylinositol	1–2	2–3

sphingosine 1-phosphate (S1P), and lysosphingomyelin. Enzymes of the sphingolipid metabolism regulate the concentrations of the sphingolipids (Fig. 7.1). They comprise sphingomyelinases (SMase), ceramidases (CDase), and sphingosine (SphK) and ceramide (CerK) kinases.

Sphingolipids intervene in the regulation of intracellular transport, cell adhesion, growth, migration, senescence, and death. They thus participate in many processes, such as inflammation and angiogenesis. Sphingolipid-mediated pathways operate according to their location within the cell, whether the pathway is triggered at an organelle or the plasma membrane, and in the latter case, at the inner or outer leaflet of the plasma membrane.

Ceramide and Ceramide 1-Phosphate

Ceramide is currently considered as a hub in sphingolipid metabolism [614]. Ceramides are synthesized via the de novo pathway[4] or from sphingomyelin by

---

[4] Ceramide is synthesized de novo from the condensation of palmitate and serine by serine palmitoyl transferase (SPT) to form 3-keto dihydrosphingosine, which is reduced to dihydrosphingosine. The latter undergoes acylation by dihydroce-

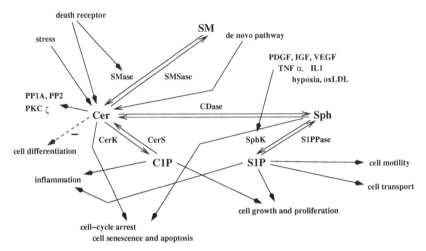

Figure 7.1. Sphingolipid metabolism, some inducers and effectors, as well as sphingolipid effects on the cell fate (Sources: [614, 615]; DAG, diacylglycerol; CerS: ceramide synthase; IGF: insulin-like growth factor; IL1: interleukin-1; oxLDL: oxidized low-density lipoprotein; PDGF: platelet-derived growth factor; PKC, protein kinase-C; PP1, PP2a: protein phosphatases; S1Pase: sphingosine 1-phosphate phosphatase; SMSase: sphingomyelin synthase; TNFα: tumor-necrosis factor-α; VEGF: vascular endothelial growth factor). Sphingomyelin (SM) is cleaved by sphingomyelinase (SMase) and generates ceramide (Cer). Ceramide and sphingosine promote apoptosis, cell cycle arrest and cellular senescence. Ceramide is phosphorylated by ceramide kinase (CerK) to form ceramide 1-phosphate (C1P), which favors cell division. Ceramide is cleaved by ceramidase (CDase) to produce sphingosine (Sph). Sphingosine is phosphorylated into sphingosine 1-phosphate (S1P) by sphingosine kinase (SphK). Lipid S1P is degraded by S1P lyase to form ethanolamine phosphate and hexadecanal. It operates via both intra- and extracellular routes to promote cell growth, proliferation, and motility. Its extracellular actions use different G-protein-coupled receptors ($S1P_1$–$S1P_5$).

activation of sphingomyelinases.[5] Ceramide also forms sphingomyelin owing to sphingomyelin synthase, and this reaction generates diacylglycerol (DAG) from phosphatidylcholine. In addition, ceramide is phosphorylated by ceramide kinase and glycosylated by specific hydrolases, glucosyl- or galactosylceramide synthases. Specific β-glucosidases and galactosidases hydrolyze glucosylceramide and galactosylceramide to regenerate ceramide. Ceramide is

---

ramide synthase (CerS) to form dihydroceramide. Ceramide is then generated by the desaturation of dihydroceramide due to desaturases.

[5] Acid sphingomyelinase is located at the outer cell membrane leaflet and assigned to the lysosome. Neutral sphingomyelinase lodges at the inner leaflet of the plasma membrane. Acid sphingomyelinase is regulated by PKCδ and oxidative stress among other agents. Neutral sphingomyelinase is activated by cytokines, such as TNFα and IL1, and various types of cell stresses.

especially targeted by ceramidases to produce sphingosine. Conversely, sphingosine is recycled to ceramide by ceramide synthases.

Ceramide mediates many stress responses. At the plasma membrane, ceramide may mediate receptor clustering and nanodomain formation. Ceramide induces cell autophagy, senescence, apoptosis, or caspase-independent cell death by limiting access to extracellular nutrients, as it downregulates nutrient transporters [616]. Ceramide also inhibits cell differentiation, growth, and proliferation. Ceramide activates protein kinase-Cζ and cathepsin-D [614]. Accumulated ceramide stimulates phosphatases, such as PP1 and PP2a. Tumor-necrosis factor-α requires ceramide to dephosphorylate PKCα and protein kinase-B among other proteins. Agent C1P activates phospholipase-A2. Ceramide 1-phosphate acts in cell transport and division, as well as inflammation. Glucosylceramide is involved in cell transfer of materials.

Sphingosine and Sphingosine 1-Phosphate

Sphingosine is phosphorylated by sphingosine kinases SK1 and SK2 to synthesize sphingosine 1-phosphate. Many growth factors acutely activate SK1 isotype. Sphingosine 1-phosphate is dephosphorylated to regenerate sphingosine owing to S1P phosphatases. Enzyme S1P lyase breaks down S1P into non-sphingolipid molecules.

Sphingosine regulates the actin cytoskeleton, cell transport, cell cycle, and apoptosis. Low nanomolar concentrations of sphingosine 1-phosphate (Vol. 2 – Chap. 3. Growth Factors) are measured in the cell, whereas higher concentrations are detected in blood, as S1P is bound to lipoproteins and albumin. Many growth factors, such as EGF, PDGF, TNFα,[6] and IL1, activate SK1, hence causing transient elevations in S1P levels, via PKC, PLD, and/or ERK [614].

Sphingosine 1-phosphate is released in the extracellular matrix through the ABC transporter ABCc1. It then targets its G-protein-coupled receptors (S1PR1–S1PR5; Vol. 3 – Chap. 7. G-Protein-Coupled Receptors) to activate phospholipase-C, small GTPases Rho and Rac, as well as the Ras–ERK and PI3K–PKB–Rac pathways [614].

High S1P1R levels are observed in endothelial cells. The SK1–S1P–S1P1R pathway regulates angiogenesis. Agent S1P not only controls endothelial cell proliferation and migration for vessel formation, but also elicits proliferation and migration of smooth muscle cells.

The circulating lipid mediator S1P targets its receptors S1P1R, S1P3R, and S1P4R on cells of lymphoid organs to regulate lymphocyte egress [617].[7]

---

[6] Kinase SK1 intervenes in various TNFα effects, such as stimulation of cyclooxygenase-2, augmented prostaglandin production, induction of adhesion molecules, and activation of NOS3.

[7] Sphingosine 1-phosphate controls T-lymphocyte egress from thymus into blood as well as from lymph node and Peyer's patches into lymph, and B-lymphocyte egress into lymph. On the other hand, S1PR agonists sequester CD4+ and

Sphingosine 1-phosphate can also act within the producing cell to promote cell growth, proliferation, and motility.

7.2.2.2 Glycerophospholipids

Structural glycerophospholipids, or phospholipids, include phosphatidylcholine (PtdCho), phosphatidylethanolamine (PtdEtn), phosphatidylserine (PtdSer), and phosphatidylinositols (PtdIns or PI), as well as phosphatidic acids.

Phosphatidylcholine abounds in cell membrane, corresponding to a large fraction of phospholipids. *Plasmalogen* provides an important quantity of choline glycerophospholipids of the myocardium. Myocardial phospholipase-A2 (iPLA2), which is mainly associated with the sarcolemma and does not require calcium, acts on plasmalogen. Activated iPLA2 produces lysoplasmenylcholine and arachidonic acid. Lysoplasmenylcholine alters calcium fluxes. Arachidonic acid decreases the conductance of gap junctions.

In addition, transarachidonic acids are produced by nitric oxide (NO)-dependent NO_2-mediated isomerization of arachidonic acid within the cell membrane. They induce time- and concentration-dependent apoptosis of endothelial cells. Transarachidonic acid effects are mediated by thrombospondin-1 [619]. *Nitrative stress* leads to microvascular degeneration and ischemia.

Glycerol-3-phosphate is acylated to form lysophosphatidic acid (LPA) that can be further acylated by transmembrane lysophosphatidic acid acyl transferase (LPAAT)[8] to form phosphatidic acid. Phosphatidic acid is a precursor for glycerophospholipids (in particular, PtdEtn and PIs). A low LPAAT activity has been assigned to endophilin and C-terminal-binding protein (CTBP)[9] that are particularly implicated in stabilization of membrane curvature at budding vesicles, but LPAAT activity can be a purification artefact [620].

Membrane anionic phospholipid phosphatidylserine contributes to the subcellular location of proteins with cationic domains. Phosphatidylserine is distributed in cytosolic leaflets of the plasma membrane, as well as endosomes and lysosomes [621]. The negative charge of phosphatidylserine directs moderately positively charged proteins to the endocytic route. Strongly charged cationic proteins that normally associate with the plasma membrane relocalize to endocytic compartments when plasmalemmal charge decreases upon calcium influx.

CD8+ T lymphocytes in lymph node [618]. Plasmatic S1P has a main hematopoietic origin, as erythrocytes are major contributors [617].

[8] Enzyme LPAATα localizes to the endoplasmic reticulum. It is ubiquitously expressed, whereas LPAATβ has a more limited tissue distribution.

[9] A.k.a. brefeldin A-ribosylated substrate (BARS).

Table 7.3. Phosphoinositide kinases and phosphatases and their genes (Source: [895]). Phosphoinositide enzyme genes can form clusters such as the OCRL–PIP5K1A–INPP5A–MTMR1 gene cluster (SAC3: Sac domain-containing inositol phosphatase-3 gene; INPP: inositol polyphosphate 5-phosphatase gene; Mtm: myotubular myopathy-associated gene [myotubularin]; Mtmr: myotubularin-related phosphatase gene Ocrl: oculocerebrorenal syndrome of Lowe phosphatase gene; PIB5PA: Phosphatidylinositol (4,5)-bisphosphate 5-phosphatase-A gene; PIxK: phosphatidylinositol x-kinase; PIPxK: phosphatidylinositol phosphate x-kinase; PTEN: phosphatase and tensin homolog deleted on chromosome ten gene; Sacm1l: suppressor of actin mutations 1-like gene; Skip: skeletal muscle and kidney-enriched inositol phosphatase gene; SYNJ: synaptojanin lipid phosphatase gene; TMEM: transmembrane protein-55A (PI(4,5)P$_2$ 4-phosphatase-2) gene; i-Pase: i-phosphatase [$i = 3, \dots, 5$]).

Enzyme	Genes
	PI Kinases
PI3K	PIK3CA/B/D, PIK3R1–PIK3R3; PIK3CG, PIK3R5–PIK3R6
	PIK3C2A/B/G,
	PI3KC3, PIK3R4
PI4K	PI4KA–PI4KB; PI4K2A–PI4K2B;
	PI4KAP1–PI4KAP2 (pseudogenes)
PIP4K	PIP4K2A/B/G
PIP5K	PIP5K1A–PIP5K1C,
	PIP5K1L1 (a.k.a. PIPSL [PIP5K1A and PSMD4-like],
	PIP5K1P3, and PSMD4P2); PIP5K3; PIP5KL1 (PIPKH)
	PI Phosphatases
3-Pase	PTEN; Mtm1, Mtmr1–Mtmr13
4-Pase	Inpp4A–B; Sacm1l; TMEM55A–TMEM55B
5-Pase	FIG4; INPP5B, INPP5D–INPP5F, INPPL1;
	Ocrl1; PIB5PA; Skip; SYNJ1–2

Phosphoinositides

Phosphoinositides, or phosphatidylinositols derive from reversible phosphorylation in 3 of the 5 hydroxyl groups of inositol. Their metabolism involves many PI kinases and phosphatases (Table 7.3). The combined activities of these enzymes yield a dynamic equilibrium between 7 distinct, interconvertible PI species. They enable phosphorylation–dephosphorylation cycles. Various isoforms of PI kinases and phosphatases localize to all cell compartments (Table 7.4).

Phosphorylation at hydroxyl positions 3, 4, and 5 of the inositol ring enables to categorize the class of phosphatidylinositols into families of mono- (PIP), bis- (PIP$_2$), and trisphosphate (PIP$_3$) derivatives (Fig. 7.2). Membrane-bound PIs hence include: (1) singly phosphorylated phosphatidylinositol, PI(4)P being much more abundant than PI(3)P and PI(5)P; (2) doubly

Table 7.4. Enzymes and their isoforms implicated in the phosphoinositide metabolism for transport and signaling (Source: [623]; PIK: phosphatidylinositol kinase; PIPK: phosphatidylinositol phosphate kinase; PTen: phosphatase and tensin homolog; SHIP: SH2-containing inositol phosphatase; VPS: vacuolar protein sorting). Distinct PI enzyme isoforms can be involved in the synthesis of a given PI in a given cell compartment according to environmental conditions (presence or absence of cell stimulation).

Type	Isoforms (catalytic subunits)
PI3K	PI3Kα–PI3Kδ (family-1)
	PI3KC2α–PI3KC2γ (family-2)
	PI3KC3 (family-3; a.k.a. VPS34)
PI4K	PI4K2α–PI4K2β
	PI4K3α–PI4K3β
PI5K	
PIP4K	PIP4Kα–PIP4Kγ
PIP5K	PIP5Kα–PIP5Kγ
3-Phosphatase	PTen1, PTen2
4-Phosphatase	Synaptojanin-1, synaptojanin-2
5-Phosphatase	Synaptojanin-1, synaptojanin-2
	SHIP1–SHIP2
	5-Phosphatase-2, 5-phosphatase-4

phosphorylated phosphatidylinositol, or bisphosphophosphatidylinositol, the concentration of $PI(4,5)P_2$ being higher than those of $PI(3,4)P_2$ and $PI(3,5)P_2$; and (3) triply phosphorylated phosphatidylinositol $(PI(3,4,5)P_3)$. The activity of PIs depends on their cellular distribution. Their activity is regulated by PI kinases, PI phosphatases, and PI transfer proteins. For example, $PI(3)P$ can be produced by the action of phosphatidylinositol 3-kinase (PI3K) on PIs or by dephosphorylation of $PI(3,4)P_2$ and $PI(3,4,5)P_3$. Cleavage by phospholipases, such as phospholipase-C and phospholipase-A2, produces signaling metabolites. Dephosphorylation mainly by 5-phosphatase switches off signaling.

Inositol polyphosphates and inositol phospholipids are metabolically and functionally interconnected. Inositol polyphosphates are involved in biosynthesis and protein phosphorylation. Inositol (1,4,5)-trisphosphate is required in intracellular calcium signaling. Inositol phospholipids are effectors in acute signaling and regulate the transport across cell membranes and within the cytosol, as well as the cytoskeleton organization.[10]

---

[10] Phosphoinositides bind to cytosolic proteins and to cytosolic domains of membrane proteins. They thus regulate the function of membrane proteins and recruit them to the membrane cytoskeletal and signaling components.

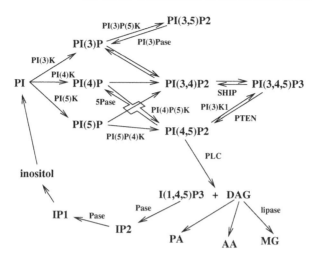

Figure 7.2. Main phosphoinositides (PI) and their relationships (IP: inositol phosphate; AA: arachidonic acid; MG: monoglyceride; PA: phosphatidic acid). Phosphatidylinositol-phosphates are phospholipids that are singly or multiply phosphorylated on 3, 4, and/or 5 positions on an inositol head group. For example, $PI(3,4,5)P_3$ can be generated from $PI(4,5)P_2$ by PI3K. Conversely, $PI(4,5)P_2$ can be produced from $PI(3,4,5)P_3$ by 3-phosphatase PTen.

Phosphoinositide Tags and their Effects

Phosphoinositides are responsible for the heterogeneity of membrane lipidic composition between membranes of cell compartments as well as within a given membrane that is needed in intracellular molecular tranfer and signal transduction. However, phosphoinositides localize to cell membranes in relatively small amounts among phospholipids (5–30% of membrane phospholipids). Nevertheless, they are involved in numerous cell-life events (Table 7.5). Phosphoinositides also reside on the cytosolic face of intracellular membranes and nucleoplasm. They particularly regulate nuclear functions, cytoskeletal dynamics, cell signaling, and membrane transport.[11]

Phosphoinositides are constitutive signals that define organelle identity. Phosphoinositide segregation according to the type of organelle membranes ensures directional transport from one cell compartment to another. Moreover, phosphorylation and dephosphorylation of phosphoinositides by kinases and phosphatases adapt their activities to the cell's demand.

The turnover of membrane phosphatidylinositol and its phosphorylated products, the phosphoinositides, occurs after tissue stimulation.[12] Each of

---

[11] Plasmalemmal phosphatidylinositol (4,5)-bisphosphate is required for endo- and exocytosis, as well as cortical cytoskeleton integrity. Both PI(3)P and $PI(3,5)P_2$ regulate membrane trafficking in both directions.

[12] Reversible phosphorylation of the inositol ring of the precursor phosphatidylinositol at positions 3, 4, and 5 generates 7 phosphoinositides.

Table 7.5. Functions of phosphoinositides (Sources: [624, 895]). Phosphoinositide regulators control the subcellular location and activation of various effector proteins that have suitable binding domains. A fraction of a given PI enzyme can relocalize upon cell stimulation.

	Location	Functions
PI(3)P	Endosomes	Endocytosis
	Plasma membrane	
PI(4)P	Golgi	Transport
	Plasma membrane	
PI(5)P	Nucleus	Apoptosis
PI(3,4)P$_2$	Plasma membrane	Signaling
		Cytoskeletal dynamics
PI(3,5)P$_2$	Late endosomes	Osmotic stress
		Signaling
PI(4,5)P$_2$	Plasma membrane	Endocytosis
	Golgi	
	Secretory granules	
	Nucleus	Cytoskeletal dynamics
PI(3,4,5)P$_3$	Plasma membrane	Signaling
		Cytoskeletal dynamics

the 7 phosphoinositides has a cellular distribution often characterized by a predominant location in membrane subsets.[13]

Phosphoinositides are precursors of intracellular signaling molecules, and thus involved in smooth muscle cell contraction and endothelial cell production of vasoactive molecules (Vol. 5 – Chaps. 8. Smooth Muscle Cells and 9. Endothelium). Phosphoinositides can specifically interact with proteins having lipid-binding domains.

Phosphoinositides serve as platforms for molecular assemblies to recruit and/or activate effectors at proper locations and times. In particular, they facilitate enzyme recruitment to the appropriate membrane region for catalysis when substrates have reached the membrane.

Phosphatidylinositol (4,5)-Bisphosphate (PI(4,5)P$_2$)

Phosphatidylinositol (4,5)-bisphosphate regulates several ion carriers, such as Na^+–Ca^{++} exchanger, ATP-sensitive $K_{IR}6.2$ channel, and G-protein-dependent inward rectifier K^+ channels (GIRKs; $K_{IR}3.1$–3.4). Furthermore, PI(4,5)P$_2$ activates enzymes and anchors proteins to the membrane.

---

[13] PI(4)P is observed in the plasma membrane, mainly in the Golgi body, and in the endoplasmic reticulum; PI(3)P in early endosomes. PI(4,5)P$_2$ and PI(3,4,5)P$_3$ are located at the plasma membrane. PI(3,4)P$_2$ is mostly found in the plasma membrane and early endocytic vesicles; PI(3,5)P$_2$ in early and late endosomes.

PI(4,5)P$_2$ is involved in cell transport because it: (1) binds to clathrin adaptors and endocytic factors such as dynamin and (2) acts on the actin cytoskeleton. The clathrin-coat assembly protein, AP180 adaptor protein,[14] and annexin-A2 that control ion channels and intracellular transport as well as, in the extracellular space, stimulate fibrinolysis and promote clearance of apoptotic cells, link to PI(4,5)P$_2$. On the other hand, phosphoinositide-sequestering proteins such as *pipmodulins*[15] buffer PI(4,5)P$_2$ and release it upon demand.

When stimulated by G-protein-coupled receptors, phospholipase-C cleaves phosphatidylinositol (4,5)-bisphosphate to manufacture the second messengers inositol (1,4,5)-trisphosphate (IP$_3$) and diacylglycerol (DAG). Diacylglycerol activates protein kinase-C. Diacylglycerol is either hydrolyzed by the diacylglycerol lipase into messenger arachidonic acid or recycled into phosphoinositide synthesis, being phosphorylated by diacylglycerol kinase to phosphatidic acid. Phospholipase-D is associated with the reverse reaction (dephosphorylation), from phosphatidic acid to diacylglycerol by phosphatidic acid phosphohydrolase. Inositol (1,4,5)-trisphosphate releases Ca^{++} from intracellular stores through inositol-trisphosphate receptors. Inositol (1,4,5)-trisphosphate can be converted into inositol by specific phosphatases.

Phosphatidylinositol (4,5)-bisphosphate of the inner leaflet of the cell membrane is synthesized from phosphoinositide by phosphoinositide 4-kinase (PI4K) and phosphoinositide 5-kinase (PI5K). Agent PI(4)P that is a precursor for PI(4,5)P$_2$ at the plasma membrane arises from the Golgi body-located PI4Ks (PI4K2α and PI4K3β) in steady state and mainly from PI4K3α in angiotensin-2-stimulated cells [625].

Phosphatidylinositol (4,5)-bisphosphate can undergo dephosphorylation by 5-phosphatase to create phosphatidylinositol 4-monophosphate. Phosphoinositide 3-kinase-1 phosphorylates phosphatidylinositol (4,5)-bisphosphate and produces phosphatidylinositol (3,4,5)-trisphosphate, which regulates small guanosine triphosphatases Rho (Vol. 4 – Chap. 8. Guanosine Triphosphatases and their Regulators).

The plasmalemmal distribution and sequestration of highly charged PIP$_2$ is affected by electrostatic interactions and Ca^{++}–calmodulin. Phosphatidylinositol (4,5)-bisphosphate and phosphoinositides phosphorylated at the 3 position have direct signaling roles. Phosphatidylinositol (4,5)-bisphosphate transduces extracellular signals, either via concentration fluctuations or its metabolites, generating 3 second messengers (diacylglycerol, inositol triphosphate, and phosphatidylinositol trisphosphate).

---

[14] A.k.a. 91-kDa synaptosomal-associated protein (SNAP91).

[15] Pipmodulins are proteic regulators of actin dynamics via interactions with PIP$_2$. They include growth-associated protein GAP43, brain-abundant, membrane-attached signal or brain acid-soluble protein BASP1 (a.k.a. 22-kDa neuronal tissue-enriched acidic or neuronal, axonal membrane protein NAP22 and CAP23), and myristoylated alanine-rich protein kinase-C substrate MARCKS.

The balance between PI(4,5)P$_2$ and PI(3,4,5)P$_3$ plays a role in cell polarity. In certain polarized cells, PI(3,4,5)P$_3$ resides exclusively at the basolateral membrane, whereas the apical membrane is enriched in PI(4,5)P$_2$. The Par3–aPKC complex intervenes in asymmetrical PI distribution by recruiting phosphatase PTen to tight junctions, thereby restricting PI(3,4,5)P$_3$ from moving across tight junctions into the apical membrane [26].

Phosphatidylinositol (3,4,5)-Trisphosphate

Phosphorylation by phosphoinositide 3-kinase produces phosphatidylinositol (3,4,5)-trisphosphate, a second messenger for the activation of protein kinase-B and phosphoinositide-dependent kinase-1, in response to growth factor stimulation.

Dephosphorylation of PI(3,4,5)P$_3$ at the 3 position by phosphatase and tensin homolog PTen switches off its signaling (Fig. 7.2). Dephosphorylation of PI(3,4,5)P$_3$ by 5-phosphatase, such as SH2-containing inositol phosphatases SHIP1 and SHIP2, produces functional PI(3,4)P$_2$ (Vol. 4 – Chap. 1. Signaling Lipids).

Phosphoinositides and Small GTPases

Small guanosine triphosphatases have a functional relationship with phosphoinositides [626]. Like phosphoinositides, small GTPases help in recruiting cytosolic proteins to specific membrane compartments.[16] Phosphoinositides regulate the recruitment of guanine nucleotide-exchange factors and GTPase-activating proteins to membranes and their activity. Several enzymes of the phosphoinositide metabolism are effectors of GTPases, inducing feedback loops. Membrane-bound GTPases form coreceptors with phosphoinositides to recruit cytosolic proteins.

Summary

Phosphoinositide metabolism and functions are summarized in Fig. 7.3.

7.2.2.3 Cholesterol

Cholesterol homeostasis is linked to the cell integrity. Cellular content and distribution of cholesterol within the cell depends on cholesterol synthesis in the endoplasmic reticulum and uptake of cholesterol ester-rich lipoproteins that are conveyed in the flowing blood.

Cholesterol Synthesis

Cholesterol synthesis involves numerous enzymes. The hydroxy-methylglutaryl-coenzyme A (HMGCoA) reductase is the rate-limiting enzyme. It is required in the conversion of HMGCoA into mevalonate (Fig 7.4).

---

[16] Ras GTPases are implicated in signaling, Rho GTPases in actin regulation, ARF, ARL, and Rab GTPases in vesicular transport.

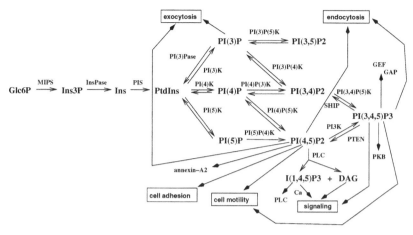

Figure 7.3. Phosphoinositide metabolism and function (Sources: [179, 626, 627]). Phosphoinositides (PI or PtdIns) form PI phosphorylated derivatives (PI^P) (polyphosphoinositides) and diverse free inositol polyphosphates (InsP). Myo-inositol (Ins) is synthesized from glucose-6-phosphate (Glc6P) by NAD^+-dependent myo-inositol-3-phosphate synthase (MIPS) that leads to ^{D}myo-inositol 3-phosphate (Ins3P), which is then dephosphorylated by inositol monophosphatase (InsPase). Certain mammalian cells produce Ins, whereas others import Ins via symporters. Phosphoinositide synthase (PIS) catalyzes PI production from Ins. Many PI kinases catalyze the addition of phosphate groups to the inositol 3, 4, and/or 5 positions. PI 3-Kinase (PI3K) synthesizes PI(3)P from PI and then PI(3)P 5-kinase (PI(3)P5K) $PI(3,5)P_2$. PI(3)P serves in secretory endosomes and $PI(3,5)P_2$ in endocytosis. $PI(4,5)P_2$ results from precursors PI(4)P produced by PI4K owing to PI(4)P5K and, to a lesser extent, PI(5)P owing to PI(5)P4K. Plasmalemmal $PI(4,5)P_2$ is involved in exo- and endocytosis. It regulates ion carriers. It is required for adaptors between the plasma membrane and actin cytoskeleton at sites of cell adhesion. Plasmalemmal second messenger $PI(3,4,5)P_3$ is synthesized by PI3K after activation of receptor Tyr kinases (PI3K1a) or G-protein-coupled receptors (PI3K1b). Conversely, $PI(3,4,5)P_3$ is dephosphorylated by 3-phosphatase PTen (phosphatase and tensin homolog on chromosome 10) into $PI(4,5)P_2$. SH2-containing inositol 5′-phosphatase (SHIP) targets $PI(4,5)P_2$. Both $PI(4,5)P_2$ and $PI(3,4,5)P_3$ in cooperation with small GTPases participate in the recruitment and activation of actin regulators at the plasma membrane to control cell shape and motility. $PI(3,4,5)P_3$ also binds endocytic clathrin adaptors. $PI(4,5)P_2$ is hydrolyzed by phosphoinositidase-C (PIC or PLC) to form second messenger $I(1,4,5)P_3$ and DAG. $I(1,4,5)P_3$ can also serve as a precursor of inositol phosphates, such as Ins-hexakisphosphate (IP6) and IP6-derived diphosphates (or pyrophosphates PPIP5 [IP7] and $(PP)_2IP4$ [IP8]). Bruton Tyr kinase (BTK), general receptor for phosphoinositides GRP1, a guanine nucleotide-exchange factor (GEF), GTPase-activating protein (GAP), and protein kinase-B (PKB) target $PI(3,4,5)P_3$ or/and $PI(3,4)P_2$. PLCδ1 binds $PI(4,5)P_2$ and $I(1,4,5)P_3$.

The membrane sterol (SREBP) cleavage-activating protein transports the sterol regulatory element-binding protein (SREBP) from the endoplasmic

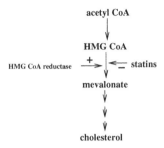

Figure 7.4. Main steps of the cholesterol synthesis. Cells synthesize cholesterol from acetyl-CoA using the mevalonate pathway. The rate-limiting enzyme of cholesterol synthesis is hydroxymethylglutaryl-CoA reductase (HMGCoAR).

reticulum to the Golgi body, thereby activating cholesterol synthesis when the cellular cholesterol level is low. Cholesterol accumulation in the endoplasmic reticulum membrane changes conformation of the sterol cleavage-activating protein so that this protein binds to insulin-induced gene product (InsIG) [628]. InsIG-bound sterol cleavage-activating protein fails to carry sterol regulatory element-binding protein to the Golgi body and hinders cholesterol synthesis. Thereby, sterol cleavage-activating protein dictates the rate of cholesterol synthesis.

Synthesized cholesterol in the endoplasmic reticulum is conveyed to the plasma membrane, mostly bypassing the Golgi body. Cholesterol moves toward the cell surface via both vesicular and non-vesicular transports.

Cholesterol Endocytosis and Exocytosis

Cholesterol uptake by cells from lipoproteins in the extracellular space involve low-density lipoprotein receptor (Fig. 7.5). Plasmalemmal LDLR binds particles that contain ApoB or ApoE, such as chylomicron remnants and very-low- (VLDL) and low-density (LDL) lipoproteins. Extracellular lipids such as those carried by LDL[17] are transported via endosomes; the LDL receptor recycles to the plasma membrane.

Lipoproteins are conveyed by clathrin-coated vesicles to endosomes. Within the lumen of late endosomes, cholesteryl esters are hydrolyzed into free cholesterol by acid lipase. Cholesterol is then recycled using tubulovesicles and multivesicular endosomes. In fact, released cholesterol joins either the cholesterol pool for recycling to the cell surface or the endoplasmic reticulum to be re-esterified and stored.

Cholesterol imported to late endosomes is sorted owing to 2 proteins in late endosomes: Niemann-Pick type-C proteins, i.e., transmembrane NPC1 and luminal NPC2.[18] Protein NPC2 is required to deliver or remove cholesterol from

---

[17] On average, LDL contains 1,500 molecules of cholesteryl esters.

[18] Mutations in the genes that encode any NPC protein causes lysosomal storage abnormality, the so-called Niemann-Pick type-C disease, which is characterized

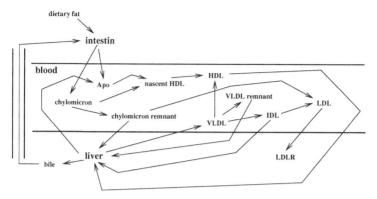

Figure 7.5. Metabolic pathways of plasma lipoproteins. Insoluble lipids are transported through the blood circulation as complexes with proteins. Dietary cholesterol is absorbed by enterocytes of the small intestine that manufacture chylomicrons, incorporating also triglycerides. Hepatocytes produce very-low-density lipoproteins (VLDL). VLDLs released by hepatocytes are processed in the circulation into intermediary (IDL) and low-density (LDL) lipoproteins. Both chylomicrons and VLDLs secreted by intestine and liver, respectively, are triglyceride-rich lipoproteins. They undergo lipolysis in the circulation and deliver fatty acids to tissues. Chylomicron remnants and about half of VLDL remnants are taken up by the liver. The remainder of VLDL remnants is further metabolized to cholesterol-rich LDLs. Some triglycerides are hydrolyzed in the circulation and new apoproteins (Apo) are added to generate chylomicron remnants that are taken up by hepatocytes. Excess cholesterol is released by cells and forms nascent then mature high-density lipoproteins (HDL). Nascent HDLs are formed in the circulation and also contain lipid-poor apolipoproteins secreted by liver and intestine and released during lipolysis of TG-rich lipoproteins. HDLs return lipids to the liver (*reverse cholesterol transport*). Cholesterol is secreted from the liver into the bile to be either reabsorbed (*enterohepatic cycle*) or excreted into feces. The brain is isolated from this circuit, but lipoproteins are used to transport lipids between cells of the nervous system. Abnormal lipoprotein concentrations in the plasma represent a risk factor of atheroma. HDLs are either pro- or anti-inflammatory. Small, dense LDLs can be more atherogenic than large, buoyant LDLs.

NPC1, as bidirectional transfer of cholesterol between NPC1 and liposomes is accelerated more than 100-fold by NPC2 [629].[19] These proteins that both bind to cholesterol are required for the cholesterol delivery to a cholesterol

---

by very slow efflux of cholesterol from late endosomes and subsequent accumulation of cholesterol and glycosphingolipids in late endosomes. Proteins NPC1 and NPC2 localize to the membrane and lumen of late endosomes, respectively. In addition, NPC1 is an oxysterol-binding protein. Both NPC1 and NPC2 are required to remove lipoprotein-derived cholesterol from endosomes and lysosomes.

[19] Protein NPC1 binds to and dissociates cholesterol slowly, whereas NPC2 links to and releases rapidly cholesterol. Consequently, NPC1 and NPC2 donors transfer cholesterol very slowly and quickly to either NPC1 or NPC2 acceptors, respectively.

efflux transporter such as ATP-binding cassette transporter ABCa1 (ABC; Vol. 3 – Chap. 4. Membrane Compound Carriers), which enables the export of cholesterol from late endosomes.[20]

Transporter ABCa1 delivers cholesterol and phospholipids to apolipoproteins [630] (Fig. 7.6). It promotes phospholipid transfer to lipid-poor apolipoprotein-A1, which can then be transformed into HDL in the blood by lysolecithin-cholesterol acyltransferase (LCAT). In addition, transporter ABCg1 on macrophages and endothelial cells supports cholesterol efflux from cells to high-density lipoproteins. In hepatocytes, HDLs bind to scavenger receptor-B1. Once cholesterol is imported, hepatocyte transporters ABCg5 and ABCg8 carry cholesterol as well as other sterols into the bile.[21]

Three cellular pathways of cholesterol transport between cells and high-density lipoproteins (HDL; Vols. 3 – Chap. 4. Membrane Compound Carriers and 5 – Chap. 1. Blood) include: (1) passive diffusion from the plasma membrane to plasma HDLs; (2) efflux via scavenger receptor-B1 (mainly in the liver and adrenal glands); and (3) active efflux via ATP-binding cassette transporters to lipid-poor apolipoproteins such as apolipoprotein-A1. The 2 first processes involve bidirectional fluxes of cholesterol, whereas the last-mentioned is associated with unidirectional outward transfer of cholesterol.

Nuclear hormone receptors (NR) — liver X (LXR [NR1h2–NR1h3]) and retinoid X (RXR [NR2b1–NR2b3]) receptors — are involved in cholesterol transport from peripheral tissues to the liver as well as cholesterol conversion into bile acids and excretion (Vol. 3 – Chap. 6. Receptors) .[22] Cholesterol overload in cells, particularly foam cells (Vol. 6 – Chap. 7. Vascular Diseases), activates liver X and retinoid X receptors, leading to cholesterol efflux, via activated receptors ABCa1 and ABCg1 as well as apolipoprotein-E, and transport to the liver for secretion in bile, using the plasma lipid transfer proteins CETP and PLTP. Transporter ABCa1 may not only act as a phospholipid translocase, but also participate in membrane rafts.

---

[20] Transporter ABCa1 is particularly expressed in the liver, kidney, adrenal glands, and intestine, as well as foam cells (transformed macrophages) of atherosclerotic lesions. It localizes to the plasma membrane, Golgi body, endosomes, and lysosomes.

[21] In the liver, cholesterol is excreted into the bile both as free cholesterol and after conversion to bile acids.

[22] Transcription of ATP-binding cassette carrier ABCa1 that transports cholesterol and phospholipids across the plasma membrane is regulated by oxysterol-activated nuclear receptor liver X receptor (LXR), a partner of retinoid X receptor (RXR) that is activated by specific oxysterols (transformed cholesterol). The activity of ABCa1 is hindered by mevalonate products [633].

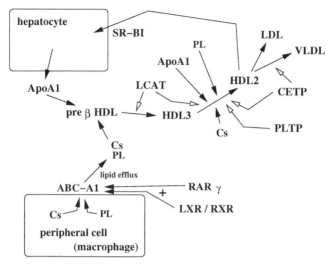

Figure 7.6. Cholesterol (Cs) and its transport via lipoproteins (Sources: [631,632]). ATP binding cassette (ABC) transporters-A1 of the cell membrane favor lipid efflux from cholesterol-loaded peripheral cells such as macrophages. They are activated by nuclear liver X (RXR or NR1h), retinoid X (RXR or NR2b), and retinoid acid receptors RARγ (NR1b3). Cholesterol and phospholipids form preβ-HDL (a minor fraction of plasma HDLs) with apolipoprotein-A1 (ApoA1) that are transformed into small high-density lipoproteins (HDL3) by lysolecithin-cholesterol acyltransferase (LCAT) and secondarily into large HDL2 by addition of Cs, phospholipids, and ApoA1. HDL2 gives birth to very-low-density lipoproteins (VLDL) by cholesterol ester transfer protein (CEPT) and low-density lipoproteins (LDL). HDL metabolism involves: (1) various subspecies of HDLs (lipid-poor preβ-HDL; small, dense HDL3; and large, light HDL2); (2) lipid components (apolipoprotein A1 [ApoA1], phospholipid [PL], cholesterol [Cs], and cholesterol ester [CE]); (3) enzymes (lecithin cholesterol acyltransferase [LCAT], cholesteryl ester transfer protein [CETP], phospholipid transfer protein [PLTP], hepatic lipase [HL], and endothelial lipase [EL]); and (4) receptors and transporters (ATP-binding cassette transporter-A1 [ABCa1] and scavenger receptor-B1 [SRb1]). HDL3 Particles have anti-atherogenic effects.

Cholesterol Metabolism

Membrane component lipids (cholesterol, sphingolipids, glycosphingolipids, and lysobisphosphatidic acid) are delivered to late endosomes and lysosomes to undergo hydrolysis.

Excess cholesterol is esterified by acylCoA-cholesterol acyltransferase in the endoplasmic reticulum. The esterified cholesterol is stored in short-term cytoplasmic lipid buffers. Cholesterol esters are indeed processed by hydrolases in late endosomes and lysosomes (de-esterification) and released for cell use.[23]

---

[23] Esterification–de-esterification cycle represents a major part of cholesterol metabolism.

7.2.3 Glucids

Membrane glucidic copulas of the external membrane layer contribute to the membrane asymmetry. Membrane glucids participate in the protein structure and stability of the membrane. They modulate the function of membrane proteins. Surface polysaccharides are directly used for cell recognition and adhesion. For example, laminin, a glycoprotein, allows adhesion of endothelial cells to collagen. The *macrophage migration-inhibitory factor* (MMIF) released by activated T lymphocytes, binds to membrane glucids of macrophages to inhibit their migration once they are on the working site.

7.2.4 Proteins

Integral membrane proteins cross the lipid bilayer (transmembrane proteins), some being able to span it, or are attached to bilayer lipids. On the other hand, peripheral membrane proteins are attached to integral membrane proteins.

Membrane proteins can be classified into the following types [634]: (1) *type-1 single-pass transmembrane* protein with a cytoplasmic C-terminus and an extracellular (or luminal for endothelium wetted surface or organelle membrane) N-terminus; (2) *type-2 single-pass transmembrane* protein with a cytoplasmic N-terminus and an extracellular or luminal C-terminus; (3) *multipass transmembrane* proteins;[24] (4) *lipid chain-anchored* membrane proteins; (5) *glycosyl-phosphatidylinositol-anchored* membrane proteins; and (6) *peripheral membrane proteins* that bind to the membrane indirectly by non-covalent interactions with other membrane proteins.

Integral membrane proteins encompass members of the 5 first types of membrane proteins of the above-mentioned classification. Numerous plasmalemmal proteins are attached to the cell membrane by a glycosyl-phosphatidylinositol anchor. These GPI-anchored proteins (gpiAP) can be removed by phospholipase-C.

Membrane proteins represent approximately 30% of the proteome. The activity of plasmalemmal proteins is regulated by switching among inactive, active, and possible intermediate states, particularly by phosphorylation. Changes in conformational dynamics between folded and unfolded protein states affect the protein stability and binding affinity.

In isolated proteins, sequence dictates structure (atomic position) that determines function. However, in membrane proteins, cellular membrane yields a mechanical medium with which proteins interact. Thickness and curvature as well as bending stiffness of membrane regions that surrounds embedded proteins influence their function. Moreover, certain lipid species are required for proper protein functioning. Conversely, membrane proteins can influence

---

[24] I.e., with many membrane-spanning or transmembrane segments.

bending modulus of lipid bilayers in a concentration-dependent manner. Protein interactions such as binding of membrane-associated proteins are able to deform membrane.

Most membrane proteins are inserted into membranes owing to N-terminal signal sequences recognized by signal recognition particle proteins. In fact, both proteins that are secreted (messengers) or inserted in the plasma membrane (sensors and receptors of the cell environment cues as well as carriers), once translated in ribosomes, have a short terminal sequence that assigns the final destination. Proteins translocate across or into membranes, either in folded or unfolded manner, using or not pre-existing *translocons*, to function. Translocon allows protein migration across and integration into membranes. Translocon contains a non-specific protein-conducting channel embedded in the membrane and its associated molecules that conveys newly synthesized proteins with the appropriate address label. *Translocase* is an enzyme devoted to protein translocation across or into cell membrane. Translocase can form a complex with ribosome [635].

Identification of the type of a membrane protein helps to determine its function. Transmembrane proteins function on both sides of the membrane. Certain proteins can transmit signals from their extra- to intracellular domains. Others transport molecules from one side to the other. Some connect the cytosleleton to the extracellular matrix. Proteins that are associated with one side of the lipid bilayer can only operate on that side. Membrane proteins, embedded in the lipidic skeleton, are implicated in several cell activities (Table 7.6).

Membrane proteins function as carriers, receptors, adhesion sites, or markers (cell recognition). Membrane carriers select ions and molecules to cross the plasma membrane from or to the cytosol, thereby creating concentration gradients and electical potential across the plasma membrane. Proteins involved in proton and electron transport form waterproof structures, whereas some transporters contain large, water-filled cavities.

The plasma membrane contains histocompatibility antigens (H-antigens), protein–carbohydrate complexes, that determine the interaction with the immunological system. The synthesis of H-antigens is controlled by the major histocompatibility complex (MHC) on chromosome 6.

The plasmalemmal prohibitin domain-containing proteins bind cholesterol and ion channels, especially those involved in mechanotransduction, such as epithelial sodium channels and transient receptor potential channels (Vol. 3 – Chap. 2. Membrane Ion Carriers). This family includes: (1) raft- and caveola-associated, integral membrane proteins flotillins (Flot1–Flot2), (2) podocin that lines podocytes of renal glomeruli, (3) type-1 and -2 prohibitins that can reside in the inner mitochondrial membrane and nucleus, (4) integral membrane stomatin that operates in a stretch-sensitive complex and participates in the regulation of ion transport, among others. The plasmalemmal prohibitin domain-containing proteins regulate the activity of associated ion channels.

Table 7.6. Activity of membrane proteins. Specialized proteins are enclosed in lipid bilayers. They can or cannot span the bilayer thickness. Membrane proteins differ by their tasks and structures, ranging from a single domain to multisubunit protein. Membrane proteins strongly interact with membrane lipids. Water can penetrate deeply into membrane proteins. Both water penetration and interactions with the lipid bilayer influence their geometry. Protein–lipid interface influences protein functioning. Mechanosensitive channels are gated by membrane stretch. Gating features of ion channels such as voltage-dependent channels can depend on membrane properties. Cooperative channel gating between neighboring proteins arises from interactions between adjacent channels via surrounding lipids. In addition, dimerization kinetics of some membrane peptides can be controlled by mechanical stress applied to membrane.

Function	Type	Effect
Mass transfer	Ion pump	Cell polarity
	Ion exchanger	Cell functioning
	Ion channel	Cell metabolism
	Vesicle (clathrin, caveolin)	Endo/exocytosis
	Gap junction (connexin)	Electrical activity coordination Molecule exchange
Transduction	Receptor	Signaling
Cell organization and adhesion	Adhesion molecules Cadherin, selectin, integrin, IgCAM	Tissue stability Signaling ECM-Csk connections Intercellular interactions Extravasation
Enzyme	Proteases	Cleavage of transmembrane proteins

7.2.4.1 Cell Adhesion Molecules

Some molecules are involved in cell organization and adhesion. Cell adhesion molecules (Sect. 7.5) are proteins that binds those of other cells or constituents of the extracellular matrix. These transmembrane receptors possess 3 domains: (1) an intracellular domain that interacts with the cytoskeleton; (2) a transmembrane domain; and (3) an extracellular domain that interacts either with other cell adhesion molecules or matrix components. Cell adhesion molecules pertain mainly to 5 classes: addressins (lymphocyte homing receptors), cadherins, immunoglobulin cell adhesion molecules, integrins, and selectins.

7.2.4.2 Signaling Receptors

Signal transduction refers to the conversion of a mechanical, physical, or chemical stimulus into a chemical signal that is transmitted inside the cell using an appropriate cascade of chemical reactions (Vols. 3 and 4). Signal transduction starts with the excitation of a plasmalemmal receptor that triggers the signaling cascade. Although some receptors reside within the cell, most receptors are transmembrane proteins. Numerous receptors respond to chemicals, such as hormones, growth factors, cytokines, and chemokines, as well as neurotransmitters and extracellular matrix components. Ligand fixation triggers synthesis of *second messengers*, such as cyclic nucleotides (cyclic adenosine monophosphate [cAMP] and cyclic guanosine monophosphate [cGMP]), phosphoinositides, etc., that are responsible for cell responses of the extracellular ligand (first messenger).

7.2.4.3 Membrane Enzymes: Regulated Intramembrane Proteolysis

Certain enzymes reside and are able to act at the cell hydrophobic–hydrophilic border as they transfer a lipophilic substrate from the membrane into their active site to catalyze its reaction with a hydrophilic compound. Proteases irreversibly cleaves peptide bonds.

In many cases, the *regulated intramembrane proteolysis* (RIP) begins with an initial cleavage (ectodomain shedding) of a single-span membrane protein substrate with type-1 or -2 orientation [636]. This ectodomain shedding occurs within the ectodomain at a peptide bond close to the transmembrane domain. Shedding results in the release of the ectodomain into the extracellular milieu. The generated membrane-bound stub undergoes a second cleavage within its transmembrane domain (intramembrane proteolysis).

Membrane-embedded proteases cleave the transmembrane segment of membrane-docked proteins. The local composition of membrane lipids can influence orientation and activity of these proteases. Ubiquitous transmembrane proteases are anchored in the plasma membrane with their catalytic site exposed to the external surface of the cell surface. They participate in extracellular proteolysis for degradation of extracellular matrix components, regulation of chemokine activity, and release of membrane-anchored cytokines, cytokine receptors, and adhesion molecules. They thus influence cell growth and motility. In particular, they contribute to the secretion of angiogenic factors [637].

Ectoenzymes localize to the external surface of the plasma membrane and thus target extracellular substrates. ADP-Ribosyltransferases, cyclases, nucleotidases, oxidases, peptidases, and proteases have catalytic domains outside the plasma membrane. In particular, they regulate leukocyte extravasation (Vol. 5 – Chap. 9. Endothelium).

On the other hand, *exoenzymes* are secreted by cells to work in the extracellular space. Exoenzymes encompass blood-coagulation factors such as thrombin, angiotensin-converting enzyme secreted by pulmonary and renal endothelial cells to generate angiotensin-2, lipoprotein lipase, which releases lipids from circulating lipoproteins, and digestive enzymes, such as salivary

Table **7.7.** Secretases (Source: [638]; ADAM: a disintegrin and metalloprotease [adamlysin]; BACE: β-amyloid precursor protein-converting enzyme [BACE1/2 a.k.a. memapsins-1/2 and aspartyl protease AsP1/2]). The γ-secretase complex (γSC) comprises 4 elementary components: presenilin (PS), nicastrin, presenilin enhancer-2 (PEn2) and anterior pharynx defective phenotype homolog APH1.

Secretase family	Secretase type
α	Adamlysins
β	BACE1/2
γ	Presenilins PS1/2, Pen2, Aph1, nicastrin (Aph2)

and pancreatic amylases, which hydrolyze dietary starch into di- and trisac-charides.

Membrane-anchored proteases lodge not only in the plasma membrane, but also in organelle membrane to release mediators inside the cytosol that can then carry out their task. Among these released fragments, some are transcription factors that then translocate to the nucleus to activate their target genes.

Neuroproteases: Illustration of Membrane Protein Turnover

In neurons, metabolism of amyloid precursor protein (APP) yields information on the turnover of a type-1 membrane protein. Turnover of amyloid precursor protein (APP) depends on intramembranal cleavage by β-secretase such as βAPP-converting enzyme (BACE), a neurosecretase for initial APP process-ing, and by γ-secretase complex (γSC), which catalyzes the final cleavage to generate amyloid-β peptide (Aβ), as well as metalloproteases of the ADAM family (α-secretase)[25] for APP turnover itself [638] (Table 7.7). Secretases are activated for regulated intramembranal proteolysis.

Intramembrane Protease Families

Intramembrane proteases comprise metallo, aspartyl, and rhomboid serine proteases (Table 7.8). Members of the family of adamlysins and matrix metalloproteases, as well as aspartate proteases β-site APP-cleaving en-zymes BACE1 and BACE2 carry out *ectodomain shedding* (cleavage step 1). *Intramembrane cleavage* (cleavage step 2) is mediated by intramembrane-cleaving proteases (iCliP), such as metallo and aspartyl proteases. Metallo

---

[25] Among members of the ADAM family, those involved in amyloid precursor protein turnover comprise ADAM8 (or CD156) to ADAM10 (or CD156c) and ADAM17 (a.k.a. (pro)tumor-necrosis factor-α-converting enzyme [TACE] and CD156b). Enzyme ADAM17 is targeted by tissue inhibitor of metalloproteases TIMP3, whereas ADAM10 is inhibited by TIMP1 and TIMP3 [638].

Table 7.8. Families of membrane proteases (P; Main source: [639]; APLP: amyloid precursor protein-like protein; HER: human epidermal growth factor receptor; SPP: signal peptide protease; TFPP: type-4 prepilin-like proteins leader peptide-processing enzyme; ZMPSte24: zinc metallopeptidase Ste24 homolog). Sheddases of the ADAM and BACE protease families are intramembrane-cleaving proteases (iCliP). Proteases ADAMs and BACE1 share same substrates, such as amyloid precursor protein (APP), selectin-P ligand (a.k.a. P-selectin glycoprotein ligand PSGL1 or CD162), neuregulin-1 (Nrg1), and β subunits of voltage-gated sodium channels (Na_V). Enzyme ADAM10 is involved in the shedding of different members of the cadherin class, once transported to adherence junctions. It also releases low-affinity immunoglobulin-E receptor FcER2 (Fc fragment of IgE receptor; a.k.a. CD23). Protease BACE1 is involved in the shedding of type-3 neuregulin-1 required for myelin sheath formation in the peripheral nervous system. Membrane-bound transcription factor peptidases site-1 and -2 (MBTPS1 and MBTPS2; a.k.a. site-1 and -2 proteases [S1P and S2P]) cleave luminal and transmembrane domains, respectively, of sterol response element-binding protein (SREBP) upon sterol deprivation as well as type-2 endoplasmic reticulum-associated transmembrane-activating transcription factor ATF6 during the unfolded protein response (UPR). Once UPR is activated, ATF6 is proteolyzed to release a cytosolic fragment that migrates to the nucleus to activate gene transcription.

Action mode	Family	Substrates	iCliP Function
MetalloP.	MBTPS2	SREBP, ATF6, ZMPSte24	Yes
	ADAM	APP, PSGL1, Nrg1, Na_V, cadherins	Yes
Serine P.	Rhomboid	EGF	Yes
Aspartyl P.	Presenilin	Notch, HER4, CSF1, CD44, cadherin, APP, APLP1/2, GluR3, TNFRSF16, IfnαR2	Yes
	SPP	Signal peptide remnants	Yes
	TFPP	Bacterial prepilin-4	No
	BACE	APP, APLP1/2, PSGL1, Nrg1, Na_V	Yes

and aspartyl proteases conduct cleavage step 2 after step 1. Intramembrane cleavage releases peptides into the extra- or intracellular space. On the other hand, rhomboid iCLiPs cleave their substrates without the requirement of a preceding ectodomain cleavage [636].

Aspartyl Proteases (Aspartic-Type Endopeptidases)

Signal peptide peptidases and their homologs — signal peptide peptidase-like proteins (SPPL2a–SPPL2c and SPPL3) — constitute a family of transmembrane aspartyl proteases. [26] Signal peptide peptidase removes membrane-anchored signal peptides from the endoplasmic reticulum membrane after an initial cleavage by signal peptidase. Substrates of SPPs and SPPLs are type-2 transmembrane proteins.[27] Type-2 membrane proteins that have experienced ectodomain shedding can subsequently become substrates for a member of the aspartyl protease iCliP subfamily of the signal peptide peptidase-like (SPPL) family or the metalloprotease membrane-bound transcription factor peptidase site-2.

β Secretase β-amyloid precursor protein-converting enzyme BACE acts on various substrates, such as P-selectin glycoprotein ligand-1, (2,6)-sialtransferase, APP-like proteins APLP1 and -2, β1 to β4 subunits of voltage-gated sodium channel, LDL receptor, and amyloid precursor protein and its derivatives [638].

The γ secretase complex (γSC) is an aspartyl protease that comprises 4 elementary components: presenilin (PS1 or PS2), nicastrin (Nct), presenilin enhancer-2 (PEn2), and anterior pharynx defective phenotype homolog APH1. These 4 components assemble together and lead to presenilin endoproteolysis to form an active γ-secretase. Enzyme γSC catalyzes intramembrane proteolysis of type-1 membrane proteins [636].

The γ-secretase complex cleaves and releases intracellular domains as potential nuclear transactivators of Notch and its ligands Delta-like proteins and Jagged, growth hormone receptor, tumor-necrosis factor receptor superfamily member TNFRSF16, interferon-αreceptor IfnαR2, receptor Tyr kinase HER4, colony-stimulating factor CSF1, LDLR-related proteins, heparan sulfate proteoglycan epican,[28] cadherins, and inositol-requiring kinase and RNase IRe1, a sensor of the unfolding protein response following endoplasmic reticulum stress. It also targets nectin-1α, β2 subunit of voltage-gated sodium channel, ionotropic glutamate receptor GluR3, amyloid precursor protein, as well as APP-like proteins APLP1 and -/2 [638].

Presenilin achieves its developmental functions either by its γ-secretase activity such as cleavage of Notch receptor into icNotch transcriptional regulator, or independently from catalysis via the PKB and Wnt pathways [636].

Presenilin-2 influences sarco(endo)plasmic reticulum Ca^{++} ATPase activity and Ca^{++} transfer between the endoplasmic reticulum and mitochondria via interaction between the endoplasmic reticulum and mitochondria [640].

---

[26] Signal peptide peptidase-3, or SPP-like-3, is a.k.a. intramembrane protease IMP2 and presenilin-like protein PSL4. Enzyme SPPL2a is also designated as IMP3 and PSL2; SPPL2b as IMP4 and PSL1; and SPPL2c as IMP5.

[27] Signal peptide peptidase-like proteins-2A (SPPL2a; a.k.a. intramembrane protease IMP3 and presenilin-like protein PSL2) and -2B (SPPL2b; a.k.a. IMP4 and PSL1) causes intramembrane cleavage of tumor-necrosis factor-αin activated dendritic cells. Both SPPL2a and SPPL2b thus contribute to the regulation of adapted and innate immunity, as TNFα cleavage releases TNFαICD that triggers the production of interleukin-12 [636].

[28] A.k.a. CD44 and phagocytic glycoprotein-1.

Rhomboids

Rhomboids are intramembrane serine proteases that are unrelated to soluble Ser proteases. Seven transmembrane domain-containing rhomboids cleave substrates in or near transmembrane domains of proteins. These integral membrane proteins typically cleave peptide hormones along the secretory pathway. Rhomboid proteases also target mitochondrial intermembrane space proteins. They are thus implicated in growth factor signaling and mitochondrial function [641].[29]

In humans, the mitochondrial intramembrane cleaving presenilin-associated rhomboid-like protease (PARL)[30] controls apoptosis via crista remodeling and cytochrome-C release after cleavage of the mitochondrial, 120-kDa, dynamin-like protein optic atrophy OpA1 [636, 642].

Rhomboids can cut the membrane-docked precursor of epidermal growth factor (Vol. 2 – Chap. 3. Growth Factors), then releasing the extracellular domain that transmits signals to surrounding cells.

Metalloproteases

Among metalloproteases, membrane-bound transcription factor peptidase site-2 (MBTPS2)[31] cleaves the cytoplasmic domain of sterol regulatory element-binding protein (SREBP) that then translocates to the nucleus to activate genes involved in cholesterol and fatty acid synthesis [643].[32]

Adamlysins[33] are Zn^{++}-dependent sheddases that cut off or shed extracellular portions of transmembrane proteins. Mammalian genomes contain

---

[29] In Drosophila, 2 essential enzymes have been identified: Star and Rhomboid-1.

[30] A.k.a. PSARL, PSARL1, and PSenIP2. It is the ortholog of mitochondrial membrane Rhomboid Rbd1p.

[31] A.k.a. site-2 protease and sterol regulatory element-binding protein (SREBP) intramembrane protease.

[32] When cholesterol abounds, SREBP is retained in the endoplasmic reticulum. Conversely, when cholesterol level decays, SREBP is cleaved and released to act as a transcription factor for genes that encode enzymes of cholesterol and fatty acid synthesis (e.g. HMGCoA synthase) as well as low-density lipoprotein receptor. Protein SREBP is cleaved and activated by 2 proteases: membrane-bound transcription factor peptidase site-1 (MBTPS1), or site-1 protease, and MBTPS2, or site-2 protease. Sterols regulate SREBP via SREBP cleavage-activating protein (SCAP) that activates MBTPS1 when sterol concentration is low. After synthesis in the endoplasmic reticulum, the SCAP–SREBP complex moves to the Golgi body. Protease MBTPS1 cleaves SREBP in the lumen of the Golgi body. N-Terminus of SREBP that remains attached to the membrane via the transmembrane segment is then cleaved by MBTPS2 on the Golgi body to release SREBP N-terminus that translocates to the nucleus.

[33] Adamlysin refers to "A disintegrin and metalloproteinase" (ADAM); a.k.a. metalloproteinases disintegrins with cysteine-rich domains (MDC).

33 Adam genes. Adamlysins shed ectodomains from protumor-necrosis factor-α, angiotensin-1-converting enzyme, Notch and its ligands Delta-like proteins and Jagged, L-selectin, epidermal growth factor receptor, interleukin-6 receptor, amyloid precursor protein-like proteins APLP1 and -2, oxytocinase, and prions, among others [638].

Membrane-Tethered Matrix Metalloproteases

Membrane-tethered matrix metalloproteases constitute a family of the subclass of MMPs, most of MMPs being secreted. Multidomain MMPs possess a prometallo, metallo, and hemopexin-like domain. The prodomain maintains the metalloprotease domain in an inactive state until its cleavage.

Matrix metalloproteases are inhibited by tissue inhibitors of metalloproteases (TIMP). Agent TIMP2 interacts with both MMP14 and soluble proMMP2 to form a ternary complex at the cell surface. Membrane-anchored reversion-inducing Cys-rich protein with Kazal motifs (RECK) inhibits MMP14 and MMP2, in adddition to MMP9 [637].

Dipeptidyl Peptidase-4

Dipeptidyl peptidase-4[34] is the founding member of the DPP4 family that encompasses DPP6 to DDP9, ^{N}acetylated α-linked acidic dipeptidases-1 to -3, attractin, seprase,[35] prolyl endopeptidase (PEP), aminopeptidase-P (APP), prolidase,[36] and thymus-specific serine protease (TSSP or PrsS16) [644].

Dipeptidyl peptidase-4 targets multiple endocrine peptides, neuropeptides, and chemokines that contain an alanine or proline at position 2. It cleaves numerous glucoregulatory peptides and food-intake regulators, such as glucagon-like peptides GLP1 and GLP2, gastric inhibitory polypeptide (GIP), gastrin-releasing peptide (GRP), vasoactive intestinal peptide (VIP), pituitary adenylate cyclase-activating peptide (PACAP), oxyntomodulin,[37] and enterostatin.[38]

---

[34] A.k.a. adenosine deaminase-binding protein and CD26.

[35] A.k.a. fibroblast activation protein-α(FAP).

[36] A.k.a. peptidase-D.

[37] Oxyntomodulin is a peptide hormone produced by oxyntic cells of the gastric (fundic) mucosa.

[38] Enterostatin is produced in the intestine by pancreatic procolipase that is activated in the intestinal lumen by trypsin to serve as a coenzyme for pancreatic lipase. Efficient absorption of dietary lipids depends on the action of pancreatic triglyceride lipase.

In addition, DPP4 degrades β-casomorphin-2 into inactive dipeptides,[39] endomorphin-2,[40] corticotropin-like intermediate peptide (CLIP),[41] among others [644].

Peptidase DPP4 contributes to the regulation of inflammation, especially the lymphocyte fate. Expression of DPP4 varies with the state of cell differentiation [644]. Ectopeptidases regulate the activity of several chemoattractants [637].

Truncation by DPP4 of neuropeptide-Y and peptide-YY (peptide Tyr–Tyr, or pancreatic peptide-YY$_{3-36}$) modifies their receptor selectivity. Therefore, they shift from vasoconstriction to growth factor activity [637].

Angiotensin-Converting Enzyme and Aminopeptidase-N

Angiotensin-converting enzyme and aminopeptidase-N convert angiotensin-1 into the vasoconstrictor angiotensin-2. Moreover, angiotensin-converting enzyme inactivates the vasodilator bradykinin.

Transmembrane Proteases and Integrins

Matrix metalloproteinases (MMP) cleave matrix molecules. Among the MMP family, 2 families exist: (1) *soluble MMPs* that are secreted as inactive prozymogens and activated by other proteases and (2) *membrane-anchored MMPs* that are expressed as active enzymes to act only in the pericellular space.

The extracellular matrix experiences a continuous remodeling that can become important under certain circumstances, such as tissue development and regeneration as well as inflammation and healing. Membrane-type mt1MMP digests not only numerous ECM constituents, such as collagens and fibronectin, but also proMMP2 and proMMP13.

---

[39] β-casomorphins are peptides derived from the digestion of milk protein.

[40] Endomorphins are endogenous opioid peptides.

[41] Pro-opiomelanocortin (POMC) is a precursor of corticotropin, or adrenocorticotropic hormone (ACTH), itself a precursor of corticotropin-like intermediate peptide (CLIP), a short neuropeptide secreted by corticotrope cells of the adenohypophysis.

Integrin subunits-α[42] and -β tethers to many partners.[43] Integrins also interact directly via their extracellular regions with mt1MMP. Guanine nucleotide-exchange factor RabIF (or RasGRF3) binds to the cytosolic region of α-integrins to trigger mt1MMP-mediated activation of MMP2 and MMP9 [645].

Transmembrane Proteases in Angiogenesis

Membrane-inserted, Zn^{++}-dependent, matrix metalloproteinase MMP14[44] and ADAM17[45] as well as 3 ectopeptidases, aminopeptidase-N (APn),[46] dipeptidyl peptidase-4 (DPP4),[47] and angiotensin-converting enzyme,[48] intervene in angiogenesis, in addition to proteolysis of extracellular matrix proteins by matrix metalloproteinases and the plasminogen activator–plasmin system [637] (Table 7.9). Transmembrane and extracellular proteases enable cell migration and can release stored angiogenic factors. Some substrates of ectopeptidases have pro-angiogenic (angiotensin-2, bradykinin, and neuropeptide-Y) or angiostatic (vasostatin and chemokines CXCL10 and CXCL11) activity [637].

Transmembrane proteases are widely distributed, in particular in all hematopoietic lineages. They participate in various functions of blood circulation (Table 7.10). Furthermore, they are involved in cell proliferation and motility (Table 7.11). Proteases MMP14, APn, DPP4 prime the MAPK pathway [637].

7.2.4.4 Constituents of Vesicular Transfer of Molecules

Molecules can cross the plasma membrane from the intra- (exocytosis) and extracellular medium (endocytosis) via fusion or budding of vesicles to

---

[42] Integrin subunits-α interact with calreticulin, guanine nucleotide-exchange factor for Rab GTPases, α integrin-binding proteins αIBP63 and αIBP80, Rab-interacting factor (RabIF; a.k.a. RasGFR3 and mammalian suppressor of secretion MSS4), and bridging integrator protein BIn1 (a.k.a. amphiphysin-2, amphiphysin-like protein, and box-dependent Myc interaction protein-1). Agent RabIF is involved in the regulation of intracellular vesicular transport. Protein BIn1 regulates membrane remodeling and trafficking during endocytosis. In striated myocytes, BIn1 localizes to transverse (T)-tubules. A muscle-specific BIn1 isoform has membrane-tubulation properties.

[43] The cytoplasmic tail of β-integrin subunits (Itgβ) interacts with α-actinin, filamin, paxillin, talin, cytohesin-1, focal adhesion kinase, integrin-linked kinase, integrin-β1-binding protein Itgβ1BP1 (or integrin cytoplasmic domain-associated protein ICAP1), Itgβ3BP (or β3-endonexin), receptor for activated protein kinase-C RACK1, WD protein interacting with integrin tails WAIT1 (or embryonic ectoderm development Polycomb protein (EED), among others.

[44] A.k.a. membrane type-1 matrix metalloproteinase (mt1MMP).

[45] A.k.a. (pro)tumor-necrosis factor-α-converting enzyme [TACE] and CD156b).

[46] A.k.a. CD13.

[47] A.k.a. CD26.

[48] A.k.a. CD143.

Table 7.9. Transmembrane proteases in angiogenesis, their endogenous substrates and distribution (Sources: [637, 644]; ACE: angiotensin-converting enzyme; ADAM: a disintegrin and metalloprotease; APn: aminopeptidase-N; DPP4: dipeptidyl peptidase-4; MMP: matrix metalloproteinase; βCM2: β-casomorphin-2; α1MG: α1-microglobulin; Bdk: bradykinin; CCLi: chemokine (C-C motif) ligand i; CCLiLj: chemokine (C-C motif) ligand i-like j; CG: chromogranin; CXCLi: chemokine (C-X-C motif) ligand i; EM2: endomorphin-2; GHRH: growth hormone-releasing hormone; GIP: gastric inhibitory polypeptide; IL: interleukin; LIF: leukemia-inhibitory factor; NK1: neurokinin-1 (a.k.a. neurokinin-A, tachykinin-1, and substance-P); NPY: neuropeptide Y; pClps: procolipase; PYY: peptide-YY; Sst: somatostatin; Tryg: trypsinogen). Chromogranin-A is the precursor of vaso-, pancrea-, cate-, and parastatin. Whereas ACE is anchored to the plasma membrane by its N-terminus, APn and DPP4 are tethered by their C-termini.

Type	Substrates	Distribution
ACE	Angiotensin-1, bradykinin, enkephalins	T lymphocyte, activated monocyte, macrophage, neuroepithelial cell, endo- and epithelial cells, fibroblast
APn	Angiotensin-1/2/3, Met-enkephalins, NK1, Sst, CCL2/4	Monocyte, T lymphocyte, neutrophil, bone-marrow stromal cell, endo- and epithelial cells, osteoclast
DPP4	CG, vasostatin, EM2, GLP1/2, GIP, GRP, PACAP, Bdk, NK1, NPY, PYY, IGF1, prolactin, GHRH, Sst, LIF, CCL5/11/22, CCL3L1, CXCL6/10/12, IL1β/2, βCM2, α1MG Tryg, pClps	B-, T lymphocyte, NK cells, macrophage, endo- and epithelial cells, fibroblast
ADAM17	Collagen-4/17, CSF1, ProTNFα, ProTGFβ, TNFRSF1b, TNFSF6/8, L-selectin, mucin-1	Monocyte, activated lymphocyte, endothelial cell, chondrocyte, myoepithelial cell, nervous system cells
MMP14	Collagens-1/2/3, fibronectin, laminin, vitronectin, fibrinogen FXII, aggregan, APP, CD44, ProMMP2/13	Monocyte, macrophage, neutrophil, endo- and epithelial cells, myoepithelial cell, fibroblast

the plasma membrane, respectively. Both endo- and exocytosis rely on similar processes, but not on the same molecular machinery. Protein insertion into target membranes depends on translocation, abundance of non-polar residues in transmembrane helices, and position of polar residues within transmembrane segments [646].

Table 7.10. Transmembrane proteases and their effect in blood circulation (Source: [637]; ↑: increase).

Type	Effect
ACE	Blood pressure (↑)
ADAM17	Migration of eosinophils, neutrophils, and monocytes
APn	Blood pressure (↑)
DPP4	Glucose tolerance

Table 7.11. Transmembrane proteases and regulation of cell proliferation and motility, in particular via cytokine release, chemokine processing, and matrix degradation (Source: [637]). Agents MMP9, interleukin-1, -6, and -8, TNFα, and TGFα are pro-angiogenic factors. Interleukin-1, -6, and -8, TGFα, TGFβ1, TNFα, and tissue factor are implicated in proliferation, differentiation, and migration of vascular endothelial cells. Tumor-necrosis factor-α promotes production of MMP9 and urokinase-type plasminogen activator (or urokinase; uPA) that degrade the extracellular matrix (ECM) as well as proteinase inhibitor (PAI) by endothelial cells. Transforming growth factor-β1 regulates MMP activity by infiltrated macrophages. Chemokines control cell migration by binding to their G-protein-coupled receptors. Ectopeptidases APn and DPP4 release chemokines. In addition, ADAM17, APn, DPP4, and MMP14 degrade the extracellular matrix. Last, but not least, transmembrane proteases can cleave cell adhesion molecules. Proteases ADAM17 and MMP14 target L-selectin and epican (or CD44), respectively.

Type	Effect
ACE	Release of TNFα, IL2/12
	Processing of CCL3L1, CXCL9/10
ADAM17	Release of TNFα, TGFβ
	ECM degradation
APn	Release of IL6/8
	Processing of CCL2/3
	ProMMP9 activation
DPP4	Release of TGFβ1, IL2/10/12, Ifnγ
	Processing of CCL5/11/22 and CXCL11/12
	ProMMP9 activation
MMP14	ProMMP2/13 activation
	ECM degradation

Protein–lipid and protein–protein interactions control the binding of signaling molecules to membrane receptors as well as endocytosis. *Clathrin* and membrane rafts are both required for compound endocytosis from the cell surface. Glycosyl-phosphatidylinositol-anchored proteins, lipoproteins, cytosolic protein Tyr kinases, small GTPases of the RAS hyperfamily, and transmembrane proteins interact with membrane rafts. Caveolins (Cav1–Cav3) are involved in receptor-independent endocytosis.

7.2.4.5 Ion Carriers and Gap Junctions

Ion pumps and channels (Vol. 3 – Chap. 3. Main Classes of Ion Channels and Pumps), as well as gap junctions (Sect. 7.7.4) coordinate the electrical activity and molecular exchanges. Ion pumps determine and maintain chemical gradients by using an energy source to move ions against their concentration gradient. Ion channels use chemical gradients to carry ions through their pores. Ion channels can be classified by their gating mode (ligand-, voltage-, mechanical stress-, temperature-, and light-gated channels).

7.2.4.6 Connexins

Connexins and pannexins constitute 2 families of proteins involved in intercellular communications (Vol. 3 – Chap. 2. Membrane Ion Carriers). Connexins mainly form gap junctions that bridge apposed cells across the extracellular space. Pannexins generate non-junctional channels that operate as juxta- and paracrine regulators owing to the release of nucleotides, like connexins.

Connexins (Cx23, Cx25, Cx26, Cx30.1–Cx30.3, Cx31, Cx31.1, Cx31.9, Cx32, Cx36–Cx37, Cx40, Cx40.1, Cx43, Cx45–Cx47, Cx50, Cx59, and Cx62, according to their molecular weights [kDa]) are encoded by 21 genes. Connexins form hemichannels — connexons — that either remain free or most often assemble to build gap junctions used for the passive diffusion of small molecules (Sect. 7.7.4). Each apposed hemichannels has 2 mechanisms of gating in response to transjunctional voltage: the fast and the slow (or loop) gate. In addition, gap junctions can be gated by intracellular H^+, Ca^{++}, post-translational modifications, and some chemical agents.

7.2.4.7 Pannexins

Pannexins (Px1–Px3) contribute not only to gap junction structure, but also form channels to release nucleotides and other small molecules. They can interact with purinergic receptors (Vol. 3 – Chap. 2. Membrane Ion Carriers). Pannexins Px1 and Px2 are widely expressed in mammalian brain. Nonjunctional channels, i.e., pannexons and some connexons, such as Cx46 and Cx50, open by membrane depolarization. However, their activation threshold depends on channel type.

During the early stage of programmed death (Vol.2 – Chap. 4. Cell Survival and Death), apoptotic cells release signals, such as nucleotides ATP and UTP, via pannexin-1 to recruit phagocytes [647]. In fact, pannexin-1 is a target of effector caspases (caspase-3 and -7).

7.2.4.8 Tetraspanins

Tetraspanins are integral membrane proteins, mediators of signal transduction, and organizers of membrane nanodomains, i.e., *tetraspanin-enriched*

Table 7.12. Distribution, partners, and effects of vascular tetraspanins (Source: [648]; EC: endothelial cell; I: integrin; SMC: smooth muscle cell).

Type	Cell type	Partners	Effects
Tspan24	SMC		
	EC	Tspan28/29	Cell–matrix adhesion
		β_1 I	Cell–cell adhesion
		ICAM1, VCAM1	Cell migration
		mt1-MMP	Angiogenesis
	Platelet	$\alpha_{2B}\beta_3$ I	Aggregation
Tspan29	SMC	β_1 I	Proliferation
			Migration
	EC	ICAM1, VCAM1, β_1 I	Diapedesis
	Platelet	$\alpha_{2B}\beta_3$ I	Aggregation
	GP1b–GP5–GP9		
Tspan30	SMC		
	EC	vWF	Vesicular transport
	Platelet	$\alpha_{2B}\beta_3$ I	Spreading

domains. Tetraspanins interact directly with cholesterol. Tetraspanin-enriched nanodomains are signaling platforms.

Tetraspanins associate with other tetraspanins as well as other plasmalemmal molecules, such as cell adhesion proteins, growth factor receptors, and members of the immunoglobulin class. Palmitoylation influences between-tetraspanin associations.

In addition, tetraspanins form molecular complexes with cytoplasmic proteins, such as syntenin-1 and signaling effectors. Tetraspanin transport is associated with the vesicular transfer of other membrane proteins.

Tetraspanins regulate cell adhesion, spreading, proliferation, and migration, as well as fusion and pericellular proteolysis. They also contribute to metastasis or viral infection. In platelets, tetraspanins interact with and regulate other platelet receptors such as $\alpha_{2B}\beta_3$ integrins.

Tetraspanins possess 4 transmembrane segments, intracellular N- and C-termini, and 2 extracellular small and large loops. In humans, the tetraspanin family includes 33 identified members.[49] Some tetraspanins are widely expressed, such as Tspan24, Tspan29, and Tspan30, especially in vascular and hematopoietic cells, whereas others such as Tspan32[50] are expressed only in hematopoietic cells [648] (Table 7.12).

---

[49] In particular, the tetraspanin family includes Tspan24 (or CD151), Tspan25 (or CD53), Tspan26 (or CD37), Tspan27 (or CD82), Tspan28 (or CD81), Tspan29 (or CD9), and Tspan30 (or CD63).

[50] A.k.a. tumor-suppressing subchromosomal transferable fragment cDNA TSSC6.

Tetraspanin-28

Widely expressed tetraspanin-28 (Tspan28)[51] is embedded in the plasma membrane, where it facilitates signaling for its partners. It is required for intercellular interactions and fusion of egg with sperm. It redistributes to immune synapses between interacting B and T lymphocytes [649].

Many proteins associate with Tspan28 in the cell membrane. Tetraspanin-28 localizes to membrane nanodomains with its partners and other members of the tetraspanin family. These nanodomains enable additional bindings with signaling molecules and the cytoskeleton. The concentration of Tspan28 varies in distinct tetraspanin-enriched nanodomains in a given cell [649]. Partners of Tspan28 change according to the cell type.

In addition to its association with plasmalemmal proteins, Tspan-28 links to cytosolic proteins. Intracellular Tspan-28 associates with actin-binding ezrin and moesin. It also interacts with 14-3-3 proteins, phosphatidylinositol 4-kinase, and activated protein kinase-C [649]. Tetraspanin-28 regulates the interaction of G-protein-coupled receptor GPR56 with G-protein subunits.

In human B lymphocytes, Tspan28 interacts with CD19, a B-cell-specific member of the Ig class, CD225, an interferon-inducible molecule, and MHC class-2 molecules [649]. It thus forms a signaling complex with CD19, CD21, and CD225 on the plasma membrane of B lymphocytes. It operates in the intracellular exocytosis and processing of CD19. In T lymphocytes, Tspan28 connects to CD4 and CD8 proteins. Engagement of Tspan28 activates $\alpha_L\beta_2$ integrin on T lymphocytes and preferentially favors maturation of T_{H2} cells [649]. Co-engagement of Tspan28 with T-cell receptor complex and CD16 activates T lymphocytes and inhibits natural killer cells, respectively.

Other Types of Tetraspanins

Tetraspanin-24

Tetraspanin-24 enhances cell motility, hence invasion and metastasis of cancer cells.

Tetraspanin-25

Tetraspanin-25 tethers to integrins. It contributes to the transduction of signals from CD2, a cell-adhesion molecule on the surface of T and natural killer cells.

Tetraspanin-26

Tetraspanin-26 complexes with integrins and other transmembrane proteins. It may play a role in interactions between T and B lymphocytes.

---

[51] A.k.a. target of the antiproliferative antibody TAPA1 and CD81.

Tetraspanin-27

Tetraspanin-27 operates as a metastasis suppressor that can be activated by P53 transcription factor. It is thus repressed in tumors. Tetraspanins-27 and -29, but not Tspan30, can inhibit Wnt signaling, as they decrease the cytoplasmic and nuclear concentration of β-catenin, in the absence of GSK-3β activity or proteasomal degradation. They cause β-catenin export via exosomes [650]. Extracellular release of β-catenin via exosomes requires E-cadherin.

Tetraspanin-29

Tetraspanin-29 participates in the control of growth, proliferation, and migration of vascular endothelial and smooth muscle cells and cardiomyocytes [648]. Tetraspanin-29 also modulates leukocyte dispedesis and promotes the stability of platelet aggregates.

Tetraspanin-30

Tetraspanin-30 resides in the membrane of Weibel-Palade bodies in platelets and endothelial cells. It is involved in vascular cell adhesion and exocytosis regulation [648]. Tetraspanin-24 is a regulator of vasculogenesis and angiogenesis [648]. It promotes cell-adhesion strengthening mediated by integrin as well as PKC- and CDC42-dependent actin cytoskeleton reorganization.

7.2.4.9 Lectins

Lectins (lectio: picking, gathering; λεγω, lego: to gather, collect) are ubiquitously produced. Most lectins have neither enzymatic nor immune functions, although some lectins can recognize carbohydrates on pathogens. Lectins bind to either a soluble carbohydrate or carbohydrate moiety of a glycoprotein or glycolipid.

Lectins contribute to the regulation of cell adhesion and glycoprotein synthesis as well as control of protein concentrations in blood. On the surface of hepatocytes, these plasmalemmal receptors may be responsible for the removal of some glycoproteins from the blood circulation.

They tether to soluble extra- and intercellular glycoproteins. They have thus been classified into ^{N}acetylglucosamine-, ^{N}acetylneuraminic acid-, fucose-, galactose and ^{N}acetylgalactosamine-, and mannose-binding lectins.

The class of lectins can be decomposed into Ca^{++}-dependent and -independent carbohydrate-binding lectin subclasses. The calcium-binding (C-type) lectin (CLec) possesses the carbohydrate-binding domain of proteins. Conversely, numerous proteins contain a C-type lectin domain and, hence, a high-affinity carbohydrate-recognition domain. C-Type lectins are classified into 17 families (Table 7.13). In particular, C-Type lectin receptors orchestrate signal transduction pathways that regulate adaptive immune responses (Vol. 3 – Chap. 11. Receptors of the Immune System).

Table 7.13. Families of C-type lectins (Sources: [651,652]; CBCP: Calxβ and CTLD domain-containing protein; CSPG: chondroitin sulfate (CS) proteoglycan; CTLD: C-type lectin-like domain; CoLec: collectin; CoLec8: conglutinin; dcAR: dendritic cell immuno-activating receptor; dcIR: denditic cell immunoreceptor [CLec4a2]; dcSIGN: dendritic cell-specific ICAM3-grabbing non-integrin [CLec4l]; dcSIGNR: dcSIGN-related [CLec4m]; DGCR: DiGeorge syndrome critical region gene product; EMBP: eosinophil major basic protein, or PRG2: proteoglycan-2, bone marrow [natural killer cell activator, eosinophil granule major basic protein]; FREM1: Fraser syndrome (FraS1)-related extracellular matrix-1 [QBRICK]; HASGPR: hepatic asialoglycoprotein receptor [CLec4H1/2]; MASGPR: macrophage asialoglycoprotein receptor [CLec10a]; MBP: mannose-binding protein; SEEC: sperm-coating glycoprotein (SCP), EGF, EGF, CTLD domain-containing protein; SftP: surfactant protein). Asialoglycoproteins have lost their terminal sialic acid residues. The exposure of the subterminal galactose residues causes a rapid clearance of the glycoproteins from the circulation in the liver via asialoglycoprotein receptors. Collectins are soluble pattern recognition receptors.

Family	Members
1	Lecticans (CSPG1–CSPG3, CSPG7)
2	Asialoglycoproteins (HASGPR1/2, MASGPR), CLec4A/C–F/K, CoLec12
	Dendritic cell C-type lectins receptors (dcSIGN, dcSIGNR, dcAR, dcIR, dectins, DLec)
3	Collectins (MBPa/c, SftPa1/a2/d, CoLec8–CoLec12)
4	Selectins-E/L/P
5	Natural killer receptors (NKRs) ([natural] killer-cell lectin-like receptors [KLRs])
6	Multi-CTLD endocytic receptors
7	Regeneration factors (Reg1α–Reg1β, Reg3α–Reg3γ, Reg4)
8	Chondrolectin, layilin (type-1 receptors)
9	Tetranectin (CLec3a/3b/11a)
10	Polycystin
11	Attractin, attractin-like protein
12	EMBP
13	DGCR2
14	Thrombomodulin
15	CLec13a (Bimlec)
16	SEEC
17	CBCP–FREM1

Calcium-Binding (C-Type) Lectins

The class of C-type lectins encompasses collectins, selectins, endocytic receptors, and C-type lectin domain-containing proteoglycans (lecticans and proteoglycan-2 and -3). Some of these proteins are secreted and others are transmembrane proteins. They can exist as monomers, but often oligomerize.

Many C-type lectins operate as endocytic receptors to deliver soluble, bound ligands to lysosomes via the clathrin-dependent delivery to early and then late endosomes. Most endocytic receptors are type-2 transmembrane proteins, whereas CLec13d[52] pertains to the type-1 transmembrane protein set. Endocytosis of liganded receptors by C-type lectins can lead to: (1) receptor degradation in phagolysosomes or (2) receptor recycling to the plasma membrane, according to the receptor and ligand type.[53]

C-Type lectins are Ca^{++}-dependent, glycan-binding proteins that share a homologous carbohydrate-recognition domain (CRD) and a C-type lectin domain (CTLD). Whereas the CRD motif of C-type lectins possesses a structural homology, the protein sequence of CTLD is variable and do not bind sugars, but connect to proteins, lipids, and inorganic molecules [652]. The C-type lectin domain is a ligand-binding motif detected in more than 1,000 proteins, C-type lectins being the major representatives of CTLD-containing proteins. The carbohydrate-recognition domain can contain up to 4 Ca^{++}-binding sites. Binding of Ca^{++} ions can stabilize the protein structure and favor ligand tethering. On the other hand, pH-induced loss of Ca^{++} destabilizes ligand-binding affinity, a necessary process before lysosomal degradation.[54]

Myeloid cells, especially dendritic cells and macrophages, express numerous C-type lectins, which belong mainly to family-2 (asialoglycoproteins), -5 (natural killer receptors), and -6 (multi-CTLD endocytic receptors).

In addition to dendritic cell immuno-activating receptor (dcAR; in mice), CLec4a2,[55] CLec4l,[56] and CLec4m,[57] type-2 receptors comprise [651, 652]: collectin-12 (CoLec12)[58] and C-type lectins CLec4e,[59] CLec4f,[60] CLec4g,[61] CLec4j,[62] CLec4k,[63] and CLec4n[64] (Table 7.14).

---

[52] A.k.a. mannose receptor-C1 and macrophage mannose receptor.

[53] Heterotetrameric asialoglycoprotein receptor (ASGPR) and mannose receptor support ligand delivery to early endosomes and receptor recycling. C-Type lectin-13B fosters delivery to late endosomes and lysosomes.

[54] Endosomes and lysosomes have an acidic luminal content induced by a proton-pumping vacuolar ATPase (vATPase).

[55] A.k.a. denditic cell immunoreceptor (dcIR).

[56] A.k.a. dendritic cell-specific intercellular adhesion molecule-3 (ICAM3)-grabbing non-integrin (dcSIGN).

[57] A.k.a. dcSIGN-related protein (dcSIGNR).

[58] A.k.a. scavenger receptor with CTLD (SRCL).

[59] A.k.a. macrophage-inducible C-type lectin (Mincle).

[60] A.k.a. Kupffer cell receptor.

[61] A.k.a. liver sinusoidal epithelial cell lectin.

[62] A.k.a. IgE Fc receptor.

[63] A.k.a. Langerhans cell-specific C-type lectin (langerin).

[64] A.k.a. CLec6a or dectin-2.

Table 7.14. Main aliases of C-type lectins.

Type	Alias and portmanteau
Family-2 receptors	
CLec4a2	dcIR
CLec4e	Mincle
CLec4k	Langerin
CLec4l	dcSIGN
CLec4m	dcSIGNR
CLec4n	CLec6a, dectin-2
Collectins (family-3)	
CoLec1	MBP, MBPc, MBL
CoLec2	MBPa
CoLec4	SftPa1
CoLec5	SftPa2
CoLec7	SftPd
CoLec8	Conglutinin
CoLec9	CoLec43
CoLec10	CoLecL1
CoLec11	CoLecK1
CoLec12	CoLecP1
Macrophage mannose receptor family	
CLec13b	DEC205
CLec13c	PLA2R1
CLec13d	MMR
CLec13e	MRC2, CD280, uPARAP, Endo180
CTLD–acidic neck-containing proteins	
CLec13a	Bimlec
PrG2	EMBP, MBP1
PrG3	EMBPH, MBP2, preproMBPH

The *regeneration factor* (Reg) family of the C-type lectin subclass comprises small, secreted proteins implicated in acute phase reactions and cell maintenance and growth.

Collectins (CoLec) are C-type lectins that currently assemble in large oligomeric complexes that contain 9 to 27 subunits. Nine types of collectins have been identified: collectin-1,[65] collectin-2,[66] the single mannose-binding protein in humans, collectin-8, or conglutinin, surfactant proteins SftPa1 (CoLec4), SftPa2 (CoLec5), and SftPd (CoLec7), as well as collectin-9 (or collectin-43), -10 (collectin-L1), -11 (or collectin-K1), and -12 (or collectin-P1). Collectin-1, -4, -5, -7 to -9, and -11 are soluble, whereas collectin-10

---

[65] A.k.a. mannose-binding protein-C (MBP or MBPc) and mannose-binding lectin (MBL).

[66] A.k.a. mannose-binding protein-A (MBPa).

and -12 are membrane proteins. Collectins are involved in innate immunity. They indeed support phagocytosis and chemotaxis and stimulate the production of cytokines and reactive oxygen species.

The *macrophage mannose receptor* (MMR) clan includes [651]: CLec13b,[67] CLec13c,[68] CLec13d,[69] and CLec13e.[70]

The *endosialin* clan comprises endosialin, CD93, thrombomodulin, and EGFR5 (CLec14a) [651].

The set of CTLD–acidic neck-containing proteins encompasses CLec13a[71] as well as proteoglycan-2 (PrG2)[72] and -3[73] (PrG3) [651].

7.2.4.10 Galectins

Galectins are secreted lectins that participate in the formation of glycoproteic lattices that yield many cell-surface functions. They are encoded by the galectin genes (LGALS1–LGALS9 and LGALS12–LGALS13). Galectin synthesis is regulated both temporally and spatially during morphogenesis, particularly in polarized epithelia. Galectins bind β-galactoside and glycosphingolipids.

Galectin-3 and -4 operate in apical membrane sorting [653]. Galectin-3 assists in the apical delivery of non-raft proteins in polarized epithelial cells. Galectin-4 associates with sulfatides to form sorting platforms for the delivery of raft proteins to the apical membrane.

Galectin-9 interacts with Forssman glycosphingolipid, thereby contributing to epithelial polarity. Forssman glycosphingolipid serves as a surface receptor that transfers galectin-9 to the Golgi body from which the protein is recycled back to the apical surface to support the apical sorting of proteins and lipids [653].

7.2.4.11 Siglecs

The immunoglobulin class comprises recognition molecules of the immune system. Immunoglobulin-type lectins exploit the structural diversity of glycans for recognition. Sialic acid-binding immunoglobulin-like lectins — siglecs —

---

[67] A.k.a. dendritic and epithelial cell receptor DEC205.

[68] A.k.a. 180-kDa phospholipase-A2 receptor PLA2R1.

[69] A.k.a. macrophage mannose receptor (MMR).

[70] A.k.a. mannose receptor MRC2, CD280, urokinase-type plasminogen activator receptor-associated protein [uPARAP], and endocytic receptor Endo180.

[71] A.k.a. dendritic and epithelial cell receptor DEC205-associated C-type lectin (Bimlec).

[72] A.k.a. eosinophil major basic protein (EMBP) and bone marrow proteoglycan MBP1.

[73] A.k.a. EMBP homolog, MBP2, and prepro-major basic protein homolog (preproMBPH).

are mostly expressed by cells of the immune system. Siglecs recognize sialic acids of pathogens.

Many siglecs have Tyr-based signaling motifs, especially immunoreceptor Tyr-based inhibitory motifs (ITIM), that operate in cell signaling and endocytosis. Siglec receptors with ITIM suppress activation signals that emanate from receptors with immunoreceptor Tyr-based activation motifs (ITAM) by recruiting tyrosine and inositol phosphatases. Siglecs thus promote cellular interactions and regulate the innate and adaptive immune systems via glycan recognition [654].

The siglec superfamily can be subdivided into 2 families. CD33-Related siglecs that are expressed mostly by the innate immune system cells (neutrophils, eosinophils, monocytes, macrophages, mastocytes, and NK and dendritic cells) share sequence similarity in their extracellular regions and most contain tyrosine-based signaling motifs in their intracellular domains. They comprise 9 siglecs and 1 siglec-like protein. They regulate cell expansion by inhibiting cell proliferation or inducing apoptosis. CD33-Related siglec ITIM recruits protein Tyr phosphatases PTPn6 and PTPn11 and suppressor of cytokine signaling SOCS3. CD33-Related siglecs that lack ITIM, such as siglec-H and -14, associate with adaptor DAP12 that contains ITAM to trigger either cell activation or inhibition. CD33-Related siglecs can also function as endocytic receptors for the clearance of sialylated antigens and/or to promote or prevent antigen presentation.

Siglec-1,[74] a macrophage-restricted siglec, siglec-2,[75] an inhibitory receptor of B lymphocytes, siglec-4,[76] and siglec-15 form a subset of distantly related substances.

7.2.4.12 Ezrin, Radixin, and Moesin (ERM Proteins)

Ezrin, radixin, and moesin (ERM proteins) participate in the organization of membrane nanodomains and regulation of signaling pathways. They indeed interact with phospholipids, transmembrane and membrane-associated cytoplasmic proteins, and the actin cytoskeleton. They are expressed in a developmental and tissue-specific manner. Many epithelial and endothelial cells synthesize predominantly ezrin and moesin, respectively [655]. Radixin is most abundantly produced by hepatocytes.

The neurofibromin-2 (NF2) gene encodes merlin, a protein closely related to ERM proteins. Merlin acts at the plasma membrane, possibly to regulate growth factor receptor availability at the cell surface.

Ezrin is activated when it binds to phosphatidylinositol (4,5)-bisphosphate. The resulting conformational change unmasks binding sites for ^{F}actin and cytoplasmic tails of membrane proteins, such as sialophorin (Spn; or CD43), epican (or CD44), intercellular adhesion molecules ICAM1 and ICAM2, Na^+–H^+

---

[74] A.k.a. sialoadhesin and CD169.
[75] A.k.a. CD22.
[76] A.k.a. myelin-associated glycoprotein (MAG).

exchanger NHE1 (or SLC9a1), syndecan-2, and β-dystrophin [655]. Binding partners of ERM proteins include ^{F}actin and the regulatory subunit of protein kinase-A, among others. Moesin and radixin may function similarly.

These interactions link signaling receptors to the cortical cytoskeleton and foster assembling of signaling complexes that, in turn, may regulate receptor transfer and coreceptor function. Tethering of ERM proteins promotes the activation of epican coreceptors such as the hepatocyte growth factor receptor.

The adaptors and SLC9a3 regulators SLC9a3R1[77] and SLC9a3R2[78] can also bind to ezrin owing to a second FERM domain. Adaptors SLC9a3R1 and SLC9a3R2 link to multiple types of membrane proteins. Consequently, ERM proteins can tether to numerous proteins directly or indirectly. Linkage of SLC9a3R1 to ERM proteins impedes that of membrane proteins [655]. Protein SLC9a3R1 can bind to partners, such as SLC9a3 (or NHE3), cystic fibrosis transmembrane conductance regulator, β2-adrenergic receptor, platelet-derived (PDGFR) and epidermal (EGFR) growth factor receptors, TBC1 domain family member TBC1D10a,[79] and PDZ domain-containing scaffold protein PDZK1. Scaffold PDZK1 can bind to SLC22a4,[80] SLC34a1,[81] and 17-kDa membrane-associated protein (MAP17, or PDZK1IP1). Similarly to ERM proteins, the conformation of SLC9a3R1 and PDZK1 is regulated by intra-molecular connections that may modulate their affinities for their ligands.

Kinase RoCK, protein kinases-C PKCα and PKCθ, NCK-interacting kinase (NIK),[82] mammalian sterile-twenty (Ste20)-like kinase MST4,[83] and lymphocyte-oriented kinase (LOK; a.k.a STK10) may phosphorylate (activate) ezrin, moesin, and radixin (Thr576, Thr558, and Thr564, respectively). Ezrin can also be phosphorylated by cyclin-dependent kinase CDK5 (Thr235) as well as epidermal growth factor receptor (Tyr145 and Tyr353). Activated ERM proteins can potentially assemble clusters of plasmalemmal proteins and link them to ^{F}actin.

ERM Proteins operate as both regulators and effectors of RhoA GTPase. They promote RhoA signaling. They regulate RhoA via RhoGEFs and RhoGDIs. Small GTPase RhoA contributes to the regulation of the cortical actin cytoskeleton via its interactions with RoCK and formins.

ERM Proteins also regulate Hedgehog signaling. They link to Patched directly or via SLC9a3R1 and SLC9a3R2 adaptor. Ezrin participates in the maintenance of junctions of the apical interface during morphogenesis that

---

[77] A.k.a. ERM-binding phosphoprotein EBP50 and Na^+–H^+ exchange regulatory cofactor NHERF1.

[78] A.k.a. NHE3 kinase-A regulatory protein [E3KARP] and NHERF2.

[79] A.k.a. 64-kDa EBP50 PDZ interactor EPI64.

[80] A.k.a. organic cation–carnitine transporter OCTN1.

[81] A.k.a. Na^+-dependent phosphate transporter-2A.

[82] A.k.a. MAP4K4; not NFκB-inducing kinase (or MAP3K14).

[83] A.k.a. MST3 and SOK1-related kinase (MASK).

Table 7.15. Examples of ion channels associated with caveolae (Source: [612]; Ca_V: voltage-gated Ca^{++} channel; HCN: hyperpolarization-activated, cyclic nucleotide-gated K^+ channel; K_V: voltage-gated K^+ channel; Na_V: voltage-gated Na^+ channel).

Channel	Cell type
$Ca_V 1.2$	Cardiomyocytes, smooth muscle cell
Connexin-43	Alveolar epithelial cell
IP_3R	Ubiquitous
HCN_4	Sinus-node cell
$K_{IR}6.1$	Smooth muscle cell
$K_V 1.5$	Ventriculomyocyte
$Na_V 1.5$	Ventriculomyocyte

involves apical expansion or reorganization. In particular, ERMs contribute to *luminogenesis* [655]. They form a free apical domain limited by tight junctions.

In immune cells, more precisely immunological synapses between antigen-presenting cells and T lymphocytes, ERM proteins can disassemble microvilli and influence the amount and location of transmembrane proteins [655]. Association of sialophorin with ERM proteins is important for immunological synapse formation.

7.2.5 Caveolae

Caveolae are specialized, invaginated membrane rafts that contribute to the regulation of vesicular transport and serve as signaling platforms. The organization and function of caveolae depend on coat proteins — caveolins (Cav1–Cav3) — and adaptors — cavins (cavin-1–cavin-4) — that promote membrane remodeling and caveolin-derived structure transfer.

Density of caveolae differs according to the cell type. They are particularly prominent in adipocytes, fibroblasts, and epithelial and vascular endothelial cells. Caveolae can form transendothelial channels and vesiculovacuolar organelles and cavicles.

Caveolae constitute invaginated endocytic and signaling platforms enriched in cholesterol and sphingolipids, in addition to cholesterol-binding caveolin. Among caveolin isoforms, caveolin-3 is expressed exclusively in myocytes.[84] Caveolae contain ion channels and gap junction proteins (Table 7.15). Proteins of the membrane-associated guanylate kinase (MAGUK) family such as Disc large homologs promote clustering of diverse regulators of ion channels [612].

Dystrophin that stretches laterally along actin filaments intervenes in assembling and maintenance links between the actin cytoskeleton and the extracellular matrix. In skeletal and cardiac myocytes, dystrophin associates

---

[84] Myocardial caveolin-3 localizes to T-tubules and sarcolemma, but not intercalated discs.

Table 7.16. Examples of constituents of caveolae (AR: adrenergic receptor; DLg: Disc large homolog; PKA: protein kinase A; PKC: protein kinase C; Src: sarcoma-associated kinase).

Type	Members
Kinases	PKA, PKC, Src
Receptors	β2AR
Adaptors	DLg1

with various proteins to form the dystrophin-associated protein complex (DAPC) that links the actin cytoskeleton to the extracellular matrix. Dystroglycans α and -β that make up the DAPC core establish a transmembrane link between laminin-2 and dystrophin. Dystrophin-associated protein complexes also contain [656]: sarcoglycans,[85] sarcospan,[86] α-dystrobrevin,[87] syntrophin,[88] syncoilin,[89] as well as caveolin-3, laminin-2, and nitric oxide synthase NOS1. Dystrophin-associated protein complexes stabilize the sarcolemma during repeated cycles of contraction and relaxation. They transmit forces generated by sarcomeres to the extracellular matrix. They are involved in cell signaling. They interact with calmodulin, GRB2 adaptor, and nitric oxide synthase NOS1 [656]. Dystrophin-associated protein complexes are anchored to caveolae [612].

7.2.6 Membrane Rafts

A membrane raft is a specialized, tiny (size order $\mathcal{O}[10\,\mathrm{nm}]$), dynamic (half-life $\mathcal{O}[100\,\mathrm{ns}]$) membrane domain with signaling and transport functions. Any activation modulates the size and stability of rafts. Membrane rafts compartmentalize the membrane of the cell and organelles into nanodomains. They are able to fix or exclude specific lipids and proteins.

Membrane rafts originate from liquid-ordered phase regions of the membrane that float in the liquid-disordered phase. Membrane rafts contain choles-

---

[85] Five transmembrane sarcoglycans (α–ε) are expressed primarily in skeletal myocytes.

[86] Sarcospan is produced predominantly in skeletal and cardiac myocytes.

[87] Alternative splicing produces 5 α-dystrobrevin isoforms. Only α-dystrobrevin-2 abounds at the sarcolemma.

[88] All the 3 syntrophin isoforms reside at the neuromuscular junction in skeletal myocyte, but only α1- and β1 isoforms lodge along the sarcolemma. Syntrophins may function as modular adaptors that recruit signaling proteins to the sarcolemma and DAPC, such as voltage-gated sodium channels, NOS1, microtubule-associated Ser/Thr kinase MAST2 (or MAST205), and P38γ (a.k.a. MAPK12, stress-activated protein kinase SAPK3, and ERK3).

[89] Syncoilin localizes in skeletal, cardiac, and smooth myocytes at the sarcolemma, Z-lines, and neuromuscular junction.

terol packed with glycosphingolipids [657], lipid-modified proteins, such as glycosyl-phosphatidylinositol (GPI)-anchored [658], palmitoylated, and myristoylated proteins.[90] Clusters of GPI-anchored proteins associate with phospholipids, in addition to cholesterol and glycosphingolipids. Membrane rafts are stabilized via interactions with the cytoskeleton. Cytoskeletal proteins, such as spectrin, actin, bands 4.1 and -4.2, are partly associated with membrane rafts.

Membrane rafts can contain phosphoinositides that are involved in cell endocytosis.[91] Manifold phosphoinositides interact with specific proteic domains using a Ca^{++}-dependent or -independent mechanism. Membrane rafts serve also for phosphatidylinositol (4,5)-biphosphate signaling.[92]

Membrane rafts are also constituted of enzymes such as members of the SRC kinase family, ion carriers, signaling receptors, among other types of proteins (Table 7.17). Raft proteins are thus involved in signaling. They can aggregate into larger platforms in response to various stimuli.

In addition, membrane rafts can contain scaffolds, such as flotillins-1[93] and -2 and stomatin. Flotillin-1 resides in tiny patches of the plasma membrane, distinct from clathrin-coated pits and caveolae, as well as in certain clathrin-independent endocytic particles with glycosyl-phosphatidylinositol anchors.

The spatial organization of signaling proteins in the cell membrane is particularly ascribed to membrane rafts. Nevertheless, clustering and trapping can also result from between-protein interactions among the relevant signaling molecules [660].

Membrane rafts participate in the regulation of various biological processes [661]. Receptors are recruited to membrane rafts that control their function. In addition, these nanoplatforms serve as intermediates of nascent cell adhesion. Membrane rafts control integrin-mediated cell adhesion, in particular $\alpha_L\beta_2$ integrin nanoclusters on all T lymphocytes as well as B lymphocytes, macrophages, and neutrophils, that colocalize with nanodomains of GPI-anchored proteins [662].

Membrane Rafts and Mechanotransduction

Mechanotransduction at least partly relies on membrane rafts. Endothelial cells detect changes in the hemodynamic stress field that are converted into a

---

[90] These post-tranalational modifications favor protein localization in membrane rafts. Sphingolipids and GPI-anchored proteins undergo transient confinements in these nanodomains.

[91] Phosphoinositides control the localization of phosphoinositide-binding proteins to targeted organelles.

[92] Distinct phospholipase-C-coupled receptors that either depend on cholesterol such as neurokinin-A receptor, or independent such as endothelin receptor, can share the same PIP_2 pool at the plasma membrane [659].

[93] Flotillin-1 is also called reggie-2.

Table 7.17. Examples of constituents of membrane rafts and associated proteins (Ca$_V$: voltage-gated Ca^{++} channel; EPCR: endothelial protein-C receptor; GluT: glucose transporter; K$_{ATP}$: ATP-sensitive K$^+$ channel; K$_{IR}$: inwardly rectifying K$^+$ channel; K$_V$: voltage-gated K$^+$ channel; Na$_V$: voltage-gated Na$^+$ channel).

Type	Examples
	Lipids
Cholesterol	
Glycosphingolipids	
Phosphoinositides	PI(4,5)P$_2$
	Proteins
Adhesion molecules	Integrin
Ion carriers	Ca$_V$, K$_{ATP}$, K$_{IR}$, K$_V$, Na$_V$
Transporters	ABCa1, GluT1
Receptors	EPCR, PAR$_1$
Enzymes	Src, PKA, PP1, PP2a
Scaffolds	Flotillin, stomatin
Cytoskeleton	Spectrin, actin, band-4.1/4.2 proteins
Miscellaneous	Thrombomodulin

sequence of mediator activation starting from various signaling molecules associated with caveolae, such as protein kinases and sphingomyelinase [663]. Increased mechanical stimuli indeed induce translocation of signaling molecules primarily to caveolae and ultimately activate the Ras–Raf–MAPK pathway (MAPK: mitogen-activated protein kinase; Vol. 4 – Chap. 5. Mitogen-Activated Protein Kinase Modules). Membrane disrupter filipin causes disassembly of caveolae, thereby precluding both signaling events at the cell surface and activation of the MAPK module.

Formation of membrane rafts can depend on integrin-dependent signaling from the extracellular matrix and reorganization of the actin cytoskeleton. Integrins (Sect. 7.5.4) and caveolin-1 (Sect. 9.3) can interact via some adaptors. Focal adhesion-associated integrins that are connected to actin microfilament terminations, such as $\alpha_5\beta_1$ (fibronectin receptor) and $\alpha_V\beta_3$ (vitronectin receptor), concentrate at the upstream edge (with respect to flow direction) of cultured bovine aortic endothelial cells to improve cell adherence on the extracellular matrix [664].[94] Phosphorylation of caveolin-1 and redistribution

---

[94] Amounts of $\alpha_5\beta_1$- and $\alpha_V\beta_3$ integrins rise and decay in response to flow, respectively. Moreover, talin and $\alpha_5\beta_1$ integrin are nearly evenly distributed along the cell surface, whereas vinculin and $\alpha_V\beta_3$ integrins tends to disappear at the downstream cell edge.

of ^{F}actin induced by hemodynamic stress happen in response to β_1 integrin and Src kinase activation [665]. Protein phosphatase-1 precludes flow-induced association of C-terminal Src kinase (CSK) with β_1 integrins and their colocalization with paxillin and phosphorylated caveolin-1. Caveolin-1 phosphorylation by Src kinase elicits CSK recruitment at integrin sites, where CSK phosphorylates myosin light chain in response to flow [666]. In addition, the extracellular matrix controls myosin light-chain phosphorylation via cytoskeletal prestress [667].

7.3 Transmembrane Voltage

Although the total ionic concentration is similar on both sides of the lipid bilayer membrane that delimits the cell, the concentration of ion species differs in the intra- and extracellular medium. The ionic gradients across the cell membrane generate a transmembrane voltage that is negative inside the cell (on the order of -100 mV). The hydrophobic part of the thin ($\sim$3 nm) lipid bilayer acts as the dielectric of the capacitor subjected to a strong electric field [668].

The electric field is sensed via charge displacements or dipole movement that induce transient electric currents. When the electric field changes, proteic sensors bear conformational changes. Charged residues (Asp, Glu, Arg, Lys, and His) and side chains with intrinsic dipole moment (Tyr) reorientate in the electric field. Plasmalemmal proteins that contain cavities in which free ions can move can undergo conformational changes (e.g., sodium–potassium pump) [668]. The charge displacement depends on the charge magnitude and local electric field strength.

The transmembrane potential regulates the function of many plasmalemmal proteins such as ion carriers (channels, transporters, and pumps) and enzymes. Voltage sensors are proteic structures that sense the transmembrane potential. Ion channels have a positively charged transmembrane sensor that moves in response to changes in transmembrane potential. Voltage-dependent G-protein-coupled receptors as well as ATP-driven ion pumps and transporters possess a voltage-sensing motif. Sensing transient currents are produced by motion of charges.

Voltage-gated Na^+, K^+, and Ca^{++} channels contain a selective ion transport pore and voltage sensors.[95] They initiate and propagate electrochemical

---

[95] Potassium channels are composed of 4 independent proteic subunits and Na^+ and Ca^{++} channels of 1 peptide with 4 homologous domains. Each subunit or domain comprises 6 transmembrane segments (S1–S6) and a pore loop between S5 and S6. The central conduction pore made of S5, S6, and the pore loop has 4 voltage sensors (S1–S4) around it. Water penetration in the K^+ channel core can increase the local electric field in the extracellular half of the protein near voltage sensors with respect to that of the bilayer. Voltage-gated proton channel has a voltage

waves of excitation. Cooperation between Na^+ channel domains at hyperpolarized potentials allows faster opening kinetics. Activation charge motion and between-domain interactions are coupled to the ion channel inactivation.

Voltage-dependent phosphatase has 4 transmembrane segments (S1–S4; voltage sensors) and a large intracellular domain for enzymatic actvity that increases on depolarization.

Voltage-dependent GPCRs and ion carriers and pumps do not possess S4 sensors. The activity of cholinergic M1 and M2 muscarinic receptors increases and decreases with depolarization, respectively. The loop between S5 and S6 is the effector of the voltage-sensor motion.

Voltage-dependent sodium–glucose cotransporter of the small intestine uses a transporter intrinsic charge to sense voltage and carried Na^+ ions to generate the gating current.

Na^+–K^+ pumps maintain ion gradients across cells using energy from ATP hydrolysis. Its electrogenic sequential transport (3 Na^+ [outflux] vs. 2 K^+) generates an outward current that makes the cell cortex potential more negative than the near-cell environment. The Na^+–K^+ pump current decreases as the transmembrane potential becomes more negative.

7.4 Membrane Nanotubes

Membrane nanotubes are thin extensions from cell membrane that transiently connect cells over long distances to facilitate intercellular communication.[96] Many cell types store excess membrane in their cell surface. Membrane extensions can then appear when the cytoskeleton extracts membrane from its reservoir. Nanotubes are either thin ($\sim$300 nm) or thick ($\sim$800 nm). Membrane nanotubes are characterized by heterogeneity in their specific features (Table 7.18).

Two different mechanisms could generate intercellular membrane nanotubes [669]. On the one hand, actin-driven protrusions from a given cell may transiently connect to a nearby cell. Closed-end nanotubes between cells then require membrane fusion to create open-ended tunnels. On the other, cells that come into contact during a sufficiently long time ($\geq$4 mn) may form nanotubes as they subsequently move apart. These 2 processes may be combined. Short extensions that connect neighboring cells may elongate as the between-cell distance increases.

Membrane nanotubes are made of ^{F}actin. Microtubules in large membrane nanotubes possibly transport materials between cells. Molecules and cell

---

sensor with 4 transmembrane segments, including positively charged S4 segment, without central conduction pore.

[96] They differ from thin, permanent projections such as microvilli and dynamic protrusions, either thin (ruffles and filopodia) or large (lamellipodia), which occur during cell migration (Vol. 2 – Chap. 6. Cell Motility).

Table 7.18. Membrane nanotube features (Source: [669]). The functioning of membrane nanotubes has mainly been demonstrated in vitro.

Cell type	Properties
Neuronal cell	Caliber 50–200 nm; length 15–60 μm ^Factin, no microtubule Duration <1 h Unidirectional actin-based vesicle transfer
Macrophage Dendritic cell	Caliber 300–800 nm Caliber 35–250 nm; length <100 μm ^Factin, microtubules in some nanotubes Duration several hours ATP-dependent vesicle and calcium flux
T lymphocyte	Caliber 180–380 nm; length 20–45 μm ^Factin, no microtubule Duration <1 h Viral protein transmission
NK cells and target cells	Length 30–140 μm ^Factin, no microtubule Duration <40 mn NK-cell receptor transfer
Cytotoxic T cells and target cells	Length ∼10 μm Major histocompatibility complex protein transport

organelles can travel between cells via membrane nanotubes, either inside or along their surface and either uni- or bidirectionally.

Membrane nanotubes can serve for intercellular signaling. Calcium fluxes have been observed through some membrane nanotubes, but not all types. In vivo, membrane nanotubes are used for surface proteins swapping between cells.

7.5 Adhesion Molecules

The coordinated expression by cells of various types of adhesions, either with the extracellular matrix or adjoining cells, regulate the cell shape, tissue structure, and cell motility. Cellular junctions maintain the structural integrity of tissues.

Adhesion molecules are plasmalemmal proteic constituents of cell adhesion sites (with apposed cells [e.g., adhesions between apposed endothelial cells] with neighboring cells [e.g., adhesions between an endothelial cell and a smooth muscle cell through the basement membrane], or with the extracellular matrix [e.g., adhesions between an endothelial cell and its basement membrane]). Adhesion molecules also serve as signaling components that control cell growth, division, differentiation, and apoptosis.

Adhesion molecules are composed of 3 domains: (1) an endodomain (or intracellular domain) that interacts with the cytoskeleton; (2) a transmembrane domain; and (3) an ectodomain (or extracellular domain) associated with either the same (homophilic binding) or other types (heterophilic binding) of adhesion molecules. *Homophilic* interactions means that similar molecules bind to one another, whereas *heterophilic* interactions refer to the attachment among different molecule types. *Homotypic* and *heterotypic* interactions are related to the binding to the same isotype or different types of a given chemical species, respectively.

Four main classes of adhesion molecules exist: cadherins, selectins, integrins, and adhesion molecules of the immunoglobulin superclass (Tables 7.19 to 7.20). The 2 first classes of substances are calcium-dependent adhesion molecules.

Mechanical forces can be transmitted across the cell surface to the cytoskeleton or fibers of the extracellular matrix by adhesion molecules (integrins, cadherins, selectins, and Ig-like cell adhesion molecules). Mechanotransduction is principally carried out by integrins that act as major mechanoreceptors. Cadherins at adherens junctions can mediate force-induced activation of mechanosensitive Ca^{++} channels in fibroblasts and increase actin polymerization primed by calcium influx [672]. In endothelial cells, vascular endothelial-cadherins that induce transient and sustained activations of RhoA GTPase control focal adhesion assembly via the Rho–RoCK pathway in response to mechanical loading [673].

Zipcode-binding proteins, the insulin-like growth factor-2 mRNA-binding proteins (IMP1–IMP3),[97] members of the RNA-binding protein hyperclass mainly located in the cytoplasm, are required for cell motility, cell adhesion, and cytoplasmic spreading.

7.5.1 Cadherins

Cadherin is a portmanteau for calcium-dependent adhesion protein, as this type of glycoproteins contain calcium-binding sites. Cadherins connect cells that express the same cadherin type, one cadherin-binding to another in the extracellular space; i.e., cadherin ectodomains undergo Ca^{++}-dependent homophilic and homotypic interactions. Moreover, the cytoplasmic domain of type-1 and -2 cadherins contains binding domains for various catenins.

---

[97] RNA localizes to the cytosol according to attached RNA-binding proteins, which dictate the RNA destination. Insulin-like growth factor-2 mRNA-binding proteins belong to the zipcode-binding protein family. The same alias (IMP) is used to designate the compound "impedes mitogenic signal propagation" also called BrCa1-associated protein (BrAP). This Ras effector modulates the sensitivity of the MAPK cascade.

Table 7.19. Calcium-independent cell adhesion molecules and related proteins (**Part 1**; Sources: [670], Wikipedia; CD: cluster of differentiation; ceaCAM: carcino-embryonic antigen-related cell adhesion molecule [or CD66a]; ICAM: intercellular adhesion molecule; lCAM: liver cell adhesion molecule; nCAM: neural cell adhesion molecule [or CD56]; ngCAM; neuron–glia cell adhesion molecule; nrCAM: neuron–glia cell adhesion molecule-related protein; obCAM: opioid-binding cell adhesion molecule or opioid-binding cell adhesion molecule-like OPCML; PECAM: platelet-endothelial cell adhesion molecule; VCAM: vascular cell adhesion molecule; CHL1: close homolog of L1CAM; CR4: complement receptor-4 or $\alpha_X\beta_2$ integrin; DCC: deleted in colorectal carcinoma [a.k.a. CRC18 and CRCR1]; LAR: leukocyte common antigen-related molecule or protein Tyr phosphatase receptor PTPRf; MAG: Myelin-associated glycoprotein [siglec-4]; NCA: non-specific cross-reacting antigen; Siglec: sialic acid-binding Ig-like lectin). The L1 family of cell adhesion molecules of the immunoglobulin superclass includes 4 members: L1CAM (a.k.a. nCAML1, L1, and CD171), CHL1, neurofascin, and nrCAM that are found on neurons (especially on axons) and glial cells. Axonal contactin-2 (a.k.a. TAG1 and axonin-1) is a ligand for neurocan and phosphacan, in addition to neural cell adhesion molecules, such as ngCAM, L1CAM, nCAM, and extracellular matrix protein tenascin-C [671].

Immunoglobulin superclass		
Neural	Immune	Miscellaneous
nCAM	ICAM1	ceaCAM
ngCAM	ICAM2	NCA
nrCAM	ICAM3	ceaCAM
L1CAM	PECAMl	VCAM
CHL1	CD4	Poliovirus receptor (CD155)
Neurofascin	LAR	obCAM
Contactin		PDGFR
Contactin-1		Muc18
Contactin-2		DCC
ICAM5		ICAM4 (erythroid cell)
MAG		
Nectins and nectin-like molecules		
Integrins		

7.5.1.1 Cadherin Class

The cadherin class includes several subclasses (Table 7.21 to 7.26). *Classical cadherins* are type-1 single-span transmembrane proteins that localize primarily to adherens junctions to confer calcium-dependent cell-cell adhesion. Most classical cadherins serve as homophilic adhesion receptors for cellular recognition and adhesion. They can transfer information inside the cell via interactions with cytoskeletal and signaling molecules.

Intercellullar junctions formed by cadherins, such as desmosomes and adherens junctions, comprise 2-dimensional arrays of trans-dimers of monomers

Table 7.20. Calcium-dependent cell adhesion molecules and related proteins (**Part 2**; Sources: [670], Wikipedia; Cdh: cadherin; CD: cluster of differentiation; mCAM: melanoma cell adhesion molecule [or CD146]; siglec: sialic acid-binding Ig-like lectin; vEC, vSMC: vascular endothelial and smooth muscle cells).

Type	Spot	Aliases
Cadherins		
E-Cadherin	Epithelial	Cdh1, CD324, L-CAM, eCdh
N-Cadherin	Neural	Cdh2, nCdh
P-Cadherin	Placental	Cdh3, pCdh
R-Cadherin	Retinal	Cdh4, rCdh
VE-Cadherin	Vascular endothelial	Cdh5, veCdh
K-Cadherin	Kidney	Cdh6, kCdh
Cadherin-7		Cdh7
Cadherin-8		Cdh8
T1-cadherin		Cdh9
T2-cadherin		Cdh10
OB-cadherin	Osteoblast	Cdh11
N-cadherin-2		Cdh12
H-cadherin	Heart	Cdh13
(T-Cadherin)		
M-Cadherin	Myotubule	Cdh15
KSP-Cadherin		Cdh16
LI-cadherin	Liver–intestine	Cdh17
Cadherin-18		Cdh18
Cadherin-19		Cdh19
Cadherin-20		Cdh20
Cadherin-23	Neurosensory epithelium	Cdh23
Selectins		
E-selectin	Endothelial cell type	CD62E
L-selectin	Lymphocyte type	CD62L
P-selectin	Platelet type	CD62P
Others		
Lymphocyte homing receptors		
Galactosyl transferase		
Siglec2		CD22
CD24	B lymphocyte, differentiating neuroblast	
CD44	Hematopoietic stem cell	
mCAM	T lymphocyte, vEC, vSMC	
CD164		

that emanate from opposing cell surfaces. On the other hand, lateral cis-

Table 7.21. Cadherins (Cdh) constitute a class of glycoproteins that serve as calcium-dependent, cell-adhesion molecules (**Part 1**; Sources: [674–676]). In vertebrates, more than 100 members of the cadherin class are grouped into several subclasses, such as classical, desmosomal, Flamingo–CELSR, Fat-type cadherins, protocadherins, and others. Type-1 classical cadherins include, in particular, E-cadherin (epithelial), also called cadherin-1, and N-cadherin (neuronal), and type-2 atypical cadherins comprise vascular endothelial (VE)-cadherin (a.k.a. cadherin-5). Cadherin genes can form clusters, e.g., the CDH7, CDH19, and CDH20 genes are clustered on chromosome 18q22-q23 [677].

Molecule	Other alias	Distribution
Type-1 cadherins		
E-Cadherin	Cdh1, L-CAM, Uvomorulin, CD324	Epithelium
M-Cadherin	Cdh15	Myotubule
N-Cadherin	Cdh2, CD325	Neural
P-Cadherin	Cdh3	Placental
R-Cadherin	Cdh4	Retinal
Type-2 cadherins		
Br-Cadherin	N-Cadherin-2 Cdh12	Brain
Ey-Cadherin	Cdh14, Cdh18 Cdh24	Eye
K-Cadherin	Cdh6	Kidney (fetal), retina, neurons
Ksp-Cadherin	Cdh16	Kidney-specific
LI-Cadherin	Cdh17	Liver, intestine
OB-Cadherin	Cdh11	Osteoblast, retina
T1-Cadherin	Cdh9	Testis (fetal), neurons, CD4+, CD8+ thymocytes
T2-Cadherin	Cdh10	Testis (postnatal), neurons, CD4+, CD8+ thymocytes
VE-Cadherin	Cdh5, CD144	Endothelium
Cadherin-7	Cdh7L1/L2, Cdh19	Neural crest, retina
Cadherin-20	Cdh7L3	Retina
Cadherin-8		Retina, neuron, fetal kidney
Cadherin-22		
Cadherin-24	Cdh11L	
Cadherin-26	VR20	
T Cadherin		
H/T-Cadherin	Cdh13	Heart, vascular cells, Neurons

interfaces between cadherins from the same cell surface are used for ordered cadherin clustering.

Cadherins mediate interactions between both cell cytoskeletons. In particular, cadherins control cell polarity and interactions between neighboring

Table 7.22. Cadherin class members. (**Part 2**) Desmosomal cadherins (CDHF: cadherin family member). The desmocollin family of the cadherin superfamily include 3 members (Dsc1–Dsc3) that, together with desmogleins (Dsg1–Dsg4) constitute desmosomes.

Type	Aliases	Distribution
	Desmocollin family	
Desmocollin-1	Dsc1, CDHF1	Widely expressed
Desmocollin-2	Dsc2, CDHF2	Myocardium, lymph node
Desmocollin-3	Dsc3, CDHF3 Dsc4	Stratified epithelia
	Desmoglein family	
Desmoglein-1	Dsg1, CDHF4	Specialized epithelia
Desmoglein-2	Dsg2, CDHF5	Ubiquitous
Desmoglein-3	Dsg3, CDHF6	Specialized epithelia
Desmoglein-4	Dsg4, CDHF13	Skin, salivary gland, testis, prostate

Table 7.23. Cadherin class members. (**Part 3**) Protocadherins. Protocadherins constitute the largest cadherin subclass that is decomposed into several families, such as α-, β-, and δ-protocadherins. Protocadherins have a weaker binding capacity than classical cadherins, but a high potential for intracellular signaling. A large set of cadherin-like transmembrane proteins that are expressed in the nervous system originate from clustered protocadherin genes. Non-clustered δ-protocadherins form a subset.

Family Type	Members, Alias(es)	Distribution, Function
α-Protocadherins	PCdhα1–PCdhα14	
β-Protocadherins	PCdhβ1–PCdhβ19	
γ-Protocadherins	PCdhγa1–PCdhγa12 PCdhγb1–PCdhγb9 PCdhγc3–PCdhγc5	
Cdh, fibroblast 3 (Fib3)	PCdhγa12, Cdh21	Fibroblast
Cdh, fibroblast 2 (Fib2)	PCdhγb4, Cdh20	Fibroblast
Cdh, fibroblast 1 (Fib1)	PCdh16, Dachsous-1, Cdh19	Fibroblast
δ-Protocadherins	PCdh1, PCdh7, PCdh8, PCdh9, PCdh10, PCdh11, PCdh17, PCdh18	Retina

Table 7.24. Cadherin class members. (**Part 4**) Protocadherins that emanate from non-clustered protocadherin genes (bhPCdh: brain, human protocadherin; OlPCdh: olfactory bulb protocadherin; PAPC: paraxial protocadherin PC42: protocadherin-42). Protocadherin-12, or vascular endothelial (VE)-cadherin-2 is a cell adhesion molecule that is produced by the vasculogenic rather than angiogenic endothelial cells as well as trophoblasts and mesangial cells [678].

Type	Aliases
Protocadherin-1	PCdh1, PC42
protocadherin-7	PCdh7, bhPCdh
protocadherin-8	PCdh8, PAPC, arcadlin
protocadherin-9	PCdh9
protocadherin-10	PCdh10, OlPCdh
protocadherin-11X	PCdh11X, PCdhX,
(X-linked)	PCdh11
protocadherin-11Y	PCdh11Y, PCdhY
(Y-linked)	PCdh22
protocadherin-12	PCdh12, veCdh2
protocadherin-17	PCdh17, PCdh68
protocadherin-18	PCdh18, PCdh68L
protocadherin-19	PCdh19
protocadherin-20	PCdh20, PCdh13

endothelial cells. The short cytoplasmic tail of cadherin interacts with catenins that are connected to cytoplasmic proteins, such as actinin and vinculin, to link the cadherin–catenin complex to the actin cytoskeleton.

Classical cadherins are classified into type-1 and -2 cadherins. Type-1 cadherins include Cdh1 to Cdh4 and Cdh15 (E-, M-, N-, P-, and R-cadherins; Table 7.21). Type-2 cadherins encompass Cdh5–Cdh12, Cdh16–Cdh17, Cdh20, Cdh22, Cdh24, and Cdh26 (i.e., Br-, Ey-, K-, Ksp-, LI-, OB-, T1-, T2-, and VE-cadherins as well as cadherins-7 to -12 and -24). Two forms of cadherin-8 exist in the central nervous system: full-length and truncated form without a cytoplasmic domain.

H-cadherin, or T-cadherin, constitutes a group of the cadherin class. It localizes to membrane rafts. It lacks the transmembrane and cytoplasmic domains; it is hence docked to the cell membrane by a glycosyl-phosphatidylinositol anchor. It serves as a LDL receptor. Binding of LDL to T-cadherin leads to the activation of extracellular signal-regulated protein kinases ERK1 and ERK2 and nuclear translocation of nuclear factor-κB. T-Cadherin overexpression in endothelial cells facilitates cell migration and formation of stress fibers.

Desmosomal cadherins are derivatives of type-1 cadherins that include *desmocollin* and *desmoglein*.[98] Desmocollins possess a higher degree of homology with the classical cadherins. Each type of desmocollin encoded by a gene

---

[98] Desmoglein refers to desmosomal glycoproteins. Desmosomes are cell adhesive junctions connected to intermediate filaments, whereas adherens junctions are

Table 7.25. Cadherin class members. (**Part 5**) Cadherin-related class members (CDHF: cadherin family member; CdhR: cadherin-related protein; CStn, ClStn: calsyntenin; Dfnb: non-syndromic (non-syndromal) deafness; Hscr: Hirschsprung's disease protein; MEN2a(b): multiple endocrine neoplasia type-2A (2B) protein; MuCdhL: mucin and cadherin-like protein; MTC: medullary thyroid cancer protein; PC-LKC: protocadherin liver, kidney, and colon; PTC: papillary thyroid carcinoma protein; ReT: rearranged during transformation; Ush1: Usher syndrome type-1 protein). Protocadherin-15 resides in the nasal ciliated epithelium as well as neurosensory epithelia of eyes and ears. ReT51 (ReT) Protein Tyr kinase receptor is expressed in various types of neurons. ReT Ligand glial cell-line-derived neurotrophic factor induces phosphorylation of Pcdhγ in motor neurons and phosphorylation of Pcdhα and Pcdhγ in sympathetic neurons [679].

Type	Aliases
CdhR1	PCdh21
CdhR2	PCdh24, PC-LKC
CdhR3	Cdh28
CdhR4	Cdh29
CdhR5	MuCdhL, MuPCdh
CdhR12	CStn1, ClStn1, alcadein-α
CdhR13	CStn2, ClStn2, alcadein-γ
CdhR14	CStn3, ClStn3, alcadein-β
CdhR15	PCdh15, Dfnb23, Ush1f
CdhR16	CDHF12, ReT, ReT51, MEN2a, MEN2b, MTC1, Hscr1, PTC
CdhR23	Cdh23, Dfnb12, Ush1d

possesses several variants (Dsc1a–Dsc1b, Dsc2a–Dsc2b, and Dsc3a–Dsc3b). Different types of desmocollins coexist in the same desmosome. Desmogleins have an extra C-terminal domain that is absent in other types of cadherins. Different isoforms of desmocollins are expressed in different regions of human epidermis.

Non-classical cadherins include Flamingo, Dachsous, and Fat that act in planar cell polarity and growth regulation, as well as usher cadherins, protocadherin-15 and cadherin-23 that regulate the formation of apical cell protrusions [759].

The protocadherin subclass is large. Its members have up to 7 extracellular Ca^{++}-binding domains, a single transmembrane region, and distinct cytoplasmic sequences. In mammals, multiple protocadherins are highly expressed in the nervous system.[99]

---

linked to actin filaments and microtubules. In addition, cadherin density in adherens junctions is lower than that in desmosomes.

[99] During brain development, protocadherins show distinct spatiotemporal expression patterns linked to other positional cues related to the development of the brain into discrete segmental and functional subdivisions [680]. They yield scaffolds of adhesive clues.

Table 7.26. Cadherin class members. (**Part 6**) Protocadherins involved in signaling pathways triggered by morphogens (CDHF: cadherin family member; CdhR: cadherin-related protein; CELSR: cadherin, EGF LAG seven-pass G-type receptor; EGFL: epidermal growth factor-like protein; Fib1: cadherin, fibroblast 1; (H)Fmi: (human) Flamingo homolog; MEGF: multiple EGF-like domain-containing protein). Adhesion G-protein-coupled receptors are plasmalemmal proteins with a long extracellular N-terminus that contain multiple domains. For example, Flamingo has EGF-like, laminin-G-like, and cadherin-like sequences in its N-terminus.

Family and types	Alias(es)	Function
Dachsous homolog		Hippo (STK3/4) and
Dchs1	Cdh19/25, CdhR6,	Wnt pathways
	PCdh16, Fib1	
Dchs2	Cdh27/J, CdhR7,	
	PCdh23, PCdhJ	
Fat homolog (Fat1)	ME5,	Hippo (STK3/4) and
(CDHF7)	CdhR8	Wnt pathways
Adhesion, G-protein-coupled receptors		
Flamingo homologs		Wnt pathway
CELSR1	CDHF9, Flamingo-1,	
	ME2, PCdhβ11	
CELSR2	CDHF10, Hfmi2, Fmi2	
	EGFL2, MEGF3	
CELSR3	CDHF11, HFmi1, Fmi1,	
	EGFL1, MEGF2,	

The subclass of protocadherins comprises the group of 7-repeat protocadherins (PCdh1, PCdh7, PCdh9, and PCdh11), 6-repeat protocadherins that include the protocadherin-G family with subfamilies Ga (12 members in humans), Gb (7 members in humans), and Gc (3 members in humans), protocadherin-3 family (15 members in humans), and CNR or protocadherin-A family (13 members in humans), in addition to other types (PCdh8, PCdh10 and PCdh16–PCdh18). Large Cadherins encompass Fat1 and Fat2, Cdh23, PCdh15, and PCdh24.

7.5.1.2 Structure–Function Relationships of Cadherins

Classical cadherins interact with catenins and the actin filament network; desmosomal cadherins with catenins and the intermediate filament meshwork; and cadherin-related neuronal receptor (CNR)-cadherins localized to the synaptic junctions with the protein Tyr kinase Fyn of the SRC family [680] (Vol. 4 – Chap. 3. Cytosolic Protein Tyrosine Kinases).

Seven transmembrane (7TM) cadherins, CNR-cadherins, and T-cadherin localize to specialized membrane rafts rich in signaling molecules. Non-classical cadherins of the 7TM cadherin group, such as cadherin EGF-like,

LAG-like, 7-pass (G-type) receptor (CELSR, Flamingo homolog), which operate as G-protein-coupled receptors, cadherin-related proteins, such as those of the Dachsous and FAT subfamilies, serve as receptors for the Hippo signaling pathway and modulators of Frizzled function in the Wnt signaling pathway.

7.5.1.3 Cadherins and Catenins

The cytoplasmic domain of cadherins binds to γ-catenin (or plakoglobin) in a mutually exclusive fashion. The juxtamembrane domain links to catenin-$\delta 1$ (or P120 catenin). Cadherins are stabilized by adaptors α- and β-catenins, as well as cortical actin filaments.

Catenins-α exist either as monomers or homodimers. α-Catenin can tether to both β-catenin or γ-catenin as well as actin filaments, either directly or via actin-binding proteins, such as vinculin and α-actinin. α-Catenin successively links to ^{F}actin and β-catenin to remodel adhesion complexes. α-Catenin monomers preferentially connect to cadherin–β-catenin complexes, whereas α-catenin dimers are attached to actin filaments. β-Catenin binds to the cytoplasmic tail of cadherins. α-Catenins transiently bound to cadherin–β-catenin complexes can dissociate from them and bind to actin.

However, cadherin–β-catenin–α-catenin complexes do not necessarily bind to actin filaments. α-Catenin homodimers hamper actin polymerization caused by actin-related protein ARP2–ARP3 complexes and Wiskott-Aldrich syndrome proteins [681]. In addition, cadherins activate small GTPase Rho that operates in remodeling of actin cytoskeleton and regulation of cadherin-mediated adhesion.

Intercellular adhesion between epithelial cells is mediated by E-cadherins Their trans-association in homophilic complexes supported by actin filaments via α- and β-catenins maintains cell polarity and ensures epithelial architecture and stability. E-Cadherin dynamics enable both tight cell adhesion and contact remodeling. E-Cadherins have 2 main pools at the surface of epithelial cells, a monomer pool that exchanges with a stabilized trans-homophilic dimer pool at contacting surfaces. Homophilic E-cadherin clusters are associated with 2 populations of ^{F}actin that have different dynamics at the zonula adherens [682]: (1) small actin patches that stabilize adherens junctions and (2) a permanently recycling pool that tethers adherens junctions to the actin cytoskeleton via α-catenins, thus allowing remodeling.

7.5.1.4 VE-Cadherins in Vascular Endothelia

Vascular endothelial (VE)-cadherins (veCdh) are specific components of intercellular adherens junctions of adjoining endothelial cells that are characterized by self-association, thereby elaborating hexamers [683]. Hexameric self-assembly prevents veCdh internalization and degradation. In other words, veCdh hexamerization maintains mature intercellular adherens junctions.

Vascular endothelial cadherins anchor adherens junctions between endothelial cells to catenin-δ1 and β-catenin. β-Catenin links VE-cadherin to the actin cytoskeleton via α-catenin. Catenin-δ1 regulates the actin cytoskeleton via Rho GTPase [684]. VE-Cadherin can interact with catenin-δ1 to mediate cell locomotion and proliferation.

Displacements of endothelial and epithelial cells in cellular layers require coordinated cell junction movements. Although cadherins form adherens junction at the cell apical region, an additional cadherin population is distributed throughout the lateral between-cell adhesion loci. VE-Cadherin motions in a basal-apical direction associated with reorganization of actin filaments occur at cell junctions between adjoining moving cells for sliding of contacting cell membranes [685].

7.5.1.5 Cadherin-Mediated Mechanotransduction

Intercellular junctions formed by local gathering of cadherins correspond to major sites of transmission of mechanical forces inside tissues. Cadherins allow cells to sense and adapt to mechanical stresses imposed by adjoining cells. The higher the rigidity of the matrix, the greater the force magnitude generated at cadherin-based intercellular contacts, the more developed the cadherin cell adhesions, and the stronger the organization of the actin cytoskeleton [686]. This adaptation to the matrix rigidity requires myosin-2 that put actin filaments under tension. Cadherin-based intercellular adhesions balance locally tensions between extra- et intracellular media.

7.5.1.6 Cadherins in Cell Migration

Migration of an isolated cell begins with protrusion of the cell membrane, followed by the formation of new adhesions at the cell front that link the actin cytoskeleton to the substratum (Vol. 2 – Chap. 6. Cell Motility). Activation of the actomyosin cytoskeleton generates traction forces that move the cell forward and disassembly of adhesions at the cell rear.

Cycles of formation, maturation, and disassembly of cell adhesions at the cell front and, with delay, rear during cell migration are coupled to actin polymerization and organization and actin–myosin contraction. The activity of both actin and myosin-2 relies on Rho GTPases and protein Tyr kinases. Adhesion dynamics are regulated by complex feedback loops supported by reciprocity between Rac and Rho activation via guanine nucleotide exchange factors and GTPase-activating proteins.

The E-cadherin–catenin complex operates in cooperation with the actin cytoskeleton and redistribution of integrin-based focal adhesions also during cell-sheet migration. E-cadherin regulates lamellipodium activity and cell migration directionality, but not cell migration rate [687].

Table 7.27. Adhesion, G-protein-coupled receptors (GPR; CELSR: cadherin, EGF-like, LAG-like, and seven-pass receptor; CIRL: calcium-independent α-latrotoxin receptor; EMR: EGF-like module containing, mucin-like, hormone receptor-like protein; Lec: lectomedin; Source: [688]). In human, 33 adhesion-GPCR species exist.

Type	Other aliases	Distribution
CD97		Leukocytes
EMR1		Eosinophils
EMR2	CD312	Myeloid cells (monocytes, macrophages, neutrophils, dendritic cells)
EMR3		Granulocytes
EMR4	GPR127, PGR16	
GPR110–GPR116		
GPR123–GPR126, GPR128		
GPR133, GPR144		
Latrophilin receptors		
Lphn1	CIRL1, Lec2	Ubiquitous
Lphn2	CIRL2, Lec1	Primarily outside of brain, (low levels in brain)
Lphn3	CIRL3, Lec3	Central nervous system
Flamingo homologs		
CELSR1	CDHF9, Flamingo-1, ME2, PCdhβ11	
CELSR2	CDHF10, Hfmi2, Fmi2 EGFL2, MEGF3	
CELSR3	CDHF11, HFmi1, Fmi1, EGFL1, MEGF2,	

7.5.2 Adhesion G-Protein-Coupled Receptors

Adhesion G-protein-coupled receptors are cell-surface molecules that possess 3 to 5 consecutive epidermal growth factor (EGF) modules linked via a mucin-like spacer to a 7-span transmembrane class-A G-protein-coupled receptor. The G-protein-coupled receptor-proteolytic site (GPS) gives rise to an autocatalytic processing of the polypeptide into an extracellular α- and a membrane-spanning β chain that associate at the plasma membrane. First identified members of the EGF-TM7 family comprise CD97 and EMR1 to EMR4 (Table 7.27). EGF-TM7 Receptors are expressed predominantly by cells of the immune system.

The prototypal member of the EGF-TM7 family is CD97. It is constitutively expressed on granulocytes and monocytes [689]. It abounds in smooth muscle cells, macrophages, and dendritic cells. Moreover, its expression is rapidly upregulated on activated T and B lymphocytes.[100] On leukocytes, it binds the complement regulatory protein CD55,[101] a glycosyl-phosphatidylinositol (GPI)-anchored membrane protein expressed on many mammalian cells. Molecule CD55 protects cells such as complement-coated erythrocytes from complement-mediated damage by regulation of the C3 convertase. In particular, CD55 prevents C3 deposition on the surface of splenocytes [690]. In addition to CD55, CD97 binds to chondroitin sulfate and angiogenic integrins. Interaction between CD55 on monocytes and CD97 on T cells leads to cell proliferation and secretion of interferon-γ [689].

A closely related molecule to CD97 is epidermal growth factor-like module containing, mucin-like, hormone receptor-like protein EMR2 that interacts with chondroitin sulphate glycosaminoglycans in an isoform-specific manner. It is expressed on monocytes, macrophages, and granulocytes. It may intervene in the migration and adhesion of myeloid cells during cell differentiation, maturation, and activation [691]. Expression of EMR2 is upregulated during differentiation and maturation of macrophages, but is downregulated during dendritic cell maturation. Alternative splicing of EMR2 and glycosylation is regulated during myeloid differentiation. Agent EMR2 regulates neutrophil responses, as it potentiates effects of numerous pro-inflammatory mediators [692]. Liganded EMR2 increases neutrophil adhesion and migration, superoxide production, and degranulation and release of proteolytic enzymes.

Expression of EMR1 is restricted to eosinophils [693].[102] Its expression overlaps that of chemokine receptor CCR3 and siglec-8.

Molecule EMR3 is another member of the EGF-TM7 family of adhesion GPCR receptors. The highest expression of EMR3 is detected granulocytes. More mature CD16+ monocytes express high levels of EMR3 [694]. CD16− Monocytes and myeloid dendritic cells are EMR3$^{dim/low}$. Lymphocytes and plasmacytoid dendritic cells are EMR3−. CD34+, CD33−, CD38− committed hematopoietic stem cells and CD34+, CD33+, CD38+ progenitors in the bone marrow express CD97 and EMR2, but not EMR3 heterodimer.

In humans, G-protein-coupled receptors GPR123 to GPR128 are additional members of the EGF-7TM family [695]. Three of these receptors form a phylogenetic cluster, 2 pertain to a cluster with GPR64 and GPR56, and one with EMR1 to EMR3.

Phylogenetic analysis of transmembrane regions TM1 to TM7 that excludes the very large G-protein-coupled receptor VLGR1 shows the existence of 8 families of adhesion G-protein-coupled receptors [696] (Table 7.28).

---

[100] Activation of T lymphocytes depends on signaling by T-cell receptors bound to major histocompatibility complexes displayed by antigen-presenting cells.

[101] A.k.a. complement decay-accelerating factor [DAF] and Cromer blood group protein (Crom).

[102] Substance EMR1 is absent on mononuclear phagocytic cells, such as monocytes, macrophages, and myeloid dendritic cells.

Table 7.28. Families of human adhesion G-protein-coupled receptors (GPR*i*; Source: [696]; BAI: brain-specific angiogenesis inhibitor; CELSR: cadherin, EGF-like, LAG-like, and seven-pass receptor; EMR: EGF-like module containing, mucin-like, hormone receptor-like protein; HE: human epididymis-specific protein; Lphn: latrophilin).

Family	Members
1	Lphn1–Lphn3
2	CD97, EMR1–EMR3, GPR127
3	GPR123–GPR125
4	CELSR1–CELSR3
5	GPR133, GPR144
6	GPR110, GPR111, GPR113, GPR115, GPR116
7	BAI1–BAI3
8	GPR56, GPR64 (HE6), GPR97, GPR112, GPR114, GPR126, GPR128

Table 7.29. Vascular selectins with 3 main types: endothelium (E)-selectins, lymphocyte (L)-selectin, and platelet (P)-selectin (Sources: [674, 700–702]).

Molecule	Ligands	Distribution
L-selectin (CD62L)	Glycosylated mucin-like molecules (Glycam-1, CD34, MadCAM1)	WBC
E-selectin (CD62E)	Sialyated mucin-like molecules (CD15s or ESL1)	EC
P-selectin (CD62P)	P-selectin glycoprotein ligand-1 (PSGL1 or CD162)	EC, TC

7.5.3 Selectins

The extracellular domain of selectins, single transmembrane polypeptides, can reach sizes of about 50 nm [697]. The selectins are expressed in endothelial cells and blood cells. They are aimed at binding 2 cell surfaces in presence of Ca^{++} (Table 7.29). They slow intravascular leukocytes before transendothelial migration (Vol. 5 – Chap. 9. Endothelium).

Three types of selectins are defined according to the cell in which they were discovered. *L-selectin*[103] is expressed on leukocytes that target activated endothelial cells. *E-selectin*[104] is produced by endothelial cells after cytokine

---

[103] A.k.a. CD62L (CD meaning cluster of differentiation with a number assigned by the international workshops on leukocyte typing) or LAM1 or Mel14.

[104] A.k.a. CD62E or ELAM1.

activation. *P-selectin*[105] is preformed and stored for rapid release in platelet granules or Weibel-Palade bodies of endothelial cells [699].[106]

Selectin–ligand complexes are able to resist detachment that results from hemodynamic force applied to leukocytes bound to vessel walls before diapedesis, particularly when they roll over the endothelium. Selectin–ligand complexes can bear forces up to the order of 100 pN. Transition from bent conformational state to the ligand-bound, extended conformation can occur with a faster time scale than force-induced rupture of selectin–ligand bond [703].

7.5.4 Integrins

Integrins are transmembrane heterodimeric glycoproteins that connect actin filaments of the cell cytoskeleton to protein constituents of the extracellular matrix (ECM). The extracellular domain of integrins can reach sizes of about 20 nm [704].

Integrins influence assembly and pattern of cytoskeletal signaling complexes. They mediate signaling to (*inside–out signaling*) or from (*outside–in signaling*) the environment.[107]

Once assembled with integrins, complexes that contain cytoskeletal and adaptor proteins as well as protein Tyr kinases initiate signaling cascades. Various molecules that bind to the cytoplasmic part of integrins include actin, α-actinin, paxilin, talin, tensin, vinculin, and zyxin [706]. In addition to these partners, 3 main other types of proteic ligands interact with integrins [707]: (1) matrix proteins, such as collagen, fibronectin, and fibrinogen; (2) circulating plasma proteins; and (3) proteins of the plasma membrane.

Activated integrins regulate cell adhesion and migration, as well as extracellular matrix assembly and mechanotransduction. They thus participate in the control of tissue development, inflammation, and immunity, as well as hemostasis and vascular permeability. Integrin activation is, indeed, implicated in anchorage-dependent cellular events, such as platelet aggregation and leukocyte transmigration. Integrins also contribute to the local regulation of

---

[105] A.k.a. CD62P or GMP140 (140-kDa granule membrane protein) or PADGEM. E-selectin and P-selectin have similar roles. P-selectin binds to P-selectin glycoprotein ligand 1 (PSGL1).

[106] Activated endothelial cells release Weibel-Palade bodies, which contain von Willebrand factor and P-selectin, to induce leukocyte rolling and platelet adhesion and aggregation. Weibel-Palade granules thus contain mediators that promote inflammation and coagulation. Prevention of Weibel-Palade body release might be a mechanism by which NO protects the vessel wall from inflammation [698].

[107] The extracellular matrix yields signals that control cell shape, migration, proliferation, differentiation, morphogenesis, and survival. These signals are derived from cryptic sites within matrix molecules (*matricryptic sites*) that are revealed after structural or conformational modifications of these molecules. *Matricryptins* refer to enzymatic fragments of matrix components that contain matricryptic sites [705].

the vasomotor tone (Vol. 5 – Chaps. 8. Smooth Muscle Cells and 9. Endothelium). Integrin interactions with matrix ligands can actually trigger signaling events and provide paths for transduction of mechanical forces across the plasma membrane. In addition to changes in cell adhesion, integrin activation can control the polarity of migrating cells.

Integrins functions are regulated by [708]: (1) the level of integrin expression,[108] (2) divalent cations,[109] (3) cellular environment,[110] and (4) affinity modulation.[111]

7.5.4.1 Integrin Structure and Types

Integrins are transmembrane heterodimeric metalloproteins made of α and β subunits. Each subunit has a large ectodomain (extracellular region), a single transmembrane segment, and, in general, a short endodomain (cytoplasmic segment). The transmembrane domains of the 2 subunits associate via 2 structural elements [709]: (1) the inner membrane clasp in the inner membrane leaflet that extends at the interface between the plasma membrane and the cell cortex and (2) the outer membrane clasp at the outer leaflet of the lipid bilayer. Therefore, perturbations at the intra- or extracellular face can destabilize the integrin transmembrane dimer.

Modulation of Integrin Configuration and Activity

Integrin activity is modulated by divalent cations, Ca^{++}, Mg^{++}, and Mn^{++}. Integrins change from a curved, compact shape in low-affinity state to an extended conformation in high-affinity state [710]. Ligand binding induces the extension of the integrin extracellular domain. Ligand binding to the metal ion-dependent adhesion site of α_L ligand-binding I domain[112] is modulated

---

[108] The transcription control of integrin expression promotes adhesion of a given type of cell and prevents others from interacting negatively such as in angiogenesis.

[109] Binding between integrin and its specific ligands requires divalent cations, such as Ca^{++}, Mg^{++}, Mn^{++}. This binding thus depends on the content of the extracellular medium. Different divalent cations can have opposite effects on a given integrin.

[110] The ligand specificities vary according to the cell type. Integrin $\alpha_2\beta_1$ on platelets binds to collagen and, on endothelial cells, to collagen and laminin. This cell type-specific binding depends on the cell environment.

[111] Integrin function can be regulated by both the number of expressed receptors and the environment.

[112] The interaction I domain exists in 9 among the 18 integrin α subunits. A Mg^{++} ion links to the domain in a metal ion-dependent adhesion site (MIDAS). Ion Mg^{++} coordinates the connection between the integrin and its ligand. Ligand binding induces a structural rearrangement in the I domain of the α subunit. Member of the immunoglobulin superclass, such as intercellular adhesion molecules ICAM1 and ICAM2, vascular cell adhesion molecule VCAM1, and mucosal vascular addressin cell adhesion molecule MAdCAM1, are recognized by some integrins.

Table 7.30. Examples of integrin types that act as collagen, fibrinogen, laminin, and/or vitronectin receptors and their aliases (CD: cluster of differentiation; CRi: complement receptor i [CR$i\alpha$–CR$i\beta$ or CRia–CRib dimer]; GP: glycoprotein; LFA: lymphocyte function-associated antigen; Mac1: macrophage-1 antigen; VLA: very late activation antigen). β_2 Subunit encoded by the ITGB2 gene corresponds to subunit β of CR3 (expressed on granulocytes, monocytes, and natural killer cells) and CR4 receptor (on monocytes, neutrophils, platelets, and B lymphocytes) for complement component C3 fragments iC3b (both; complement component C3 is cleaved into C3a and C3b; C3b is broken down progressively to inactive derivative of C3b [iC3b], then C3c and C3dg) and C3dg (CR4).

Integrin	Aliases
$\beta_1\alpha_1$	VLA1, CD29–CD49a
$\beta_1\alpha_2$	VLA2, CD29–CD49b, GP2a–GP1a
$\beta_1\alpha_3$	VLA3, CD29–CD49c
$\beta_1\alpha_4$	VLA4, CD29–CD49d
$\beta_1\alpha_5$	VLA5, CD29–CD49e
$\beta_1\alpha_6$	VLA6, CD29–CD49f
$\beta_2\alpha_L$	CD18–CD11a, LFA1
$\beta_2\alpha_M$	CD18–CD11b, CR3, Mac1
$\beta_2\alpha_X$	CD18–CD11c, CR4, GP150/95,
$\beta_2\alpha_D$	CD18–CD11d
$\beta_3\alpha_{2B}$	CD61–CD41, GP3a–GP2b
$\beta_3\alpha_V$	CD61–CD51

by the β_2 I domain. When integrin extends, its β_2 cytoplasmic and transmembrane domains separate, enabling signal transmission across the plasma membrane. Furthermore, integrin cytoplasmic domains can then cooperate with cytoskeletal components to trigger cell spreading.

Integrin Types and Composition

Various integrins combine different kinds of α and β subunits (Tables 7.30 and 7.31). In mammals, 24 canonical integrins are produced by combinations of 18 α subunits (α_1–α_{11}, α_D, α_E, α_L, α_M, α_V, α_W, and α_X) and 8 β subunits (β_1–β_8). In addition, some subunits possess differentially spliced variants. Integrins are classified into subclasses according to their β subunits that are associated with one or more α subunits [708].

Integrin in Cells of the Vasculature

Vascular cells express several members of the integrin class. In platelets, α_{2B} integrin subunit encoded by the ITGA2B gene joins with β_3 to form a fibronectin receptor (glycoprotein GP2b–GP3a complex) that operates in

Table 7.31. Vascular integrins (Sources: [674, 701, 702, 711]; Cn: collagen; FN: fibronectin; Fng: fibrinogen; FX: clotting factor X; ICAM: intercellular adhesion molecule; Ln: laminin; MAdCAM: mucosal vascular addressin cell adhesion molecule; PECAM: platelet endothelial cell adhesion molecule; Tsp: thrombospondin; VCAM: vascular cell adhesion molecule; VN: vitronectin; vWF: von Willenbrand factor; CMC: cardiomyocyte; EC: endothelial cell; Eφ: eosinophil; FB: fibroblast; Lφ: lymphocyte; Mφ: macrophage; Mo: monocyte; Nφ: neutrophil; SMC: smooth muscle cell; TC: thrombocyte (platelet); WBC: white blood cell).

Molecule	Ligands	Distribution
$\alpha_1\beta_1$	Ln, Cn	NK-, B-, T Lφ, FB, EC
$\alpha_2\beta_1$	Ln, Cn	NK-, B-, T Lφ, TC, FB, EC
$\alpha_3\beta_1$	Ln, Cn, FN	T Lφ, FB, EC
$\alpha_4\beta_1$	VCAM1, ICAM1, FN	NK-, B-, T Lφ, Eφ, Mo, EC
$\alpha_4\beta_7$	FN, VCAM1, MadCAM	NK-, B-, T Lφ
$\alpha_5\beta_1$	FN	B-, T Lφ, TC, FB, EC
$\alpha_6\beta_1$	Ln	WBC, FB, EC
$\alpha_6\beta_4$	Ln	FB, EC
$\alpha_7\beta_1$	Ln	CMC, FB, EC
$\alpha_{11}\beta_1$	Cn	CMC, SMC
$\alpha_{2B}\beta_3$	FN, VN, vWF, Tsp	TC
$\alpha_L\beta_2$	ICAMs	Nφ, Mo, Mφ, T-, B-, and NK Lφ
$\alpha_M\beta_2$	ICAMs, FX, Fng	NK-, B Lφ, Nφ, Mφ
$\alpha_X\beta_2$	Fng	B Lφ, Mφ
$\alpha_V\beta_3$	FN, VN, Cn, PECAM1, vWF, Fng, Tsp	B-, T Lφ, Mo, EC
$\alpha_V\beta_5$	FN, VN, Fng	FB, Mo, Mφ
$\alpha_{IEL}\beta_7$	E-cadherin	T Lφ

coagulation. These cell fragments also contain other types of α subunits, such as α_1 to α_{11} and α_V integrins. Thrombocytes produce β_1 integrin subunit that form $\alpha_2\beta_1$, $\alpha_4\beta_1$, and $\alpha_5\beta_1$ integrins.[113] Platelet β_3 integrin subunit binds to fibrinogen during clotting.[114]

Integrins β_2 (α_L, α_M, α_X, α_D) are leukocyte-specific receptors, involved in leukocyte adhesion and transmural migration, granulocyte aggregation, T-lymphocyte killing, and T helper cell response [712]. In leukocytes, $\alpha_L\beta_2$,[115]

---

[113] $\alpha_4\beta_1$ Integrin is also called very late antigen-4 (VLA4) or CD49d–CD29 complex. Integrins $\alpha_2\beta_1$ and $\alpha_5\beta_1$ are also termed glycoproteic complexes GP1a–GP2a and GP1c–GP2a, respectively.

[114] Platelet $\alpha_{2B}\beta_3$ integrin recognizes fibrinogen, in the presence of agonists (ADP, thrombin, and thromboxan-A2), fibronectin, vitronectin, von Willebrand factor, and thrombospondin.

[115] Integrin $\alpha_L\beta_2$ is also refered to as leukocyte function-associated antigen-1 (LFA1) and CD11a–CD18 complex. This integrin is indeed a heterodimer formed by subunits α_L (CD11a) and β_2 (CD18). It binds to intercellular adhesion molecule-1 (ICAM1) on endothelial cells to promote leukocyte adhesion and transmural

$\alpha_M\beta_2{}^{116}$ and $\alpha_X\beta_2$ integrins[117] contain the same β_2 (CD18) subunit. Leukocytes possess $\alpha_1\beta_2$ and $\alpha_M\beta_2$ integrins that bind to endothelial cells.

Integrin β_3 subunit (connected to α_{2B} and α_V integrin) are observed in various cells (platelets, endothelial cells, monocytes, smooth muscle cells, etc.). Other integrins include β_4 (α_6), β_5 (α_V), β_6 (α_V), β_7 (α_4 and α_E), β_8 (α_V), among others.

Integrin β_2

Subunit β_2 heterodimerizes with: α_L (or CD11a), α_M (or CD11b), α_X (or CD11c), and α_D (or CD11d). β_2 Integrins participate in innate and adaptive immune responses, especially leukocyte diapedesis and recruitment to inflammatory sites and secondary lymphoid organs, pathogen recognition, phagocytosis, and antibody- or cell-mediated cytotoxicity [714].

In normal conditions, β_2 integrins are inactive with a bent conformation. They thus prevent adhesion of circulating leukocytes. They switch within seconds from a low- to high-affinity conformation in response to extracellular agonists (inside–out activation) [714]. Binding of cytosolic talin or kindlin-3 also primes β_2 integrin activation.

In fact, β_2 integrin ectodomain interacts with more than 30 extracellular and plasmalemmal ligands [714]: (1) soluble proteins, such as chemokines, growth factors, fibrinogen, inactive product of the complement cleavage fragment C3b, coagulation factor X, elastase-2;[118] (2) counter-receptors, such as ICAM1 to ICAM4, JAM3, platelet receptor of von Willebrand factor, glycoprotein GP1bα (or CD42b), monocyte differentiation antigen CD14, and MAPK overlapping kinase (MOK);[119] and (3) inhibitors, such as developmentally-regulated endothelial cell locus DEL1 and neutrophil

---

migration. Once bound and activated by ICAM1, $\alpha_L\beta_2$ integrin triggers abrupt, firm adhesion of rolling lymphocytes [713]. It is expressed on plasma membranes of all the leukocyte types. It is involved in migration of leukocytes across vessel walls both during normal lymphocyte recirculation through lymph nodes and in the response of myeloid and lymphocytic cells to inflammation. This integrin is also required in the interaction between cytotoxic T cells and their target cells, and in the formation and stabilization of the immunological synapse between T lymphocytes and antigen-presenting cells during the immune response. Last but not least, ligand binding to the extracellular domain of $\alpha_L\beta_2$ integrin triggers intracellular pathways that influence the cell fate (signaling toward intracellular medium). In addition, leukocyte behavior shifts from adhesion to migration and conversely owing to $\alpha_L\beta_2$ integrin (signaling toward extracellular medium).

[116] Integrin $\alpha_M\beta_2$ is also termed CD11b–CD18, Mac1, and CR3.

[117] Integrin $\alpha_X\beta_2$ is also termed CD11c–CD18 or CR4.

[118] A.k.a. neutrophil elastase.

[119] A.k.a. renal tumor antigen-1 and receptor for advance glycation end products RAGE1.

inhibitory factor. Integrins β_2 also interact with plasmalemmal receptors, such as immune receptor FcγR3[120] and urokinase receptor.

Integrin α_2

Subunit α_2 connects to β_1 subunit to form the functional receptor $\alpha_2\beta_1$ integrin on activated T lymphocytes (very late activation antigen VLA2), platelets (collagen receptor glycoproteic dimer GP1a–GP2a).[121] It is particularly expressed in airway epithelial cells as well as dendritic cells, neutrophils, natural killer cells, monocytes, and mastocytes [715].

In humans, $\alpha_2\beta_1$ integrin is widely expressed on mesenchymal, epithelial, and endothelial cells as well as platelets. It contributes to the control of cell differentiation, proliferation, and apoptosis, as well as remodeling of the extracellular matrix, as it regulates the synthesis of collagens and matrix metalloproteinases. It regulates the activity of kinases PI3K, PKB, P38MAPK, and ERK, cyclin-dependent kinase inhibitor CKI1b, phosphatase PP2a, Runt-related transcription factor Runx2, small GTPase Rac1, and guanine nucleotide-exchange factor Vav2 [715]. Furthermore, it cooperates with other integrins, other receptors of matrix components such as syndecans, platelet glycoprotein GP6, discoidin domain receptors, as well as growth factor receptors (EGFR, HGFR, and VEGFR).

Integrin $\alpha_2\beta_1$ interacts with E-cadherin, collagens, especially fibril-forming collagen types 1 to 3, 5, and 9, as well as fibrils formed by collagen-1, -4, -6 to -10, and -16, laminins, decorin, endorepellin, tenascin-C, chondroadherin, matrix metalloproteinase MMP1 (or collagenase-1), and members of the collectin family (C1q complement protein, mannose-binding lectin, and surfactant protein-A) [715]. It binds to intracellular compounds, such as intermediate filament constituent vimentin, ^{F}actin, and calreticulin [715]. Small GTPase Rab21 controls its endosomal transfer [715].

Integrin α_4

Subunit α_4 builds a heterodimer with a β_1[122] or β_7 chain. It interacts with vascular (VCAM1) and mucosal vascular addressin (MAdCAM1) cell adhesion molecule. α_4 Integrin intervenes in the motility of hematopoietic stem cells, eosinophils, mastocytes, and T and B lymphocytes [716]. It is also involved in the recruitment of bone marrow-derived endothelial progenitors to the neovasculature.

$\alpha_4\beta_1$ Integrin operates in leukocyte diapedesis due to its interaction with junctional adhesion molecule JAM2 on endothelial cells. During T-cell development, $\alpha_4\beta_1$ integrin is required in selection and migration of T lymphocytes

---

[120] I.e., crystallizable fragment (Fc) receptor of antibody IgG.
[121] A.k.a. collagen-binding extracellular matrix receptor-2 (ECMR2).
[122] Very late antigen VLA4.

in the thymus [716]. It also promotes the localization of B lymphocytes in the bone marrow and germinal centers, mediates B-cell retention in the marginal zones of spleen, and participates in cell compartmentation in peripheral lymphoid tissues. In addition, it acts as a costimulator on T lymphocytes and regulates differentiation of T_{H1} and T_{H2} helper T cells.

Counter-receptors of $\alpha_4\beta_1$ integrin include VCAM1, JAM2, ADAM28 of the adamlysin family of transmembrane metalloproteases, fibronectin, osteopontin, thrombospondin, and propolypeptide of von Willebrand factor [716].

$\alpha_4\beta_7$ Integrin is expressed by a small subset of circulating memory T cells that preferentially localize to the gut-associated lymphoid tissue, owing to MAdCAM1 [716].

Integrin α_{11}

α_{11} Subunit constitutes $\alpha_{11}\beta_1$ collagen receptor. This integrin enhances polymerization of collagen-1 and -3 [717]. $\alpha_{11}\beta_1$ Integrin preferentially targets fibrillar collagens (type 1 to 3) than mesh-forming (collagen-4) and beaded filament-building (type 6) collagens. It operates in specialized subsets of mesenchymal cells. It can impede fibroblast chemotaxis in response to PDGF [717].

Integrin α_E

α_E Subunit[123] assembles with β_7 chain. α_E Subunit is expressed by T lymphocytes in mucosal tissues, cytolytic T lymphocytes, E-selectin+, CD4+ memory T cells, CD4+ effector T cells, and CD4+, CD25+ regulatory T cells, as well as subsets of mucosal dendritic cells and bronchoalveolar and activated mucosal mastocytes [718].

Integrin α_L

α_L Subunit (or CD11a) associates with β_2 subunit (or CD18) to form the functional leukocyte function-associated antigen LFA1. It is expressed on the cell surface of all leukocytes. It is involved in transmigration of leukocytes across blood vessel walls, interaction between cytotoxic T cells and their target cells, and formation and stabilization of immunological synapse between T lymphocytes and antigen-presenting cells [719]. It participates in cell growth, differentiation, migration, and apoptosis.

Its ligands and counter-receptors include intercellular adhesion molecules ICAM1 to ICAM5, junctional adhesion molecule JAM1 on endothelial cells, as well as various intracellular molecules, such as cytoskeletal components, adaptors, signaling mediators, and transcription factors (α-actinin, filamin, talin, cytohesin-1 to -3, focal adhesion- and integrin-linked kinase, receptor for

---

[123] A.k.a. CD103 and mucosal lymphocyte-1 antigen.

activated C kinase RACK1, Jun activation domain-binding protein JAB1 (or signalosome subunit-5), regulator of adhesion and cell polarization enriched in lymphoid tissues (RAPL) [719].

Integrin α_M

α_M Subunit[124] is expressed on the surface of leukocytes. It dimerizes with β_2 subunit to generate $\alpha_M\beta_2$ integrin.[125] It is implicated in inflammation and immunity, especially leukocyte adhesion, migration, activation, phagocytosis, degranulation, superoxide release, as well as development and apoptosis, in addition to blood coagulation and clot dissolution [720].

$\alpha_M\beta_2$ Integrin can bind to [720]: (1) extracellular matrix proteins (collagen-1, -2, -4, -6, and -14, fibronectin, laminin, and vitronectin) (2) cell-adhesion molecules (ICAM1, -2, and -4, GP1bα, JAM3, E-selectin, and C-type lectin CLec4l); (3) proteins of blood coagulation and clot dissolution (coagulation factor X, fibrinogen, high-molecular-mass kininogen, plasminogen, urokinase) as well as other plasma proteins (angiostatin, complement factors I and H, catalase, elastase, myeloperoxidase, proMMP9); (4) plasmalemmal receptors (B-cell receptor-associated protein BCAP31, FcϵR2, FcγR3b, MAPK overlapping kinase [MOK], thymocyte GPI-anchored cell-surface antigen Thy1 [or CD90], uPAR, and TNFRSF5); (5) intracellular ligands (talin, radixin, IGF-binding protein-8, and cyclin-A2); as well as (6) various components of pathogens.

Integrin α_V

α_V Subunit forms a complete integrin by heterodimerization with 5 β subunits (β_1, β_3, β_5, β_6, and β_8). A metal ion is bound when the integrin is ligated (Mg^{++}) or unligated (Ca^{++}) [721].[126] Ion Mn^{++} stimulates integrin activation, as it imitates ligand binding.

Ligand types depend on β subunit type. Irrespective to β subunit, α_V integrin tethers to latency-associated peptide (proregion of transforming growth factor-β), fibronectin, and vitronectin. Associated with any β subunit except β_8, α_V integrin links to osteopontin; with β_1, β_3, or β_6, to fibrinogen; with β_1 or β_3 to L1 cell adhesion molecule (L1CAM);[127] with β_3 or β_6 to A disintegrin and metalloproteinase domain-containing proteins (ADAM); with β_1, β_3, or β_8 to laminin; with β_3 or β_5 to osteonectin, periostin, and canstatin [721].[128] Furthermore, $\alpha_V\beta_3$ Integrin binds to von Willebrand

---

[124] A.k.a. CD11b and complement receptor CR3a.

[125] A.k.a. macrophage antigen Mac1.

[126] Divalent cations Ca^{++} and Mg^{++} inhibit and promote ligand binding, respectively.

[127] A.k.a. neural cell adhesion molecule L1 and CD171.

[128] A.k.a. collagen-4 chain-α2.

factor, thrombospondin, fibrillin, tenascin, PECAM1, ICAM4, MMP2 and MMP9, fibroblast growth factors FGF1 and FGF2, urokinase, urokinase receptor, angiostatin, plasmin, cardiotoxin, prothrombin, and developmentally-regulated endothelial cell locus DEL1, among others.

α_V Integrin possesses numerous plasmalemmal partners, such as insulin receptor substrate IRS1, insulin-like growth factor IGF1, and transforming TβR2, vascular endothelial VEGFR2, platelet-derived (PDGF), fibroblast FGFR3, and hepatocyte growth factor receptor (HGFR), as well as insulin (IR), urokinase (uPAR) and low-density lipoprotein (LDLR) receptor, in addition to caveolin and integrin-associated protein (IAP or CD47).

Like other types of integrins, α_V integrin connects to multiple intracellular proteins, such as actin-binding proteins (filamin, kindlins, and talin), calreticulin, receptor for activated protein kinase-C RACK1, dual Ig domain-containing adhesion molecule (DICAM), and adherens junction constituent nectins.

Activated α_V integrin operates in embryogenesis, especially neuro-, myo-, and vasculogenesis, as well as angiogenesis,[129] wound healing,[130] and inflammation, among other normal functions, as well as participates in atherosclerosis and tumorigenesis [721]. α_V Subunit is expressed in the myocardium, vascular endothelium (with β_1, β_3, β_5, and β_8), and bronchi [721]. $\alpha_V\beta_3$ Integrin is also expressed on resting T lymphocytes as well as activated natural killer T lymphocytes [721].

7.5.4.2 Integrin Activation

At rest, interactions between transmembrane domains of integrin subunits maintain a low-affinity, inactive state. Integrin activation then requires disruption of these transmembrane interactions.

Integrin ectodomains possess several types of configurations [709]: (1) bent closed conformations; (2) intermediate extended conformations with a closed head-piece; and (3) extended open conformations. These configurations that result from rotation relative to bonds without bond break and creation might correspond to low-affinity, activated, and activated, ligand-occupied integrin conformers, respectively.

Interactions of integrin cytoplasmic domains with each other or cytoplasmic proteins create allosteric rearrangements of integrins that support activation. Changes in conformation of a given integrin heterodimer and clustering

---

[129] $\alpha_V\beta_5$ Integrin binds to developmentally-regulated endothelial cell locus DEL1 (a.k.a. EGF-like repeats and discoidin-I-like domain-containing protein EDIL3), an integrin-binding pro-angiogenic factor that activates an angiogenic program. This program stimulates the expression of $\alpha_V\beta_3$ integrin and urokinase-type plasminogen activator receptor [721].

[130] During wound healing, $\alpha_V\beta_3$ integrin is upregulated in platelets, endothelial cells, macrophages, and fibroblasts [721]. Production of $\alpha_V\beta_5$ and $\alpha_V\beta_6$ integrins also rises.

of heterodimers into oligomers influence the binding of ligands, via variations in receptor affinity and the elevated number of possible chemical bonds (receptor valency) that results from integrin clustering.

Integrin clustering can be generated by the recruitment of proteic complexes to cytoplasmic domains of integrins as well as binding of extracellular multivalent ligands to ectodomains [709]. In addition, binding of multimeric ligands may support integrin clustering seeds, as it can modify the nature of the bond between integrin and ligand upon conformational changes.

Integrin activation comprises triggering, intermediate signaling events with changes in interaction between transmembrane domains of α and β subunits, and, finally, interaction of integrins with cytoplasmic regulators that changes binding affinity of integrins for their ligands.

7.5.4.3 Bidirectional Integrin Signaling

Integrins signal in the 2 directions with different consequences. During *inside–out signaling*, an intracellular activator, such as talin or kindlins, binds to integrin and causes a conformational change that increases its affinity for extracellular ligands. Inside–out signaling controls adhesion strength, as it enables strong interactions between integrins and extracellular matrix proteins that allow integrins to transmit forces required for cell migration and matrix remodeling [709].

Integrins also transmit information into cells from their environment. Integrins can actually be activated directly by extracellular factors. During *outside–in signaling*, the binding of integrins to their extracellular ligands changes the integrin conformation and promotes integrin clustering. The combination of these 2 events primes intracellular signals to control cell polarity and cytoskeletal rearrangement, as well as gene expression for cell survival and proliferation [709].

The 2 unidirectional signalings, in fact, are often closely linked, as ligand binding associated with outside–in signaling stimulates integrins and, conversely, integrin activation increases ligand binding for inside–out signaling.

Cytosolic Regulators

Final intracellular steps of integrin activation in outside–in signaling involve the binding of talins and/or kindlins [709].[131] Conversely, during inside–out

---

[131] Kindlin-1 deficiency that results from a mutation in the KIND1 gene causes the rare congenital Kindler or Weary-Kindler syndrome (a.k.a. bullous acrokeratotic poikiloderma of Kindler and Weary, congenital poikiloderma with blisters and keratoses, congenital poikiloderma with bullae and progressive cutaneous atrophy, hereditary acrokeratotic poikiloderma, hyperkeratosis–hyperpigmentation syndrome, and acrokeratotic poikiloderma). The Kindler syndrome is an autosomal recessive genodermatosis characterized by epithelial cell dysfunction with a skin that has a strong tendency to blister as well as gasterointestinal manifestations.

signaling, integrin cytoplasmic domains trigger conformational changes. Talins and kindlins are the triggering factors.

Talin binding destabilizes the interaction between the transmembrane regions of integrin subunits, as talin-1 or its paralog talin-2 disrupts electrostatic interaction of α with β subunit. In addition, talin is able to modify the tilt angle of the integrin transmembrane domain [709]. In addition to its activation of integrins, cytoskeletal protein talin links integrins to filamentous actin and actin-binding proteins, hence linking the actin cytoskeleton to the extracellular matrix. Talins are necessary, but not sufficient, for integrin activation.

Kindlins (kindlin-1–kindlin-3) cooperate with talins to regulate the activation of specific integrins. In the absence of talin, kindlins do not exhibit any stimulatory effect [709]. Therefore, kindlins act as coactivators for talin of integrins.

Small GTPases Rap1a and Rap1b stimulate integrin activation. Protein RasGRP1[132] is a Rap1 guanine nucleotide-exchange factor (Rap1GEF) in hematopoietic cells that activate Rap1 GTPase. Active Rap1 binds to Rap1-interacting adaptor molecule (RIAM)[133] to form the Rap1–RIAM–talin complex. The Rap1–RIAM–talin complex enables talin recruitment to integrin and subsequent integrin activation [709].

Adaptor RIAM activates talin-1. In lymphoid cells, RIAM supports integrin-mediated cell adhesion. Moreover, RIAM interacts with Enabled homolog and vasoactive stimulatory phosphoprotein (Ena–VASP)-related proteins that can promote actin polymerization and then form ^{F}actin. Whereas RIAM abounds in hematopoietic cells, Ras-associated (RalGDS/AF6) and pleckstrin homology domains-containing protein RAPH1[134] is a paralog in fibroblasts and other cells. The latter is a binding protein of Ena–VASP homology proteins that localizes at tips of lamellipodia and filopodia.

Other activators and inhibitors specifically target integrin subunits. Ca^{++}-and integrin-binding protein CIB1[135] is an α_{2B} integrin tail-binding protein. It antagonizes talin-1 binding, thereby functioning as an inhibitor of integrin activation [709].

Regulator for cell adhesion and polarization enriched in lymphoid tissues (RAPL)[136] is a Rap1-binding protein that specifically connects to $\alpha_L\beta_2$ integrin [709].

Activated hRas stimulates extracellular signal-regulated kinases ERK1 and ERK2 and suppresses integrin activation in many cell types [709].

Type-1 phosphatidylinositol 4-phosphate 5-kinase PI(4)P5K1γ synthesizes PI(4,5)P$_2$. The latter regulates interactions between scaffolding and signaling

---

[132] A.k.a. Ca^{++}- and DAG-regulated CalDagGEF1.

[133] A.k.a. amyloid β (A4) precursor protein-binding, family-B, member 1-interacting protein APBb1IP.

[134] A.k.a. lamellipodin (Lpd).

[135] A.k.a. kinase-interacting protein KIP1.

[136] A.k.a. new Ras effector NORE1 and Ras association domain-containing family protein RASSF5.

proteins at focal adhesions such as interaction of vinculin with actin and talin. Talin colocalizes with integrins in cell–matrix junctions and links focal adhesions to stress fibers, as it connects integrins to ^{F}actin. Kinase PI(4)P5K1γ associates with talin; the resulting complex is recruited to focal adhesions and both components are activated. Talin is able to bind both PI(4)P5K1γ and β-integrin cytoplasmic domain, using the same binding site. Therefore, PI(4)P5K1γ competes with β-integrin tails for binding to talin-1. In other words, talin binding to integrin is inhibited by PI(4)P5K1γ [722].

Phosphorylation of integrin tails by SRC family kinases triggered by ligand binding to integrins reduces the affinity of integrin for talin-1 and promotes the interaction of integrin with docking protein DOK1 that impedes integrin activation [709].

Integrin cytoplasmic domain-associated protein ICAP1, or integrin β1-binding protein Itgβ1BP1, can bind to $β_{1A}$ integrin and restrict talin-1 binding to integrin, hence limiting integrin activation [709].

Actin-linking filamin-A is another inhibitor of integrin activation, as it blocks binding of the other actin-linking talin-1 to β-integrin cytoplasmic tails. On the other hand, cytoskeletal adaptor migfilin targets the same filamin-binding site that is used by β integrins. Migfilin interaction thus dissociates filamin from integrin and promotes the talin–integrin binding [723].

7.5.4.4 Integrin Adhesome

Integrin adhesome integrates the entire set of interactions involved in integrin-mediated adhesion and signaling. Cells indeed sense multiple environmental signals via cell–matrix adhesion complexes. Cells are able to detect differences in adhesive ligand density. Moreover, they respond to mechanical stress and surface rigidity. At least 156 known components (151 proteins, 4 lipids, and calcium ions)[137] intervene, including 90 substances that reside within

---

[137] Six sets can be defined. (1) The actin–integrin set (actin, 11 actin regulators, 46 adaptors, adhesion molecule-associated proteins such as syndecan, and transmembrane molecules, mainly integrins) connects the adhesome to actin. Talin, tensin, plectin, filamin, and α-actinin especially bridge integrin to actin. (2) The set of Ser/Thr kinases and phosphatases that have the following substrates: Tyr kinases and phosphatases, adaptors, actin regulators, RacGEFs, calpain, and integrins. Protein Ser/Thr kinases are phosphorylated by protein Ser/Thr kinases and dephosphorylated by protein Ser/Thr phosphatases. Serine/threonine phosphatases are not regulated by protein Ser/Thr kinases. Serine/threonine kinases are also regulated by Tyr kinases, lipids, and Rho GTPase. (3) The set of Tyr kinases and phosphatases that target various substrates (e.g., adaptors, adhesion molecules, actin regulators, Ser/Thr and phosphoinositide kinases, and GTPase regulators) and are targeted by their regulators. (4) The set of GTPases of the RHO superfamily with their effectors (phosphoinositide kinases and Ser/Thr kinases), their Ser/Thr kinase regulators, GTPase-activating proteins, guanine nucleotide-exchange factors, and regulators of GAPs and GEFs.

adhesion sites and 66 peripheral molecules that interact with intrinsic components to influence cell fate [724].

Adhesome contains structural and functional nodes and module adaptors. In addition to adhesion molecules, adhesome comprises cytoskeletal components, actin-binding proteins, cell-membrane receptors, Ser/Thr protein kinases and phosphatases, Tyr protein kinases and phosphatases, phosphoinositide kinases and phosphatases, GTPases, GTPase-activating proteins, and guanine nucleotide-exchange factors (Tables 7.32 and 7.33).

Some adhesome proteic components such as integrins can interact with more than 20 molecules and correspond to prominent hubs of the adhesome network. About half of the links between distinct adhesome components can be switched on and off by signaling effectors.

Various proteins link integrins to the cytoskeleton, such as tensin and filamin. Among these proteins, some have several binding sites; therefore crosslinking actin filaments. They include α-actinin, fimbrin, and ezrin–radixin–moesin. Last-mentioned partners coordinate signals triggered by cytokine-induced adhesion molecules with extracellular adhesive functions [725]. Moreover, ERMs are markers of RoCK kinase activity.

Integrins switch between active and inactive conformations. In the inactive state, integrins have a low affinity for ligands. Signaling events induce a conformational change with exposure of the ligand-binding site. Ligand–integrin binding connects the actin cytoskeleton to the extracellular matrix, via the assembly of a proteic complex, and interacts with signaling pathways. Mechanical stresses modify the conformational state of integrins that form new connections with their specific matrix ligands and cytosolic molecules to relay mechanical stress signaling to intracellular pathways and increase cell binding [726].

Integrin-Linked Kinase

Integrin-linked kinase (ILK) is a cytoplasmic integrin effector that regulates ^{F}actin rearrangements. Integrin-linked kinase, isoforms of particularly interesting Cys-His-rich protein (PINCH), and parvin form the *IPP complex* in the cytosol [727].[138] The IPP complex is recruited at focal adhesions with

---

GTPase-activating proteins are primarily regulated by Tyr kinases and guanine nucleotide-exchange factors by protein Ser/Thr kinases. (5) The set of phosphoinositides that are regulated by phosphoinositide kinases and phosphatases, which are controlled by GTPases and Tyr and Ser/Thr kinases. They regulate adaptor proteins (e.g., talin and vinculin), actin-bundling and -crosslinking proteins (e.g., filamin and α-actinin), as well as Ser/Thr kinases. (6) The set of protease calpain (isoform-1 and -2) and ubiquitin ligase CBL that degrade adhesome proteins. Calpain that is regulated by calcium and Ser/Thr phosphorylation mostly targets adaptors and actin regulators, as well as 2 Tyr kinases and Tyr phosphatases.

[138] Integrin-linked kinase binds to: (1) PINCH and ILK-associated phosphatase on its ankyrin domain; (2) possibly phosphatidylinositol (3,4,5)-trisphosphate on its

Table 7.32. Components of the integrin adhesome (**Part 1**; Source: [724]).

Component type	Genes
Adhesion molecules	CEACAM1, ENG, ITGA5, ITGB1, SDC4, THY1
Receptor	CD47, INSR, KTN1, LAYN, LRP1, MAPK8IP3, PLAUR, SIRPA
Messenger	Calcium, DAG, IP$_3$
Adaptor	AbI1, CRKL, DMN, FHL2, Gab1, GNB2L1, Grb2/7, IRS1, LIMS1, LPP, LPXN, NCK2, NUDT16L1, PPFIA1, PXN, SH3KBP1, Shc1, SORBS1/2/3, TES, TGFB1I1, TRIP6, TSPAN1
Cytoskeleton elements	ACTB, BCAR1, CRK, CSRP1, FBLIM1, JUB, Kα1, NEDD9, NF2, PALLD, PARVA/B, PLEC1, PlekHc1, SDCBP, TLN1, TNS1, VCL, VIL2, VIM, ZYX
Actin-bundling proteins	ACTN1, FLNA, KEAP1, MARCKS
Actin-binding proteins	ARPC2, CFL1, CTTN, ENAH, LASP1, NEXN, PFN1, VASP
Vesicular components	CAV1
Lipid	PIP$_2$, PIP$_3$
Ion channels	KCNH2, PKD1, TRPM7
Exchangers	SLC9A1
Chaperones	CALR, HSPA2, HSPB1
Transcription factors	ITGB3BP, STAT3
RNA metabolism	PABPC1, RAVER1

paxillin, vinculin, and focal adhesion kinases and builds an interface between integrins on the one hand and the actin cytoskeleton and signaling pathways on the other. The IPP complex participates in the regulation of the migration

---

plekstrin homology domain; (3) parvin, paxillin, kindlin-2, β_1 and β_3 integrins, protein kinase-B, phosphatidylinositol 3-kinase-dependent kinase-1 on its kinase domain. Kindlin-2 binds to migfilin, which links to filamin and interacts with filamentous actin and integrins. Adaptor PINCH1 binds to receptor Tyr kinases as well as Ras-suppressor protein RSu1, a negative regulator of Jun N-terminal kinase, and to receptor Tyr-kinase adaptor NCK2. α-Parvin binds to ^{F}actin, and kinase TESK1, which phosphorylates cofilin. β-Parvin binds to ^{F}actin, α-actinin, and guanine nucleotide-exchange factor RhoGEF6. Agent RhoGEF6 binds to the Rac1/CDC42 effector PAK1 and calpain-4, which cleaves talin. When talin is recruited to the plasma membrane and activated in association with phosphatidylinositol phosphate kinase-1γ (PIPK1γ), it binds to β integrins.

Table 7.33. Components of the integrin adhesome (**Part 2**; Source: [724]).

Component type	Molecules
Phospholipase	PLD1
Tyr kinase	ABL1, Fyn, Lyn, FAK1/2, Src, Syk
Ser/Thr kinase	ILK, LIMK1, MAPK1/8, PDPK1, PKB1, PAK1, PDPK1, PRKACA, PRKCA, RhoK1
Dual kinase	TESK1
PI kinase	PIK3CA, PIP5K1C, PLCG1, PPP2CA
Tyr phosphatase	CSK, PTPn1/2/6/11/12, PTPRa/f/h/o
Ser/Thr phosphatase	ILKAP, PPM1m
PI phosphatase	INPP5D, INPPL1, PTen
GTPase	Dynamin-2, ARF1, hRas, Rac1, RhoA
GAP	RhoGAP5/24/26, DDEF1/2, GIT1, RASA1, RICS
GEF	RhoGEF6, BCAR3, DOCK1, ElMo1, RAPGEF1, SOS1, TIAM1, Trio, Vav1
Ubiquitin complex	CBL
Protease	CapN1
Metalloproteinase	ADAM12

of endothelial cells and cardiomyocytes, leukocyte recruitment, and platelet aggregation.

Parvins

The parvin family of actin-binding, integrin-linked kinase-binding, focal-adhesion resident adaptors include: α-,[139] β-,[140] and γ-parvin that are encoded by the PARVA, PARVB, and PARVG genes, respectively. α-Parvin is nearly ubiquitously expressed, β-parvin is preferentially expressed in the myocardium and skeletal muscles, and γ-parvin in lymphoid tissues [728].

α-Parvin, but not β-parvin, impedes Rac activation and lamellipodium formation [729]. β-Parvin binds to and inhibits integrin-linked kinase [730]. It also contributes to the linkage between integrins and the cytoskeleton in combination with ILK [731]. β-Parvin interacts with α-actinin [731]. It mediates

---

[139] A.k.a. actopaxin, matrix remodeling-associated protein-2, and calponin homology [calponin-like] integrin-linked kinase-binding protein (CHILKBP).

[140] A.k.a. affixin.

integrin signaling for reorganization of ^{F}actin induced by the initial cell–matrix interaction. It activates small GTPases CDC42 and Rac via CDC42/Rac1-specific guanine nucleotide-exchanging factor RhoGEF6 at tips of lamellipodia in motile cells [732]. β-Parvin tethers to dysferlin at the sarcolemma of skeletal myocytes [733].[141] β-Parvin is inhibited by α-parvin [729]. β-Parvin promotes apoptosis, whereas the ILK–α-parvin complex protects cells from apoptosis [729].

Talin

Talin that binds the cytoplasmic tail of β integrins mediates integrin activation to link the extracellular matrix, actin cytoskeleton, and signaling networks.[142] Integrin activation involves a conformational change in integrin dimer to raise its affinity to the matrix. In mammals, 2 redundant talin isoforms exist (talin-1 and -2). Both are needed to form focal adhesions and stable cell spreading.

Kindlins

Kindlins are cytoplasmic components of cell–matrix adhesions that bind to cytoplasmic tails of $β_1$ and $β_3$ integrins and cooperate with talin in integrin activation. The kindlin family include 3 known members that mediate protein interactions (kindlin-1–kindlin-3).[143] Integrin-binding kindlin-2 and -3 are required for maximal integrin activation. Kindlins interact with other focal-adhesion proteins, such as migfilin and integrin-linked kinase, to promote cytoskeletal reorganization.

Kindlin-1 and -2 are widely expressed in tissues, but kindlin-3 expression is restricted to hematopoietic cells. Kindlin-3 is particularly involved in integrin activation in leukocytes and platelets. Kindlin-1 is phosphorylated at least

---

[141] Dysferlin is a sarcolemmal protein that is implicated in calcium-dependent membrane repair.

[142] Talin binds to ^{F}actin with lower affinity at higher pH. Talin contains several actin-binding sites, but a single domain shows pH-dependency [734]. In addition to talin, other proteins, such as cofilin, huntingtin-interacting protein HIP1, and villin have pH-dependent actin binding.

[143] Because kindlins contain a FERM (band-4.1, ezrin, radixin and moesin) domain, they are named fermitin family homologs. Kindlin-1 (Kind1) is also called kindlerin, fermitin family homolog Fermt1 and Unc112-related protein URP1. Membrane-associated, intracellular kindlin-2 (Kind2) that colocalizes with integrin at cell–matrix adhesion complexes is an actin-extracellular-matrix linker. It is also designated as mitogen-inducible gene product MIG2, pleckstrin homology domain-containing family C member PlekHc1, Fermt2, and Uncoordinated Unc112. Kindlin -3 (Kind3) is also named MIG2-like protein, Fermt3, and URP2.

partly by casein kinase-2 that is involved in cytoskeleton regulation [735]. Integrin coactivator kindlin-2 binds to migfilin[144] and recruits it to cell–matrix adhesions. However, migfilin is not essential for kindlin-2 localization to adhesion sites. Migfilin presence in cellular adhesion complexes facilitates the accumulation of migfilin-binding proteins, such as vasodilator-stimulated phosphoprotein (VASP), which is involved in filamentous actin formation, and filamin. On the other hand, the recruitment of kindlin-2 and integrin-linked kinase in cell–matrix adhesions depends on both molecules. Kindlin-2 binds to β_3 integrins for synergistic enhancement of talin-induced integrin activation that is needed for cell spreading and cell–matrix adhesion [736, 737].

Activated CDC42-Associated Kinase

Cytosolic protein Tyr kinase Activated CDC42-associated kinase (ACK)[145] is ubiquitously expressed. Kinase ACK is phosphorylated (activated) by activated integrins linked to cell adhesion on fibronectin by growth factors, such as epidermal and platelet-derived growth factors. Kinase ACK is an early transducer of extracellular stimuli. It interacts with adaptors and CDC42 guanosine triphosphatase. It binds to active $CDC42^{GTP}$ and inhibits both its intrinsic and GTPase-activating protein (GAP)-stimulated GTPase activity. It can also play a role in vesicle dynamics.

7.5.4.5 Leukocyte Integrins

Integrins are used during the migration and functioning of leukocytes toward and in inflammatory sites. β_2 Integrins are restricted to leukocytes. They intervene in leukocyte adhesion. They include 4 heterodimeric transmembrane glycoproteins that share the same β_2 subunit paired to one among 4 distinct α subunits [738, 739]: (1) $\alpha_L\beta_2$;[146] (2) $\alpha_M\beta_2$;[147] (3) $\alpha_X\beta_2$;[148] and (4) $\alpha_D\beta_2$.[149]

---

[144] A.k.a. filamin-binding LIM protein FBLIM1 and mitogen-inducible gene product MIG2-interacting protein.

[145] A.k.a. Tyr non-receptor kinase TNK2 (Vol. 4 – Chap. 3. Cytosolic Protein Tyrosine Kinases).

[146] A.k.a. the CD11a–CD18 dimer and lymphocyte function-associated antigen LFA1. Integrins $\alpha_L\beta_2$ exist on early progenitors of all myeloid and erythroid cells.

[147] A.k.a. the CD11b–CD18 dimer, complement receptor CR3, and macrophage-1 antigen Mac1. Integrin $\alpha_M\beta_2$ is highly expressed on antigen-presenting cells, such as monocytes, macrophages, and dendritic cells, as well as cells of the innate immune response, such as granulocytes and natural killer cells.

[148] A.k.a. the CD11c–CD18 dimer and glycoproteic heterodimer GP150–GP195. Like $\alpha_M\beta_2$ integrin, $\alpha_X\beta_2$ integrin is normally expressed only on monocytes, macrophages, and granulocytes.

[149] A.k.a. the CD11d–CD18 dimer.

Toll-like receptors (Vol. 3 – Chap. 11. Receptors of the Immune System) activated by microbial components prime the innate immune response with the production of type-1 interferons and pro-inflammatory cytokines. Because excessive cytokine production damages host tissues and can even provoke a septic shock, certain mechanisms hamper TLR signaling, in particular via spleen Tyr kinase (SYK). Stimulation of various Toll-like receptors in macrophages causes the activation of α_X integrin via phosphatidylinositol 3-kinase and Ras association domain-containing family member RASSF5,[150] an effector of the integrin-associated GTPase Rap [740]. In turn, α_X integrin triggers via immunoreceptor Tyr-based activation motif (ITAM)-containing protein the Src–SYK pathway. Kinase SYK phosphorylates TLR adaptors MyD88 and TRIF that are then ubiquitinated by the ubiquitin ligase CBL for degradation. Therefore, α_X integrin ignites a negative feedback, as it suppresses MyD88- and TRIF-dependent inflammatory responses.

7.5.4.6 Integrins and Mechanotransduction

Mechanical forces activate the adhesion receptor $\alpha_5\beta_1$ integrin. The latter and $\alpha_V\beta_3$ integrins connect the cell cytoskeleton to fibronectin in the extracellular matrix. Integrin $\alpha_5\beta_1$ binds to fibronectin via specific crosslinkers that determines, in the presence and absence of mechanical tension, 2 types of binding states, relaxed and stretched [741]. $\alpha_5\beta_1$ Integrins indeed switch from a relaxed to a tensioned state in response to myosin-2-generated cytoskeletal force. Fibronectin-binding regions (RGD domain) are committed in relaxed form, whereas both RGD and synergy sites are engaged upon tension for a stronger bond.

The proportion of $\alpha_5\beta_1$ integrins in stretched bound state depends on the extracellular matrix stiffness rather than the amount of $\alpha_5\beta_1$ integrins bound to fibronectin. Conversion of bound state of $\alpha_5\beta_1$ integrin from relaxed to stretched state generates downstream signaling, as quantity of phosphorylated focal adhesion kinase correlates with stretched state. Hence, $\alpha_5\beta_1$ integrin has mechanosensing and mechanoresponsive properties.

Fibronectin clusters recruit $\alpha_5\beta_1$ integrin to increase the cell's ability to withstand high applied mechanical loads and maintain adhesion strength [742]. Moreover, mechanical forces favor interaction between force-bearing cytoskeletal protein talin and its partner vinculin [743]. Talin that links $\alpha_5\beta_1$ integrin to the cytoskeleton recruits additional focal adhesion protein vinculin upon stretch, as talin contains multiple vinculin-binding sites, thereby allowing cytoskeleton reorganization. Cells are indeed able to adjust their cytoskeletal organization to minute changes in their immediate surroundings, especially via integrin-based adhesomes that recognize not only the chemical nature of their extracellular neighborhood, but also its structure (ligand spacing) and

---

[150] A.k.a. regulator for cell adhesion and polarization enriched in lymphoid tissues (RAPL).

Table 7.34. Vascular cell adhesion molecules of the immunoglobulin superclass (IgCAM; Sources: [674, 701, 702]; ALCAM: activated leukocyte cell adhesion molecule; Bsg: basigin, or extracellular matrix metalloproteinase inducer [EMMPrIn]; ICAM: intercellular adhesion molecule; PECAM: platelet endothelial cell adhesion molecule; siglec: sialic acid-binding Ig-like lectin; VCAM: vascular cell adhesion molecule). CD44 involved in homing and Indian blood group system is a receptor for hyaluronic acid as well as osteopontin, collagens, and matrix metalloproteinases. Its sialyfucosylated glycoforms such as HCELL on human hematopoietic stem cells bind P-, L-, and E-selectins.

Molecule	Ligands	Distribution
ALCAM (CD166)	CD6, CD166	WBC
Bsg (CD147)		WBC, RBC, TC, EC
Siglec-2 (CD22)	CD45	B Lφ
CD44	Ankyrin, FN, etc.	Lφ
ICAM1 (CD54)	$\alpha_4\beta_1$, $\alpha_L\beta_2$, $\alpha_M\beta_2$	WBC, EC
ICAM2 (CD102)	$\alpha_L\beta_2$	Lφ, EC, Mo
ICAM3 (CD50)	$\alpha_L\beta_2$	WBC
PECAM (CD31)	CD31, $\alpha_V\beta_3$	WBC, EC
VCAM1 (CD106)	$\alpha_4\beta_1$, $\alpha_4\beta_7$	Mo, EC

physical characteristics. Stretching of cytoskeletal and extracellular molecules can expose binding sites for adhesion complex components. Mechanotransduction then happens, at least partly, by protein binding after exposure of buried binding sites in the talin–vinculin complex.

On the other hand, less stable links via $\alpha_V\beta_3$ integrins enable mechanotransduction and cause reinforcement of integrin–cytoskeleton linkages via talins that provoke formation of $\alpha_V\beta_3$ integrin cluster and tether these integrin clusters to the cytoskeleton. Fast rate of binding and unbinding of $\alpha_V\beta_3$ integrins may allow these molecules to continuously monitor force and react. Force sensors may recruit additional integrins and other cytoskeletal proteins. Downstream signaling involves SRC family kinases via activated phosphatases, such as non-receptor protein Tyr phosphatase PTPn1b and PTPRa receptor Tyr phosphatases.

7.5.5 Immunoglobulin-Like Cell Adhesion Molecules

Certain members of the immunoglobulin (Ig) superclass — the Ig cell adhesion molecules (IgCAM) — are involved in calcium-independent intercellular binding (Table 7.34).

Intercellular adhesion molecules (ICAM) are expressed on activated endothelial cells, being the ligand for integrins expressed by leukocytes. ICAM1 (or CD54) causes reversible adhesion, stabilizing leukocytes for possible extravasation.

Platelet–endothelial cell adhesion molecule PECAM1 (or CD31) belongs to leukocytes, platelets, and intercellular junctions of endothelial cells.

Vascular cell adhesion molecule VCAM1, once bound to $\alpha_4\beta_1$ integrin, induces firm adhesion of leukocytes on endothelium.

7.5.5.1 Junctional Adhesion Molecules

Junctional adhesion molecules (JAM) belong to the set of Ca^{++}-independent immunoglobulin-like cell adhesion molecules of the immunoglobulin superclass. Junctional adhesion molecules are expressed by endothelial cells, leukocytes, and platelets. Junctional adhesion molecules support both homo- and heterophilic interactions.[151]

Junctional adhesion molecules include JAM1 to JAM4 and JAM-like protein [744].[152] Protein JAM1 localizes to the apical region of tight junctions, where it links cingulin, occludin, protein zonula occludens-1, etc. Molecule JAM1 lodges on endothelial, circulating, and antigen-presenting cells. It can interact with $\alpha_L\beta_2$ integrins. It is highly expressed in the brain vasculature, where it stabilizes cell junctions, thereby reducing the vascular permeability.

Molecules JAM2 and JAM3 are expressed by endothelial cells. Molecule JAM2 can interact with $\alpha_4\beta_1$ integrins. Protein JAM3 is also expressed by platelets, monocytes, natural killer and dendritic cells, B lymphocytes, and a subset of T lymphocytes. It is strongly expressed in lymph nodes where it destabilizes cadherin junctions, thus increasing the vascular permeability. It is also found in high endothelial venules. It binds to leukocyte $\alpha_M\beta_2$ and $\alpha_X\beta_2$ integrins to link thrombocytes to leukocytes.

7.5.5.2 Nectins and Nectin-like Molecules

Nectins (nectin-1–nectin-4) and nectin-like molecules (NecL1–NecL5) are ubiquitous immunoglobulin-like cell adhesion molecules with homo- and heterophilic links between family members for intercellular communication. The family of nectin and nectin-like molecules comprises several identified members with splice variants (Table 7.35). Nectin or nectin-like proteins in a given cell first form dimers that secondarily associate with dimers of apposing cells.[153] They are involved not only in intercellular adhesions, but also in cell differentiation, polarization, proliferation, survival, and motion.

---

[151] Homophilic interaction of JAM1 is used for platelet adhesion to the endothelium [744]. Molecule JAM3 has a higher affinity for heterodimerization with JAM2 than for homodimerization. Interaction of JAM3 also occurs with integrins during platelet adhesion to leukocytes and extravasation of leukocytes.

[152] Junctional adhesion molecules JAM1 to JAM3 are also named JAMa to JAMc.

[153] Heterophilic interactions occur between nectin-1 and -3, nectin-2 and -3, and nectin-1 and -4. Heterophilic nectin interaction between adjoining cells is stronger than homophilic interactions. Extracellular regions of NecL4 and NecL5 only form heterophilic interactions between neighboring cells, whereas other NecLs generate both homophilic and heterophilic interactions.

Table 7.35. Nectins and nectin-like (NecL) molecules (Source: [745]).

Molecule	Effect
Nectin-1	Intercellular adhesion
Nectin-2	Intercellular adhesion
Nectin-3	Intercellular adhesion
Nectin-4	Intercellular adhesion
NecL1	Intercellular adhesion
	(specific to nervous tissue)
	Axoglial interactions, Schwann cell differentiation, and myelination
NecL2	Intercellular adhesion
	(basolateral membrane of epithelial cells)
	Tumor suppression
	Regulation of immune defense
NecL3	
NecL4	Intercellular adhesion
	(neural tissue)
	Axoglial interactions, Schwann cell differentiation, and myelination
NecL5	Cell proliferation and motion

Nectins and nectin-like proteins form heterophilic interactions between adjoining cells (especially cytotoxic T and natural killer cells) with other Ig-CAMs, such as CD96 (or Tactile) and CD226,[154] as well as class-1 MHC-restricted T-cell-associated molecule.

Nectins initiate the formation of adherens junctions (zonula adherens) before recruiting cadherins that stabilize these cell junctions. The formation of adherens junctions induces a reorganization of the actin cytoskeleton.

Nectins interact with ^{F}actin-binding protein afadin.[155] Afadin then links nectins to the actin cytoskeleton.[156] Nectin-1 and -3 also bind cell polarity protein partitioning defective Par3 [745]. The nectin–afadin complex is required in the formation of cadherin-based adherens junctions[157] and claudin-based tight junctions.

Integrins of cell–matrix adhesions operates in cell proliferation and migration. $\alpha_V\beta_3$ Integrin that is activated and connected to the actin cytoskeleton

---

[154] A.k.a. T-lineage-specific activation antigen TLiSA1 and DNAX accessory molecule DNAM1.

[155] Nectins bind afadin, but not nectin-like molecules.

[156] Afadin also recruits α-catenin that is either free or associated with cadherins via β-catenin.

[157] Homophilic linkages among cadherins can stabilize nectin-based intercellular contacts to form adherens junctions. Afadin binds ^{F}actin-binding protein ponsin that connects the nectin–afadin and cadherin–catenin complexes. The recruitment of cadherin–catenin complex to nectin-based adhesion sites is favored by nectin-induced activation of Rac and CDC42 by several filamentous actin-binding proteins.

by talin operates in nectin-induced formation of adherens junctions.[158] $\alpha_V\beta_3$
Integrin is afterward inactivated to maintain adherens junctions by the nectin
stimulation of phosphatase PTPRm that dephosphorylates phosphatidylinositol phosphate kinases-1γ.[159]

Nectin-3 associates with platelet-derived growth factor receptor at intercellular adhesion sites and favors cell survival. The phosphoinositide 3-kinase
signaling pathway downstream from the PDGF receptor could be involved in
the nectin-3- and afadin-mediated suppression of apoptosis [745].

Nectin-like proteins NecL1 and NecL2 cooperate with scaffold proteins
membrane-associated guanylyl kinase MAGUK and Band4.1. Protein NecL5
links to dynein light-chain component TCTEX1 [745].[160] It also forms a
complex with $\alpha_V\beta_3$ integrin and platelet-derived growth factor receptor to
promote cell proliferation and migration.[161] However, nectin-3 interacts with
NecL5 to reduce its level at adhesion sites and prevent cell proliferation and
migration. Protein NecL5 favors cell proliferation by enhancing growth factor
pathways.[162]

7.5.6 Claudins

Claudins (Cldn1–Cldn24) are important components of the tight junction
that form a branching and anastomosing network of strands of plasma membranes (Sect. 7.7.3). They establish a paracellular barrier and regulate the
transfer of molecules, especially small substances such as ions, in the intercellular space between endo- or epithelial cells. They build tight junctions with

---

[158] Nectin-1 and -3, but not nectin-2, link to $\alpha_V\beta_3$ integrin that, once activated,
primes nectin-induced stimulation of Src kinase.

[159] Phosphatidylinositol phosphate kinases-1γ produces phosphatidylinositol (4,5)-
bisphosphate that enhances binding of talin to $\alpha_V\beta_3$ integrin.

[160] Protein NecL5 could then regulate microtubule reorientation during cell motion.

[161] Protein NecL5 is located at the leading edge of moving cells. On the other hand,
nectins favor intercellular adhesion. Afadin at the leading edge of migrating cell
regulates PDGF signaling during the formation of the leading edge. Both Src and
Rap1 are activated at the leading edge by the NecL5–PDGFR–$\alpha_V\beta_3$ integrin
complex. Activated Rap1 activates (phosphorylates) Cdc42GEF FRG, thereby
CDC42. Lamellipodia are formed due to Rac, filopodia owing to CDC42, and
focal complexes by the actions of Rac and CDC42. Focal complexes are transformed into focal adhesions by inactivation of CDC42 and Rac and activation
of Rho. Activated Rap1 leads to activation of Rac, inactivation of RhoA, and
recruitment of afadin to the leading edge.

[162] Activation by growth factors of cell proliferation uses effector module Ras–Raf–
ERK (Vol. 4 – Chap. 5. Mitogen-Activated Protein Kinase Modules). Protein
NecL5 interacts with sprouty-2, an inhibitor of cell proliferation signaling, to
prevent sprouty-2 phosphorylation by Src in response to growth factors, so that
NecL5 prolongs cell proliferation signaling.

2 other types of transmembrane proteins, occludins and junctional adhesion molecules.

Claudins constitute a family of the tetraspanin superfamily. These transmembrane proteins possess 4 transmembrane domains with 2 extracellular loops and cytoplasmic short N- and relatively longer C-termini. Assembly of claudins into tight junctions necessitates zonula occludens scaffold proteins ZO1 or ZO2. These scaffolds interact with both claudins and tight-junction peripheral proteins, such as cingulin and partitioning defective proteins Par3 and Par6.

Heteromers formed by claudin-16 and -19 are required not only for assembling of tight junctions, but also their proper functioning, i.e., the generation of cation-selective paracellular channels [746].[163] In tight junctions of the epithelium of the thick ascending limb (Vol. 2 – Chap. 1. Remote Control Cells), depletion in claudin-16 and -19 does not perturb the epithelial barrier function, but cripple ion selectivity.

7.6 Focal Adhesions

Focal adhesions (2–10 µm long, 250–500 nm wide, 10–15 nm gap; 40–50 per cell) are cell–matrix junctions that are also defined as focal adhesion plaques or focal contacts. However, they differ from hemidesmosomes.

Focal adhesions are complexes of clustered integrins and associated proteins that link components of the extracellular matrix, such as fibronectin, collagen, laminin, and vitronectin, to the cell cytoskeleton and mediate cell adhesion [747]. Actin filaments, rather than intermediate filaments, attach to specific membrane proteins.

Integrin-based focal adhesions serve as interfaces for mechanical force transduction between cells and their environment. In addition to focal adhesions, some plasmalemmal receptors, the majority of which are not connected to focal adhesions, participate in mechanotransduction.

7.6.1 Focal Adhesion Components

The components of focal adhesions can be grouped into several functional sets of focal adhesion proteins (FAP) [748]: (1) transmembrane integrins and syndecans that possess heparan- and chondroitin sulfate chains;[164] (2) cytoskeletal proteins, with direct or indirect association with actin, without

---

[163] In humans, impaired claudin-16 and -19 cause the inherited renal disorder familial hypomagnesemia with hypercalciuria and nephrocalcinosis caused by mutations in the CLDN16 and CLDN19 genes, i.e., a chronic renal wasting of calcium and magnesium.

[164] Syndecans constitute a family with 4 members regrouped into 2 subfamilies with syndecan-1 and -3 on the one hand and syndecan-2 and -4 on the other. Heparan- and chondroitin sulfate chains interact with numerous ligands, such as fibroblast

enzymatic activity, such as α-actinin, α-parvin, paxillin, talin, tensin, and vinculin; (3) protein Tyr kinases, such as members of the SRC family, focal adhesion kinases FAK1 and FAK2, C-terminal Src kinase (CSK), and Abelson leukemia viral proto-oncogene product (Abl kinase); (4) protein Ser/Thr kinases, such as integrin-linked kinase (ILK), P21-activated kinase (PAK), and protein kinases-C (PKC); (5) modulators of small guanosine triphosphatases, such as GTPase-activating protein ASAP1 (ArfGAP with SH3, ankyrin repeat, and PH domains) and RhoGAP10 and -26; (6) protein Tyr phosphatases, such as receptor PTPRf and cytosolic PTPn11 phosphatases; and (7) other enzymes, such as phosphatidylinositol 3-kinase (PI3K) and calpain-2. However, all these components are not necessarily constitutively located in all focal adhesions [747].

Talin and vinculin are cytoskeletal proteins that bind integrins and actin cytoskeleton [749]. *Talin* possesses actin-binding and vinculin-binding sites. It also binds to β integrin cytoplasmic tails [750]. *Vinculin* binds ^{F}actin and may crosslink talin and actin, thereby stabilizing the interaction [751]. Talin, vinculin, and paxillin regulate the formation of focal adhesions and stress fibers.

Integrins and syndecans are 2 adhesion molecule types that synergistically signal to control formation of focal adhesions [752]. Cytoplasmic domain of syndecan-4 binds and activates protein kinase-Cα to regulate integrin activation in addition to signals downstream from both adhesion molecule types. Syndecan-4 regulates both Rac1 and RhoA GTPases via PKCα kinase.

Small adhesion plaques (30–40 nm) contain integrins and some associated proteins. Focal complexes are nascent, transient, integrin-containing adhesion spots (size ~100 nm) that are composed of multiple molecules.[165] Focal complexes experience mechanical forces during cell migration. The formation of mature focal adhesions requires myosin-2A activity [753], whereas myosin-2B is needed to form stable actin filament bundles and adhesions.

Components of focal adhesions, such as Breast cancer anti-estrogen resistance docking protein BCAR1[166] and fibronectin, can undergo force-mediated conformational changes and become active. A force-induced conformational change in scaffold BCAR1 enables its phosphorylation by Src kinase.

Mechanosensors that are associated with focal adhesions include mechanosensitive Ca^{++} channels. Forces developed by stress fibers can thus induce a local Ca^{++} influx near focal adhesions [753].

(FGF), transforming (TGFβ), and vascular endothelial (VEGF) growth factors, fibronectin, and antithrombin-1. Interactions between some syndecans and fibronectin can be modulated by extracellular tenascin-C.

[165] Zyxin is a distinctive marker, as it localizes to focal adhesions, but not focal complexes.

[166] A.k.a. CAS and P130CAS.

7.6.2 Focal Adhesion Functions

Focal adhesions regulate actin assembly and conversely cytoskeletal forces control assembly and maturation of cell adhesions (bidirectional interactions). Focal adhesions participate in actin nucleation, as actin subunits are predominantly incorporated at the membrane-associated end of actin filaments [753]. Integrin adhesions determine the spatial organization of the actin cytoskeleton. Conversely, cytoskeleton-generated forces affect the initiation, growth, and maturation of focal adhesions.

Cells are actually able to adjust their cytoskeletal organization, and hence their shape and motility, to minute changes in their immediate surroundings. Stress fibers connected to focal adhesions grow by incorporation of mainly new components at the focal adhesion–stress fiber interface. The Diaphanous family of formins operate in the actin-nucleating function of focal adhesions and stress fiber formation.

Mechanosensitive integrin-based adhesion complexes that form focal adhesions are tightly tethered to the actin cytoskeleton. They contain the cellular machinery that recognizes the chemical nature of the extracellular medium. Combination of integrins can be controlled by the presence of distinct extracellular molecules, such as fibronectin and vitronectin, to yield adaptability. The focal adhesion machinery can also detect the physical features of the extracellular matrix, e.g., its rigidity and anisotropy.

Protein complexes that generate integrin-mediated adhesions are called *adhesomes.* Adhesomes not only link the extracellular matrix to the actin cytoskeleton, but also recruit signaling molecules to trigger adhesion-dependent cell process. Chemo- and mechanosensitive adhesome can contain up to 160 intrinsic constituents and transiently associated components. Adhesome components include several actin regulators, adaptor proteins that directly or indirectly link actin to integrins, and signaling molecules (kinases, phosphatases, and G proteins and their regulators) [753]. Integrin receptors are operational elements, as they serves as cell adhesion-dependent sensors and actuators that provide an output.

Dynamical connections between the adhesome and actin cytoskeleton that regulate each other is composed of: (1) receptor module (integrin and its coreceptors such as syndecan) that associates the cell with its extracellular matrix; (2) actin-linking module (e.g., talins, tensin, integrin-linked kinase, and vinculin);[167] (3) actin-polymerizing module (e.g., zyxin and formins); and (4) signaling module (e.g., focal adhesion kinase and scaffold BCAR1) [753].

A feedback between adhesome and actin cytoskeleton allows efficient sensing and suitable reaction. Focal adhesions are sensitive to regulators of both actin polymerization and myosin-2-mediated contractility. Stimulated signaling module of the chemo- and mechanosensitive adhesome hastens guanine

---

[167] Vinculin binding to talin triggers clustering of activated integrins and their linkage to actin.

Table 7.36. Cellular junction features.

Junction	Function	Attachment	Intermembrane space (nm)	Associated molecules
Desmosome	Cell–cell anchor	Cadherin	25–35	Intermediate filaments
Hemidesmosome	Cell–BM adhesion	Integrin	25–35	Intermediate filaments
Adhesion belt	Cell–cell adhesion	Cadherin Integrin	10–25	Actin filaments
Focal contact	Cell–ECM adhesion	Cadherin Integrin	20–25	Actin filaments
Tight junction	Occlusion of between-cell space	Membrane-joining strands	<1	Membrane proteins
Gap junction	Cell–cell communication	Connecting channel	2–3	Connexin

nucleotide-exchange factors and GTPase-activating proteins to activate or inactivate small GTPases. These small GTPases, especially Rho and Rac, are major regulators of actin cytoskeleton functioning. Thereupon, small GTPases affect actin polymerization and actomyosin contractility via cytoskeleton-regulating proteins to modulate force generation by the cytoskeleton. Conversely, the cytoskeleton signals to adhesome via actin-linking module.

Focal adhesion kinase in cooperation with Src kinase can bind, phosphorylate, and activate both guanine nucleotide-exchange factors (RhoGEF) and GTPase-activating proteins (RhoGAP). Focal adhesion kinase also cooperates with several other kinases and phosphatases of focal adhesions, such as Fyn kinase, receptor-type protein Tyr phosphatase PTPRa,[168] and phosphatase PTPn11, in the regulation of the mechanosensory activity of focal adhesions. Integrin-linked kinase with particularly interesting new Cys–His protein (PINCH) and parvin form a ternary complex that also participates in integrin tethering to the actin cytoskeleton and regulation of actin dynamics.

7.7 Cellular Junctions

Cellular junctions are tiny specialized regions of the plasma membrane. Several histological and functional categories include: (1) impermeable junctions that maintain an internal area chemically distinct from surroundings; (2) adhering junctions that reinforce tissue integrity; and (3) communicating junctions for exchange of nutrients and messengers with the environment (Fig. 7.7, Table 7.36).

---

[168] A.k.a. RPTPα.

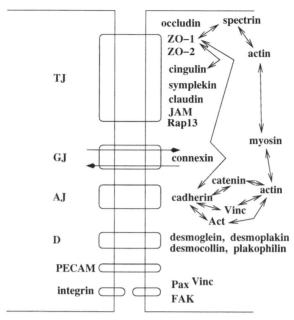

Figure 7.7. Endothelial cell junctions (Source: [754]). Tight junction (TJ) is formed by occludin and peripheral membrane proteins, such as cingulin and small GTPase Rap13. Zonula occludens protein ZO1 binds to spectrin and cingulin. It crosslinks catenin–cadherin complexes. Cadherins (ve-, p-, and nCdhs) are found in adherens junction (AJ), as well as clusters of VE-cadherins (predominant cadherin) that bind to actin. Cadherins bind to the cytoskeleton via catenins such as γ-catenin (plako-globin). β-Catenins bind to α-actinin (Act), vinculin (Vinc), and actin as well as VE-cadherin with γ-catenin. Tight junction protein occludin and adherens junction proteins β-catenin and VE-cadherin colocalize in endothelial cells. Platelet endothe-lial cell adhesion molecule (PECAM1) maintains junction integrity. Connexins form gap junctions (GJ). Integrins bridge the cells. Desmosomes (D) are made of desmo-plakins. Vinculin is located along junctions.

Within the junctions, membrane proteins have specific configurations. In-tegrins and cadherins are associated with signaling molecules, leading either to cell proliferation or apoptosis (Sect. 7.5). Cell junctions can remodel.

7.7.1 Desmosomes and Hemidesmosomes

Desmosomes (δεσμωμα: bond; σωμα: body), or *macula adherens*, are large bundles that anchor cells to its neighboring cells. This type of junctional com-plex localizes to the lateral sides of plasma membranes as randomly arranged adhering spots. Desmosomes contribute to the resistance against mechanical forces.

7.7.1.1 Types of Desmosomes

The intercellular space is filled with filaments that bridge not only membranes, but also cytoskeletons of adjacent cells. Two main desmosome types exist: belt and spot desmosomes. *Belt desmosomes* contain actin filament susceptible to contraction in presence of ATP, Ca^{++}, Mg^{++}, to close the gap during cell apoptosis. In fact, the belt desmosome corresponds to the adherens junction (Sect. 7.7.2). *Spot desmosomes* contain filaments and transmembrane linkers that connect cytoplasmic networks of tonofilament bundles for mechanical coupling between apposed cells.

7.7.1.2 Hemidesmosomes

Hemidesmosomes are small proteic complexes in the basal layer of epithelia and basal cell surface of endothelial cells that form a monolayer. Hemidesmosomes resemble half of a desmosome. Hemidesmosomes connect the elements of the cellular cytoskeleton to the underlying basement membrane. They do not link adjacent cells. Hemidesmosomes also serve as anchoring sites for tonofilament bundles.

Hemidesmosomes, as adhesion loci of cells to basement membranes, allow tissue mechanical integration via links between collagen-7 and integrins. They act in signal transduction via integrin and regulate cellular activities. Cells subjected to mechanical stresses have numerous spot desmosomes and hemidesmosomes, which limit cell distensibility and distribute stresses among layer cells and to the underlying tissues to avoid disruption.

7.7.1.3 Constituents and Partners

Desmosomes contain 2 families of desmosomal cadherins: *desmocollins* and *desmogleins*. Both desmosomal cadherin families possess several molecular types that are specific to differentiation status and cell type. These transmembrane proteins close the space between apposed epithelial cells (homophilic binding).

Plakophilin (Pkp) localizes to desmosomes. Plakophilins (Pkp1–Pkp4) pertain to the *Armadillo family*, or catenin family, with catenin-β, -γ, and -δ1 and -2, as well as catenin-δ1-related proteins, Armadillo repeat gene deleted in velocardiofacial syndrome protein (ARVCF), adenomatous polyposis coli protein, and Rap1 guanine nucleotide-dissociation stimulator-1 (Rap1GDS1; Table 7.37).[169] Armadillo family proteins couple cadherins to cytoskeletal filaments and regulate junction assembly. They intervene in interactions between the cytoskeleton and plasma membrane. In addition, they can be involved in signal transduction.

---

[169] A.k.a. stimulatory GDP–GTP exchange protein and Rap1–GDP dissociation stimulator-1.

Table 7.37. Cell adhesion sites and the cytoskeleton (Source: [755, 756]; APC: adenomatous polyposis coli protein; ARVCF: armadillo repeat gene deleted in velo-cardiofacial syndrome). Both adherens junctions and desmosomes couple the extra-cellular matrix to the cytoskeleton via transmembrane cadherins. Desmosomes and adherens junctions anchor intermediate and actin filaments to the plasma membrane, respectively. Catenin-γ (Ctnnγ) is also called plakoglobin; catenin-δ1 P120-catenin (P120Ctn), cadherin-associated protein, and P120 cadherin-associated Src substrate (P120CAS); catenin-δ2 neural plakophilin-related Armadillo repeat protein, δ-catenin, neurojungin (NPRAP). Proteins of the Armadillo family couple cadherins to cytoskeletal filaments and regulate the assembling of cell adhesion sites.

Adhesion contituent	Localization and interactions
Desmosome	
Desmosomal cadherins	
Desmoglein	Intermediate filaments
Desmocollin	Intermediate filaments
Desmoplakin (cytoskeletal linker)	Intermediate filaments (vimentin, keratin-1, desmin) Plakophilin-1/2, catenin-γ, cadherin Desmoplakin (homodimer)
Armadillo family – Catenin family	
APC	Catenin-α/β
ARCVF	Adherens junction
Catenin-α	Adherens junction Actin filament, cadherin
Catenin-β	Adherens junction Actin filament, cadherin
Catenin-γ	Adherens junction, desmosome Intermediate filament, desmoglein, classical cadherins
Catenin-δ1	Adherens junction Actin filament, cadherin
Catenin-δ2	Adherens junction
Plakophilins-1/2/3	Desmosome Intermediate filament
Plakophilins-4	Adherens junction, desmosome Actin and intermediate filaments
Adherens junctions	
Classical cadherins (E/N/VE-Cdh)	Actin filaments
Catenin-α/β	Actin filaments

Plakophilin 2 (Pkp2) deficiency can induce arrhythmogenic right ventricular cardiomyopathy,[170] ventricular tachyarrhythmias, and sudden death [767].

---

[170] Arrhythmogenic right ventricular cardiomyopathies are associated with fibro-fatty replacement of cardiomyocytes.

Plakophilin-4 (Pkp4) is observed in the desmosomes and adherens junctions of endothelia.

Dystonin,[171] a member of the plakin family of adhesion plaque proteins, possesses several isoforms (Dst1–Dst5 and Dst8). They anchor intermediate filaments to the actin cytoskeleton and hemidesmosomes.

7.7.2 Zonula Adherens or Adherens Junction

Zonula adherens (zonula: little girdle), or adherens junction, forms an adhesion belt that links adjacent epithelial or endothelial cells.[172] Adherens junctions are also called belt desmosomes. Adherens junctions are formed by opposite dense plaques at each apposed cell.

The intercellular space is filled with molecular spokes that connect both membranes. This adhesion belt separates the apical and basolateral membranes of each polarized cell. Adherens junctions are involved in dynamical tension sensing, force transduction, and signaling.

7.7.2.1 Fascia Adherens

A similar cell junction in non-epithelial cells is the fascia adherens (fascia: band). This broad, stabilizing, ribbon-like structure does not completely encircle the cell. In particular, this intercellular junction localizes to intercalated discs of cardiomyocytes. It participates in the transmission of contraction forces.

7.7.2.2 Constituents and Partners

Cadherins and actin filaments are the main components of these adhesion plaques. This cellular bridging depends on calcium [757]. Cadherin–catenin complexes constitute the core of this type of intercellular adhesion. In paticular, adherens junctions contain E-cadherins. The cytoplasmic tail of cadherin binds to catenins. Crosstalk between cadherin–catenin clusters and actin regulators directs the assembly of adherens junctions.

Filamentous actin stabilizes E-cadherin–β-catenin–α-catenin complexes. Localization of E-cadherins is stabilized via the recruitment of actin filaments either by E-cadherin or independently of E-cadherin [758].[173]

---

[171] A.k.a. hemidesmosomal plaque protein and bullous pemphigoid antigen BPAg1.

[172] Epithelial and endothelial cells are able to tightly join via intercellular adhesion, often with an asymmetric architecture of protein and lipid constituents of the plasma membrane, leading to an apical–basal polarity.

[173] Synaptotagmin-like protein bitesize is mainly required to organize ^{F}actin in the apical junctional region. Bitesize binds to phosphatidylinositol (4,5)-bisphosphate, partitioning defective protein Par3, and ^{F}actin-binding protein moesin.

Table 7.38. Adherens junction and its constituents (Source: [759]). α-Catenin binds to several actin partners, such as α-actinin, afadin, formin, LIM domain and actin-binding protein LIMA1, vinculin, zonula occludens protein ZO1, which may mediate interactions between α-catenin and ^{F}actin. Catenin-δ1 links cadherin to microtubules via kinesin or adaptor PH domain-containing protein PlekHa7 that connects catenin-δ1 to calmodulin-regulated spectrin-associated protein CamSAP3, or Nezha protein. Protein CamSAP3 links to microtubule minus-ends and tethers them to the zonula adherens.

Component	Partners	Role
Cadherin	β-Catenin	Connection to microtubules
		Cadherin exocytosis
	α-Catenin	Connection to actin cytoskeleton
		Connection to microtubules
	Catenin-δ1	Connection to microtubules
		Inhibition of cadherin endocytosis
Nectin	Afadin	Connection to actin cytoskeleton

Cadherin–catenin complexes link not only to actin, but also microtubules to organize epithelial cells. Moreover, catenins link to exo- and endocytic machinery. On the other hand, regulated endocytosis of cadherin–catenin clusters facilitates the remodeling of adherens junctions. The effects of the regulation of adherens junctions by the cytoskeleton and constituent transfer on overall tissue structure depends on whether a subset or the entire set of intercellular contacts is involved [759].

In addition to cadherins, another type of transmembrane protein, nectins of the class of immunoglobulin cell adhesion molecules lodges in adherens junctions. Nectins form homo- or heterophilic interactions with other nectins. Nectins bind to the cytoplasmic adaptor afadin (or AF6) that links to actin. Therefore, adherens junctions are composed of at least 2 types of proteic complex: cadherin–catenin and afadin–nectin complexes (Table 7.38).

7.7.2.3 Adherens Junctions and the Cytoskeleton

At adherens junctions, the cadherin ring couples with a contiguous ring of actin filaments. Therefore, cadherin-based adherens junctions link cytoskeletal proteins of a given cell to the cytoskeleton of its neighboring cells as well as to proteins of the extracellular matrix. Adherens junction is a third type of anchoring junction with its cytoskeletal protein anchors and transmembrane linker proteins.[174]

Actin filaments connect to the adherens junctions via catenins. Cadherin–catenin complex organizes adherens junctions. Three types of catenins exist:

---

[174] Actin filaments are linked to adherens junctions rather than intermediate filaments that are mainly associated with desmosomes and hemidesmosomes.

α-, β-, and γ-catenin. Catenins-β and -γ bind directly to cadherins, whereas α-catenin links β- or γ-catenin and then the cadherin–catenin complex to the actin cytoskeleton.

The cadherin–catenin complex ties to ^{F}actin via actin-binding protein LIM domain and actin-binding protein-1 (LIMA1)[175] and α-catenin that interacts with actin [760]. Protein LIMA1 hinders actin depolymerization, and hence stabilizes the associated bundle of cortical actin filaments. Small GTPase Rho controls the tension of these actin fibers. α-Catenin also interacts with vinculin that can be involved in myosin-6-dependent junction formation.

Different isoforms of myosin-2 are implicated in adherens junction assembly, functioning, and remodeling. In response to signals, adherens junctions form, change their size, move, and degrade. Adherens junctions sense forces from different directions and respond. They regulate and are regulated by the cytoskeleton. Myosin-2-generated tension unfolds α-catenin, thereby unmasking a binding site for vinculin that binds to ^{F}actin [761]. Vinculin recruitment to α-catenin increases the amount of ^{F}actin in adherens junction that, in turn, attracts more cadherins and α-catenins. Among the 3 myosin-2 isoforms in mammalian cells, 2 isoforms – myosins-2A and -2B – that have different duty ratios[176] localize to adherens junctions [762].

Localization to adherens junctions of myosin-2A and -2B depends on signaling by Rho and Rap1, respectively. In addition, junctional residence of myosin-2A requires E-cadherin adhesion and myosin light-chain kinase. Myosins-2A and -2B lodge in distinct regions of adherens junctions. Myosin-2A is required for cadherin clustering and proper adhesion. Myosin-2B permanently controls the distribution of E-cadherin along the length of adherens junction and the normal concentration of ^{F}actin. Myosin-2A may control the initial stage of adherens junction assembly, whereas myosin-2B that can resist to sustained tension promotes adherens junction maturation and adaptation. In other words, myosin-2A functions as a cortical organizer to concentrate E-cadherin in adherens junctions, whereas myosin-2B prevents fragmentation of adherens junctions.[177]

7.7.2.4 Catenins as Constituents and Sensors

Catenins are involved not only in cell adhesion but also in signaling. Because distinct forms of catenins exist, adhesion and signaling are not necessarily coupled [763].

---

[175] A.k.a. Epithelial protein lost in neoplasm (EPLIN).

[176] The duty ratio is a measure of binding stability (time of binding) to ^{F}actin during the ATPase cycle. Myosin-2A duty ratio is 3 times smaller than that of myosin-2B.

[177] Loss of myosin-2B, but not myosin-2A, does not affect integrity of the apical actin ring, but reduces catenin-dependent vinculin recruitment and ^{F}actin accumulation at adherens junctions. On the other hand, excess myosin-2B can compensate for loss of myosin-2A, but not vice versa [762].

At adherens junctions, α-catenin can serve as a mechanosensor that converts tension generated by myosin-2 and actin as well as traction exerted by the cell environment into a chemical signal. α-Catenin linked to the actin cytoskeleton recruits vinculin. Tension indeed stretches α-catenin, thereby exposing its vinculin-binding site. Vinculin then tethers to α-catenin in adherens junctions when tension is exerted by the actin cytoskeleton [771].

β-Catenins interact with E-cadherin and α-catenin. They anchor the cadherin complex to the actin cytoskeleton. β-Catenins are mediators of Wnt signaling (Vol. 3 – Chap. 10. Morphogen Receptors).[178] β-Catenin activity is controlled by binding partners.

γ-Catenin binds E-cadherin, α-catenin, and transcription factors. It is also involved in cell adhesion as well as Wnt signaling. Catenins actually are not only components of the adherens junction complex, but also translocate to the nucleus to regulate gene expression.

7.7.2.5 Formation, Remodeling, and Degradation of Adherens Junctions

Cadherin–catenin clusters promote local actin polymerization to assembling adherens junctions [759]. The actin-related protein ARP2–ARP3 complex, formins, and members of the Enabled homolog and vasoactive stimulatory phosphoprotein (Ena–VASP) family localize specifically to cadherin–catenin clusters at early intercellular contact sites. Cadherin–catenin clusters signal via small GTPases Rac and Rho. Guanine nucleotide-exchange factor (RacGEF) T lymphoma invasion and metastasis-inducing protein TIAM1 is recruited to adherens junctions [759]. Small GTPase Rac can then attract WASP-family verprolin homolog WAVe2 that acts with WAVe1 to activate the ARP2–ARP3 complex. On the other hand, Rho activity that promotes actomyosin filament contraction is blocked until adherens junctions are strong enough to resist. Small GTPase Rac inhibits Rho via GTPase-activating protein RhoGAP35. Crosstalk between Rac and Rho contributes to the reconfiguration of actin during adherens junction assembly.

Formation and dissolution of adherens junctions is regulated in coordination with actin cytoskeleton by Rho GTPases. Abl kinases Abl1 and Abl2[179] control the formation and maintenance of cadherin-mediated intercellular adhesion via the Rho–RoCK–myosin pathway [768]. Kinase Abl operates as a

---

[178] The Wnt signaling pathway depends (conventional axis) or not (unconventional axis) on β-catenin [765]. Diversin functions as a molecular switch between these 2 types of Wnt pathways. Inversin (Inv) can associate with Disheveled. It inhibits β-catenin-dependent Wnt signaling upstream of the β-catenin degradation complex [766]. Flow over ciliated MDCK cells, ciliae being the sensing element of the urinary tract, increases the expression of inversin and slightly reduces β-catenin levels.

[179] Cytosolic Tyr kinases Abl1 and Abl2 are also called Abl and Arg, respectively (Vol. 4 – Chap. 3. Cytosolic Protein Tyr Kinases).

regulatory link between cadherin–catenin complex and actin cytoskeleton, as it activates Rac GTPases and inhibits Rho GTPases.[180] On the other hand, Abl2 promotes adherens junction formation via a CRK–CRKL-dependent axis in cells that have weak intercellular junctions. Kinase Abl activated by cadherin commitment phosphorylates the CRK–CRKL complex. Phosphorylation of CRK by Abl activates Rac possibly by either recruiting CRK-binding RacGEF dedicator of cytokinesis DOCK1 or phosphorylating GEF SOS1.[181]

In neurons, Abl kinases bind to a proteic complex that contains Robo receptor,[182] N-cadherin, β-catenin, and adaptor Cables-1.[183] Once associated with this complex, Abl kinase phosphorylates β-catenin that loses its affinity for N-cadherin. Within the Robo–N-cadherin complex, Cables binds to Robo-associated Abl kinase and cadherin-associated β-catenin [769].

Once the adherens junction is formed, Rho can act via RoCK kinase to trigger contraction of stress fibers to strengthen the adhesive belt. Moreover, Rho regulates adherens junctions via Diaphanous homolog Dia1.

Remodeling of adherens junction results at least partly from endocytosis. Cadherins are continuously internalized and can be recycled to adherens junctions. Clathrin adaptor β-arrestin interacts with phosphorylated VE-cadherin. The clathrin adaptor proteic complex AP2 also contributes to VE-cadherin endocytosis using actin and microtubule tracks. Ubiquitin ligase Casitas B-lineage lymphoma (CBL)-transforming sequence-like protein CBLL1[184] can trigger cadherin endocytosis and degradation [759]. Small GTPase CDC42 promotes dynamin-dependent scission of endocytic vesicles that contain adherens junction constituents using 2 CDC42 effectors: CDC42-interacting protein CIP4 and Par6 [759]. On the other hand, ArfGAP1,[185] an ArfGAP for ADP-ribosylation factor ARF6, counteracts this endocytosis.

7.7.2.6 Adherens Junctions and Microtubules

Microtubule plus-ends repeatedly grow and shrink. Adherens junctions stabilize microtubule plus-ends, as microtubule dynamics are slowed in the presence of the adhesive belt. Microtubule plus-ends associate with plus-end-tracking proteins (+TP). Among these +TPs, cytoplasmic linker protein CLiP170 localizes to cadherin–catenin clusters [759].

---

[180] Loss in Abl kinases impairs Rac activation induced by cadherin engagement.

[181] In epithelial sheets, Abl kinases hamper Rho GTPases. Consequently, myosin regulatory light-chain phosphorylation, stress fiber formation, and actomyosin contractility are impeded, thereby avoiding disruption of adherens junctions.

[182] Axon repulsive guidance signal Slit binds to its Robo receptor and inactivates N-cadherin.

[183] Cables is a substrate for both cyclin-dependent kinase CDK5 and Abl kinase. Enzyme CDK5 can link to the regulatory subunit P35 that interacts with β-catenin.

[184] A.k.a. Hakai and RING finger protein RNF188.

[185] A.k.a. stromal membrane-associated GTPase-activating protein SMAP1.

Furthermore, the minus-end-directed nanomotor dynein also resides at young and mature adherens junctions [759]. Dynein can also interact directly with β-catenins as well as eukaryotic translation initiation factor-3 subunit eIF3k that is linked to adherens junctions via actin.

In addition, the stabilizer of apical zonula adherens CamSAP3 interacts with microtubule minus-ends. It links to kinesin-14B family member KIFC3 and catenin-δ1-binding protein PlekHa7, another stabilizer of apical zonula adherens. It connects to catenin-δ1 to anchor microtubule minus-ends in adherens junctions. Moreover, CamSAP3 promotes microtubule polymerization at these sites. Nanomotor kinesin-1 can interact with catenin-δ1 to target adherens junction proteins to the plasma membrane for assembly of the adhesive belt. Besides, connexin-43 and aquaporin-3 are recruited to the lateral plasma membrane by microtubules linked to adherens junctions [759].

7.7.2.7 α-Catenin Mechanotransduction in Adherens Junctions

E-Cadherin forms homophilic complexes and clusters at adherens junctions of intercellular interfaces. α-Catenin links transiently to E-cadherin and ^{F}actin. It shuttles between E-cadherin–β-catenin complexes and ^{F}actin. It indeed cannot bind simultaneously to E-cadherin–β-catenin complex and actin filaments. α-Catenin also tethers to actin partners, such as vinculin, α-actinin, zonula occludens protein ZO1, formin-1, and afadin.

Adherens junctions contain mechanosensors that can detect pulling forces from adjacent cells and generate opposing forces to maintain tissue integrity. Forces generated by the actomyosin cytoskeleton are transmitted directly to adjacent cells via adherens junctions that are connected to actin–myosin filaments such as stress fibers.

Vinculin participates in force-dependent reinforcement of focal adhesions upon integrin linkage to the extracellular matrix. Forces applied on a focal adhesion strengthen the association between integrin and actin via actin-binding vinculin and talin. In addition, stretched talin triggers vinculin binding [770].

Vinculin is also recruited via α-catenin to adherens junctions when they are subjected to tension exerted by actomyosin filaments (*force-dependent association*) [771]. Tension of actomyosin filaments reinforces mechanical coupling of actin to E-cadherin through α-catenin. Myosin-2 activity promotes the recruitment of vinculin to adherens junctions. In addition, α-catenin bound to E-cadherin–β-catenin complexes might unmask the vinculin-binding site of α-catenin. Intracellular contractile actomyosin forces can then be transmitted to adherens junctions mostly by E-cadherins as well as α-catenin and vinculin that mediate the mechanical coupling between E-cadherin and ^{F}actin.

7.7.3 Tight Junctions or Zonula Occludens

Tight junctions (TJ), or zonula occludens, form protein junctions that leave tiny between-cell spaces (<1 nm).[186] Membranes of adjacent cells fuse at a belt.[187] Two rows of integral membrane proteins, each adjacent membrane contributing for one row, holds the 2 membranes close together by reciprocal contacts. Sealing strands that form attachment lines between adjacent cells are organized into a flexible network that modifies the junction tightness according to local physiological needs and maintains the cell sealing under stress.

7.7.3.1 Functions

Tight junctions attach apposed cells together. Tight junctions thus selectively modulate paracellular permeability. Therefore, epithelia are classified as tight or leaky whether tight junctions exist and prevent water and solute movement or not.

Tight junctions divide the plasma membrane of epithelial cells into distinct apical and basolateral regions with different protein and lipid compositions.

7.7.3.2 Constituents and Partners

Several proteins contribute to the formation of tight junctions: *cingulin, claudin,*[188] *junctional adhesion molecules, occludin,*[189] *symplekin, zonula occludens proteins* (ZO),[190] among others. E-cadherin is specifically required for tight junction formation and is involved in signaling rather than cell contact [774].

Zonulin,[191] the inactive precursor of haptoglobin-2 (preHp2), modulates intercellular tight junctions, hence reversibly regulating paracellular uptake routes, i.e., epithelial permeability [775].[192]

---

[186] The tight junction dimensions average 27.4 nm in width and 1.1 μm in length.

[187] The membranes are not cemented together. Such intimate contact blocks molecule passage through the obliterated intercellular space.

[188] Claudin assembly modes increase the diversity of the structure and functions of tight junction strands [772].

[189] Occludin brings opposite external cell leaflets into contact.

[190] There are several types of zonula occludens proteins (ZO1–ZO3) [773].

[191] Intestinal permeation modulator single chain zonulin (but not its cleaved mature form) enables transactivation of EGF receptors from activated proteinase-activated receptor PAR_2 that augments epithelial permeability. Proteolysis of zonulin into its α and β subunits neutralizes its ability to activate EGFR and increase intestinal permeability.

[192] Haptoglobin (Hp) is a tetramer with 2 α and 2 β chains that originate from a common precursor that is cleaved during protein synthesis. The haptoglobin gene (HP) has 2 alleles (HP1 and HP2), hence 3 genotypes exist (HP1-1, HP2-1, and HP2-2). Haptoglobin binds to free plasma hemoglobin released from ery-

In polarized epithelial cells, phosphatidylinositol (3,4,5)-trisphosphate that is stably localized at the basolateral plasma membrane regulates the formation of this segment of the cell surface [776].

Small GTPase RhoA regulates the tight junction assembly, hence cell polarity. Partitioning defective protein Par6 interacts with transforming growth factor-β receptors (Par6 is a substrate of TβR2) and ubiquitin ligase SMURF1 that degrades RhoA, leading to a loss in tight junctions [777].

A cell apicobasal polarity complex located in tight junctions of any epithelial cell consists of partitioning defective proteins Par3 and Par6 as well as atypical protein kinase-C (aPKC) and guanine nucleotide-exchange factor for small GTPase Rac1 TIAM1. Another proteic complex that associates Par3 and Par6 with VE-cadherin, but lacks aPKC, exists in adherens junctions of endothelial cells [778].

In the endothelium, integral membrane proteins junctional adhesion molecules, occludins and claudins that localize to tight junctions control the vascular permeability. They are also involved in angiogenesis in association with $\alpha_V\beta_3$ integrins as well as in inflammation, as they are required for leukocyte recruitment and transendothelial diapedesis.

Junctional adhesion molecules redistribute from the cell junctions to the apical surface of stimulated endothelial cells. Nectins recruit cadherins to form adherens junctions, and afterward, claudins, occludin, and junctional adhesion molecules, owing to peripheral membrane proteins (among which are many actin filament-binding proteins), to the apical side of adherens junctions to build tight junctions.

Protein kinases and phosphatases either localize to tight junctions or interact with tight junction constituents. Several protein kinase-C isoforms regulate function as well as assembly and maintenance of tight junctions.

Occludin interacts with signaling mediator kinases, such as regulatory subunit PI3KR1 of phosphoinositide 3-kinase and atypical protein kinase-C. Kinase PI3K regulates activation of small GTPase Rac1 that controls lamellipodium formation. During wound healing and cell migration, after birth as well as during embryogenesis, occludin accumulates on the leading edge membrane to regulate directional cell migration and reorientation of the microtubule organization center [779]. Occludin keeps polarity proteins at the leading edge, as it recruits and connects to the aPKC–Par3–PATJ polarity complex and regulates actin organization and lamellipodium formation.

---

throcytes. Mature haptoglobin (plasma heterodimeric glycoprotein; 100–300 mg per 100 ml) scavenges free hemoglobin to inhibit its oxidative activity. The haptoglobin–hemoglobin complex is processed by the reticuloendothelial system (mostly the spleen). Haptoglobin that arises from different genotypes binds hemoglobin with different affinities (Hp2-2 being the weakest binder). Haptoglobin pertains to the family of mannose-binding lectin-associated Ser protease (MASP) that also includes plasminogen-related growth factors, such as epidermal and hepatocyte growth factor, which can affect tight junction integrity.

Moreover, occludin phosphorylation (Tyr473) allows the recruitment of regulatory subunit of PI3K (PI3KR1, i.e., P85α) to the leading edge.

7.7.4 Gap Junctions (Nexus or Macula Communicans)

Gap junctions, *nexus*,[193] or macula communicans,[194] build between-cell channels, which bridge adjacent membranes and thus connect the cytosol of neighboring cells (bore of 1.2–2 nm).

7.7.4.1 Channel Function

These intercellular proteic channels enable low-molecular-weight molecules $(< \sim 1\,kDa)$, small signaling molecules, such as second messengers inositol (1,4,5)-trisphosphate and cyclic adenosine monophosphate, as well as ions, to diffuse between neighboring cells (metabolic and signaling coupling).[195] These clusters of membrane channels transmit electrochemical current between adjoining cells for coordinated activity (electrical coupling).

Conformational changes close the channel. Gap junction permeability is controlled by multiple factors: calcium concentration mediated by calmodulin, pH, membrane potential, protein phosphorylation, etc.

7.7.4.2 Constituents and Partners

At a gap junction, each cell provides a hemichannel, or *connexon*, that docks with a hemichannel in the apposed cell. Connexon is an hexameric array of *connexins*. Connexon is an homohexamer or or heterohexamer according to available connexin types. Pairs of connexons form an intercellular membrane channel.

Connexon–containing vesicles are transported to the cell surface and fuse with the cell membrane. Inserted hemichannels diffuse in the membrane until they dock with a hemichannel in an apposed membrane to form an intercellular channel.

Various connexins are involved in gap junctions. Connexins, encoded by a gene family, are commonly named by their molecular mass (Cx·). Different connexins assemble to form junctions, which differ in channel conductance, gating, permeability depending on both size and charge, as well as temporal and spatial patterns of expression. Another type of proteins — pannexins — can form gap junctions.

---

[193] Nexus: binding together, fastening, joining, interlacing, entwining, and clasping.

[194] Macula: spot, mark.

[195] Gap junctions mediate between-cell transfer of small molecules like amino acids, nucleotides, vitamins, hormones, cAMP, glucids, etc.

Connexins have many partner proteins, such as Src kinases, zonula occludens-1, adhesion junction proteins (N-cadherin, β- and γ-catenins, plakophilin-2, and desmoplakin), tubulin, and caveolins [780]. Binding to tubulin and its associated nanomotors allows connexin transport to the correct destination at the cell surface. N-cadherin is implicated in Cx43 trafficking and assembly into functional gap junctions. Protein ZO1 participates in the regulation of gap junction size. Plakophilin-2 modulates the activity of β-catenin and Cx43 intracellular distribution.

7.7.4.3 Gap Junction in the Cardiomyocyte

Gap junctions of the cardiomyocyte mainly consist of 2 connexons made from 6 connexins. Many types of connexins generate heterotypic (heterogeneous connexons, each made from a single connexin type), heteromeric (connexons made from different connexins), and homotypic junctions.

Gap-junctional coupling between the same and different types of cardiomyocytes (nodal, atrial, and ventricular, as well as epi-, endo-, and midmyocardial; Vol. 5 – Chap. 5. Cardiomyocytes) depends on many factors, such as the amount and types of expressed connexin, proportion of each assembled connexin, size and distribution of gap junctions, and the gating of gap junctions [780]. Gap junctions in the heart wall allow propagation of ion fluxes between cardiomyocytes that govern the heart rhythm. Most gap junctions are located in longitudinal (end-to-end) intercalated discs. Intercalated discs also contain force-transmitting adherens junctions that are mostly located in lateral (side-to-side) intercalated discs, and desmosomes.

Many connexin types are expressed in the heart, mainly connexin-43, -40, and -45 that are expressed in distinct combinations and relative quantities in different cardiomyocyte subsets. Myocytes of the sinoatrial and atrioventricular nodes have small, dispersed gap junctions composed of Cx45. Downstream from the His bundle, nodal cells prominently express Cx40, but distal cells abundantly produce Cx43, although Cx45 is continuously synthesized from the atrioventricular node to the ends of the Purkinje fibers. Ventricular myocytes are interconnected by clusters of Cx43-containing gap junctions.

Phosphorylation of connexin-43 in gap junctions by protein kinase-C reduces the permeability to large hydrophilic solutes (ATP, cAMP, IP$_3$, NAD$^+$), but not to ions. Connexin-43 phosphorylation thus does not affect the propagation of the electrochemical wave.

7.7.5 Epithelial Cell Junctions and Polarity

Epithelial cells exhibit an asymmetric organization of junctional complexes. Strong intercellular adhesion is generated at adherens junctions by cadherins. Tight junctions create a diffusion barrier for soluble molecules in intercellular space and for lipids and proteins in the plasma membrane, thereby maintaining distinct compositions of apical and basolateral plasmalemmal regions.

Transmembrane proteins of the junctional adhesion molecule and nectin families are involved in the recruitment of the partitioning defective complexes to primordial adhesions during formation of epithelia. Dimers and clusters of E-cadherins activate small GTPases CDC42 and Rac1, respectively [595]. Nectins stimulate both CDC42 and Rac1. Adherens and tight junctions requires atypical protein kinase-C (aPKC) activated by interactions of small GTPase Rho and the partitioning defective complex (Sect. 6.10). Binding of $CDC42^{GTP}$ or $Rac1^{GTP}$ to Par6 allows Par6 to activate aPKC. In addition, Rac-specific guanine nucleotide-exchange factor T-cell-lymphoma invasion and metastasis TIAM1 interacts with Par3 for Rac1 activation that hastens aPKC. Activated atypical protein kinase-C phosphorylates its substrates for cell polarization and epithelium maturation.

Conversely, transforming growth factor-β yields epithelial–mesenchymal transition via diverse mechanisms: (1) Par6 phosphorylation by ligand-activated TGFβ receptor-2 that, in turn, activates ubiquitin ligase SMURF1 for RhoA proteasomal degradation and loss of the apical actomyosin ring; (2) Par3 downregulation that causes cytoplasmic retention of aPKC and Par6, E-cadherin repression, and loss of adherens and tight junctions; and (3) activation of transcriptional repressor Snail that inhibits Crumbs-3 expression [595]. Moreover, the loss of RhoA activity in the cell basal region induces the disassembly of basal microtubules, disruption of integrin-mediated adhesion, and subsequent degradation of the basement membrane.

Protein Ser/Thr kinases microtubule affinity-regulating kinases MARK1 to MARK4 constitute a kinase family as well as are members of the partitioning defective (Par) family of proteins. Several substrates of MARK1 (or Par1 polarity kinase) have been identified, such as Par1 and Par3.[196]

Atypical protein kinase-C (PKCζ or PKCι) phosphorylates MARK1[197] as well as other members of the MARK family to regulate their subcellular location (relocalization from the membrane into the cytosol), 14-3-3 binding, and kinase activity. Recruitment of MARK2 in and removal from lateral membranes maintain the cell apicobasal polarity. The partitioning defective complex antagonizes MARK1, as atypical PKC of the Par3–Par6–aPKC complex phosphorylates MARK1 to remove it from the plasma membrane. In addition, novel PKC activates protein kinase-D that phosphorylates MARK2 to support the binding to 14-3-3 protein and, hence, impede its membrane association [858].

---

[196] Kinase MARK1 phosphorylates Par3 to exclude it from lateral membranes of epithelial cells.

[197] At Thr595.

8

Extracellular Matrix

The extracellular matrix (ECM) provides a scaffold for appropriate three-dimensional cell assembly. It supports tissue formation and remodeling as well as cell functions, such as division, differentiation, and apoptosis. The extracellular matrix produced by cells serves as a binding medium.

The composition, structure, and function of the extracellular matrix varies between different tissues. It particularly abounds in connective tissues. The extracellular matrix is mainly synthesized by fibroblasts and smooth muscle cells.

Imaging techniques have been developed for the structural analysis of soft tissues, in particular to study the crimp of collagen fibers within bundles [781]. Protein content (*matrixome*) of tissue-customized extracellular matrix has been identified using computational screening of secreted protein genes and functional assays [782].

Cells interact and communicate with other cells and the extracellular matrix. Signaling from the extracellular matrix influences regulation of gene expression. As a reservoir of signaling molecules, the extracellular matrix modulates activities of extracellular soluble (e.g., growth factors) and insoluble (e.g., adhesion molecules) factors. It also transmits physical stimuli.[1]

8.1 Structural Proteins

The extracellular matrix is composed of 3 molecule types: (1) *structural proteins*, i.e., abundant, insoluble, glycoproteic collagens (Col)[2] and

---

[1] Transforming growth factor-β regulates the deposition of proteins of the extracellular matrix and their interactions with the cells using integrins. It stimulates the synthesis of collagens and fibronectin, as well as the balance between matrix metalloproteinases and tissue inhibitors of matrix metalloproteinases. Protein TGFβ and its SMAD effectors are involved in the contraction of collagen gels by fibroblasts.

[2] In humans, 29 species have been identified (κολλαω: to glue).

Table 8.1. Glycoproteins and proteoglycans of the extracellular matrix are relatively slightly and strongly glycosylated, respectively. Glycan refers to a poly- or oligosaccharide of a glycoconjugate, such as a glycoprotein and proteoglycan. Glycosaminoglycans include heparin, heparan-, dermatan-, chondroitin-, and keratin sulfates. Serglycin that is also called hematopoietic cell granule proteoglycan is stored in secretory granules of many hematopoietic cells. Fibromodulin interacts with collagen-1 and -2 fibrils and sequesters transforming growth factor-β. Betaglycan is the transforming growth factor-βreceptor TβR3.

	Glycoproteins
Extracellular Matrix	Fibrillins (Fbn1–Fbn4), insoluble fibronectin, fibulins (Fbln1–Fbln7), vitronectin, tenascins (TnC/N/R/W/XA/XB; TnM1–TnM4)
Basement membrane	Entactin (or nidogen-1), osteonidogen (or nidogen-2), laminins (α–β–γ trimers)

	Proteoglycans
Leu-rich	Decorin, biglycan, fibromodulin
Lectican	Aggrecan, brevican, neurocan, versican
Basement membrane	Agrin, entactin (or nidogen-1), osteonidogen, perlecan
Cell surface	Syndecans, betaglycan (or TβR3), epican (CD44), glypicans (Gpc1–Gpc6)
Intracellular	Serglycin

non-glycosylated proteins elastins that yield flexibility (Eln);[3] (2) *specialized glycoproteins*, such as fibrillin (Fbn), fibronectin (FN), laminin (Lam); and (3) heavily glycosylated *proteoglycans* (PoG),[4] such as chondroitin (CSPG1–CSPG6), heparan (HSPG),[5] and keratan sulfate proteoglycans, as well as hyaluronic acid that build the ground substance (Table 8.1).[6]

Without anchorage to the extracellular matrix, the cells cannot survive. Fibronectins attach cells to the extracellular matrix, except in the basal lamina[7] of endothelia, which involve laminin as adhesive molecule. Large, dimeric fibronectin forms fibrils that indeed provide numerous binding sites for plasmalemmal and matrix proteins, such as integrins as well as collagen, heparin,

---

[3] ελασις: a driving away, banishing; ελαστρεω: to drive, push.

[4] The alias PG is used for prostaglandins.

[5] Heparan (ηπαρ: liver) sulfate proteoglycans regulate the distribution of the extracellular morphogens, such as Hedgehog and Wnt (Vol. 3 – Chap. 10. Morphogen Receptors), as well as bone morphogenetic proteins [783] (Vols. 2 – Chap. 3. Growth Factors and 3 – Chap. 8. Receptor Kinases).

[6] An altered content of hyaluronan in the arterial wall disturbs the wall functions. Hyaluronan overproduction in the aorta is associated with a thinning of elastic lamellae [784]. It can lead to increased wall stiffness and promote atherosclerosis.

[7] Lamina: plate, stands for a thin layer of connective tissue.

and fibrin. Binding site exposure caused by mechanical forces on the extracellular matrix imposed by cells, i.e., by resulting stretch exerted on fibronectin, is due to fibronectin unfolding rather than conformational changes [785].

The cell life depends on the microstructure of the extracellular matrix. Fibroblasts grown in high-fibril-density substrates have decreased length-to-height ratios, increased surface areas, and a greater number of projections [786]. Fibroblasts in low-fibril-density matrices reorganize their extracellular matrix to a greater extent. Fibroblast proliferation is enhanced in a low-fibril-density extracellular matrix. Furthermore, β_1 integrins are localized according to the local strain field and matrix remodeling events.

8.1.1 Proteoglycans

Like other glycoconjugates, proteoglycans play numerous roles. The gel-like ground substance has an important water-binding capacity that amplifies the volume occupied by the macromolecules.

8.1.1.1 Proteoglycans and Glycosaminoglycans

Proteoglycans are composed of a protein to which are attached glycosaminoglycans (GAG; or mucopolysaccharides [MPS]). Glycosaminoglycans with their negative charges act as buffer for excess cations. They are made of repeated disaccharide units that contains: (1) a modified glucid, either ^{N}acetylgalactosamine (GalNAc) or ^{N}acetylglucosamine (GlcNAc) and (2) a uronic acid, either glucuronate, or glucuronic acid (GlcUA), or iduronate, or iduronic acid (IdoUA; Table 8.2). Glycosaminoglycan chains are added to a proteic core in the Golgi body and modified by sulfations and epimerizations before transport to the plasma membrane.

Glycosaminoglycans defined by the presence of amino sugars and other carbohydrates in a polymeric form include (Table 8.2): (1) chondroitin sulfate (CS) with a disaccharide unit modified with ester-linked sulfate at certain positions ($[4GlcUA\beta 1$–$3GalNAc\beta 1]^n$); (2) dermatan sulfate (DS), a modified form of chondroitin sulfate in which a part of the ^{D}glucuronate residues are epimerized to ^{L}iduronates ($[4IdoUA\beta 1$–$3GalNAc\beta 1]^n$); (3) heparan sulfate (HS) characterized by a disaccharide unit with ^{N}- and ^{O}sulfate esters at various positions ($[4GlcUA\beta 1(IdoUA\alpha 1)$–$4GlcNAc\alpha 1]^n$); (4) heparin, a type of heparan sulfate that has the highest amount of iduronic acid and ^{N}- and ^{O}sulfate residues; (5) hyaluronan, or hyaluronic acid (hyaloid [vitreous] and uronic acid), the single glycosaminoglycan synthesized at the plasma membrane and composed of a disaccharide unit that is neither sulfated nor modified by epimerization of the glucuronic acid moiety to iduronic acid ($[4GlcUA\beta 1$–$4GlcNAc\beta 1]^n$); (6) keratan sulfate (KS), a poly^{N}acetyllactosamine with sulfate esters at C6 of ^{N}acetylglucosamine and galactose residues ($[3Gal\beta 1$–$4GlcNAc\beta 1]^n$).

Table 8.2. Glycosaminoglycans are abundant heteropolysaccharides (Source: [787]). Glycosaminoglycans are highly negatively charged, long, unbranched polysaccharides with a variably sulfated, repeating disaccharide unit. The disaccharide unit contains a modified glucid ([N]acetylgalactosamine [GalNAc] or [N]acetylglucosamine [GlcNAc]; 2S, 4S, 6S: 2-, 4-, 6-sulfate) and a uronic acid (glucuronate [glucuronic acid (GlcUA)] or iduronate [iduronic acid (IdoUA)]). Heparin and closely related heparan sulfates, as well as hyaluronates and chondroitin, dermatan, and keratan sulfates belong to the glycosaminoglycan superclass of carbohydrates. Heparin is stored within secretory granules of basophils and mastocytes. This anticoagulant prevents the formation of clots and their extension, but does not break down clots. Chemokine CXCL4, also called platelet factor PF4 and oncostatin-A, is released from α-granules of activated platelets and binds to and antagonizes heparin. This chemoattractant for monocytes, neutrophils, and fibroblasts, connects to the chemokine receptor CXCR3b. The cell-surface glycoprotein CD44, a determinant for the Indian blood group system and a marker for effector memory T lymphocytes, is a receptor for hyaluronic acid that also interacts with collagens, osteopontin, and matrix metalloproteinases.

Type	Location	Composition
Chondroitin sulfate	Heart valves, bone, cartilage	GlcUA, GalNAc(4S) or (6S)
Dermatan sulfate	Blood vessels, heart valves, skin	IdoUA (sulfated), GalNAc(4S)
Heparan sulfate	Basement membranes, cell surfaces	GlcUA or IdoUA(2S), GlcNAc(6S)
Heparin	Intracellular granules	GlcUA or IdoUA(2S), GlcNAc(6S)
Hyaluronan	Synovial fluid, vitreous humor, loose connective tissue	GlcUA, GlcNAc
Keratan sulfate	Bone, cartilage, cornea	Gal(6S), GlcNAc(6S)

The names of proteoglycans derive from associated glycosaminoglycan chains (Tables 8.3 to 8.4). For example, heparan sulfate proteoglycans are composed of a protein core and heparan sulfate glycosaminoglycan chains. Like glycoproteins, most proteoglycans contain also [N]- and [O]glycans, but the glycosaminoglycan chains are much larger than other types of glycans. The properties of the glycosaminoglycans thus tend to dominate.

In addition, some proteoglycans contain a single glycosaminoglycan chain (e.g., decorin), whereas others have more than 100 chains (e.g., aggrecan). A given proteoglycan often exhibits differences in the number and size of attached glycosaminoglycans according to the cell type, when it resides in different cell types.

Table 8.3. Proteoglycans and glycosaminoglycans (**Part 1**). The small leucine-rich proteoglycan family (SLRP) includes biglycan, decorin, and keratan sulfate proteoglycans, such as fibromodulin, keratocan, lumican, and mimecan (a.k.a. osteoglycin [OGn] and osteoinductive factor [OIF]). Lecticans constitute a family of chondroitin sulfate proteoglycans that includes aggrecan, brevican, neurocan, and versican. They act as linkers of carbohydrate and protein ligands in the extracellular matrix. Proteoglycans can be categorized according to the nature of their glycosaminoglycan chains and size. Perlecan and testicans (Ticn and sometimes Tic or Tes), or Sparc–osteonectin, Cwcv and Kazal-like domain-containing proteoglycans (SPOCK), which are Ca^{++}-binding molecules, contain both chondroitin (CS) and heparan (HS) sulfate chains. Biglycan, decorin, and versican possess both chondroitin (CS) and dermatan (DS) sulfate chains.

Type	Other alias (name)	Size
Chondroitin sulfate proteoglycans		
(extracellular, sulfated)		
CSPG1	Acan (aggrecan)	Large
CSPG2	Vcan (versican)	Large
CSPG3	Ncan (neurocan)	Large
CSPG4	NG2	Large
CSPG5	NGc (neuroglycan-C)	
CSPG6	Bam (bamacan)	
CSPG7	Bcan (brevican)	Medium
Dermatan sulfate proteoglycans		
DSPG1	Dcn (decorin)	Small
DSPG2	BGn (biglycan)	Small
DSPG3	EPyc (epiphycan)	
CS–HS		
SPOCK1	Ticn1	Small
SPOCK2	Ticn2	Small
SPOCK3	Ticn3	Small
Keratan sulfate proteoglycans		
(extracellular, sulfated)		
Lum (lumican)		Small
FMod (fibromodulin)		Small
Kera, Ktn (keratocan)		Small
OGn (osteoglycin)	OIF (mimecan)	Small
OMd (osteomodulin)		Small

8.1.1.2 Types of Proteoglycans

Cells synthesize proteoglycans and store them in secretory granules, insert them into the plasma membrane, or secrete them into the extracellular matrix. Cells produce diverse species of proteoglycans (Tables 8.3 to 8.8).

Table 8.4. Proteoglycans and glycosaminoglycans (**Part 2**; ADAMTS: a disintegrin and metalloproteinase with thrombospondin motif; HGF: hepatocyte growth factor; IGF: insulin-like growth factor; IGFR: IGF receptor; PTn: pleiotrophin; TDGF: teratocarcinoma-derived growth factor; TFPI: tissue factor pathway inhibitor). Hyaluronan is the single known unsulfated proteoglycans. Several enzymes cleave and release the extracellular domain of syndecans in response to various stimuli. Members of the glypican family of heparan sulfate proteoglycans are attached to the plasma membrane via a glycosylphosphatidylinositol anchor. Glypican-1 may act as a coreceptor for FGFR1 targeted by fibroblast growth factor FGF2. Glypican-3 is a coreceptor for heparin-binding proteins and their corresponding receptors (e.g., HGF and TNFRSF11b). It is first processed into a 65-kDa core protein and further cleaved by a furin-like convertase into a 40-kDa protein that modulates Wnt signaling.

Type	Other name and alias	Binding partners
Heparan sulfate proteoglycans (extracellular, sulfated)		
	Syndecans	
Sdc1	CD138	Collagen1–6, laminin-1/2, FGF2, HGF, midkine, PTn, antithrombin-3, cathepsin-G, CD44, ADAMTS4, MMP7, elastase, CCL5/7/11/17, CXCL1
Sdc2, HSPG1	Fibroglycan, HSPG core protein	FGF2, HBEGF, VEGF, gmCSF
Sdc3	N-syndecan (ScdN)	PTn
Sdc4	Amphiglycan, ryudocan	
	Perlecan	
HSPG2	Plc, Prcan	
	Glypicans	
Gpc1		TDGF1
Gpc2	Cerebroglycan	Midkine
Gpc3	OCI5, MXR7	TFPI1, FGF2, Wnt IGF2, IGF1R
Gpc4	K-glypican	FGF1/2, HGF, Wnt
Gpc5		
Gpc6		
Heparin (intracellular, sulfated)		
Hyaluronic acid (extracellular, unsulfated)		

Proteoglycans of secretory granules may regulate the sequestration of positively charged components, such as proteases and bioactive amines, via the

Table 8.5. Examples of heparan sulfate proteoglycans and their tissue distribution (Source: [788]). Agrin is involved in the aggregation of acetylcholine receptors during synaptogenesis. It binds to the receptor Tyr kinase muscle specific kinase, dystroglycan, and laminin. It also participates in the renal filtration, enabling the retention of anionic molecules within the vasculature.

Proteoglycan	Localization
Secreted – Basement membrane proteoglycans	
Perlecan	Basement membrane, cartilage
Agrin	Neuromuscular junction,
	alveolar and glomerular basement membrane
Collagen-18	Neuromuscular junction
Membrane bound	
Syndecan-1–4	Epithelial cell, fibroblast
β-glycan	Fibroblast
Glycosyl-phosphatidylinositol-anchored	
Glypican-1–6	Epithelial cell, fibroblast
Intracellular granule	
Serglycin	Endothelial, endocrine, hematopoietic cells,
	mastocyte

interaction with negatively charged glycosaminoglycans. Among membrane proteoglycans, some belong to type-1 single-pass transmembrane protein (i.e., with a single membrane-spanning domain, a cytoplasmic C-terminus, and an extracellular N-terminus) or are glycosyl-phosphatidylinositol anchored to the cell surface.

8.1.1.3 Functions of Proteoglycans

Proteoglycans have many functions, as they interact with numerous substances [788]. Interstitial proteoglycans, especially those that contain chondroitin and dermatan sulfate chains form hydrated matrices, which fill the intercellular space. Proteoglycans of the basement membrane regulate its permeability. They also serve as a support for epithelial cells. Proteoglycans contribute to coagulation, immunity, and wound repair. In the extracellular matrix, proteoglycans bind growth factors, morphogens, cytokines, and chemokines, thereby acting as stores. Proteoglycans can act as receptors for proteases and protease inhibitors, thus regulating their activity. Membrane proteoglycans can operate as coreceptors for growth factor receptors, and as endocytic receptors for ligand clearance. They can cooperate with cell adhesion molecules to foster cell attachment and intercellular interactions.

Table 8.6. Examples of chondroitin sulfate proteoglycans (CSPG) and their tissue distribution (Source: [788]). Members of the aggrecan family of proteoglycans are C-type lectins. Small Leu-rich proteoglycans (SLRPG) stabilize and organize collagen fibers. Decorin sequesters transforming growth factor-β. Leucine and proline-enriched proteoglycan is also known as leprecan-1 (Lepre1), phosphacan as protein Tyr phosphatase receptor PTPRz or PTPζ. Three PTPRz splice variants exist in the central nervous system: full-length [PTPRz$_L$] and short [PTPRz$_S$] versions and matrix form. Leprecan is a basement membrane CSPG, whereas the 3 other main proteoglycans of the basement membrane (perlecan, agrin, and collagen-18) are HSPGs.

Proteoglycan	Localization
Aggrecan family – lecticans	
Aggrecan	Cartilage
Brevican	Brain
Neurocan	Brain
Versican	Connective tissue, aorta, brain
Small leucine-rich proteoglycans	
Decorin	Connective tissue
Biglycan	Connective tissue
Miscellaneous	
Leprecan	Basement membrane
Collagen-9 α2 chain	Cartilage; vitreous humor
Phosphacan	Brain (membrane bound)
Thrombomodulin	Endothelial cells (membrane bound)
CD44	Leukocytes (membrane bound)
CSPG4	Neural cells (membrane bound)
	Glial progenitors. chondroblasts, myoblasts, endothelial cells of the brain
Serglycin	Myeloid cells (intracellular granules)

8.1.1.4 Heparan Sulfate Proteoglycans

Proteoglycans with heparan sulfate chains are ubiquitously expressed at cell surfaces and in the extracellular matrix. Heparan sulfate chains bind to various types of extracellular proteins, such as growth factors (e.g., fibroblast and vascular endothelial growth factor) and morphogens (e.g., Hedgehog and Wnt), as well as constituents of the extracellular matrix. Heparan sulfate proteoglycans also act as receptors for triglyceride-rich lipoproteins in the liver to incorporate lipoproteins and control lipid metabolism.

Heparan sulfate chains are covalently bound to Ser residues in core proteins via a tetrasaccharide linkage. Assembly and polymerization of heparan sulfate chains requires numerous modifications by exostosin Ext1–Ext2 polymerase

Table 8.7. Examples of keratan sulfate proteoglycans (KSPG) and their tissue distribution (Source: [788]; SV2 or SV2a: synaptic vesicle glycoprotein-2A). Claustrin inhibits neural cell adhesion and neurite outgrowth.

Proteoglycan	Localization
Small leucine-rich proteoglycans (secreted)	
Fibromodulin	Broadly expressed
Keratocan	Wide distribution, sulfated only in cornea
Lumican	Broadly expressed
Mimecan	Wide distribution, sulfated only in cornea
Others	
Aggrecan	Cartilage (secreted)
Claustrin	Central nervous system (membrane bound)
SV2	synaptic vesicles (membrane bound)

Table 8.8. Examples of types of proteoglycans and matrices produced by selected cell types (Source: [787]; CS, DS, HS: chondroitin, dermatan, heparan sulfate). Laminin receptors comprise integrins and dystroglycan.

Producing cells	Collagen	Cell–matrix anchor	GAG	Receptor
Fibroblasts	Col1	Fibronectin	CS, DS	Integrin
	Col3	Fibronectin	HS, heparin	Integrin
	Col5	Fibronectin	HS, heparin	Integrin
	Col6	Fibronectin	HS	Integrin
Epithelial cells	Col3	Fibronectin	HS, heparin	Integrin
	Col4	Laminin	HS, heparin	Laminin receptors
Endothelial cells	Col4	Laminin	HS, heparin	Laminin receptors

complexes.[8] Transmembrane glycosyltransferases Ext1 and Ext2 cooperate for initiation and elongation of heparan sulfate chains and exostosin-like protein ExtL3 regulates the chain length [789].

Heparan sulfate proteoglycans can be divided into 3 main different families according to protein backbones: (1) transmembrane syndecans (the largest amount), (2) glycosylphosphatidylinositol-linked glypicans, and (3) secreted perlecans (the smallest amount).

Syndecans have an extended extracellular domain, which can sense applied shear, and a cytoplasmic sequence for signal transmission, possibly linked to

---

[8] The gene family EXT, which comprises 5 identified members, encodes polymerases with saccharide transferase activity responsible for heparan sulfate synthesis.

Table 8.9. Binding partners of sulfated glycosaminoglycans (Source: [788]; PAI: plasminogen activator inhibitor; TFPI: tissue factor pathway inhibitor; tPA: tissue plasminogen activator). Serine peptidase (protease) inhibitors serpin-A5, -C1, and -D1 are also called protein-C inhibitor or plasminogen activator inhibitor-3, antithrombin (or antithrombin-3), and heparin cofactor-2, respectively. Serpin-C1 in plasma inactivates thrombin. Serpin-D1 also inhibits thrombin, acting as a cofactor for heparin and dermatan sulfate.

Activity	Molecules
Cell–matrix interactions	Collagen, fibronectin, laminin, vitronectin, thrombospondin, tenascin
Coagulation, fibrinolysis	Thrombin, TFPI, tPA, PAI1, serpin-A5/C1/D1
Lipolysis	Lipoprotein lipase, hepatic lipase, apolipoprotein-A5/B/E
Inflammation	Interleukin-2/7/8, chemokines CCL4, CXCL12, TNFα, superoxide dismutase, L/P-selectins, microbial adhesins
Growth factors, morphogens	BMP, FGF, FGFR, HGF, TGFβ, VEGF, Hedgehog, Wnt

G-protein-coupled receptors, especially those bound with endothelial nitric oxide synthase, and to cytoskeletal components such as actin.

Syndecan-1, the most represented heparan sulfate proteoglycan in the vascular endothelium, is also involved in inflammation, as it binds chemokines (chemotactic cytokines; Table 8.4; Vol.2 – Chap. 6. Cell Motility).

The glypican family includes 6 known members. Glypicans regulate several developmental pathways (bone morphogenetic protein, fibroblast growth factor, Hedgehog, and Wnt).

Glypican-3 stimulates Wnt signaling [790].[9] On the other hand, glypican-3 core protein binds to Sonic Hedgehog and inhibits Hh signaling.

Heparan sulfate proteoglycans contribute to the formation of gradients of diffusible morphogens, such as Sonic Hedgehog and bone morphogenetic proteins, as well as growth factors, such as vascular endothelial and fibroblast growth factors.

8.1.1.5 Syndecan and Mechanotransduction

Syndecan-4, a heparan sulfate proteoglycans positioned at the cell surface, can support cell attachment and spreading even in the absence of binding

---

[9] Loss-of-function mutations of glypican-3 cause X-linked Simpson-Golabi-Behmel syndrome. Elevated glypican-3 level in serum is a marker of hepatocellular carcinoma.

Table 8.10. Laminin heterotrimers assembled from α, β, and γ chains.

Chain	Type	Gene
α chain	Lamα1–Lamα5	LAMA1–LAMA5
β chain	Lamβ1–Lamβ4	LAMB1–LAMB4
γ chain	Lamγ1–Lamγ3	LAMC1–LAMC3

of integrins to extracellular matrix components [791]. Integrin receptors are major constituents of focal adhesion complexes that also include paxillin, vinculin, and talin, among others. Focal adhesion complexes form a structural signaling connection between the extracellular matrix and the cytoskeleton. Plasmalemmal integrins are involved in mechanical responses. They then control cell structure and influence cell adhesion and migration.

Syndecans, like integrins, cluster at *focal adhesion complexes*. Syndecans, like integrins, have binding affinities for both the extracellular matrix and the actin cytoskeleton. Syndecans can provoke the activation of the Rho and Rac pathways.

Transmembrane syndecan-4 actually indirectly attaches to components of focal adhesion complexes such as paxillin that are connected to the actin cytoskeleton. It can recruit components of focal adhesion complexes to sites of syndecan-4-specific cellular attachments. Upon mechanical stimulation via syndecan-4, syndecan-4-supported adhesion loci cause phosphorylation of extracellular signal-regulated kinases [791]. This mechanotransductive signaling depends on the actin cytoskeleton.

8.1.2 Laminin

Laminin is a major glycoprotein of the basal lamina. Laminin promotes cell adhesion, migration, growth, and differentiation. Laminin is composed of 3 chains (A, B1, and B2), 3 arms that bind to other laminins, and one arm that is anchored to the cell membrane via integrins, dystroglycan, and Lutheran blood group glycoprotein. Fifteen laminins have been identified. Each laminin molecule is a heterotrimer assembled from α, β, and γ chains (Table 8.10).

Laminins form networks that associate with collagen-4 meshworks via entactin/nidogen and perlecan. Dysfunctional laminin-2 causes muscular dystrophy.

8.1.3 Fibronectins

Cell anchorage and migration involve another glycoprotein (∼5% carbohydrate), *fibronectin* (FN).[10] These connecting elements are fixed to heparan

---

[10] Necto, nectere: to connect, bind, link.

sulfate proteoglycans, fibrin, collagen, and elastin of the extracellular matrix on the one hand and the cell membrane on the other hand via integrins [793]. Fibronectin yields an important substrate for cell adhesion. Fibronectin links to actin filaments, hence the cytoskeleton.

Plasma fibronectin is synthesized in the liver by hepatocytes. Soluble fibronectin is conveyed in plasma. Insoluble form is a large complex of crosslinked units. There are several fibronectin isoforms that are the product of a single gene. Fibronectin is composed of 2 similar polypeptide chains attached by disulfide bonds that are folded into a series of 5 or 6 functional units with binding sites.

Fibronectin acts in clotting and healing. Clotting factor XIII bridges fibronectin to fibrin. Fibronectin enhances platelet adhesion. Activated platelets have specific fibronectin adhesion sites composed of 2 glycoproteins. Fibroblasts also have such sites, but with distinct affinity glycoproteins. Whereas platelets must be strongly moored whatever the stress field (strong affinity), fibroblasts travel through the fibronectin network (weak affinity). Fibronectin also promotes chemotaxis [794] and activates integrin signaling [795]. Fibronectin has several sites that can bind to integrins [796].

$\alpha(1,3)$-Mannosyltransferase apoptosis-linked gene-2 (ALG2)-interacting protein X (ALIX) is a ubiquitous, cytoplasmic adaptor involved in endosomal sorting and an ^{F}actin-binding protein for actin cytoskeleton assembly. Adaptor ALIX acts on both sides of the plasma membrane. A fraction of ALIX is secreted and localizes extracellularly to regulate integrin-mediated cell adhesions via $\alpha_5\beta_1$ integrins and fibronectin matrix assembly [797].

8.2 Basement Membrane

The basement membrane (BM)[11] is a specialized, dense extracellular matrix sheet at the interface between the connective tissue and epithelia or endothelia. It acts as a barrier to the passive diffusion of large molecules (size > 50 nm), but some cells can cross the basement membrane.

Both the basement membrane and the extracellular matrice are major regulators of embryo- and fetogenesis as well as tissue integrity. The basement membrane influences the functions of contacting cells (regulation of cell shape, gene expression, cell proliferation, migration, and apoptosis) [798].

Basement membrane contains laminins, members of the entactin/nidogen family, collagen-4, glycoproteins, and proteoglycans [799,800]. Laminin receptors are integrins and *dystroglycan*.[12] The laminin type affects receptor interaction. Laminin and collagen-4 networks are linked by *nidogens* [801]. This

---

[11] Two terms — basement membrane and basal lamina — are often used as synonyms. However, the basal lamina is a dense layer of the basement membrane.

[12] Dystroglycan associates with other proteins into complexes that are linked to the cytoskeleton.

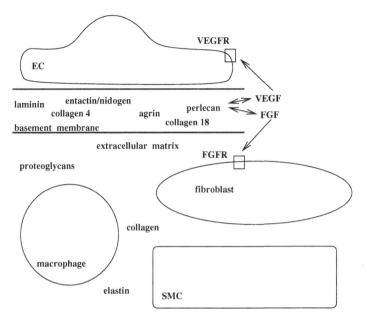

Figure 8.1. The endothelial cell (EC) lying on the basement membrane. The in-terstitial matrix contains cells (resident macrophages, fibroblasts, smooth muscle cells [SMC], pericytes), as well as fibers and proteoglycans. Heparan sulfate proteo-glycans of the basement membrane such as perlecan form complexes with growth factors, here the vascular endothelial (VEGF) and fibroblast (FGF) growth factors. The growth factors can then bind to their respective receptors VEGFR and FGFR. Perlecan can induce growth factor–receptor interactions and angiogenesis.

combined network connects to other glycoproteins and proteoglycans that act as both structural elements and receptor ligands. Proteoglycans bind, via their charged carbohydrate chains, soluble molecules, such as growth factors and ions, and regulate the diffusion of macromolecules.

The synthesis of the basement membrane is spatially and temporally re-gulated by neighboring cells. Regulation mediated by receptors on the cell membrane depends on the distribution of ligands and receptors [802,803]. The basement membrane binds various growth factors [804] (Fig. 8.1). This bind-ing regulates the accessibility of these ligands via retention or release from the basement membrane. Binding to basement membrane components can alter the local ion composition, with proteoglycans linking divalent cations [805].

The basement membrane has manifold structural roles. It gives a stable surface for endothelium firm anchorage that protects from shearing and de-tachment. Basement membrane acts as a selective barrier for macromolecular transfer.

8.3 Interstitial Matrix

The interstitial matrix[13] influences the functions of contacting cells. It has a fibrillar structure, with a large amount of collagens. The structure of the interstitial matrix depends mainly on the type of fibrils as well as type and amount of proteoglycans. Proteoglycans not only act as gel formers by water adsorption, but also have structural and cellular interactions [806]. Osmolarity of interstitial fluid affects the repulsive forces of the negatively charged GAGs, causing collapse or inflation, which can affect fiber stretching and folding.

Collagen and elastin control the rheology of the connective tissue. The ground substance with glycosoaminoglycans stabilizes the fiber network. Proteoglycans control the level of hydration of connective tissues, and thus partially determine the physical properties of connective tissues.

8.3.1 Elastin

The first type of major fibers of the extracellular matrix is the *elastin fiber* (ElnF), with elastin and fibrillin. Elastic fibers (thickness 0.5–1 μm) can ramify and form elastic networks and fenestrated membranes.

Tropoelastin is a precursor of elastic fibers synthesized by fibroblasts and smooth muscle cells and secreted in the extracellular space. These cells form a microfibrillar glycoprotein scaffold with fibrillin on which are deposited tropoelastin monomers. Microfibrils are grouped in alignment in infoldings of the cell surface. Tropoelastin bridges, as well as connections between elastin and collagen, need lysyloxydase.

Desmosine crosslinks elastin to form elastin fibers [807]. Elastin makes long-chain molecules tied together in a 3D network by scattered crosslinks with a degree of order influenced by its environment. Elastin also binds to cells via *elastonectin*.[14] G-protein-coupled receptor of elastin recognizes also laminin. Ligand-bound receptor induces Ca^{++} influx. Subsequent efflux is much slower with aging.

Elastic fibers hence consist of 2 components: microfibrils and polymerized elastin. Microfibrils (10–12 nm) in the extracellular matrix are composed of fibrillin-1 and -2, and associated proteins, such as latent transforming growth factor-β-binding protein[15] (LTBP1–LTBP4) and microfibril-associated glycoproteins. Microfibrils can form aggregates devoid of elastin.

---

[13] The interstitial matrix is the extracellular matrix part without basement membrane. The interstitial matrix, with the cells, forms the connective tissue.

[14] Two different types of interactions exist between cells and elastin [808]: (1) the adhesion of cells to insoluble fibrous elastin via elastonectin and (2) the binding of cells to soluble elastin-derived peptides via elastin receptors.

[15] Latent transforming growth factor-β-binding proteins (LTBP) not only are structural components of the extracellular matrix, but also (except LTBP2) a binding partner of the precursor of transforming growth factor TGFβ (TGFβ–LAP or small latent TGFβ complex) and modulate TGFβ activity.

Deposition and crosslinking of elastin along microfibrils depend on fibrillin-1, fibulin-5, and latent TGFβ-binding protein LTBP2 [809].

Elastin fibers are the most elastic biomaterials, at least up to a stretch ratio of 1.6 [810], the loading and unloading cycles being nearly superimposed. Strained elastic fibers produce recoil forces. The elastic modulus of elastin fibers ranges from 0.1 to 1.2 MPa [811–813].[16]

8.3.2 Fibrillins

Fibrillin is a large glycoprotein that serves in the formation of elastic fibers. Fibrillin is secreted by fibroblasts and incorporated into regular microfibrils with a periodic structure. Fibrillins align in parallel with crosslinks to form head-to-tail arrays. They construct staggered arrangements. In the unstressed state, they adopt a complex folded organization that can bear extensions [814].

Fibrillin-rich microfibrils are thin, filamentous components of connective tissues. These microfibrils yields a scaffold for elastin deposition. In elastic tissues, such as aorta wall and lung parenchyma, preformed bundles of microfibrils form a template for tropoelastin deposition during elastic fiber formation. They are retained as an outer mantle of mature elastic fibers [814]. The human genome contains at least 3 known fibrillins (Fbn1–Fbn3).

Matrix microfibril-associated glycoproteins, such as microfibril-associated glycoproteins MAGP1 and MAGP2, latent transforming growth factor-binding proteins, and chondroitin sulfate proteoglycans, colocalize with microfibrils [814]. During elastic fiber formation, microfibrils that interact with tropoelastin are connected to lysyl oxidase, elastin, and elastin-microfibril interface proteins, such as fibulin-2 and emilin.

Fibrillin-1 forms a sheath surrounding amorphous elastin.[17] Fibrillin-2 intervenes in early elastogenesis.[18] Unlike fibrillin-1 and -2, fibrillin-3 localizes mainly in the brain [815]. Like fibrillin-2, fibrillin-3 reaches its highest concentration in fetuses, especially in developing skeletal elements, skeletal muscle, lung, kidney, and skin. Its concentration decays at low levels in postnatal tissues.

8.3.3 Fibulins

Fibulins (Fbln1–Fbln7) constitute a class of secreted calcium-binding glycoproteins that are encoded by 7 genes. They interact with several matrix proteins, such as fibrillin, fibronectin, laminins, proteoglycans, and tropoelastin.

---

[16] A value of elastic modulus of elastin from 0.4 to 0.6 MPa is often considered.

[17] Mutations in the FBN1 gene are responsible for the Marfan syndrome.

[18] Mutations in the FBN2 gene cause congenital contractural arachnodactyly (Beals-Hecht syndrome).

Table 8.11. Class of fibulins (Fbln or Fibl; ARMD: age-related macular degeneration; AxPC1: ataxia, posterior column 1, with retinitis pigmentosa; DANCE: developmental arteries and neural crest EGF-like protein; DHRD: Doyne honeycomb retinal dystrophy (macular degeneration) protein; EFEMP: EGF domain-containing fibulin-like extracellular matrix protein). Fibulins regulate organ shape with growth factors and stromal cells. They are involved in cell proliferation, migration, differentiation, and survival.

Type	Alias	
Fbln1	Fibl1	Blood coagulation
Fbln2	Fibl2	Differentiation of organ structures
Fbln3	Fibl3, EFEMP1, DHRD	Indirect angiogenic effect
Fbln4	Fibl4, EFEMP2	Elastic fiber assembly
Fbln5	Fibl5, DANCE	Promotion of adhesion of endothelial cells
		Atherosclerosis, restenosis
Fbln6	Hemicentin-1 (Hmcn1)	Anchorage of mechanosensory neurons,
	Fibl6, ARMD1, AXPC1	Exocrine gland architecture
Fbln7	Fibl7	Exocrine gland architecture,
		tooth development

Fibulin-1 links to elastic fibers. It lodges particularly in basement membranes. It is also present in blood, where it binds to fibrinogen. It is incorporated in fibrin clots. Fibulin-1 suppresses fibronectin-mediated adhesion and motility [816].

Fibulin-2 is involved in microfibril and elastic fiber structuring [817]. It can participate in clusters made of nidogen-1 and -2, fibronectin, and heparan sulfate proteoglycans such as perlecan in the basement membrane via its immunoglobulin-like domains [818].

Fibulin-3 stimulates production of vascular endothelial growth factor, hence angiogenesis [819]. Moreover, fibulin-3 hinders tumor cell apoptosis.

Fibulin-4 is an elastic fiber-associated protein that is required for formation of elastic fibers, especially in arteries, lungs, and skin. It regulates the tropoelastin expression [820].

Fibulin-5 abounds in large vessels and cardiac valves during embryo- and fetogenesis as well as in many adult tissues that contain abundant elastic fibers such as the aorta. This calcium-dependent, elastin-binding protein localizes to the surface of elastic fibers. Fibulin-5 acts as a scaffold protein that links elastic fibers to cell integrins [821]. Fibulin-5 is a ligand for $\alpha_V\beta_3$, $\alpha_V\beta_5$, and $\alpha_9\beta_1$ integrins [822].[19]

Several fibulin-5-binding proteins exist, such as elastin, emilin, superoxide dismutase, among others. Fibulin-5 interacts with fibrillin-1. Fibrillin-1

---

[19] Fibulin-5-deficient mice have disorganized elastic fibers and fragmented elastin. Consequently, aortas in these mice are tortuous and softer. Fibulin-5 expression is reinduced in injured vessels and atherosclerosis, in intimal smooth muscle and endothelial cells [823].

microfibrils with fibulin-5 deposits can support the maturation of elastic fibers. Latent transforming growth factor-β-binding protein LTBP2 located in elastin-associated microfibrils interacts with fibulin-5 and can promote deposition of fibulin-5 onto fibrillin-1 microfibrils to form elastic fibers.

8.3.4 Collagens

Collagens are structural proteins that constitute the second major type of fibers of the extracellular matrix.

8.3.4.1 Collagen Production

Procollagens are synthesized by fibroblasts and vascular smooth muscle cells and then cleaved to collagen after secretion from the cell. Extracellular procollagen is combined to tropocollagen that spontaneously self-assembles into collagen fibrils [824]. Collagen is a load-bearing material that deforms differently in different directions and exhibits viscoelastic behavior. Collagen is stiff (high elastic modulus).[20]

8.3.4.2 Collagen Fibrils

Collagens form fibrils characterized by triplet of helical chains coiled around each other and stabilized by covalent crosslinks.[21] The triple helix domain is common for all collagens; the heterogeneity resides in the assembling mode and in the resulting structure. Collagen fibrils are more resistant to collagen proteolysis than collagen monomers.

Certain microfibrils (length ∼ 300 nm), which contain 5 rows of basic collagen molecule forming a helix, are packed together to form a fibril (caliber 50–200 nm), crosslinks providing stability.[22] Collagen fibrils can vary in thickness and in association with each other.[23]

---

[20] Collagen is about a thousand times stiffer than elastin. Collagen stiffness depends on fiber arrangement.

[21] Helical structure is made from 3 separate collagen chains by intertwining a coiled structure. Possible periodic striated aspects of collagen fibrils are due to their trellis arrangements.

[22] Neighboring collagen molecules are arranged to form a right-handed twist. The resulting microfibril, the basic building block of the collagen fibril, interdigitates with adjoining microfibrils. The quasihexagonal arrangement of collagens is continuous. Each microfibril contains at least 2 to 3 intermicrofibrillar crosslinkages, and one intramicrofibrillar bond [825]. Decorin regulates collagen-1 fibrillogenesis.

[23] Collagen-3, -5, and -11 regulate fibril size by incorporation into fibril-1 and 2.

8.3.4.3 Collagen Fibers

Long tropocollagens (length ~ 300 nm; bore ~ 1.5 nm) self-assemble in the extracellular milieu, constitute a staggered array, and form collagen fibrils (length ~ 1 µm) that aggregate and bundle into collagen fibers (ColF; length ~ 10 µm). Collagen design maximizes the strength with large energy dissipation during deformation.

Due to their configuration, collagen fibers can be stretched by strong stresses up to 16.3 kPa. The mechanical properties of blood vessels depend on interactions between elastic and collagenous elements. Fibroblasts remodel collagen to fit tissue to its environment requirements, in particular to regulate the cell growth and tissue development.

8.3.4.4 Collagen Types

The collagen subclasses encompass: (1) hexagonal-like network-building collagens (types 8 and 10); (2) beaded filament collagens (type 6); (3) anchoring fibril collagens (type 7); (4) transmembrane collagens (types 13 and -17); and (5) multiplexin collagens (types 15 and -18; Tables 8.12 and 8.13).

Collagen-1, the main constituent of arterial wall, is a tensile fiber (thickness 1–20 µm) that usually forms wavy bundles, which do not ramify. It is located between elastic lamellae of elastic arteries. Trimeric collagen-1 can favor activation of plasmalemmal proteases for cell migration and adhesion. Collagen-1 degradation is initiated by interstitial collagenases of the matrix metalloproteinase subclass such as MMP1, which can be associated with the plasma membrane of the migrating cell.

Collagen-3 strengthens walls of blood vessels. Its ultrastructure is similar to collagen-1, but its composition differs, with higher molecular stability. Collagen-3 determines vessel compliance [826].

Collagen-4 provides the basal lamina of endothelia and filter for blood capillaries.

Transmembrane collagen-13 is found in intercalated discs of cardiomyocytes.

8.3.4.5 Collagen Environment

The collagen is surrounded by extensible glycoproteins and proteoglycans. Proteoglycans are attached to collagen. The rheological properties of pure collagen are thus difficult to assess. Collagen fibers and fiber bundles have sheaths with collagen-3 and -4.

8.3.4.6 Collagen Remodeling

Collagen remodeling by fibroblasts has a crucial role in organizing tissue structures. Fibroblast lamellipodia extend along the collagen fibers, and work

Table 8.12. Types of collagens (**Part 1**; Main source: Wikipedia). The collagen class is composed of different subclasses (FACIT: fibril-associated collagen with interrupted triple helices that is both a collagen and proteoglycan; MACIT: membrane-associated collagen with interrupted triple helices that is a transmembrane collagen).

Type	Gene(s)	Location
	Fibrillar collagens	
1	COL1A1–COL1A2	Connective tissue, vessel wall
2	COL2A1	Cartilage, vitreous humor of eyes
3	COL3A1	Reticular fibers
		Distensible connective tissue
5	COL5A1–COL4A3	All tissues
11	COL11A1–COL11A2	Cartilage
24	COL24A1	
27	COL27A1	
	FACIT Collagens	
9	COL9A1–COL9A3	Cartilage
12	COL12A1	Interactions with collagen-1 and -3,
		decorin and glycosaminoglycans,
14	COL14A1	
16	COL16A1	
19	COL19A1	
20	COL20A1	
21	COL21A1	
	MACIT Collagens	
13	COL13A1	Intercalated discs of cardiomyocytes
		Interactions with $\alpha_1\beta_1$ integrin,
		fibronectin, nidogen, and perlecan
17	COL17A1	
23	COL23A1	
25	COL25A1	

on collagen by a mechanism requiring $\alpha_2\beta_1$ integrins as well as myosin-2B assembly and contraction in lamellipodia [827]. Fibronectin provides a separate structural network in interstitial matrices, which interacts with cells.

Animal models of essential hypertension show that the number of connections between elastin and smooth muscle cell increases and fenestrations of the internal elastic lamina are smaller [828]. Elastin and collagen not only intervene in the vessel wall rheology, via their mechanical properties, density, and spatial organization, but also control the function of smooth muscle cells.

8.3.4.7 Collagen Prolyl 4-Hydroxylases

Collagen prolyl 4-hydroxylases (CP4H) are $\alpha^2\beta^2$ tetramers that intervene in the synthesis of collagens. β Subunit is identical to protein disulfide

Table 8.13. Types of collagens (**Part 2**; Main source: Wikipedia).

Type	Gene(s)	Location
	Basal lamina collagen	
4	COL3A1–COL3A6	Basement membrane
	Type-6 collagen	
6	COL6A1–COL6A3	All tissues
	Anchoring fibrils	
7	COL7A	Dermis–epidermis interface
	Multiplexins	
15	COL15A1	
18	COL18A1	Cleavage into endostatin
	Network	
4	COL3A1–COL3A6	Basement membrane
7	COL7A	Dermis–epidermis interface
10	COL10A1	Cartilage
	Miscellaneous	
8	COL8A1–COL8A2	Endothelia
22	COL22A1	
26	COL26A1	
28	COL28A1	
29	COL29A1	Epidermis

isomerase [829]. Several isoforms of the catalytic α subunit exist ($\alpha 1$–$\alpha 3$). In humans, 3 known types of collagen prolyl 4-hydroxylases correspond to CP4H1 ($[\alpha_1]^2\beta^2$), CP4H2 ($[\alpha_2]^2\beta^2$), and CP4H3 ($[\alpha_3]^2\beta^2$). Transcription factor P53 activates the genes that encode collagen prolyl 4-hydroxylase.

Type-1 collagen prolyl 4-hydroxylase is the main form in most cell tissues [829]. Type-2 is a major kind in endo- and epithelial cells, as well as chondrocytes and osteoblasts. Type-3 collagen prolyl 4-hydroxylase is more restricted and its concentration lower, with its the highest levels in the liver as well as placenta and fetal skin.

8.3.5 Tenascins

The tenascin (Ten) class of matrix, cell-adhesion glycoproteins (Table 8.14) contributes to vasculogenesis and various normal and pathological processes of mature life, such as wound healing and vascular diseases [830].

In association with matrix proteins and plasmalemmal receptors such as integrins, tenascins have opposite cellular functions, according to the mode of presentation and cell types and differentiation states. Tenascins are regulated by growth factors, vasoactive peptides, matrix proteins, and mechanical factors [830].

Table 8.14. Class of tenascins (Ten). These matrix, cell-adhesion glycoproteins act via integrins, cell-adhesion molecules of the immunoglobulin class (IgCAMs), protein Tyr phosphatase receptor PTPRz1, annexin-2, fibronectin, and lecticans. The tenascin class includes TenC, TenR, TenW, TenX, and TenY. Tenascins can have anti-adhesive effects. Tenascin-C can promote integrin-dependent protein clustering at focal adhesions, as it interacts with RTKs and activates ERK. Like TenC, TenX (or flexilin) is expressed in the heart as well as vascular smooth muscle cells. TenY is coexpressed with TenC in lungs.

Tenascin	Alias	Function
TenC	Cytotactin	Tissue development and remodeling (vasculogenesis, wound healing)
TenR	Restrictin, janusin	Development of the nervous system
TenW		Development of the nervous system
TenX	Flexilin, hexabrachion	Wound healing
TenY		

Tenascin-C interacts with integrins, collagens, proteoglycans, and fibronectin. It behaves either as an adhesive or anti-adhesive protein. It binds to annexin-2, a plasmalemmal receptor of endothelial cells [831]. The TenC gene expression is mechanosensitive.

8.3.6 Lysyl Oxidase

Lysyl oxidase (LOx) is a copper-dependent amine oxidase that initiates crosslinking of collagen and elastin. It participates in the maintenance of the extracellular matrix. Four LOx-like enzymes (LOxL1–LOxL4) have been identified.

Lysyl oxidase is synthesized as pre-proLOx that is transformed into pro-Lox. This proenzyme is secreted into the extracellular medium, where it is processed by bone morphogenetic protein BMP1 and other procollagen C-proteinases to release mature form and propeptide.

In the vascular wall, LOx is expressed in fibroblasts and vascular endothelial and smooth muscle cells. Transforming growth factor-β increases LOx expression and activity in cultured smooth muscle cells and lung fibroblasts. It upregulates the synthesis of proteoglycans, collagen-1 and -3, and fibronectin in smooth muscle cells [832]. Platelet-derived growth factor augments LOx production in smooth muscle cells. Granulocyte–macrophage colony-stimulating factor (CSF2) elevates both LOx and BMP1 expression in these cells.

Lysyl oxidase is a chemoattractant for vascular smooth muscle cells and monocytes. It may participate in vascular remodeling in cardiovascular diseases [832].[24]

---

[24] Lysyl oxidase downregulation is a feature of endothelial dysfunction as observed in earlier stages of atherosclerotic plaque formation. Low-density lipoproteins

8.3.7 Matrix Metalloproteinases

Proteolytic degradation and remodeling of the extracellular matrix is controlled by matrix metalloproteinases (MMP) and the *tissue inhibitors of metalloproteinases* (TIMP). Matrix metalloproteinases are endopeptidases that depend on metal ions as cofactors.

Two main MMP superfamilies can be defined: (1) secreted, soluble MMPs and (2) membrane-type MMPs (mtMMP; Sect. 7.2.4.3). Secreted MMPs are composed of an auto-inhibitory propeptide, furin cleavage site, catalytic domain with Zn^{++}- and Ca^{++}-binding motifs as well as a PRCGVPD sequence that prevents access of substrate into the catalytic site, and a flexible hinge region with a hemopexin domain that can bind proteins, such as factor Xa, thrombin, and heparin. Membrane-type MMPs contain either a transmembrane domain, such as mt1MMP to mt3MMP and mt5MMP, or are glycosylphosphatidylinositol-anchored, such as mt4MMP and mt6MMP. In addition, mtMMPs can activate some soluble MMPs. In particular, mt1MMP cleaves proMMP2 to produce MMP2.

The subclass of matrix metalloproteinases (MMP1 to MMP28) comprises several infraclasses (Table 8.15):[25] (1) collagenases, such as MMP1, MMP8, and MMP13; (2) gelatinases, such as MMP2 and MMP9, or gelatinases-A and -B, that digest denatured collagen (gelatin) and other matrix components such as those of basement membranes, i.e., intact collagen-4; (3) stromelysins, such as MMP3, MMP10, and MMP11; (4) matrilysins, such as MMP7 and MMP26; (5) metalloelastases such as zinc-dependent MMP12; and (6) membrane-type MMPs (mt1MMP–mt8MMP).

Matrix metalloproteinases are very slow enzymes that most often target many substrates (Table 8.16). Their substrates include extracellular matrix proteins as well as numerous bioactive molecules, plasmalemmal receptors, apoptotic ligands such as TNFSF6 that can be released, and chemo- and cytokines that can be activated or inactivated. They are secreted as inactive zymogens with a propeptide domain, which is removed for enzyme activation. These latent enzymes can be activated by reactive oxygen species and proteinases, among other factors.

disturb endothelial permeability. Moreover, high LDL concentration reduces LOx activity in endothelial cells. Interferon-γ in ruptured atherosclerotic plaques reduces LOx expression and activity in vascular smooth muscle cells. Homocysteine thiolactone (but not homocysteine) supresses LOx activity. Pro-inflammatory cytokines such as tumor-necrosis factor-α reduce LOx expression in endothelial cells. Conversely, LOx upregulation could induce neointimal thickening. In addition, high incidence of aortic aneurysms, aortic tortuosity, and wall fissuration with extended fragmentation of elastic fibers, disruption of smooth muscle cell contact, discontinuities of both internal elastic lamina and lamellae, and detachment of endothelial cells from the basement membrane are observed in LOx-deficient mice.

[25] Gelatinases, matrilysins (MMP7), and macrophage elastase (MMP12) that overexpressed in aortic aneurysms (Vol. 6 – Chap. 7. Vascular Diseases) degrade elastin.

Table 8.15. Types of matrix metalloproteinases and their aliases (mt$i$MMP: membrane-type matrix metalloproteinase-i; cMMP: chicken matrix metalloproteinase; caMMP: cysteine array matrix metalloproteinase; endometase: endometrial tumor-derived metalloproteinase; RASI: rheumatoid arthritis synovium inflamed; xMMP: human ortholog for Xenopus laevis matrix metalloproteinase).

Type	Alias
	Secreted, soluble MMPs
MMP1	Collagenase-1, interstitial collagenase
	Tissue collagenase, fibroblast-type collagenase
MMP8	Neutrophil collagenase
MMP13	Collagenase-3
MMP18	Collagenase-4
MMP2	Gelatinase-A, 72-kDa type-4 collagenase
MMP9	Gelatinase-B, 92-kDa gelatinase
MMP3	Stromelysin-1
MMP10	Stromelysin-2, transin-2
MMP11	Stromelysin-3
MMP19	RASI1, stromelysin-4
MMP7	Matrilysin-1
MMP26	Matrilysin-2, endometase
MMP12	Macrophage metalloelastase
MMP20	Enamelysin
MMP21	xMMP
MMP22	
MMP27	MMP22, cMMP
MMP28	Epilysin
	Membrane-associated MMPs
MMP14	mt1MMP
MMP15	mt2MMP
MMP16	mt3MMP
MMP17	mt4MMP
MMP23a	caMMP
MMP23b	
MMP24	mt5MMP
MMP25	mt6MMP, MMP20a

8.3.7.1 Regulation of MMP Synthesis and Activity

Matrix metalloproteinases are regulated at transcriptional, post-transcriptional, and post-translational levels, as well as via their localization. In addition, they are controlled by tissue inhibitors of metalloproteinases.

Table 8.16. Examples of matrix metalloproteinases (MMP) and their substrates (Source: [833]).

Type	Substrates
MMP1	Collagen-1/2/3/7/8/11,
	glycoproteins (entactin), proteoglycans (aggrecan),
	L-selectin,
	IL1β,
	MMP2, MMP9
MMP2	Collagen-4/5/6/10, elastin,
MMP3	Collagen-3/4/5/9, laminin, elastin,
	fibronectin, entactin, aggrecan, perlecan, decorin,
	plasminogen, IL1β,
	MMP2/TIMP2, MMP7, MMP8, MMP9, MMP13
MMP7	Collagen-4/10, elastin, fibronectin, laminin,
	aggrecan, decorin,
	transferrin, plasminogen,
	β4 integrin,
	MMP1, MMP2, MMP9/TIMP1
MMP8	Collagen-1/2/3/5/7/8/10, fibronectin
	aggrecan
MMP9	Collagen-4/5/7/10/14, elastin, fibronectin,
	entactin, aggrecan,
	plasminogen, IL1β,
MMP10	Collagen-3/5, elastin,
	aggrecan,
	MMP1, MMP8
MMP12	Collagen-4, elastin,
	fibronectin, vitronectin, laminin, entactin,
	fibrinogen, fibrin, plasminogen
MMP13	Collagen-1/2/3/4/9/10/14, fibronectin,
	aggrecan, perlecan
	plasminogen,
	MMP9
MMP14	Collagen-1/3,
	fibronectin, laminin, vitronectin, entactin,
	aggrecan, perlecan, decorin,
	MMP2, MMP13
MMP15	Fibronectin, entactin, laminin, aggrecan, perlecan,
	MMP2
MMP16	Collagen-3,
	fibronectin,
	MMP2
MMP18	Collagen-1
MMP19	Collagen-1

In cardiac cells, during hypoxia, MMP2 production is upregulated by angiotensin-2, endothelin-1, estrogen, melatonin, epidermal (EGF), fibroblast (FGF2), and platelet-derived (PDGF) growth factor, as well as interleukin-1β and -6, and tumor-necrosis factor-α [834].[26]

In blood, α2-macroglobulin[27] and thrombospondin-1 and -2 can bind to some MMPs and prevent their activity [834]. Complexes of α2-macroglobulin and MMP (MMP2 and MMP9) are removed by scavenger receptor-mediated endocytosis for an irreversible MMP clearance.

Full-length matrix metalloproteinases can undergo post-translational modifications, such as S-glutathiolation, S-nitrosylation, and phosphorylation. Peroxynitrite activates MMP1, -8, -9, -10, and -28 via a S-glutathiolation of the propeptide, whereas it inactivates tissue inhibitor of metalloproteinases TIMP1 and TIMP4 [834]. Phosphorylation of MMP2 by protein kinase-C reduces its activity, whereas MMP2 dephosphorylation enhances its activity.

Zymogen proMMP2 is activated by the combined action of MMP14 (or mt1MMP) and TIMP2.[28] Zymogen proMMP2 can also be activated by oxidative stress without proteolytic removal of the propeptide domain to produce an active, full-length enzyme.[29]

Many molecules act on matrix metalloproteinases. Nitric oxide inhibits MMP2 via the repressor activating transcription factor ATF3, but stimulates MMP13. Inducible cyclooxygenase Cox2 activated by cyclic strain upregulates MMP9 and MMP14 expression. Prostaglandin PGJ2 upregulates MMP1 production.

Tissue inhibitors of metalloproteinases are synthesized by endothelial and smooth muscle cells as well as fibroblasts, among other cells. Activity of TIMPs of endothelial cells is sensitive to stresses. According to its active terminal domain, TIMP2 either activates MMP2 or inhibits MMPs. Interaction between MMP14 and TIMP2 is required to activate proMMP2. The fibrinolytic system, also sensitive to stresses, which includes plasminogen activators (tPA and uPA) and inhibitors (PAI1 and PAI2),[30] regulates the conversion of plasminogen into plasmin. Plasmin participates in MMP activation.

Cyclic strains and shear stresses regulate the expression of matrix metalloproteinases by endothelial and vascular smooth muscle cells, as well as the

---

[26] Transcription of MMP2 is regulated by transcription factors, coactivators, and corepressors, such as transcription factor SPI1, Activating proteins AP1 and AP2, P53, Y-box-binding protein YBP1, and ETS1 [834].

[27] α2-Macroglobulin, encoded by the A2M gene, is a large plasma protein produced by the liver. It is able to inactivate numerous proteinases, such as serine-, cysteine-, aspartic-, and metalloproteinases. It prevents both blood coagulation by inhibiting thrombin and fibrinolysis by inhibiting plasmin and kallikrein.

[28] Full-length MMP2 is obviously heavier than proteolytically cleaved, enzymatically active MMP2 (72- and 64-kDa, respectively) [834].

[29] Oxidative stress causes S-glutathiolation of the PRCGVPD domain, hence relieving its shielding on the catalytic sequence.

[30] Endothelial PAI1 serves also as an antimigratory molecule.

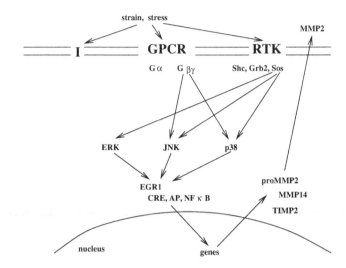

Figure 8.2. Strain promotes the synthesis of matrix metalloproteinase MMP2 in endothelial cells. Cyclic strains are detected by G-protein-coupled receptors (GPCR), receptor Tyr kinases (RTK), and integrins (I). Signaling cascades activate ERK1 and -2, P38MAPK, and JNK, then transcription factors, early growth response protein EGR1, cAMP response element (CRE), activator protein AP1, and nuclear factor-κB (NFκB), leading to the production of MMPs and TIMPs (Source: [833]).

production of proteins of the basement membrane and the interstitial matrix (Fig. 8.2).

Extracellular matrix metalloproteinase inducer increases both soluble and membrane MMPs.[31] Synthesis of MMP2 as well as of collagen-1 and -3 is controlled by nuclear factor-κB. Collagen accumulation that is observed during pressure overload-mediated cardiac remodeling can be highly repressed by proteasome inhibition [836]. Proteasome inhibition indeed suppresses NFκB expression, which rises upon pressure overload.

8.3.7.2 Extra- and Intracellular Effects

Matrix metalloproteinases exert not only extracellular effects, but also operate intracellularly. Zinc-dependent matrix metalloproteinase MMP2 proteolyzes extracellular matrix proteins to particularly remodel the collagen matrix as well as specific proteins within the cardiomyocyte, once activated by oxidative stress [834]. Matrix metalloproteinase-22 that is ubiquitously expressed in cells of the cardiovascular system, such as endothelial cells, cardiomyocytes, and fibroblasts, degrades the cytoskeletal constituent α-actinin.

---

[31] Sustained, persistent myocardial expression of this transmembrane protein can cause myocardial remodeling with fibrosis in aging mice [835].

In cardiomyocytes, MMP2 localizes to the nucleus, sarcomeres, cytoskeleton, mitochondria, and caveolae, where it can exert proteolytic cleavage. Caveolin-1 and -3 impede MMP2 activity [834]. In the sarcomere, MMP2 colocalizes with troponin-I, myosin light chain-1, α-actinin, and TIMP4 that regulates its activity. On the cytoskeleton, MMP2 can degrade α-actinin and desmin, but not spectrin [834]. Both MMP2 and MMP3 possess a nuclear localization sequence.[32]

Matrix metalloproteinases can have many functions, in addition to extracellular matrix degradation. Matrix metalloproteinase membrane type-1 can regulate the development of the microvasculature via platelet-derived growth factor-B (PDGFb) and its receptor PDGFRβ [837]. Factor PDGFb regulates the function of intramural cells. Enzyme mt1MMP is associated with the PDGFb–PDGFRβ signaling.

Matrix metalloproteinases alter the endothelial barrier by proteolysis of cell junctions. However, in normal conditions, MMP stimulation is counterbalanced by barrier upregulation by cyclic stresses. Plasmalemmal adamlysins ADAM10 and ADAM15 of the A disintegrin and metalloprotease (ADAM) family also disassemble cell junctions. Endothelial MMP9 is involved in cyclic strain-induced angiogenesis and endothelial MMP2 in hypoxia-generated angiogenesis and migration of smooth muscle cells [833].

8.3.7.3 MMPs in Tumors and Arterial Wall Diseases

Several types of proteinases, such as plasminogen activators, serine elastases, and cathepsins, expressed in atheromatous plaques (Vol. 6 – Chap. 7. Vascular Diseases), contribute to the degeneration of the vessel wall. Matrix metalloproteinases are involved in the evolution of atherosclerosis and aneurysms. Matrix metalloproteinases are naturally inhibited by TIMPs. Inhibitor TIMP3 is a constituent of the extracellular matrix and some basement membranes. Tissue inhibitors of metalloproteinases also promotes cell growth and proliferation (Vol. 2 – Chap. 2. Cell Growth and Proliferation).

The extracellular matrix is much stiffer in tumors than normal tissues, due to a high content in collagen fibers. Stiff extracellular matrices enhance Rho GTPase activity [838]. Matrix metalloproteinases frequently have high levels in tumors. They are involved not only in local cancer invasion by degrading the structural components of the extracellular matrix, but also in the release of cell-bound inactive precursors of growth factors and activate survival factors. They may also participate in cancer initiation, especially in the epithelial–mesenchymal transition.[33]

---

[32] Full-length MMP3 remains in the cytosol, whereas its truncated, active form translocates to the nucleus [834].

[33] In mammary glands, MMP3 acts via Rac1b, which raises the production of reactive oxygen species, particularly mitochondrial superoxide, leading to DNA damage and genomic instability, and eventually to tumorigenesis. Agents MMP3,

8.3.8 Periostin

Periostin is an extracellular matrix molecule of the fasciclin family secreted by fibroblasts that acts in cell adhesion, growth, and migration. Periostin is mainly expressed in collagen-rich fibrous connective tissues that are subjected to permanent mechanical stress, such as heart valves, tendons, perichondrium, cornea, and periodontal ligament. Periostin interacts with collagen-1 and regulates collagen-1 fibrillogenesis and collagen crosslinking [840].

In the heart, periostin is expressed at very early stages of embryogenesis, but in normal adult myocardium, it is only detected in valves. During normal growth of neonatal heart, cardiomyocytes undergo hypertrophy in response to various signals such as changes in blood pressure. The number of fibroblasts increase and connective tissue is formed. Coordination of postnatal signals and cell interactions allows heart adaptation to adult function. Periostin contributes to cardiomyocyte–fibroblast interactions during cardiac remodeling, as it promotes fibroblast adherence to cardiomyocytes [841].

Periostin is upregulated after environmental changes or heart injury. Periostin associates with matrix regulators, such as transforming growth factor-β, tenascin, and fibronectin.[34] Periostin-promoted cell migration involves α_V integrin, focal adhesion kinase, and protein kinase-B.[35]

8.3.9 Thrombospondins

All 5 members (Tsp1–Tsp5) of the thrombospondin family bind numerous calcium ions. They may then sequester calcium ions at the cell surface or in some regions of the extracellular matrix, where Ca^{++} can be mobilized during matrix remodeling or cell signaling [844]. All the Tsp isotypes promote synaptogenesis during postnatal development of the central nervous system as well as in adults [844]. They may also function in synapse organization.

Thrombospondin-1

Thrombospondin-1 (Tsp1) is a secreted glycoprotein that is constitutively produced to operate in the extracellular environment, where its concentration is regulated by the rates of synthesis and clearance. Its synthesis is regulated by PTen phosphatase as well as P53 and SMAD4 transcription factors [844]. Thrombospondin-1 is widely expressed during embryo- and fetogenesis.

The turnover of Tsp1 is faster than that of other matrix proteins. The clearance of Tsp1 is mediated by an interaction with proteoglycans and LDL

Rac1b, and H_2O_2 enhance the expression of Snail, a transcription repressor of E-cadherin [839].

[34] Periostin is a TGFβ-responding factor that operates in cardiac healing after acute myocardial infarction [842].

[35] Periostin participates in myocardial infarction-induced fibrosis as well as pressure overload-induced cardiac hypertrophy and fibrosis [843].

receptor-related proteins [844]. Proteins bound to Tsp1, such as matrix metalloproteinases and vascular endothelial growth factor, are taken up along with Tsp1 [844].

Thrombospondin-1 is a multimer that functions during organogenesis and tissue remodeling, as it regulates the activity of various proteases in the extracellular matrix. It binds to and inhibits the activity of plasmin, urokinase plasminogen activator, neutrophil elastase, cathepsin-G, and matrix metalloproteinases [844]. The inhibition of MMP9 contributes to the Tsp1 anti-angiogenic activity [844].

Conversely, thrombospondin-1 is a substrate for numerous proteases, such as plasmin, cathepsin-G, leukocyte elastases, and A disintegrin and metalloproteinase with thrombospondin motifs ADAMTS1. The cleavage of Tsp1 by ADAMTS1 releases its pro-angiogenic N-terminus and an anti-angiogenic fragment [844].

Thrombospondin-1 also links to extracellular matrix proteins, such as fibrinogen, fibronectin, and collagens. In addition, Tsp1 interacts with vascular endothelial (VEGF), fibroblast (FGF2), hepatocyte (HGF), and platelet-derived (PDGF) growth factor, as well as insulin-like growth factor-binding protein-5, and histidine-rich glycoprotein located in plasma and platelets (an inhibitor of fibrinolysis and antagonist of inhibitors of blood coagulation), osteonectin, tumor-necrosis factor-α-stimulated protein TSG6, tissue factor pathway inhibitor, proteoglycans, and glycosaminoglycans [844].

Thrombospondin-1 connects to various membrane proteins, such as integrins, proteoglycans, calreticulin, transforming growth factor-β, and thrombospondin receptors CD36[36] and CD47. Owing to these interactions, Tsp1 promotes intercellular interactions. Thrombospondin-1 binds to β_1, $\alpha_{2B}\beta_3$, and $\alpha_V\beta_3$ integrins, among others [844].

Thrombospondin receptor CD36 is a member of the scavenger receptor superclass that binds oxidized low-density lipoprotein, fatty acids, myristic acid, and anionic phospholipids. Thrombospondin-1 prevents CD36 fatty acid translocase activity. Interaction of Tsp1 with CD36 is involved in platelet aggregation, angiogenesis inhibition, as well as binding of platelets to monocytes and apoptotic neutrophils to macrophages [844].

Cluster of differentiation CD47[37] operates as a receptor for both Tsp1 and its counter-receptor, the transmembrane glycoproteic enzyme Signal-

---

[36] A.k.a. glycoprotein GP4 or GP3b.

[37] Cluster of differentiation CD47, also named integrin-associated protein (IAP) and ovarian carcinoma antigen OA3, is a protein associated with the Rhesus complex on erythrocytes. Agent CD47 indeed tethers to several integrins, such as $\alpha_2\beta_1$, $\alpha_4\beta_1$, $\alpha_V\beta_3$, and $\alpha_{2B}\beta_3$ integrins. Cyclooxygenase-2 regulation is mediated by association of CD47 with collagen-binding $\alpha_2\beta_1$ integrin [845]. $\alpha_{2B}\beta_3$ Integrin forms a complex with CD47 and cytosolic protein Tyr kinases Src, focal adhesion, and spleen Tyr kinase [845]. $\alpha_V\beta_3$ Integrin also complexes with CD47 and Gα_i subunit of the heterotrimeric G protein as well as kinases Src and Lyn, and non-receptor Tyr phosphatase PTPn11.

regulatory protein-α (SiRPα), or protein Tyr phosphatase non-receptor substrate PTPnS1.[38] Molecule PTPnS1 impedes receptor Tyr kinase-triggered signaling.

Peptide Tsp1 can activate integrins via CD47. In addition, Tsp1 binding of CD47 potently impedes the nitric oxide–soluble guanylate cyclase–cyclic guanosine monophosphate–cGMP-dependent protein kinase (NO–sGK–cGMP–PKG) signaling in vascular cells. As nitric oxide is a vasodilator, the Tsp1–CD47 complex disturbs the local regulation of blood flow. Besides, Tsp1, together with CD47, promotes apoptosis of endothelial cells and fibroblasts [845].

Thrombospondin receptor CD47 also mediates the interaction of Tsp1 with neutrophils [844]. Subunit Gα$_i$ of heterotrimeric G protein can associate with integrin-associated protein CD47. Consequently, CD47 ligation by Tsp1 that can favor GTP loading of Gi reduces the cytoplasmic cAMP concentration, hence protein kinase-A signaling, especially in platelets and T lymphocytes [845].[39]

Binding of Tsp1 to CD47 primes several other actions, such as histamine release from mastocytes, activation-independent agglutination of platelets, inhibition of development of naive T lymphocytes into type-1 helper T effectors, repression of FGF2-stimulated angiogenesis, dissociation of CD47 from SiRPα in vascular smooth muscle cells, enhancement of IGF1 receptor signaling, and downregulation of pro-inflammatory interleukin-12 and tumor-necrosis factor-α production by monocyte-derived dendritic cells [845].

Calreticulin localizes in the endoplasmic reticulum and plasma membrane, where it acts as a receptor for Tsp1. The interaction of Tsp1 with the complex of calreticulin and low-density lipoprotein receptor-related protein LRP1: (1) provokes the disassembly of focal adhesions via focal adhesion kinase and small GTPase RhoA and (2) activates protein kinase-B. Thrombospondin-1 binds to 2 other members of the low-density lipoprotein receptor family: very-low-density lipoprotein receptor and apolipoprotein-E receptor-2.

---

[38] Agent CD47 is a type-1 integral membrane protein that is composed of a glycosylated extracellular immunoglobulin variable-like domain, 5 transmembrane segments, and an alternatively spliced cytoplasmic C-terminus [845]. Four CD47 splice variants have been detected, from the shortest form 1 to the longest form 4. Isoform 2 of CD47 is the most widely expressed and is the dominant form in all circulating and immune cells in humans. Molecule CD47 acts as a bidirectional receptor in cellular signaling. Binding of CD47 to SiRPα initiates a signaling cascade in SiRPα-expressing cells that limits activity of phagocytic cells and regulates insulin-like growth factor-1 receptor signaling in vascular smooth muscle cells. Reverse signaling refers to SiRP binding that modifies CD47 signaling [845]. Ligation by physiological ligands Tsp1, SiRPα, and SiRPβ regulates CD47 signaling.

[39] In T lymphocytes, CD47 connection to Tsp1 causes phosphorylation of extracellular signal-regulated protein kinase, a member of the mitogen-activated protein kinase module. In smooth muscle cells, Tsp1 precludes ERK phosphorylation, but activates Jun N-terminal kinases [845].

Thrombospondin-1 is expressed by T lymphocytes, dendritic cells, and macrophages. It is upregulated in response to TGFβ in the 2 last-mentioned cell types. Linkage of Tsp1 to $\alpha_4\beta_1$ integrin promotes T-cell adhesion and chemotaxis [844]. Interaction of Tsp1 with CD47 on dendritic cells decreases cytokine production. This interaction also prevents early T-cell activation, but promotes the conversion of memory T cells into regulatory T cells [844]. The Tsp1–TGFβ pathway regulates the helper 17 T-cell response. Thrombospondin-1 activates the latent TGFβ complex, especially during wound healing and immune response.[40] It also tethers to latency-associated peptide to control monocyte movement [844].

Thrombospondin-1 suppresses nitric oxide activity and angiogenesis via its engagement with CD36 and CD47. It regulates thrombosis by interacting with von Willebrand factor, nitric oxide, and platelet integrins [844]. On the one hand, Tsp1 binds von Willebrand factor and protect it from cleavage by ADAMTS13. On the other hand, Tsp1 and von Willebrand factor both bind platelet glycoproteic complex GP1b–GP5–GP9. Thrombospondin-1 can hence partially inhibit platelet adhesion and rolling on von Willebrand factor.

By interacting with CD36, Tsp1 inhibits endothelial cell migration via SRC family kinase Fyn and P38MAPK, causes apoptosis via caspase-3 and tumor-necrosis factor axis components TNFSF6 and TNFRSF1a or death receptors [844]. It promotes VEGF clearance and precludes VEGFR2 activation via association of CD36 with β_1 integrins. It hampers the production of proangiogenic matrix proteins. Moreover, it can suppress PKB activation in response to vascular endothelial growth factor and endothelial cell proliferation via P53 and CKI1a as well as VLDLR and the PKB–MAPK pathway [844].

8.4 Integrins in Matrix Remodeling

At least 20 integrins serve as receptors for matrix constituents, such as laminins, collagens, fibronectin, osteopontin, tenascins, and vitronectin (Table 8.17).

Plasmalemmal integrins contribute to the organization and maintenance of the extracellular matrix. They tether the extracellular matrix to the cytoskeleton. They operate as sensors and transducers of applied mechanical forces. Actin microfilaments and, to a lesser extent, microtubules modulate the time-dependent stiffening of focal adhesions that experience mechanical stresses exerted by the environment [847].

In cells such as vascular smooth muscle cells, focal adhesions are involved in interactions between integrins and the extracellular matrix that are

---

[40] Factors TGFβs are synthesized as precursors. Afterward, TGFβ homodimer complexed to latency-associated peptide forms the small latent TGFβ complex. This complex remains in the cell until its binding to latent TGFβ-binding protein to build large latent TGFβ complex that is secreted to the extracellular matrix.

Table 8.17. Matrix constituents and their cellular receptors, the integrins. (Source: [846]).

Matrix constituent	Cell-surface receptors (mostly integrins [$\alpha_i \beta_j$ dimers])
	Basal lamina
Collagen-4	$\alpha_1\beta_1$, $\alpha_2\beta_1$, $\alpha_3\beta_1$, $\alpha_V\beta_3$
Laminins	$\alpha_1\beta_1$–$\alpha_3\beta_1$, $\alpha_6\beta_1$, $\alpha_6\beta_4$, $\alpha_7\beta_1$, $\alpha_V\beta_3$
	Dystroglycan
Perlecan	Dystroglycan
Nidogen	$\alpha_3\beta_1$
Collagen-1/3	$\alpha_1\beta_1$, $\alpha_2\beta_1$, $\alpha_V\beta_3$
Fibrillin-1	$\alpha_V\beta_3$
Fibronectin	$\alpha_1\beta_1$, $\alpha_5\beta_1$, $\alpha_8\beta_1$, $\alpha_V\beta_3$
Nephronectin	$\alpha_8\beta_1$
Osteopontin	$\alpha_8\beta_1$, $\alpha_V\beta_3$
Tenascin-C	$\alpha_8\beta_1$, $\alpha_V\beta_3$
Vitronectin	$\alpha_V\beta_3$

responsible for mechanotransduction. Fibronectin connects to collagen and proteoglycans on the one hand and $\alpha_3\beta_1$, $\alpha_5\beta_1$, $\alpha_V\beta_1$, $\alpha_V\beta_6$, and $\alpha_{2B}\beta_3$ integrins on the other. However, $\alpha_5\beta_1$ integrin only serves as fibronectin receptor in all studied tissues. Proliferation of fibroblasts from anterior cruciate ligament is not influenced by the type of culture subtrate (collagen-1, laminin, elastin, and fibronectin), but that of fibroblasts from medial collateral ligament is faster on fibronectin-based matrix with respect to collagen [848].

In addition, integrin endocytosis and recycling influences the deposition and reorganization of matrix constituents such as fibronectin. In myofibroblasts, caveolin-dependent internalization of $\alpha_5\beta_1$ integrin is required for endocytosis and turnover of matrix fibronectin [849]. In endothelial cells, endocytosis and subsequent recycling of active $\alpha_5\beta_1$ integrins occurs primarily at plasmalemmal regions that are connected to fibronectin-abundant matrix regions.

Neuropilin-1, independently of its receptor function for vascular endothelial growth factor isoform VEGFa$_{165}$ and semaphorin-3A, promotes adhesion of endothelial cells to fibronectin via endocytosis $\alpha_5\beta_1$ integrins in Rab5+ early endosomes [850].[41]

---

[41] Neuropilin-1 is a coreceptor for vascular endothelial growth factor isoform VEGFa$_{165}$ and semaphorin-3A. It is also required for endothelial response to VEGFa$_{121}$ isoform that does not bind Nrp1. Neuropilin-1 binds to homomultimeric endocytic adaptor GAIP-interacting protein C-terminus GIPC1 and interacts with $\alpha_5\beta_1$ integrin at adhesion sites for integrin endocytosis. Agent GIPC1 indeed also interacts with $\alpha_5\beta_1$ integrin and nanomotor myosin-6 to support endocytosis.

$\alpha_5\beta_1$ Integrin is initially recruited to focal adhesions that contain $\alpha_V\beta_3$ integrin. $\alpha_5\beta_1$ Integrin then leaves these focal adhesions, moves, and stimulates fibronectin fibrillogenesis in other sites that connect to actin stress fibers for stress-adaptative focal adhesion redistribution [851]. In endothelial cells, VEGF-mediated fibronectin polymerization and deposition of a fibrillar fibronectin matrix results from Rab4-dependent recycling of $\alpha_V\beta_3$ integrin from early endosomes to the plasma membrane [852].[42] $\alpha_V\beta_3$ Integrin activates Rho GTPase, thereby causing cell contraction.

Initially, fibronectin monomers associate primarily with plasmalemmal $\alpha_5\beta_1$ integrins. Fibronectin polymerization then proceeds as strains that result from Rho-mediated contractions of actin–myosin stress fibers transmitted by $\alpha_5\beta_1$ integrins exhibit cryptic sites of fibronectin N-termini.

Binding of fibronectin to $\alpha_5\beta_1$ integrins is influenced by numerous factors [853]. Platelet-derived growth factor homodimer PDGFbb and lysophosphatidic acid reduces and raises integrin–fibronectin linkage, respectively. Moreover, after engagement, the force required to disrupt integrin–fibronectin binding increases with augmented contact time.

Cardiac hypertrophy is associated with remodeling of the extracellular matrix. In particular, fibronectin and collagen move during cardiac hypertrophy. Collagen and fibronectin receptors, the α_1 and α_5 integrins, redistribute from a diffuse to sarcomeric banding pattern [854]. Furthermore, the synthesis rate of integrins increases during cardiac hypertrophy. Protein phosphatase-2A dephosphorylates β_1 integrin and promotes cardiac hypertrophy.

8.5 Cell–Matrix Mechanical Coupling

Forces are applied to the cell by the cell environment, and conversely forces are produced by the cell and transmitted via the cytoskeleton and the adhesion sites to the surrounding medium. The interface between the cell and extracellular space is characterized by plasmalemmal plaques composed of integrins, adaptors, and anchor proteins. Numerous stress-induced responses of cells, particularly endothelial cells, which must withdraw shear stress and stretch exerted by the flowing blood on its wetted surface, require integrins.

Focal adhesions sense loadings undergone by the cell. The size of focal adhesions increases when any loading is applied to the cell, the number of molecular interactions rising due to conformational changes of proteins involved in adhesion sites or molecule recruitment. Conversely, the size of focal adhesions decreases when loading is removed [855].

Stretch-dependent growth of cell adhesion sites is mediated by small GTPase Rho via RoCK kinases and Diaphanous Dia1. RoCK kinase activates myosin-2. Agent Dia1 primes focal adhesion assembly and regulates

---

[42] Integrin recycling regulated by Rab4a GTPase is required for VEGFR1-induced branching angiogenesis.

microtubule dynamics. Microtubules stop focal adhesion growth and suppress cellular contractibility via microtubule-associated motors such as kinesin and microtubule-associated signaling molecules such as RhoGEF2. Guanine nucleotide-exchange factor RhoGEF2 stimulates Rho GTPase when it is not bound to microtubules.

The simple thermoelastic theory for the equilibrium state of adhesion is based upon the competition between binding tendency of adhesion molecules (enthalpy gain) and mobility tendency (entropy gain) [856]. A balance of these contributions results in a contact zone with partial cell adhesion, with bound and free partners such as integrins. In the absence of force, mobile integrins migrate into the contact zone. The probability of bond formation thereby increases. Applied forces affect binding and unbinding rates in the contact zone and augment adhesion energy density.

Using reflection interference contrast microscopy to identify adhesion complexes and applying forces (4 ± 2 pN) to membrane-bound paramagnetic beads by magnetic tweezers, the occurrence of force-induced adhesion strengthening is shown to be caused by the thermoelastic response to deformation of the plasma membrane [856]. Secondarily, the regulated actin cytoskeleton couples to adhesion domains and stiffens. Cytoskeletal remodeling due to external mechanical forces happens over a time scale of minutes. At shorter time scales, the cell response is dominated by the physical properties of the plasma membrane. In between-cell contact domains before adhesion, binding partners are mobile, whereas in cell–matrix, the counterpartner is fixed. Membrane deformation is associated with bond formation that depends on adhesion partner mobility. Cell adhesion and resistance to deadhesion are enhanced when both binding partners are mobile rather than when a single one diffuses in the membrane. Cell mechanosensing conducts a response such that adhesion structures grow and strengthen under force. Cell mechanosensing appears to be enhanced in cell–cell adhesion rather than in cell–matrix adhesion.

Extracellular strains and stresses are transmitted to cells. They can be sensed by appropriate molecules and converted into chemical signals either at the cell surface and possibly, when they exert on cell–cell and cell–matrix adhesion regions, within the cell, as they can be relayed by cytoskeleton filaments. Cytoskeletal elements permanently remodel for cell adaptation. However, a sufficiently stable, prestressed[43] cell cytoskeleton connects cell units and withstands any applied force. In a cell subjected to a stress field, the cytoskeleton reorientates and cell organelles change their positions.

Mechanotransduction is a mechanism by which applied mechanical forces are translated into chemical signals. Mechanosensitive receptors trigger signaling cascades that initiate: (1) release of stored active molecules for adaptation to stress and strain and (2) gene expression. Newly synthesized and released stored molecules can then cause intracrine regulation or are secreted outside the cell to prime local control (auto-, juxta-, and paracrine regulation).

---

[43] Myosin-2 is able to generate prestress in the cytoskeleton [857].

Nucleoskeleton that withstands mechanical loads deforms in response to applied forces. These loads can also stretch nuclear-pore complexes that control nuclear import and export. The nucleoskeleton can transmit stress and strain from the nuclear envelope receptors down to chromatin network, the organization of which influences DNA accessibility. Moreover, it can modulate the activity of transcription and splicing factors as well as self-assembly of regulatory complexes involved in gene regulation [857].

9

Intracellular Transport

Directional transport of cytoplasmic cargos (mRNA, proteins, lipids, and organelles), achieves translocation of signaling mediators, establishes cell polarity, and supports cell division and differentiation. Cells possess an adequate machinery to transport substances across membranes and through the cytosol between various organelles and the cell surface.

Intracellular molecular transport relies on a pathway, i.e., an organized set of intermediate chemical reactions that can be represented by a series of discrete events from the uptake of cargos into a membrane patch, formation of cargo-containing vesicle initiated by a progressive invagination of a membrane nanodomain, subsequent detachment, transfer down to vesicle fusion with target membrane, to cargo release at its destination, whatever the function to be fulfilled (Table 9.1).

Molecular transfer occurs in both directions between cell compartments. Bidirectional fluxes encompass import fluxes that follow uptake of materials

Table 9.1. Different steps of intracellular transfer of molecules (t[v]SNARE: target [vesicle] membrane soluble ^{N}ethylmaleimide-sensitive factor [NSF] attachment protein [SNAP] receptors). Many mechanisms ensure cargo sorting and destination compartment identity.

Stages	Comments
Molecular assembling	Formation of coat and adaptor complexes; cargo sorting
Membrane budding	Curved patches by membrane bending-generating proteins
Membrane fission	Dynamin-induced scission of vesicles
Mass transfer	Traveling using nanomotors on cytoskeletal tracks
Membrane tethering	Establishment of contact with the correct target membrane (loose connection); tethering components include long coiled-coil proteins and large complexes
Membrane fusion	vSNARE–tSNARE interaction
Material release	

from the environment (endocytosis) and export fluxes (exocytosis) that deliver to the plasma membrane products necessary for cell fate or secreting synthesized compounds to cell surroundings, such as matrix constituents or auto-, juxta-, and paracrine regulators, as well as substances that are conveyed away in blood and lymph, such as endocrine factors. Cholesterol ingress and egress are illustrated in Fig. 9.1.

9.1 Transport Features

Each transport stage comprises simultaneous processes. For example, early stages of vesicle formation include cargo recruitment, membrane bending, and clathrin polymerization, which happen synchronously. Various involved proteic and lipidic interactors form the so-called *cellular transport interactome* [860].

9.1.1 Mass Conservation at Cellular Membranes

Endo- and exocytosis must be coupled to ensure a constant cell surface area. Internalization of extracellular molecules with plasmalemmal and recruited cytosolic components is balanced by delivery of new lipids and proteins to the plasma membrane. Conversely, inhibited secretion impedes uptake.

9.1.2 Transport Tubules and Vesicles

The macromolecular transport network comprises a set of interconnected tubulovesicular membranous elements. Early endosomes actually include tubular components that serve as sorting platforms. These tubular components encompass: (1) retromer-labeled endosome-to-trans-Golgi network transport carriers (ETC) and (2) tubular sorting endosome (TSE) or tubular endosomal network (TEN). Early-to-late endosomal transition occurs by maturation during which endosomes progressively incorporate a rising number of internal vesicles to form the so-called *multivesicular bodies* (MVB), or multivesicular endosomes (MVE).

   In the Golgi body that is characterized by a high rate of exchange among Golgi cisternae, cargo moves between 2 different Golgi membrane domains. Golgi enzymes and transmembrane cargo are actually partitioned between 2 distinct lipid phases; transport intermediates bud from regions that are enriched in export domains, whereas processing domains are enriched in Golgi enzymes [861]. Cargos exit the Golgi body at an exponential rate proportional to their total Golgi concentration without lag or transit time, unlike regular progression in Golgi cisternae. Incoming cargos rapidly mix with ready ones. Furthermore, a Golgi-based signaling network is activated by transport to monitor and balance transport rates into and out of the Golgi body [862].

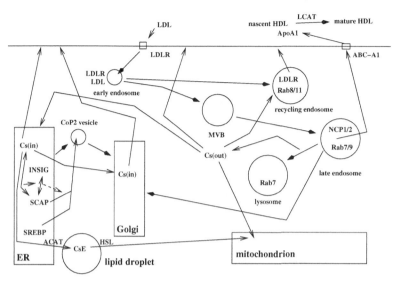

Figure 9.1. Cholesterol intracellular transfer that partially involves endosomes and several Rab GTPases illustrates exo- and endocytosis (Source: [859]). Low-density lipoprotein receptors (LDLR) bind LDLs that are then imported in clathrin-coated pits for delivery to early sorting endosomes. These receptors are recycled back to the plasma membrane (PM), whereas LDLs are delivered to late (LE) and multivesicular endosomes (MVB) for hydrolysis by acid lipase. Free cholesterol exits endosomes to be incorporated in other cell compartments. Cholesterol partners include sterol regulatory element-binding protein (SREBP) cleavage-activating protein (SCAP), Niemann-Pick-C protein (NPC), oxysterol-binding protein (OSBP), and OSBP-related proteins (ORP). Proteins NPC1 and NPC2 in MVBs and cathepsin-D in LEs favor endosomal egress of cholesterol into the cytosol for delivery to membranes of recycling endosomes, endoplasmic reticulum (ER), and mitochondria, as well as PM. Protein ORP1 promotes LE displacement toward the minus-end of microtubules. Synthesized cholesterol into ER is carried to PM, via or mostly bypassing the Golgi body. Excess cholesterol in ER is esterified by acylCoA cholesterol acyltransferase (ACAT) and stored in lipid droplets. Cholesterol derived from lipid droplets by hydrolysis of cholesteryl esters due to hormone-sensitive lipase (HSL) as well as other sources is used by mitochondria for steroidogenesis in steroidogenic cells. When sterol level is low, SREBP in ER is carried to and processed into the Golgi body. The activated fragment is then conveyed to the nucleus for transcription. Sterols inhibit the ER–Golgi transport of SREBP, as SREBP escort SCAP complexes with ER anchor insulin-induced gene product (InsIG). Retention protein InsIG prevents SCAP recruitment to CoP2+ vesicles and SREBP–SCAP transport to the Golgi body. Cholesterol efflux from cell to apoprotein-A1 involves ATP-binding cassette transporters ABCa1 to generate nascent high-density lipoproteins (HDL). Lecithin–cholesterol acyl transferase (LCAT) in the interstitial space esterifies cholesterol and produces globular HDLs. Transporter ABCg1 in macrophages cooperates with ABCa1 by adding lipids from other cells to nascent HDLs.

Proteic chaperones leave the endoplasmic reticulum and reach the Golgi body to bind to KDEL receptor and activate Src kinases.

Multivesicular bodies with their intraluminal vesicles and also, sometimes, onion-like spires of internal membrane correspond to an endocytosis stage between early, peripheral, Rab5+ endosomes and the terminal, lysosomal compartment. Multivesicular endosomes are therefore able to construct carrier membranes in- and outwardly. On the one hand, nascent multivesicular bodies sort plasmalemmal proteins and lipids within emanating spherical or tubular carriers oriented toward the surrounding cytosol. These tubules carry cargos to other intracellular destinations, such as the trans-Golgi network and recycling endosomes. On the other, inward deformation of the encapsulating endosomal membrane and internal packaging of selected cargos into limiting membranes generate intraluminal vesicles. These intraluminal vesicles partition plasmalemmal molecules into the multivesicular body lumen. This spatial remodeling isolates internalized receptors from the surrounding cytosol, thereby terminating signal transmission and preparing receptors for destruction by the lysosomal compartment. Endosomal-sorting complexes required for transport (ESCRT; Sect. 9.1.13.9) assemble on the endosome membrane to produce intraluminal vesicles, hence coordinating cargo concentration.

9.1.3 Molecular Transport and Cell Signaling

Molecule transports regulate not only the composition of the plasma membrane, but also the interaction between the cell and its environment. The set of interconnected tubulovesicular transport elements regulate various sorting and signaling events. Molecular internalization and intracellular transport actually participate in cell signaling (Sect. 9.5).

A common consequence of endocytosis is signaling termination at the plasma membrane. On the other hand, signal transduction triggered at the plasma membrane can continue in a different manner inside the cell from endosomes.

Exo- and endocytosis modulate the presence of receptors in the plasma membrane and, hence, regulate the assembly of plasmalemmal signaling platforms. They can also control the recruitment of effectors at the plasma membrane or intermediate stations along the endocytic route. Ligand availability can also be controlled by endocytosis (e.g., ligands of the Notch receptor family; Vol. 3 – Chap. 10. Morphogen Receptors).

In summary, signaling: (1) can continue from endosomal receptors using another pathway; (2) can be initiated secondarily from the plasma membrane after receptor recycling, after a more or less long delay, whether recycling takes a fast or slow route;[1] or (3) can end when endocytosis leads to lysosomal degradation of transported receptors.

---

[1] Endocytosis can also contribute to resensitization of inactivated, internalized receptors. G-protein-coupled receptors (Vol. 3 – Chap. 7. G-Protein-Coupled Receptors) that are phosphorylated by GPCR kinases bind to β-arrestins that

9.1.4 Node and Hub-Based Modeling

Intracellular transfer pathways can be described as compartments that exchange their membranes and cargos in an organized sequence of events between cell constituents. Material fluxes are strongly controlled owing to various sources of signaling, especially signaling molecules in trafficking membranes.

The transport pathways can be modeled by a set of nodes and hubs at each stage of molecular interactions. A node corresponds to a chemical that interacts with a small set of substances to execute a given function after experiencing a modification, such as phosphorylation by a kinase, in eventual association with its adaptor(s) and/or activator(s). A hub represents a molecule that interacts with many partners at different times or locations (dynamical hubs) or simultaneously (static hubs), thereby connecting several pathways. Transport pathways can contain common modules with defined functionalities that mostly comprise nodes and static hubs.

9.1.5 Membrane Curvature

Membrane curvature (Sect. 4.3.5) contributes to the generation of vesicular or tubular structure in the plasma membrane. Endophilin with its membrane-binding BAR domain is required in the formation of highly curved membrane patches. These domains can interact with dynamin and actin regulatory proteins to form narrow tubular invaginations of the plasma membrane.

Proteins that coordinate membrane curvature determine initial sites of molecule internalization, as they recruit the endocytic machinery for formation of clathrin-coated vesicles. Membrane-sculpting FCH domain-only proteins FCHO1 and FCHO2 that form crescent-shaped dimers are recruited to sites of clathrin-coated pits [863]. Their recruitment precedes that of the coat protein clathrin and the adaptor protein AP2 complex. Moreover, they act as nucleators of clathrin-coated pits that direct clathrin and AP2 association and vesicle budding. Both FCHO1 and FCHO2 interact with 2 components of the vesicular transfer machinery: epidermal growth factor receptor (EGFR) pathway substrate EPS15 and intersectin-1 that recruit AP2 and clathrin to initiate maturation of clathrin-coated pits. The AP2 complex and clathrin then bind to cargos destined for intracellular transfer by clathrin-coated vesicles.

9.1.6 Membrane Deformation and Scission

During molecule transfer, vesicles or tubules bud from a donor membrane and then detach. Proteins that produce membrane curvature and fission

---

prevent Gs-subunit recruitment (Vol. 4 – Chap. 8. Guanosine Triphosphatases) and terminate signaling. β-Arrestins indeed also bind to clathrin, thereby targeting receptors for clathrin-mediated endocytosis. Yet, in early endosomes, GPCRs can be dephosphorylated by PP2a phosphatase to be recycled to and resensitized at the cell surface.

generally remain located on the cytosolic side of the vesicle during budding. Coat proteins, such as clathrin and CoP1 or CoP2 complexes, polymerize on the external side of vesicles. Proteins involved in membrane deformation that localize on the luminal side of the nascent transport shuttle are carried with it [864]. Afterward, dynamin GTPase provokes the scission of vesicles.

The formation of multivesicular endosomes starts with the generation of invaginations in endosomes that give rise to intraluminal vesicles. The ESCRT complexes (Sect. 9.1.13.9) are involved in morphogenesis of intraluminal vesicles as well as sorting of ubiquitinated cargos into these intraluminal vesicles. They lodges on the cytosolic side of endosomes, which corresponds to future luminal edge of intraluminal vesicles [864].

9.1.7 Membrane Fusion

Molecule endocytosis and secretion is mainly achieved by transport of vesicles between donor and acceptor membranes. Cargos are delivered once donor and acceptor membranes fuse. In addition, although transport between cellular membranes is mostly due to spherical vesicles (diameter 60–100 nm), tubules and larger transport structures exist, either from membrane budding or vesicle merging (e.g., fusion between endosomes and/or lysosomes to form multivesicular endosomes).

Membrane fusion is the usual process in cell endo- and exocytosis. It is regulated by many proteins, such as membrane-anchored soluble ^{N}ethylmaleimide-sensitive factor attachment protein (SNAP) receptors (SNARE) and protein phosphatases.[2] Conformational changes of these proteins at least may assist in the formation of a fusion neck delineating a pore between the 2 membranes in contact. A fusion neck can be formed quickly (time scale of order $\mathcal{O}[100\,ns]$) [867]. Three families of SNARE proteins exist: (1) vesicle-associated membrane proteins (VAMP); (2) syntaxins located on target membranes; and (3) synaptosome-associated proteins (SnAP). These proteins interact to form a SNARE complex.

9.1.8 Nucleocytoplasmic Transport

The double lipid bilayer of the nuclear envelope that separates the nuclear content from other intracellular compartments must be crossed to fullfil

---

[2] For example, cytosolic complexin interacts with the SNARE complex of synaptic vesicles and cell membranes to regulate neurotransmitter release. Complexin accessory helix competes with synaptobrevin (a.k.a. vesicle-associated membrane protein [VAMP]) for the same binding site in the SNARE complex. Complexin binds to VAMP and then acts as a fusion inhibitor [865]. On the other hand, its N-terminus serves as a fusion enhancer [866]. Complexin impedes membrane fusion to avoid spontaneous fusion by SNARE proteins, thereby allowing appropriate secretion of neurotransmitters and other mediators, once fusion inhibition is relieved. The dual function of complexin can both suppress spontaneous fusion and activates fast calcium-mediated fusion.

basic cellular processes. The nucleocytoplasmic transport yields bidirectional transfer of RNAs and proteins.

The nucleocytoplasmic transport regulates RNA processing, protein synthesis, signal transduction, cell division cycle, and circadian rhythm [868].[3] Controlled molecular synthesis and signaling takes place in both the nucleoplasm and cytoplasm and hence necessitate molecule exchange between these 2 compartments.[4]

Numerous channels and carriers reside in cellular membrane that enable the translocation of ions and small molecules. Specialized proteins maintain pores in the nuclear envelope, which are continuous with the lumen of the endoplasmic reticulum.[5] Molecule flux across the lipid bilayer occurs via large nuclear-pore complexes (caliber $\sim$10 nm) that span the nuclear membrane (Sect. 4.4.4.2).

9.1.8.1 Nuclear-Pore Complexes and Carriers

Nuclear-pore complexes can carry large cargos (caliber $\sim$25 nm, weight 1–3 MDa), such as ribosomal subunits and proteasomal precursors. Ions and small molecules with a characteristic size less than about 5 nm ($< \sim$40 kD) can diffuse across the nuclear pore in few minutes [868]. Cargos larger than the passive diffusion limit ($\sim$50 kDa) need specific transport receptors to cross pores. These specific transport receptors recognize nuclear transport signals on proteins. *Nuclear localization signal* (NLS) is a sequence observed in many proteins. Nuclear import or export carriers recognize cargo and shuttle back and forth through the nuclear pores [868] (Table 9.2).

Nuclear-pore complexes are composed of *nucleoporins*. Peripheral nucleoporins operate as specific assembly and disassembly sites for import and export complexes.

The largest class of transport receptors comprise members of the family of karyopherins-β (KaPβ), or importins-β (transportins, importins, and exportins). In other words, most of the importins and exportins are members of the karyopherin-β family. Transportins such as importin-13 can transport cargo in both directions. All other carriers serve as importins or exportins.

---

[3] DNA replication and transcription necessitate nuclear in- and outflux of molecules. Nuclear import of histones is required during S phase of the cell cycle to package newly replicated DNA.

[4] Transcription and translation are spatially separated by the nuclear membrane. These processes thus need nucleocytoplasmic transport, especially import of transcription factors and export of ribosomal subunits and mRNA transcripts for ribosome assembly and protein translation in the cytoplasm.

[5] The pore size is large enough to allow the passage of macromolecules, such as ribosomes (that contain about 75 different proteins) and nucleic acids, especially mRNA.

Table 9.2. Carriers of the nucleocytoplasmic transport (Sources: [868, 869]; CAS: cellular apoptosis susceptibility protein; NTF: nuclear transport factor; TAP: tip-associated protein). Nuclear-pore complexes are composed of nucleoporins (NuP) that operate as soluble carriers and accessory proteins such as nucleoporin-interacting proteins (NIC). Nucleoporin domains collapse when nuclear transport receptors act. Most of the proteins that carry cargo through nuclear-pore complexes are karyopherins (KaP) that carry cargo directly or via adaptors. Certain molecules such as ribosomal protein-L23A (RPl23a) are imported by multiple carriers. Molecules imported into the nucleus are divided into different sets according to their nuclear import signals: (1) proteins with a nuclear localization signal motif that requires importin-α–β heterodimer and (2) proteic complexes such as spliceosomal U small nuclear ribonucleoprotein that needs 2 copies of importins-β.

Molecule	Function
	Carriers
	Importins
Importin-β	Import of importin-α, XRIPα, snurportin, ribosomal proteins
Importin-5	Import of ribosomal proteins
Importin-7	Import of ribosomal proteins, histone H1
Importin-11	Import of ubiquitin conjugase subunit UbcM2
Transportin-1	Import of mRNA-binding proteins, ribosomal proteins
Transportin-SR	Import of mRNA-binding proteins
Hmtr10	Import of ribosomal proteins, histones H2A/B
NTF2	Import of Ran
	Exportins
Exportin-1	Export of proteins with nuclear export signal motifs, snurportin-1
Exportin-4	Export of eIF5A
Exportin-5	Export of phosphorylated proteins Import of RPA
Exportin-t	Export of tRNAs
CAS	Export of importin-α
	Non-karyopherin carriers
TAP	Export of pre-mRNAs
Calreticulin	Export of steroid receptors

Some karyopherins are importins, as they import cargos into the nucleus.[6] Import karyopherins are heterodimers of importin-α and -β. Importin-α binds the NLS-containing cargo in the cytoplasm. Importin-β docks the cargo at the cytoplasmic side of the nuclear-pore complex and mediates transfer through

---

[6] Nuclear import of proteins that possess a nuclear localization signal requires the import receptors.

the nuclear-pore complex. On the other hand, exportins remove cargos from the nucleus.

Importin-β carries cargo either directly or via adaptor. Most KaPβs bind cargos directly, but KaPβ1 commonly uses adaptors (or transport factors) such as KaPα isoforms to recognize cargo. In fact, adaptors snurportin and importin-α bind to importin-β.

However, transport factors are not all related to the karyopherin family. In addition, proteins such as β-catenins interact directly with nuclear-pore complexes and translocate across nuclear-pore complexes without soluble factors.

Different importin types can have opposite effects: import of transcriptional regulator Snail homolog of E-cadherin, which depends on importin-β, is prevented by importin-α3 and -α5.

Nuclear RNA export factor NxF1 and Transporter-associated with antigen processing (TAP) mediates export of messenger RNA that possesses the *constitutive transport element* (CTE). Protein NxF1 heterodimerizes with NTF2-like export factor NxT1 and binds nucleoporins. More precisely, TAP is an export factor for pre-mRNA.[7] Small protein NxT1 is a cofactor for TAP.

Another carrier is calreticulin. Cytoplasmic calreticulin binds to the glucocorticoid receptor and other nuclear receptors. In the absence of cortisol, glucocorticoid receptor is cytoplasmic. Ligand binding exposes nuclear localization signal motifs and allows translocation to the nucleus. DNA-Binding domain of glucocorticoid receptor serves also as a calreticulin-binding site. Therefore, calreticulin helps to switch off the gene expression after ligand dissociation by blocking DNA rebinding and by promoting export from the nucleus [868]. Calreticulin could either act as an export carrier for steroid receptors or transport cofactor that promotes cargo loading onto karyopherin.

Transport across the nuclear pore can be supported by a concentration gradient. On the other hand, small GTPase Ran can be required for cargo accumulation in a compartment against a concentration gradient. Ran^{GTP} promotes the assembly of export complexes and the disassembly of import complexes.

Proteins 14-3-3 can operate as cytoplasmic retention factors for certain phosphorylated proteins. β-Catenins can be sequestered by specific transcription factors in the nucleus. In addition, modified cargo (e.g., phosphorylated) in one compartment cannot interact with the carrier.

Binding of nuclear-pore complexes and nuclear transport factors, such as nucleoporins, karyopherins, and Ran guanine nucleotide-exchange factor, with genes can regulate transcription.[8] Transcription activators and their

---

[7] Pre-mRNAs are assembled into ribonucleoproteic complexes before nuclear export for cytoplasmic processing. Different pre-mRNA types recruit different ribonucleoproteins. Some ribonucleoproteins are removed prior to nuclear export. Others remain bound during translocation before splicing and are re-imported into the nucleus.

[8] In Saccharomyces cerevisiae, most nucleoporins and karyopherins preferentially associate with highly transcribed genes and genes that possess Rap1-

coactivators recruit, at least in some species, nucleoporins to nuclear cortex platform for full transcriptional induction.[9]

9.1.8.2 Nuclear Import and Export Signals

Importin-α binds to *nuclear localization signal*, an import motif of the structure of proteins. The NLS sequence can be processed or masked to impede the protein transfer into the nucleus.[10] Exportin targets *nuclear export signal* (NES).[11] The NES sequence can also undergo post-translational modifications and be masked.[12] Nuclear export and localization signals can require formation of a complex that controls nucleocytoplasmic transport.

Many transport receptors regulate a set of cargos. Each transport receptor recognizes a specific subset of NLSs and/or NESs, and thereby governs the transport of selected cargos. Nucleocytoplasmic transport can be controlled by modulation of transport receptors. In addition, KaPβ abundance as well as its cognate cargo level vary. Moreover, different transport receptor types can compete for a limited number of NPC binding sites.

Most terminal signaling effectors contain a nuclear localization signal that binds to importin-α with which they translocate into the nucleus. In addition, extracellular signal-related kinase ERK2, mitogen-activated protein kinase kinase MA2PK1, and SMAD3, which lack NLS,contain a 3-amino acid domain (Ser-Pro-Ser [SPS] for ERK2 and Thr-Pro-Thr [TPT] for MAP2K1 and SMAD3) that, upon phosphorylation, binds to importin-7 for nuclear import, thereby serving as a general *nuclear translocation signal* for NLS-lacking proteins [872]. Other regulators that shuttle between the cytolasm and nucleus and possess a nuclear translocation signal include SMAD2 and -4, JNK2, PKB, STAT4, APC, P53, among others.

---

binding sites, whereas RanGEF preferentially links to transcriptionally inactive genes [870]. Transcriptional activation of the GAL genes causes their association with nuclear-pore proteins, relocation to the nuclear cortex, and loss of RanGEF linkage.

[9] In yeast, Rap1 coactivators Gcr1 and Gcr2 bridge nuclear-pore complex and transcriptional machinery via NuP84 [871].

[10] Phosphorylation by PKB of nuclear localization signal of forkhead transcription factor FoxO4 prevents its nuclear import. Agent IκB masks nuclear localization signal of NFκB for cytoplasmic sequestration.

[11] Protein Tyr phosphatase CDC25 that regulates the cell cycle possesses both nuclear localization signal and nuclear export signal sequences. Its dominant nuclear localization signal motif is shielded when it is phosphorylated and then attached to 14-3-3 protein; consequently, CDC25 accumulates in the cytoplasm.

[12] Transcription factor NFAT contains nuclear export signal and one or 2 nuclear localization signal motifs depending on NFAT isoform. Calcineurin (PP3) causes NFAT nuclear accumulation by shielding its nuclear export signal. Protein kinase-A favors 14-3-3–NFAT binding that masks NFAT nuclear localization signal motif, hence causing cytoplasmic sequestration. In response to cellular stress, P53 homotetramerizes and is retained in the nucleus by NES masking, hence prevented to interact with exportin-1 (a.k.a. Crm1).

9.1.8.3 Transport Directionality

Nuclear-pore carriers bind the small GTPase Ran. Transport directionality for KaPβ family members is, indeed, achieved via Ran. Monomeric GTPase Ran is characterized by cell compartmentation with predominantly nuclear Ran^{GTP} and cytosolic Ran^{GDP}. Its regulator, the guanine nucleotide-exchange factor regulator of chromosome condensation RCC1 and GTPase-activating protein RanGAP are restricted to the nucleus and cytoplasm, respectively [873].[13]

Karyopherins bind to nucleoporins and form complexes with Ran^{GTP}. Binding to Ran^{GTP} excludes cargo attachment and vice versa (mutual exclusion). The affinity of transportin-1 for Ran^{GTP} (∼100 nmol) is 10,000-fold higher than that for Ran^{GDP}, so that transportin-1 loads cargo in the cytosol, where Ran^{GTP} concentration is very low, and releases it inside the nucleus, where Ran^{GTP} level is high (10–15 μmol) [868].

Protein RanGEF is a constitutively nuclear protein, connected to chromatin. Lamin-A influences Ran distribution between the nucleus and the cytoplasm. Active Ran^{GTP} promotes the dissociation of import carriers from their cargo and the assembly of export carriers with cargo. The RanGAP activity is suppressed by karyopherins [874]. In the presence of Ran, cargo moves through the nuclear-pore complex. Importin subunits then dissociate, as Ran^{GTP} binds importin-β and displaces importin-α. Importin-α enters the nucleoplasm with its cargo. Importin-β remains at the pore.

Ran-Binding proteins that shuttle rapidly across nuclear pores have a high affinity for Ran^{GTP}. They stimulate RanGAP activity about 10-fold. In particular, RanBP1 interacts with the Ran^{GTP}–karyopherin complex and facilitates RanGAP binding to Ran^{GTP}. Each Ran-dependent transport cycle delivers at least one Ran to the cytoplasm, which returns to the nucleus to recover its GTP-bound state.

Once NLS-containing cargo has reached the nucleoplasm, importin-α returns to the cytoplasm. Cellular apoptosis susceptibility protein (CAS) mediates the re-export of cargo-free importin-α, only in the presence of Ran^{GTP} that forms the importin-α–CAS–Ran^{GTP} complex [875]. Importin-α is released from this complex in the cytoplasm by the combined action of RanBP1 and RanGAP1 regulators (Table 9.3).

Ran-Binding protein RanBP3, an effector of the Ras–ERK signaling pathway and Ran regulator, coordinates the nucleocytoplasmic transport of many proteins and RNAs. It is phosphorylated via the Ras–ERK pathway by ribosomal S6 kinase as well as via the PI3K pathway by protein kinase-B [876]. The Ras–ERK and PI3K–PKB pathways modulate Ran gradient and hence Ran gradient-dependent nuclear transport. Phosphorylated RanBP3 increases the affinity of RanBP3 for Ran and efficiency of nuclear import.

---

[13] Kinetic coefficients are approximately equal to 1.5×10^{-5}/s and 1.8×10^{-5}/s for endogenous GTPase activity and conversely.

Table 9.3. Regulators of nucleocytoplasmic transport (Sources: [868, 869]; NTF: nuclear transport factor; NxF1: nuclear RNA export factor; NxT1: NTF2-like export factor; NuP: nucleoporin; RanBP: Ran-binding protein; RanGAP: Ran GTPase-activating protein; RCC: regulator of chromosome condensation).

Molecule	Function
RCC1	Guanine nucleotide-exchange factor (RanGEF)
RanGAP	GTPase-activating protein
RanBP1	RanGAP coactivator
RanBP2	Binding of sumoylated RanGAP
RanBP3	Activation of RanGEF
NuP50	Binding of karyopherins
NxT1	Cofactor for NxF1 adaptor

9.1.8.4 Nucleocytoplasmic Transport and Body's Development

The nucleocytoplasmic transport relays developmental signals inside the cell. Among about 30 different types of proteins that compose the nuclear-pore complex in mice, the expression of nucleoporin NuP133 is restricted to some cell types and developmental stages to modulate the activity of the nuclear-pore complex [877].

Coordinated switching between KaPα subtypes can be aimed at regulating cell differentiation [878]. Temporal expression of KaPα can modulate nuclear import of transcription factors according to the tissue development stage, the developmental progression requiring the activity of a given set of transcription factors at a given stage.

9.1.9 Exocytosis

Endocytosis is the uptake of nutrients and vesicle-mediated internalization of plasmalemmal lipids and associated proteins, especially receptors.[14] Exocytosis corresponds to the transport of newly synthesized proteins from the endoplasmic reticulum, through the Golgi body, to the plasma membrane.

Secretion by neuroendocrine or endocrine cells is a relatively slow, but persistent process.[15] Spatiotemporal organization of the signal calcium inside the cell determines its specific role. Buffers bind to calcium that triggers secretion and thus delay the secretion. Unlike neurons (Vol. 2 – Chap. 1. Remote Control Cells), in neuroendocrine cells, vesicles and channels are separated by a distance of 100 to 300 nm for long-time scale secretion.

---

[14] At the end of the nineteenth century, Metchnikoff observed particle ingestion by cells.

[15] Secretion by chromaffin and pancreatic β cells persists for dozens of milliseconds after a short pulse ceases.

Most of the proteins must cross at least one biological membrane to reach its correct location. Protein translocation is carried out into 3 steps: (1) substrate recognition, (2) delivery to the destination membrane, and (3) translocation into or across a membrane. Following substrate recognition, 2 delivery pathways serve for protein translocation to target cellular membrane between cell compartment or extracellular medium in the case of cell export [879].

In the *post-translational delivery pathway*, ATP-dependent factors bind to newly synthesized proteins during the entire transit until membrane translocation is initiated.[16] The GTP-dependent *co-translational delivery pathway* that is mediated by GTPases of signal recognition particles couples protein synthesis to membrane transport.

Proteins destined for secretion or the plasma membrane are synthesized and inserted in the endoplasmic reticulum. In the endoplasmic reticulum, proteins fold properly owing to chaperones such as protein disulfide isomerase. However, misfolding can occur. Misfolded proteins may be retained in the endoplasmic reticulum and carried into the cytosol for degradation by the proteasome.

Certain proteins are secreted owing to a specific stimulation using different export routes.[17] Molecule transport is easier through transmembrane channels and pores, molecules interacting with their carriers. Specific binding sites can exist inside the channel, apparently leading to an increase in transmembrane transport.

9.1.10 Ectocytosis

Vesicles serve to transport substances not only inside the cell, but also to transfer molecules between adjacent or remote cells. Vesicle shedding from the plasma membrane for intercellular communication occurs during cell division, apoptosis, coagulation, immunomodulation, inflammation, and tumorigenesis.

According to their nature, size, and origin, these vesicles are called ectosomes, exosomes, or microparticles. Vesiculation toward the extracellular space in activated secretory or transformed cells enables the release of materials either from plasma membrane (ectosomes) or endosomes that reach the cell surface (exosomes). Composition of ectosomes differs from original plasma membrane because of material sorting during ectosome formation. Exosomes are derived from multivesicular bodies that undergo exocytosis.

Transfer of vesicular cargo between cells often involves interaction between phosphatidylserine of vesicular surface and plasma membrane. Extracellular

---

[16] Two ATP-dependent delivery routes to the endoplasmic reticulum exist. The first route requires the HSP40–HSP70 chaperone complex. The second pathway is mediated by cytosolic ATPase ASNA.

[17] They include pro-angiogenic lectins FGF1 and FGF2 (expelled by heat shock and shear stress, respectively), interleukin-1β (secreted by activated monocytes), macrophage migration inhibitory factor (released by monocytes in the presence of bacterial lipopolysaccharides), and galectin (during differentiation).

vesicles can contain lipids, proteins, nucleic acids, receptors (e.g., chemokine and transferrin receptors as well as receptor Tyr kinases). Cargo uptake implicates various molecules (e.g., kidney injury molecule KIM1,[18] phosphatidylserine receptor T-cell immunoglobulin- and mucin-containing molecule TIM4, brain-specific angiogenesis inhibitor BAI1, and P-selectin glycoprotein ligand PSGL1).

Vesicle-mediated intercellular exchange can be blocked by annexin-5. Ectocytosis is an ubiquitous, rapid process that can involve cytosolic Ca^{++} as well as activated protein kinases, such as protein kinase-C and extracellular signal-regulated protein kinase.

Vesicles released from tumor cells can stimulate angiogenesis. Vesicles that contain epidermal growth factor receptor can actually be secreted from cancer cells to interact with endothelial cells. Once these EGFR-containing vesicles have been incorporated by endothelial cells, EGFR effectors MAPK and PKB are activated. Furthermore, endothelial cells express vascular endothelial growth factor that can then activate VEGFR2 receptor (autocrine regulation) [880].

9.1.11 Endocytosis

Endocytosis refers to internalization of molecules, such as nutrients and other extracellular molecules, as well as plasmalemmal receptors, from the cell surface, together with plasmalemmal proteins and lipids, into membrane compartments and then vesicules for intracellular transfer. Therefore, molecules destined for endocytosis that are embedded into the plasma membrane are internalized with segments of the plasma membrane. Transcytosis from apical to basal membranes is an endocytosis followed by an exocytosis.

Endocytosis regulates nutrient uptake, cell-surface lipid and protein homeostasis, synaptic transmission, immune defense, and signaling from plasmalemmal receptors, i.e., hormone and growth factor signaling. Endocytosis is regulated by extracellular molecular signals. On the other hand, endocytosis acts as an organizer of signaling circuits by tuning space and time resolution of signals via internalization of plasmalemmal receptors that can either terminate signaling or change the signaling mode. Plasmalemmal proteins (e.g., transporters, receptors, and ion carriers) that undergo endocytosis are first modified at the plasma membrane by kinases and/or ubiquitin ligases.

9.1.11.1 Internalization Types

Endocytosis includes various internalization types (Fig. 9.2; Table 9.4). *Phagocytosis* (φαγειν: to eat) refers to uptake, especially by macrophages, of big particles (size >500 nm) into plasmalemmal distortions initiated by

---

[18] A.k.a. TIM3.

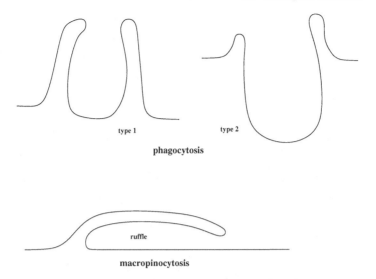

Figure 9.2. Microscopic mechanisms of endocytosis. (**Top**) Phagocytosis from plasma membrane protrusions (**left**) or invagination associated with slight protrusions and ingress of soluble effectors (**right**). (**Bottom**) Ruffle-based macropinocytosis.

receptors that recognize the particle; these distortions encircle the particle and give rise to phagosomes.[19]

Macropinocytosis refers to uptake of a fraction of the extracellular medium, sometimes together with particles, using large membrane invaginations that depends on growth factor receptor signaling.[20]

Pinocytosis (πινω: to drink) refers to uptake of liquid and certain macromolecules at a scale smaller than 200 nm. Fluid-phase pinocytosis deals with dilution of molecules in the absorbed medium. Adsorptive pinocytosis refers to adsorption of substance on the membrane with or without the receptor.

Phagocytosis of apoptotic cells by phagocytic cells participates in tissue remodeling as well as regulation of inflammation and immune responses. During apoptotic cell clearance, plasmalemmal, apoptotic, and phagocytic receptors of the cell death abnormal CeD1 homolog family (i.e., mammalian CeD1 ortholog Multiple EGF-like domain-containing protein MEGF10 that accumulates at the contact region during engulfment of apoptotic cells [882]) are involved in recognition of apoptotic cell corpses, transduction of engulfment

---

[19] Two types of phagocytosis comprise [881]: (1) engulfment of Fc receptor-bound immunoglobulin IgG-opsonized particles from protrusions of the plasma membrane (phagocytic cup) and (2) complement receptor CR3-mediated ingestion via plasma membrane, such as in the case of C3bi-coated particles.

[20] Macropinosomes (size 0.2–10 μm) that originate from plasma membrane ruffles are controlled by epidermal growth factor and its receptor, Rac1 GTPase, PI(4,5)P$_2$, WASPs, ARP2–ARP3 complex, and P21-activated kinase PAK1 [881].

Table 9.4. Mechanisms of endocytosis (Source: [881]; AP: [clathrin-associated] adaptor proteic complex; ARF: ADP-ribosylation factor; ARP: actin-related protein; CTBP: C-terminal-binding protein [a.k.a. brefeldin-A-ribosylated substrate (BARS)]; EPS: epidermal growth factor receptor pathway substrate; GAP: GTPase-activating protein; GEF: guanine nucleotide-exchange factor; GBF: Golgi-specific brefeldin-A-resistant GEF; PAK: p21-activated kinase; PI3K: phosphatidylinositol 3-kinase; PTRF: RNA polymerase-1 and transcript release factor; SDPR: serum deprivation protein response; WASP: Wiskott-Aldrich syndrome protein). Phagosomal membranes are composed of lysosomal proteins, lysosomal-associated membrane proteins LAMP1 (CD107a), LAMP2 (CD107a or LGP110), mitochondrial voltage-dependent anion channel VDAC1, several small GTPases of the RAB superfamily (Rab2, -3c, -5, -7, -11, and -14), as well as apoptosis-linked gene 2-interacting protein-X (ALIX) and flotillin-1. Glycosyl-phosphatidylinositol-anchored protein (gpiAP)-enriched early endosomal compartment (GEEC) may result from fusion of uncoated tubulovesicular clathrin-independent carriers (ClIC).

Process	Small GTPases	Other partners
	Microscale endocytosis	
Phagocytosis	CDC42, Rac1, RhoA	Actin, ARP2–ARP3, nWASP, PI3K, $PI(4,5)P_2$
Macropinocytosis	CDC42, Rac1	Actin, PI3K, PAK1, Ras, Src, CTBP1, dynamin
	Nanoscale endocytosis	
Clathrin-based endocytosis	Rab5	AP2, Eps15, epsin, amphiphysin, dynamin
Caveola-based endocytosis		Actin, PTRF, Src, SDPR, dynamin
IL2Rβ endocytosis	RhoA, Rac1	PAK1/2, dynamin
GEEC	ARF1, CDC42	Actin, Cholesterol, GBF1 (Arf1GEF), RhoGAP10/26
Flotillin-based endocytosis		Flotillins-1/2
ARF6-dependent endocytosis	ARF6	

signals, and initiation of the maturation of phagosomes that contain apoptotic cell corpses.[21] The retromer complexes allow retrograde transport of transmembrane cargos from the endosome to the trans-Golgi network [884].

These processes depend on actin-mediated remodeling of the plasma membrane. *Endocytic pits*, or membrane indentations, and vesicles that arise from

---

[21] In Caenorhabditis elegans, engulfment and phagosome maturation require the sequential enrichment on phagocytic membranes of phagocytic receptor CeD1, large GTPase dynamin Dyn1, phosphatidylinositol 3-phosphate, and small GTPase Rab7, as well as the incorporation of endosomes and lysosomes to phagosomes [883].

Table 9.5. Molecular transfer via membrane rafts or coating (Source: [609]).

Process type	Mechanism
Raft-mediated	Incorporation of cargos in a raft
	Membrane bending
	Formation of raft transport container
Coat-mediated	Cargo sorting by adaptors
	(recognition of cargo sorting signals)
	Assembling of coat proteins
	Membrane bending and subsequent budding
	Vesicle fission

membrane invaginations correspond to micropinosomes, macropinosomes, and phagosomes. Coated vesicles lose their coat when they form endosomes (larger vesicles) that can then mature and follow various paths.

Early endosomes follow recycling routes to the plasma membrane or degradative route when late and multivesicular endosomes interact by fusion with lysosomes. Lysosomes are vesicles with hydrolytic enzymes that digest molecules conveyed by endocytosis. Lysosomes recycle both produced and incorporated molecules.

9.1.11.2 Intracellular Vesicle-Mediated Transport Modes

Regulated molecular endocytosis can take multiple routes, such as Rab4-dependent, fast and Rab8/11-mediated, slow recycling routes (endocytosis–exocytosis cycle) and Rab7-dependent, degradative route [885]. Receptor ubiquitination initiates the degradative route.[22]

These distinct vesicular endocytic mechanisms that are programmed at the cell surface coexist and operate concurrently in a single cell. Cargo specificity and endocytosis path depend on physical and chemical properties of cargo molecules, specific coat, coat-associated protein assembly, regulatory molecules, and plasmalemmal parameters.

The 2 major vesicular transport modes include clathrin and membrane-raft endocytosis (Table 9.5). Many forms of non-clathrin-mediated endocytosis exist according to the composition of the starting plasmalemmal nanodomains.

Clathrin-mediated endocytosis is involved in the internalization and recycling of receptors by endocytic vesicles coated with the protein clathrin. Clathrin-mediated receptor endocytosis can promote signaling, whereas caveolar internalization is required for receptor turnover.

A given molecule can be internalized by different mechanisms in different cell types (e.g., caveola-mediated pathway in given cell types and

---

[22] Ubiquitinated receptors are recognized by many endosomal-sorting complex required for transport.

RhoA-dependent mechanism in other types) or switch pathways in a single cell type under different conditions.

Many endocytosis pathways exist according to involved adaptors, such as arrestin, arrestin-related transport adaptor (ART), endocytic adaptor low-density lipoprotein receptor adaptor protein LDLRAP1,[23] CBL-interacting protein of 85 kDa (CIN85), and epsin,[24] that bind to the cytoplasmic tail of the transmembrane protein. Receptor endocytosis can require interaction with ubiquitinated arrestin or arrestin-like adaptors. Adaptor proteic complex AP2 links to clathrin-coated vesicles.[25]

The endocytic apparatus includes early, recycling, and late endosomes, as well as multivesicular bodies and lysosomes to sort, process, and eventually degrade internalized cargos. Endocytosis can thus terminate signaling by degrading activated receptors. Vesicles are delivered to target membrane, using molecular motors along microtubules or actin filaments.

Membrane proteins are translocated into early endosomes. Endosomal cargos are sorted to distinct destinations: (1) some cargos such as nutrient receptors are recycled back to the plasma membrane, either directly or via recycling endosome; (2) others such as receptors for lysosomal enzymes travel to the trans-Golgi network; (3) ubiquitinated membrane proteins such as activated growth factor receptors are sorted into vesicles that target multivesicular endosomes and then lysosomes.

Internalized material is transported from the cell cortex by early endosomes. Clathrin-dependent endocytosis of ligand-bound receptors starts with a clathrin-coated pit that bears scission to form a clathrin-coated vesicle that uncoats and fuses with early endosome. Caveolin-dependent endocytosis of circulating proteins leads to a caveosome that also fuses with early endosomes. Clathrin-, caveolin-, dynamin-independent, ARF6-regulated vesicle that internalizes integrins, major histocompatibility complex molecules, and glycosyl-phosphatidylinositol-anchored proteins fuses with early endosomes. Substances and plasmalemmal proteins either are recycled back to the plasma membrane by recycling endosomes to be reused or conveyed by late endosomes and subsequently by lysosomes for degradation. Ubiquitination causes endosome fusion into multivesicular bodies.

---

[23] A.k.a. autosomal recessive hypercholesterolaemia protein (ARH).

[24] Endocytic adaptor epsin should not be confused with actin-bundling espin.

[25] Ligand binding to its cognate receptor accelerates the recruitment of this receptor to clathrin in clathrin-coated pits via adaptors, such as AP2 or β-arrestins. More than 50 different proteins reside in clathrin-coated pits [885]. Clathrin then polymerizes and provokes pit invagination to form a vesicle. In clathrin-coated pits, actin filaments created by nWASP activation of ARP2 and ARP3 enlarges invagination. The created vesicle is then released into the cytoplasm after scission of the clathrin-coated pit from the plasma membrane owing to large GTPase dynamin and actin.

Early endosomes move from the cell periphery to the cell core, where they can mature.[26] However, endosomes do not necessarily mature. The mature endosomes fuse with late endosomes and can be targeted for degradation.

Molecule internalization can also be done via structures that contain gpi-APs and fluid-phase markers [886].[27] Corresponding carriers fuse to form tubular early endocytic compartments or *gpiAP-enriched endosomal compartments* (GEEC). Internalization, after binding to the plasma membrane, can require membrane rafts and caveosome-like structures (devoided of caveolin) [887].[28]

Endosomes are also implicated in abscission during the cytokinesis stage of the cell cycle[29] as well as cell polarization[30] and migration [888].[31]

Endocytosis and signaling are strongly coupled. Removal of receptors from the plasma membrane and subsequent internalization determines the signaling competence of receptors. Receptors are then either recycled to the cell membrane or transported to lysosomes for degradation. Reversible protein changes, such as phosphorylation and ubiquitination, drive this process. In the endosome, ubiquitin serves as a molecular tag on trafficking cargos. Furthermore, signaling can continue from endosomes, which then operate as signaling stations, using another mode of signal transduction. Endosomes hence contribute to signal diversification and specificity. The endocytic route thus determines the bulk signaling output.

9.1.12 Endocytosis Events

Endocytosis begins at the plasma membrane once agonists have bound their corresponding plasmalemmal receptors. Receptors are concentrated in

---

[26] Small GTPase Rab5 characterizes an early endosome. During endosome maturation, Rab5 is replaced with Rab7 GTPase.

[27] This pathway is not only caveola- and clathrin-independent, but also dynamin- and ARF6-independent.

[28] This pathway is caveola-, clathrin-, dynamin-2-, and ARF6-independent, but cholesterol-, and Tyr kinase-dependent.

[29] During late telophase, Rab11+, Rab35+ endosomes accumulate in the midbody, as they interact with the vesicle-tethering exocyst complex and proteic platform BRUCE, or baculoviral IAP repeat-containing protein BIRC6, that regulates the delivery of endosomes to the midbody and serves as a ubiquitin ligase [888].

[30] Endosomes participate in asymmetrical localization of partitioning defective proteins. Moreover, endosomal-sorting complexes required for transport are involved in cell polarity.

[31] Rab5+ Endosomes carry Rac1 that can fuse with RacGEF TIAM1-loaded endosomes to activate Rac1. ARF6+ Vesicles that contain activated Rac1 then bud off this compartment and run to the ruffling cell zone. In addition, C-type lectin-13E (a.k.a. mannose receptor MRC2 and endocytic receptor Endo180) that generates RoCK-mediated adhesion disassembly can use endosomes to travel to the cell rear at focal adhesions, where it helps in assembling Rho, RoCK1, and MLC2 to form a contractile complex [888].

membrane nanodomains in association with a network of endocytic adaptor proteins. Ligand–receptor interactions often lead to aggregation of numerous ligand–receptor complexes at a site where the membrane begins to invaginate, followed by cell absorption and intracellular transport. Bulging primes vesicle formation. Membrane budding is initiated by protrusion of plasmalemmal patches with carrier proteins and their extracellular ligands. Membrane coat proteins, such as clathrins, assemble and capture the ligand-bound receptors. An initial inward protrusion of the plasma membrane generates a coated invagination.

Actin polymerization regulators as well as actin assembly and stabilization promoters associated with the coat stimulate actin filament formation. These actin filaments drive slow inward movement of the coat. Scission molecules separate the budded membrane from the adjoining plasma membrane to form a coated vesicle. The vesicle moves rapidly into the cell, being coupled to the cytoskeleton for efficient delivery to the target organelle, and the coat components are removed for another cycle of vesicle formation. When it reaches its target organelle, the vesicle is tethered to the target membrane before fusing with it. In summary, vesicle history can be divided into several successive steps: initiation, assembling, budding, detachment, uncoating, displacement, followed at destination determined by sorting by vesicle membrane budding, docking, and fusion.

9.1.13 Transport Factors

Intracellular transport comprises several steps: selection of motors, uptake of cargo, targeting and control of directionality, displacement, and finally release of the cargo at its destination. Substrate transport within the cell hence requires a cluster of subtances: assembling and docking proteins, coat components, adaptors, effectors, and regulators that target, tether, and cut or fuse membranes on the one hand and load, carry, and unload cargo on the other.

Cargo size, shape, and surface features can influence endocytosis [889]. Cells are able to incorporate non-spherical particles with size of 3 μm. Rodlike shape favors cellular internalization.

Cellular cargo is packaged into a vesicle of given size that is surrounded by a coat made of scaffold and adaptor protein complexes. The cytoskeleton is involved in the spatial distribution, motility, and morphology of transport compartments. Intracellular cargos are usually transported as motor-driven translocation complexes to their sites of destination. Nanomotors contribute to the uptake, transport, and release of cargo.

Most internalized substances are delivered to early endosomes that derive from the plasma membrane via: (1) clathrin- or caveolin-coated vesicles and (2) tubular intermediates, the clathrin- and dynamin-independent carriers (CLIC). Macropinocytosis and phagocytosis (actin-dependent uptake of fluid and large particles, respectively) are different types of endocytosis that involve internalization of relatively large (>1 μm) membrane patches.

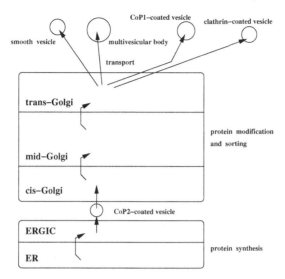

Figure 9.3. The Golgi body receives proteins via transition vesicles from the endoplasmic reticululm–Golgi intermediate compartments (ERGIC), a smooth region of the rough endoplasmic reticulum. The Golgi stack modifies proteins within its 3 compartments. The proteins are sorted for destination sites using different types of vesicles.

9.1.13.1 Coat Proteins

Intracellular transport is carried out after vesicle formation that relies on aggregation of specialized membrane-bound proteins into a coat able to curve cellular membranes. Turnover of coat components occurs in different conditions: (1) quiescent membranes; (2) coat protein aggregates into elementary coatbuilding units after energy-consuming desorption of coat proteins; (3) vesicles produced upon presence of cargo and variation in membrane tension; and (4) cytosolic coat component recycling compartment [890]. Membrane-bound elementary coat-building units polymerize into coat structures that locally bend the membrane and recruit cargo molecules. As the coat expands, the coated domain of the membrane invaginates to form a nearly spherical vesicle containing cargo. The coat components soon disassemble to participate in the formation of a new vesicle.

Vesicle movement between cell compartments involves various types of containers that are composed of coat protein complex (CoP1 and CoP2; Fig. 9.3), clathrin, and caveolin during the different steps of endocytosis and exocytosis. These coats can also be linked to adaptors to ensure efficient substrate selection and coordination with vesicle formation.

Coat proteins are heptamers with β, γ, δ, and ζ subunits. The coat protein complex CoP2 is composed of: (1) SAR1 GTPase, (2) GTPase-activating proteins, the Sec23 and Sec24 homologs, of the endoplasmic reticulum–Golgi body transfer[32] and (3) Sec13 and Sec31 homologs that have a structural role in cargo transport.

Caveolea

Caveolae are able to internalize molecules quickly and to transport them across thin cells at a rate on the order of minute. Transport vesicles can be classified according to the components of the protein coat and transported molecules. Anterograde and retrograde transport is mediated by distinct sets of cytosolic coat proteins CoP1 and CoP2. Hence, 3 major kinds of coating proteins are involved in the transport between intracellular membranes and the plasma membrane: (1) the clathrin–adaptin complexes for endocytosis and vesicular transfer between the Golgi body, lysosomes, and endosomes; (2) the CoP1 complex for intra-Golgi and Golgi body-to-endoplasmic reticulum displacement; and (3) the CoP2 complex for vesicular motion from the endoplasmic reticulum to the Golgi body.

Several transport pathways use a diverse set of clathrin-independent cargo transporters. The latter are involved in the regulation of cell growth and development [891]. They include the caveolar pathway and fluid-phase endocytosis. Clathrin-independent endocytosis of specific cargos can be classified according to involvement of dynamin and small GTPases CDC42, RhoA, and ARF6. Cargo selection depends on initial processing such as protein-based modifications (e.g., ubiquitination) and lipid-based mechanisms (e.g., clustering of lipid-tethered proteins).

Certain substances can be internalized by clathrin-, caveolar- and CDC42-dependent processes, as they can be managed using different modalities and membrane patch organization.

Clathrin

Clathrin intervenes as a transporter for importing materials. For example, in nerve terminals, clathrin-coated vesicles participate in retrieval of synaptic vesicles following exocytosis. Clathrin also serves in sorting proteins that are destined for the basolateral membrane from the trans-Golgi network, thus contributing to the cell polarity setting [892]. Adaptor protein AP4 is involved in protein delivery to the basolateral plasma membrane, but it does not interact with clathrin. The clathrin adaptor complex AP2 strongly impedes endocytosis, but not basolateral protein export from the trans-Golgi network. Fast endocytosis and recycling of basolateral receptors, such as transferrin receptor and low-density lipoprotein receptor, require clathrin tetrameric AP1b adaptor. Some basolateral proteins, such as CD147 and E-cadherin, necessitate clathrin, but not AP1b for their basolateral localization. Consequently, other clathrin adaptors participate in basolateral polarity.

---

[32] Sec23 is a SAR1-specific GTPase-activating protein. Sec24 selects transported substances.

9.1.13.2 Kinesins

Endosomes rapidly move along cytoskeletal filaments to exchange proteins and lipids between different cell compartments. The motor-driven transport between the trans-Golgi network and Rab4+ perinuclear endosomal compartments relies on AP1 adaptor complex and γ-ear-containing, ARF-binding proteins (GGA). Gadkin[33] that is associated with the trans-Golgi network and its derived endosomes form a complex with clathrin adaptor AP1 and nanomotor kinesin KIF5 (Sect. 6.4.3.3), hence linking TGN-derived endosomes to the microtubule network [893]. AP1-binding membrane protein gadkin then acts as a receptor for KIF5 on TGN-derived endosomes. Rab4+, Gadkin+ endosomes that are conveyed by KIF5-mediated mechanism can be distinguished from recycling endosomal tubules that experience myosin-5b- and KIF3b-dependent transport.

9.1.13.3 Dynamin

Large GTPase dynamin severs clathrin-coated vesicles from the plasma membrane and participates in transport mediated by caveolae. Dynamin-dependent processes use dynamin for pinching and scission. However, some transport pathways are dynamin-independent.

Dynamin-dependent endocytic pathways include mechanisms that depend on: clathrin, caveolin, and GTPase RhoA (especially in immune cells and fibroblasts) [891]. Dynamin-independent endocytosis (non-clathrin, non-caveolar uptake) copes with small GTPase CDC42 and/or ARF6. The CDC42 pathway is cholesterol-sensitive and coupled to actin polymerization. It is able to internalize lipid-linked proteins such as GPI-anchored proteins. ARF6 acts in actin remodeling and stimulates the formation of phosphatidylinositol (4,5)-bisphosphate by activating phosphatidylinositol 4-phosphate 5-kinase. Clathrin-independent ARF6-regulated pathway can involve protein-based mechanisms of cargo sorting (cytoplasmic tail sequence), but not necessarily. Endocytosis mediated by ARF6 and pathway regulated by CDC42 can experience crossregulation, as an excitation of the former (possibly by G proteins) represses the latter.

9.1.13.4 Flotillins

The integral membrane proteins flotillin-1 and -2 may select lipid cargo for dynamin-independent endocytosis. Dietary absorption that includes biliary cholesterol reabsorption is a major source of cholesterol. Free cholesterol is taken up via vesicular endocytosis with Niemann-Pick-C1-like protein NPC1L1 [894]. Flotillins of membrane rafts associate with NPC1L1 to form cholesterol-enriched membrane nanodomains before acting as carriers.

---

[33] γ1-adaptin and kinesin interactor, a.k.a. γ1-adaptin brefeldin-A resistance protein (γBAR).

Table 9.6. Location of phosphoinositides (Source: [895]). Phosphoinositides organize membrane nanodomains, control membrane sorting and deformation, and integrate functions involved in intracellular transport (cytoskeleton dynamics, cell signaling, lipid metabolism, and energy control). For example, phosphatidylinositol 3,5-bisphosphate ($PI(3,5)P_2$) that is ubiquitously expressed, but at low abundance, is necessary for retrograde membrane trafficking from lysosomes and late endosomes to the Golgi body, among other functions. Cellular concentrations of phosphoinositides are regulated.

Molecule	Location
	Phosphoinositides
$PI(3)P$	Early endosome, sorting endosomes, multivesicular body/late endosome
$PI(4)P$	Trans-Golgi network, TGN–plasmalemmal transport vesicle
$PI(3,5)P_2$	Early endosome, multivesicular body/late endosome
$PI(4,5)P_2$	Plasma membrane, Golgi body, early endosome, recycling endosome
$PI(3,4,5)P_3$	Plasma membrane

9.1.13.5 Phosphoinositides

Phosphoinositides are strongly implicated in endocytosis, as they control assembly–disassembly cycles of complexes involved in membrane trafficking. Phosphoinositides are quickly phosphorylated and dephosphorylated by many phosphoinositide kinases and phosphatases in different intracellular compartments to regulate their spatial and temporal distribution (Tables 9.6 to 9.8).

A given phosphoinositide species can produce alternative metabolites. Conversely, distinct pathways leads to a single product among all different possible end PI species owing to the distribution of PI-metabolizing enzymes and organelle-specific assembly of enzymatic complexes that associate PI kinases, PI phosphatases, and regulators. However, a given PI species can have a distinct fate in different subcellular locations. Small GTPases operate as recruiters and timers for PI enzymes. Certain small GTPases can also coordinate the activity of multiple PI enzymes.

Phosphatidylinositol 3-phosphate is found on the surface of endosomes, where it can recruit proteins. Among such proteins, certain ones coordinate the activity of small GTPases Rab, some are involved in ubiquitination for subsequent lysosomal degradation, and others are required for multivesicular body formation. Phosphatidylinositol 3-phosphate or its metabolite $PI(3,5)P_2$ in endosomes is required for maturation of the transport of certain receptors assigned to recycling or degradation, at the microtubule-associated tubularization stage [896].

Phosphatidylinositol 4-phosphate and $PI(4,5)P_2$, therefore PI 4-kinases and $PI(4)P$ 5-kinases, are involved in transport of hormones and neurotransmitters by immature secretory granules from the trans-Golgi network.

Table 9.7. Location of phosphoinositide kinases (Source: [895]).

Molecule	Location
PI3Kα	Plasma membrane, early endosome
PI3KC2α	Trans-Golgi network
PI3KC3	Plasma membrane,
	early endosome, pre-autophagosome, phagosome,
	trans-Golgi network
PI4Kα	Endoplasmic reticulum
PI4Kβ	Golgi body
PI4K2α	Plasma membrane, trans-Golgi network,
	early endosome, multivesicular body/late endosome
PI4K2β	Plasma membrane, early endosome
PI5K1α	Plasma membrane
PIP5K1α	Plasma membrane, early and recycling endosomes
PIP5K1β	Plasma membrane, early and recycling endosomes
PIP5K1γ	Plasma membrane, early and recycling endosomes
PIP5K3	Early and late endosomes, multivesicular body

Table 9.8. Location of phosphoinositide phosphatases (Source: [895]; INPP: inositol polyphosphate 5-phosphatase; MTM1: myotubularin-1; MTMR4: myotubularin-related protein-4; OCRL: oculocerebrorenal syndrome of Lowe protein; PTen: phosphatase and tensin homolog deleted on chromosome-10; SACM1L: suppressor of actin mutations 1-like; FIG4 homolog is also known as KIAA0274; SKIP: skeletal muscle and kidney enriched inositol phosphatase; Synj1: synaptojanin-1).

Molecule	Location
	3-Phosphatases
MTM1	Early endosome
MTMR4	Early endosome
PTen	Phagosome
	4-Phosphatases
INPP4a	Early endosome
SACM1L	Endoplasmic reticulum
TMEM55a/b	Late endosome, multivesicular body, lysosome
	5-Phosphatases
FIG4	Late endosome, multivesicular body
INPP5b	Early endosome
INPP5d	Phagosome
OCRL	Plasma membrane, early endosome
SKIP	Plasma membrane, endoplasmic reticulum
SYNJ1/2	Early endosome

Immature secretory granules change in size and composition to form mature secretory granules owing to membrane remodeling by adaptor protein-1-containing clathrin coats. Adaptor AP1 is recruited to immature secretory granules by small GTPase ARF1. Mature secretory granules neither bind ARF1 nor have AP1-containing clathrin coats [897]. Activity of PI4K on immature secretory granules is not regulated by ARF1.

Phosphatidylinositol (4,5)-bisphosphate is required in cell transport, especially in clathrin-dependent endocytosis of plasmalemmal proteins, such as nutrients and growth factor receptors. It is synthesized by phosphatidylinositol 4-phosphate 5-kinase in association with clathrin adaptor protein AP2 complex. It also serves as a substrate for the synthesis of phosphatidylinositol (3,4,5)-trisphosphate.

Phosphoinositides control protein recruitment and activation, as well as assembly of molecular complexes involved in membrane bending and fission as well as vesicle motion, tethering, and fusion (Tables 9.9 and 9.10). Phosphoinositide–protein interactions usually occur at relatively low affinities and thus require commitment of adaptors, cofactors, and small GTPases.

9.1.13.6 Cytoskeleton, Associated Nanomotors, and other Partners

Actin Cytoskeleton

The actin cytoskeleton intervenes during early endocytosis, and acts in endosome dynamics and distribution. Filamentous actin and the ARP2–ARP3 complex accumulate on endosomes. The ARP2–ARP3 complex is actually involved in the regulation of endocytosis.

WASP and SCAR Homolog

The ARP2–ARP3 activator WASP and SCAR homolog (WASH) acts as a nucleation-promoting factor. As it activates the ARP2–ARP3 complex, it promotes actin polymerization and cytoskeletal reorganization for endocytic vesicle scission and endosomal sorting [898, 899]. It is required in retrograde transport, in particular of cation-independent mannose 6-phosphate receptors, from endosomes to the Golgi body.

Protein WASH localizes to early endosome domains enriched in ARP2–ARP3, F-actin, and retromer components, where it interacts with tubulin [898]. In T lymphocytes, WASH resides in endosomes that polarize toward the immunological synapse during T-cell activation [898].

Protein WASH functions as a constituent of WASH proteic complexes. It links to Family with sequence similarity FAM21, among the 7 components of the WASH complex (e.g., strumpellin [or SPG8], WASH complex subunit-7 SWIP, and coiled-coil domain-containing subunit CCDC53) that comprises a capping protein heterodimer [899]. Protein WASH also interacts with dynamin. Structurally similar to the WASH complex, the WAVe complex impedes WAVe activity [900].

Table 9.9. Phosphoinositide partners for intracellular transport (**Part 1**; Source: [895]; AP: clathrin adaptor protein; AP180: 180-kDa assembly protein; ARF: ADP ribosylation factor; CLINT: clathrin-interacting protein localized in the trans-Golgi network; Dab2: Disabled homolog-2; GAP: GTPase-activating protein; GGA: Golgi-localized γ-ear-containing ARF-binding protein, GEF: guanine nucleotide-exchange factor LDLRAP1: endocytic adaptor low-density lipoprotein receptor adaptor protein [a.k.a. autosomal recessive hypercholesterolemia adaptor protein (ARH) that binds to clathrin, AP2, and low-density lipoprotein receptor to promote LDLR clustering into clathrin-coated pits and endocytosis]). The spatial and temporal control of PI phosphorylation and dephosphorylation by locally recruited PI kinases and phosphatases enables multiple, dynamical PI functions in intracellular transport. Phosphoinositides interact with small GTPases (Rab and ARF), coat proteins, adaptors, etc. Phosphoinositides operate at the interface between cellular membranes and the cytoskeleton (microfilaments and microtubules) to prime membrane deformation and fission, as well as vesicle displacement, anchoring, and fusion to the target membrane. Local actin polymerization that accompanies the internalization of materials destined for intracellular transport requires phosphatidylinositol (4,5)-bisphosphate, small GTPase CDC42, ARP2–ARP3 complex, and other proteins. Endocytic accessory proteins, such as Huntingtin-interacting protein HIP1 and HIP1-related protein HIP1R that is an actin-binding component of clathrin-coated pits and vesicles, connects clathrin with actin filaments and their nanomotors myosin-1 and -6. Hepatocyte growth factor-regulated Tyr kinase substrate (HRS) regulates the sorting of ligand-stimulated and unstimulated growth factor receptors in early endosomes, as it can serve as an adaptor between ubiquitinated receptors and clathrin.

PI	Targets
	Coat and adaptor proteins
$PI(3)P$	HRS
$PI(4)P$	AP1γ, epsin-R, CLINT
$PI(3,4)P_2$	HIP1/HIP1R
$PI(3,5)P_2$	HIP1/HIP1R
$PI(4,5)P_2$	AP2α/μ, AP3,AP180/CALM, CLINT, Dab2,
	Epsin1–3, β-arrestins, HIP1–HIP1R, LDLRAP1
$PI(3,4,5)P_3$	CLINT
PIP_4	GGA
	Small GTPase accessory proteins
$PI(3,4)P_2$	ArfGAP, ArfGEF
$PI(3,5)P_2$	ArfGAP
$PI(4,5)P_2$	ArfGAP, ArfGEF, RabGAP, RabGEF
$PI(3,4,5)P_3$	ArfGAP, ArfGEF

Protein WASH that generates an actin network on restricted domains of sorting and recycling endosomes contributes to the recycling of transferrin [899]. It preferentially connects to Rab4+ or Rab11+ early and recycling endosomes rather than Rab7+ endosomes of the degradative pathway.

Table 9.10. Phosphoinositide partners for intracellular transport (**Part 2**; Source: [895]; EEA: early endosomal antigen; ESCRT: endosomal-sorting complexes required for transport; KIF: microtubule motor of kinesin family; SCAMP: secretory carrier membrane protein; SNAP: soluble ^{N}ethylmaleimide-sensitive attachment protein; SNx: Sorting nexin; Synj1: synaptojanin-1; VAMP: vesicle-associated membrane protein). Phosphoinositides act on microtubule-associated nanomotors such as kinesin family protein KIF16b for endosome motility. Phosphoinositides also control sphingolipid synthesis via sphingolipid-transfer proteins (ceramide-transfer protein CERT and 4 (four)-phosphate-adaptor protein FAPP2). Phosphoinositides integrate signaling and membrane trafficking, as they mediate the effects of plasmalemmal receptor-initiated signaling cascades on membrane trafficking.

PI	Targets
Sorting proteins	
PI(3)P	SNx1–3, SNx9–10, SNx13, SNx16–17; ESCRT2–3
PI(4)P	SNx2, SNx5, SNx13
PI(5)P	SNx2, SNx5, SNx13
PI(3,4)P$_2$	SNx9; ESCRT2
PI(3,5)P$_2$	SNx1, SNx5, SNx9, SNx13; ESCRT3
PI(4,5)P$_2$	SNx9
PI(3,4,5)P$_3$	SNx1
Fission proteins	
PI(4,5)P$_2$	Dynamin-1–3, amphiphysin-2
Fusion proteins	
PI(3)P	EEA1, rabenosyn-5, VAMP8
PI(4,5)P$_2$	Exophilin-4, synaptogamin, SNAP25/syntaxin-1A, SCAMP2
Cytoskeleton-associated nanomotors and regulators	
PI(3)P	KIF16b
PI(4,5)P$_2$	KIF1a, myosin-1C and -6
PI(3,4,5)P$_3$	KIF13b/16b, myosin-10, dynein, dynactin

In summary, WASH participates in the endosomal recycling and retrograde transport.

Myosin Nanomotors

Most plus-end-directed myosins form stable dimers and interact constitutively with their cargo complexes. However, myosin-6 that is fully functional only within translocation complex moves toward the minus-end of actin. Transport myosins are made of 2 major domains: (1) a motor domain that binds ^{F}actin and (2) a tail domain for cargo binding. Myosin-1, -5 to -7, and -10 are involved in cellular transport. Processive myosins, such as myosin-5, can remain attached throughout the sequential steps of their activity along actin filament.

Myosin-5 is a 2-headed (dimeric)[34] motor protein responsible for intracellular transport on actin filaments and positioning of certain vesicles (secretory vesicles, lysosomes, and peroxisomes) and organelles.[35] Distant cargo-binding domains regulate cargo type (cargo-specific receptors), attachment–detachment, and destination [904]. Myosin-5 has functional similarities with kinesin and differs from myosin-2 activity. Myosin-5 functioning is mediated by Rab GTPases, which bind to myosin either directly or via adaptors. The light chain domain serves as a mechanical lever arm, due to tilting of the myosin neck. The 2 heads maintain association with actin during displacement. At each step, the trailing head becomes the leading head. Another orientational change occurs without producing significant motion along actin. The lever arms probe their local environment prior to executing a full step [905]. Myosin-5 binds to calcium–calmodulin-dependent protein kinase-2 to activate the kinase activity. Enzyme CamK2 phosphorylates myosin-5 to regulate the organelle transport.

Cell organelles are carried along microtubules by kinesin, then locally along actin by myosin-5A. Myosin-5A can travel according to a spiral path along actin. At intersecting filaments (either actin or microtubules), the nanomotor adapts to transport its cargo to the cell periphery [906]. Myosin-5A not only easily maneuvers at actin filament intersections, but also when encountering a microtubule, steps onto the microtubule and begins a one-dimensional diffusive search for a kinesin. At filament intersection, myosin-5 rapidly diffuses along microtubule mainly due to electrostatic interaction (in vitro estimated diffusion constant $\sim0.3\,\mu m^2/s$, maximal velocity $\sim3\,\mu m/s$).

Myosin-6, the single nanomotor that moves along actin filaments toward the minus-end, participates in endocytosis. It is recruited to clathrin-coated pits and uncoated endocytic vesicles, as well as liposomes in the presence of calcium. Myosin-6 links clathrin-coated pits via the adaptor protein Dab2. Adaptor synectin[36] used in internalization of plasmalemmal receptors is involved in the association of myosin-6 with uncoated endocytic vesicles [907].

---

[34] Myosin-5 Myo4p binds its cargo via 2 distinct binding domains [901]. Isolated Myo4p is monomeric. Dimerization of Myo4p that occurs within the nanomotor-translocation complex promotes stabilization of myosin–cargo complexes.

[35] At nanomolar calcium concentrations, myosin remains folded, whereas at micromolar calcium levels, ATPase is activated and myosin unfolds [902]. Myosin-5 hydrolyzes ATP to move toward the plus-end of actin filaments. Strain between the 2 heads of a moving myosin-5 can modulate the ADP release rate. Release of ADP is rate limiting in myosin kinetics. In addition, the forward motion rate of the nanomotor depends on ATP concentration, whereas load-dependent backstepping can occur in the absence of ATP, as ATP binding is inhibited by molecular loading. This mechanical asymmetry can be explained by the strength of actin binding of a nanomotor head that is modulated by the lever arm conformation, myosin-5 functioning as a ratchet [903].

[36] Synectin is also called the GAIP interacting protein-C terminus (GIPC).

Annexins

Annexin-2 is a ubiquitous Ca^{++}-binding protein for actin-dependent vesicle transport.[37] Annexin-2 is present both as a cytosolic monomer and heterotetramer at the outer cytoskeleton linked to the plasma membrane, interacting with receptors and ion channels. Annexin-2 is located in cell regions undergoing actin remodeling. It indeed reduces the polymerization rate of actin monomers in a dose-dependent manner [908]. Annexin-2 inhibits filament elongation at the barbed ends.

CyFIP

Actin dynamics are involved in the generation of clathrin- and adaptor protein AP1-coated transport carriers that connect the trans-Golgi network to endocytosis. ADP-ribosylation factor ARF1 synchronizes the association of clathrin–AP1 coats and a clathrin heavy chain-binding protein, the cytoplasmic fragile-X mental retardation (FMR)-interacting protein CyFIP2[38] at the trans-Golgi network [909]. Upon ARF1 priming, Rac1 and RhoGEF7 activate this complex to enable nWASP- and ARP2–ARP3-dependent actin polymerization at membranes, thus promoting tubule formation.

Microtubules

Microtubules are required for the motility of early endosomes. Kinesins act as ATPases to translocate cargos along microtubules. Many kinesins have long extensions and flexible parts between segments of these extensions. Nanomotors often work in sets to collectively transport materials, as they form clusters bound to cargos or act as crosslinkers in interconnected filament networks. Kinesin sets can propel a microtubule. The average number of nanomotors bound to a microtubule determines the transport velocity. When nanomotors are coupled to form a processive set, each nanomotor has to slow down for coordinated transport. Collective microtubule transport in fact requires mutual nanomotor inhibition.

Kinesin-1 drives the movement of cell organelles, such as rough endoplasmic reticulum, Golgi body, mitochondria, and secretory and endocytic vesicles. Kinesin-1 contains 2 motor heavy chains (KHC) and 2 light chains (KLC). Kinesin light chains interact with multiple cargo molecules. Many different KLC1 splicing isoforms exist, but a given kinesin-1 has a single KLC isoform, and thus a given targeted cargo. In the absence of load, kinesin-1 is able to make multiple consecutive steps along a microtubule. Kinesin-1 extension opposite to motor domain allows loose mechanical coupling between nanomotors that is needed for efficient transport by ensembles of processive

---

[37] Annexin-2 is also required for the formation of tight junctions.
[38] A.k.a. SRA2 and PIR121.

nanomotors [910]. Kinesin-1 dimers with and without extensions behave similarly as single molecules, but not as a coupled ensemble.

Kinesin-8 and -13 also control microtubule polymerization dynamics, as they are microtubule depolymerases. Kinesin-17, a member of the kinesin-2 family, is regulated by phosphorylation by calcium–calmodulin-dependent protein kinase-2 that disrupts the kinesin-17–Mint1 complex and causes the release of transported cargo from microtubules [911].

Cytoplasmic microtubule-activated ATPase dynein is the primary nanomotor for transport of vesicles, organelles, proteins, and RNA cargos from the cell periphery toward the nucleus along the microtubule cytoskeleton.[39] Among microtubule-associated proteins, dynactin moves progressively along microtubules in the absence of molecular motors. Dynein requires dynactin for functioning. The skating domain of dynactin is used by motor dynein to maintain longer interactions with microtubules during each binding [912]. Tight binding prevents premature dissociation of the dynein–dynactin–cargo complex from the microtubule. Moreover, dynactin maintains a binding with the microtubule, but does not impede the dynein velocity along the microtubule.

Motor complexes, such as sets of kinesins and dyneins, operate with better efficiency than single molecules [913]. Multimotor transport depends on the architecture of the molecule assembly and its ability to bind multiple microtubule sites. Complexes formed by dynein, its activator dynactin, and actin-related protein ARP1, moves in the 2 directions (toward plus and minus microtubule ends) along microtubules [914]. The direction switch for backward excursions allows flexible navigation through a crowded cellular environment.

9.1.13.7 Soluble NSF Attachment Protein Receptors

Soluble ^{N}ethylmaleimide-sensitive factor (NSF) attachment protein receptors (SNARE) constitute a large superfamily with more than 60 members. They primarily mediate vesicle fusion. Members of the SNARE superfamily belong to the *vesicle-associated membrane protein* (VAMP) and *syntaxin* families.[40]

Members of the SNARE superfamily can be divided into 2 main categories: (1) vesicle (vSNARE) that are incorporated into membranes of transport vesicles and (2) target (tSNARE) SNAREs that localize in membranes of target cell compartments.

Recent classification based on the structural features of SNARE proteins divide them into SNAREs^R and SNAREs^Q whether an arginine (R) and glutamine (Q) residue contribute to the formation of the zero-ionic layer in

---

[39] Dynein is a minus-end-directed microtubule nanomotor, which can be controlled either by regulation of its subunits or regulation via accessory proteins, such as dynactin. Nuclear distribution genes encode dynein chains and dynactin components.

[40] Neurotransmitter exocytosis involves syntaxins, soluble ^{N}ethylmaleimide-sensitive factor attachment proteins (SNAP), and SNAREs.

the assembled core SNARE complex [915]. Proteins SNARE^R and SNARE^Q tether together to strongly anchor vesicles on target membranes.

A particular SNARE^R is synaptobrevin (or vesicle-associated membrane protein [VAMP]) in the synaptic vesicles (Table 9.11). Proteins of the SNARE^Q class comprise syntaxin and synaptosomal-associated protein SNAP25.

The tSNARE members of the syntaxin-1 family are SNAREs^{Q_A}; those of the SNAP25 family with a SNARE motif similar to that of SNAP25 N- and C-terminus SNARE^{Q_B} and SNARE^{Q_C}, respectively. In addition, some tSNAREs possess both ^{Q_B} and ^{Q_C} domains (SNARE$^{Q_B Q_C}$), such as SNAP23, -25, and -29. On the other hand, vSNAREs of the synaptobrevin (VAMP) family correspond to SNARE^R.

Pathways of cargo transfer include: (1) direct recycling of receptors to the plasma membrane; (2) transport of elongated tubulovesicular structures to recycling endosomes; (3) delivery of vesicles to the Golgi body using the retromer complex; and (4) maturation of early endosomes into late endosomes and assembly into multivesicular bodies that then fuse with lysosomes.

Cargo type directs the endocytosis path. Whereas transferrin is a marker of the recycling axis,[41] low-density lipoprotein is a marker for the degradative pathway. They are thus differentially sorted within early endosomes. All carrier vesicles initially experience sorting and membrane budding, then, after their transfer, docking and fusion at destination.

The recognition between vesicle and target membranes requires the interaction of vesicle coat and target membrane proteins. Hence, vSNAREs link to tSNAREs. Assembly of vSNARE–tSNARE (SNARE^R–SNARE^Q) complexes is modulated by Rab GTPases [916].

Vesicle sorting and budding as well as docking and fusion use the same factors [917]. Early endosomes represent major sorting platforms in cells. They receive cargos from fusion with recently internalized as well as carrier vesicles from the Golgi body.

Molecules SNAREs can interact with coat proteins for vesicle sorting. Sorting and budding are insensitive to clathrin coats, coatomer protein complex CoP1,[42] dynamin, and actin, but are sensitive to retromer subunits.[43]

These processes require Rab5 and its effectors, such as class-3 phosphatidylinositol 3-kinase PI3KC3 (vacuolar protein sorting VPS34) and docking factor Early endosomal auto-antigen EEA1, as well as fusion cofactor, the homohexameric AAA ATPase ^{N}ethylmaleimide-sensitive factor (NSF), and soluble NSF attachment receptors. Protein EEA1 that cooperates with NSF,

---

[41] Transferrin recycles via the formation of small vesicles from endosomes,

[42] The coatomer complex coats transport vesicles: CoP1 those of the retrograde transport from the trans-Golgi network to the cis-Golgi compartment and endoplasmic reticulum, CoP2 those of the anterograde transport from the endoplasmic reticulum to the cis-Golgi network, and clathrin and its associated adaptins endosomes from the plasma membrane and trans-Golgi network to lysosomes.

[43] The retromer complex recycles transmembrane receptors from endosomes to the trans-Golgi network.

Table 9.11. Cellular distribution of SNARE protein family members (Source: [915]; ER: endoplasmic reticulum; TGN: trans-Golgi network; BET1: blocked early in transport homolog 1 for endoplasmic reticulum-to-cis-Golgi network vesicular transport [a.k.a. Golgi vesicular membrane-trafficking protein P18]; GoSR1: Golgi SNAP receptor complex member [a.k.a. Golgi vSNARE GoS28]; Sec22: secretory pathway vesicle trafficking protein homolog; SNAP: soluble ^{N}ethylmaleimide-sensitive factor-attachment protein; Stx: syntaxin; VAMP: vesicle-associated membrane protein; VTI1: vesicle transport through interaction with tSNARE homolog 1). Membrin is also called Golgi SNAP receptor complex member GoSR2.

Locus	SNARE class	SNARE types
Rough ER	$SNARE^{QA}$	Stx5/11/18
	$SNARE^{QC}$	BET1
	vSNARE	Sec22b
Smooth ER	$SNARE^{QA}$	Stx17
Golgi body	$SNARE^{QA}$	Stx5
	$SNARE^{QB}$	Membrin (GoSR2), GoSR1
	$SNARE^{QBQC}$	SNAP29
	vSNARE	Sec22b
TGN	$SNARE^{QA}$	Stx11/16
	$SNARE^{QB}$	VTI1a/1b
	$SNARE^{QC}$	Stx6/10
Early endosome	$SNARE^{QA}$	Stx7/13
	$SNARE^{QC}$	Stx8
	vSNARE	VAMP3/8
Late endosome	$SNARE^{QA}$	Stx7
	$SNARE^{QC}$	Stx8
	vSNARE	VAMP7/8
Dense core endosome	$SNARE^{QA}$	Stx3
	$SNARE^{QBQC}$	SNAP23
	vSNARE	VAMP2
Clathrin-coated vesicle	vSNARE	VAMP1/2
Lysosome	$SNARE^{QA}$	Stx7
	$SNARE^{QC}$	Stx8
Synaptic vesicle	vSNARE	VAMP1/2
Plasma membrane	$SNARE^{QA}$	Stx1–Stx4
	$SNARE^{QBQC}$	SNAP23/25
	vSNARE	VAMP6

Rab5, and endosomal SNAREs, such as syntaxin-6 or -13, participates in budding even more than in fusion. The fusion cofactor NSF ATPase that disassembles SNAREs and docking factor EEA1 are also required for sorting and budding of recycling vesicles.

Molecular events associated with the priming, tethering, and docking steps precede vesicle fusion. They can differ according to the vesicle type, e.g., endosome and secretory vesicle. Sterols, phosphoinositides, 3 SNARE^{Q}s, and an SNARE^R together form a bundle to fuse 2 membranes.

Syntaxin-binding protein StxBP1 promotes the stability of the 3 SNARE^Q complexes. Small GTPase Rab5 coordinates the establishment of a specific, local, lipidic platform and recruitment of a proteic complex that orchestrates the homotypic endosome fusion.

The component SNARE^R of fusion reactions is dispensable for the initial docking step, whereas SNARE^Q is required [918]. Yet, SNARE^R remains essential for the downstream fusion reaction. Moreover, other factors than SNARE^R may carry out the initial docking and bind to the 3 SNARE^Q complex such as synaptotagmin-1, the partner of the 3 SNARE^Q complex in secretory vesicles.

Arachidonic acid released from phospholipid membranes by phospholipases and their metabolites are involved in the regulation of synaptic transmission. It is a common lipid regulator of SNAREs and α-synuclein, a cytosolic protein that abounds in mature nerve terminals.[44] Arachidonic acid indeed upregulates syntaxin and enhances its engagement with the SNARE complex.

9.1.13.8 Sorting Nexins

Sorting nexins control the formation of proteic complexes involved in endosomal sorting and signaling. Thirty-three sorting nexins (SNx1–SNx33) have been identified. The SNx-BAR subset comprises 12 sorting nexins with BAR (Bin, amphiphysin, Rvs) domain that forms a rigid curved structure. They also contain SNx-PX phosphoinositide-binding domains, the variation of which allows corresponding proteins to target distinct endosomal compartments [921].[45] However, variation in specificity and affinity of SNx-BARs for distinct phosphoinositides is insufficient to characterize vesicles and tubules.

---

[44] α-Synuclein sequesters arachidonic acid, thereby preventing SNARE activation and, in particular, fatty acid-mediated exocytosis [919]. On the other hand, α-synuclein directly interacts with the SNARE protein synaptobrevin and assists the SNARE complex assembly [920]. Presynaptic nerve terminals release neurotransmitters repeatedly, often at high frequency. Repeated release requires cycles of SNARE complex assembly and disassembly. α-Synuclein connects to VAMP2 to elicit SNARE complex assembly.

[45] Sorting nexin-1 binding to phosphatidylinositol (3,5)-bisphosphate targets high-curvature domains of early endosomes. SNx9 binding to phosphatidylinositol (4,5)-bisphosphate targets high-curvature domains of the plasma membrane. SNx9 also binds activator nWASP of the actin-related protein-2/3 complex to link the actin cytoskeleton to the membrane and drive membrane remodeling and tubulation, as well as to couple the transport element to molecular motors for displacement. SNx9 associates with dynamin for membrane scission. Moreover, SNx9 is a binding partner of clathrin and adaptor protein AP2 for clathrin-mediated endocytosis. SNx18 in endosomes colocalizes and interacts

The formation of SNx-BAR oligomers restricted to a given compartment of the transport system could add another level of specificity.

Sorting nexins with BAR domain stabilize the formation of tubules. They also serve as a scaffold for the formation of endosomal-sorting complexes. The *retromer* involved in transcytosis comprises a membrane-bound coat of sorting nexin complexes (SNx1–SNx2 and SNx5–SNx6). Once formed, the SNx-BAR coat acts as a docking site for cargo-selective, vacuolar protein sorting-associated proteins VPS5 and VPS17 that are linked to the cargo-selective VPS26–VPS29–VPS35 complex [921].[46]

Sorting nexins bind to molecular motors, such as kinesin and dynein. They also couple endosomal sorting to signaling. SNx27 not only regulates the endosomal sorting of cargo such as G-protein-gated potassium channels (Kir3), but also scaffolds the formation of signaling complexes by binding to diacylglycerol kinase-ζ and cytohesin-associated scaffolding protein (CASP) that controls signaling via ARF GTPase.

As sorting nexins bind phosphatidylinositol 3-monophosphate, they are associated with PI(3)P-enriched elements of the early endocytic network to operate in endocytosis, endosomal sorting, and signaling [921]. After internalization, receptors enter early endosomes, where they are segregated into separate paths by sorting.

9.1.13.9 Endosomal-Sorting Complex Required for Transport Complexes

Endosomal-sorting complex required for transport complexes operate in endocytosis and receptor signaling termination. The sequential action of 5 distinct endosomal-sorting complexes required for transport promotes lysosomal degradation of plasmalemmal receptors via multivesicular bodies. Multivesicular bodies are formed when cargo-rich patches of the limiting membrane of endosomes bud inward and are then cleaved to yield cargo-carrying intraluminal vesicles. Internalized receptors and other cargos destined for lysosomal degradation are ubiquitinated and sorted by the ESCRT complexes. Ubiquitin serves as a sorting motif for most, but not all, conveyed plasmalemmal proteins into intraluminal vesicles.

Seventeen known vacuolar protein sorting (VPS) molecules assemble into 5 complexes [922]: ESCRT0[47] to ESCRT3, and VPS4. These 5 complexes are transiently recruited from the cytoplasm to endosomes. Their sequential

---

with dynamin-2 and adaptor protein AP1. It binds PI(4,5)P_2 and colocalizes also with phosphofurin acidic-cluster-sorting protein-1 for retrograde endosomal sorting.

[46] Protein VPS5 operates as a sorting nexin. Proteins VPS26, VPS29, and VPS35 are all components of the retromer complex.

[47] A.k.a. hepatocyte growth factor-regulated Tyr kinase substrate–signal-transducing adaptor molecule (HRS–STAM) complex. Adaptors STAM1 and STAM2 that are regulators of receptor signaling and transfer interact directly with HRS that mediates the endocytic sorting of ubiquitinated membrane proteins.

action directs the sorting of ubiquitinated transmembrane proteic cargos and the inward budding of vesicles of the multivesicular body into the lumen of the endosome. The complexes ESCRT0 to ESCRT3 operate early in the endosomal compartment. They are recruited to endosomes as stable heteromers. Their distinct ubiquitin-interacting modules collect and concentrate ubiquitinated transmembrane proteins.

The transient ESCRT0 complex is made of VPS27. The transient ES-CRT1 complex is composed of VSP23, VSP28, and VSP37. The transient ESCRT2 complex consists of VPS22, 2 VPS25, and VPS36. The transient ESCRT3 complex is constituted by VPS2, VPS20, VPS24, and VPS32. Both human AAA ATPase paralogs VPS4a and VPS4b also associate with endosomes. They are recruited by VPS2 to disassemble ESCRT3 before vesicle formation from multivesicular bodies.

Generation of multivesicular bodies involves 2 coupled, but separate phases [923]: a first step of membrane invagination linked to cargo selection and a second final step with intraluminal vesicle fission. Inward budding and involution of surface membrane (away from the cytosol) is then followed by membrane scission.

Initial complexes ESCRT0, ESCRT1, and ESCRT2 are involved in cargo grouping and ubiquitin recognition (Table 9.12). The ESCRT0 complex forms domains of clustered cargos, owing to its ubiquitin-binding domain, but does not deform membranes [924].[48] The ESCRT1 and ESCRT2 complexes deform synergistically the membrane into buds, in which cargo is confined. These complexes localize to the cytosolic side of the bud neck, stabilizes the bud neck, and recruit ESCRT0-ubiquitin domains to the buds as well as ESCRT3 subunits VPS20 and VPS32a to VPS32c[49] to the neck, but not the lumen, of the bud.[50] The bud can then be cleaved by ESCRT3 subunits to form intraluminal vesicles after scission from the cytoplasmic side of the bud without ESCRT consumption [924]. Intraluminal vesicles thus contain cargos, but not ESCRT complexes.

Fusion of multivesicular bodies with lysosomes delivers cargos into lysosomes, where they are degraded. The length of ESCRT3 oligomers controls the size of multivesicular bodies and contributes to the regulation of ESCRT3 assembly by ESCRT2 [922]. The first step of ESCRT3 assembly is mediated by VPS20 that assembles VPS32 oligomers. The ESCRT2 subunit VPS25

---

[48] Although ESCRT0, -1, and -2 contain a Ub-binding domain (UBD), only ES-CRT0 is able to form cargo clusters.

[49] Protein VPS32a is also known as charged multivesicular body protein ChMP4a, human sucrose non-fermenting SNF7-1, and SNF7 homolog associated with ALIX SHAX2; VPS32b as CHMP4b, SNF7-2, and SHAX1; and VPS32c as CHMP4c, SNF7-3, and SHAX3.

[50] The ESCRT3 complex that aims at extruding sorted cargos into the endosomal lumen is able to form long, ring-like, filaments that can remain attached to the endosomal membrane, as well as tubular structures, membrane evaginations, and budded intraluminal vesicles.

Table 9.12. Role of the ESCRT complexes.

Type	Function
ESCRT0	Clustering of ubiquitinated cargos
ESCRT1	Formation of membrane invaginations
ESCRT2	Formation of membrane invaginations
ESCRT3	Vesicle scission (genesis of intraluminal vesicles)

provokes a conformational change that converts inactive monomeric VPS20 into active nucleator for VPS32 oligomerization.[51]

The ESCRT complexes enable inward vesiculation and lysosomal targeting. However, all internal vesicles of multivesicular bodies are not destined for destruction. Some transmembrane proteins actually move out of intraluminal vesicles. The ESCRT1 accessory protein programmed cell death-6-interacting protein PdCD6IP[52] that interacts with VPS37c assigns transport cellular destinations other than lysosomes, possibly owing to repeated cycles of intraluminal vesicle scission and fusion for fine-tune sorting and signaling in maturing endosomes [923].

Lysobisphosphatidic acid (LBPA), or bis(monoacylglycero)phosphate, a common, minor constituent of all cells, binds to PDCD6IP and may act as a regulator. It can promote the formation of intraluminal vesicles. However, it is mostly restricted to late Rab7+ mature multivesicular bodies [923]. On the other hand, secretory carrier membrane protein SCaMP3, a polyubiquitinated tetraspanin that binds to ESCRT0 and ESCRT1 complexes, resides on maturing endosomes. It may participate in back fusion with ESCRT1 component PdCD6IP.

9.1.13.10 G Proteins and Small Guanosine Triphosphatases

Subunit Gα of guanine nucleotide-binding proteins (Gαβγ trimers) localize to cellular membranes, particularly those of the Golgi body. G Proteins are involved in the regulation of the CoP1 transport and transport between the trans-Golgi network and plasma membrane. Several endocytic routes of specialized cargos that do or do not use clathrin require small guanosine triphosphatases, or monomeric GTPases (Table 9.13).

The growth of coated pits and the secretion of coated vesicles result from a balance between polymerization and depolymerization of coat components. Assembly and disassembly of coat protein complexes is associated with activation and inactivation of GTPase ARF1 for CoP1 and SAR1 for CoP2.

---

[51] Each ESCRT2 complex contains 2 VPS25 that synergistically prime assembly of a functional ESCRT3 complex for cargo sequestration and vesicle formation during sorting to multivesicular body.

[52] A.k.a. apoptosis-linked gene-2 (ALG2)-interacting protein-X (ALIX).

Table 9.13. Role of small GTPases in clathrin-independent endocytosis (Source: [891], ARF6: ADP-ribosylation factor-6; Cav1: caveolin-1; CavCV: caveolin-coated vesicles; CLIC: clathrin- and dynamin-independent carriers; EEA1: early endosomal antigen-1; ETRC: Early tubular recycling compartment; GEEC: gpiAP-enriched early endosomal compartments; gpiAP: glycosyl-phosphatidylinositol-anchored protein; IgER: immunoglobulin E receptor; IL2R: interleukin-2 receptor; MHC-1: major histocompatibility complex-1; PI3K: phosphatidylinositol 3-kinase; PKCα: protein kinase-Cα).

GTPase	Cargo	Primary carrier and acceptor compartment	Other components
RhoA	γC-cytokine receptor, IL2Rβ, IgER	Uncoated vesicles Early endosome	Dynamin
CDC42	gpiAPs	CLICs GEECs	Flotillin-1, Rab5, PI3K, EEA1
ARF6	β1 integrins, E-cadherin, gpiAPs, MHC-1, IL2Rα	Unknown ETRC	Flotillin-1, Rab11, Rab22, PI3K
Rab5	Glycosphingolipids, gpiAPs β1 integrins,	CavCV CavCV Early endosome	Cav1, dynamin Cav1, dynamin PKCα, Src

Upon activation, GTPases bind to the membrane and recruit elementary coat-building units (coatomer complexes) that then polymerize into coats. GTPase inactivation leads to its unbinding from the membrane and to coatomer disassembly. Competition between growth and unbinding can produce stable coats of given area, as new membrane-bound elementary coat-building units polymerize at the coat periphery, whereas others within the coat disassemble and are expelled to the cytosol [890].

ARF GTPases

ADP-ribosylation factors, especially GTPases ARF1 and ARF6, participate in multiple intracellular transport events. They control carrier genesis. They indeed regulate membrane budding, cargo sorting, and vesicle fission.

Small GTPase ARF1 acts in vesicle formation during early and late exocytosis. It recruits the AP3 complex and CoP1 to endosomes. Small GTPase ARF6 regulates the cortical actin cytoskeleton.

In addition, members of the ARF superfamily regulate phosphoinositide 4-kinases and PI(4)P 5-kinases. Both ARF1 and ARF6 stimulate PIP5K [925]. Whereas PI4K on immature secretory granules is not regulated by ARF1, ARF1 recruits and activates PI4K3β in the Golgi body for synthesis of PI(4)P and then PI(4,5)P$_2$ [926].

Monomeric GTPase ARF1 is also involved in endocytosis of gpiAPs and fluid phase that occurs via CDC42-regulated pinocytosis and requires neither dynamin nor coating proteins caveolin and clathrin. Activated, cholesterol-sensitive CDC42 recruits the actin-polymerization machinery to specific plasmalemmal loci. Small GTPase ARF1 binds and recruits RhoGAP10 to the plasma membrane that inactivates CDC42, thereby modulating endocytosis [927].

Small GTPase ARF6 and the ArfGEF activator — ADP-ribosylation factor nucleotide site opener (ARNO) — operate in vesicle-coat formation and actin cytoskeletal remodeling. Endosomes and lysosomes have an acidic luminal content induced by a proton-pumping vacuolar ATPase (vATPase). Both ARF6 and ARNO are recruited from the cytosol to endosomes owing to intra-endosomal acidification [928].[53] Hence, endosomal acidification promotes transport between early and late endosomes.

Guanine nucleotide-exchange factor BRAG2 [54] partially colocalizes with ARF6 to the cell periphery. It activates ARF6 for protein transport near the plasma membrane, including receptor recycling.[55]

Rab GTPases

Small GTPases of the RAB superfamily implicated in intracellular transport constitutes the largest set of the RAS hyperfamily of small GTPases.[56] They regulate the dynamics of vesicular carriers. They participate in many stages of the vesicular transfer of molecules inside the cell from vesicle birth and scission from the donor membrane to docking and fusion with acceptor compartments. Proteins of the RAB superfamily form foci for the assembly of proteic complexes on various vesicle membranes to sort and regulate the cargo transfer.

Small Rab GTPases undergo a membrane insertion and extraction cycle, which is partially coupled to their activation–inactivation cycle. In the absence of efficient intrinsic guanine nucleotide exchange, they interact with their effectors via guanine nucleotide-exchange factors and GTPase-activating proteins, which promote the cyclic assembly and disassembly of Rab-containing complexes, respectively. Membrane-bound, Rab^{GTP} can bind to their specific effectors. In addition, GDP-dissociation inhibitor binds to and sequesters Rab enzymes in the cytosol. On the other hand, GDI displacement factor (GDF) enables membrane attachment of Rab proteins.[57]

---

[53] Small GTPase ARF1 is not recruited by an acidification-dependent mechanism.

[54] A.k.a. guanine nucleotide-exchange activator ArfGEP100.

[55] It also interacts with α-catenin, a regulator of adherens junctions and actin cytoskeleton remodeling, thereby contributing to cell adhesion and migration.

[56] The greater the plasma membrane complexity, the larger the number of Rab proteins.

[57] Regulators GDFs are integral membrane proteins that remove GDI from Rab for its activation.

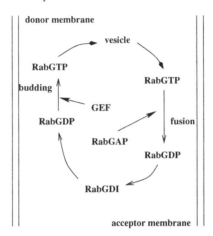

Figure 9.4. Regulation of vesicle transport by Rab GTPases (Source: [929]).

Small Rab GTPases regulate the 4 steps in vesicular transport to achieve specificity and directionality in intracellular transport: (1) membrane budding,[58] (2) delivery,[59] (3) tethering,[60] and (4) fusion[61] with the membrane of the target compartment (Fig. 9.4).

Effectors of Rab GTPases couple membranes to the cytoskeleton, recruit dyneins, kinesins, and myosins, initiate vesicle docking, and mediate membrane fusion. Active Rabs are incorporated into specific membrane nanodomains by effector- and GEF-mediated positive feedback loops [930]. Active Rabs can then recruit additional effectors into these nanodomains for cell transport. The RabGEF effector complexes stabilize activated Rab GTPases on membranes. After inactivation by specific GTPase-activating proteins, Rab^{GDP} can be extracted from the membrane by GDI and recycled back to the cytosol.

Small Rab GTPases regulate the dynamical assembly and disassembly of proteic scaffolds implicated in vesicle motions. Hubs regulated by Rab GTPases are building blocks for vesicle formation based on Rab GTPases that

---

[58] During carrying vesicle formation, the suitable transport and fusion molecular complex must be incorporated into the vesicle before scission from the donor membrane, involving Rab proteins.

[59] Vesicles are transported through the cytoplasm toward target membranes by using either actin-dependent motors such as myosins or microtubule-dependent motors, such as kinesins or dyneins. These nanomotors or motor adaptors are Rab effectors. Smell Rab GTPases often interact with nanomotors via intermediary proteins.

[60] Vesicle tethering brings the vesicle and target membrane into close proximity for merging.

[61] Smell Rab GTPases affect vesicle fusion via soluble ^{N}ethylmaleimide-sensitive factor attachment protein receptors.

function as coding systems to regulate the dynamics of specialized plasmalemmal patches, the so-called *membrome* [931].

Small GTPase Rab5 regulates endocytic membrane dynamics from the vesicle formation to the endosome fusion. Clathrin-coated vesicles possess Rab5 exchange-factor activity. Protein RabGAP5 affects endosomal Rab5 effector Early endosome antigen EEA1 and blocks endocytosis via endosomes [929]. In addition, Rab5 is involved in phagocytosis, especially of apoptotic cells [932]. Activity of Rab5 on phagosome membranes begins to increase on disassembly of actin coat-encapsulating phagosomes. However, Rab5 activity changes during the engulfment of apoptotic thymocytes. Activation of Rab5 is continuous or repetitive for up to 10 mn and ends before the collapse of engulfed apoptotic cells. GTPase-activating protein and VPS9 domain-containing protein GAPVD1[62] precludes Rab5 activation at engulfment sites. Protein GAPVD1 interacts with microtubule-associated protein of the RP/EB family MAPRE1[63] that allows microtubules to deliver GAPVD1 to phagosomes. In addition, Rab5 acts as a controller and coordinator of phosphoinositide kinases and phosphatases in endocytosis, as it activates PI3K3, PI3Kβ1, 4-phosphatase-1, and 5-phosphatases inositol polyphosphate-5-phosphatases InPP5b[64] and InPP5f [933].[65]

Small, Golgi-associated GTPase Rab6 and myosin-2 regulate the fission of Rab6+ vesicles from the Golgi body. More precisely, 2 splice variants of Rab6 — Rab6a and Rab6a'— that are attached to tubulovesicular structures that bud from the Golgi body interact with the non-processive nanomotor myosin-2[934]. Proteins Rab6 control the generation of tubulovesicular carriers at the Golgi body as well intra-Golgi transfer, Golgi body-to-endoplasmic reticulum, trans-Golgi network-to-plasma membrane, and endosome-to-Golgi body transport.[66] Isoform Rab6a does not compensate for the lack of Rab6a', and conversely. These 2 splice variants cooperate in membrane fission without redundancy. In addition, kinesin KIF5b may affect tubule extension before fission of Rab6+ vesicles by actomyosin filaments.

Small Rab GTPases and soluble ^{N}ethylmaleimide-sensitive factor attachment protein receptors (docking and fusion plasmalemmal proteins)[67] direct

---

[62] A.k.a. Rab5-activating protein RAP6 and GAPex5.

[63] A.k.a. microtubule plus-end-binding protein EB1. The RP/EB family includes retinitis pigmentosa protein RP1, a photoreceptor microtubule-associated protein (MAPRE2 and EB2) and end-binding proteins.

[64] A.k.a. type-2 inositol (1,4,5)-trisphosphate 5-phosphatase.

[65] A.k.a. oculocerebrorenal syndrome of Lowe protein (OCRL).

[66] Various types of Rab GTPases and myosins are involved in the transport from the trans-Golgi network to the plasma membrane, such as Rab6 and -8 and myosin-2, -6, and -18a.

[67] Proteins SNAREs are important components of proteic complexes that act as lipid mergers for membrane fusion between transfer vesicles and target compartments. In particular, SNAREs form nanoclusters that are heterogeneously distributed in the plasma membrane to enable the docking of secretory vesicles.

the transport of vesicles involved in endo- and exocytosis to their single destinations, SNAREs being required for membrane recognition and fusion. Docking and fusion mediated by SNARE complexes are regulated by specific effectors that either promote or prevent SNARE assembly. Several types of SNARE-binding proteins exist, such as *synaptotagmins*, which may regulate SNARE-mediated fusion, and *complexins*. These accessory proteins affect SNARE sorting and location. Proteins SNAREs are also regulated by kinases and phosphatases as well as other signaling proteins.

Transport protein particles (TraPP) are multimeric guanine nucleotide-exchange factors for Rab1 GTPase.[68] The TraPP complexes regulate endoplasmic reticulum–Golgi and intra-Golgi body transfer. The 3 forms of TraPP complexes contain a common core of subunits and distinct subunits that determine their localization [936]. Agent TraPP1 tethers to coated vesicles that run from the endoplasmic reticulum to Golgi body and TraPP2 to coated vesicles within the Golgi body. Protein TraPP3 is involved in autophagy.

Diverse combinations of Rab GTPases with members of the myosin-5 subclass (Myo5a–Myo5c) control distinct cargo transfer routes. Small GTPase Rab8a can link to all 3 Myo5 isoforms. Small GTPase Rab11a interacts with Myo5a and Myo5b. Myosin-5B associates with Rab8a, Rab10, and Rab11a during recycling to the plasma membrane of proteins internalized by clathrin-dependent or -independent endocytosis. Myosin-5B ensures the apical membrane transfer with both Rab8a and Rab11a in epithelial cells [937]. On the other hand, in non-polarized cells, myosin-5B cooperates with Rab11a, but not Rab8a. In particular, in non-polarized cells, the recycling of the transferrin receptor relies on Myo5b and Rab11a. In polarized cells, the apical recycling

---

The stability of these nanodomains depends on sterols such as cholesterol. The SNARE superfamily comprises 36 members in humans [935]. Many SNAREs reside predominantly or selectively in given cell compartments. Syntaxins-1, -2, -4, SNAP23, and SNAP25 are located at the plasma membrane, syntaxin-5, vesicle-associated membrane protein VAMP4 (or synaptobrevin-4) in the membrane of the Golgi body. SNARE proteins involved in transport between 2 cell compartments are found in both compartment membranes, as well as in the associated vesicles. SNARE proteins clamp the membranes together and initiate fusion. Merged opposing membranes form an aqueous pore that connects the distal leaflets of the membranes in contact. After membrane fusion, SNARE complexes are disassembled and SNARE proteins are recycled owing to AAA ATPases (ATPases associated with various cellular activities), ^{N}ethylmaleimide-sensitive factor (NSF), and soluble NSF attachment proteins (SNAP cofactors).

[68] The contact of a vesicle and its target membrane is ensured by long coiled-coil proteins and large complexes [936]. Long coiled-coil proteins include early endosome antigen EEA1, golgins, and USO1 homolog (a.k.a. general vesicular transport factor P115, transcytosis-associated protein [TAP], and vesicle-docking protein [VDP]). Large complexes encompass exocyst, TraPP, conserved oligomeric Golgi protein (COG), homotypic fusion and vacuole protein sorting (HoPS), class-C core vacuole–endosome tethering (CorVET), Dependence on Sly1 (DSl1), and Golgi-associated retrograde protein (GARP).

of the transferrin receptor depends on Myo5b and Rab11a. The apical transfer is controlled by Rab11a that recruits and activates Rab8a on subapical vesicles via the RabGEF Rabin8.

Ras GTPases

Small GTPases Ras localize with their partners to the Golgi body. Here, they are regulated with modes of activation and deactivation that differ from those at the plasma membrane.

SAR GTPases

Small GTPase secretion-associated and Ras-related protein SAR1, a member of the RAS hyperfamily of small GTPases, recruits: (1) CoP2 vesicle to the membranes of the endoplasmic reticulum;[69] (2) ARF1, CoP1, and clathrin vesicles to the Golgi body; and (3) plasmalemmal ARF6, phosphatidylinositol (4,5)-bisphosphate kinase, ARL1, and ARFRP1 to the trans-Golgi network. GTPase ARF1 involved in CoP1 coat assembly and recycling from Golgi body to the endoplasmic reticulum links with Rab1b, which modulates CoP1 recruitment to compartments of Golgi body.[70]

9.1.13.11 Protein Kinases

The cAMP–PKA pathway acts on many trafficking steps [939]. The cAMP–PKA pathway includes protein kinase-A, adenylyl cyclase, which synthesizes cAMP, adenylyl cyclase activators, such as Gα protein, phosphodiesterases which hydrolyze cAMP, and PKA-anchoring scaffold proteins. Protein kinase-A is actived when the catalytic and regulatory subunits dissociate in the presence of cAMP. The entire set of components is located on the plasma membrane. Enzyme PKA recruits ARF1 to the Golgi body. It phosphorylates certain SNAREs.

Protein kinase-C is recruited into cellular membranes for transport regulation. It is required in the transport of glucosaminoglycans from the Golgi body to the plasma membrane. It is involved in carrier membrane fission from the trans-Golgi network.

---

[69] Transport from the endoplasmic reticulum to the Golgi body requires assembly of the CoP2 complex at endoplasmic reticulum outlets. Cytosolic small CoP2-coated vesicles contain SNARE Sec22b necessary for downstream fusion [938].

[70] Protein ARF1 also associates with Rab11a, dynamin-2, clathrin light chain-A, clathrin-binding protein HRS/HG, HRS/HG-interacting protein TSG101, and TSG101-interacting protein EAP30. Enzyme ARF1 can form clusters with Rab1b and -C, GEF TraPPC4, Rab1 tether GRaSP55 (or 55-kDa Golgi reassembly stacking protein-2 [GoRaSP2]), mannose 6-phosphate receptor-binding protein M6PRBP1 (or cargo selection protein TIP47), and CoP2 components Sec13R and Sec24C.

9.1.13.12 Ubiquitination

Ubiquitination is a post-translational modification that corresponds to addition of protein *ubiquitin* to a substrate. Intracellular transport, like many other cell processes, uses ubiquitin label.

Ubiquitin[71] (Ub) is involved in signal transduction and endocytosis, among other cellular processes [940, 941]. It acts on protein–protein interactions implicated in membrane fusion. Monoubiquitination causes endocytosis of plasmalemmal receptors by clathrin-dependent or -independent mechanisms, according to stimulation intensity and ubiquitination status.[72] It is also required for cellular transport associated with endosomal sorting.

Proper cell signaling and metabolism require transport of membrane proteins to lysosomes via multivesicular endosomes for destruction. Lysine-48-linked polyubiquitination serves as a tag for proteasomal degradation, whereas multiple monoubiquitination and Lys-63-linked polyubiquitination mediate protein degradation in the lysosome.

Monoubiquitination targets plasmalemmal receptors and associated proteins, as well as endocytic accessory Ub-binding proteins.[73] Recruitment of Ub ligase can lead to receptor endocytosis and hence termination of signaling.[74]

Endosomal-sorting complexes required for transport (ESCRT0–ESCRT3) with several accessory components, PI(3)P, and AAA ATPase vacuolar protein sorting-associated protein VPS4, as well as deubiquitinating enzymes and programmed cell death 6-interacting protein (PdCD6IP or ALIX) sort ubiquitinated cargos into invaginations of endosome membranes and catalyze

---

[71] Ubiquitin is activated by an ATP-dependent ubiquitin-activating (E1) enzyme, attached to target proteins by ubiquitin-conjugating (E2) enzyme in association with a ubiquitin–protein (E3) ligase to eventually induce their degradation by a 26S proteasome. Ubiquitin-like proteins also control the activity of proteins. Ubiquitin-like proteins are activated by specific Ub activases and are transferred to their substrates by Ub conjugases.

[72] Slight stimulations lead to clathrin-dependent endocytosis. Strong stimulations increase the level of receptor ubiquitination and favor endocytosis supported by membrane raft or caveolae.

[73] Multiple monoubiquitination of epidermal and platelet-derived growth factor receptors yields tags that ensure receptor endocytosis and degradation, whereas monoubiquitination of accessory proteins could regulate their function in endosome [942]. Ubiquitination of some receptor Tyr kinases such as epidermal growth factor receptor is mediated by Ub ligase CBL [891]. Receptor EGFR can be internalized using clathrin-mediated and -independent pathways. Inositol trisphosphate receptors of the endoplasmic reticulum membrane are ubiquitinated and degraded upon elevation of both Ca^{++} and IP_3 intracellular levels.

[74] Not only GPCRs, but also their associated proteins are ubiquitinated for proteasomal degradation [943]. Degradation of $G\alpha$ and $G\beta\gamma$ subunits yields another mechanism for fine-tuning cell signaling. In addition, activated $\beta2$-adrenoceptors induce ubiquitination and proteasomal degradation of GRK2.

abscission of endosomal invaginations to form intraluminal vesicles that contain tagged cargos [944].

Cargo partitioning into intraluminal vesicles of multivesicular endosomes usually involves protein ubiquitination as well as hepatocyte growth factor-regulated Tyr-kinase substrate and endosomal-sorting complexes required for transport. The former recruit not only ubiquitinated receptors but also ES-CRT complexes.[75] However, intraluminal vesicle sorting in fibril formation does not require HRS [945].

Ubiquitination is a reversible process, as deubiquitinases are involved in endocytosis. Deubiquitinases include: (1) isopeptidases or Ub C-terminal hydrolases (UCH); (2) ubiquitin specific proteases (USP);[76] (3) otubains of the OTU family (OTUB1–OTUB2); (4) Machado-Joseph disease (MJD; or type-3 spinocerebellar ataxia SCA3) proteases;[77] and (5) Zn^{++}-dependent JAMM metalloproteases.[78]

9.1.13.13 Arrestins

β-Arrestins intervene in the desensitization of G-protein-coupled signaling. They also form an endocytic complex that can initiate a G-protein-independent transmission of signals. In particular, this complex regulates extracellular signal-regulated protein kinases involved in various signaling pathways. The cytoplasmic tail of activated transmembrane receptors is phosphorylated to create binding sites for arrestins, which preclude interaction of receptors with G proteins and, hence, further signaling from activated receptors. Arrestins also promote receptor internalization, as they act as adaptors for endocytic proteins clathrin and clathrin-associated adaptor protein AP2 complex.

The arrestin family of adaptor or scaffold proteins include 2 photoreceptor-specific visual arrestins (arrestin-1 and -4), 2 ubiquitous β-arrestins, and

---

[75] Transport of chemokine receptor CXCR4 to lysosomes involve HRS that colocalizes with ubiquitinated proteins on clathrin-coated endosomes to prevent its recycling [943]. Ubiquitin ligase Itch (a.k.a. Itchy homolog, atrophin-1-interacting protein, and NFE2-associated polypeptide NAPP1) ubiquitinate both CXCR4 and HRS in endosomes.

[76] Deubiquitinase USP4 prevents proteasomal degradation of newly synthesized intracellular A_{2A} adenosine receptors [943].

[77] The founding member of the MJD family is ataxin-3. The N-terminal Josephin domain exists in many other proteins. Ataxin-3 interacts with the type-2 AAA ATPase valosin-containing protein that acts in membrane fusion, endoplasmic reticulum-associated degradation, and regulation of the NFκB pathway. It also interacts with the repair protein RAD23 homolog, a shuttle for the transfer of ubiquitinated proteins to the proteasome.

[78] JAMM: JAB1–MPN–Mov34 metalloenzyme; MPN: Mpr1p–Pad1p–N-terminal domain. The JAMM motif in regulatory particle number-11 (Rpn11) and G-protein pathway suppressor (or COP9 signalosome) component CSn5 support the isopeptidase activity intrinsic to the proteasome and signalosome, respectively [946].

arrestin-related proteins that couple to transmembrane proteins (GPCRs and others) for endocytosis [947].

β-Arrestins assemble proteic complexes upon GPCR stimulation that trigger receptor signaling, endocytosis, and degradation [948]. β-Arrestins target not only classical GPCRs, but also atypical GPCRs, such as Frizzled and Smoothened, as well as receptor Tyr kinases, cytokine receptors, and nicotinic (ionotropic) cholinergic receptors (ligand-gated ion channels).

The set of arrestin-related proteins comprises 6 known arrestin-related transport adaptors (ART), α-arrestins) and 4 identified VPS26 members, components of the retromer complex.

α-Arrestins are associated with membranes, whereas β-arrestins are generally cytoplasmic in unstimulated cells. Members of the ART class can have high affinity for given plasmalemmal proteins (receptors and transporters).

Membrane proteins are often ubiquitinated during arrestin-mediated endocytosis, as β-arrestins can operate as ubiquitin ligase adaptor. β-Arrestin–GPCR complexes can be transient or stable according to the GPCR class [943]. Activation of class-A receptors induces a transient ubiquitination. β-Arrestins are rapidly deubiquitinated and dissociate from receptor that is internalizing for fast recycling. Activation of class-B receptors generates a sustained β-arrestin ubiquitination and formation of stable endocytic complexes. Persistent ubiquitination of β-arrestin is required not only to form stable, endocytic, signaling β-arrestin–activated GPCR complexes, but also to recruit extracellular signal-related kinase on these signalosomes.[79] β-Arrestin–GPCR signalosomes that contain ERK are compartmentalized to perinuclear endosomes.

Arrestins, arrestin-related transport adaptors, and membrane proteins (receptors and carriers) contain PY-binding motifs that recruit ubiquitin ligase [949]. Arrestins not only operate as adaptors in receptor ubiquitination, but also can undergo ubiquitination. Arrestin ubiquitination stabilizes its binding to activated receptors. Ubiquitination of receptors can also acts as a signal for endocytosis down to lysosomal degradation.[80]

9.1.13.14 ATPases Associated with Diverse Cellular Activities

Members of the superclass of AAA ATPases regulate the transport of proteins. In particular, AAA ATPase vacuolar protein sorting-associated pro-

---

[79] Prolonged β-arrestin ubiquitination following angiotensin-2 stimulation leads to endosome-localized AT_{1A}–β-arrestin–ERK complex.

[80] β-Arrestin-1 escorts Ub ligase DM2 to bind insulin-like growth factor-1 receptor for receptor ubiquitination and degradation [943]. β-Arrestin-2 acts as a Ub ligase adaptor for β2-adrenoceptors and vasopressin V_2 receptors that are polyubiquitinated in response to ligand binding. β-Arrestin-2 also interacts with Ub ligases DM2 and NEDD4 for ubiquitination and lysosomal degradation of endocytozed β2-adrenoceptors [950].

tein VPS4 is necessary for delivery of ubiquitinated CXCR4 to late endosomes [943].

9.1.13.15 Cell Surface Molecules

Integrins

Integrins, which are involved in cell anchorage (Sect. 7.5), contribute to the
control of the transfer of other receptors, such as vascular endothelial growth
factor receptor VEGFR2 and epidermal growth factor EGFR1. Integrins also
regulate Rho GTPases. In particular, intracellular integrin transfer directs
Rho signaling pattern during cytokinesis and cell migration. They control the
translocation of Rac^{GTP} to the plasma membrane, before Rac binding to effectors.[81] Integrin-regulated Rac-binding sites are within cholesterol-enriched
membrane nanodomains. Integrins prevent the internalization of components
of these plasmalemmal nanodomains, and hence control Rac signaling. This
internalization is mediated by dynamin-2 and caveolin-1 [951].

Integrin endo- and exocytosis can contribute to a bulk motion across the
cell such as during mitosis or support a spatially restricted displacement such
as during cell migration. Endo- and exocytosis can be localized spatially to
concentrate receptors and restrict signaling to given subcellular regions.

Endocytosis of many integrins occurs by clathrin-dependent and -indepen-
dent mechanisms (Table 9.14). A given integrin can be internalized by different
processes. Once internalized using clathrin- or caveolin-dependent mode, integrins travel via early endosomes, where they are sorted to late endosomes and
lysosomes for degradation or, most often, recycled back to the plasma membrane. Integrin recycling use one of 2 pathways that are spatially and temporally distinct: (1) quick Rab4-dependent recycling or (2) slower Rab11- and/or
ARF6-dependent recycling via the perinuclear compartment (Table 9.15).

The mode of intracellular transfer of integrins influences their function.
Endosomes mediate redistribution of integrins during the end of mitosis. Integrin transfer intervenes in cell signaling, especially that linked with integrin-
associated growth factor receptors.

Spatial restriction of integrins can modulate Rho signaling and affect recycling of other receptors, such as EGFR and VEGFR [952]. Integrins regulate
Rac signaling, as it control endocytosis and recycling of membrane rafts.[82]

---

[81] Integrins increase the membrane affinity for Rac, thus leading to RhoGDI dissociation and effector coupling.

[82] In adherent cells, integrin engagement with fibronectin retains phosphorylated
caveolin in focal adhesions, keeps active Rac at the cell surface, and prevents
endocytosis of membrane rafts. When cells detach and integrins lose their matrix
ligands, phosphorylated caveolin is released from focal adhesions and recruited to
caveolae that trigger endocytosis of membrane rafts and subsequent transport
along microtubules to the perinuclear region. Cell reattachment reverses this
process. Activated ARF6 then promotes the rapid recycling of membrane rafts,
hence recruitment of active Rac to the plasma membrane.

Table 9.14. Integrin endocytosis (Source: [952]).

Integrin	Adaptors and partners
	Clathrin-dependent pathway
$\alpha_V\beta_3$	Numb
$\alpha_V\beta_6$	HCLS1-associated protein-X1
$\alpha_V\beta_5$	Clathrin
$\alpha_1\beta_1$	Disabled-2 (clathrin adaptor)
$\alpha_2\beta_1$	Disabled-2
$\alpha_3\beta_1$	Adaptor protein-2 via Tspan24, Disabled-2, adaptor protein-2 via L1CAM (or CD171)
$\alpha_5\beta_1$	Adaptor protein-2 via Tspan24, Numb, NPXY-Binding motif-containing proteins, neuropilin-1, myosin-6, GAIP C-terminus-interacting protein GIPC1
$\alpha_6\beta_1$	Adaptor protein-2 via Tspan24
	Clathrin-independent pathway
$\alpha_2\beta_1$	Rab21
$\alpha_5\beta_1$	Rab21
	Caveolin-dependent pathway
$\alpha\beta_1$	Protein kinase-Cα

Table 9.15. Small, monomeric GTPases of the RAB superfamily and integrin recycling (Source: [952]; ACAP: ArfGAP with coiled coil, Ankyrin repeat, and PH domain-containing protein; ARF: ADP-ribosylation factor; EHD: EH domain-containing protein; RabIP4: Rab4-interacting protein; RCP, Rab-coupling protein; VAMP: vesicle-associated membrane protein).

Integrin	GTPase	Associated proteins
$\alpha\beta_1$	Rab11	PKCϵ
	ARF6	ACAP1
$\alpha_2\beta_1$	Rab21	
$\alpha_5\beta_1$	Rab11	RCP, EHD1, VAMP3
	Rab25	
	Rab21	
$\alpha_V\beta_3$	Rab4	PKD1, RabIP4

Cell migration relies on a balance between Rac and RhoA GTPases. Small GTPase Rac stabilizes the extension of a broad, flat lamellipodium at the cell front. Small GTPase RhoA favors contraction of actin–myosin filaments that limits lamellipodial activity to the cell front and helps to retract the rear of the cell. Integrin $\alpha_V\beta_3$ supports Rac [952]. Integrin $\alpha_5\beta_1$ stimulates RhoA

GTPase. Also, the mode of integrin transfer can influence their signaling to Rho GTPases.[83]

Proteoglycans

Proteoglycans close to the external face of the plasma membrane participate in cell transport. The fibroblast growth factor FGF2 is secreted to transmit angiogenic signals via a ternary complex with FGF receptors and extracellular heparan sulfate proteoglycans [953].

9.1.14 Coupled Transport of Proteins and Lipids

The 3 most common types of lipids in cellular membranes are glycerolipids (e.g., phosphatidylinositol and its derivatives), sterols, and sphingolipids. The latter act in signaling pathways and formation of membrane rafts. Glucosylceramide, a precursor for more complex glycosphingolipids, is formed by the addition of glucose to ceramide. Although ceramide is produced in the endoplasmic reticulum, like most cell lipids, glycosphingolipids of cell membranes are synthesized in the Golgi body. Glycosphingolipid synthesis depends on glucosylceramide-transfer protein 4 (Four)-phosphate adaptor FAPP2 that binds phosphatidylinositol 4-phosphate and ARF1 [954]. Protein FAPP2 also regulates cellular transfer of proteins and glucosylceramide from proximal to distal Golgi compartments. Glucosylceramide is required for protein transport out of the distal compartment. Moreover, FAPP2 mediates the backward transport of glucosylceramide from the Golgi body to the endoplasmic reticulum, where it is processed into complex glycosphingolipids. Proteins FAPP1 and FAPP2 also operate in transport from the trans-Golgi network to the plasma membrane and control cilium formation in polarized cells.

Lipid-binding proteins couple the lipid synthesis to the anterograde transport of these lipids to the plasma membrane. The plasma membrane–endosome–trans-Golgi network and the early Golgi body–endoplasmic reticulum routes are connected by intense bidirectional transfer. However the respective membranes have a distinct lipid composition, as the former is highly enriched in glycosphingolipids, sphingomyelin, and cholesterol compared with the latter. These lipids bind lipid-binding proteins such as FAPPs, CERT, and oxysterol-binding protein OSBP1 that target the trans-Golgi network, as they are recruited by PI(4)P and ARF1.

---

[83] During cell migration, when recycled $\alpha_V\beta_3$ integrin can engage matrix ligands, Rab11 family-interacting protein Rab11FIP1 associates with β_3 integrins. Dual effector Rab11FIP3 for Rab11 and ARF6 controls endosomal transfer that carries RhoGEF2 to the cleavage furrow during late telophase. Activated RhoA may then activate Rho effectors Diaphanous homolog-1 and RoCK and citron kinase in this region where integrins concentrate to assemble and contract the splitting ring.

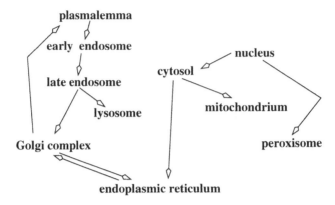

Figure 9.5. Transport between the main cell constituents.

9.1.15 Transport Sorting

Once membrane-bound vesicles enter the endosomal system, they are sorted for delivery to their specific destinations. Incorporation into the intraluminal vesicles of multivesicular bodies involves different mechanisms. Transport sorting is controlled by numerous molecules. Ubiquitin, used for degradation, is also a sorting tag on endocytic cargos and a regulatory switch on endocytic adaptor proteins.

Hepatocyte growth factor-regulated Tyr kinase substrate is an early endosomal protein that contributes to the regulation of tranfer of growth factor-receptor complexes through early endosomes. Protein HRS operates also in postendocytic sorting of other plasmalemmal receptors such as protease-activated receptor PAR_2 [955]. It also controls lysosomal transfer and recycling of calcitonin receptor-like receptor (CRLR or CalcRL), a non-ubiquitinated receptor that travels to lysosomes after sustained activation and recycles after transient activation. It thus intervenes in both degradative and recycling pathways.

Sorting nexins control the formation of proteic complexes involved in endosomal sorting and signaling [921]. Transferrin receptor is recycled back to the plasma membrane owing to sorting nexin-4 via the tubular endosomal network. Epidermal growth factor receptor is sorted into early endosomes that mature into late endosomes, which fuse with lysosomes for EGFR degradation. Cation-independent mannose 6-phosphate receptor is sorted for retrieval back to the trans-Golgi network through both early and late endosomal pathways.

Endosomal-sorting complexes required for transport ESCRT1 to ESCRT3 (Sect. 9.1.13.9) are necessary for transporting proteins in vesicles [956]. Another mechanism that does not require direct ubiquitination exists for certain substances. The lipid-driven process of multivesicular body formation involves lipid lysobisphosphatidic acid [945]. Furthermore, protein processing can require a passage by intraluminal vesicles of multivesicular bodies.

The intracellular transport of substances deals not only with the flux of molecules from the extracellular spaces, but also with the transfer of compounds manufactured by the cell toward the targeted (intra- or extracellular) medium (Fig. 9.5). In both cases, the cell membrane must be crossed. The cell secretion from a donor organelle requires the budding of transport vesicles, their displacement, and the fusion of the vesicles with an acceptor membrane, once the vesicles loaded with a specific compound have recognized the appropriate target. Large GTPase dynamin is recruited to coated pits before vesicle budding [957]. Dynamin binds to inositol lipids, constraining the membrane into tubular shapes. They form necks of vesicles and promote their scission from the membrane via GTP hydrolysis.

9.1.16 Endocytic Recycling

Endocytic uptake and recycling are coordinated to control the composition of the plasma membrane. Endocytic recycling comprises clathrin-dependent (CDE) and -independent (CIE) endocytosis (Table 9.16).

Clathrin-independent endocytosis and recycling include tubular carriers [958]. Specialized actin-driven clathrin-independent endocytosis includes macropinocytosis and phagocytosis.

Recycling endosomes participate in temporal and spatial regulation of cytokinesis, cell adhesion, and tissue morphogenesis, as well as learning and memory. Various agents, particularly monomeric GTPases (Table 9.17), adaptors, cytoskeletal constituents, and nanomotors (Table 9.18), regulate endocytic recycling.

9.1.16.1 Recycling Pathways

A *fast recycling* back to the plasma membrane from earlier stages of endocytosis is regulated by small GTPases of the RAB superfamily, especially Rab35. The *slow recycling* corresponds to the transport of cargos from early endosomes to the juxtanuclear endocytic recycling compartment (JNERC) and, then, to the plasma membrane. Small GTPase Rab35 is also involved in ARF6+ and EHD1+ tubular recycling endosomes that carry cargos back to the plasma membrane from the juxtanuclear endocytic recycling compartment in clathrin-independent endocytosis.

The *juxtanuclear endocytic recycling compartment* often localizes near the microtubule organizing center and Golgi body, but not in polarized cells.

Early endosomes mature into multivesicular bodies or extend tubules, lose Rab5, and acquire Rab11 to form the Rab11+, EHD1+, tubular, juxtanuclear, endocytic recycling compartment [958]. Protein EHD4 is needed for transfer from early endosomes to the juxtanuclear endocytic recycling compartment and late endosomes. Protein EHD3 that binds to 2 Rab effectors, rabenosyn-5

Table 9.16. Clathrin-dependent and -independent endocytosis and its factors (Source: [958]). Both routes can use the same factors (EEA: early endosome antigen; EHD: C-terminal epidermal growth factor receptor substrate-15 homology domain-containing protein; PdCD6IP: programmed cell death protein-6-interacting protein [a.k.a. ALIX: apoptosis-linked gene 2-interacting protein-X1]; Rab11FIP: Rab11 family-interacting protein; SNx: sorting nexin). Some cargos that do not take recycling itineraries are selected in early endosomes by endosomal-sorting complex required for transport (ESCRT) to travel into late endosomes and multivesicular bodies. Clathrin-dependent cargos can recycle back to the cell surface through a rapid recycling route that requires Rab4 and Rab35 GTPases. Both CDE and CIE can lead to juxtanuclear endocytic recycling compartment (JNERC). From ERC, recycling of both clathrin-dependent and -independent types of cargo requires Rab11 GTPase. Recycling of clathrin-independent cargos involves the generation of distinctive Rab8- and Rab22a-dependent tubules. Some clathrin-dependent cargos might also recycle via Rab8- and Rab22a-dependent tubules. In the cell periphery, tubules break up into vesicles before fusing with the plasma membrane under the control of partitioning defective protein Par3 and CDC42 and subsequently ADP-ribosylation factor ARF6, as well as Rab11 and cortical actin. Tubular recycling endosomes are more or less associated with the clathrin-independent endocytosis.

Transfer vesicle or tubule	Mediators
Early endosome	Rab5, EEA1 Reception of endocytic vesicles (clathrin-dependent and -independent endocytosis)
Rapid recycling endosome	Rab4, Rab35 (clathrin-dependent endocytosis)
Recycling endosome	Rab11 (clathrin-dependent and -independent endocytosis)
	Clathrin-independent endocytosis
Juxtanuclear endocytic recycling compartment	Rab10, Rab22a, Rab11FIP5, Rab11FIP2–EHD3, Rab11FIP3–ARF6, Rab11FIP3–Rab11 Dynein, SNx4
Tubule-originated cortical recycling vesicle	Rab11, ARF6, CDC42 actin, Par3
Tubular recycling endosome	Rab8, Rab22a, Rab10, Rab11, Rab35, CDC42 EHD1–PdCD6IP, EHD1–Rab11FIP2, Par3

and Rab11 family-interacting protein Rab11FIP2, may link Rab5+ early endosomes to Rab11+ juxtanuclear endocytic recycling compartment. Small GTPase Rab22a and Rab11FIP5 are involved in the movement of cargos from early endosomes to the juxtanuclear endocytic recycling compartment [958]. Both

Table 9.17. Small GTPases in endocytic recycling, their location, and interactors, especially regulators (Source: [958]; CIE: clathrin-independent endocytosis; JNERC: juxtanuclear endocytic recycling compartment; ACAP: ArfGAP with coiled-coil, Ank repeat, and PH domain-containing protein; ARF: ADP-ribosylation factor; ASAP: ArfGAP with SH3 domain, ANK repeat, and PH domain-containing protein; Exo: exocyst subunit; GAP: GTPase-activating protein; GEF: guanine nucleotide-exchange factor; Rab11FIP: Rab11 family-interacting protein). Jun N-Terminal kinase-interacting proteins JIP3 (or MAPK8IP3) and JIP4 (or MAPK8IP4) bind to ARF6, kinesin light chain KLC1, and dynactin. Connecdenn (or DENN domain-containing receptor-mediated endocytosis protein RME4) interacts with Rab35 and AP2clathrin-associated adaptor proteic complex. Rabenosyn-5 interacts with Rab4 and Rab5, epidermal growth factor receptor substrate-15 homology domain-containing proteins EHD1 and EHD3, as well as vacuolar protein sorting-associated protein VPS45 that segregates intracellular molecules into distinct organelles. The recycling of syndecan-1 together with fibroblast growth factor receptor requires ARF6, $PI(4,5)P_2$ produced by ARF6-mediated activation of phosphatidylinositol 4-phosphate 5-kinase (PI(4)P5K), and syntenin that binds to syndecans. Small GTPase CDC42 colocalizes with EHD1+ endosomes. The CDC42–Par6–Par3–aPKC complex regulates cell adhesion and the maintenance of E-cadherin in adherens junctions.

GTPases	Localization	Interactors
ARF6	CIE cortical vesicle	Phospholipase-D2, PI(4)P5K
		ExoC5, JIP3/4, Arf6GAPs ASAP1 and ACAP1
		Arf6GEF (or ArfGEP100)
CDC42	CIE cortical vesicle	actin
Rab4	Early endosome	Rabenosyn-5
Rab5	Early endosome	Rabenosyn-5
Rab8	Recycling tubule	
Rab10	Early endosome	
	JNERC	
	Recycling tubule	
Rab11	JNERC	Rab11FIP2/3/5
	Recycling tubule	
	Recycling endosome	
Rab22a	Recycling tubule	
Rab35	Early endosome	Rab35GAP (TBC1D10C or EP164C)
	Recycling tubule	Connecdenn
Rac1		
Ras		

Rab11FIP3 and ArfGAP ASAP1 contribute to the juxtanuclear positioning of the endocytic recycling compartment. The AP2 complex participates in recycling of class-1 MHC proteins and β integrins. Hence, clathrin-dependent and -independent endocytosis (Table 9.19) are connected [958].

Table 9.18. Cytoskeletal constituents, nanomotors, and adaptors in endocytic recycling (Source: [958]; aPKC: atypical protein kinase C; EHD: C-terminal epidermal growth factor receptor substrate-15 homology domain-containing protein; ERK: extracellular signal-regulated kinase; Par: partitioning defective protein; PdCD6IP: PdCD 6-interacting protein). Dynactin (Dctn), or dynein activator complex (Dctn1–Dctn6 and actin-related proteins ActR1a and ActR3c), binds to dynein and kinesin-2 and links them to vesicular carriers to be transported along microtubules. Sorting nexin-4 (SNx4) interacts with nanomotor dynein through linker protein WW domain-containing protein WWC1. Inwardly rectifying potassium channel $K_{IR}3.4$ binds to ARF6 guanine nucleotide-exchange factor EFA6 to activate ARF6 and raise its plasmalemmal density.

Protein	Function
Clathrin	Vesicle coat protein
Dynactin	Dynein activator
Dynein	Microtubule-based, minus-end-directed nanomotor
Myosin-5b	Actin-based, barbed end-directed nanomotor
Myosin-6	Actin-based, pointed end-directed nanomotor
Kinesin	Microtubule-based, plus-end-directed nanomotor
EHD	ATPase
ERK	Phosphorylation
aPKC	Cell polarity
Src	Phosphorylation
Syndapin	Membrane tubulation and actin remodeling
PdCD6IP	Regulation of multivesicular bodies and recycling
Par3/6	Cell polarity
Myoferlin	
SNx4	Sorting for juxtanuclear endocytic recycling compartment (avoid degradative endosomal compartment)
Mucolipin-2	ARF6 Activation
$K_{IR}3.4$	ARF6 Activation

9.1.16.2 Small GTPases in Recycling Axes

Several recycling routes from the juxtanuclear endocytic recycling compartment back to the plasma membrane probably exist: tubular and vesicular recycling endosomes. Both types require Rab11 GTPase. Tubular recycling endosomes are, more or less, associated with the clathrin-independent endocytosis [958].

Many proteins contribute to the final stage of slow recycling such as ARF6 that activates phospholipase-D2 to produce phosphatidic acid and diacylglycerol implicated in recycling as well as PI(4)P5K kinase to generate $PI(4,5)P_2$ for the recruitment of proteins involved in vesicle formation, membrane fusion, and cortical actin polymerization [958]. Several factors, such as ERK and Src kinases, small GTPases Rac and Ras, cation channel mucolipin-2,

Table 9.19. Examples of proteins transported in clathrin-independent endocytosis (Source: [958]; AR: adrenergic receptor; CD44: hyaluronan receptor [Indian blood group]; CD55: complement decay-accelerating factor [Cromer blood group]; CD59: protectin, or complement regulatory protein, a membrane inhibitor of the complement membrane attack complex; CD147: basigin, or extracellular matrix metalloproteinase inducer [EMMPrIn], a type-1 integral membrane receptor, which binds cyclophilins CyPa and CyPb and some integrins as well as monocarboxylate transporters; GluT: glucose transporter; ICAM: intercellular adhesion molecule; mGluR: metabotropic glutamate receptor; M_3: type-3 acetylcholine muscarinic receptor; MHC: major histocompatibility complex). Molecule CD1a, like CD1b and CD1c (i.e., group-1 CD1), related to the class-1 MHC molecules, are produced in antigen-presentig cells.

Cargo	Function
β2AR	Signaling
β integrins	Cell–matrix adhesion and interaction
E-Cadherin	Between-cell junction
CD1a	Immune response
CD44	Cell–matrix interaction
CD55, CD59	Protection against complement
CD147	Cell–matrix interaction
GluT1	Glucose transporter (SLC2a1)
ICAM1	Cell–matrix adhesion and interaction
K_{IR}3.4	KCNJ5-encoded K^+ channel (or GIRK4)
LAT1	L-type amino acid transporter (or SLC7a5)
M_3, mGluR7	Signaling
Class-1 MHC	Immune response
Class-2 MHC	Immune response
Mucolipin-2	Cation channel
SLC3a2	Amino acid transport
Syndecan-1	Cell–matrix interaction

and K_{IR}3.4 channel (or GIRK4) can regulate the activation of ARF6 GTPase. The ARF6 effectors Jun N-terminal kinase-interacting protein JIP3 and JIP4 (or MAPK8IP3 and MAPK8IP4) are scaffold proteins that bind to plus-end-directed nanomotor kinesin and minus-end-directed nanomotor dynein, thereby controlling the direction of transport along microtubules [958].

Small GTPase Rab22a is implicated not only in transfer from early endosome to endocytic recycling compartment, but also in the genesis of recycling tubules back to the plasma membrane. Small GTPase Rab8 also resides on tubular endosomes, where it may promote cortical actin-driven protrusions of the plasma membrane [958].

Blood vessel–epicardial substance (BVES) is a widely expressed, dimeric, transmembrane protein, especially on all 3 myocyte types (smooth, skeletal, and cardiac) and epithelial cells. It influences cell adhesion and motility. It interacts with RhoGEF25[84] that activates CDC42 and Rac1 to initiate formation of lamellipodia and filopodia. It also contributes to the regulation of vesicular transport. It directly interacts with ubiquitous vesicle-associated membrane protein VAMP3 that binds syntaxin-4 in the basolateral region of epithelial cells to tether vesicles and specifically recycles transferrin and integrins via recycling endosomes [959].

9.1.16.3 Neurotransmitter Recycling

Neurotransmitters experience multiple rounds of exo–endocytosis. The clathrin adaptor proteic complex AP1, especially its adaptin-σ1b subunit,[85] participates in the regeneration of neurotransmitter-filled presynaptic vesicles, once its proteic and lipidic constituents have been reinternalized [960].

Neurons use many recycling pathways to reuptake released materials, as they must respond to various stimulus frequencies over extended periods of time and thus maintain a functional pool of synaptic vesicle.

(1) "Kiss and run" mode corresponds to the release of neurotransmitters from the vesicle lumen that forms a transient pore that communicates with the extracellular space. This secretion is followed by a simple reversal of exocytosis at the active zone, as the vesicle membrane never fully fuses with the plasma membrane. Once the pore closed, the vesicle returns to the functional vesicle pool to be refilled with transmitters.

(2) Secretion vesicles fully fuse with the membrane prior to reuptake. In the latter case, 2 other internalization processes can then be used. (2.1) AP2–clathrin-mediated endocytosis operates with slow kinetics that limit exo–endocytosis cycles during intense stimulation. (2.2) Quicker, non-specific bulk endocytosis of a substantial amount of materials in the vicinity of the release site also involves recruitment of a clathrin coat and subsequent fission of the vesicle from the plasma membrane by dynamin. Endophilin and synaptojanin then uncoat the vesicle that is brought back to the functional vesicle pool. Bulk endocytosis with adequate sorting owing to AP1 can be used to maintain neurotransmission during high-level activity.

9.1.17 Receptor Endocytosis

Most receptors and ligands follow mainly one of available routes from the early endosome [961]. (1) The *intake path* can lead to late endosomes and lysosomes. It is required for nutrient uptake and receptor downregulation. (2) The *recycling route* transfers unloaded receptors from early endosomes back to the plasma membrane. (3) An additional *retrograde transport* connects

---

[84] A.k.a. Rho guanine nucleotide-exchange factor-T (GEFT).

[85] Adaptor protein complexes (AP1–AP4) mediate different types of vesicle formation. Adaptor protein complex AP1 contains 4 adaptins (β1, γ1, μ1, and σ1). Three isoforms of the σ1 subunit that binds cargos exist (σ1a–σ1c).

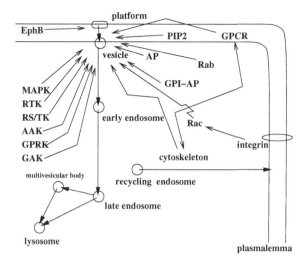

Figure 9.6. Examples of receptors, kinases, and pathways involved in endocytosis.

early endosomes to the trans-Golgi network, Golgi body, and endoplasmic reticulum.

9.1.17.1 Cofactors and Regulators

The activity and recycling of membrane receptors need cofactors and regulators (Fig. 9.6). The low-density lipoprotein receptor-related protein (LRP) is a multiligand receptor. Sorting nexin-17 (SNx17) regulates the cell membrane level of LRP. It interacts with phosphatidylinositol 3-phosphate and binds to LRP, promoting LRP recycling in the early endosomes [962].

After ligand binding and phosphorylation of β-adrenergic receptors, recruitment of β-arrestin and adaptor proteic complex AP2 to the activated receptor allow the link of the receptor to clathrin-coated pits for internalization.[86] Plasmalemmal phosphoinositide $PI(4,5)P_2$ is required for budding of clathrin-coated vesicles. Phosphoinositide 3-kinase, which phosphorylates both proteins and lipids, is required in the endocytosis associated with the β-adrenergic receptor. Both protein and lipid kinase activities are involved in receptor endocytosis. Tropomyosin, a PI3K substrate, acts in the internalization of β-adrenergic receptors [963].

9.1.17.2 Endosomes and Signaling

Signaling can be triggered at the plasma membrane or after receptor internalization from endocytic compartments [964]. Multiple plasmalemmal receptors are ubiquitinated when they are stimulated by extracellular ligands.

---

[86] The adaptor complex AP2 is always associated with clathrin on the plasma membrane.

Ubiquitination regulates receptor signaling by modulating the magnitude and duration of the signal. Ubiquitinated proteins are able to interact with other proteins, and, thus drive the different steps of the signaling pathway. Reciprocally, signaling regulates ubiquitination. Monoubiquitination can be required for receptor activation, receptor endocytosis, and lysosomal degradation.

Ligand binding to receptor Tyr kinases activates the receptor to initiate the signaling cascade. The signaling lifetime is regulated, activated RTKs triggering negative feedback loops that cause receptor endocytosis and lysosomal degradation. This event requires the formation of large protein complexes that remove activated receptors from the plasma membrane. Ubiquitin ligase CBL downregulate plasmalemmal receptors by multiple monoubiquitination, and then act as endocytic adaptors for subsequent lysosomal degradation.[87] Action of CBL is modulated by interactions with regulators (CIN85 and RhoGEF7), which promote or preclude receptor endocytosis [965].[88]

Kinases associated with the G-protein-coupled-receptor signaling pathways control clathrin-mediated endocytosis and regulate actin and tubulin cytoskeleton [966]. Kinases acting on the endocytosis are involved in cell adhesion. Each set of kinases exerts opposite effects on the 2 endocytic mechanisms (caveolae and membrane rafts).

Other types of raft-mediated molecular transfer can be used for endocytosis [967]. Cells use various pathways to control plasmalemmal receptors, i.e., cell signaling features and receptor turnover. *Dynamin* is involved in cell transport, especially in apical transport to the plasma membrane and in receptor-mediated endocytosis. C-Terminal-binding protein-1 controls transport mechanisms distinct from the dynamin-required endocytosis, such as the basolateral transport from the Golgi body to the plasma membrane [968].

9.1.18 Enzymatic Control

Phosphorylation–dephosphorylation cycles control the coated-vesicle transport (Fig. 9.6). Numerous kinases are involved in clathrin- and raft-mediated transport with endosomes and caveosomes, respectively [966]. Each endocytic route is regulated by a specific kinase group. Within this group, some kinases act directly, whereas other kinases modulate the endocytic path.

Certain kinases exert opposite effects on the 2 main endocytosis types for coordination between endocytic routes (Ca^{++} channels with kinase domains, Ca^{++}–calmodulin-activated kinases, effectors of CDC42 and Rac1,

---

[87] Ubiquitin ligase CBL forms with other proteins an interactome inside the cell that controls cell proliferation, survival, and motion. It is located in focal adhesions with actin linkers and small GTPases.

[88] The adaptor protein CIN85 binds to different molecules involved in the regulation of receptor endocytosis. Protein RhoGEF7 binds to P21-activated kinase.

Wnt signaling components, and receptor Tyr kinases and Ser/Thr kinases, as well as MAPKs.[89]

9.1.19 Alternative Splicing of Transport Proteins

Like other proteins, proteins involved in intracellular transfer of molecules can arise from alternative splicing of their cognate pre-messenger RNAs. This alternative splicing is specific to the cell type or not (Chap. 5). In epithelial cells, epithelial splicing regulatory proteins targets pre-mRNAs that encode proteins of the vesicular transfer (Table 9.20).

9.2 Clathrin-Mediated Transport

Clathrin-coated vesicles participate in the intracellular transfer between various compartments in both exo- and endocytosis. Clathrin forms a scaffold around vesicles. This scaffold of the vesicle membrane is linked to cargos by clathrin adaptors.

Clathrin-mediated endocytosis packages and internalizes selected cargos from the plasma membrane into transport vesicles, which fuse with early endosomes. Selected cargos encompass nutrients, hormones, growth factors, receptors, low-density lipoproteins, and ferritin, as well as antigens and pathogens.

To ensure a rapid uptake (within minutes) from the cell surface, a dedicated sorting machinery gathers and allocates diverse sets of membrane proteins into selected membrane patches that form vesicles. Endocytosis that dynamically moves transmembrane proteins and their bound ligands as well as membrane lipids off the surface can be constitutive or temporally regulated [970].

Regulated endocytosis often involves addition of a reversible post-translational sorting tag. Sorting signals associated with some types of cargos are recognized and decoded by different cargo-selective clathrin adaptors such as AP2 heterotetramer. Monomeric clathrin-associated sorting proteins (ClASP) that typically bind to lipids, AP2 complex, and clathrin can sort signals into assembling vesicles [970].

---

[89] Metabolic kinases, mediators of the TOR signaling pathway, G-protein-coupled receptor kinases, regulators of the cytoskeleton are involved in clathrin-mediated endocytosis. Regulators of the integrin-dependent focal adhesion, collagen receptor Tyr kinase MRC2, and members of cell adhesion-dependent signaling regulate raft trafficking via focal adhesions. Factors of cortical actin depolymerization and polymerization, associated with CDC42- and PIP_2-dependent signaling act on endocytosis mediated by membrane rafts and caveolae. Members of the ARK–PRK family of kinases, such as cyclin-G-associated kinase and adaptin-associated kinase-1 are required in endocytosis, especially as they participate in clathrin-coated vesicles [969].

Table 9.20. Set of proteins of the vesicular transport, the pre-mRNAs of which undergo an alternative splicing regulated by epithelial splicing regulatory proteins (ESRP; Source: [68]). These proteins include: (1) constituents of the octameric exocyst involved in targeting and tethering of post-Golgi vesicles to the plasma membrane prior to vesicle fusion as well as recruitment of Rho GTPase; (2) RalA-specific activating guanine nucleotide-exchange factor (GEF) with PH domain and SH3-binding motif RalGPS2, small GTPase RalA being required for exocyst assembly and function; (3) Components of the SNARE complexe that mediate vesicle fusion with the plasma membrane; and (4) Rho and Rab regulators and effectors (AKAP: A-kinase anchoring protein; CDC42BP: CDC42-binding protein kinase; CDC42SE: CDC42 small effector; DAAM: Disheveled-associated activator of morphogenesis; ECT: epithelial cell transforming sequence; ERC: ELKS–Rab6-interacting–CAST family; ExoC: exocyst complex component; FAM: family with sequence similarity; GAP: GTPase-activating protein; MPRIP: myosin phosphatase Rho-interacting protein; Myo: myosin; Plxn: plexin; RabEP: rabaptin, Rab-binding effector protein; RabEPK: Rab9 effector protein with kelch motifs; Rtkn: rhotekin; SRGAP: Slit–Robo Rho GTPase activating protein; Stx: syntaxin; StxBP: syntaxin-binding proteins; TBC1D: Tre2–USP6, BUB2, CDC16 (TBC1) domain family member [Rab-GAP]; TSC: tuberous sclerosis).

Set	Members
Excyst	ExoC1/7
Ral GTPase	RalGPS2
SNARE	Stx2/3, StxBP2/5/6
RHO and its regulators and effectors	RhoT1, RhoGAP10–12/17, CDC42BPα/β CDC42SE1, Myo9b, AKAP13, DAAM1, ECT2, FAM13b1, MPRIP, PlxnB2, Rtkn2, SRGAP2
RAB regulators and effectors	Rab6a/18/28, RabEP1, RabEPK, ERC1, TBC1D13, TSC2

The clathrin-dependent route is responsive to activated receptors and ion channels to ensure a rapid clearance of tagged cargos off the plasma membrane. Concentrated receptors linked by coat proteins form *clathrin-coated pits*.

The relocalization of activated receptors (G-protein-coupled β2-adrenergic receptor, epidermal growth factor receptor, low-density lipoprotein receptor, etc.) after their removal from the plasma membrane allows: (1) interactions of receptors with other proteins in the endosome; (2) recycling of receptors back to the plasma membrane; and (3) degradation, with subsequent signaling shutdown, by directing receptors to lysosomes.

9.2.1 Clathrin

Clathrin is a membrane constituent. Clathrin can form structures of variable curvature. However, clathrin coats show an invariant local pattern of contacts that appear to stabilize the lattice and then control assembly and uncoating [971].

Clathrin-coated vesicles have 3 layers: (1) an outer clathrin layer, clathrin forming a scaffold on the vesicle surface; (2) a middle layer with various clathrin adaptors and assembling regulators; and (3) an inner layer with transported substances.

The clathrin triskelion consists of 3 clathrin heavy and light chains. It interacts with ligands via its N-terminal domain. During coated vesicle formation, clathrin becomes concentrated by interacting with both adaptor proteic complexes and accessory proteins. Concentrated clathrin self-polymerizes to form an interaction surface composed of about 200 binding domains that simultaneously tether to adaptor and accessory proteins [860].

Clathrin causes membrane deformation for budding and can contribute to the activity of membrane-bending accessory proteins, serving as a flexible scaffold for adaptors and accessory proteins [972].

9.2.2 Transport and Recycling

Clathrin-mediated endocytosis follows a set of transient and localized events. The first step is the binding of extracellular cargo molecules to specific plasmalemmal receptors. Receptors and other membrane proteins destined for endocytosis are then sequestered by intracellular adaptors.

The formation of clathrin-coated vesicles at the plasma membrane begins with the recruitment of coated vesicle adaptors, the clathrin adaptor complex-2, accessory proteins,[90] and clathrin to $PI(4,5)P_2$-enriched plasmalemmal nanodomains.

Once the endocytic coat is formed at the plasma membrane, clathrin drives coated pit invagination. Invagination and scission of clathrin-coated pits are tightly coupled. The motion of clathrin-coated pits away from the plasma membrane begins before membrane scission [973]. Cortactin recruitment occurs at the site of clathrin-coated pits before scission with a peak recruitment coincident with scission.

Membrane bending is generated during vesicle budding, owing to amphiphysin, endophilin, epsin, and SAR1 GTPase.

Cytoplasmic *endophilins* (A1–A3; B1–B2)[91] bind to membrane scissor dynamin and phosphoinositide phosphatase synaptojanin, a clathrin uncoater. They contain a BAR domain that senses and generates membrane curvature.

Epsin binds to $PI(4,5)P_2$, promotes membrane curvature, and stimulates clathrin polymerization, thereby stabilizing the generated curvature. The

---

[90] Accessory proteins, such as clathrin assembly lymphoid myeloid leukemia protein, epsin, and eps15–eps15R coordinate the coat assembly.

[91] Endophilin-A1 is also called SH3P4, SH3GL2, and EEN-B1; endophilin-A2, SH3P8 and SH3GL1; endophilin-A3, SH3P13 and SH3GL3; endophilin-B1, SH3GLB1; and endophilin-B2, Bif1, SH3GLB2, and EEN.

coated membrane invaginates and subsequently pinches off to form a vesicle.

Actin polymerization is tightly coupled to vesicle budding, from the plasmalemmal invagination to form a coated pit to the vesicle scission. A transient burst of actin polymerization supports molecule internalization. Actin polymerization at the neck of constricted coated vesicles pushes coated pits and nascent coated vesicles away from the plasma membrane into the cytosol.[92] Plasmalemmal myosins are recruited at the onset of actin polymerization. The ARP2–ARP3 complex, nWASP, and cortactin participate in actin organization and endocytosis. The ARP2–ARP3 complex is itself regulated by many interacting proteins.[93] Dynamin coordinates actin filament assembly at endocytic sites.

Once inside the cytosol, the clathrin coat of newly formed endocytic vesicles disassembles and clathrin is rapidly released for subsequent use. The naked vesicle fuses with endosomal membranes of other uncoated vesicles and forms early endosomes. Some ligands are separated from their receptors. Cargos can be recycled back to the plasma membrane or transported to late endosomes and lysosomes for degradation.

Such a mechanism involves microtubules. In some cases, actin can associate with motile endosomes. Early endosomes can lead to tubules and even to vacuoles, characterized by a large fluid content. Sorted vesicles and other particles move along the cytoskeleton to fuse with acceptor membrane. Membrane proteins can be either recycled to the plasma membrane or retained in multivesicular endosome before being delivered to lysosomes.

9.2.3 Mediators of Clathrin-Based Transport

Clathrin-coated vesicle formation and uncoating (duration of 30–90 s) requires interactions of numerous proteins. Various adaptors allow sorting of cargos into assembling clathrin-coated pits on a membrane. Once the membrane invaginated and then elongated, a vesicle forms. To deform the membrane and form a tubular bud and then a vesicle, the membrane elastic resistance forces must be balanced by forces that can be generated by actin polymerization and action of myosin-1 [975]. These active forces are also important in scission.[94] Membrane constriction leads to coated vesicle budding. The released vesicle quickly loses its clathrin coat.

---

[92] Actin polymerization at the plasma membrane either drives the protrusion of lamellipodia and filopodia at the leading edge of migrating cells or intracellular invaginations for endocytosis.

[93] Part of actin cytoskeleton-regulatory complex Pan1 regulates the ARP2–ARP3 complex involved in actin structuring at endocytic sites. It contains a Wiskott-Aldrich syndrome protein homology-2 (WH2)-like motif necessary for ARP2–ARP3 activation. The Pan1 WH2-like motif binds to ^{F}actin. ^{F}Actin binding to Pan1 is required prior to ARP2–ARP3-mediated nucleation [974].

[94] Dynamin does not always intervene.

Various proteins regulate coat formation and disassembly. Vesicle uncoating implicates cytosolic ATPase HSC70, which is recruited to clathrin-coated vesicles by its cofactor auxilin.[95] The timing of auxilin recruitment determines the onset of uncoating. A low level of auxilin is transiently incorporated during the growth of coated pits. However, immediately after vesicle budding, accumulation of a specific phosphoinositide recruits a sufficient amount of auxilin to trigger uncoating (second and major burst of auxilin recruitment) [976]. Auxilin-2 is involved in AP1 binding.

Clathrin adaptors include amphiphysin-1 and -2, clathrin-associated adaptor proteic complexes (AP1–AP4), which recognize different types of molecules, β-arrestin-1 and -2, 91-kDa synaptosomal-associated protein SnAP91,[96] low-density lipoprotein receptor adaptor protein LDLRAP1, Disabled-2,[97] epsin-1 to -3 and -R,[98] Huntington-interacting protein HIP1[99] and its related protein HIP1R, sorting nexin Snx9, Golgi-localized γ-ear-containing ARF-binding proteins (GGA1–GGA3),[100] and hepatocyte growth factor-regulated Tyr kinase substrate[101] (Tables 9.21 and 9.22). These peripheral membrane proteins are recruited either via GTPases ARF, ARL, Rab, and SAR, or via phosphorylated phosphatidylinositols, these 2 mechanisms being interlinked.

Adaptor protein complex AP2 is composed of 4 subunits (α, $\beta2$, $\mu2$, $\sigma2$). Distinct domains bind cargo and accessory proteins. It is stabilized by clustering molecules such as Eps15.

Adaptin binds clathrin of clathrin-coated vesicles to the membrane. Adaptor protein complexes bind to clathrin to stimulate the formation of clathrin coat. The phospholipid bilayer via PIP_2 acts in the clathrin coat formation. After endocytosis, the membrane components in the endosome are either recycled back to the plasma membrane after dissociation from the ligand, or are degraded in the endosome fused with the lysosome.

α-*Adaptin-associated kinase* AAK1, located in clathrin-coated pits, promotes AP2 conformational change to bind the cargo [979]. Conversely, *synaptojanin*, which degrades AP2-recruiting $PI(4,5)P_2$, promotes AP2 dissociation

---

[95] Two auxilin variants exist: brain-specific auxilin-1 and ubiquitous auxilin-2, also called cyclin-G-associated kinase (GAK).

[96] A.k.a. clathrin coat assembly protein AP180 and clathrin assembly lymphoid–myeloid leukemia protein (CALM). It binds to $PI(4,5)P_2$.

[97] Proteins LDLRAP1 and Dab2 are involved in the uptake of LDL receptors.

[98] Epsin-R is involved in transport from trans-Golgi network to endosomes. The epsin family includes epsin-1, SnAP91, and HIP1–HIP1R, characterized by a $PI(4,5)P_2$-binding domain.

[99] Huntington-interacting protein HIP1 can promote with AP2 clathrin recruitment and assembly.

[100] Golgi-localized γ-ear-containing ARF-binding proteins transport substances between the trans-Golgi network and endosomes.

[101] Protein HRS incorporates targeted molecules into multivesicular bodies from vesicles.

Table 9.21. Clathrin adaptors and membrane constituents (Sources: [977, 978]; CHC: clathrin heavy chain; CLC: clathrin light chain; SnAP: synaptosomal-associated protein [SnAP91 a.k.a. AP180]).

Clathrin adaptor	Membrane constituent	Binding partners
AP1	PI(4)P	CHC
AP2	PI(4,5)P$_2$	CHC, epsin, auxilin, Synaptojanin, amphilysin
SnAP91	PI(4,5)P$_2$	Clathrin, AP2, PLD
Epsin	PI(4,5)P$_2$	Clathrin, AP2, intersectin
Arrestin	PI(4,5)P$_2$	
Amphilysin	PI(4,5)P$_2$	Clathrin, AP2, endophilin, PLD

Table 9.22. Clathrin-coated vesicle components and their interactions (Source: [977]; GGA: Golgi-localized γ-adaptin ear-containing Arf-binding protein).

Protein	Interacting partners
β-Arrestin	Clathrin, APs, GPCRs
Dynamin	Amphilysin, endophilin, intersectin PIs, G proteins, cortactin, profilin GRB2, Src, PI3K, PLCγ
Auxilin	Clathrin, APs, HSC70
Synergin	APs, GGAs
Endophilin	Synaptojanin, amphilysin, dynamin
Ankyrin	Clathrin, spectrin, vimentin, tubulin (Ank1) Na$^+$ channel (Ank2) K$^+$ ATPase, IP$_3$R, RyR, PKC (Ank3)
GGA	ARF, GGAs

from the plasma membrane and hinders the formation of clathrin-coated vesicles.

Epsins and *amphiphysins* are implicated in membrane deformation. Clathrin-associated sorting proteins and arrestins recruit target molecules. Clathrin coats bind to auxilin-1 and HSC70, creating a distortion of the clathrin coat, destabilization, and uncoating [980].

The *EphB–Ephrin-B* complex regulates clathrin-mediated endocytosis, acting on protein interactions and phosphoinositide turnover via phosphorylation of synaptojanin-1,[102] a phosphatidylinositol phosphatase, which is required for clathrin-mediated endocytosis [981]. Activation of receptor of EphB2 and subsequent synaptojanin-1 Tyr phosphorylation has 2 distinct effects on endocytosis. The inhibition of the interaction of synaptojanin-1

---

[102] More precisely, EphB2-mediated phosphorylation of synaptojanin-1 inhibits the interaction of synaptojanin 1 with the endophilin and lowers the phosphatase activity of synaptojanin-1.

with endophilin promotes the early phase of clathrin-mediated endocytosis, whereas it hinders the late phase of endocytosis. The vesicle uncoating requires interactions with endophilin.

9.3 Caveolin-Mediated Transport

Caveolin-mediated transport represents another kind of transport used by cells. Nanoindentations in the plasma membrane of many cell types, especially endothelial cells, can trap molecules. These membrane-lined recesses can extend from the inner to outer membrane layer. They are involved not only in cellular transport, but also in signal transduction, as well as in cell displacement.

9.3.1 Caveolin and Caveolae

Caveolin, an integral membrane protein, builds caveolae [982]. Caveolae are invaginations (size 50–100 nm) of the plasma membrane that are enriched in cholesterol, glycosphingolipids, and caveolins, as well as signaling and clustered GPI-anchored proteins. Caveolae can switch from stable nanodomains to carriers.

Three types of caveolae may exist: (1) caveolae strictly speaking; (2) glycosphingolipid-enriched membranes (GEM);[103] and (3) polyphosphoinositol-rich (PIP$_2$-enriched) rafts.

The caveolin family comprises 3 members (Cav1–Cav3). Caveolin-1 lodges in endothelia in particular. Caveolin-2 colocalizes with Cav1. Small GTPase Rab13 forms clusters with Cav1 and Cav2. Caveolin-3 is expressed in myocytes, such as smooth muscle cells and cardiomyocytes.[104] Caveolin-1, and Cav3 in myocytes, interacts also with integrins and other focal adhesion constituents.

Caveolin forms oligomers that are able to connect to cholesterol and sphingolipids. Caveolar cargos comprise lipids, proteins, and pathogen lipid-anchored proteins. Aminopeptidase-P abounds in caveolae of endothelial cells.

Caveolae localize to numerous cell types, in particular, endothelial and smooth muscle cells, cardiomyocytes, fibroblasts, and adipocytes. Caveolae are involved in lipid regulation in adipocytes and other cell types. Caveolae are particularly involved in cholesterol homeostasis.

Caveolae are involved in endocytosis, transcytosis, and exocytosis. Caveolar transport is as rapid as that with clathrin-coated vesicles. However, they are not involved in clathrin-independent intracellular transport under basal conditions [983].

---

[103] Glycosphingolipid-enriched membranes are caveolae without caveolin-1.

[104] Caveolin-3 is essential for the formation of caveolae in cardiomyocytes, but caveolin-1 is also important in these cells.

Caveolae also function as signaling platforms at the plasma membrane. They can sense membrane physical and chemical changes. Caveolae may intervene in mechanosensation in certain cell types. The high density of caveolae in endothelial cells that are sensitive to changes in mechanical stresses is associated with increased phosphorylation of Cav1 [984]. Moreover, caveolin is connected to integrins and Gq protomers. Caveolae can also sense membrane tension in smooth muscle cells.

Transfer of ion carriers with signaling molecules enables the control of cell excitability. In cardiomyocytes, several ion channels, pumps, and exchangers localize to caveolae, where their activity is modulated, such as voltage-dependent Na^+, K^+ ($K_V1.5$), and Ca^{++} channels, ATP-sensitive K^+ channels ($K_{IR}6.2$ and SUR2a), Na^+–K^+ ATPase, and Na^+–Ca^{++} exchangers. In vascular smooth muscle cells, K_{ATP} channels ($K_{IR}6.1$ and SUR2b) also reside in caveolae.

9.3.2 Involved Molecules in Caveolin-Mediated Transport

Molecular transfer mediated by caveolae employs caveolin+ vesicles (*cavicles* and *caveosomes*). The caveosome does not contain classical markers of early endosomes. The motility of caveosomes at the cell cortex depends on cortical actin filaments. On the other hand, the cytoplasmic motion of caveosomes depends on the microtubule network. The caveosome route is regulated by cytosolic Tyr kinase and protein kinase-C [985]. Kinase Src phosphorylates Cav1 and dynamin.

9.3.2.1 Cavins

All 4 cavins (cavin-1–cavin-4) are recruited to caveolae [986]. Cavin-1[105] participates in the formation of caveolae [987]. It promotes the recruitment and local accumulation of cavin-2, and vice versa. At the plasma membrane, caveolin-1 forms stable oligomers of approximately 150 monomers that can be rendered highly dynamical by cavin-2 to form flexible coats for changes in caveolar size and shape with possible tubulation.

Cavin-2[106] primes Cav1+ membrane tubules and regulates the morphology and dynamics of caveolae [986]. However, cavins-1 and -2 have distinct role. Cavin-2 may favor malleability of caveolae.

Cavin-3[107] regulates endocytosis of caveolae. Cavin-4[108] is specific to muscles. It limits budding and intracellular transfer of Cav1+ vesicles from the plasma membrane, but does not influence Cav1 concentration.

---

[105] A.k.a. polymerase-1 and transcript release factor (PTRF).

[106] A.k.a. serum deprivation protein response (SDPR).

[107] A.k.a. serum deprivation response (SDR) factor-related gene product that binds to C-kinase (SRBC) and protein kinase-Cδ-binding protein (PrKCδBP).

[108] A.k.a. muscle restricted coiled-coiled protein (MuRC).

9.3.2.2 Other Mediators

Endocytosis mediated by caveolae involves SNARE proteins. Caveolar endocytosis is regulated by syntaxin-6, a target SNARE [988]. Syntaxin-6 is also required for the delivery of GM1 ganglioside and GPI-anchored protein from the Golgi body to the plasma membrane.

Large GTPase dynamin acts not only in clathrin-coated vesicle budding, but also in fission of caveolae from the plasma membrane. Phospholipase-D2 functions as a GTPase-activating protein for dynamin. GTP hydrolysis leads to tubule twisting and elongation and then tubule breakage and release [989].

Caveolin-1 can act as a CDC42 guanine nucleotide dissociation inhibitor (CDC42GDI). In pancreatic β cells, reduced Cav1 level activates CDC42 and increases CDC42-dependent insulin release [891].

9.4 Polarized Transport

In polarized cells, intracellular transport polarizes during and after the establishment of cell polarity. The generation and maintenance of apical and basolateral surfaces in polarized cells requires suitable protein sorting that relies on recognition of sorting motifs of the protein structure and/or glycolipid anchors to determine the destination. The spatial restriction of signaling is achieved via an asymmetrical distribution or redistribution of membrane nanopatches or signaling molecules. This process is often mediated by exo–endocytosis cycles [885]. These cycles can redirect and confine signaling mediators to specialized apical or basal regions of the plasma membrane of polarized epithelial cells.

Intracellular transport matches the polarized distribution via coordinated interaction of 3 molecular complexes [990]: (1) protein-sorting signals that are recognized and segregated by cytoplasmic decoding adaptor complexes to regulate transport toward appropriate plasmalemmal domains; (2) signaling complexes and scaffolds associated with the cytosolic face of the plasma membrane that define plasmalemmal domains to which proteins are delivered; and (3) cell adhesion complexes that influence target plasmalemmal domains.

In polarized as well as non-polarized cells, sorting occurs at the trans-Golgi network. In polarized cells, apical and basolateral vesicles are separately transported to their final destinations. Apical-destined vesicles that carry raft-dependent cargos differ from those that transport raft-independent cargos. Basolateral-destined vesicles can be classified into several groups according to their cargos. Distinct molecules participate in apical and basolateral transport [991] (Table 9.23).

Protein transport to the plasma membrane involves the sequential transfer between different membrane compartments (endoplasmic reticulum, Golgi body, endosomes, and plasma membrane). Transport is mediated by vesicles

Table 9.23. Molecules of polarized transport (Source: [991]). Membrane rafts, i.e., glycolipid and cholesterol-containing membrane nanodomains, where glycosylphosphatidylinositol (GPI)-anchored proteins and glycosylated proteins accumulate, are essential for apical transport. Raft-associated members of the MAL proteolipid family contribute to direct the polarized transfer. Agent MAL1 is a highly hydrophobic, non-glycosylated integral membrane protein. Phosphoinositide-binding specific pleckstrin homology domain-containing protein-A8 (PlekHa8 or FAPP2) that binds to ARF and PIP_2 is involved in apical sorting. Minus-ends of microtubules are directed to apical plasma membrane, but directions of microtubules are mixed below the apical plasma membrane. Plus-ends of microtubules localize to the distal ends of axons. In neurons, transport from the cell body to axon terminal is mediated by kinesin family (KIF) nanomotors and in the opposite direction by dynein. In dendrites, microtubules have both polarities; some plus-ends point outwardly and others inwardly. Small GTPase Rab8 is involved in localization of apical proteins. Annexin-13b is necessary for apical tethering. Syntaxin-3 and soluble NSF attachment protein SNAP23 serve as target SNAP receptors (tSNAREs) and vesicle-associated membrane protein VAMP7 as vesicle SNARE (vSNARE) in apical transport. Syntaxin-4 acts as tSNARE and VAMP3 as vSNARE in basolateral transport (StxBP: syntaxin-binding protein).

Apical transport	Basolateral transport
GPI signals	Sorting signals based on
Glycosylation	Tyr and diLeu motifs
Galectin linking	Clathrin
MAL proteolipids	Adaptors AP1b and AP4
PlekHa8	Protein kinase-D2
Rab8	Rab13
Syntaxin-3	Syntaxin-4
StxBP2	StxBP3a
VAMP7	VAMP3
SNAP23	
KIF5c	

that bud from original compartment and fuse with the arrival station.[109] Polarity-sorting motifs and adaptors are required for the polarized delivery of newly synthesized and recycling receptors such as low-density lipoprotein receptors that are assigned to the basolateral membrane of epithelial cells.

In epithelial cells, ankyrin–spectrin scaffolding complexes that bind ion carriers, receptors, and cell-adhesion proteins are differentially located in the

---

[109] Coatomer protein complex-1 is required in Golgi–endoplasmic reticulum and intra-Golgi transport, coatomer protein complex-2 in endoplasmic reticulum–Golgi transport with SNARE, and the clathrin-associated adaptor proteic complex with vesicle-tethering and SNARE complexes between the 3 major sites of protein sorting in exo- and endocytosis: Golgi body, plasma membrane, and endosomes. Adaptor proteic complexes recognize and bind specific amino acid motifs of the cytoplasmic domain of membrane proteins and cluster these proteins into patches.

basolateral membrane, thereby allowing local accumulation of membrane proteins. Like tight junctions that act as a fence to prevent the free diffusion of proteins, diffusion barriers are represented by cytoskeletal scaffolds that selectively recruit high concentrations of membrane proteins and impede the displacement of freely diffusing membrane proteins.

Signaling complexes, such as Crumbs-3, partitioning defective complex (Par), and Scribble,[110] associate with the cytoplasmic surface of the plasma membrane around sites of cell adhesion to demarcate different plasma membrane regions to guide the establishment of cell polarity.

Phosphoinositides bind scaffold proteins required for substance delivery, thereby contributing to cell polarity-directed transport. Moreover, monomeric GTPases that regulate actin cytoskeletal dynamics operate in cell polarization. Crosstalk exists between Rap1 and various Rho and members of polarity complexes to induce spatially restricted cytoskeletal remodeling required for cell polarization in different cellular contexts [992].

Recognition of sorting cues happens in the trans-Golgi network, where segregation of apical and basolateral cargo takes place. Some cargos are conveyed to the plasma membrane via endocytic intermediate compartments for proper delivery. Newly synthesized apical proteic markers are conveyed by different populations of endosomes [993]. These distinct delivery routes of apical cargos differ from trafficking path used by newly synthesized basolateral markers. There are indeed: (1) distinct apical and basolateral early endosomes; (2) common recycling endosome that is able to segregate apical and basolateral cargo; and (3) a Rab11+, immunoglobulin-A+, apical, recycling endosome[111] that regulates apically directed transport. A fraction of newly synthesized basolateral proteins transits via transferrin+ common recycling endosome. Actin-dependent transport selective for membrane raft-associated apical proteins such as GPI-anchored proteins and raft-independent process of apical proteins such as glycan-bound proteins take different endocytic intermediates. The latter uses apical recycling endosome, whereas raft-associated apical proteins could travel via an apical endocytic compartment different from apical early endosomes and apical recycling endosomes.[112]

Integrins prevent local non-clathrin-mediated endocytosis of membrane rafts (that serve as anchors for Rac). On the other hand, clathrin-mediated, recycling endocytosis maintains active Rac near sites of integrin-mediated signaling [885]. ADP-ribosylation factor ARF6 is involved in this recycling endocytosis. This process enables the redelivery to the plasma membrane not only of Rac and integrins, but also of membrane rafts.

---

[110] Crumbs-3 and Scribble complexes participate in the identity of apical and basolateral regions of the plasma membrane, respectively.

[111] Apical recycling endosomes can be decomposed into Rab11a- and Rab11b-positive apical recycling compartments.

[112] Glycosylphosphatidyl inositol-anchored proteins are conveyed into the specific tubular early endocytic compartment that is called gpiAP-enriched endosomal compartments (GEEC).

9.5 Signaling Mediated by Endosomes

Binding of an extracellular ligand to its cognate plasmalemmal receptor initiates a cascade of reactions that begins at the plasma membrane. Once, activated, many receptors bear endocytosis. Ligand-bound receptor are internalized via the endosomal tubulovesicular network that extends throughout the cytoplasm. Afterward, receptors are either degraded in lysosomes and proteasomes or recycled back to the cell surface where they can rebind extracellular ligands to sustain signaling.

Yet, endosomes not only serve for recycling or degradation of plasmalemmal receptors, but also as signaling mediators. Activated receptors can indeed accumulate in endosomes. Signaling can then be initiated and terminated from endosomal receptors. Moreover, certain signaling components are exclusively located in endosomes, as receptors that continue to transmit signals from endosomes that differ from those sent from the plasma membrane. Receptor signaling from endosome membranes is regulated by ligand availability, receptor coupling to signaling effectors (Tables 9.24 and 9.25), and subcellular location of signaling mediators [994].

Endosomes yield a moving source of signals propelled by microtubular motors. Endosomal signaling thus avoids the slow diffusion in the cytosol of signaling mediators and their possible deactivation during their displacement. The endosomal signaling platform transmits signals close to the next site involved in the reaction cascade such as the nucleus.[113]

Endosomes initiate their own signaling types that cannot be delivered from the plasma membrane. These signals actually rely on lipids or proteins that are unique to endosomes, as endosome enriched in some signaling mediators recruited by these constituents can be involved in specific signaling pathways.

Phosphatidylinositol 3-phosphate that is locally and specifically synthesized enables endosome-specific assembly of signaling complexes, as it links to proteins such as early endosomal antigen-1 [885].[114]

---

[113] In neurons, neurotrophins produced by postsynaptic cells target presynaptic neurotrophic Tyr kinase receptor in axonal termini. The signal then travels to the neuron body to promote gene transcription [885]. This movement can be efficiently achieved by the retrograde endosomal transport of activated receptors.

[114] The FYVE (Fab-1 [vacuolar membrane 1-phosphatidylinositol 3-phosphate 5-kinase that generates $PI(3,5)P_2$ for vacuolar sorting and governs yeast vacuole homeostasis; known as PIP(5)K3 and PIKFYVE in mammals], metallodependent hydrolase YotB, vesicle transport protein Vac1, and early endosome antigen EEA1) domain enables the recruitment of proteins involved in membrane trafficking and cell signaling to phosphatidylinositol 3-phosphate-containing membranes. Internalized TβR receptor interacts with the PI(3)P-binding protein SMAD anchor for receptor activation (SARA or zinc finger FYVE domain-containing protein ZFYVE9) and phosphorylates the SARA-associated molecule SMAD2, thereby promoting the dissociation of SMAD2 and its interaction with SMAD4 to elicit a transcriptional response [885]. PI(3)P-binding endofin (or ZFYVE16) can interact with TβR and SMAD4 and facilitate the formation of the SMAD2–SMAD4 complex.

Table 9.24. Receptor signaling from endosomes. (**Part 1**) G-Protein-coupled receptors (Source: [994]; AT1aR: angiotensin-2 type-1A receptor; β2AR: β2-adrenoceptor; CCR: chemokine CC-motif receptor; CXCR: chemokine CXC-motif receptor; D2R: dopamine 2 receptor; ERK: extracellular signal-regulated protein kinase; JNK: Jun N-terminal kinase; KOR: κ-opioid receptor; LPAR: lisophosphatidic acid receptor; NK1R, neurokinin-1 receptor; PAR2: protease-activated receptor-2; PI3K: phosphatidylinositol 3-kinase; PKB: protein kinase-B; PP2a: protein phosphatase-2A; V2R: vasopressin receptor [V_2]).

Receptor	Endosomal signaling mediators	Outcomes (general)
AT1aR	ERK1/2, JNK3	Cell survival and proliferation Protein synthesis
CCR7	ERK1/2	Homing and transfer of dendritic cells and lymphocytes to secondary lymphoid tissues
CXCR4	P38MAPK	Macrophage chemotaxis, lymphocyte homing, hematopoiesis, metastasis
D2R	PP2a–PKB	Dopamine-dependent behavior
KOR	P38MAPK	Growth, K^+ conductance elevation, Ca^{++} conductance reduction, intracellular Ca^{++} mobilization
LPAR	PI3K–PKB	Cell proliferation and survival
NK1R	ERK1/2	Cell survival and proliferation
PAR2	ERK1/2, PI3K	Rearrangement of cytoskeleton and tight junctions
V2R	ERK1/2	Aquaporin density increase
β2AR	Src, ERK1/2	Cardioprotection

In Rab7+ late endosomes, endosomal membrane-raft adaptor chromosome-11 open reading frame-59 (C11ORF59)[115] anchors the MAP2K1–ERK module [996]. The MAPK scaffold protein MAPKSP1[116] connects to endosomal adaptor protein MAPKSP1-associated protein MAPKSP1AP[117] that is necessary and sufficient to localize MAPKSP1 to endosomes. The newly assembled

---

[115] A.k.a. membrane-raft adaptor protein P18 and Ragulator complex protein P27[KIP1] (cyclin-dependent kinase inhibitor CKI1b)-releasing factor from RhoA (P27RFRho). The Ragulator (MAPKSP1–RoBlD3–C11ORF59) complex interacts with Rag GTPase and recruits it to lysosomes. The Ragulator-Rag complex that targets TORC1 to the lysosomal surface is necessary for TORC1 activation by amino acids [995].

[116] A.k.a.mitogen-activated protein kinase kinase-1-interacting protein MAP2K1IP1 and MP1.

[117] A.k.a. roadblock domain-containing protein RoBlD3, endosomal adaptor protein P14, EndAP, late endosomal–lysosomal MP1-interacting protein, and mitogen-activated protein-binding protein-interacting protein (MAPBPIP).

Table 9.25. Receptor signaling from endosomes. (**Part 2**) Receptor Tyr kinases and Toll-like receptors (Source: [994]; EGFR: epidermal growth factor receptor; ERK: extracellular signal-regulated protein kinase; IR [InsR]: insulin receptor; IRF: interferon-regulatory protein; MyD88, myeloid differentiation primary response gene product-88; NTRK: neurotrophic Tyr receptor kinase; PDGFR: platelet-derived growth factor receptor; PI3K: phosphatidylinositol 3-kinase; PLC: phospholipase-C; PKB: protein kinase-B; TRAM–TRIF: Toll-receptor-associated molecule–Toll-receptor-associated activator of interferon; VEGFR: vascular endothelial growth factor receptor; VSMC: vascular smooth muscle cell).

Receptor	Endosomal signaling mediators	Outcomes (general)
Receptor tyrosine kinases		
EGFR	ERK1/2, PI3K–PKB	Cell proliferation, tumor growth
InsR (IR)	ERK1/2, PI3K	VSMC proliferation and hypertrophy
NTRK1	ERK1/2, ERK5, P38MAPK, PI3K–PKB, PLCγ1	Neurite outgrowth Modulation of neuronal architecture Neural remodeling
PDGFR	ERK1/2, PI3K–PKB	Cell migration, proliferation, and protection
VEGFR2	ERK1/2	Cell growth, chemotaxis
Toll-like receptors		
TLR3	Src Kinase	Immune response
TLR4	TRAM–TRIF	Immune response
TLR9	MyD88-IRF7	Immune response

MAPKSP1–MAPKSP1AP scaffold complex recruits C11ORF59 to create a signaling platform for the extracellular signal-regulated kinase [997].

9.5.1 Endosomal Receptor Tyrosine Kinases

Endocytosis allows full activity of receptor Tyr kinases, such as platelet-derived (PDGFR) and vascular endothelial (VEGFR2) growth factor receptors. In hepatocytes, epidermal growth factor causes accumulation of activated EGFR and its signaling effectors (GRB2, SHC, and SOS) in early endosomes [994]. Receptor EGFR is phosphorylated in endosomes. It is dephosphorylated (inactivated) before entering into lysosomes.

In endosomes that contain EGFR, 2 Rab5 effectors are recruited, the adaptors proteins containing phosphotyrosine-interaction domain, pH domain, and Leu-zipper motif APPL1 and APPL2 [885]. These adaptors lead to PKB activation and substrate selection by PKB, as they activate the PKB–GSK3β axis. APPL-enriched endosomes correspond to early stages of the early endosome. Their maturation is controlled by the localized production of phosphoinositides. During their maturation, PI(3)P is generated and APPLs are shed and replaced by PI(3)P-binding proteins such as EEA1.

In adipocytes, endosomal insulin receptors that are more strongly phosphorylated than plasmalemmal insulin receptors and insulin receptor substrate IRS1 are able to connect to endosomal membranes.

In nociceptive neurons (and pheochromocytoma cells), nerve growth factor elicits accumulation of activated NTRK1 receptors, phospholipase-Cγ1, and components of the mitogen-activated protein kinase module and PI3K signaling pathways in endosomes.

Nerve growth factors released from cells to promote survival of tissue-innervating neurons can signal from axon terminals. These signals must be sustained to travel long distances to the soma. Retrograde transport by signaling endosomes transmits NGF signals. Furthermore, NGF signaling from endosomes causes sustained MAPK activation, whereas from the plasma membrane, NGF activates Ras transiently [994]. The site of receptor activation can influence the type of signaling axis. Activation of the receptor neurotrophin Tyr kinase NTRK1 in axon terminals of dorsal root ganglia neurons leads to MAPK6 (ERK5) stimulation and translocation to the nucleus to activate CREB for neuronal survival. This signaling pattern also promotes activation of transcription factor myocyte enhancer factor MEF2d. On the other hand, NTRK1 activation at the cell body signals via ERK1, ERK2, and ERK5. This signaling pattern does not elicit MEF2d activation. Therefore, retrograde endosomal signaling and direct stimulation at the soma have different outcomes.

Endocytosis of NTRK and EGFR differs and causes opposite signaling effects. Neurotrophins and NGF, but not other growth factors, promote neuronal differentiation and survival. Nerve growth factor provokes NTRK endocytosis via Rac GTPase and transfer protein Pincher [994]. Receptor NTRK then accumulates in Rab5+ immature multivesicular bodies that lack late endosome protein Rab7. On the other hand, EGF stimulates clathrin-dependent endocytosis of EGFR into Rab5+ endosomes, but Rab5 is rapidly exchanged for Rab7 that leads to transition to late endosomes and lysosomes.

Transport of hepatocyte growth factor receptor from peripheral to perinuclear endosomes promotes nuclear accumulation of STAT3 transcription factor [994]. Displacement of EGFR between peripheral and perinuclear locations also affects EGFR degradation and MAPK activation.

9.5.2 Endosomal G-Protein-Coupled Receptors

β-Arrestins (Sect. 9.1.13.13) not only desensitize guanine nucleotide-binding (G)-protein-coupled receptors at the plasma membrane, but also operate during GPCR endocytosis and endosomal signaling from these receptors inside the cytoplasm. Some G-protein-coupled receptors such as angiotensin-2 receptor AT_{1A} indeed remain associated with β-arrestins after endocytosis.

β-Arrestins uncouple receptors from G proteins and couple them to clathrin and AP2 adaptor complex for desensitization and endocytosis on the one hand, and recruit signaling mediators to activated receptors at both plasma and endosome membranes on the other.

β-Arrestins recruit mitogen-activated protein kinases to endosomes.[118] β-Arrestins act as scaffolds that complex with MAP3K4, MAP3K5, and Jun N-terminal kinase JNK3 [998]. They thus stably anchor MAPKs and direct the spatial distribution of activity of this MAPK module under the control of a GPCR.

Interactions of GPCRs with β-arrestins depend on the extent of GPCR phosphorylation by G-protein-coupled receptor kinases. Class-A GPCRs, such as β2ARs and α1bARs, have few phosphorylation sites and transiently interact with β-arrestin-1 and -2 (higher affinity for β-arrestin-2), mostly at the plasma membrane. Class-B GPCRs (e.g., AT_{1A}, NK_1, PAR_2, and V_2) are phosphorylated at multiple sites and link to both β-arrestin-1 and -2 with high affinity for prolonged periods at both plasma and endosome membranes. Class-C GPCRs (e.g., B_2 bradykinin receptor) are internalized with β-arrestins into endosomes followed by rapid dissociation of β-arrestins upon ligand removal [994].

Intensity of β-arrestin-induced MAPK signaling depends on the affinity of the receptor for β-arrestins that itself relies on the type of G-protein-coupled receptor kinase. Phosphorylation of AT_{1A} by GRK5 and GRK6 activates ERK1 and ERK2, whereas phosphorylation by GRK2 and GRK3 prevents signaling.

Different receptor ligands can lead to distinct signaling patterns. Chemokine receptor CCR7 has 2 ligands, CCL19 and CCL21, that activate G proteins and cause β-arrestin-2-dependent ERK activation. However, only CCL19 provokes GRK3- and GRK6-mediated phosphorylation of CCR7, its redistribution to endosomes, and desensitization. On the other hand, CCL21 only activates GRK6 [994].

Endothelin-converting enzyme ECE1 is a membrane-associated metalloendopeptidase that shuttles between plasma and endosome membranes. It degrades substance-P, calcitonin gene-related peptide, and somatostatin in endosomes to disrupt the peptide receptor–β-arrestin complex, thereby releasing into the cytoplasm β-arrestin and receptors (neurokinin-1 receptor, calcitonin gene-related peptide receptor, and somatostatin receptor-2A, respectively) that can recycle. Inhibition of ECE1 causes retention of the substance-P–NK_1–β-arrestin–MAP2K1–ERK1/2 complex in endosomes and sustained activation of ERK1 and ERK2 [994].

9.5.3 Endosomal Toll-Like Receptors

Endosomes serve as signaling platforms for Toll-like receptors in innate immunity. In dendritic cells stimulated by double-stranded RNAs, TLR3 and

---

[118] The MAPK cascades (ERK, JNK, P38MAPK) are stimulated by β2-adrenoceptors, neurokinin receptor NK_1, protease-activated receptor PAR_2, angiotensin-2 receptor AT_{1A}, and vasopressin receptor V_2 [994].

Src kinase accumulate in endosomes that contain double-stranded RNA. Receptor TLR4 for lipopolysaccharides from bacterial walls activates 2 pathways: the TIRAP-MyD88[119] and the TRAM–TRIF endosomal pathway[120] that cause cytokine and interferon activity, respectively [994]. In plasmacytoid dendritic cells, CpG oligodeoxynucleotide binds to TLR9 in endosomes to induce interferon activity, as it leads to activation of transcription factor interferon-regulatory protein IRF7 via adaptor MyD88.[121]

9.5.4 Endosomal Signaling and Proteolysis

Notch signaling (Vol. 3 – Chap. 10. Morphogen Receptors) requires its proteolysis at an extracellular site by metalloproteases followed by intramembrane cleavage of the membrane-anchored intracellular domain by γ-secretase to release the Notch intracellular domain that translocates to the nucleus to regulate gene transcription. Early endosomes are required for efficient γ-secretase-mediated cleavage of Notch [994].

The acidic pH of endosomes influences signaling. Notch activation requires its cleavage by γ-secretase at both the plasma membrane and on endosomes. As γ-secretase reaches its peak activity at pH 7.8, Notch is efficiently processed during its endosomal transit. In addition, Notch can be activated in the absence of ligand by ubiquitin ligase Deltex that promotes Notch endocytosis, but precludes the route toward the multivesicular body [885].[122]

9.5.5 Endosomal Signaling and Ubiquitination and Deubiquitination

Ligand-induced ubiquitination of G-protein-coupled receptors and receptor Tyr kinases at the plasma membrane and possibly in endosomes controls the receptor fate. In particular, receptor ubiquitination supports its lysosomal sorting.

Ubiquitin–protein ligase CBL ubiquitinates epidermal (EGFR) and platelet-derived (PDGFR) growth factor as well as protease-activated PAR$_2$ receptors.[123]

---

[119] TIRAP: Toll-interleukin-1 receptor domain-containing adaptor; MyD88: myeloid differentiation primary response gene product-88.

[120] TRAM: Toll-receptor-associated molecule and TRIF: Toll-receptor-associated activator of interferon.

[121] In conventional dendritic cells, CpG oligodeoxynucleotide is rapidly degraded in lysosomes.

[122] Ubiquitin ligases of the Deltex (DTX) family act as Notch signaling modifiers. Members of the human DTX family (DTX1–DTX4 and Deltex-3-like protein DTX3L) that homo- and heterodimerize are capable of self-ubiquitination.

[123] Protease-activated receptor PAR$_2$, chemokine receptor CXCR4, platelet-activating factor receptor PAFR are monoubiquitinated.

Ubiquitin ligase Itch[124] targets chemokine receptor CXCR4 as well as δ-opioid receptor. Ubiquitination of CXCR4 receptor by Ub ligase Itch impedes CXCR4 retention in endosomes [943].

Ubiquitin ligase NEDD4 links to β2–adrenergic receptoradrenoceptors.[125] Ligands of β2ARs also promote interaction between β-arrestin-2 and Ub ligase DM2 that ubiquitinates β-arrestin-2.

Conversely, endosomal deubiquitinating proteases (DUBs), such as associated molecule with SH3 domain (AMSH)[126] and ubiquitin-specific protease-Y (UBPY or USP8) control EGFR deubiquitination [994].[127] Like CBL that promotes EGFR lysosomal degradation, USP8 favors lysosomal sorting of EGFR, whereas AMSH elicits EGFR recycling.

β-Arrestin ubiquitination and deubiquitination regulates the association of β-arrestins with GPCRs and β-arrestin-dependent activation of ERK1 and ERK2 enzymes.[128]

9.6 Endocytosis and Cell Division

9.6.1 Symmetrical Cell Division

Proteins involved in endocytosis, such as clathrin, dynamin, endocytic adaptor low-density lipoprotein receptor adaptor protein LDLRAP1 and small GTPase Rab6a also intervene in mitosis. They associate with centrosomes or mitotic spindle [885]. Clathrin-mediated endocytosis works throughout mitosis, but with a less potent recycling. The recycling pathway recovers at the last stage of mitosis.

During mitosis, endocytic proteins bind to centrosomes and kinetochores involved in chromosome segregation that form part of the mitotic spindle.[129] Clathrin heavy chain localizes to the kinetochore fibers of the spindle and contributes to adequate chromosome alignment [885]. Small GTPase Rab6a

---

[124] Ubiquitin ligase Itch that is located in endosomes is a binding partner for endophilin that operates in clathrin-mediated endocytosis. Ubiquitin ligase Itch is required for the ubiquitin-dependent endocytosis of plasmalemmal proteins.

[125] Mutations in PY-binding motif of the epithelial sodium channel in kidneys prevent its binding to NEDD4 Ub ligase. Subsequent increased level of the epithelial sodium channel at the plasma membrane causes excessive reabsorption of Na^+ and water that leads to familial hypertension (Liddle's syndrome).

[126] A.k.a. endosome-associated ubiquitin isopeptidase and STAM binding protein.

[127] Both AMSH and USP8 deubiquitinate δ-opioid receptor and PAR_2 for lysosomal transport and degradation. β2-Adrenoceptors are deubiquitinated by USP33 and USP20 that promote recycling from late endosomes.

[128] In particular, USP33 deubiquitinates β-arrestin-2, thereby destabilizing the interaction of β-arrestin-2 with V_2 vasopressin receptor and attenuating activation of ERK1 and ERK2 enzymes.

[129] Kinetochore is the protein structure on chromosomes to which spindle fibers attach during the cell division to pull the chromosomes apart.

is also recruited to the kinetochores and cooperates with the spindle checkpoint protein Mitotic arrest-deficient MAD2 to ensure that attachment of spindle microtubules to kinetochores at metaphase. Large GTPase dynamin participates in centrosome cohesion. Low-density lipoprotein receptor adaptor protein LDLRAP1 that binds to the nanomotor dynein localizes to the centrosome during interphase and kinetochores and spindle poles during mitosis.

Components of recycling endosomes, such as Rab11, Rab35, and ARF6, and components of the secretory machinery such as SNAREs cooperate with microtubule plus-end-directed nanomotors to orient the recycling of endosomes toward the transient midbody of the cell, where homotypic vesicle-fusion events generate distinct membrane subcompartments with defined lipid compositions. The latter act as signaling platforms for abscission (separation of daughter cells at the end of cytokinesis) [885].

9.6.2 Cell Division into Stem and Progenitor Cells

Asymmetrical division yields a stem and a progenitor cell. Whereas stem cells are quiescent, although able to self-renew, progenitor cells proliferate and, eventually, differentiate. Endosomes obey to an asymmetrical partitioning during cell division of hematopoietic stem cells. Endosomes with Zinc finger FYVE domain-containing protein ZFYVE9,[130] an anchoring protein involved in TGFβ signaling, may be involved in this process.

Transcription factor P53 is controlled by endocytic protein NUMB homolog. It regulates intracellular transfer of molecules. Endocytic protein NUMB, which antagonizes Notch, prevents the ubiquitination of P53 by DM2, thereby precluding P53 degradation [885].[131] Endocytic control exerted on P53 affects the mode of cell division, either asymmetrical or symmetrical. Conversely, P53 regulates the endosomal compartment, particularly the recycling pathways.

9.7 Endocytosis and Gene Transcription

Endocytosis can deliver cargos to the inner nuclear membrane. Delivery of transcriptional mediators pertains to the overall endocytosis role. Membrane-anchored growth factors, such as precursors of amphiregulin and heparin-binding EGF-like factor, are carried to the inner nuclear membrane using endocytosis machinery upon suitable signaling [885].

Endocytic proteins can shuttle between the cytoplasm and nucleus, where they modulate gene expression [885]. Endocytic proteins bind either transcriptional regulators or chromatin-remodeling complexes. Various endosomal

---

[130] A.k.a. SMAD anchor for receptor activation (SARA).

[131] In dividing mammary stem cells, NUMB partitions the daughter cell that adopts the stem-cell fate.

proteins translocate to the nucleus and act as coregulators of transcription. They tether to transcription factors to modulate their stability or activity. Yet, endocytic proteins can act independently of their role in endocytosis. In particular, the endocytic and nuclear functions of Huntingtin interacting protein HIP1 seem to be mutually exclusive [885].

Among endocytic proteins, endocytic adaptor and nucleic acid binder β-arrestin operates as a nuclear messenger. Proteins APPL1 and APPL2 as well as the ESCRT3 complex bind to chromatin-remodeling complexes. Clathrin heavy chain links to the P53-responsive promoter and stabilizes the association between P53 and P300 histone acetyltransferase, thereby enhancing P53-dependent transactivation. The transcriptional activity of clathrin heavy chain does not require its trimerization domain that is indispensable for its endocytic coat-protein function.

9.8 Exosomes

Cells can communicate via exosomes (size 40–100 nm). Exosomes are intraluminal vesicles in multivesicular endosomes that are released in the extracellular milieu when these endosomes fuse with the plasma membrane. Exosomes can then be captured by surrounding cells and internalized.

Exosome-mediated communication enables the donor cell to genetically reprogram the recipient cell [885]. Exosomes can convey genes. Mastocytes release exosomes that contain more than 1,000 mRNA species and more than 100 types of microRNAs. These 2 types of RNAs can reprogram a cell when taken up. Nevertheless, only a subset of mRNAs that are transcribed in a cell is found in exosomes. The endocytic machinery in the donor cell not only contributes to the generation of exosomes, but also RNA sorting into these vesicles, especially via the ESCRT2 complex.

Non-coding regulatory microRNAs of cellular and viral origin that repress the translation of mRNAs into proteins can be secreted via exosomes [999]. These exosomes protect miRs from degradation by ribonucleases. MicroRNA-mediated gene silencing can then result from intercellular communications.

Transcription factor P53 controls the release of exosomes. The P53-inducible transmembrane protein and metalloreductase Six transmembrane prostate protein STEAP3[132] facilitates the secretion of exosomes that contain Translationally controlled tumor protein [1000] (TCTP).[133] This secreted protein participates in inflammatory responses by promoting the release of histamine. Another P53-regulated gene product, Charged multivesicular body protein CHMP4c[134] is a subunit of the ESCRT3 complex that can increase the rate of exosome production.

---

[132] A.k.a. tumor suppressor-activated pathway protein TSAP6.

[133] A.k.a. histamine-releasing factor.

[134] A.k.a. Snf7 homolog associated with Alix SHAx3. Charged multivesicular body proteins ChMP4a and CHMP4b are also called SHAx2 and SHAx1, respectively.

Transcription factor P53 controls the activity of the gene that encodes caveolin-1. Hence, P53 can regulate non-clathrin-mediated endocytosis of plasmalemmal receptors such as EGFR [885]. It also governs the recycling of integrins and EGFR via the recycling protein Rab11 family-interacting protein Rab11FIP1[135] Transcription factor P53 also binds to clathrin.

Exosomes released by tumor cells can deliver RNAs, angiogenic factors, and oncoproteins to surrounding normal cells, thus promoting tumor growth. Cancer cells can then induce genetic reprogramming of adjacent cells [885].

9.9 Lipid-Transfer Proteins

The cell as well as its organelles are identified by membranes. Cell organelles ensure segregation between diverse metabolisms and exchanges between these compartments. The lipid composition labels cellular membranes. The plasma membrane characterized by its transverse asymmetry in lipid composition is enriched in sphingolipids and sterols. On the other hand, the membrane of the endoplasmic reticulum that possesses a transversally symmetrical lipid distribution contains low concentrations of both lipid types.

The distribution of proteins among cellular organelles often results from specific sorting motifs in proteins as well as post-translational modifications and/or conformational changes. On the other hand, lipids that lack locus-segragation motifs travel within the cytosol between membranes of the cell by non-vesicular lipid transport mechanisms. These processes are supported by lipid-transfer proteins, in addition to slow, free diffusion[136] and vesicular lipid transport [165].[137] Membrane lipidic composition, curvature, and fluidity influence the kinetics of the transfer.[138]

Endosomes carry proteins along both exo- and endocytic routes, using chemical energy and the cytoskeleton. Lipids are constituents of transport vesicles. Hence, vesicular tranfer allows lipid exchange. Nevertheless, lipid

---

[135] A.k.a.Rab-coupling protein (RCP).

[136] Spontaneous movement of lipid monomers between membranes is extremely slow. The rate of spontaneous lipid exchange is correlated with aqueous-phase solubility [165]. Lysophosphatidylcholine that has a single acyl chain exchanges more rapidly than phosphatidylcholine. Cholesterol that is more soluble than phosphatidylcholine travels much faster. Three types of spontaneous lipid exchange exist [165]: (1) aqueous diffusion that depends on concentration in donor membrane; (2) membrane collision that depends on concentration in both donor and acceptor membranes; and (3) activated collision associated with lipid extension from the bilayer.

[137] In dynamical membranes of the cell, lipids experience lateral diffusion and transbilayer flip-flop, i.e., motion between the 2 leaflets of the membrane bilayer [165]. Lateral diffusion (in the membrane plane) occurs at speeds of 0.1 to 1 μm/s. Transbilayer flip-flop is spontaneous or mediated by proteins, such as flippases and translocases.

[138] High transfer occurs generally in more fluid or highly curved membranes.

transport exist in the absence of endosomes. In addition, lipid transfer is enhanced in regions with a small cytosolic gap (10–20 nm) between membranes [165].[139]

Most lipid-transfer proteins can bind lipid monomers in a hydrophobic pocket and transfer hydrophobic lipids through an aqueous phase [165]. Soluble lipid-transfer proteins modulate the lipid composition of membranes, as they promote lipid transport between membranes according to membrane environment. Furthermore, they regulate vesicular transfer of materials, signal transduction, and lipid metabolism via lipid extraction and delivery, in addition to the controlled access of lipid-binding proteins to membrane lipids. Lipid-transfer proteins operate also as lipid level sensors and/or lipid-presenting proteins [165].

According to their lipid-binding specificity and transfer capability, lipid-transfer proteins can be grouped into 3 main classes [165]: phospholipid-, sterol-, and sphingolipid-transfer proteins. *Non-specific lipid* (NSLTP), *phosphatidylcholine* (PCTP), and *phosphatidylinositol* (PITP) transfer proteins compose the 3 principal mammalian subclasses of phospholipid-transfer proteins.

Non-specific lipid-transfer proteins carry all common diacyl phospholipids, glycolipids, and cholesterol.

Ceramide-transfer protein (CerT) has a high specificity for natural ceramides. Both CeRT and phosphatidylinositol 4-phosphate adaptor FAPP2[140] are sphingolipid-transfer proteins. *Steroidogenic acute regulator* (StAR) is a cholesterol-transfer protein. *Oxysterol-binding protein* (OSBP) and OSBP-related proteins (ORPs) are sterol sensor and/or sterol-transfer proteins. Protein ORP2 can act as a sterol transporter between the endoplasmic reticulum and plasma membrane.

According to the structure of the single lipid-transfer domain, several families of lipid-transfer proteins have been defined [165]: *GLTP family* of glycolipid transfer; *PITP family* of phosphatidylinositol and phosphatidylcholine transfer; *OSBP–ORP family* of oxysterol-binding; *SCP2 family* (SCP2: sterol carrier protein-2) of non-specific lipid transfer; and *START family* of steroidogenic acute regulator-related lipid transfer proteins.

---

[139] Small between-membrane gaps can be observed between mitochondria and either the endoplasmic reticulum or the plasma membrane as well as the endoplasmic reticulum and the Golgi body, peroxisomes, lipid droplets, late endosomes, and lysosomes [165]. They are involved in the regulation of cellular lipid and Ca^{++} homeostasis.

[140] A.k.a. pleckstrin homology (PH) domain-containing PlekHa8.

9.10 Modeling of the Cytoskeleton-Based Molecular Transport

Intracellular transport relies on proteic motors, such as dyneins, kinesins, and myosins, that transduce chemical energy into mechanical energy. Models of molecular transport in cells using nanomotors along constituents of the cytoskeleton are based on partial differential equations.

A 2-head kinesin nanomotor runs along a microtubule. One head responds to a potential and moves, then the molecule changes of conformation and the second head moves and binds to a new site. A system of evolution equations for the partial probabilities ($\Pi = \{\Pi_1, \Pi_2\}$) of active heads with potentials Ψ_1 and Ψ_2 and conformational change coefficients Ξ_1 and Ξ_2 has been proposed by [1001]:

$$\frac{\partial \Pi_i}{\partial t} = \frac{\partial}{\partial x}(\kappa \frac{\partial \Pi_i}{\partial x} + \Psi_i' \Pi_i) - \Xi_i \Pi_i + \Xi_j \Pi_j, \qquad (9.1)$$
$$i = 1, 2, \ i \neq j,$$
$$\text{in } \Omega = (0, 1), t > 0,$$

$$\kappa \frac{\partial \Pi_i}{\partial x} = -\Psi_i' \Pi_i, \quad i = 1, 2, \text{ on } \Gamma, \ t > 0,$$

$$\Pi_i(x, 0) = \Pi_i^0 \geq 0 \quad i = 1, 2, \text{ in } \Omega,$$

$$\int_\Omega (\Pi_1 + \Pi_2) \, dx = 1.$$

A framework for molecules displacement in an asymmetrical potential Ψ proposed by Hastings, Kinderlehrer, and Kowalczyk is the Fokker-Planck-Kolmogorov eigenfunction equation [1002]:

$$-\epsilon \frac{\partial^2 u}{\partial x^2} - \left(\Psi'(x)u\right)_x = 0, \quad 0 < x < 1$$
$$-\epsilon \frac{\partial u}{\partial x} + \Psi'(x)u(x) = 0, \quad \text{at } x = 0, 1, \qquad (9.2)$$

i.e., the flux equals zero at cytoskeletal filament inlet and outlet. This model is not appropriate. Eigenvalue problems have been devised in which molecules can reach 2 conformations. Initial-value problems such as the flashing rachets are more difficult to handle, but are more suitable, using the same boundary condition (zero flux at cytoskeletal filament ends):

$$\frac{\partial u}{\partial t} - \epsilon \frac{\partial^2 u}{\partial x^2} - \left(\Psi'(\frac{t}{\epsilon}, \frac{x}{\epsilon})u\right)_x = 0, \quad 0 < x < 1$$
$$-\epsilon \frac{\partial u}{\partial x} + \Psi'(\frac{t}{\epsilon}, \frac{x}{\epsilon})u(t, x) = 0, \quad \text{at } x = 0, 1 \qquad (9.3)$$
$$u(t, x) : \ \epsilon\text{-periodic in time.}$$

10

Concluding Remarks

> " *La curiosité dynamise l'esprit humain.* *[Curiosity potentiates the human spirit.]* " *(G. Bachelard, 1884–1962).*

The tier architecture of any living system is characterized by its communication means and regulation procedures. It enables integration of environmental changes to adapt. Multiple molecules interact to create the adaptable activity of the cells, tissues, organs, and body. Any integrative model then incorporates a set of models developed at distinct length scales and includes characteristic times to efficiently describe the structure-function relationships of the explored physiological system.

The present volume of the series of books devoted to Circulatory and Ventilatory Systems provides the information needed to understand and possibly model the organization of cells that are attached to neighboring cells of the same or different type (e.g., a vascular endothelial cell to an endothelial or smooth muscle cell) to form, in particular, organ coating and glands, as well as any interacting tissues such as the vascular endothelium and muscular layer (media) to locally control blood flow.

Biological systems — from the molecular level to the physiological apparatus — are characterized by their complicated structure, variable nature, and complex behavior. The processing of the signals that control the activity of transcription factors and the expression of genes to direct cell decision (differentiation, growth, proliferation, or death), organization of metabolism, and cellular communication for coordinated action in a tissue, relies on non-linear dynamics that control spatial distribution and clustering of molecular species at a given time. Fast protein modifications that result from protein interactions in the cytoplasm propagate signals and lead to slow transcription and translation. Volume 2 is devoted to cell life from birth to death, as well as the local, regional, and remote control that regulates various events of the cell fate.

The control on a given cell is carried out by molecules synthesized by and secreted from the cell to act on the cell itself (intra- [without release] and autocrine control), its neighbors (juxta- and paracrine control), or distant cells (endocrine control). Once it has reached its cell target, any messenger triggers a signal transduction axis characterized by a cascade of chemical reactions from the cell surface to the cytoplasm, and possibly the nucleoplasm, that allows the cell to adapt. Therefore, cells appropriately respond in a controlled, coherent manner to external stimuli (adaptation robustness). Specific responses with their respective intracellular biochemical reaction cascades can be generated over a wide range of parameter variation. Signals are transduced by information processing networks that are characterized by signal transduction complexity and between-pathway connectivity. Mediators of signaling pathways that govern the cell behavior are detailed in Volumes 3 and 4.

Biomathematical modeling can be developed at different scales, from the signal transduction pathway (nanoscopic scale) to the cell reaction (microscopic scale) and tissue adaptation and remodeling (mesoscopic scale). The combined response of the cellular and tissular constituents of walls of conduits of the circulatory and ventilatory systems launches a reaction to attune the organ to the environmental condition at a given time. The structure and function of specialized cells and tissues are described in Volume 5.

The wetted surface of any segment or organ of the cardiovascular system is covered by the endothelium, which constitutes the interface between the flowing blood and the deformable solid wall. The endothelium is a layer of connected and anchorage-dependent cells. The endothelium has several functions: it controls molecule exchange between the blood and the vessel wall and perfused tissues; regulates flowing cell adhesion on the blood vessel wall and extravasation, especially for immune defense; controls the blood coagulation and thrombolysis; and regulates the vasomotor tone and proliferation of vascular smooth muscle cells via the release of several compounds. It is required in angiogenesis. Endothelial cells detect hemodynamic stresses via mechanosensors.

The blood vessel wall is a living tissue that quickly reacts to loads applied to it by the flowing blood. In any segment of a blood vessel, the endothelial and smooth muscle cells sense the large-amplitude space and time variations in small-magnitude wall shear stress and larger wall stretch generated by the blood pressure. These cells respond within a short time scale (from seconds to hours) to adapt the vessel caliber according to the loading, especially when changes exceed the limits of the usual stress range. This regulatory mechanism is much swifter than nervous and hormonal control. The mechanotransduction pathways determine the local vasomotor tone and subsequently the lumen bore of the reacting blood vessel.

Future explorations of blood and air flows in the cardiovascular and respiratory systems will thus require the development of models that couple different length and time scales. Similarly to physiology, biomechanics focus on the macroscopic scale. Like molecular biology, biomathematics address the nano- and microscopic scales that should be coupled to macroscopic events to take into account the fundamental features of living cells and tissues that sense, react, and adapt to applied loadings. Simulation coupling platforms will be associated with high-performance computing.

References

Introduction

1. Jacob F (1970) La logique du vivant, une histoire de l'hérédité [The logic of life: a history of heredity]. Gallimard, Paris
2. Lucretius (1997) De rerum natura (Book 5-322) [On the Nature of Things]. Garnier Flammarion, Paris

Volume I. Chap. 1. Cells and Tissues

3. Rothpearl A, Varma AO, Goodman K (1988) Radiographic measures of hyperinflation in clinical emphysema. Discrimination of patients from controls and relationship to physiologic and mechanical lung function. Chest 94:907–913
4. Bates DV, Christie RV (1964) Respiratory Function in Disease, WB Saunders, Philadelphia and London
5. Hwang NS, Varghese S, Lee HJ, Zhang Z, Ye Z, Bae J, Cheng L, Elisseeff J (2008) In vivo commitment and functional tissue regeneration using human embryonic stem cell-derived mesenchymal cells. Proceedings of the National Academy of Sciences of the United States of America 105:20641–20646
6. Yang J, Chai L, Fowles TC, Alipio Z, Xu D, Fink LM, Ward DC, Ma Y (2008) Genome-wide analysis reveals Sall4 to be a major regulator of pluripotency in murine-embryonic stem cells. Proceedings of the National Academy of Sciences of the United States of America 105:19756–19761
7. Kim JB, Sebastiano V, Wu G, Araúzo-Bravo MJ, Sasse P, Gentile L, Ko K, Ruau D, Ehrich M, van den Boom D, Meyer J, Hübner K, Bernemann C, Ortmeier C, Zenke M, Fleischmann BK, Zaehres H, Schöler HR (2009) Oct4-induced pluripotency in adult neural stem cells. Cell 136:411–419
8. Heng JC, Feng B, Han J, Jiang J, Kraus P, Ng JH, Orlov YL, Huss M, Yang L, Lufkin T, Lim B, Ng HH (2010) The nuclear receptor Nr5a2 can replace Oct4 in the reprogramming of murine somatic cells to pluripotent cells. Cell Stem Cell 6:167–174

9. Jones DL, Wagers AJ (2008) No place like home: anatomy and function of the stem cell niche. Nature Reviews – Molecular Cell Biology 9:11–21

10. Reuter G, Cavalli G (2009) Epigenetics and the control of multicellularity. EMBO Reports 10:25–29

11. Inada A, Nienaber C, Katsuta H, Fujitani Y, Levine J, Morita R, Sharma A, Bonner-Weir S (2008) Carbonic anhydrase II-positive pancreatic cells are progenitors for both endocrine and exocrine pancreas after birth. Proceedings of the National Academy of Sciences of the United States of America 105:19915–19919

12. Peerani R, Rao BM, Bauwens C, Yin T, Wood GA, Nagy A, Kumacheva E, Zandstra PW (2007) Niche-mediated control of human embryonic stem cell self-renewal and differentiation. EMBO Journal 26:4744–4755

13. International Stem Cell Initiative Adewumi O, Aflatoonian B, Ahrlund-Richter L, Amit M, Andrews PW, Beighton G, Bello PA, Benvenisty N, Berry LS, Bevan S, Blum B, Brooking J, Chen KG, Choo ABH, Churchill GA, Corbel M, Damjanov I, Draper JS, Dvorak P, Emanuelsson K, Fleck RA, Ford A, Gertow K, Gertsenstein M, Gokhale PJ, Hamilton RS, Hampl A, Healy LE, Hovatta O, Hyllner J, Imreh MP, Itskovitz-Eldor J, Jackson J, Johnson JL, Jones M, Kee K, King BL, Knowles BB, Lako M, Lebrin F, Mallon BS, Manning D, Mayshar Y, Mckay RDG, Michalska AE, Mikkola M, Mileikovsky M, Minger SL, Moore HD, Mummery CL, Nagy A, Nakatsuji N, O'Brien CM, Oh SKW, Olsson C, Otonkoski T, Park KY, Passier R, Patel H, Patel M, Pedersen R, Pera MF, Piekarczyk MS, Pera RAR, Reubinoff BE, Robins AJ, Rossant J, Rugg-Gunn P, Schulz TC, Semb H, Sherrer ES, Siemen H, Stacey GN, Stojkovic M, Suemori H, Szatkiewicz J, Turetsky T, Tuuri T, van den Brink S, Vintersten K, Vuoristo S, Ward D, Weaver TA, Young LA, Zhang W (2007) Characterization of human embryonic stem cell lines. Nature – Biotechnology 25:803–816

14. Adamo L, Naveiras O, Wenzel PL, McKinney-Freeman S, Mack PJ, Gracia-Sancho J, Suchy-Dicey A, Yoshimoto M, Lensch MW, Yoder MC, García-Cardeña G, Daley GQ (2009) Biomechanical forces promote embryonic haematopoiesis. Nature 459:1131–1135

15. North TE, Goessling W, Peeters M, Li P, Ceol C, Lord AM, Weber GJ, Harris J, Cutting CC, Huang P, Dzierzak E, Zon LI (2009) Hematopoietic stem cell development is dependent on blood flow. Cell 137:736–748

16. Bearzi C, Rota M, Hosoda T, Tillmanns J, Nascimbene A, De Angelis A, Yasuzawa-Amano S, Trofimova I, Siggins RW, LeCapitaine N, Cascapera S, Beltrami AP, D'Alessandro DA, Zias E, Quaini F, Urbanek K, Michler RE, Bolli R, Kajstura J, Leri A, Anversa P (2007) Human cardiac stem cells. Proceedings of the National Academy of Sciences of the United States of America 104:14068–14073

17. Sainz J, Al Haj Zen A, Caligiuri G, Demerens C, Urbain D, Lemitre M, Lafont A (2006) Isolation of "side population" progenitor cells from healthy arteries of adult mice. Arteriosclerosis, Thrombosis, and Vascular Biology 26:281–286

18. Hobert O (2008) Gene regulation by transcription factors and microRNAs. Science 319:1785–1786

19. Bowness P, Caplan S, Edidin M (2009) MHC molecules lead many lives. EMBO Reports 10:30–34

20. Chidgey AP, Layton D, Trounson A, Boyd RL (2008) Tolerance strategies for stem-cell-based therapies. Nature 453:330–337

21. Vogel V, Sheetz M (2006) Local force and geometry sensing regulate cell functions. Nature Reviews – Molecular Cell Biology 7:265–275

22. Shilo BZ (2010) The numbers behind morphogenesis. EMBO Reports 11:243–245

23. Bateman JF, Boot-Handford RP, Lamandé SR (2009) Genetic diseases of connective tissues: cellular and extracellular effects of ECM mutations. Nature Reviews – Genetics 10:173–183

24. Lecuit T, Lenne PF (2007) Cell surface mechanics and the control of cell shape, tissue patterns and morphogenesis. Nature Reviews – Molecular Cell Biology 8:633–644

25. Overholtzer M, Brugge JS (2008) The cell biology of cell-in-cell structures. Nature Reviews – Molecular Cell Biology 9:796–809

26. Bryant DM, Mostov KE (2008) From cells to organs: building polarized tissue. Nature Reviews – Molecular Cell Biology 9:887–901

27. Trepat X, Deng L, An SS, Navajas D, Tschumperlin DJ, Gerthoffer WYT, Butler JP, Fredberg JJ (2007) Universal physical responses to stretch in the living cell. Nature 447:592–595

28. Patel AB, Gibson WT, Gibson MC, Nagpal R (2009) Modeling and inferring cleavage patterns in proliferating epithelia. PLoS Computational Biology 5:e1000412

29. Maniotis AJ, Bojanowski K, Ingber DE (1997) Mechanical continuity and reversible chromosome disassembly within intact genomes removed from living cells. Journal of Cellular Biochemistry 65:114–130

30. Wozniak MA, Chen CS (2009) Mechanotransduction in development: a growing role for contractility. Nature Reviews – Molecular Cell Biology 10:34–43

31. Hahn C, Schwartz MA (2009) Mechanotransduction in vascular physiology and atherogenesis. Nature Reviews – Molecular Cell Biology 10:53–62

32. Leeson TS, Leeson CR (1970) Histology. Saunders, Philadelphia

33. Bloom W, Fawcett DW, Jensh RP (2002) Bloom & Fawcett's concise histology, Arnold, London

34. Poirier J, Ribadeau Dumas J-L, Catala M, André J-M, Gherardi R, Bernaudin J-F (2002) Histologie – Les tissus. Masson, Paris

35. Kishioka C, Okamoto K, Kim J, Rubin BK (2001) Regulation of secretion from mucous and serous cells in the excised ferret trachea. Respiration Physiology 126:163–171

36. Mitchell B, Jacobs R, Li J, Chien S, Kintner C (2007) A positive feedback mechanism governs the polarity and motion of motile cilia. Nature 447:97–101

37. Roberts WG, Palade GE (1995) Increased microvascular permeability and endothelial fenestration induced by vascular endothelial growth factor. Journal of Cell Science 108:2369–2379

38. Bearer EL, Orci L (1985) Endothelial fenestral diaphragms: a quick-freeze, deep-etch study. Journal of Cell Biology 100:418–428

39. Stan RV (2004) Multiple PV1 dimers reside in the same stomatal or fenestral diaphragm. American Journal of Physiology – Heart and Circulatory Physiology 286:H1347–H1353

40. Movata HZ, Fernando NVP (1964) The fine structure of the terminal vascular bedstar, open: IV. The venules and their perivascular cells (pericytes, adventitial cells). Experimental and Molecular Pathology 3:98–114

41. Iolov A, Kane A, Bourgault Y, Owens R, Fortin A (2011) A finite element method for a microstructure-based model of blood. International Journal for Numerical Methods in Biomedical Engineering (to appear)

42. Thiriet M, Graham JMR, Issa RI (1992) A pulsatile developing flow in a bend. Journal de Physique III 2:995–1013

43. Oates C (2008) Cardiovascular Haemodynamics and Doppler Waveforms Explained, Cambridge University Press, Cambridge, UK

44. Daley PJ, Sagar KB, Wann LS (1985) Doppler echocardiographic measurement of flow velocity in the ascending aorta during supine and upright exercise. British Heart Journal 54:562–567

45. Sukernik MR, West O, Lawal O, Chittivelu B, Henderson R, Sherzoy AA, Vanderbush EJ, Francis CK (1996) Hemodynamic correlates of spontaneous echo contrast in the descending aorta. American Journal of Cardiology 77:184–186

46. Jáuregui-Renaud K, Hermosillo JA, Jardón JL, Márquez MF, Kostine A, Silva MA, Cárdenas M (2005) Cerebral blood flow during supine rest and the first minute of head-up tilt in patients with orthostatic intolerance. Europace 7:460–464

47. Pott F, van Lieshout JJ, Ide K, Madsen P, Secher NH (2000) Middle cerebral artery blood velocity during a valsalva maneuver in the standing position. Journal of Applied Physiology 88:1545–1550

48. Kochanowicz J, Turek G, Rutkowski R, Mariak Z, Szydlik P, Lyson T, Krejza J (2009) Normal reference values of ratios of blood flow velocities in internal carotid artery to those in common carotid artery using Doppler sonography. Journal of Clinical Ultrasound 37:208–211

49. Matsue H, Sawa Y, Matsumiya G, Matsuda H, Hamada S (2005) Mid-term results of freestyle aortic stentless bioprosthetic valve: clinical impact of quantitative analysis of in-vivo three-dimensional flow velocity profile by magnetic resonance imaging. Journal of Heart Valve Disease 14:630–636

50. Gilmore ED, Hudson C, Nrusimhadevara RK, Harvey PT, Mandelcorn M, Lam WC, Devenyi RG (2007) Retinal arteriolar diameter, blood velocity, and blood flow response to an isocapnic hyperoxic provocation in early sight-threatening diabetic retinopathy. Investigative Ophthalmology and Visual Science 48:1744–1750

51. Aorta, Arch Vessel, and Great Veins (2007) In Naidich DP, Webb WR, Müller NL, Vlahos I, Krinsky GA (eds) Computed Tomography And Magnetic Resonance of the Thorax, Lippincott Williams & Wilkins, Philadelphia, PA

52. Mao SS, Ahmadi N, Shah B, Beckmann D, Chen A, Ngo L, Flores FR, Gao YL, Budoff MJ (2008) Normal thoracic aorta diameter on cardiac computed tomography in healthy asymptomatic adults: impact of age and gender. Academic Radiology 15:827–834

53. Sligar SG (2010) glimpsing the critical intermediate in cytochrome P450 oxidations. Science 330:924–925

54. Rittle J, Green MT (2010) Cytochrome p450 compound I: capture, characterization, and C-h bond activation kinetics. Science 330: 933-937

55. Valvano JW (2006) Bioheat Transfer. In Webster JG (Ed.) Encyclopedia of Medical Devices and Instrumentation, 2nd edition, Wiley-Interscience, Hoboken, NJ

56. DuBois D, DuBois EF (1916) A formula to estimate the approximate surface area if height and weight be known. Archives of Internal Medicine 17:863–871

57. Wang Y, Moss J, Thisted R (1992) Predictors of body surface area. Journal of Clinical Anesthesia 4:4–10

58. Met – Metabolic Rate, The Engineering ToolBox, www.engineeringtoolbox.com/met-metabolic-rate-d_733.html

59. Koehler KR. College Physics for Students of Biology and Chemistry. Thermodynamics. 27. Heat Flow (www.rwc.uc.edu/koehler/biophys.2ed/heat.html)

60. Woodcock JP (1976) Physical properties of blood and their influence on blood-flow measurement. Reports on Progress in Physics 39:65

61. Jiang SC, Ma N, Li HJ, Zhang XX (2002) Effects of thermal properties and geometrical dimensions on skin burn injuries. Burns 28:713–717

62. Mendlowitz M (1948) The Specific Heat of Human Blood. Science 107:97–98

63. Nahirnyak VM, Yoon SW, Holland CK (2006) Acousto-mechanical and thermal properties of clotted blood. Journal of the Acoustical Society of America 119:3766–3772

64. Holmes KR. Thermal conductivity of biomaterials (www.ece.utexas.edu/ṽalvano/research/Thermal.pdf)

65. Valvano JW. Bioheat Transfer. IV. Tissue Thermal Transport Properties. http://users.ece.utexas.edu/ṽalvano/research/jwv.pdf

66. Gregory PA, Bert AG, Paterson EL, Barry SC, Tsykin A, Farshid G, Vadas MA, Khew-Goodall Y, Goodall J (2008) The miR-200 family and miR-205 regulate epithelial to mesenchymal transition by targeting ZEB1 and SIP1. Nature – Cell Biology 10:593–601

67. Park SM, Gaur AB, Lengyel E, Peter ME (2008) The miR-200 family determines the epithelial phenotype of cancer cells by targeting the E-cadherin repressors ZEB1 and ZEB2. Genes and Development 22:894–907

68. Warzecha CC, Jiang P, Amirikian K, Dittmar KA, Lu H, Shen S, Guo W, Xing Y, Carstens RP (2010) An ESRP-regulated splicing programme is abrogated during the epithelial-mesenchymal transition. EMBO Journal 29:3286–3300

69. Nakaya Y, Kuroda S, Katagiri YT, Kaibuchi K, Takahashi Y (2004) Mesenchymal-epithelial transition during somitic segmentation is regulated by differential roles of Cdc42 and Rac1. Developmental Cell 7:425–438

70. Hader C, Marlier A, Cantley L (2010) Mesenchymal-epithelial transition in epithelial response to injury: the role of Foxc2. Oncogene 29:1031–1040

Volume I. Chap. 2. Cells of the Blood Circulation

71. Long X, Creemers EE, Wang DZ, Olson EN, Miano JM (2007) Myocardin is a bifunctional switch for smooth versus skeletal muscle differentiation. Proceedings of the National Academy of Sciences of the United States of America 104:16570–16575

72. Simper D, Stalboerger PG, Panetta CJ, Wang S, Caplice NM (2002) Smooth muscle progenitor cells in human blood. Circulation 106:1199–1204

73. Hu Y, Davison F, Ludewig B, Erdel M, Mayr M, Url M, Dietrich H, Xu Q (2002) Smooth muscle cells in transplant atherosclerotic lesions are originated from recipients, but not bone marrow progenitor cells. Circulation 106:1834–1839

74. Schwartz SM, Campbell GR, Campbell JH (1986) Replication of smooth muscle cells in vascular disease. Circulation Research 58:427–444

75. Hu Y, Zhang Z, Torsney E, Afzal AR, Davison F, Metzler B, Xu Q (2004) Abundant progenitor cells in the adventitia contribute to atherosclerosis of vein grafts in ApoE-deficient mice. Journal of Clinical Investigation 113:1258–1265

76. Zhang X, Adamson RH, Curry FE, Weinbaum S (2008) Transient regulation of transport by pericytes in venular microvessels via trapped microdomains. Proceedings of the National Academy of Sciences of the United States of America 105:1374–1379

77. Howson KM, Aplin AC, Gelati M, Alessandri G, Parati EA, Nicosia RF (2005) The postnatal rat aorta contains pericyte progenitor cells that form spheroidal colonies in suspension culture. American Journal of Physiology – Cell Physiology 289:C1396–C1407

78. Lewitzky M, Yamanaka S (2007) Reprogramming somatic cells towards pluripotency by defined factors. Current Opinion in Biotechnology 18:467–473

79. Takahashi K, Tanabe K, Ohnuki M, Narita M, Ichisaka T, Tomoda K, Yamanaka S (2007) Induction of pluripotent stem cells from adult human fibroblasts by defined factors. Cell 131:861–872

80. Viswanathan SR, Daley GQ, Gregory RI (2008) Selective blockade of microRNA processing by Lin28. Science 320:97–100

81. Bogdanov KY, Vinogradova TM, Lakatta EG (2001) Sinoatrial nodal cell ryanodine receptor and Na^+-Ca^{2+} exchanger. Molecular partners in pacemaker regulation. Circulation Research 88:1254

82. Weihe E, Kalmbach P (1978) Ultrastructure of capillaries in the conduction system of the heart in various mammals. Cell and Tissue Research 192:77–87

83. Bell JJ, Bhandoola A (2008) The earliest thymic progenitors for T cells possess myeloid lineage potential. Nature 452:764–767

84. Wada H, Masuda K, Satoh R, Kakugawa K, Ikawa T, Katsura Y, Kawamoto H (2008) Adult T-cell progenitors retain myeloid potential. Nature 452:768–772

85. Eshghi S, Vogelezang MG, Hynes RO, Griffith LG, Lodish HF (2007) $\alpha_4\beta_1$ integrin and erythropoietin mediate temporally distinct steps in erythropoiesis: integrins in red cell development. Journal of Cell Biology 177:871–880

86. Ji P, Jayapal SR, Lodish HF (2008) Enucleation of cultured mouse fetal erythroblasts requires Rac GTPases and mDia2. Nature – Cell Biology 10:314–321

87. Nemazee D (2006) Receptor editing in lymphocyte development and central tolerance. Nature Reviews – Immunology 6:728–740

88. Melichar HJ, Narayan K, Der SD, Hiraoka Y, Gardiol N, Jeannet G, Held W, Chambers CA, Kang J (2007) Regulation of $\gamma\delta$ versus $\alpha\beta$ T lymphocyte differentiation by the transcription factor SOX13. Science 315:230–233

89. He X, Kappes DJ (2006) CD4/CD8 lineage commitment: light at the end of the tunnel? Current Opinion in Immunology 8:135–142

90. Kovalovsky D, Uche OU, Eladad S, Hobbs RM, Yi W, Alonzo E, Chua K, Eidson M, Kim HJ, Im JS, Pandolfi PP, Sant'Angelo DB (2008) The BTB-zinc finger transcriptional regulator PLZF controls the development of invariant natural killer T cell effector functions. Nature – Immunology 9:1055–1064

91. Boyden LM, Lewis JM, Barbee SD, Bas A, Girardi M, Hayday AC, Tigelaar RE, Lifton RP (2008) Skint1, the prototype of a newly identified immunoglobulin superfamily gene cluster, positively selects epidermal $\gamma\delta$ T cells. Nature – Genetics 40:656–662

92. Medeiros RB, Burbach BJ, Mueller KL, Srivastava R, Moon JJ, Highfill S, Peterson EJ, Shimizu Y (2007) Regulation of NF-κB activation in T cells via association of the adapter proteins ADAP and CARMA1. Science 316:754–758

93. Harrington LE, Janowski KM, Oliver JR, Zajac AJ, Weaver CT (2008) Memory CD4 T cells emerge from effector T-cell progenitors. Nature 452:356–360

94. Lieberman J (2003) The ABCs of granule-mediated cytotoxicity: new weapons in the arsenal. Nature Reviews – Immunology 3:361–370

95. Vivier E, Tomasello E, Baratin M, Walzer T, Ugolini S (2008) Functions of natural killer cells. Nature – Immunology 9:503–510

96. Vignali DAA, Collison LW, Workman CJ (2008) How regulatory T cells work. Nature Reviews – Immunology 8:523–532

97. Lund JM, Hsing L, Pham TT, Rudensky AY (2008) Coordination of early protective immunity to viral infection by regulatory T cells. Science 320:1220–1224

98. Loder F, Mutschler B, Ray RJ, Paige CJ, Sideras P, Torres R, Lamers MC, Carsetti R (1999) B cell development in the spleen takes place in discrete steps and is determined by the quality of B cell receptor-derived signals. Journal of Experimental Medicine 90:75–89

99. Moser K, Tokoyoda K, Radbruch A, MacLennan I, Manz RA (2006) Stromal niches, plasma cell differentiation and survival. Current Opinion in Immunology 18:265–270

100. Cariappa A, Mazo IB, Chase C, Shi HN, Liu H, Li Q, Rose H, Leung H, Cherayil BJ, Russell P, von Andrian U, Pillai S (2005) Perisinusoidal B cells in the bone marrow participate in T-independent responses to blood-borne microbes. Immunity 23:397-407

101. Sapoznikov A, Pewzner-Jung Y, Kalchenko V, Krauthgamer R, Shachar I, Jung S (2008) Perivascular clusters of dendritic cells provide critical survival signals to B cells in bone marrow niches. Nature – Immunology 9:388–395

102. Weintz G, Olsen JV, Frühauf K, Niedzielska M, Amit I, Jantsch J, Mages J, Frech C, Dölken L, Mann M, Lang R (2010) The phosphoproteome of toll-like receptor-activated macrophages. Molecular Systems Biology 6:371

103. Tufet M (2008) Macrophages: re-educating macrophages. Nature Reviews – Immunology 8:488

104. Hagemann T, Lawrence T, McNeish I, Charles KA, Kulbe H, Thompson RG, Robinson SC, Balkwill FR (2008) "Re-educating" tumor-associated macrophages by targeting NF-kappaB. Journal of Experimental Medicine 205:1261–1268

105. Fong CHY, Bebien M, Didier LA, Nebauer R, Hussell T, Broide D, Karin M, Lawrence T (2008) An antiinflammatory role for IKKβ through the inhibition of "classical" macrophage activation. Journal of Experimental Medicine 205:1269–1276

106. Feng R, Desbordes SC, Xie H, Sanchez Tillo E, Pixley F, Stanley ER, Graf T (2008) PU.1 and C/EBPα/β convert fibroblasts into macrophage-like cells. Proceedings of the National Academy of Sciences of the United States of America 105:6057–6062

107. Krishnamoorthy N, Oriss TB, Paglia M, Fei M, Yarlagadda M, Vanhaesebroeck B, Ray A, Ray P (2008) Activation of c-Kit in dendritic cells regulates T helper cell differentiation and allergic asthma. Nature – Medicine 14:565–573

108. Gilliet M, Cao W, Liu YJ (2008) Plasmacytoid dendritic cells: sensing nucleic acids in viral infection and autoimmune diseases. Nature Reviews – Immunology 8:594–606

109. Galli SJ, Grimbaldeston M, Tsai M (2008) Immunomodulatory mast cells: negative, as well as positive, regulators of immunity. Nature Reviews – Immunology 8:478–486

Volume I. Chap. 3. Cells of the Ventilatory Tract

110. Mitchell B, Jacobs R, Li J, Chien S, Kintner C (2007) A positive feedback mechanism governs the polarity and motion of motile cilia. Nature 447:97–101

111. Jeffery PK (1998) The development of large and small airways. American Journal of Respiratory and Critical Care Medicine 157:S174–S180

112. Bartlett JA, Fischer AJ, McCray PB (2008) Innate immune functions of the airway epithelium. Contributions to Microbiology 15:147–163

113. Laube DM, Yim S, Ryan LK, Kisich KO, Diamond G (2006) Antimicrobial peptides in the airway. Current Topics in Microbiology and Immunology 306:153–182

114. Boers JE, Ambergen AW, Thunnissen FB (1998) Number and proliferation of basal and parabasal cells in normal human airway epithelium. American Journal of Respiratory and Critical Care Medicine 157:2000–2006

115. Boers JE, Ambergen AW, Thunnissen FB (1999) Number and proliferation of Clara cells in normal human airway epithelium. American Journal of Respiratory and Critical Care Medicine 159:1585–1591

116. Serafini SM, Wanner A, Michaelson ED (1976) Mucociliary transport in central and intermediate size airways: effect of aminophyllin. Bulletin Européen de Physiopathologie Respiratoire

117. Ghabrial AS, Krasnow MA (2006) Social interactions among epithelial cells during tracheal branching morphogenesis. Nature 441:746–749

118. Cutz E (1987) Cytomorphology and differentiation of airway epithelium in developing human lung. In: McDowell EM (ed), Lung carcinomas. Churchill Livingstone, Edinburgh

119. De Paepe ME (2005) Lung growth and development. In: Churg AM, Myers JL, Tazelaar HD, Wright JL (eds), Thurlbeck's Pathology of the Lung, third edition, Thieme, New York

120. Bucher U, Reid L (1961) Development of the mucus-secreting elements in human lung. Thorax 16:219–225

121. Cutz E (1982) Neuroendocrine cells of the lung: an overview of morphologic characteristics and development. Experimental Lung Research 3:185–208

122. Pandit L, Kolodziejska KE, Zeng S, Eissa NT (2009) The physiologic aggresome mediates cellular inactivation of iNOS. Proceedings of the National Academy of Sciences of the United States of America 106:1211–1215

123. Xu W, Koeck T, Lara AR, Neumann D, DiFilippo FP, Koo M, Janocha AJ, Masri FA, Arroliga AC, Jennings C, Dweik RA, Tuder RM, Stuehr DJ, Erzurum SC (2007) Alterations of cellular bioenergetics in pulmonary artery endothelial cells. Proceedings of the National Academy of Sciences of the United States of America 104:1342–1347

124. Delplanque A, Coraux C, Tirouvanziam R, Khazaal I, Puchelle E, Ambros P, Gaillard D, Péault B (2000) Epithelial stem cell-mediated development of the human respiratory mucosa in SCID (severe combined immunodeficiency) mice. Journal of Cell Science 113:767–778

125. Jetten AM (1991) Growth and differentiation factors in tracheobronchial epithelium. American Journal of Physiology – Lung Cellular and Molecular Physiology 260:L361–L373

126. Stiles AD, D'Ercole AJ (1990) The insulin-like growth factors and the lung. American Journal of Physiology – Lung Cellular and Molecular Physiology 3:93–100

127. Wolbach SB, Howe PR (1926) Tissue changes following deprivation of fat-soluble A vitamin. Journal of Experimental Medicine 42:753–781

128. Hong KU, Reynolds SD, Watkins S, Fuchs E, Stripp BR (2004) Basal cells are a multipotent progenitor capable of renewing the bronchial epithelium. American Journal of Pathology 164:577–588

129. Reynolds SD, Giangreco A, Power JHT, Stripp BR (2000) Neuroepithelial bodies of pulmonary airways serve as a reservoir of progenitor cells capable of epithelial regeneration. American Journal of Pathology 156:269-278

130. Rock JR, Onaitis MW, Rawlins EL, Lu Y, Clark CP, Xue Y, Randell SH, Hogan BL (2009) Basal cells as stem cells of the mouse trachea and human airway epithelium. Proceedings of the National Academy of Sciences of the United States of America 106:12771–12775

131. Ling TY, Kuo MD, Yu AL, Huang YH, Wy TJ, Lin YC, Chen SH, Yu J (2006) Identification of pulmonary Oct-4+ stem/progenitor cells and demonstration of their susceptibility to SARS coronavirus (SARS-CoV) infection in vitro. Proceedings of the National Academy of Sciences of the United States of America 103:9530-9535

132. Wang D, Haviland DL, Burns AR, Zsigmond E, Wetsel RA (2007) A pure population of lung alveolar epithelial type II cells derived from human embryonic stem cells. Proceedings of the National Academy of Sciences of the United States of America 104:4449–4454

133. Kaliner MA (1991) Human nasal respiratory secretions and host defense. American Review of Respiratory Diseases 144, S52–S56

134. Mercer RR, Russell ML, Roggli VL, Crapo JD (1994) Cell number and distribution in human and rat airways. American Journal of Respiratory Cell and Molecular Biology 10:613–624

135. Evans MJ, Plopper CG (1988) The role of basal cells in adhesion of columnar epithelium to airway basement membrane. American Review of Respiratory Diseases 138:481–483

136. Laoukili J, Perret E, Willems T, Minty A, Parthoens E, Houcine O, Coste A, Jorissen M, Marano F, Caput D, Tournier F (2001) IL-13 alters mucociliary differentiation and ciliary beating of human respiratory epithelial cells. Journal of Clinical Investigation 108:1817–1824

137. Farquhar MG, Palade GE (1963) Junctional complexes in various epithelia. Journal of Cell Biology 17:375-412

138. Yoshisue H, Puddicombe SM, Wilson SJ, Haitchi HM, Powell RM, Wilson DI, Pandit A, Berger AE, Davies DE, Holgate ST, Holloway JW (2004) Characterization of ciliated bronchial epithelium 1, a ciliated cell–associated gene induced during mucociliary differentiation. American Journal of Respiratory Cell and Molecular Biology 31:491–500

139. Lansley AB (1993) Mucociliary clearance and drug delivery via the respiratory tract. Advanced Drug Delivery Reviews 11:299–327

140. Foliguet B, Puchelle E (1986) Apical structure of human respiratory cilia, Bulletin Européen de Physio-pathologie Respiratoire 22:43–47

141. Kim CS, Rodriguez CR, Eldridge MA, Sackner MA (1986) Criteria for mucus transport in the airways by two-phase gas-liquid flow mechanism. Journal of Applied Physiology 60:901–907

142. Kim CS, Greene MA, Sankaran S, Sackner MA (1986) Mucus transport in the airways by two-phase gas-liquid flow mechanism: continuous flow model. Journal of Applied Physiology 60:908–917

143. Plopper CG, Mariassy AT, Wilson DW, Alley JL, Nishio SJ, Nettesheim P (1983) Comparison of nonciliated tracheal epithelial cells in six mammalian species: ultrastructure and population densities. Experimental Lung Research 5:281-294

144. Willems LN, Kramps JA, Jeffery PK, Dijkman JH (1988) Detection of antileukoprotease in the developing fetal lung. Thorax 43: 784–786

145. De Water R, Willems LN, Van Muijen GN, Franken C, Fransen JA, Dijkman JH, Kramps JA (1986) Ultrastructural localization of bronchial antileukoprotease in central and peripheral airways by a gold-labeling technique using monoclonal antibodies. American Review of Respiratory Diseases 133:882–890

146. Dunnill MS, Massarella GR, Anderson JA (1969) A comparison of the quantitative anatomy of the bronchi in normal subjects, in status asthmaticus, in chronic bronchitis, and in emphysema. Thorax 24:176–179

147. Rogers DF (2003) The airway goblet cell. International Journal of Biochemistry and Cell Biology 35:1–6

148. Devereux TR, Domin BA, Philpot RM (1989) Xenobiotic metabolism by isolated pulmonary cells. Pharmacology and Therapeutics 41:243–256

149. Li Y, Martin LD, Spizz G, Adler KB (2001) MARCKS protein is a key molecule regulating mucin secretion by human airway epithelial cells in vitro. Journal of Biological Chemistry 276:40982–40990

150. Koch H, Hofmann K, Brose N (2000) Definition of Munc13-homology-domains and characterization of a novel ubiquitously expressed Munc13 isoform. Biochemical Journal 349:247–253

151. Boers JE, den Brok JL, Koudstaal J, Arends JW, Thunnissen FB (1996) Number and proliferation of neuroendocrine cells in normal human airway epithelium. American Journal of Respiratory and Critical Care Medicine 154:758–763

152. Allen KM, Haworth SG (1989) Cytoskeletal features of immature pulmonary vascular smooth muscle cells and the influence of pulmonary hypertension on normal human development. Journal of Pathology 158:311–317

153. Bucher U, Reid L (1961) Development of the intrasegmental bronchial tree: the pattern of branching and development of cartilage at various stages of intra-uterine life. Thorax 16:207–218

154. Wright JL, Thurlbeck WM (2005) Quantitative Anatomy of the Lung. In: Churg AM, Myers JL, Tazelaar HD, Wright JL (eds), Thurlbeck's Pathology of the Lung, third edition, Thieme, New York

155. Thiriet M, Douguet D, Bonnet JC, Canonne C, Hatzfeld C (1979) Influence du mélange He–O_2 sur la mixique dans les bronchopneumopathies obstructives chroniques. [Influence of a He–O_2 mixture on gas mixing in chronic obstructive lung diseases.] Bulletin Européen de Physiopathologie Respiratoire 15:1053–1068

156. Maina JN, West JB (2005) Thin and strong! The bioengineering dilemma in the structural and functional design of the blood–gas barrier. Physiological Reviews 85:811–844

157. Massaro GD, Massaro D (1973) Hyperoxia: a stereologic ultrastructural examination of its influence on cytoplasmic components of the pulmonary granular pneumocyte. Journal of Clinical Investigation 52:566–570

158. Fehrenbach H (2001) Alveolar epithelial type II cell: defender of the alveolus revisited. Respiratory Research 2:33–52

159. Sirianni FE, Chu FS, Walker DC (2003) Human alveolar wall fibroblasts directly link epithelial type 2 cells to capillary endothelium. American Journal of Respiratory and Critical Care Medicine 168:1532–1537

160. Lipscomb MF, Pollard AM, Yates JL (1993) A role for TGF-beta in the suppression by murine bronchoalveolar cells of lung dendritic cell initiated immune responses Regional Immunology 5:151-157

161. Thepen T, Kraal G, Holt PG (1994) The role of alveolar macrophages in regulation of lung inflammation. Annals of the New York Academy of Sciences 725:200-206

162. Lambrecht BN (2006) Alveolar macrophage in the driver's seat. Immunity 24:366–368

163. Takabayshi K, Corr M, Hayashi T, Redecke V, Beck L, Guiney D, Sheppard D, Raz E (2006) Induction of a homeostatic circuit in lung tissue by microbial compounds. Immunity 24, 475–487

Volume I. Chap. 5. Cell Structure and Function

164. Alberts B, Bray D, Lewis J, Raff M, Roberts K, Watson JD (2002) Molecular biology of the cell, 4th ed. Garland Science, New York

165. Lev S (2010) Non-vesicular lipid transport by lipid-transfer proteins and beyond. Nature Reviews – Molecular Cell Biology 11:739–750

166. Levy Y, Onuchic JN (2006) Water mediation in protein folding and molecular recognition. Annual Review of Biophysics and Biomolecular Structure 35:389–415

167. Makarov V, Pettitt BM, Feig M (2002) Solvation and hydration of proteins and nucleic acids: a theoretical view of simulation and experiment. Accounts of Chemical Research 35:376–384

168. Raschke TM (2006) Water structure and interactions with protein surfaces. Current Opinion in Structural Biology 16:152–159

169. Ball P (2008) Water as an active constituent in cell biology. Chemical Reviews 108:74–108

170. Persson E, Halle B (2008) Cell water dynamics on multiple time scales. Proceedings of the National Academy of Sciences of the United States of America 105:6266–6271

171. Rafelski SM, Marshall WF (2008) Building the cell: design principles of cellular architecture. Nature Reviews – Molecular Cell Biology 9:593–602

892. Deborde S, Perret E, Gravotta D, Deora A, Salvarezza S, Schreiner R, Rodriguez-Boulan E (2008) Clathrin is a key regulator of basolateral polarity. Nature 452:719–723

173. Mitrossilis D, Fouchard J, Guiroy A, Desprat N, Rodriguez N, Fabry B, Asnacios A (2009) Single-cell response to stiffness exhibits muscle-like behavior. Proceedings of the National Academy of Sciences of the United States of America 106:18243–18248

174. Mitrossilis D, Fouchard J, Pereira D, Postic F, Richert A, Saint-Jean M, Asnacios A (2010) Real-time single-cell response to stiffness. Proceedings of the National Academy of Sciences of the United States of America 107:16518–16523

175. Bergmann A, Steller H (2010) Apoptosis, stem cells, and tissue regeneration. Science Signaling 3:re8

176. Gullingsrud J, Schulten K (2004) Lipid bilayer pressure profiles and mechanosensitive channel gating. Biophysical Journal 86:3496–3509

177. Blood PD, Ayton GS, Voth GA (2005) Probing the molecular-scale lipid bilayer response to shear flow using nonequilibrium molecular dynamics. Journal of Physical Chemistry. B, Condensed Matter, Materials, Surfaces, Interfaces & Biophysical Chemistry 109:18673–18679

178. Kraft ML, Weber PK, Longo ML, Hutcheon ID, Boxer SG (2006) Phase separation of lipid membranes analyzed with high-resolution secondary ion mass spectrometry. Science

179. Lemmon MA (2008) Membrane recognition by phospholipid-binding domains. Nature Reviews – Molecular Cell Biology 9:99–111

180. van Meer G, Voelker DR, Feigenson GW (2008) Membrane lipids: where they are and how they behave. Nature Reviews – Molecular Cell Biology 9:112–124

181. Ikonen E (2008) Cellular cholesterol trafficking and compartmentalization. Nature Reviews – Molecular Cell Biology 9:125–138

182. Gonen T, Cheng Y, Sliz P, Hiroaki Y, Fujiyoshi Y, Harrison SC, Walz T (2005) Lipid-protein interactions in double-layered two-dimensional AQP0 crystals. Nature 438:633–638

183. Behnia R, Munro S (2005) Organelle identity and the signposts for membrane traffic. Nature 438:597–604

184. Weirich CS, Erzberger JP, Barral Y (2008) The septin family of GTPases: architecture and dynamics. Nature Reviews – Molecular Cell Biology 9:478–489

185. Lindsey R, Momany M (2006) Septin localization across kingdoms: three themes with variations. Current Opinion in Microbiology 9:559–565

186. Barral Y (2010) Septins at the nexus. Science 329:1289–1290

187. Kim SK, Shindo A, Park TJ, Oh EC, Ghosh S, Gray RS, Lewis RA, Johnson CA, Attie-Bittach T, Katsanis N, Wallingford JB (2010) Planar cell polarity acts through septins to control collective cell movement and ciliogenesis. Science 329:1337–1340

188. Hu Q, Milenkovic L, Jin H, Scott MP, Nachury MV, Spiliotis ET, Nelson WJ (2010) A septin diffusion barrier at the base of the primary cilium maintains ciliary membrane protein distribution. Science 329:436–439

189. Zimmerberg J, Kozlov MM (2006) How proteins produce cellular membrane curvature. Nature Reviews – Molecular Cell Biology 7:9–19

190. Reynwar BJ, Illya G, Harmandaris VA, Müller MM, Kremer K, Deserno M (2007) Aggregation and vesiculation of membrane proteins by curvature-mediated interactions. Nature 447:461–464

191. Blood PD, Voth GA (2006) Direct observation of Bin/amphiphysin/Rvs (BAR) domain-induced membrane curvature by means of molecular dynamics simulations. Proceedings of the National Academy of Sciences of the United States of America 103:15068–15072

192. Hatzakis NS, Bhatia VK, Larsen J, Madsen KL, Bolinger PY, Kunding AH, Castillo J, Gether U, Hedegård P, Stamou D (2009) How curved membranes recruit amphipathic helices and protein anchoring motifs. Nature – Chemical Biology 5:835–841

193. Bhatia VK, Madsen KL, Bolinger PY, Kunding A, Hedegård P, Gether U, Stamou D (2009) Amphipathic motifs in BAR domains are essential for membrane curvature sensing. EMBO Journal 28:3303–3314

194. Suetsugu S (2010) The proposed functions of membrane curvatures mediated by the BAR domain superfamily proteins. Journal of Biochemistry 148:1–12

195. Alliegro MA, Henry JJ, Alliegro MC (2010) Rediscovery of the nucleolinus, a dynamic RNA-rich organelle associated with the nucleolus, spindle, and centrosomes. Proceedings of the National Academy of Sciences of the United States of America 107:13718–13723

196. Strambio-De-Castillia C, Niepel M, Rout MP (2010) The nuclear pore complex: bridging nuclear transport and gene regulation. Nature Reviews – Molecular Cell Biology 11:490–501

197. Akhtar A, Gasser SM (2007) The nuclear envelope and transcriptional control. Nature Reviews – Genetics 8:507–517

198. Guenther MG, Young RA (2010) Repressive transcription. Science 329:150–151

199. Gamble MJ, Frizzell KM, Yang C, Krishnakumar R, Kraus WL (2010) The histone variant macroH2A1 marks repressed autosomal chromatin, but protects a subset of its target genes from silencing. Genes and Development 24:21–32

200. Sarkies P, Reams C, Simpson LJ, Sale JE (2010) Epigenetic instability due to defective replication of structured DNA. Molecular Cell 40:703–713

201. Verzi MP, Shin H, He HH, Sulahian R, Meyer CA, Montgomery RK, Fleet JC, Brown M, Liu XS, Shivdasani RA (2010) Differentiation-specific histone modifications reveal dynamic chromatin interactions and partners for the intestinal transcription factor CDX2. Developmental Cell 19:713–726

202. Bernad R, Sánchez P, Losada A (2009) Epigenetic specification of centromeres by CENP-A. Experimental Cell Research 315:3233–3241

203. Bloom K, Joglekar A (2010) Towards building a chromosome segregation machine. Nature 463:446–456

204. Azzalin CM, Reichenbach P, Khoriauli L, Giulotto E, Lingner J (2007) Telomeric repeat-containing RNA and RNA surveillance factors at mammalian chromosome ends. Science 318:798–801

205. Gard S, Light W, Xiong B, Bose T, McNairn AJ, Harris B, Fleharty B, Seidel C, Brickner JH, Gerton JL (2009) Cohesinopathy mutations disrupt the subnuclear organization of chromatin. Journal of Cell Biology 187:455–462

206. Berger SL (2007) The complex language of chromatin regulation during transcription. Nature 447:407–412

207. Zhou B, Margariti A, Zeng L, Xu Q (2011) Role of histone deacetylases in vascular cell homeostasis and arteriosclerosis. Cardiovascular Research 90:413–420

208. Chang S, Young BD, Li S, Qi X, Richardson JA, Olson EN (2006) Histone deacetylase 7 maintains vascular integrity by repressing matrix metalloproteinase 10. Cell 126:321–334

209. Trivedi CM, Luo Y, Yin Z, Zhang M, Zhu W, Wang T, Floss T, Goettlicher M, Ruiz Noppinger P, Wurst W, Ferrari VA, Abrams CS, Gruber PJ, Epstein JA (2007) Hdac2 regulates the cardiac hypertrophic response by modulating Gsk3beta activity. Nature – Medicine 13:324–331

210. Wee HJ, Voon DC, Bae SC, Ito Y (2008) PEBP2β/CBFβ-dependent phosphorylation of RUNX1 and p300 by HIPK2: implications for leukemogenesis. Blood 112:3777–3787

211. Aikawa Y, Nguyen LA, Isono K, Takakura N, Tagata Y, Schmitz ML, Koseki H, Kitabayashi I (2006) Roles of HIPK1 and HIPK2 in AML1- and p300-dependent transcription, hematopoiesis and blood vessel formation. EMBO Journal 25:3955–3965

212. Lu A, Zougman A, Pudelko M, Bebenek M, Ziółkowski P, Mann M, Wisniewski JR (2009) Mapping of lysine monomethylation of linker histones in human breast and its cancer. Journal of Proteome Research 8:4207–4215

213. Franz H, Mosch K, Soeroes S, Urlaub H, Fischle W (2009) Multimerization and H3K9me3 binding are required for CDYL1b heterochromatin association. Journal of Biological Chemistry 284:35049–35059

214. Bender A, Hajieva P, Moosmann B (2008) Adaptive antioxidant methionine accumulation in respiratory chain complexes explains the use of a deviant genetic code in mitochondria. Proceedings of the National Academy of Sciences of the United States of America 105:16496–16501

215. Misteli T (2009) Self-organization in the genome. Proceedings of the National Academy of Sciences of the United States of America 106:6885–6886

216. Rajapakse I, Perlman MD, Scalzo D, Kooperberg C, Groudine M, Kosak ST (2009) The emergence of lineage-specific chromosomal topologies from coordinate gene regulation. Proceedings of the National Academy of Sciences of the United States of America 106: 6679–6684

217. Alexander RP, Fang G, Rozowsky J, Snyder M, Gerstein MB (2010) Annotating non-coding regions of the genome. Nature Reviews – Genetics 11:559–571

218. Caron H, van Schaik B, van der Mee M, Baas F, Riggins G, van Sluis P, Hermus MC, van Asperen R, Boon K, Voûte PA, Heisterkamp S, van Kampen A, Versteeg R (2001) The human transcriptome map: clustering of highly expressed genes in chromosomal domains. Science 291:1289–1292

219. Versteeg R, van Schaik BD, van Batenburg MF, Roos M, Monajemi R, Caron H, Bussemaker HJ, van Kampen AH (2003) The human transcriptome map reveals extremes in gene density, intron length, GC content, and repeat pattern for domains of highly and weakly expressed genes. Genome Research 13:1998–2004

220. Feschotte C (2008) Transposable elements and the evolution of regulatory networks Nature Reviews – Genetics 9:397–405

221. Berretta J, Morillon A (2009) Pervasive transcription constitutes a new level of eukaryotic genome regulation. EMBO Reports 10:973–982

222. Thomas MC, Chiang CM (2006) The general transcription machinery and general cofactors. Critical Reviews in Biochemistry and Molecular Biology 41:105–178

223. Belasco JG (2010) All things must pass: contrasts and commonalities in eukaryotic and bacterial mRNA decay. Nature Reviews – Molecular Cell Biology 11:467–478

224. Goodrich JA, Tjian R (2010) Unexpected roles for core promoter recognition factors in cell-type-specific transcription and gene regulation. Nature Reviews – Genetics 11:549–558

225. Durier S, Fassot C, Laurent S, Boutouyrie P, Couetil J-P, Fine E, Lacolley P, Dzau VJ, Pratt RE (2003) Physiological genomics of human arteries: quantitative relationship between gene expression and arterial stiffness. Circulation 108:1845–1851

226. Yamashita T, Honda M, Takatori H, Nishino R, Hoshino N, Kaneko S (2004) Genome-wide transcriptome mapping analysis identifies organ-specific gene expression patterns along human chromosomes. Genomics 84:867–875

227. Xu M, Long C, Chen X, Huang C, Chen S, Zhu B (2010) Partitioning of histone H3-H4 tetramers during DNA replication-dependent chromatin assembly. Science 328:94–98

228. Bahar R, Hartmann CH, Rodriguez KA, Denny AD, Busuttil RA, Dollé MET, Calder RB, Chisholm GB, Pollock BH, Klein CA, Vijg J (2006) Increased cell-to-cell variation in gene expression in ageing mouse heart. Nature 441:1011–1014

229. Longhese M, Bonetti D, Manfrini N, Clerici M (2010) Mechanisms and regulation of DNA end resection. EMBO Journal

230. Izumi N, Yamashita A, Iwamatsu A, Kurata R, Nakamura H, Saari B, Hirano H, Anderson P, Ohno S (2010) AAA+ proteins RUVBL1 and RUVBL2 coordinate PIKK activity and function in nonsense-mediated mRNA decay. Science Signaling 3:ra27

231. Liu T, Ghosal G, Yuan J, Chen J, Huang J (2010) FAN1 acts with FANCI-FANCD2 to promote DNA interstrand cross-link repair. Science 329:693–696

232. Naim V, Rosselli P (2009) The FANC pathway and BLM collaborate during mitosis to prevent micro-nucleation and chromosome abnormalities. Nature – Cell Biology 11:761–768

233. Pace P, Mosedale G, Hodskinson M, Rosado IV, Sivasubramaniam M, Patel KJ (2010) Ku70 corrupts DNA repair in the absence of the Fanconi anemia pathway. Science 329:219–223

234. Jackson SP, Bartek J (2009) The DNA-damage response in human biology and disease. Nature 461:1071–1078

235. Petrini JHJ (2007) A touching response to damage. Science 316:1138–1139

236. Misteli T, Soutoglou E (2009) The emerging role of nuclear architecture in DNA repair and genome maintenance. Nature Reviews – Molecular Cell Biology 10:243–254

237. Galanty Y, Belotserkovskaya R, Coates J, Polo S, Miller KM, Jackson SP (2009) Mammalian SUMO E3-ligases PIAS1 and PIAS4 promote responses to DNA double-strand breaks. Nature 462:935–939

238. Huen MSY, Sy SMH, Chen J (2010) BRCA1 and its toolbox for the maintenance of genome integrity. Nature Reviews – Molecular Cell Biology 11:138–148

239. Morris JR, Boutell C, Keppler M, Densham R, Weekes D, Alamshah A, Butler L, Galanty Y, Pangon L, Kiuchi T, Ng T, Solomon E (2009) The SUMO modification pathway is involved in the BRCA1 response to genotoxic stress. Nature 462:886–890

240. Ip SCY, Rass U, Blanco MG, Flynn HR, Skehel JM, West SC (2008) Identification of Holliday junction resolvases from humans and yeast. Nature 456:357–361

241. Chen Z, Yang H, Pavletich NP (2008) Mechanism of homologous recombination from the RecA–ssDNA/dsDNA structures. Nature 453:489–484

870. Casolari JM, Brown CR, Komili S, West J, Hieronymus H, Silver PA (2004) Genome-wide localization of the nuclear transport machinery couples transcriptional status and nuclear organization. Cell 117:427–439

871. Menon BB, Sarma NJ, Pasula S, Deminoff SJ, Willis KA, Barbara KE, Andrews B, Santangelo GM (2005) Reverse recruitment: the Nup84 nuclear pore subcomplex mediates Rap1/Gcr1/Gcr2 transcriptional activation. Proceedings of the National Academy of Sciences of the United States of America 102:5749–5754

244. Saitoh N, Spahr CS, Patterson SD, Bubulya P, Neuwald AF, Spector DL (2004) Proteomic analysis of interchromatin granule clusters. Molecular Biology of the Cell 15:3876–3890

245. Boisvert FM, van Koningsbruggen S, Navascues J, Lamond AI (2007) The multifunctional nucleolus. Nature Reviews – Molecular Cell Biology 8:574–585

246. Stark LA, Taliansky M (2009) Old and new faces of the nucleolus. EMBO Reports 10:35–40

247. Stewart CL, Roux KJ, Burke B (2007) Blurring the boundary: the nuclear envelope extends its reach. Science 318:1408–1412

248. Güttinger S, Laurell E, Kutay U (2009) Orchestrating nuclear envelope disassembly and reassembly during mitosis. Nature Reviews – Molecular Cell Biology 10:178–191

249. Ooshio T, Irie K, Morimoto K, Fukuhara A, Imai T, Takai Y (2004) Involvement of LMO7 in the association of two cell-cell adhesion molecules, nectin and E-cadherin, through afadin and α-actinin in epithelial cells. Journal of Biological Chemistry 279:31365–31373

250. Holaska JM, Rais-Bahrami S, Wilson KL (2006) Lmo7 is an emerin-binding protein that regulates the transcription of emerin and many other muscle-relevant genes. Human Molecular Genetics 15:3459–3472

251. Bengtsson L (2007) What MAN1 does to the Smads. TGFβ/BMP signaling and the nuclear envelope. FEBS Journal 274:1374–1382

252. Burke B, Ellenberg J (2002) Remodelling the walls of the nucleus. Nature Reviews – Molecular Cell Biology 3:487–497

253. Busson S, Dujardin D, Moreau A, Dompierre J, De Mey JR (1998) Dynein and dynactin are localized to astral microtubules and at cortical sites in mitotic epithelial cells. Current Biology 8:541–544

254. Prunuske AJ, Liu J, Elgort S, Joseph J, Dasso M, Ullman KS (2006) Nuclear envelope breakdown is coordinated by both Nup358/RanBP2 and Nup153, two nucleoporins with zinc finger modules. Molecular Biology of the Cell 17:760–769

255. Joseph J, Tan SH, Karpova TS, McNally JG, Dasso M (2002) SUMO-1 targets RanGAP1 to kinetochores and mitotic spindles. Journal of Cell Biology 156:595–602

256. Kaiser TE, Intine RV, Dundr M (2008) De novo formation of a subnuclear body. Science 322:1713–1717

257. Trinkle-Mulcahy L, Lamond AI (2007) Toward a high-resolution view of nuclear dynamics. Science 318:1402–1407

258. Marzluff WF, Wagner EJ, Duronio RJ (2008) Metabolism and regulation of canonical histone mRNAs: life without a poly(A) tail. Nature Reviews – Genetics 9:843–854

259. Gruenbaum Y, Margalit A, Goldman RD, Shumaker DK, Wilson KL (2005) The nuclear lamina comes of age. Nature Reviews – Molecular Cell Biology 6:21–31

260. Scaffidi P, Misteli T (2008) Lamin A-dependent misregulation of adult stem cells associated with accelerated ageing. Nature – Cell Biology 10:452–459

261. Gerace L, Blobel G (1980) The nuclear envelope lamina is reversibly depolymerized during mitosis. Cell 19:277–287

262. Tsai MY, Wang S, Heidinger JM, Shumaker DK, Adam SA, Goldman RD, Zheng Y (2006) A mitotic lamin B matrix induced by RanGTP required for spindle assembly. Science 311:1887–1893

263. Zhang Q, Skepper JN, Yang F, Davies JD, Hegyi L, Roberts RG, Weissberg PL, Ellis JA, Shanahan CM (2001) Nesprins: a novel family of spectrin-repeat-containing proteins that localize to the nuclear membrane in multiple tissues. Journal of Cell Science 114:4485–4498.

264. Gottschalk AJ, Conaway RC, Conaway JW (2008) New clues to actin function in chromatin regulation. Nature – Structural & Molecular Biology 15:432–433

265. Vartiainen MK, Guettler S, Larijani B, Treisman R (2007) Nuclear actin regulates dynamic subcellular localization and activity of the SRF cofactor MAL. Science 316:1749–1752

266. Gilchrist A, Au CE, Hiding J, Bell AW, Fernandez-Rodriguez J, Lesimple S, Nagaya H, Roy L, Gosline SJ, Hallett M, Paiement J, Kearney RE, Nilsson T, Bergeron JJ (2006) Quantitative proteomics analysis of the secretory pathway. Cell 127:1265–1281

267. Missiaen L, Van Acker K, Van Baelen K, Raeymaekers L, Wuytack F, Parys JB, De Smedt H, Vanoevelen J, Dode L, Rizzuto R, Callewaert G (2004) Calcium release from the Golgi apparatus and the endoplasmic reticulum in HeLa cells stably expressing targeted aequorin to these compartments. Cell Calcium 36:479–487

268. Diao A, Frost L, Morohashi Y, Lowe M (2008) Coordination of golgin tethering and SNARE assembly: GM130 binds syntaxin 5 in a p115-regulated manner. Journal of Biological Chemistry 283:6957–6967

269. Allan BB, Moyer BD, Balch WE (2000) Rab1 recruitment of p115 into a cis-SNARE complex: programming budding COPII vesicles for fusion. Science 289:444–448

270. Malsam J, Satoh A, Pelletier L, Warren G (2005) Golgin tethers define subpopulations of COPI vesicles. Science 307:1095-1098

271. Drin G, Morello V, Casella JF, Gounon P, Antonny B (2008) Asymmetric tethering of flat and curved lipid membranes by a golgin. Science 320:670–673

272. De Matteis MA, Luini A (2008) Exiting the Golgi complex. Nature Reviews – Molecular Cell Biology 9:273–284

273. Bonifacino JS, Traub LM (2003) Signals for sorting of transmembrane proteins to endosomes and lysosomes. Annual Review of Biochemistry 72:395–447

274. Hayashi M, Imanaka-Yoshida K, Yoshida T, Wood M, Fearns C, Tatake RJ, Lee JD (2005) A crucial role of mitochondrial Hsp40 in preventing dilated cardiomyopathy. Nature – Medicine 12:128–132

275. Breckenridge DG, Stojanovic M, Marcellus RC, Shore GC (2003) Caspase cleavage product of BAP31 induces mitochondrial fission through endoplasmic reticulum calcium signals, enhancing cytochrome c release to the cytosol. Journal of Cell Biology 160:1115–1127

276. Cribbs JT, Strack S (2007) Reversible phosphorylation of Drp1 by cyclic AMP-dependent protein kinase and calcineurin regulates mitochondrial fission and cell death. EMBO Reports 8:939–944

277. Fan W, Waymire KG, Narula N, Li P, Rocher C, Coskun PE, Vannan MA, Narula J, MacGregor GR, Wallace DC (2008) A mouse model of mitochondrial disease reveals germline selection against severe mtDNA mutations. Science 319:958–962

278. Meinecke M, Wagner R, Kovermann P, Guiard B, Mick DU, Hutu DP, Voos W, Truscott KN, Chacinska A, Pfanner N, Rehling P (2006) Tim50 maintains the permeability barrier of the mitochondrial inner membrane. Science 312:1523–1526

279. Kutik S, Stojanovski D, Becker L, Becker T, Meinecke M, Krüger V, Prinz C, Meisinger C, Guiard B, Wagner R, Pfanner N, Wiedemann N (2008) Dissecting membrane insertion of mitochondrial beta-barrel proteins. Cell 132:1011–1024

280. Rostovtseva TK, Sheldon KL, Hassanzadeh E, Monge C, Saks V, Bezrukov SM, Sackett DL (2008) Tubulin binding blocks mitochondrial voltage-dependent anion channel and regulates respiration. Proceedings of the National Academy of Sciences of the United States of America 105:18746–18751

281. Pinton P, Rimessi A, Marchi S, Orsini F, Migliaccio E, Giorgio M, Contursi C, Minucci S, Mantovani F, Wieckowski MR, Del Sal G, Pelicci PG, Rizzuto R (2007) Protein kinase Cβ and prolyl isomerase 1 regulate mitochondrial effects of the life-span determinant p66Shc. Science 315:659–663

282. Wagner BK, Kitami T, Gilbert TJ, Peck D, Ramanathan A, Schreiber SL, Golub TR, Mootha VK (2008) Large-scale chemical dissection of mitochondrial function. Nature – Biotechnology 26:343–351

283. Woodson JD, Chory J (2008) Coordination of gene expression between organellar and nuclear genomes. Nature Reviews – Genetics 9:383–395

284. Junge W, Sielaff H, Engelbrecht S (2009) Torque generation and elastic power transmission in the rotary FOF1-ATPase. Nature 459:364–370

285. Cairns CB, Walther J, Harken AH, Banerjee A (1998) Mitochondrial oxidative phosphorylation thermodynamic efficiencies reflect physiological organ roles, The American Journal of Physiology – Regulatory, Integrative and Comparative Physiology 274:1376–1383

286. Belevich I, Verkhovsky MI, Wikstrom M (2006) Proton-coupled electron transfer drives the proton pump of cytochrome c oxidase. Nature 440:829–832

287. Kussmaul L, Hirst J (2006) The mechanism of superoxide production by NADH:ubiquinone oxidoreductase (complex I) from bovine heart mitochondria. Proceedings of the National Academy of Sciences of the United States of America 103:7607–7612

288. Wilson DF (1994) Factors affecting the rate and energetics of mitochondrial oxidative phosphorylation. Medecine & Science in Sports & Exercise 26:37-43

289. Hallows WC, Lee S, Denu JM (2006) Sirtuins deacetylate and activate mammalian acetyl-CoA synthetases. Proceedings of the National Academy of Sciences of the United States of America 103:10230–10235

290. Hui STY, Andres AM, Miller AK, Spann AJ, Potter DW, Post NM, Chen AZ, Sachithanantham S, Jung DY, Kim JK, Davis RA (2008) Txnip balances metabolic and growth signaling via PTEN disulfide reduction. Proceedings of the National Academy of Sciences of the United States of America 105:3921–3926

291. Grahame Hardie D (2007) AMP-activated/SNF1 protein kinases: conserved guardians of cellular energy. Nature Reviews – Molecular Cell Biology 8:774–785

292. Xiao B, Heath R, Saiu P, Leiper FC, Leone P, Jing C, Walker PA, Haire L, Eccleston JF, Davis CT, Martin SR, Carling D, Gamblin SJ (2007) Structural basis for AMP binding to mammalian AMP-activated protein kinase. Nature 449:496–500

293. Minokoshi Y, Kim YB, Peroni OD, Fryer LG, Müller C, Carling D, Kahn BB (2002) Leptin stimulates fatty-acid oxidation by activating AMP-activated protein kinase. Nature 415:339–343

294. Yamauchi T, Kamon J, Minokoshi Y, Ito Y, Waki H, Uchida S, Yamashita S, Noda M, Kita S, Ueki K, Eto K, Akanuma Y, Froguel P, Foufelle F, Ferre P, Carling D, Kimura S, Nagai R, Kahn BB, Kadowaki T (2002) Adiponectin stimulates glucose utilization and fatty-acid oxidation by activating AMP-activated protein kinase. Nature – Medicine 8:1288–1295

295. Weigert C, Düfer M, Simon P, Debre E, Runge H, Brodbeck K, Häring HU, Schleicher ED (2007) Upregulation of IL-6 mRNA by IL-6 in skeletal muscle cells: role of IL-6 mRNA stabilization and Ca2+-dependent mechanisms. American Journal of Physiology – Cell Physiology 293:C1139–C1147

296. Watt MJ, Dzamko N, Thomas WG, Rose-John S, Ernst M, Carling D, Kemp BE, Febbraio MA, Steinberg GR (2006) CNTF reverses obesity-induced insulin resistance by activating skeletal muscle AMPK. Nature – Medicine 12:541–548

297. Cantó C, Gerhart-Hines Z, Feige JN, Lagouge M, Noriega L, Milne JC, Elliott PJ, Puigserver P, Auwerx J (2009) AMPK regulates energy expenditure by modulating NAD^+ metabolism and SIRT1 activity. Nature 458:1056–1060

298. Bungard D, Fuerth BJ, Zeng PY, Faubert B, Maas NL, Viollet B, Carling D, Thompson CB, Jones RG, Berger SL (2010) Signaling kinase AMPK activates stress-promoted transcription via histone H2B phosphorylation. Science 329:1201–1205

299. Delgoffe GM, Kole TP, Cotter RJ, Powell JD (2009) Enhanced interaction between Hsp90 and raptor regulates mTOR signaling upon T cell activation. Molecular Immunology 46):2694–2698

300. Qian SB, Zhang X, Sun J, Bennink JR, Yewdell JW, Patterson C (2010) mTORC1 links protein quality and quantity control by sensing chaperone availability. Journal of Biological Chemistry 285:27385–27395

301. Dzhindzhev NS, Yu QD, Weiskopf K, Tzolovsky G, Cunha-Ferreira I, Riparbelli M, Rodrigues-Martins A, Bettencourt-Dias M, Callaini G, Glover DM (2010) Asterless is a scaffold for the onset of centriole assembly. Nature 467:714–718

302. Goetz SC, Anderson KV (2010) The primary cilium: a signalling centre during vertebrate development. Nature Reviews – Genetics 11:331–344

303. Boehlke C, Kotsis F, Patel V, Braeg S, Voelker H, Bredt S, Beyer T, Janusch H, Hamann C, Gödel M, Müller K, Herbst M, Hornung M, Doerken M, Köttgen M, Nitschke R, Igarashi P, Walz G, Kuehn EW (2010) Primary cilia regulate mTORC1 activity and cell size through Lkb1. Nature – Cell Biology 12:1115–1122

304. Wiczer BM, Kalender A, Thomas G (2010) Bending the path to TOR. Nature – Cell Biology 12:1031–1033

305. Davey MG, James J, Paton IR, Burt DW, Tickle C (2007) Analysis of talpid3 and wild-type chicken embryos reveals roles for Hedgehog signalling in development of the limb bud vasculature. Developmental Biology 301:155–165

306. Kim J, Lee JE, Heynen-Genel S, Suyama E, Ono K, Lee KY, Ideker T, Aza-Blanc P, Gleeson JG (2010) Functional genomic screen for modulators of ciliogenesis and cilium length. Nature 464:1048–1051

307. Knödler A, Feng S, Zhang J, Zhang X, Das A, Peränen J, Guo W (2010) Coordination of Rab8 and Rab11 in primary ciliogenesis. Proceedings of the National Academy of Sciences of the United States of America 107:6346–6351

308. Dishinger JF, Kee HL, Jenkins PM, Fan S, Hurd TW, Hammond JW, Truong YN, Margolis B, Martens JR, Verhey KJ (2010) Ciliary entry of the kinesin-2 motor KIF17 is regulated by importin- 2 and RanGTP. Nature – Cell Biology

309. Fan S, Fogg V, Wang Q, Chen XW, Liu CJ, Margolis B (2007) A novel Crumbs3 isoform regulates cell division and ciliogenesis via importin β interactions. Journal of Cell Biology 178:387–398

310. Zhang MZ, Mai W, Li C, Cho SY, Hao C, Moeckel G, Zhao R, Kim I, Wang J, Xiong H, Wang H, Sato Y, Wu Y, Nakanuma Y, Lilova M, Pei Y, Harris RC, Li S, Coffey RJ, Sun L, Wu D, Chen XZ, Breyer MD, Zhao ZJ, McKanna JA, Wu G (2004) PKHD1 protein encoded by the gene for autosomal recessive polycystic kidney disease associates with basal bodies and primary cilia in renal epithelial cells. Proceedings of the National Academy of Sciences of the United States of America 101:2311–2316

311. Guirao B, Meunier A, Mortaud S, Aguilar A, Corsi JM, Strehl L, Hirota Y, Desoeuvre A, Boutin C, Han YG, Mirzadeh Z, Cremer H, Montcouquiol M, Sawamoto K, Spassky N (2010) Coupling between hydrodynamic forces and planar cell polarity orients mammalian motile cilia. Nature – Cell Biology 12:341–350

312. Ikegami K, Sato S, Nakamura K, Ostrowski LE, Setou M (2010) Tubulin polyglutamylation is essential for airway ciliary function through the regulation of beating asymmetry. Proceedings of the National Academy of Sciences of the United States of America 107:10490–10495

313. Martin S, Parton RG (2006) Lipid droplets: a unified view of a dynamic organelle. Nature Reviews – Molecular Cell Biology 7:373–378

314. Luzio JP, Pryor PR, Bright NA (2007) Lysosomes: fusion and function. Nature Reviews – Molecular Cell Biology 8:622–632

315. Laporte D, Salin B, Daignan-Fornier B, Sagot I (2008) Reversible cytoplasmic localization of the proteasome in quiescent yeast cells. Journal of Cell Biology 181:737-745

316. Anderson P, Kedersha N (2006) RNA granules. Journal of Cell Biology 172:803–808

317. Kedersha N, Anderson P (2002) Stress granules: sites of mRNA triage that regulate mRNA stability and translatability. Biochemical Society Transactions 30:963–969

318. Tsai NP, Ho PC, Wei LN (2008) Regulation of stress granule dynamics by Grb7 and FAK signalling pathway. EMBO Journal 27:715–726

319. Arimoto K, Fukuda H, Imajoh-Ohmi S, Saito H, Takekawa M (2008) Formation of stress granules inhibits apoptosis by suppressing stress-responsive MAPK pathways. Nature – Cell Biology 10:1324–1332

Volume I. Chap. 6. Protein Synthesis

320. Radisky DC, Stallings-Mann M, Hirai Y, Bissell MJ (2009) Single proteins might have dual but related functions in intracellular and extracellular microenvironments. Nature Reviews – Molecular Cell Biology 10:228–234

321. Bonasio R, Tu S, Reinberg D (2010) Molecular signals of epigenetic states. Science 330:612–616

322. Morimoto RI, Driessen AJ, Hegde RS, Langer T (2011) The life of proteins: the good, the mostly good and the ugly. Nature – Structural and Molecular Biology 18:1–4

323. Karpova TS, Kim MJ, Spriet C, Nalley K, Stasevich TJ, Kherrouche Z, Heliot L, McNally JG (2008) Concurrent fast and slow cycling of a transcriptional activator at an endogenous promoter. Science 319:466–469

324. Mirny LA (2010) Nucleosome-mediated cooperativity between transcription factors. Proceedings of the National Academy of Sciences of the United States of America 107:22534–22539

325. Reik W, Walter J (2001) Genomic imprinting: parental influence on the genome. Nature Reviews – Genetics 2:21–32

326. Jirtle RL, Skinner MK (2007) Environmental epigenomics and disease susceptibility. Nature Reviews – Genetics 8:253–262

327. Staudt AC, Wenkel S (2011) Regulation of protein function by "microProteins". EMBO Reports 12:35–42

328. Ray PS, Jia J, Yao P, Majumder M, Hatzoglou M, Fox PL (2009) A stress-responsive RNA switch regulates VEGFA expression. Nature 457:915–919

329. Muñoz IG, Yébenes H, Zhou M, Mesa P, Serna M, Park AY, Bragado-Nilsson E, Beloso A, de Cárcer G, Malumbres M, Robinson CV, Valpuesta JM, Montoya G (2011) Crystal structure of the open conformation of the mammalian chaperonin CCT in complex with tubulin. Nature – Structural and Molecular Biology 18:14–19

330. Tyedmers J, Mogk A, Bukau B (2010) Cellular strategies for controlling protein aggregation. Nature Reviews – Molecular Cell Biology 11:777–788

331. Wandinger SK, Richter K, Buchner J (2008) The Hsp90 chaperone machinery. Journal of Biological Chemistry 283:18473–18477

332. Li J, Richter K, Buchner J (2011) Mixed Hsp90-cochaperone complexes are important for the progression of the reaction cycle. Nature – Structural and Molecular Biology 18:61–66

333. Simpson NE, Lambert WM, Watkins R, Giashuddin S, Huang SJ, Oxelmark E, Arju R, Hochman T, Goldberg JD, Schneider RJ, Reiz LF, Soares FA, Logan SK, Garabedian MJ (2010) High levels of Hsp90 cochaperone p23 promote tumor progression and poor prognosis in breast cancer by increasing lymph node metastases and drug resistance. Cancer Research 70:8446–8456

334. Karagöz GE, Duarte AM, Ippel H, Uetrecht C, Sinnige T, van Rosmalen M, Hausmann J, Heck AJ, Boelens R, Rüdiger SG (2011) N-terminal domain of human Hsp90 triggers binding to the cochaperone p23. Proceedings of the National Academy of Sciences of the United States of America

335. Holt SE, Aisner DL, Baur J, Tesmer VM, Dy M, Ouellette M, Trager JB, Morin GB, Toft DO, Shay JW, Wright WE, White MA (1999) Functional requirement of p23 and Hsp90 in telomerase complexes. Genes and Development 1999 13:817–826

336. Affymetrix/Cold Spring Harbor Laboratory ENCODE Transcriptome Project (2009) Post-transcriptional processing generates a diversity of $5'$-modified long and short RNAs. Nature 457:1028–1032

337. Czerwoniec A, Dunin-Horkawicz S, Purta E, Kaminska KH, Kasprzak JM, Bujnicki JM, Grosjean H, Rother K (2009) MODOMICS: a database of RNA modification pathways. Nucleic Acids Research 37:D118–D121

338. Meier UT (2011) Pseudouridylation goes regulatory. EMBO Journal 30:3–4

339. Wu G, Xiao M, Yang C, Yu YT (2011) U2 snRNA is inducibly pseudouridylated at novel sites by Pus7p and snR81 RNP. EMBO Journal 30:79–89

340. Preker P, Nielsen J, Kammler S, Lykke-Andersen S, Christensen MS, Mapendano CK, Schierup MH, Jensen TH (2008) RNA exosome depletion reveals transcription upstream of active human promoters. Science 322:1851–1854

341. Seila AC, Calabrese JM, Levine SS, Yeo GW, Rahl PB, Flynn RA, Young RA, Sharp PA (2008) Divergent transcription from active promoters. Science 322:1849–1851

342. Core LJ, Waterfall JJ, Lis JT (2008) Nascent RNA sequencing reveals widespread pausing and divergent initiation at human promoters. Science 322:1845–1848

343. Tsai MC, Manor O, Wan Y, Mosammaparast N, Wang JK, Lan F, Shi Y, Segal E, Chang HY (2010) Long noncoding RNA as modular scaffold of histone modification complexes. Science 329):689–693

344. Rinn JL, Kertesz M, Wang JK, Squazzo SL, Xu X, Brugmann SA, Goodnough LH, Helms JA, Farnham PJ, Segal E, Chang HY (2007) Functional demarcation of active and silent chromatin domains in human HOX loci by noncoding RNAs. Cell 129:1311–1323

345. Guang S, Bochner AF, Burkhart KB, Burton N, Pavelec DM, Kennedy S (2010) Small regulatory RNAs inhibit RNA polymerase II during the elongation phase of transcription. Nature 465:1097–1101

346. Kaikkonen MU, Lam MTY, Glass CK (2011) Non-coding RNAs as regulators of gene expression and epigenetics. Cardiovascular Research 90:430–440

347. Zamore PD (2010) Somatic piRNA biogenesis. EMBO Journal 29:3219–3221

348. Olivieri D, Sykora MM, Sachidanandam R, Mechtler K, Brennecke J (2010) An in vivo RNAi assay identifies major genetic and cellular requirements for primary piRNA biogenesis in Drosophila. EMBO Journal 29:3301–3317

349. Ghildiyal M, Zamore PD (2009) Small silencing RNAs: an expanding universe. Nature Reviews – Genetics 10:94-108

350. Aravin AA, Sachidanandam R, Girard A, Fejes-Toth K, Hannon GJ (2007) Developmentally regulated piRNA clusters implicate MILI in transposon control. Science 316:744–747

351. Yang N, Kazazian HH (2006) L1 retrotransposition is suppressed by endogenously encoded small interfering RNAs in human cultured cells. Nature – Structural and Molecular Biology 13:763–771

352. Nilsen TW (2008) Endo-siRNAs: yet another layer of complexity in RNA silencing. Nature – Structural and Molecular Biology 15:546–548

353. Lee Y, Kim M, Han J, Yeom KH, Lee S, Baek SH, Kim VN (2004) MicroRNA genes are transcribed by RNA polymerase II. EMBO Journal 23:4051–4060

354. Filipowicz W, Bhattacharyya SN, Sonenberg N (2008) Mechanisms of posttranscriptional regulation by microRNAs: are the answers in sight? Nature Reviews – Genetics 102–114

355. Morlando M, Ballarino M, Gromak N, Pagano F, Bozzoni I, Proudfoot NJ (2008) Primary microRNA transcripts are processed co-transcriptionally. Nature – Structural and Molecular Biology 15:902–909

356. Davis-Dusenbery BN, Hata A (2010) Mechanisms of control of microRNA biogenesis. Journal of Biochemistry 148:381–392

357. Suzuki HI, Miyazono K (2011) Emerging complexity of microRNA generation cascades. Journal of Biochemistry 149:15–25

358. Trabucchi M, Briata P, Garcia-Mayoral MF, Haase AD, Filipowicz W, Ramos A, Gherzi R, Rosenfeld MG (2009) The RNA-binding protein KSRP promotes the biogenesis of a subset of microRNAs. Nature 459:1010–1014

359. Stewart M (2009) Nuclear export of small RNAs. Science 326:1195–1196

360. Dorner S, Eulalio A, Huntzinger E, Izaurralde E (2007) Symposium on microRNAs and siRNAs: biological functions and mechanisms. EMBO Reports 8:723–729

361. Selbach M, Schwanhäusser B, Thierfelder N, Fang Z, Khanin R, Rajewsky N (2008) Widespread changes in protein synthesis induced by microRNAs. Nature 455:58–63

362. Baek D, Villén J, Shin C, Camargo FD, Gygi SP, Bartel DP (2008) The impact of microRNAs on protein output. Nature 455:64–71

363. Flynt AS, Lai EC (2008) Biological principles of microRNA-mediated regulation: shared themes amid diversity. Nature Reviews – Genetics 9:831–842

364. Mortensen RD, Serra M, Steitz JA, Vasudevan S (2011) Posttranscriptional activation of gene expression in Xenopus laevis oocytes by microRNA-protein complexes (microRNPs). Proceedings of the National Academy of Sciences of the United States of America 108:8281–8286

365. Ramachandran V, Chen X (2008) Degradation of microRNAs by a family of exoribonucleases in arabidopsis. Science 321:1490–1492

366. Zhang Y, Liu D, Chen X, Li J, Li L, Bian Z, Sun F, Lu J, Yin Y, Cai X, Sun Q, Wang K, Ba Y, Wang Q, Wang D, Yang J, Liu P, Xu T, Yan Q, Zhang J, Zen K, Zhang CY (2010) Secreted monocytic miR-150 enhances targeted endothelial cell migration. Molecular Cell 39:133–144

367. Thum T, Catalucci D, Bauersachs J (2008) MicroRNAs: novel regulators in cardiac development and disease. Cardiovascular Research 79:562–570

368. Hutvagner G, Simard MJ (2008) Argonaute proteins: key players in RNA silencing. Nature Reviews – Molecular Cell Biology 9:22–32

369. Cifuentes D, Xue H, Taylor DW, Patnode H, Mishima Y, Cheloufi S, Ma E, Mane S, Hannon GJ, Lawson ND, Wolfe SA, Giraldez AJ (2010) A novel miRNA processing pathway independent of Dicer requires Argonaute2 catalytic activity. Science 328:1694–1698

370. Cheloufi S, Dos Santos CO, Chong MMW, Hannon GJ (2010) A dicer-independent miRNA biogenesis pathway that requires Ago catalysis. Nature 465:584–589

371. Krebs AR, Demmers J, Karmodiya K, Chang NC, Chang AC, Tora L (2010) ATAC and Mediator coactivators form a stable complex and regulate a set of non-coding RNA genes. EMBO Reports 11:541–547

372. Nilsen TW, Graveley BR (2010) Expansion of the eukaryotic proteome by alternative splicing Nature 463:457–463

373. Stamm S (2002) Signals and their transduction pathways regulating alternative splicing: a new dimension of the human genome. Human Molecular Genetics 11:2409–2416

374. Blaustein M, Pelisch F, Coso OA, Bissell MJ, Kornblihtt AR, Srebrow A (2004) Mammary epithelial-mesenchymal interaction regulates fibronectin alternative splicing via phosphatidylinositol 3-kinase. Journal of Biological Chemistry 279:21029–21037

375. Bessonov S, Anokhina M, Will CL, Urlaub H, Lührman R (2008) Isolation of an active step I spliceosome and composition of its RNP core. Nature 452:846–850

376. Tarn WY, Lee KR, Cheng SC (1993) Yeast precursor mRNA processing protein PRP19 associates with the spliceosome concomitant with or just after dissociation of U4 small nuclear RNA. Proceedings of the National Academy of Sciences of the United States of America 90:10821–10825

377. Keren H, Lev-Maor G, Ast G (2010) Alternative splicing and evolution: diversification, exon definition and function. Nature Reviews – Genetics 11:345–355

378. Chanfreau GF (2010) A dual role for RNA splicing signals. EMBO Reports 11:720–721

379. Chen M, Manley JL (2009) Mechanisms of alternative splicing regulation: insights from molecular and genomics approaches. Nature Reviews – Molecular Cell Biology 10:741–754

380. Lin S, Coutinho-Mansfield G, Wang D, Pandit S, Fu XD (2008) The splicing factor SC35 has an active role in transcriptional elongation. Nature – Structural and Molecular Biology 15:819–826

381. Abelson J (2008) Is the spliceosome a ribonucleoprotein enzyme? Nature – Structural and Molecular Biology 15:1235–1237

382. Bellare P, Small EC, Huang X, Wohlschlegel JA, Staley JP, Sontheimer EJ (2008) A role for ubiquitin in the spliceosome assembly pathway. Nature – Structural and Molecular Biology 15:444–451

383. Pessa HKJ, Will CL, Meng X, Schneider C, Watkins NJ, Perälä N, Nymark M, Turunen JJ, Lührmann R, Frilander MJ (2008) Minor spliceosome components are predominantly localized in the nucleus. Proceedings of the National Academy of Sciences of the United States of America 105:8655-8660

384. Jackson RJ, Hellen CUT, Pestova TV (2010) The mechanism of eukaryotic translation initiation and principles of its regulation. Nature Reviews – Molecular Cell Biology 11:113–127

385. Gandin V, Miluzio A, Barbieri AM, Beugnet A, Kiyokawa H, Marchisio PC, Biffo S (2008) Eukaryotic initiation factor 6 is rate-limiting in translation, growth and transformation. Nature 455:684–688

386. Isken O, Kim YK, Hosoda N, Mayeur GL, Hershey JW, Maquat LE (2008) Upf1 phosphorylation triggers translational repression during nonsense-mediated mRNA decay. Cell 133:314–327

387. Ma XM, Yoon SO, Richardson CJ, Jülich K, Blenis J (2008) SKAR links pre-mRNA splicing to mTOR/S6K1-mediated enhanced translation efficiency of spliced mRNAs. Cell 133:303–313

388. Heck JW, Cheung SK, Hampton RY (2010) Cytoplasmic protein quality control degradation mediated by parallel actions of the E3 ubiquitin ligases Ubr1 and San1. Proceedings of the National Academy of Sciences of the United States of America 107:1106–1111

389. Bengtson MH, Joazeiro CAP (2010) Role of a ribosome-associated E3 ubiquitin ligase in protein quality control. Nature 467:470–473

390. Zaher HS, Green R (2009) Quality control by the ribosome following peptide bond formation. Nature 457:161–166

391. Shyu AB, Wilkinson MF, van Hoof A (2008) Messenger RNA regulation: to translate or to degrade. EMBO Journal 27:471–481

392. Okiyoneda T, Barrière H, Bagdány M, Rabeh WM, Du K, Höhfeld J, Young JC, Lukacs GL (2010) Peripheral protein quality control removes unfolded CFTR from the plasma membrane. Science 329:805–810

393. Hutt D, Balch WE (2010) The proteome in balance. Science 329:766–767

394. Tcherkezian J, Brittis PA, Thomas F, Roux PP, Flanagan JG (2010) Transmembrane receptor DCC associates with protein synthesis machinery and regulates translation. Cell 141:632–644

395. Pechmann S, Levy ED, Tartaglia GG, Vendruscolo M (2009) Physicochemical principles that regulate the competition between functional and dysfunctional association of proteins. Proceedings of the National Academy of Sciences of the United States of America 106:10159–10164

396. Kim SJ, Born B, Havenith M, Gruebele M (2008) Real-time detection of protein-water dynamics upon protein folding by Terahertz absorption. Angewandte Chemie 47:6486–6489

397. Itoh K, Sasai M (2008) Cooperativity, connectivity, and folding pathways of multidomain proteins. Proceedings of the National Academy of Sciences of the United States of America 105:13865–13870

398. Papoian GA (2008) Proteins with weakly funneled energy landscapes challenge the classical structure-function paradigm. Proceedings of the National Academy of Sciences of the United States of America 105:14237–14238

399. Aragón T, van Anken E, Pincus D, Serafimova I, Korennykh AV, Rubio CA, Walter P (2009) Messenger RNA targeting to endoplasmic reticulum stress signalling sites. Nature 457:736–740

400. Korennykh AV, Egea PF, Korostelev AA, Finer-Moore J, Zhang C, Shokat KM, Stroud RM, Walter P (2009) The unfolded protein response signals through high-order assembly of Ire1. Nature 457:687–693

401. Kaufman RJ (1999) Stress signaling from the lumen of the endoplasmic reticulum: coordination of gene transcriptional and translational controls. Genes and Development 13:1211-1233

402. Park SW, Zhou Y, Lee J, Lu A, Sun C, Chung J, Ueki K, Ozcan U (2010) The regulatory subunits of PI3K, p85α and p85β, interact with XBP-1 and increase its nuclear translocation. Nature – Medicine 16:429–437

403. Holmberg CI, Tran SE, Eriksson JE, Sistonen L (2002) Multisite phosphorylation provides sophisticated regulation of transcription factors. Trends in Biochemical Sciences 27:619–627

404. McBride AE, Silver PA (2001) State of the Arg: protein methylation at arginine comes of age. Cell 106:5–8

405. Huang J, Berger SL (2008) The emerging field of dynamic lysine methylation of non-histone proteins. Current Opinion in Genetics and Development 18:152–158

406. Jansson M, Durant ST, Cho EC, Sheahan S, Edelmann M, Kessler B, La Thangue NB (2008) Arginine methylation regulates the p53 response. Nature – Cell Biology 10:1431–1439

407. Stamler JS, Hess DT (2010) Nascent nitrosylases. Nature – Cell Biology 12:1024–1026

408. Kornberg MD, Sen N, Hara MR, Juluri KR, Nguyen JV, Snowman AM, Law L, Hester LD, Snyder SH (2010) GAPDH mediates nitrosylation of nuclear proteins. Nature – Cell Biology 12:1094–1100

409. Janke C, Rogowski K, van Dijk J (2008) Polyglutamylation: a fine-regulator of protein function? EMBO Reports 9:636–641

410. Moore KL (2003) The biology and enzymology of protein tyrosine O-sulfation. Journal of Biological Chemistry 278:24243–24246

411. Ravid T, Hochstrasser M (2008) Diversity of degradation signals in the ubiquitin–proteasome system. Nature Reviews – Molecular Cell Biology 9:679–689

412. Ikeda F, Dikic I (2008) Atypical ubiquitin chains: new molecular signals. EMBO Reports 9:536–542

413. Rotin D, Kumar S (2009) Physiological functions of the HECT family of ubiquitin ligases. Nature Reviews – Molecular Cell Biology 10:398–409

414. Sun SC (2008) Deubiquitylation and regulation of the immune response. Nature Reviews – Immunology 8:501–511

415. Rabut G, Peter M (2008) Function and regulation of protein neddylation. EMBO Reports 9:969–976

416. Pedrioli PGA, Leidel S, Hofmann K (2008) Urm1 at the crossroad of modifications. EMBO Reports 9:1196–1202

417. Kinoshita T, Fujita M, Maeda Y (2008) Biosynthesis, remodelling and functions of mammalian GPI-anchored proteins: recent progress. Journal of Biochemistry 144:287–294

418. Laczy B, Hill BG, Wang K, Paterson AJ, White CR, Xing D, Chen YF, Darley-Usmar V, Oparil S, Chatham JC (2009) Protein O-GlcNAcylation: a new signaling paradigm for the cardiovascular system. American Journal of Physiology – Heart and Circulatory Physiology 296:H13–H28

419. Golks A, Guerini D (2008) The O-linked N-acetylglucosamine modification in cellular signalling and the immune system. EMBO Reports 9:748–753

420. Wang Z, Gucek M, Hart GW (2008) Cross-talk between GlcNAcylation and phosphorylation: Site-specific phosphorylation dynamics in response to globally elevated O-GlcNAc. Proceedings of the National Academy of Sciences of the United States of America 105:13793–13798

421. Marsh SA, Chatham JC (2011) The paradoxical world of protein O-GlcNAcylation: a novel effector of cardiovascular (dys)function. Cardiovascular Research 89:487–488

422. Lima VV, Giachini FR, Carneiro FS, Carvalho MH, Fortes ZB, Webb RC, Tostes RC (2011) O-GlcNAcylation contributes to the vascular effects of ET-1 via activation of the RhoA/Rho-kinase pathway. Cardiovascular Research 89:614–622

423. Hakmé A, Wong HK, Dantzer F, Schreiber V (2008) The expanding field of poly(ADP-ribosyl)ation reactions. EMBO Reports 9:1094–1100

424. Miyagi T, Wada T, Yamaguchi K, Hata K, Shiozaki K (2008) Plasma membrane-associated sialidase as a crucial regulator of transmembrane signalling. Journal of Biochemistry 144:279–285

425. Guo Z, Wu YW, Das D, Delon C, Cramer J, Yu S, Thuns S, Lupilova N, Waldmann H, Brunsveld L, Goody RS, Alexandrov K, Blankenfeldt W (2008) Structures of RabGGTase-substrate/product complexes provide insights into the evolution of protein prenylation. EMBO Journal 27:2444–2456

Volume I. Chap. 11. Cell Cytoskeleton

426. Fletcher DA, Mullins RD (2010) Cell mechanics and the cytoskeleton. Nature 463:485–492

427. Detournay E, Cheng AHD (1993) Fundamentals of Poroelasticity. In Fairhurst C (ed.) Comprehensive Rock Engineering: Principles, Practice and Projects, Vol. II, Analysis and Design Method, Pergamon Press, New York

428. Cowin SC (1999) Bone poroelasticity. Journal of Biomechanics 32:217–238

429. Miyake K, McNeil PL, Suzuki K, Tsunoda R, Sugai N (2001) An actin barrier to resealing. Journal of Cell Science 114:3487–3494

430. de Brabander MJ (1993) Le cytosquelette et la vie cellulaire. La Recherche 145:810–820

431. Ingber DE (2006) Cellular mechanotransduction: putting all the pieces together again. FASEB Journal 20:811–827

432. Matthews BD, Overby DR, Mannix R, Ingber DE (2006) Cellular adaptation to mechanical stress: role of integrins, Rho, cytoskeletal tension, and mechanosensitive ion channels. Journal of Cell Science 119, 508–518

433. Hu S, Chen J, Fabry B, Numaguchi Y, Gouldstone A, Ingber DE, Fredberg JJ, Butler JP, Wang N (2003) Intracellular stress tomography reveals stress focusing and structural anisotropy in cytoskeleton of living cells. American Journal of Physiology – Cell Physiology 285:C1082–C1090

434. Meyer CJ, Alenghat FJ, Rim P, Fong JH, Fabry B, Ingber DE (2000) Mechanical control of cyclic AMP signalling and gene transcription through integrins. Nature – Cell Biology 2:666-668

435. Na S, Collin O, Chowdhury F, Tay B, Ouyang M, Wang Y, Wang N (2008) Rapid signal transduction in living cells is a unique feature of mechanotransduction. Proceedings of the National Academy of Sciences of the United States of America 105:6626–6631

436. Otey CA, Rachlin A, Moza M, Arneman D, Carpen O (2005) The palladin/myotilin/myopalladin family of actin-associated scaffolds. International Review of Cytology 246:31–58

437. Ivetic A, Ridley AJ (2004) Ezrin/radixin/moesin proteins and Rho GTPase signalling in leucocytes. Immunology 112:165–176

438. Matsuoka Y, Li X, Bennett V (2000) Adducin: structure, function and regulation. Cellular and Molecular Life Sciences 57:884–895

439. Soldati T, Schliwa M (2006) Powering membrane traffic in endocytosis and recycling. Nature Reviews – Molecular Cell Biology 7:897–908

440. Taylor DL, Wang YL (1978) Molecular cytochemistry: incorporation of fluorescently labeled actin into living cells. Proceedings of the National Academy of Sciences of the United States of America 75:857–861

441. Pavalko FM, Otey CA (1994) Role of adhesion molecule cytoplasmic domains in mediating interactions with the cytoskeleton. Proceedings of the Society for Experimental Biology and Medicine 205:282–293

442. de Lanerolle P, Johnson T, Hofmann WA (2005) Actin and myosin I in the nucleus: what next? Nature – Structural and Molecular Biology 12:742–746

443. Zhao K, Wang W, Rando OJ, Xue Y, Swiderek K, Kuo A, Crabtree GR (1998) Rapid and phosphoinositol-dependent binding of the SWI/SNF-like BAF complex to chromatin after T lymphocyte receptor signaling. Cell 95:625–636

444. Pestic-Dragovich L, Stojiljkovic L, Philimonenko AA, Nowak G, Ke Y, Settlage RE, Shabanowitz J, Hunt DF, Hozak P, de Lanerolle P (2000) A myosin I isoform in the nucleus. Science 290:337–341

445. Gross SR, Kinzy TG (2005) Translation elongation factor 1A is essential for regulation of the actin cytoskeleton and cell morphology. Nature – Structural and Molecular Biology 12:772–778

446. Belyantseva IA, Perrin BJ, Sonnemann KJ, Zhu M, Stepanyan R, McGee J, Frolenkov GI, Walsh EJ, Friderici KH, Friedman TB, Ervasti JM (2009) γ-Actin is required for cytoskeletal maintenance but not development. Proceedings of the National Academy of Sciences of the United States of America 106:9703–9708

447. Weygand-Durasevic I, Ibba M (2010) New roles for codon usage. Science 329:1473–1474

448. Zhang F, Saha S, Shabalina SA, Kashina A (2010) Differential arginylation of actin isoforms is regulated by coding sequence–dependent degradation. Science 329:1534–1537

449. Onfelt B, Nedvetzki S, Benninger RK, Purbhoo MA, Sowinski S, Hume AN, Seabra MC, Neil MA, French PM, Davis DM (2006) Structurally distinct membrane nanotubes between human macrophages support long-distance vesicular traffic or surfing of bacteria. Journal of Immunology 177:8476-8483

450. Stossel TP (1993) On the crawling of animal cells. Science 260:1086–1094

451. Chhabra ES, Higgs HN (2007) The many faces of actin: matching assembly factors with cellular structures. Nature – Cell Biology 9:1110–1121

452. Parekh SH, Chaudhuri O, Theriot JA, Fletcher DA (2005) Loading history determines the velocity of actin-network growth. Nature – Cell Biology 7:1219–1223

453. Chesarone MA, DuPage AG, Goode BL (2010) Unleashing formins to remodel the actin and microtubule cytoskeletons. Nature Reviews – Molecular Cell Biology 11:62–74

454. Campellone KG, Welch MD (2010) A nucleator arms race: cellular control of actin assembly. Nature Reviews – Molecular Cell Biology 11:237–251

455. Oda T, Iwasa M, Aihara T, Maéda Y, Narit A (2009) The nature of the globular- to fibrous-actin transition. Nature 457:441–445

456. Olson EN, Nordheim A (2010) Linking actin dynamics and gene transcription to drive cellular motile functions. Nature Reviews – Molecular Cell Biology 11:353-365

457. Kuwahara K, Barrientos T, Pipes GC, Li S, Olson EN (2005) Muscle-specific signaling mechanism that links actin dynamics to serum response factor. Molecular and Cellular Biology 25:3173-3181

458. Barrientos T, Frank D, Kuwahara K, Bezprozvannaya S, Pipes GC, Bassel-Duby R, Richardson JA, Katus HA, Olson EN, Frey N (2007) Two novel members of the ABLIM [?] protein family, ABLIM-2 and -3, associate with STARS and directly bind F-actin. Journal of Biological Chemistry 282:8393–8403

459. Petit MM, Meulemans SM, Van de Ven WJ (2003) The focal adhesion and nuclear targeting capacity of the LIM-containing lipoma-preferred partner (LPP) protein. Journal of Biological Chemistry 278:2157–2168

460. Gorenne I, Nakamoto RK, Phelps CP, Beckerle MC, Somlyo AV, Somlyo AP (2003) LPP, a LIM protein highly expressed in smooth muscle. American Journal of Physiology – Cell Physiology 285:C674–C685

461. Schreiber V, Moog-Lutz C, Régnier CH, Chenard MP, Boeuf H, Vonesch JL, Tomasetto C, Rio MC (1998) Lasp-1, a novel type of actin-binding protein accumulating in cell membrane extensions. Molecular Medicine 4:675–687

462. Achard V, Martiel JL, Michelot A, Guérin C, Reymann AC, Blanchoin L, Boujemaa-Paterski R (2010) A "primer"-based mechanism underlies branched actin filament network formation and motility. Current Biology 20:423–428

463. Chereau D, Boczkowska M, Skwarek-Maruszewska A, Fujiwara I, Hayes DB, Rebowski G, Lappalainen P, Pollard TD, Dominguez R (2008) Leiomodin is an actin filament nucleator in muscle cells. Science 320:239–243

464. Takeya R, Taniguchi K, Narumiya S, Sumimoto H (2008) The mammalian formin FHOD1 is activated through phosphorylation by ROCK and mediates thrombin-induced stress fibre formation in endothelial cells. The EMBO Journal 27:618–628

465. Lincoln TM (2007) Myosin phosphatase regulatory pathways: different functions or redundant functions? Circulation Research 100:10–12

466. Wagner B,Tharmann R, Haase I, Fischer M, Bausch AR (2006) Cytoskeletal polymer networks: The molecular structure of cross-linkers determines macroscopic properties. Proceedings of the National Academy of Sciences of the United States of America 103:13974–13978

467. Kueh HY, Brieher WM, Mitchison TJ (2008) Dynamic stabilization of actin filaments. Proceedings of the National Academy of Sciences of the United States of America 105:16531–16536

468. Ma Y, Machesky LM (2010) Fascin UCSD-Nature Molecule Pages, UCSD-Nature Signaling Gateway (www.signaling-gateway.org)

469. Lin CS, Shen W, Chen ZP, Tu YH, Matsudaira P (1994) Identification of I-plastin, a human fimbrin isoform expressed in intestine and kidney. Molecular and Cellular Biology 14:2457–2467

470. Delanote V, Vandekerckhove J, Gettemans J (2005) Plastins: versatile modulators of actin organization in (patho)physiological cellular processes. Acta Pharmacologica Sinica 26:769–779

471. Pollard TD, Blanchoin L, Mullins RD (2001) Actin dynamics. Journal of Cell Science 114:3577–3579

472. Cao H, Weller S, Orth JD, Chen J, Huang B, Chen JL, Stamnes M, McNiven MA (2005) Actin and Arf1-dependent recruitment of a cortactin-dynamin complex to the Golgi regulates post-Golgi transport. Nature – Cell Biology 7:483–492.

473. Hernandez-Valladares M, Kim T, Kannan B, Tung A, Aguda AH, Larsson M, Cooper JA, Robinson RC (2010) Structural characterization of a capping protein interaction motif defines a family of actin filament regulators. Nature – Structural and Molecular Biology 17:497-503

474. McCullough BR, Blanchoin L, Martiel JL, De la Cruz EM (2008) Cofilin increases the bending flexibility of actin filaments: implications for severing and cell mechanics. Journal of Molecular Biology 381:550–558

475. Pfaendtner J, De La Cruz EM, Voth GA (2010) Actin filament remodeling by actin depolymerization factor/cofilin. Proceedings of the National Academy of Sciences of the United States of America 107:7299–7304

476. Gohla A, Birkenfeld J, Bokoch GM (2004) Chronophin, a novel HAD-type serine protein phosphatase, regulates cofilin-dependent actin dynamics. Nature – Cell Biology 7:21–29

477. Han L, Stope M, López de Jesús M, Oude Weernink PA, Urban M, Wieland T, Rosskopf D, Mizuno K, Jakobs KH, Schmidt M (2007) Direct stimulation of receptor-controlled phospholipase D1 by phospho-cofilin. EMBO Journal 26:4189–4202

478. Soosairajah J, Maiti S, Wiggan O, Sarmiere P, Moussi N, Sarcevic B, Sampath R, Bamburg JR, Bernard O (2005) Interplay between components of a novel LIM kinase-slingshot phosphatase complex regulates cofilin. EMBO Journal 24:473–486

479. Hansen SD, Mullins RD (2010) VASP is a processive actin polymerase that requires monomeric actin for barbed end association. Journal of Cell Biology 191:571–584

480. Nolen BJ, Tomasevic N, Russell A, Pierce DW, Jia Z, McCormick CD, Hartman J, Sakowicz R, Pollard TD (2009) Characterization of two classes of small molecule inhibitors of Arp2/3 complex. Nature 460:1031–1034

481. Thrasher AJ, Burns SO (2010) WASP: a key immunological multitasker. Nature Reviews – Immunology 10:182–192

482. Gomez TS, Billadeau DD (2010) WASH UCSD-Nature Molecule Pages, UCSD-Nature Signaling Gateway (www.signaling-gateway.org)

483. Yu X, Machesky LM (2010) N-WASP UCSD-Nature Molecule Pages, UCSD-Nature Signaling Gateway (www.signaling-gateway.org)

484. Takano K, Watanabe-Takano H, Suetsugu S, Kurita S, Tsujita K, Kimura S, Karatsu T, Takenawa T, Endo T (2010) Nebulin and N-WASP cooperate to cause IGF-1-induced sarcomeric actin filament formation. Science 330:1536–1540

485. Pollard TD (2011) Formin tip tracking. Science 331:39–41

486. Mizuno H, Higashida C, Yuan Y, Ishizaki T, Narumiya S, Watanabe N (2011) Rotational movement of the formin mDia1 along the double helical strand of an actin filament. Science 331:80–83

487. Sun H, Schlondorff JS, Brown EJ, Higgs HN, Pollak MR (2011) Rho activation of mDia formins is modulated by an interaction with inverted formin 2 (INF2). Proceedings of the National Academy of Sciences of the United States of America 108:2933–2938

488. Quinlan ME, Heuser JE, Kerkhoff E, Mullins DR (2005) Drosophila Spire is an actin nucleation factor. Nature 433:382–388

489. Lin YC, Redmond L (2008) CaMKIIβ binding to stable F-actin in vivo regulates F-actin filament stability. Proceedings of the National Academy of Sciences of the United States of America 105:15791–15796

490. Machacek M, Hodgson L, Welch C, Elliott H, Pertz O, Nalbant P, Abell A, Johnson GL, Hahn KM, Danuser G (2009) Coordination of Rho GTPase activities during cell protrusion. Nature 461:99–103

491. Buday L, Wunderlich L, Tamás P (2002) The Nck family of adapter proteins: regulators of actin cytoskeleton. Cellular Signalling 14:723–731

492. Li W, Fan J, Woodley DT (2001) Nck/Dock: an adapter between cell surface receptors and the actin cytoskeleton. Oncogene 20:6403–6417

493. Rohatgi R, Nollau P, Ho HY, Kirschner MW, Mayer BJ (2001) Nck and phosphatidylinositol 4,5-bisphosphate synergistically activate actin polymerization through the N-WASP–Arp2/3 pathway. Journal of Biological Chemistry 276:26448–26452

494. Baumgartner M, Sillman AL, Blackwood EM, Srivastava J, Madson N, Schilling JW, Wright JH, Barber DL (2006) The Nck-interacting kinase phosphorylates ERM proteins for formation of lamellipodium by growth factors. Proceedings of the National Academy of Sciences of the United States of America 103:13391–13396

495. Gu C, Yaddanapudi S, Weins A, Osborn T, Reiser J, Pollak M, Hartwig J, Sever S (2010) Direct dynamin–actin interactions regulate the actin cytoskeleton. EMBO Journal 29:3593–3606

496. Roux A, Plastino J (2010) Actin takes its hat off to dynamin. EMBO Journal 29:3591–3592

497. Delettre C, Lenaers G, Griffoin JM, Gigarel N, Lorenzo C, Belenguer P, Pelloquin L, Grosgeorge J, Turc-Carel C, Perret E, Astarie-Dequeker C, Lasquellec L, Arnaud B, Ducommun B, Kaplan J, Hamel CP (2000) Nuclear gene

OPA1, encoding a mitochondrial dynamin-related protein, is mutated in dominant optic atrophy. Nature – Genetics 26:207–210

498. Hodge T, Cope MJTV (2000) A myosin family tree. Journal of Cell Science 113:3353–3354 (Myosin homepage: www.mrc-lmb.cam.ac.uk/myosin)

499. Berg JS, Powell BC, Cheney RE (2001) A millennial myosin census. Molecular Biology of the Cell 12:780–794

500. Wang A, Ma X, Conti MA, Liu C, Kawamoto S, Adelstein RS (2010) Nonmuscle myosin II isoform and domain specificity during early mouse development. Proceedings of the National Academy of Sciences of the United States of America 107:14645–14650

501. Brawley CM, Rock RS (2009) Unconventional myosin traffic in cells reveals a selective actin cytoskeleton. Proceedings of the National Academy of Sciences of the United States of America 106:9685–9690

502. Laakso JM, Lewis JH, Shuman H, Ostap EM (2008) Myosin I can act as a molecular force sensor. Science 321:133–136

503. Laakso JM, Lewis JH, Shuman H, Ostap EM (2010) Control of myosin-I force sensing by alternative splicing. Proceedings of the National Academy of Sciences of the United States of America 107:698–702

504. Milton DL, Schneck AN, Ziech DA, Ba M, Facemyer KC, Halayko AJ, Baker JE, Gerthoffer WT, Cremo CR (2011) Direct evidence for functional smooth muscle myosin II in the 10S self-inhibited monomeric conformation in airway smooth muscle cells. Proceedings of the National Academy of Sciences of the United States of America 108:1421–1426

505. Resnicow DI, Deacon JC, Warrick HM, Spudich JA, Leinwand LA (2010) Functional diversity among a family of human skeletal muscle myosin motors. Proceedings of the National Academy of Sciences of the United States of America 107:1053–1058

506. Kovács M, Thirumurugan K, Knight PJ, Sellers JR (2007) Load-dependent mechanism of nonmuscle myosin 2. Proceedings of the National Academy of Sciences of the United States of America 104:9994–9999

507. Dietz ML, Bernaciak TM, Vendetti F, Kielec JM, Hildebrand JD (2006) Differential actin-dependent localization modulates the evolutionarily conserved activity of Shroom family proteins. Journal of Biological Chemistry 281:20542–20554

508. Walcott S, Sun SX (2010) A mechanical model of actin stress fiber formation and substrate elasticity sensing in adherent cells. Proceedings of the National Academy of Sciences of the United States of America 107:7757–7762

509. Rauzi M, Lenne PF, Lecuit T (2010) Planar polarized actomyosin contractile flows control epithelial junction remodelling. Nature 468:1110–1114

510. Wilson CA, Tsuchida MA, Allen GM, Barnhart EL, Applegate KT, Yam PT, Ji L, Keren K, Danuser G, Theriot JA (2010) Myosin II contributes to cell-scale actin network treadmilling through network disassembly. Nature 465:373–377

511. Craig EM, Linke H (2009) Mechanochemical model for myosin V. Proceedings of the National Academy of Sciences of the United States of America 106:18261–18266

512. Phichith D, Travaglia M, Yang Z, Liu X, Zong AB, Safer D, Sweeney HL (2009) Cargo binding induces dimerization of myosin VI. Proceedings of the National Academy of Sciences of the United States of America 106:17320–17324

513. Sweeney HL, Park H, Zong AB, Yang Z, Selvin PR, Rosenfeld SS (2007) How myosin VI coordinates its heads during processive movement. EMBO Journal 26:2682–2692

514. Spudich JA, Sivaramakrishnan S (2010) Myosin VI: an innovative motor that challenged the swinging lever arm hypothesis. Nature Reviews – Molecular Cell Biology 11:128–137

515. Reifenberger JG, Toprak E, Kim H, Safer D, Sweeney HL, Selvin PR (2009) Myosin VI undergoes a 180 degrees power stroke implying an uncoupling of the front lever arm. Proceedings of the National Academy of Sciences of the United States of America 106:18255–18260

516. Sun Y, Schroeder HW 3rd, Beausang JF, Homma K, Ikebe M, Goldman YE (2007) Myosin VI walks "wiggly" on actin with large and variable tilting. Molecular Cell 28:954–964

517. Sun Y, Schroeder HW 3rd, Beausang JF, Homma K, Ikebe M, Goldman YE (2010) Myosin VI lever arm rotation: fixed or variable? Proceedings of the National Academy of Sciences of the United States of America 107:E63

518. Dunn AR, Chuan P, Bryant Z, Spudich JA (2010) Contribution of the myosin VI tail domain to processive stepping and intramolecular tension sensing. Proceedings of the National Academy of Sciences of the United States of America 107:7746–7750

519. Breshears LM, Wessels D, Soll DR, Titus MA (2010) An unconventional myosin required for cell polarization and chemotaxis. Proceedings of the National Academy of Sciences of the United States of America 107:6918–6923

520. Sakai T, Umeki N, Ikebe R, Ikebe M. (2011) Cargo binding activates myosin VIIA motor function in cells. Proceedings of the National Academy of Sciences of the United States of America 108:7028–7033

521. Hanley PJ, Xu Y, Kronlage M, Grobe K, Schön P, Song J, Sorokin L, Schwab A, Bähler M (2010) Motorized RhoGAP myosin IXb (Myo9b) controls cell shape and motility. Proceedings of the National Academy of Sciences of the United States of America 107:12145–12150

522. Sun Y, Sato O, Ruhnow F, Arsenault ME, Ikebe M, Goldman YE (2010) Single-molecule stepping and structural dynamics of myosin X. Nature – Structural and Molecular Biology 17:485–491

523. Zoghbi ME, Woodhead JL, Moss RL, Craig R (2008) Three-dimensional structure of vertebrate cardiac muscle myosin filaments. Proceedings of the National Academy of Sciences of the United States of America 105:2386–2390

524. Glover DM, Gonzalez C, Raff JW (1993) The centrosome. Scientific American 268:62–68

525. Kodani A, Sütterlin C (2009) A new function for an old organelle: microtubule nucleation at the Golgi apparatus. EMBO Journal 28:995–996

526. Efimov A, Kharitonov A, Efimova N, Loncarek J, Miller PM, Andreyeva N, Gleeson P, Galjart N, Maia AR, McLeod IX, Yates JR, Maiato H, Khodjakov A, Akhmanova A, Kaverina I (2007) Asymmetric CLASP-dependent nucleation of noncentrosomal microtubules at the trans-Golgi network. Developmental Cell 12:917–930

527. Rivero S, Cardenas J, Bornens M, Rios RM (2009) Microtubule nucleation at the cis-side of the Golgi apparatus requires AKAP450 and GM130. EMBO Journal 28:1016–1028

528. Pizon V, Gerba F, Diaz CC, Karsenti E (2005) Microtubule-dependent transport and organization of sarcomeric myosin during skeletal muscle differentiation. EMBO Journal 24:3781–3792

529. Dimitrov A, Quesnoit M, Moutel S, Cantaloube I, Poüs C, Perez F (2008) Detection of GTP-tubulin conformation in vivo reveals a role for GTP remnants in microtubule rescues. Science 322:1353–1356

530. Rice LM, Montabana EA, Agard DA (2008) The lattice as allosteric effector: structural studies of $\alpha\beta$- and γ-tubulin clarify the role of GTP in microtubule assembly. Proceedings of the National Academy of Sciences of the United States of America 105:5378–5383

531. Wang HW, Nogales E (2005) Nucleotide-dependent bending flexibility of tubulin regulates microtubule assembly. Nature 435:911–915

532. Akhmanova A, Steinmetz MO (2008) Tracking the ends: a dynamic protein network controls the fate of microtubule tips. Nature Reviews – Molecular Cell Biology 9:309–322

533. Allen RD (1987) The microtubule as an intracellular engine. Scientific American 356:42–49

534. Kuli IM, Brown AEX, Kim H, Kural C, Blehm B, Selvin PR, Nelson PC, Gelfand VI (2008) The role of microtubule movement in bidirectional organelle transport. Proceedings of the National Academy of Sciences of the United States of America 105:10011–10016

535. Ali MY, Lu H, Bookwalter CS, Warshaw DM, Trybus KM (2008) Myosin V and Kinesin act as tethers to enhance each others' processivity. Proceedings of the National Academy of Sciences of the United States of America 105:4691–4696

536. Frederick RL, Shaw JM (2007) Moving mitochondria: establishing distribution of an essential organelle. Traffic 8:1668–1675

537. Fransson S, Ruusala A, Aspenström P (2006) The atypical Rho GTPases Miro-1 and Miro-2 have essential roles in mitochondrial trafficking. Biochemical and Biophysical Research Communications 344:500–510

538. Laib JA, Marin JA, Bloodgood RA, Guilford WH (2009) The reciprocal coordination and mechanics of molecular motors in living cells. Proceedings of the National Academy of Sciences of the United States of America 106:3190–3195

539. Shima T, Kon T, Imamula K, Ohkura R, Sutoh K (2006) Two modes of microtubule sliding driven by cytoplasmic dynein. Proceedings of the National Academy of Sciences of the United States of America 103:17736–17740

540. Hanson PI, Whiteheart SW (2005) AAA+ proteins: have engine, will work. Nature Reviews – Molecular Cell Biology 6:519–529

541. Tanenbaum ME, Akhmanova A, Medema RH (2010) Dynein at the nuclear envelope. EMBO Reports 11:649

542. Li J, Lee WL, Cooper JA (2005) NudEL targets dynein to microtubule ends through LIS1. Nature – Cell Biology 7:686–690

543. Kardon JR, Vale RD (2009) Regulators of the cytoplasmic dynein motor. Nature Reviews – Molecular Cell Biology 10:854–865

544. Hirokawa N, Noda Y, Tanaka Y, Niwa S (2009) Kinesin superfamily motor proteins and intracellular transport. Nature Reviews – Molecular Cell Biology 10:682–696

545. Miki H, Setou M, Kaneshiro K, Hirokawa N (2001) All kinesin superfamily protein, KIF, genes in mouse and human. Proceedings of the National Academy of Sciences of the United States of America 98:7004–7011

546. Wu G, Zhou L, Khidr L, Guo XE, Kim W, Lee YM, Krasieva T, Chen PL (2008) A novel role of the chromokinesin Kif4A in DNA damage response. Cell Cycle 7:2013–2020

547. Mazumdar M, Misteli T (2005) Chromokinesins: multitalented players in mitosis. Trends in Cell Biology 15:349–355

548. Nangaku M, Sato-Yoshitake R, Okada Y, Noda Y, Takemura R, Yamazaki H, Hirokawa N (1994) KIF1B, a novel microtubule plus end-directed monomeric motor protein for transport of mitochondria. Cell 79:1209–1220

549. Yamazaki H, Nakata T, Okada Y, Hirokawa N (1996) Cloning and characterization of KAP3: a novel kinesin superfamily-associated protein of KIF3A/3B. Proceedings of the National Academy of Sciences of the United States of America 93:8443–8448

550. Lawrence CJ, Dawe RK, Christie KR, Cleveland DW, Dawson SC, Endow SA, Goldstein LS, Goodson HV, Hirokawa N, Howard J, Malmberg RL, McIntosh JR, Miki H, Mitchison TJ, Okada Y, Reddy AS, Saxton WM, Schliwa M, Scholey JM, Vale RD, Walczak CE, Wordeman L (2004) A standardized kinesin nomenclature. Journal of Cell Biology 167:19–22

551. Goldstein LS (2001) Kinesin molecular motors: transport pathways, receptors, and human disease. Proceedings of the National Academy of Sciences of the United States of America 98:6999–7003

552. Kuo SC, Sheetz MP (1993) Force of single kinesin molecules measured with optical tweezers. Science 260:232–234

553. Seitz A, Surrey T (2006) Processive movement of single kinesins on crowded microtubules visualized using quantum dots. The EMBO Journal 25:267–277

554. Mori T, Vale RD, Tomishige M (2007) How kinesin waits between steps. Nature 450:750–754

555. Jolly AL, Kim H, Srinivasan D, Lakonishok M, Larson AG, Gelfand VI (2010) Kinesin-1 heavy chain mediates microtubule sliding to drive changes in cell shape. Proceedings of the National Academy of Sciences of the United States of America 107:12151–12156

556. Peters C, Brejc K, Belmont L, Bodey AJ, Lee Y, Yu M, Guo J, Sakowicz R, Hartman J, Moores CA (2010) Insight into the molecular mechanism of the multitasking kinesin-8 motor. EMBO Journal 29:3437–3447

557. Helenius J, Brouhard G, Kalaidzidis Y, Diez S, Howard J (2006) The depoly-merizing kinesin MCAK uses lattice diffusion to rapidly target microtubule ends. Nature 441:115–119

558. Hoepfner S, Severin F, Cabezas A Habermann B, Runge A, Gillooly D, Sten-mark H, Zerial M (2005) Modulation of receptor recycling and degradation by the endosomal kinesin KIF16B. Cell 121:437–450

559. Dixit R, Ross JL, Goldman YE, Holzbaur ELF (2008) Differential regulation of dynein and kinesin motor proteins by Tau. Science 319:1086–1089

560. Herrmann H, Bar H, Kreplak L, Strelkov SV, Aebi U (2007) Intermediate fil-aments: from cell architecture to nanomechanics. Nature Reviews – Molecular Cell Biology 8:562–573

561. Schweizer J, Bowden PE, Coulombe PA, Langbein L, Lane EB, Magin TM, Maltais L, Omary MB, Parry DA, Rogers MA, Wright MW (2006) New consensus nomenclature for mammalian keratins. Journal of Cell Biology 174:169–174

562. Srougi MC, Burridge K (2009) Undressing a cellular corset: septins exposed. Nature – Cell Biology 11:9–10

563. Tooley AJ, Gilden J, Jacobelli J, Beemiller P, Trimble WS, Kinoshita M, Krummel MF (2008) Amoeboid T lymphocytes require the septin cytoskele-ton for cortical integrity and persistent motility. Nature – Cell Biology 11:17–26

747. Zamir E, Geiger B (2001) Molecular complexity and dynamics of cell-matrix adhesions. Journal of Cell Science 114:3583–3590

565. Homan SM, Martinez R, Benware A, LaFlamme SE (2002) Regulation of the association of alpha 6 beta 4 with vimentin intermediate filaments in endothelial cells. Experimental Cell Research 281:107–114

566. Hu S, Chen J, Butler JP, Wang N (2005) Prestress mediates force propaga-tion into the nucleus. Biochemical and Biophysical Research Communications 329:423–428

567. Bode J, Goetze S, Heng H, Krawetz SA, Benham C (2003) From DNA struc-ture to gene expression: mediators of nuclear compartmentalization and dy-namics. Chromosome Research 11:435–445

568. Itano N, Okamoto S, Zhang D, Lipton SA, Ruoslahti E (2003) Cell spreading controls endoplasmic and nuclear calcium: a physical gene regulation pathway from the cell surface to the nucleus. Proceedings of the National Academy of Sciences of the United States of America 100:5181–5186

569. Lee KK, Haraguchi T, Lee RS, Koujin T, Hiraoka Y, Wilson KL (2001) Dis-tinct functional domains in emerin bind lamin A and DNA-bridging protein BAF. Journal of Cell Science 114:4567–4573

570. Dreuillet C, Tillit J, Kress M, Ernoult-Lange M (2002) In vivo and in vitro interaction between human transcription factor MOK2 and nuclear lamin A/C. Nucleic Acids Research 30:4634–4642

571. Holaska JM (2008) Emerin and the nuclear lamina in muscle and cardiac disease. Circulation Research 103:16–23

572. Georgatos SD, Blobel G (1987) Lamin B constitutes an intermediate filament attachment site at the nuclear envelope. Journal of Cell Biology 105:117–125

573. Haque F, Lloyd DJ, Smallwood DT, Dent CL, Shanahan CM, Fry AM, Trembath RC, Shackleton S (2006) SUN1 interacts with nuclear lamin A and cytoplasmic nesprins to provide a physical connection between the nuclear lamina and the cytoskeleton. Molecular and Cellular Biology 26:3738–3751

574. Lammerding J, Fong LG, Ji JY, Reue K, Stewart CL, Young SG, Lee RT (2006) Lamins A and C but not lamin B1 regulate nuclear mechanics. Journal of Biological Chemistry 281:25768–25780

575. Wilhelmsen K, Litjens SH, Kuikman I, Tshimbalanga N, Janssen H, van den Bout I, Raymond K, Sonnenberg A (2005) Nesprin-3, a novel outer nuclear membrane protein, associates with the cytoskeletal linker protein plectin. Journal of Cell Biology 171:799–810

576. Ketema M, Wilhelmsen K, Kuikman I, Janssen H, Hodzic D, Sonnenberg A (2007) Requirements for the localization of nesprin-3 at the nuclear envelope and its interaction with plectin. Journal of Cell Science 120:3384–3394

577. Roux KJ, Crisp ML, Liu Q, Kim D, Kozlov S, Stewart CL, Burke B (2009) Nesprin 4 is an outer nuclear membrane protein that can induce kinesin-mediated cell polarization. Proceedings of the National Academy of Sciences of the United States of America 106:2194–2199

578. Liu Q, Pante N, Misteli T, Elsagga M, Crisp M, Hodzic D, Burke B, Roux KJ (2007) Functional association of Sun1 with nuclear pore complexes. Journal of Cell Biology 178:785–798

579. Hodzic DM, Yeater DB, Bengtsson L, Otto H, Stahl PD (2004) Sun2 is a novel mammalian inner nuclear membrane protein. Journal of Biological Chemistry 279:25805–25812

580. Crisp M, Liu Q, Roux K, Rattner JB, Shanahan C, Burke B, Stahl PD, Hodzic D (2006) Coupling of the nucleus and cytoplasm: role of the LINC complex. Journal of Cell Biology 172:41–53

581. Salpingidou G, Smertenko A, Hausmanowa-Petrucewicz I, Hussey PJ, Hutchison CJ (2007) A novel role for the nuclear membrane protein emerin in association of the centrosome to the outer nuclear membrane. Journal of Cell Biology 178:897–904

582. Holaska JM, Rais-Bahrami S, Wilson KL (2006) Lmo7 is an emerin-binding protein that regulates the transcription of emerin and many other muscle-relevant genes. Human Molecular Genetics 15:3459-3472

583. Holaska JM, Lee KK, Kowalski AK, Wilson KL (2003) Transcriptional repressor germ cell-less (GCL) and barrier to autointegration factor (BAF) compete for binding to emerin in vitro. Journal of Biological Chemistry 278:6969–6975

584. Haraguchi T, Holaska JM, Yamane M, Koujin T, Hashiguchi N, Mori C, Wilson KL, Hiraoka Y (2004) Emerin binding to Btf, a death-promoting transcriptional repressor, is disrupted by a missense mutation that causes Emery-Dreifuss muscular dystrophy. European Journal of Biochemistry 271:1035–1045

585. Liu H, Lu ZG, Miki Y, Yoshida K (2007) Protein kinase Cδ induces transcription of the TP53 tumor suppressor gene by controlling death-promoting factor Btf in the apoptotic response to DNA damage. Molecular and Cellular Biology 27:8480–8491

586. Wilkinson FL, Holaska JM, Zhang Z, Sharma A, Manilal S, Holt I, Stamm S, Wilson KL, Morris GE (2003) Emerin interacts in vitro with the splicing-associated factor, YT521-B. European Journal of Biochemistry 270:2459–2466

587. Holaska JM, Wilson KL (2007) An emerin "proteome": purification of distinct emerin-containing complexes from HeLa cells suggests molecular basis for diverse roles including gene regulation, mRNA splicing, signaling, mechanosensing, and nuclear architecture. Biochemistry 46:8897–8908

588. Lattanzi G, Cenni V, Marmiroli S, Capanni C, Mattioli E, Merlini L, Squarzoni S, Maraldi NM (2003) Association of emerin with nuclear and cytoplasmic actin is regulated in differentiating myoblasts. Biochemical and Biophysical Research Communications 303:764–770

589. Ye J, Zhao J, Hoffmann-Rohrer U, Grummt I (2008) Nuclear myosin I acts in concert with polymeric actin to drive RNA polymerase I transcription. Genes and Development 22:322–330

590. Pederson T (2008) As functional nuclear actin comes into view, is it globular, filamentous, or both? Journal of Cell Biology 180:1061–1064

591. Vreugde S, Ferrai C, Miluzio A, Hauben E, Marchisio PC, Crippa MP, Bussi M, Biffo S (2006) Nuclear myosin VI enhances RNA polymerase II-dependent transcription. Molecular Cell 23:749–755

592. Holaska JM, Kowalski AK, Wilson KL (2004) Emerin caps the pointed end of actin filaments: evidence for an actin cortical network at the nuclear inner membrane. PLoS Biology 2:E231

593. Li R, Gundersen GG (2008) Beyond polymer polarity: how the cytoskeleton builds a polarized cell. Nature Reviews – Molecular Cell Biology 9:860–873

594. Bornens M (2008) Organelle positioning and cell polarity. Nature Reviews – Molecular Cell Biology 9:874-886

595. Iden S, Collard JG (2008) Crosstalk between small GTPases and polarity proteins in cell polarization. Nature Reviews – Molecular Cell Biology 9:846–859

596. Deshpande VS, McMeeking RM, Evans AG (2006) A bio-chemo-mechanical model for cell contractility. Proceedings of the National Academy of Sciences of the United States of America 103:14015–14020

597. Civelekoglu G, Edelstein-Keshet L (1994) Modelling the dynamics of F-actin in the cell. Bulletin of Mathematical Biology 56:587–616

598. Dembo M (1986) The mechanics of motility in dissociated cytoplasm. Biophysical Journal 50:1165–1183

599. Storm C, Pastore JJ, MacKintosh FC, Lubensky TC, Janmey PA (2005) Nonlinear elasticity in biological gels. Nature 435:191–194

600. Ingber DE, Madri JA, Jamieson JD (1981) Role of basal lamina in neoplastic disorganization of tissue architecture. Proceedings of the National Academy of Sciences of the United States of America 78:3901–3905

601. Fuller RB (1961) Tensegrity. Portfolio and ART News Annual 4:112–127

602. Stamenovic D, Fredberg JJ, Wang N, Butler JP, Ingber DE (1996) A microstructural approach to cytoskeletal mechanics based on tensegrity. Journal of Theoretical Biology 181:125–136

603. Wang N, Butler JP, Ingber DE (1993) Mechanotransduction across the cell surface and through the cytoskeleton. Science 260:1124–1127

604. Wendling S, Oddou C, Isabey D (1999) Stiffening response of a cellular tensegrity model. Journal of Theoretical Biology 196:309–325

605. Canadas P, Laurent VM, Oddou C, Isabey D, Wendling S (2002) A cellular tensegrity model to analyse the structural viscoelasticity of the cytoskeleton. Journal of Theoretical Biology 218:155–173

Volume I. Chap. 12. Plasma Membrane

606. Groves JT, Kuriyan J (2010) Molecular mechanisms in signal transduction at the membrane. Nature – Structural and Molecular Biology 17:659–665

607. Nakada C, Ritchie K, Oba Y, Nakamura M, Hotta Y, Iino R, Kasai RS, Yamaguchi K, Fujiwara T, Kusumi A (2003) Accumulation of anchored proteins forms membrane diffusion barriers during neuronal polarization. Nature – Cell Biology 5:626–632

608. Marguet D, Lenne PF, Rigneault H, He HT (2006) Dynamics in the plasma membrane: how to combine fluidity and order. EMBO Journal 25:3446–3457

609. Simons K, Gerl MJ (2010) Revitalizing membrane rafts: new tools and insights. Nature Reviews – Molecular Cell Biology 11:688–699

610. Lillemeier BF, Pfeiffer JR, Surviladze Z, Wilson BS, Davis MM (2006) Plasma membrane-associated proteins are clustered into islands attached to the cytoskeleton. Proceedings of the National Academy of Sciences of the United States of America 103:18992-18997

961. Johannes L, Lamaze C (2002) Clathrin-dependent or not: is it still the question? Traffic 3:443

612. Maguy A, Hebert TE, Nattel S (2006) Involvement of lipid rafts and caveolae in cardiac ion channel function. Cardiovascular Research 69:798–807

613. Tourte M (1998) Introduction à la biologie moléculaire [Introduction to molecular biology]. Diderot Editeur, Paris

614. Hannun YA, Obeid LM (2008) Principles of bioactive lipid signalling: lessons from sphingolipids. Nature Reviews – Molecular Cell Biology 9:139–150

615. Wymann MP, Schneiter R (2008) Lipid signalling in disease. Nature Reviews – Molecular Cell Biology 9:162–176

616. Guenther GG, Peralta ER, Rosales KR, Wong SY, Siskind LJ, Edinger AL (2008) Ceramide starves cells to death by downregulating nutrient transporter proteins. Proceedings of the National Academy of Sciences of the United States of America 105:17402–17407

617. Pappu R, Schwab SR, Cornelissen I, Pereira JP, Regard JB, Xu Y, Camerer E, Zheng YW, Huang Y, Cyster JG, Coughlin SR (2007) Promotion of lymphocyte egress into blood and lymph by distinct sources of sphingosine-1-phosphate. Science 316:295–298

618. Rosen H, Sanna G, Alfonso C (2003) Egress: a receptor-regulated step in lymphocyte trafficking. Immunological Reviews 195:160–177

619. Kermorvant-Duchemin E, Sennlaub F, Sirinyan M, Brault S, Andelfinger G, Kooli A, Germain S, Ong H, d'Orleans-Juste P, Gobeil F, Zhu T, Boisvert C, Hardy P, Falck JR, Balazy M, Chemtob S (2005) Trans-arachidonic acids generated during nitrative stress induce a thrombospondin-1–dependent microvascular degeneration. Nature – Medicine 11:1339–1345

620. Gallop JL, Butler JG, McMahon HT (2005) Endophilin and CtBP/BARS are not acyl transferases in endocytosis or Golgi fission. Nature 438:675–678

621. Yeung T, Gilbert GE, Shi J, Silvius J, Kapus A, Grinstein S (2008) Membrane phosphatidylserine regulates surface charge and protein localization. Science 319:210–213

895. Vicinanza M, D'Angelo G, Di Campli A, De Matteis AM (2008) Function and dysfunction of the PI system in membrane trafficking. EMBO Journal 27:2457–2470

623. Kraus M, Haucke V (2007) Phosphoinositide-metabolizing enzymes at the interface between membrane traffic and cell signalling. EMBO Reports 8:241–246

624. Rusten TE, Stenmark H (2006) Analyzing phosphoinositides and their interacting proteins. Nature – Methods 3:251–258

625. Balla A, Kim YJ, Varnai P, Szentpetery Z, Knight Z, Shokat KM, Balla T (2008) Maintenance of hormone-sensitive phosphoinositide pools in the plasma membrane requires phosphatidylinositol 4-kinase IIIα. Molecular Biology of the Cell 19:711–721

626. Di Paolo G, De Camilli P (2006) Phosphoinositides in cell regulation and membrane dynamics. Nature 443:651–657

627. Michell RH (2008) Inositol derivatives: evolution and functions. Nature Reviews – Molecular Cell Biology 9:151-161

628. Feramisco JD, Radhakrishnan A, Ikeda Y, Reitz J, Brown MS, Goldstein JL (2005) Intramembrane aspartic acid in SCAP protein governs cholesterol-induced conformational change. Proceedings of the National Academy of Sciences of the United States of America 102:3242–3247

629. Infante RE, Wang ML, Radhakrishnan A, Kwon HJ, Brown MS, Goldstein JL (2008) NPC2 facilitates bidirectional transfer of cholesterol between NPC1 and lipid bilayers, a step in cholesterol egress from lysosomes. Proceedings of the National Academy of Sciences of the United States of America 105:15287–15292

630. Tall AR, Costet P, Luo Y (2000) "Orphans" meet cholesterol. Nature – Medicine 6:1104–1105

631. Linsel-Nitschke P, Tall AR (2005) HDL as a target in the treatment of atherosclerotic cardiovascular disease. Nature Reviews – Drug Discovery 4:193–205

632. Kontush A, Chapman MJ (2006) Antiatherogenic small, dense HDL guardian angel of the arterial wall? Nature – Clinical Practice Cardiovascular Medicine 3:144–153

633. Tall AR, Costet P, Wang N (2002) Regulation and mechanisms of macrophage cholesterol efflux. Journal of Clinical Investigation 110:899–904

634. Chou KC, Elrod DW (1999) Prediction of membrane protein types and subcellular locations. Proteins: Structure, Function, and Bioinformatics 34:137–153

635. Mitra K, Schaffitzel C, Shaikh T, Tama F, Jenni S, Brooks CL, Ban N, Frank J (2005) Structure of the E. coli protein-conducting channel bound to a translating ribosome. Nature 438:318–324

636. Lichtenthaler SF, Steiner H (2007) Sheddases and intramembrane-cleaving proteases: RIPpers of the membrane. EMBO Reports 8:537–541

637. Bauvois B (2004) Transmembrane proteases in cell growth and invasion: new contributors to angiogenesis? Oncogene 23:317–329

638. Marks N, Berg MJ (2008) Neurosecretases provide strategies to treat sporadic and familial Alzheimer disorders. Neurochemistry International 52:184–215

639. Wolfe MS. γ-Secretase and presenilin. NCBI – Bookshelf – Madame Curie Bioscience Database – Neurodegenerative Disease (www.ncbi.nlm.nih.gov/bookshelf). Landes Bioscience and Springer Science+Business Media

640. Zampese E, Fasolato C, Kipanyula MJ, Bortolozzi M, Pozzan T, Pizzo P (2011) Presenilin 2 modulates endoplasmic reticulum (ER)-mitochondria interactions and Ca^{2+} cross-talk. Proceedings of the National Academy of Sciences of the United States of America 108:2777–2782

641. Freeman M (2008) Rhomboid proteases and their biological functions. Annual Review of Genetics 42:191–210

642. Wolfe MS (2009) Intramembrane-cleaving proteases. Journal of Biological Chemistry 284:13969–13973

643. Ye J, Davé UP, Grishin NV, Goldstein JL, Brown MS (2000) Asparagine-proline sequence within membrane-spanning segment of SREBP triggers intramembrane cleavage by site-2 protease. Proceedings of the National Academy of Sciences of the United States of America 97:5123–5128

644. Drucker DJ (2007) Dipeptidyl peptidase-4 inhibition and the treatment of type 2 diabetes: preclinical biology and mechanisms of action. Diabetes Care 30:1335–1343 (www.medscape.com/viewarticle/559076)

645. Knoblauch A, Will C, Goncharenko G, Ludwig S, Wixler V (2007) The binding of Mss4 to α-integrin subunits regulates matrix metalloproteinase activation and fibronectin remodeling. FASEB Journal 21:497–510

646. Hessa T, Kim H, Bihlmaier K, Lundin C, Boekel J, Andersson H, Nilsson I, White SH, von Heijne G (2005) Recognition of transmembrane helices by the endoplasmic reticulum translocon. Nature 433:377–381

647. Chekeni FB, Elliott MR, Sandilos JK, Walk SF, Kinchen JM, Lazarowski ER, Armstrong AJ, Penuela S, Laird DW, Salvesen GS, Isakson BE, Bayliss DA, Ravichandran K (2010) Pannexin 1 channels mediate "find-me" signal release and membrane permeability during apoptosis. Nature 467:863–867

648. Zhang F, Kotha J, Jennings LK, Zhang XA (2009) Tetraspanins and vascular functions. Cardiovascular Research 83:7–15

649. Levy S (2010) CD81. UCSD-Nature Molecule Pages, UCSD-Nature Signaling Gateway (www.signaling-gateway.org)

650. Chairoungdua A, Smith DL, Pochard P, Hull M, Caplan MJ (2010) Exosome release of β-catenin: a novel mechanism that antagonizes Wnt signaling. Journal of Cell Biology 190:1079–1091

651. Drickamer K (2006) Database of human and mouse proteins containing CTLDs (www.imperial.ac.uk/research/animallectins/ctld/mammals/human-vmousedata.html)

652. Cummings RD, McEver RP (2009) C-type Lectins. In Varki A, Cummings RD, Esko JD, Freeze HH, Stanley P, Bertozzi CR, Hart GW, and Etzler ME (eds) Essentials of Glycobiology, Cold Spring Harbor Laboratory Press, New York (www.ncbi.nlm.nih.gov/books/NBK1943)

653. Mishra R, Grzybek M, Niki T, Hirashima M, Simons K (2010) Galectin-9 trafficking regulates apical-basal polarity in Madin-Darby canine kidney epithelial cells. Proceedings of the National Academy of Sciences of the United States of America 107:17633–17638

654. Crocker PR, Paulson JC, Varki A (2007) Siglecs and their roles in the immune system. Nature Reviews – Immunology 7:255–266

655. Fehon RG, McClatchey AI, Bretscher A (2010) Organizing the cell cortex: the role of ERM proteins. Nature Reviews – Molecular Cell Biology 11:276–287

656. Ehmsen J, Poon E, Davies K (2002) The dystrophin-associated protein complex. Journal of Cell Science 115:2801–2803

657. Le Roy C, Wrana JL (2005) Clathrin- and non-clathrin-mediated endocytic regulation of cell signalling. Nature Reviews – Molecular Cell Biology 6:112–126

658. Lenne PF, Wawrezinieck L, Conchonaud F, Wurtz O, Boned A, Guo XJ, Rigneault H, He HT, Marguet D (2006) Dynamic molecular confinement in the plasma membrane by microdomains and the cytoskeleton meshwork. EMBO Journal 25:3245–3256

659. van Rheenen J, Achame EM, Janssen H, Calafat J, Jalink K (2005) PIP2 signaling in lipid domains: a critical re-evaluation. EMBO Journal 24:1664–1673

660. Nichols B. (2005) Cell biology: without a raft. Nature 436: 638–639

661. Simons K, Toomre D (2000) Lipid rafts and signal transduction. Nature Reviews – Molecular Cell Biology 1:31–39

662. van Zanten TS, Cambi A, Koopman M, Joosten B, Figdor CG, Garcia-Parajo MF (2009) Hotspots of GPI-anchored proteins and integrin nanoclusters function as nucleation sites for cell adhesion. Proceedings of the National Academy of Sciences of the United States of America 106:18557–18562

663. Rizzo V, Sung A, Oh P, Schnitzer JE (1998) Rapid mechanotransduction in situ at the luminal cell surface of vascular endothelium and its caveolae. Journal of Biological Chemistry 273:26323–26329

664. Girard PR, Nerem RM (1995) Shear stress modulates endothelial cell morphology and F-actin organization through the regulation of focal adhesion-associated proteins. Journal of Cellular Physiology 163:179–193

665. Sun RJ, Muller S, Zhuang FY, Stoltz JF, Wang X (2003) Caveolin-1 redistribution in human endothelial cells induced by laminar flow and cytokine. Biorheology 40:31–39

666. Radel C, Rizzo V (2005) Integrin mechanotransduction stimulates caveolin-1 phosphorylation and recruitment of Csk to mediate actin reorganization. American Journal of Physiology – Heart and Circulatory Physiology 288:H936–H945

667. Polte TR, Eichler GS, Wang N, Ingber DE (2004) Extracellular matrix controls myosin light chain phosphorylation and cell contractility through modulation of cell shape and cytoskeletal prestress. American Journal of Physiology – Cell Physiology 286:C518–C528

668. Bezanilla F (2008) How membrane proteins sense voltage. Nature Reviews – Molecular Cell Biology 9:323–332

669. Davis D, Sowinski S (2008) Membrane nanotubes: dynamic long-distance connections between animal cells. Nature Reviews – Molecular Cell Biology 9:431–436

670. Edelman GM, Crossin KL (1991) Cell adhesion molecules: implications for a molecular histology. Annual Review of Biochemistry 60:155–190

671. Milev P, Maurel P, Häring M, Margolis RK, Margolis RU (1996) TAG-1/axonin-1 is a high-affinity ligand of neurocan, phosphacan/protein-tyrosine phosphatase-ζ/β, and N-CAM. Journal of Biological Chemistry 271:15716–15723

672. Ko KS, Arora PD, McCulloch CA (2001) Cadherins mediate intercellular mechanical signaling in fibroblasts by activation of stretch-sensitive calcium-permeable channels. Journal of Biological Chemistry 276:35967–35977

673. Nelson CM, Pirone DM, Tan JL, Chen CS (2004) Vascular endothelial-cadherin regulates cytoskeletal tension, cell spreading, and focal adhesions by stimulating RhoA. Molecular Biology of the Cell 15:2943–2953

674. Pestronk A (2005) Adhesion molecules (http://www.neuro.wustl.edu/neuromuscular/lab/adhesion.htm)

675. St. Edward's University's Computer Science. School of Natural Sciences. Cadherin Proteins (www.cs.stedwards.edu/chem/Chemistry/CHEM43/CHEM43/Cadherins/FUNCTION.html)

676. Hoffmann R, Valencia A (2004) A gene network for navigating the literature. Nature – Genetics 36:664 (Information Hyperlinked over Proteins www.ihop-net.org/) [676].

677. Kools P, Van Imschoot G, van Roy F (2000) Characterization of three novel human cadherin genes (CDH7, CDH19, and CDH20) clustered on chromosome 18q22-q23 and with high homology to chicken cadherin-7. Genomics 68:283–295

678. Rampon C, Prandini MH, Bouillot S, Pointu H, Tillet E, Frank R, Vernet M, Huber P (2005) Protocadherin 12 (VE-cadherin 2) is expressed in endothelial, trophoblast, and mesangial cells. Experimental Cell Research 302:48–60

679. Schalm SS, Ballif BA, Buchanan SM, Phillips GR, Maniatis T (2010) Phosphorylation of protocadherin proteins by the receptor tyrosine kinase Ret. Proceedings of the National Academy of Sciences of the United States of America 107:13894–13899

680. Angst BD, Marcozzi C, Magee AI (2001) The cadherin superfamily: diversity in form and function. Journal of Cell Science 114:629–641

681. Drees F, Pokutta S, Yamada S, Nelson WJ, Weis WI (2005) Alpha-catenin is a molecular switch that binds E-cadherin-beta-catenin and regulates actin-filament assembly. Cell 123: 903–915

682. Cavey M, Rauzi M, Lenne PF, Lecuit T (2008) A two-tiered mechanism for stabilization and immobilization of E-cadherin. Nature 453:751–756

683. Bibert S, Ayari H, Riveline D, Concord E, Hermant B, Vernet T, Gulino-Debrac D (2008) Establishment of cell–cell junctions depends on the oligomeric states of VE-cadherin. Journal of Biochemistry 143:821–832

684. Norena NK, Liua BP, Burridgea K, Krefta B (2000) p120 catenin regulates the actin cytoskeleton via Rho family GTPases. Journal of Cell Biology, 150:567–580

685. Kametani Y, Takeichi M (2006) Basal-to-apical cadherin flow at cell junctions. Nature – Cell Biology 9:92–98

686. Ladoux B, Anon E, Lambert M, Rabodzey A, Hersen P, Buguin A, Silberzan P, Mège RM (2010) Strength dependence of cadherin-mediated adhesions. Biophysical Journal 98:534–542

687. Borghi N, Lowndes M, Maruthamuthu V, Gardel ML, Nelson WJ (2010) Regulation of cell motile behavior by crosstalk between cadherin- and integrin-mediated adhesions. Proceedings of the National Academy of Sciences of the United States of America 107:13324–13329

688. Südhof TC (2001) α-Latrotoxin and its receptors: neurexins and CIRL/latrophilins. Annual Review of Neuroscience 24:933–962

689. Abbott RJ, Spendlove I, Roversi P, Fitzgibbon H, Knott V, Teriete P, McDonnell JM, Handford PA, Lea SM (2007) Structural and functional characterization of a novel T cell receptor co-regulatory protein complex, CD97-CD55. Journal of Biological Chemistry 282:22023–22032

690. van Denderen BJ, Pearse MJ, Katerelos M, Nottle MB, Du ZT, Aminian A, Adam WR, Shenoy-Scaria A, Lublin DM, Shinkel TA, d'Apice AJ (1996) Expression of functional decay-accelerating factor (CD55) in transgenic mice protects against human complement-mediated attack. Transplantation 61:582–588

691. Chang GW, Davies JQ, Stacey M, Yona S, Bowdish DM, Hamann J, Chen TC, Lin CY, Gordon S, Lin HH (2007) CD312, the human adhesion-GPCR EMR2, is differentially expressed during differentiation, maturation, and activation of myeloid cells. Biochemical and Biophysical Research Communications 353:133–138

692. Yona S, Lin HH, Dri P, Davies JQ, Hayhoe RP, Lewis SM, Heinsbroek SE, Brown KA, Perretti M, Hamann J, Treacher DF, Gordon S, Stacey M (2008) Ligation of the adhesion-GPCR EMR2 regulates human neutrophil function. FASEB Journal 22:741–751

693. Hamann J, Koning N, Pouwels W, Ulfman LH, van Eijk M, Stacey M, Lin HH, Gordon S, Kwakkenbos MJ (2007) EMR1, the human homolog of F4/80, is an eosinophil-specific receptor. European Journal of Immunology 37:2797–2802

694. Matmati M, Pouwels W, van Bruggen R, Jansen M, Hoek RM, Verhoeven AJ, Hamann J (2007) The human EGF-TM7 receptor EMR3 is a marker for mature granulocytes. Journal of Leukocyte Biology 81:440–448

695. Fredriksson R, Gloriam DE, Höglund PJ, Lagerström MC, Schiöth HB (2003) There exist at least 30 human G-protein-coupled receptors with long Ser/Thr-rich N-termini. Biochemical and Biophysical Research Communications 301:725–734

696. Bjarnadóttir TK, Fredriksson R, Höglund PJ, Gloriam DE, Lagerström MC, Schiöth HB (2004) The human and mouse repertoire of the adhesion family of G-protein-coupled receptors. Genomics 84:23–33

697. Ushiyama S, Laue TM, Moore KL, Erickson HP, McEver RP (1993) Structural and functional characterization of monomeric soluble P-selectin and comparison with membrane P-selectin. Journal of Biological Chemistry 268:15229–15237

698. Qian Z, Gelzer-Bell R, Yang Sx SX, Cao W, Ohnishi T, Wasowska BA, Hruban RH, Rodriguez ER, Baldwin WM 3rd, Lowenstein CJ (2001) Inducible nitric oxide synthase inhibition of Weibel-Palade body release in cardiac transplant rejection. Circulation 104:2369–2375

699. Wagner DD (1993) The Weibel–Palade body: the storage granule for von Willebrand factor and P-selectin. Thrombosis and Haemostasis 70:105–110

700. Hulín I (2005) Inflammation (nic.sav.sk/logos/books/scientific/node4.html)

701. Huston DP (1997) The biology of the immune system. The Journal of the American Medical Association 278:1804–1814

702. Hynes RO (1999) Cell adhesion: old and new questions. Trends in Genetics 15:M33–M37

703. Springer TA (2009) Structural basis for selectin mechanochemistry. Proceedings of the National Academy of Sciences of the United States of America 106:91–96

704. Parise LV, Phillips DR (1985) Reconstitution of the purified platelet fibrinogen receptor. Fibrinogen binding propertie of the glycoprotein IIb-IIIa complex. Journal of Biological Chemistry 260:10698–10707

705. Davis GE, Bayless KJ, Davis MJ, Meininger GA (2000) Regulation of tissue injury responses by the exposure of matricryptic sites within extracellular matrix molecules. American Journal of Pathology 156:1489–1498

706. Horwitz AF (1997) Integrins and health. Scientific American 276:68–75

707. Hynes RO (1992) Integrins: versatility, modulation, and signaling in cell adhesion. Cell 69:11–25

708. Smyth SS, Joneckis CC, Parise LV (1993) Regulation of vascular integrins. Blood 181:2827–2843

709. Shattil SJ, Kim C, Ginsberg MH (2010) The final steps of integrin activation: the end game. Nature Reviews – Molecular Cell Biology 11:288–300

710. Chen JF, Yang W, Kim M, Carman CV, Springer TA (2006) Regulation of outside-in signaling and affinity by the $\beta_2 I$ domain of integrin $\alpha_L \beta_2$. Proceedings of the National Academy of Sciences of the United States of America 103:13062–13067

711. Koster J (2005) Integrins. http://integrins.hypermart.net/ligand.htm

712. Larson RS, Springer TA (1990) Structure and function of leukocyte integrins. Immunological Reviews 114:181–217

713. Shamri R, Grabovsky V, Gauguet J-M, Feigelson S, Manevich E, Kolanus W, Robinson MK, Staunton DE, von Andrian UH, Alon R (2005) Lymphocyte arrest requires instantaneous induction of an extended LFA-1 conformation mediated by endothelium-bound chemokines. Nature – Immunology 6:497–506

714. Arnaout A (2009) Integrin β_2. UCSD-Nature Molecule Pages, UCSD-Nature Signaling Gateway (www.signaling-gateway.org)

715. Heino J (2009) Integrin α_2. UCSD-Nature Molecule Pages, UCSD-Nature Signaling Gateway (www.signaling-gateway.org)

716. Mittelbrunn M, Cabanas C, Sanchez-Madrid F (2006) Integrin α_4. UCSD-Nature Molecule Pages, UCSD-Nature Signaling Gateway (www.signaling-gateway.org)

717. Gullberg DE (2005) Integrin α_{11}. UCSD-Nature Molecule Pages, UCSD-Nature Signaling Gateway (www.signaling-gateway.org)

718. El-Asady R, Hadley GA (2006) Integrin α_E. UCSD-Nature Molecule Pages, UCSD-Nature Signaling Gateway (www.signaling-gateway.org)

719. Cabanas C, Mittelbrunn M, Sanchez-Madrid F (2008) Integrin α_L. UCSD-Nature Molecule Pages, UCSD-Nature Signaling Gateway (www.signaling-gateway.org)

720. Sadhu C (2007) Integrin α_M UCSD-Nature Molecule Pages, UCSD-Nature Signaling Gateway (www.signaling-gateway.org)

721. Kerr BA, Byzova TV (2010) Integrin α_V. UCSD-Nature Molecule Pages, UCSD-Nature Signaling Gateway (www.signaling-gateway.org)

722. Barsukov IL, Prescot A, Bate N, Patel B, Floyd DN, Bhanji N, Bagshaw CR, Letinic K, Di Paolo G, De Camilli P, Roberts GC, Critchley DR (2003) Phosphatidylinositol phosphate kinase type 1γ and $\beta1$-integrin cytoplasmic domain bind to the same region in the talin FERM domain. Journal of Biological Chemistry 278:31202–31209

723. Ithychanda SS, Das M, Ma YQ, Ding K, Wang X, Gupta S, Wu C, Plow EF, Qin J (2009) Migfilin, a molecular switch in regulation of integrin activation. Journal of Biological Chemistry 284:4713–4722

724. Zaidel-Bar R, Itzkovitz S, Ma'ayan A, Iyengar R, Geiger B (2007) Functional atlas of the integrin adhesome. Nature – Cell Biology 9:858–867 (www.adhesome.org)

725. Kluger MS (2004) Vascular endothelial cell adhesion and signaling during leukocyte recruitment. Advances in Dermatology 20:163–201

726. Jalali S, del Pozo MA, Chen K, Miao H, Li Y, Schwartz MA, Shyy JY, Chien S (2001) Integrin-mediated mechanotransduction requires its dynamic interaction with specific extracellular matrix (ECM) ligands. Proceedings of the National Academy of Sciences of the United States of America 98:1042–1046

727. Legate KR, Montanez E, Kudlacek O, Fassler R (2006) ILK, PINCH and parvin: the tIPP of integrin signalling Nature Reviews – Molecular Cell Biology 7:20–31

728. Korenbaum E, Olski TM, Noegel AA (2001) Genomic organization and expression profile of the parvin family of focal adhesion proteins in mice and humans. Gene 279:69-79

729. Zhang Y, Chen K, Tu Y, Wu C (2004) Distinct roles of two structurally closely related focal adhesion proteins, α-parvins and β-parvins, in regulation of cell morphology and survival. Journal of Biological Chemistry 279:41695–41705

730. Mongroo PS, Johnstone CN, Naruszewicz I, Leung-Hagesteijn C, Sung RK, Carnio L, Rustgi AK, Hannigan GE (2004) β-Parvin inhibits integrin-linked kinase signaling and is downregulated in breast cancer. Oncogene 23:8959–8970

731. Yamaji S, Suzuki A, Kanamori H, Mishima W, Yoshimi R, Takasaki H, Takabayashi M, Fujimaki K, Fujisawa S, Ohno S, Ishigatsubo Y (2004) Affixin interacts with α-actinin and mediates integrin signaling for reorganization of F-actin induced by initial cell-substrate interaction. Journal of Cell Biology 165:539–551

732. Mishima W, Suzuki A, Yamaji S, Yoshimi R, Ueda A, Kaneko T, Tanaka J, Miwa Y, Ohno S, Ishigatsubo Y (2004) The first CH domain of affixin activates Cdc42 and Rac1 through αPIX, a Cdc42/Rac1-specific guanine nucleotide exchanging factor. Genes to Cells 9:193–204

733. Matsuda C, Kameyama K, Tagawa K, Ogawa M, Suzuki A, Yamaji S, Okamoto H, Nishino I, Hayashi YK (2005) Dysferlin interacts with affixin (β-parvin) at the sarcolemma. Journal of Neuropathology and Experimental Neurology 64:334–340

734. Srivastava J, Barreiro G, Groscurth S, Gingras AR, Goult BT, Critchley DR, Kelly MJS, Jacobson MP, Barber DL (2008) Structural model and functional significance of pH-dependent talin–actin binding for focal adhesion remodeling. Proceedings of the National Academy of Sciences of the United States of America 105:14436–1444

735. Larjava H, Plow EF, Wu C (2008) Kindlins: essential regulators of integrin signalling and cell–matrix adhesion. EMBO Reports 9:1203–1208

736. Montanez E, Ussar S, Schifferer M, Bösl M, Zent R, Moser M, Fässler R (2008) Kindlin-2 controls bidirectional signaling of integrins. Genes and Development 22:1325–1330

737. Ma YQ, Qin J, Wu C, Plow EF (2008) Kindlin-2 (Mig-2): a co-activator of β3 integrins. Journal of Cell Biology 181:439–446

738. Arnaout MA (1990) Structure and function of the leukocyte adhesion molecules CD11/CD18. Blood 75:1037–1050

739. Mazzone A, Ricevuti G (1995) Leukocyte CD11/CD18 integrins: biological and clinical relevance. Haematologica 80:161–175

740. Han C, Jin J, Xu S, Liu H, Li N, Cao X (2010) Integrin CD11b negatively regulates TLR-triggered inflammatory responses by activating Syk and promoting degradation of MyD88 and TRIF via Cbl-b. Nature – Immunology 11:734–742

741. Friedland JC, Lee MH, Boettiger D (2009) Mechanically activated integrin switch controls $\alpha_5\beta_1$ function. Science 323:642–644

742. Roca-Cusachs P, Gauthier NC, del Rio A, Sheetz MP (2009) Clustering of $\alpha_5\beta_1$ integrins determines adhesion strength whereas $\alpha_v\beta_3$ and talin enable mechanotransduction. Proceedings of the National Academy of Sciences of the United States of America 106:16245-16250

743. del Rio A, Perez-Jimenez R, Liu R, Roca-Cusachs P, Fernandez JM, Sheetz MP (2009) Stretching single talin rod molecules activates vinculin binding. Science 323:638-641

744. Weber C, Fraemohs L, Dejana E (2007) The role of junctional adhesion molecules in vascular inflammation. Nature Reviews – Immunology 7:467-477

745. Takai Y, Miyoshi J, Ikeda W, Ogita H (2008) Nectins and nectin-like molecules: roles in contact inhibition of cell movement and proliferation. Nature Reviews – Molecular Cell Biology 9:603-615

746. Hou J, Renigunta A, Gomes AS, Hou M, Paul DL, Waldegger S, Goodenough DA (2009) Claudin-16 and claudin-19 interaction is required for their assembly into tight junctions and for renal reabsorption of magnesium. Proceedings of the National Academy of Sciences of the United States of America 106:15350-15355

747. Zamir E, Geiger B (2001) Components of cell-matrix adhesions. Journal of Cell Science 114:3577-3579

748. Clark EA, Brugge JS (1990) Integrins and signal transduction pathways: the road taken. Science 268:233-239

749. Critchley DR (2000) Focal adhesions – the cytoskeletal connection. Current Opinion in Cell Biology 12:133-139

750. Calderwood DA, Zent R, Grant R, Rees DJ, Hynes RO, Ginsberg MH (1999) The Talin head domain binds to integrin beta subunit cytoplasmic tails and regulates integrin activation. Journal of Biological Chemistry 274:28071-28074

751. Gilmore AP, Wood C, Ohanian V, Jackson P, Patel B, Rees DJ, Hynes RO, Critchley DR (1993) The cytoskeletal protein talin contains at least two distinct vinculin binding domains. Journal of Cell Biology 122:337-347

752. Morgan MR, Humphries MJ, Bass MD (2007) Synergistic control of cell adhesion by integrins and syndecans. Nature Reviews – Molecular Cell Biology 8:957-969

753. Geiger B, Spatz JP, Bershadsky AD (2009) Environmental sensing through focal adhesions. Nature Reviews – Molecular Cell Biology 10:21-33

754. Michel CC, Curry FE (1999) Microvascular permeability. Physiological Reviews 79:703-761

755. Setzer SV, Calkins CC, Garner J, Summers S, Green KJ, Kowalczyk AP (2004) Comparative analysis of armadillo family proteins in the regulation of a431 epithelial cell junction assembly, adhesion and migration. Journal of Investigative Dermatology 123:426-433

756. Hatzfeld M (2007) Plakophilins: Multifunctional proteins or just regulators of desmosomal adhesion? Biochimica et Biophysica Acta 1773:69-77

757. Yap AS, Brieher WM, Gumbiner BM (1997) Molecular and functional analysis of cadherin-based adherens junctions. Annual Review of Cell and Developmental Biology 13:119–146

758. Pilot F, Philippe JM, Lemmers C, Lecuit T (2006) Spatial control of actin organization at adherens junctions by a synaptotagmin-like protein. Nature 442:580–584

759. Harris TJC, Tepass U (2010) Adherens junctions: from molecules to morphogenesis. Nature Reviews – Molecular Cell Biology 11:502–514

760. Abe K, Takeichi M (2008) EPLIN mediates linkage of the cadherin–catenin complex to F-actin and stabilizes the circumferential actin belt. Proceedings of the National Academy of Sciences of the United States of America 105:13–19

761. Sandquist JC, Bement WM (2010) Hold on tightly, let go lightly: myosin functions at adherens junctions. Nature – Cell Biology 12:633–635

762. Smutny M, Cox HL, Leerberg JM, Kovacs EM, Conti MA, Ferguson C, Hamilton NA, Parton RG, Adelstein RS, Yap AS (2010) Myosin II isoforms identify distinct functional modules that support integrity of the epithelial zonula adherens. Nature – Cell Biology 12:696–702

763. Gottardi CJ, Gumbiner BM (2004) Distinct molecular forms of β-catenin are targeted to adhesive or transcriptional complexes. Journal of Cell Biology 167:339–349

771. Yonemura S, Wada Y, Watanabe T, Nagafuchi A, Shibata M (2010) α-Catenin as a tension transducer that induces adherens junction development. Nature – Cell Biology 12:533–542

765. Germino GG (2005) Linking cilia to Wnts. Nature – Genetics 37:455–457

766. Simons M, Gloy J, Ganner A, Bullerkotte A, Bashkurov M, Krönig C, Schermer B, Benzing T, Cabello OA, Jenny A, Mlodzik M, Polok B, Driever W, Obara T, Walz G (2005) Inversin, the gene product mutated in nephronophthisis type II, functions as a molecular switch between Wnt signaling pathways. Nature – Genetics 37:537–543

767. Gerull B, Heuser A, Wichter T, Paul M, Basson CT, McDermott DA, Lerman BB, Markowitz SM, Ellinor PT, MacRae CA, Peters S, Grossmann KS, Drenckhahn J, Michely B, Sasse-Klaassen S, Birchmeier W, Dietz R, Breithardt G, Schulze-Bahr E, Thierfelder L (2004) Mutations in the desmosomal protein plakophilin-2 are common in arrhythmogenic right ventricular cardiomyopathy. Nature – Genetics 36:1162–1164

768. Zandy NL, Playford M, Pendergast AM (2007) Abl tyrosine kinases regulate cell–cell adhesion through Rho GTPases. Proceedings of the National Academy of Sciences of the United States of America 104:17686–17691

769. Rhee J, Buchan T, Zukerberg L, Lilien J, Balsamo J (2007) Cables links Robobound Abl kinase to N-cadherin-bound β-catenin to mediate Slit-induced modulation of adhesion and transcription. Nature – Cell Biology 9:883–892

770. Lecuit T (2010) α-Catenin mechanosensing for adherens junctions. Nature – Cell Biology 12:522–524

771. Yonemura S, Wada Y, Watanabe T, Nagafuchi A, Shibata M (2010) alpha-Catenin as a tension transducer that induces adherens junction development. Nature – Cell Biology 12: 533–542

772. Furuse M, Sasaki H, Tsukita S (1999) Manner of interaction of heterogeneous claudin species within and between tight junction strands. Journal of Cell Biology 147:891–903

773. Haskins J, Gu L, Wittchen ES, Hibbard J, Stevenson BR (1998) ZO-3, a novel member of the MAGUK protein family found at the tight junction, interacts with ZO-1 and occludin. Journal of Cell Biology 141:199–208

774. Tunggal JA, Helfrich I, Schmitz A, Schwarz H, Ganzel D, Fromm M, Kemler R, Krieg T, Niessen CM (2005) E-cadherin is essential for in vivo epidermal barrier function by regulating tight junctions. EMBO Journal 24:1146–1156

775. Tripathi A, Lammers KM, Goldblum S, Shea-Donohue T, Netzel-Arnett S, Buzza MS, Antalis TM, Vogel SN, Zhao A, Yang S, Arrietta MC, Meddings JB, Fasano A (2009) Identification of human zonulin, a physiological modulator of tight junctions, as prehaptoglobin-2. Proceedings of the National Academy of Sciences of the United States of America 106:16799–16804

776. Gassama-Diagne A, Yu W, ter Beest M, Martin-Belmonte F, Kierbel A, Engel J, Mostov K (2006) Phosphatidylinositol-3,4,5-trisphosphate regulates the formation of the basolateral plasma membrane in epithelial cells. Nature – Cell Biology 8:963–970

777. Ozdamar B, Bose R, Barrios-Rodiles M, Wang HR, Zhang Y, Wrana JL (2005) Regulation of the polarity protein Par6 by TGFbeta receptors controls epithelial cell plasticity. Science 307:1603–1609

778. Iden S, Rehder D, August B, Suzuki A, Wolburg-Buchholz K, Wolburg H, Ohno S, Behrens J, Vestweber D, Ebnet K (2006) A distinct PAR complex associates physically with VE-cadherin in vertebrate endothelial cells. EMBO Reports 7:1239–1246

779. Du D, Xu F, Yu L, Zhang C, Lu X, Yuan H, Huang Q, Zhang F, Bao H, Jia L, Wu X, Zhu X, Zhang X, Zhang Z, Chen Z (2010) The tight junction protein, occludin, regulates the directional migration of epithelial cells. Developmental Cell 18:52–63

780. Severs NJ, Bruce AF, Dupont E, Rothery S (2008) Remodelling of gap junctions and connexin expression in diseased myocardium. Cardiovascular Research 80:9–19

Volume I. Chap. 13. Extracellular Matrix

781. Elbischger PJ, Bischof H, Holzapfel GA, Regitnig P (2005) Computer vision analysis of collagen fiber bundles in the adventitia of human blood vessels. Studies in Health Technology and Informatics, Suri JS, Yuan C, Wilson DL, Laxminarayan S (eds), In Plaque Imaging: Pixel to Molecular Level 113:97–129

782. Manabe RI, Tsutsui K, Yamada T, Kimura M, Nakano I, Shimono C, Sanzen N, Furutani Y, Fukuda T, Oguri Y, Shimamoto K, Kiyozumi D, Sato Y,

Sado Y, Senoo H, Yamashina S, Fukuda S, Kawai J, Sugiura N, Kimata K, Hayashizaki Y, Sekiguchi K (2008) Transcriptome-based systematic identification of extracellular matrix proteins. Proceedings of the National Academy of Sciences of the United States of America 105:12849–12854

783. Häcker U, Nybakken K, Perrimon N (2005) Heparan sulphate proteoglycans: the sweet side of development. Nature Reviews – Molecular Cell Biology 6:530–541

784. Chai S, Chai Q, Danielsen CC, Hjorth P, Nyengaard JR, Ledet T, Yamaguchi Y, Rasmussen LM, Wogensen L (2005) Overexpression of hyaluronan in the tunica media promotes the development of atherosclerosis. Circulation Research 96:583–591

785. Smith ML, Gourdon D, Little WC, Kubow KE, Eguiluz RA, Luna-Morris S, Vogel V (2007) Force-induced unfolding of fibronectin in the extracellular matrix of living cells. PLoS Biology 5:e268

786. Pizzo AM, Kokini K, Vaughn LC, Waisner BZ, Voytik-Harbin SL (2005) Extracellular matrix (ECM) microstructural composition regulates local cell-ECM biomechanics and fundamental fibroblast behavior: a multidimensional perspective. Journal of Applied Physiology 98:1909-1921.

787. King MW (1996–2011) The Medical Biochemistry Page (http://themedicalbiochemistrypage.org/)

788. Esko JD, Kimata K, Lindahl U (2009) Proteoglycans and Sulfated Glycosaminoglycans. In Varki A, Cummings RD, Esko JD, Freeze HH, Stanley P, Bertozzi CR, Hart GW, and Etzler ME (eds) Essentials of Glycobiology, Cold Spring Harbor Laboratory Press, New York (www.ncbi.nlm.nih.gov/books/NBK1900)

789. Nadanaka S, Kitagawa H (2008) Heparan sulphate biosynthesis and disease. Journal of Biochemistry 144:7–14

790. Yan D, Lin X (2008) Opposing roles for glypicans in Hedgehog signalling. Nature – Cell Biology 10:761–763

791. Bellin RM, Kubicek JD, Frigault MJ, Kamien AJ, Steward RL Jr, Barnes HM, Digiacomo MB, Duncan LJ, Edgerly CK, Morse EM, Park CY, Fredberg JJ, Cheng CM, LeDuc PR (2009) Defining the role of syndecan-4 in mechanotransduction using surface-modification approaches. Proceedings of the National Academy of Sciences of the United States of America 106:22102–22107

792. Perumal S, Antipova O, Orgel JPRO (2008) Collagen fibril architecture, domain organization, and triple-helical conformation govern its proteolysis. Proceedings of the National Academy of Sciences of the United States of America 105:2824–2829

793. Hynes RO (1986) Fibronectins. Scientific American 254:42–51

794. Clark RA, Wikner NE, Doherty DE, Norris, DA (1988) Cryptic chemotactic activity of fibronectin for human monocytes resides in the 120-kDa fibroblastic cell-binding fragment, The Journal of Biological Chemistry 263:12115–12123

795. Giancotti FG, Ruoslahti E (1999) Integrin signaling. Science 285:1028–1032

796. Yamada KM (1999) Cell adhesion molecules. In: Creighton TE (ed) Encyclopedia of molecular biology. John Wiley & Sons Inc, New York

797. Pan S, Wang R, Zhou X, Corvera J, Kloc M, Sifers R, Gallick GE, Lin SH, Kuang J (2008) Extracellular Alix regulates integrin-mediated cell adhesions and extracellular matrix assembly. EMBO Journal 27:2077–2090

798. Aumailley M, Gayraud B (1998) Structure and biological activity of the extracellular matrix. Journal of Molecular Medecine 76:253–265

799. Timpl R, Brown JC (1996) Supramolecular assembly of basement membranes. Bioessays 18:123–132

800. Erickson AC, Couchman JR (2000) Still more complexity in mammalian basement membranes, The Journal of Histochemistry and Cytochemistry 48:1291–1306

801. Yurchenco PD, O'Rear JJ (1994) Basal lamina assembly. Current Opinion in Cell Biology 6:674–681

802. Colognato H, Winkelmann DA, Yurchenco PD (1999) Laminin polymerization induces a receptor-cytoskeleton network. Journal of Cell Biology 145:619–631

803. Henry MD, Satz JS, Brakebusch C, Costell M, Gustafsson E, Fassler R, Campbell KP (2001) Distinct roles for dystroglycan, β1 and perlecan in cell surface laminin organization. Journal of Cell Science 114:1137–1144

804. Gohring W, Sasaki T, Heldin CH, Timpl R (1998) Mapping of the binding of platelet-derived growth factor to distinct domains of the basement membrane proteins BM-40 and perlecan and distinction from the BM-40 collagen-binding epitope. European Journal of Biochemistry 255:60–66

805. Lerner L, Torchia DA (1986) A multinuclear NMR study of the interactions of cations with proteoglycans, heparin, and ficoll. Journal of Biological Chemistry 261:12706–12714

806. Iozzo RV (1998) Matrix proteoglycans: from molecular design to cellular function. Annual Review of Biochemistry 67:609–652

807. Robert L (1994) Élasticité des tissus et vieillissement [Tissue elasticity and aging]. Pour la Science 201:56–62

808. Robert L, Jacob MP, Fülop T, Timar J, Hornebeck W (1989) Elastonectin and the elastin receptor. Pathologie et Biologie (Paris) 37:736–741

809. Hirai M, Horiguchi M, Ohbayashi T, Kita T, Chien KR, Nakamura T (2007) Latent TGF-beta-binding protein 2 binds to DANCE/fibulin-5 and regulates elastic fiber assembly. EMBO Journal 26:3283–3295

810. Fung YC (1981) Biomechanics. Springer, New York

811. Aaron BB, Gosline JM (1981) Elastin as a random-network elastomer: a mechanical and optical analysis of single elastin fibers. Biopolymers 20:1247–1260

812. Canfield TR, Dobrin PB (1987) Static elastic properties of blood vessels. In: Skalak R, Chien S (eds) Handbook of bioengineering. McGraw-Hill, New York

813. Viidik A (1987) Properties of tendons and ligaments. In: Skalak R, Chien S (eds) Handbook of bioengineering. McGraw-Hill, New York

814. Kielty CM, Baldock C, Lee D, Rock MJ, Ashworth JL, Shuttleworth CA (2002) Fibrillin: from microfibril assembly to biomechanical function. Philosophical Transactions of the Royal Society – London – B Biological Sciences 357:207-217

815. Corson GM, Charbonneau NL, Keene DR, Sakai LY (2004) Differential expression of fibrillin-3 adds to microfibril variety in human and avian, but not rodent, connective tissues. Genomics 83:461–472

816. Twal WO, Czirok A, Hegedus B, Knaak C, Chintalapudi MR, Okagawa H, Sugi Y, Argraves WS (2001) Fibulin-1 suppression of fibronectin-regulated cell adhesion and motility. Journal of Cell Science 114:4587–4598

817. Hunzelmann N, Nischt R, Brenneisen P, Eickert A, Krieg T (2001) Increased deposition of fibulin-2 in solar elastosis and its colocalization with elastic fibres. British Journal of Dermatology 145:217–222

818. Hopf M, Göhring W, Mann K, Timpl R (2001) Mapping of binding sites for nidogens, fibulin-2, fibronectin and heparin to different IG modules of perlecan. Journal of Molecular Biology 311:529–541

819. Seeliger H, Camaj P, Ischenko I, Kleespies A, De Toni EN, Thieme SE, Blum H, Assmann G, Jauch KW, Bruns CJ (2009) EFEMP1 expression promotes in vivo tumor growth in human pancreatic adenocarcinoma. Molecular Cancer Research 7:189–198

820. Chen Q, Zhang T, Roshetsky JF, Ouyang Z, Essers J, Fan C, Wang Q, Hinek A, Plow EF, Dicorleto PE (2009) Fibulin-4 regulates expression of the tropoelastin gene and consequent elastic-fibre formation by human fibroblasts. Biochemical Journal 423:79–89

821. Yanagisawa H, Davis EC, Starcher BC, Ouchi T, Yanagisawa M, Richardson JA, Olson EN (2002) Fibulin-5 is an elastin-binding protein essential for elastic fibre development in vivo. Nature 415:168–171

822. Nakamura T, Lozano PR, Ikeda Y, Iwanaga Y, Hinek A, Minamisawa S, Cheng CF, Kobuke K, Dalton N, Takada Y, Tashiro K, Ross J, Honjo T, Chien KR (2002) Fibulin-5/DANCE is essential for elastogenesis in vivo. Nature 415:171–175

823. Nakamura T, Ruiz-Lozano P, Lindner V, Yabe D, Taniwaki M, Furukawa Y, Kobuke K, Tashiro K, Lu Z, Andon NL, Schaub R, Matsumori A, Sasayama S, Chien KR, Honjo T (1999) DANCE, a novel secreted RGD protein expressed in developing, atherosclerotic, and balloon-injured arteries. Journal of Biological Chemistry 274:22476–22483

824. Yurchenco PD, Furthmayr H (1984) Self-assembly of basement membrane collagen. Biochemistry 23:1839–1850

825. Orgel JPRO, Irving TC, Miller A, Wess TJ (2006) Microfibrillar structure of type I collagen in situ. Proceedings of the National Academy of Sciences of the United States of America 103:9001–9005

826. Viidik A, Danielson CC, Oxlund H (1982) On fundamental and phenomenological models, structure and mechanical properties of collagen, elastin and glycosaminoglycan complexes. Biorheology 19:437–451

827. Meshel AS, Wei Q, Adelstein RS, Sheetz MP (2005) Basic mechanism of three-dimensional collagen fibre transport by fibroblasts. Nature – Cell Biology 7:157–164

828. Laurent S, Boutouyrie P, Lacolley P (2005) Structural and genetic bases of arterial stiffness. Hypertension 45:1050–1055

829. Hieta R (2003) Prolyl 4-hydroxylase. Structural and functional characterization of the peptide-substrate-binding domain of the human enzyme, and cloning and characterization of a plant enzyme with unique properties. PhD Thesis, Department of Medical Biochemistry and Molecular Biology, Faculty of Medicine, University of Oulu (herkules.oulu.fi/isbn9514271793/html/index.html)

830. Jones FS, Jones PL (2000) The tenascin family of ECM glycoproteins: structure, function, and regulation during embryonic development and tissue remodeling. Developmental Dynamics 218:235–259.

831. Chung CY, Murphy-Ullrich JE, Erickson HP (1996) Mitogenesis, cell migration, and loss of focal adhesions induced by tenascin-C interacting with its cell surface receptor, annexin II. Molecular Biology of the Cell 7:883–892

832. Rodríguez C, Martínez-González J, Raposo B, Alcudia JF, Guadall A, Badimon L (2008) Regulation of lysyl oxidase in vascular cells: lysyl oxidase as a new player in cardiovascular diseases. Cardiovascular Research 79:7–13

833. Cummins PM, von Offenberg Sweeney N, Killeen MT, Birney YA, Redmond EM, Cahill PA (2007) Cyclic strain-mediated matrix metalloproteinase regulation within the vascular endothelium: a force to be reckoned with. American Journal of Physiology – Heart and Circulatory Physiology 292:H28–H42

834. Kandasamy AD, Chow AK, Ali MA, Schulz R (2010) Matrix metalloproteinase-2 and myocardial oxidative stress injury: beyond the matrix. Cardiovascular Research 85:413–423

835. Zavadzkas JA, Plyler RA, Bouges S, Koval CN, Rivers WT, Beck CU, Chang EI, Stroud RE, Mukherjee R, Spinale FG (2008) Cardiac-restricted overexpression of extracellular matrix metalloproteinase inducer causes myocardial remodeling and dysfunction in aging mice. American Journal of Physiology – Heart and Circulatory Physiology 295:H1394–H1402

836. Hedhli N, Lizano P, Hong C, Fritzky LF, Dhar SK, Liu H, Tian Y, Gao S, Madura K, Vatner SF, Depre C (2008) Proteasome inhibition decreases cardiac remodeling after initiation of pressure overload. American Journal of Physiology – Heart and Circulatory Physiology 295:H1385–H1393

837. Lehti K (2005) An MT1-MMP-PDGF receptor-beta axis regulates mural cell investmen of the microvasculature. Genes & Development 19:979–991

838. Paszek MJ, Zahir N, Johnson KR, Lakins JN, Rozenberg GI, Gefen A, Reinhart-King CA, Margulies SS, Dembo M, Boettiger D, Hammer DA, Weaver VM (2005) Tensional homeostasis and the malignant phenotype. Cancer Cell 8:241–254

839. Radisky DC, Levy DD, Littlepage LE, Liu H, Nelson CM, Fata JE, Leake D, Godden EL, Albertson DG, Nieto MA, Werb Z, Bissell MJ (2005) Rac1b and reactive oxygen species mediate MMP-3-induced EMT and genomic instability. Nature 436:123-127

840. Norris RA, Damon B, Mironov V, Kasyanov V, Ramamurthi A, Moreno-Rodriguez R, Trusk T, Potts JD, Goodwin RL, Davis J, Hoffman S, Wen X, Sugi Y, Kern CB, Mjaatvedt CH, Turner DK, Oka T, Conway SJ, Molkentin JD, Forgacs G, Markwald RR (2007) Periostin regulates collagen fibrillogenesis and the biomechanical properties of connective tissues. Journal of Cellular Biochemistry 101:695–711

841. Oka T, Xu J, Kaiser RA, Melendez J, Hambleton M, Sargent MA, Lorts A, Brunskill EW, Dorn GW, Conway SJ, Aronow BJ, Robbins J, Molkentin JD (2007) Genetic manipulation of periostin expression reveals a role in cardiac hypertrophy and ventricular remodeling. Circulation Research 101:313–321

842. Shimazaki M, Nakamura K, Kii I, Kashima T, Amizuka N, Li M, Saito M, Fukuda K, Nishiyama T, Kitajima S, Saga Y, Fukayama M, Sata M, Kudo A (2008) Periostin is essential for cardiac healing after acute myocardial infarction. Journal of Experimental Medicine 205:295-303

843. Ruiz-Ortega M, Rodríguez-Vita J, Sanchez-Lopez E, Carvajal G, Egido J (2007) TGF-βsignaling in vascular fibrosis. Cardiovascular Research 74:196–206

844. Lawler J (2010) Thrombospondin 1. UCSD-Nature Molecule Pages, UCSD-Nature Signaling Gateway (www.signaling-gateway.org)

845. Frazier WA, Isenberg JS, Kaur S, Roberts DD (2010) CD47. UCSD-Nature Molecule Pages, UCSD-Nature Signaling Gateway (www.signaling-gateway.org)

846. Kanwar YS, Wada J, Lin S, Danesh FR, Chugh SS, Yang Q, Banerjee T, Lomasney JW (2004) Update of extracellular matrix, its receptors, and cell adhesion molecules in mammalian nephrogenesis. American Journal of Physiology – Renal Physiology 286:F202–F215

847. Bordeleau F, Bessard J, Sheng Y, Marceau N (2008) Keratin contribution to cellular mechanical stress response at focal adhesions as assayed by laser tweezers. Biochemistry and Cell Biology 86:352–359

848. Hannafin JA, Attia EA, Henshaw R, Warren RF, Bhargava MM (2006) Effect of cyclic strain and plating matrix on cell proliferation and integrin expression by ligament fibroblasts. Journal of Orthopaedic Research 24:149–158

849. Shi F, Sottile J (2008) Caveolin-1-dependent β_1 integrin endocytosis is a critical regulator of fibronectin turnover. Journal of Cell Science 121:2360–2371

850. Valdembri D, Caswell PT, Anderson KI, Schwarz JP, König I, Astanina E, Caccavari F, Norman JC, Humphries MJ, Bussolino F, Serini G (2009) Neuropilin-1/GIPC1 signaling regulates alpha5beta1 integrin traffic and function in endothelial cells. PLoS Biology 7:e25

851. Huveneers S, Truong H, Fässler R, Sonnenberg A, Danen EH (2008) Binding of soluble fibronectin to integrin $\alpha_5\beta_1$ – link to focal adhesion redistribution and contractile shape. Journal of Cell Science 121:2452–2462

852. Jones MC, Caswell PT, Moran-Jones K, Roberts M, Barry ST, Gampel A, Mellor H, Norman JC (2009) VEGFR1 (Flt1) regulates Rab4 recycling to control fibronectin polymerization and endothelial vessel branching. Traffic 10:754–766

853. Sun Z, Martinez-Lemus LA, Trache A, Trzeciakowski JP, Davis GE, Pohl U, Meininger GA (2005) Mechanical properties of the interaction between fibronectin and $\alpha_5\beta_1$-integrin on vascular smooth muscle cells studied using atomic force microscopy. American Journal of Physiology – Heart and Circulatory Physiology 289:H2526–H2535

854. Kim DJ, Park SH, Lim CS, Chun JS, Kim JK, Song WK (2003) Cellular localization of integrin isoforms in phenylephrine-induced hypertrophic cardiac myocytes. Cell Biochemistry and Function 21:41–48

855. Geiger B, Bershadsky A (2002) Exploring the neighborhood: adhesion-coupled cell mechanosensors. Cell 110:139–142

856. Smith AS, Sengupta K, Goennenwein S, Seifert U, Sackmann E (2008) Force-induced growth of adhesion domains is controlled by receptor mobility. Proceedings of the National Academy of Sciences of the United States of America 105:6906–6911

857. Wang N, Tytell JD, Ingber DE (2009) Mechanotransduction at a distance: mechanically coupling the extracellular matrix with the nucleus. Nature Reviews – Molecular Cell Biology 10:75–82

858. Watkins JL, Lewandowski KT, Meek SE, Storz P, Toker A, Piwnica-Worms H (2008) Phosphorylation of the Par-1 polarity kinase by protein kinase D regulates 14-3-3 binding and membrane association. Proceedings of the National Academy of Sciences of the United States of America 105:18378–18383

Volume I. Chap. 14. Intracellular Transport

859. Ikonen E, Jansen M (2008) Cellular sterol trafficking and metabolism: spotlight on structure. Current Opinion in Cell Biology 20:371–377

860. Schmid EM, McMahon HT (2007) Integrating molecular and network biology to decode endocytosis. Nature 448:883–888

861. Patterson GH, Hirschberg K, Polishchuk RS, Gerlich D, Phair RD, Lippincott-Schwartz J (2008) Transport through the Golgi apparatus by rapid partitioning within a two-phase membrane system. Cell 133:1055–1067

862. Pulvirenti T, Giannotta M, Capestrano M, Capitani M, Pisanu A, Polishchuk RS, San Pietro E, Beznoussenko GV, Mironov AA, Turacchio G, Hsu VW, Sallese M, Luini A (2008) A traffic-activated Golgi-based signalling circuit coordinates the secretory pathway. Nature – Cell Biology 10:912–922

863. Henne WM, Boucrot E, Meinecke M, Evergren E, Vallis Y, Mittal R, McMahon HT (2010) FCHo proteins are nucleators of clathrin-mediated endocytosis. Science 328:1281–1284

864. Bassereau P (2010) Division of labour in ESCRT complexes. Nature – Cell Biology 12:422–423

865. Giraudo CG, Garcia-Diaz A, Eng WS, Chen Y, Hendrickson WA, Melia TJ, Rothman JE (2009) Alternative zippering as an on-off switch for SNARE-mediated fusion. Science 323:512–516

866. Maximov A, Tang J, Yang X, Pang ZP, Südhof TC (2009) Complexin controls the force transfer from SNARE complexes to membranes in fusion. Science 323:516–521

867. Haluska CK, Riske KA, Marchi-Artzner V, Lehn JM, Lipowsky R, Dimova R (2006) Time scales of membrane fusion revealed by direct imaging of vesicle fusion with high temporal resolution. Proceedings of the National Academy of Sciences of the United States of America 103:15841–15846

868. Macara IG (2001) Transport into and out of the nucleus. Microbiology and Molecular Biology Reviews 65:570–594

869. Fornerod M, Clarke PR (2008) To the centre of the volcano. Workshop on the Mechanisms of Nucleocytoplasmic Transport. EMBO Reports 9:419–424

870. Casolari JM, Brown CR, Komili S, West J, Hieronymus H, Silver PA (2004) Genome-wide localization of the nuclear transport machinery couples transcriptional status and nuclear organization. Cell 117:427–439

871. Menon BB, Sarma NJ, Pasula S, Deminoff SJ, Willis KA, Barbara KE, Andrews B, Santangelo GM (2005) Reverse recruitment: the Nup84 nuclear pore subcomplex mediates Rap1/Gcr1/Gcr2 transcriptional activation. Proceedings of the National Academy of Sciences of the United States of America 102:5749–5754

872. Chuderland D, Konson A, Seger R (2008) Identification and characterization of a general nuclear translocation signal in signaling proteins. Molecular Cell 31:850–861

873. Görlich D, Panté N, Kutay U, Aebi U, Bischoff FR (1996) Identification of different roles for RanGDP and RanGTP in nuclear protein import. EMBO Journal 15:5584–5594

874. Floer M, Blobel G (1996) The nuclear transport factor karyopherin β binds stoichiometrically to Ran-GTP and inhibits the Ran GTPase activating protein. Journal of Biological Chemistry 271:5313–5316

875. Kutay U, Bischoff FR, Kostka S, Kraft R, Görlich D (1997) Export of importin α from the nucleus is mediated by a specific nuclear transport factor. Cell 90:1061–1071

876. Yoon SO, Shin S, Liu Y, Ballif BA, Woo MS, Gygi SP, Blenis J (2008) Ran-binding protein 3 phosphorylation links the Ras and PI3-kinase pathways to nucleocytoplasmic transport. Molecular Cell 29:362–375

877. Lupu F, Alves A, Anderson K, Doye V, Lacy E (2008) Nuclear pore composition regulates neural stem/progenitor cell differentiation in the mouse embryo. Developmental Cell 14:831–842

878. Terry LJ, Shows EB, Wente SR (2007) Crossing the nuclear envelope: hierarchical regulation of nucleocytoplasmic transport. Science 318:1412–1416

879. Cross BCS, Sinning I, Luirink J, High S (2009) Delivering proteins for export from the cytosol. Nature Reviews – Molecular Cell Biology 10:255–264

880. Al-Nedawi K, Meehan B, Kerbel RS, Allison AC, Rak J (2009) Endothelial expression of autocrine VEGF upon the uptake of tumor-derived microvesicles containing oncogenic EGFR. Proceedings of the National Academy of Sciences of the United States of America 106:3794–3799

881. Kumari S, Mg S, Mayor S (2010) Endocytosis unplugged: multiple ways to enter the cell. Cell Research 20:256-275

882. Suzuki E, Nakayama M (2007) The mammalian Ced-1 ortholog MEGF10/KIAA1780 displays a novel adhesion pattern. Experimental Cell Research 313:2451–2464

883. Yu X, Lu N, Zhou Z (2008) Phagocytic receptor CED-1 initiates a signaling pathway for degrading engulfed apoptotic cells. PLoS Biology 6:e61

884. Chen D, Xiao H, Zhang K, Wang B, Gao Z, Jian Y, Qi X, Sun J, Miao L, Yang C (2010) Retromer is required for apoptotic cell clearance by phagocytic receptor recycling. Science 327:1261–1264

885. Scita G, Di Fiore PP (2010) The endocytic matrix. Nature 463:464–473

886. Kirkham M, Fujita A, Chadda R, Nixon SJ, Kurzchalia TV, Sharma DK, Pagano RE, Hancock JF, Mayor S, Parton RG (2005) Ultrastructural identification of uncoated caveolin-independent early endocytic vehicles. The Journal of Cell Biology 168:465–476

887. Damm EM, Pelkmans L, Kartenbeck J, Mezzacasa A, Kurzchalia T, Helenius A (2005) Clathrin- and caveolin-1-independent endocytosis: entry of simian virus 40 into cells devoid of caveolae. The Journal of Cell Biology 168:477–488

888. Gould GW, Lippincott-Schwartz J (2009) New roles for endosomes: from vesicular carriers to multi-purpose platforms. Nature Reviews – Molecular Cell Biology 10:287–292

889. Gratton SEA, Ropp PA, Pohlhaus PD, Luft JC, Madden VJ, Napier ME, DeSimone JM (2008) The effect of particle design on cellular internalization pathways. Proceedings of the National Academy of Sciences of the United States of America 105:11613–11618

890. Foret L, Sens P (2008) Kinetic regulation of coated vesicle secretion. Proceedings of the National Academy of Sciences of the United States of America 105:14763–14768

891. Mayor S, Pagano RE (2007) Pathways of clathrin-independent endocytosis. Nature Reviews – Molecular Cell Biology 8:603–612

892. Deborde S, Perret E, Gravotta D, Deora A, Salvarezza S, Schreiner R, Rodriguez-Boulan E (2008) Clathrin is a key regulator of basolateral polarity. Nature 452:719–723

893. Schmidt MR, Maritzen T, Kukhtina V, Higman VA, Doglio L, Barak NN, Strauss H, Oschkinat H, Dotti CG, Haucke V (2009) Regulation of endosomal membrane traffic by a Gadkin/AP-1/kinesin KIF5 complex. Proceedings of the National Academy of Sciences of the United States of America 106:15344–15349

894. Ge L, Qi W, Wang LJ, Miao HH, Qu YX, Li BL, Song BL (2011) Flotillins play an essential role in Niemann-Pick C1-like 1-mediated cholesterol uptake. Proceedings of the National Academy of Sciences of the United States of America 108:551–556

895. Vicinanza M, D'Angelo G, Di Campli A, De Matteis MA (2008) Function and dysfunction of the PI system in membrane trafficking. EMBO Journal 27:2457–2470

896. Fili N, Calleja V, Woscholski R, Parker PJ, Larijani B (2006) Compartmental signal modulation: Endosomal phosphatidylinositol 3-phosphate controls endosome morphology and selective cargo sorting. Proceedings of the National Academy of Sciences of the United States of America 103:15473–15478

897. Panaretou C, Tooze SA (2002) Regulation and recruitment of phosphatidylinositol 4-kinase on immature secretory granules is independent of ADP-ribosylation factor 1. Biochemical Journal 363:289–295

898. Gomez TS, Billadeau DD (2009) A FAM21-containing WASH complex regulates retromer-dependent sorting. Developmental Cell 17:699–711

899. Derivery E, Sousa C, Gautier JJ, Lombard B, Loew D, Gautreau A (2009) The Arp2/3 activator WASH controls the fission of endosomes through a large multiprotein complex. Developmental Cell 17:712–723

900. Jia D, Gomez TS, Metlagel Z, Umetani J, Otwinowski Z, Rosen MK, Billadeau DD (2010) WASH and WAVE actin regulators of the Wiskott-Aldrich syndrome protein (WASP) family are controlled by analogous structurally related complexes. Proceedings of the National Academy of Sciences of the United States of America 107:10442-10447

901. Heuck A, Du TG, Jellbauer S, Richter K, Kruse C, Jaklin S, Müller M, Buchner J, Jansen RP, Niessing D (2007) Monomeric myosin V uses two binding regions for the assembly of stable translocation complexes. Proceedings of the National Academy of Sciences of the United States of America 104:19778–19783

902. Thirumurugan K, Sakamoto T, Hammer JA, Sellers JR, Knight PJ (2006) The cargo-binding domain regulates structure and activity of myosin 5. Nature 442:212–215

903. Gebhardt JCM, Clemen AEM, Jaud J, Rief M (2006) Myosin-V is a mechanical ratchet. Proceedings of the National Academy of Sciences of the United States of America 103:8680–8685

904. Pashkova N, Jin Y, Ramaswamy S, Weisman LS (2006) Structural basis for myosin V discrimination between distinct cargoes. EMBO Journal 25:693–700

905. Syed S, Snyder GE, Franzini-Armstrong C, Selvin PR, Goldman YE (2006) Adaptability of myosin V studied by simultaneous detection of position and orientation. EMBO Journal 25:1795–1803

906. Ali MY, Krementsova EB, Kennedy GG, Mahaffy R, Pollard TD, Trybus KM, Warshaw DM (2007) Myosin Va maneuvers through actin intersections and diffuses along microtubules. Proceedings of the National Academy of Sciences of the United States of America 104:4332–4336

907. Naccache SN, Hasson T, Horowitz A (2006) Binding of internalized receptors to the PDZ domain of GIPC/synectin recruits myosin VI to endocytic vesicles. Proceedings of the National Academy of Sciences of the United States of America 103:12735–12740

908. Hayes MJ, Shao D, Bailly M, Moss SE (2006) Regulation of actin dynamics by annexin 2. EMBO Journal 25:1816–1826

909. Anitei M, Stange C, Parshina I, Baust T, Schenck A, Raposo G, Kirchhausen T, Hoflack B (2010) Protein complexes containing CYFIP/Sra/PIR121 coordinate Arf1 and Rac1 signalling during clathrin–AP-1-coated carrier biogenesis at the TGN. Nature – Cell Biology 12:330–340

910. Bieling P, Telley IA, Piehler J, Surrey T (2008) Processive kinesins require loose mechanical coupling for efficient collective motility. EMBO Reports 9:1121–1127

911. Guillaud L, Wong R, Hirokawa N (2008) Disruption of KIF17-Mint1 interaction by CaMKII-dependent phosphorylation: a molecular model of kinesin-cargo release. Nature – Cell Biology 10:19–29

912. Culver-Hanlon TL, Lex SA, Stephens AD, Quintyne NJ, King SJ (2006) A microtubule-binding domain in dynactin increases dynein processivity by skating along microtubules. Nature – Cell Biology 8:264–270

913. Diehl MR, Zhang K, Lee HJ, Tirrell DA (2006) Engineering cooperativity in biomotor-protein assemblies. Science 311:1468–1471

914. Ross JL, Wallace K, Shuman H, Goldman YE, Holzbaur ELF (2006) Processive bidirectional motion of dynein–dynactin complexes in vitro. Nature – Cell Biology 8:562–570

915. Bock JB, Matern HT, Peden AA, Scheller RH (2001) A genomic perspective on membrane compartment organization. Nature 409:839–841

916. Lupashin VV, Waters MG (1997) t-SNARE activation through transient interaction with a rab-like guanosine triphosphatase. Science 276:1255–1258

917. Barysch SV, Aggarwal S, Jahn R, Rizzoli SO (2009) Sorting in early endosomes reveals connections to docking- and fusion-associated factors. Proceedings of the National Academy of Sciences of the United States of America 106:9697–9702

918. Stroupe C, Hickey CM, Mima J, Burfeind AS, Wickner W (2009) Minimal membrane docking requirements revealed by reconstitution of Rab GTPase-dependent membrane fusion from purified components. Proceedings of the National Academy of Sciences of the United States of America 106:17626–17633

919. Darios F, Ruipérez V, López I, Villanueva J, Gutierrez LM, Davletov B (2010) α-Synuclein sequesters arachidonic acid to modulate SNARE-mediated exocytosis. EMBO Reports 11:528–533

920. Burré J, Sharma M, Tsetsenis T, Buchman V, Etherton MR, Südhof TC (2010) α-Synuclein promotes SNARE-complex assembly in vivo and in vitro. Science 329:1663–1667

921. Cullen PJ (2008) Endosomal sorting and signalling: an emerging role for sorting nexins. Nature Reviews – Molecular Cell Biology 9:574–582

922. Teis D, Saksena S, Judson BL, Emr SD (2010) ESCRT-II coordinates the assembly of ESCRT-III filaments for cargo sorting and multivesicular body vesicle formation. EMBO Journal 29:871–883

923. Traub LM (2010) The reverse logic of multivesicular endosomes. EMBO Reports 11:79–81

924. Wollert T, Hurley JH (2010) Molecular mechanism of multivesicular body biogenesis by ESCRT complexes. Nature 464:864–869

925. Oude Weernink PA, Schmidt M, Jakobs KH (2004) Regulation and cellular roles of phosphoinositide 5-kinases. European Journal of Pharmacology 500:87–99

926. Godi A, Pertile P, Meyers R, Marra P, Di Tullio G, Iurisci C, Luini A, Corda D, De Matteis MA (1999) ARF mediates recruitment of PtdIns-4-OH kinase-beta and stimulates synthesis of PtdIns(4,5)P2 on the Golgi complex. Nature – Cell Biology 1:280–287

927. Kumari S, Mayor S (2008) ARF1 is directly involved in dynamin-independent endocytosis. Nature – Cell Biology 10:30–41

928. Hurtado-Lorenzo A, Skinner M, El Annan J, Futai M, Sun-Wada GH, Bourgoin S, Casanova J, Wildeman A, Bechoua S, Ausiello DA, Brown D, Marshansky V (2006) V-ATPase interacts with ARNO and Arf6 in early endosomes and regulates the protein degradative pathway. Nature – Cell Biology 8:124–136

929. Haas AK, Fuchs E, Kopajtich R, Barr FA (2005) A GTPase-activating protein controls Rab5 function in endocytic trafficking. Nature – Cell Biology 7:887–893

930. Grosshans BL, Ortiz D, Novick P (2006) Rabs and their effectors: achieving specificity in membrane traffic. Proceedings of the National Academy of Sciences of the United States of America 103:11821–11827

931. Gurkan C, Lapp H, Alory C, Su AI, Hogenesch JB, Balch WE (2005) Large-scale profiling of Rab GTPase trafficking networks: the membrome. Molecular Biology of the Cell 16:3847–3864

932. Kitano M, Nakaya M, Nakamura T, Nagata S, Matsuda M (2008) Imaging of Rab5 activity identifies essential regulators for phagosome maturation. Nature 453:241–245

933. Shin HW, Hayashi M, Christoforidis S, Lacas-Gervais S, Hoepfner S, Wenk MR, Modregger J, Uttenweiler-Joseph S, Wilm M, Nystuen A, Frankel WN, Solimena M, De Camilli P, Zerial M (2005) An enzymatic cascade of Rab5 effectors regulates phosphoinositide turnover in the endocytic pathway. Journal of Cell Biology 170:607–618

934. Miserey-Lenkei S, Chalancon G, Bardin S, Formstecher E, Goud B, Echard A (2010) Rab and actomyosin-dependent fission of transport vesicles at the Golgi complex. Nature – Cell Biology 12:645–654

935. Jahn R, Scheller RH (2006) SNAREs? engines for membrane fusion. Nature Reviews – Molecular Cell Biology 7:631–643

936. Barrowman J, Bhandari D, Reinisch K, Ferro-Novick S (2010) TRAPP complexes in membrane traffic: convergence through a common Rab. Nature Reviews – Molecular Cell Biology 11:759–763

937. Roland JT, Bryant DM, Datta A, Itzen A, Mostov KE, Goldenring JR (2011) Rab GTPase-Myo5B complexes control membrane recycling and epithelial polarization. Proceedings of the National Academy of Sciences of the United States of America 108:2789–2794

938. Zeuschner D, Geerts WJC, van Donselaar E, Humbel BM, Slot JW, Koster AJ, Klumperman J (2006) Immuno-electron tomography of ER exit sites reveals the existence of free COPII-coated transport carriers. Nature – Cell Biology 8:377–383

939. Sallese M, Pulvirenti T, Luini A (2006) The physiology of membrane transport and endomembrane-based signalling. EMBO Journal 25:2663–2673

940. Welchman RL, Gordon C, Mayer RJ (2005) Ubiquitin and ubiquitin-like proteins as multifunctional signals. Nature Reviews – Molecular Cell Biology 6:599–609

941. Wilkinson KD, Ventii KH, Friedrich KL, Mullally JE (2005) The ubiquitin signal: assembly, recognition and termination. Symposium on ubiquitin and signaling. EMBO Reports 6:815–820

942. Haglund K, Di Fiore PP, Dikic I (2003) Distinct monoubiquitin signals in receptor endocytosis. Trends in Biochemical Sciences 28:598–603

943. Shenoy SK (2007) Seven-transmembrane receptors and ubiquitination. Circulation Research 100:1142–1154

670   References

944. Raiborg C, Stenmark H (2009) The ESCRT machinery in endosomal sorting of ubiquitylated membrane proteins. Nature 458:445–452

945. Theos AC, Truschel ST, Tenza D, Hurbain I, Harper DC, Berson JF, Thomas PC, Raposo G, Marks MS (2006) A lumenal domain-dependent pathway for sorting to intralumenal vesicles of multivesicular endosomes involved in organelle morphogenesis. Developmental Cell 10:343–354

946. Ambroggio XI, Rees DC, Deshaies RJ (2004) JAMM: a metalloprotease-like zinc site in the proteasome and signalosome. PLoS Biology 2:E2

947. Mittal R, McMahon HT (2009) Arrestins as adaptors for ubiquitination in endocytosis and sorting. EMBO Reports 10:41–43

948. Lefkowitz RJ, Rajagopal K, Whalen EJ (2006) New roles for β-arrestins in cell signaling: not just for seven-transmembrane receptors. Molecular Cell 24:643–652

949. Lin CH, MacGurn JA, Chu T, Stefan CJ, Emr SD (2008) Arrestin-related ubiquitin ligase adaptors regulate endocytosis and protein turnover at the cell surface. Cell 135:714–725

950. Shenoy SK, Xiao K, Venkataramanan V, Snyder PM, Freedman NJ, Weissman AM (2008) Nedd4 mediates agonist-dependent ubiquitination, lysosomal targeting, and degradation of the β2-adrenergic receptor. Journal of Biological Chemistry 283:22166–22176

951. del Pozo MA, Balasubramanian N, Alderson NB, Kiosses WB, Grande-García A, Anderson RGW, Schwartz MA (2005) Phospho-caveolin-1 mediates integrin-regulated membrane domain internalization. Nature – Cell Biology 7:901–908

952. Caswell PT, Vadrevu S, Norman JC (2009) Integrins: masters and slaves of endocytic transport. Nature Reviews – Molecular Cell Biology 10:843–853

953. Zehe C, Engling A, Wegehingel S, Schäfer T, Nickel W (2006) Cell-surface heparan sulfate proteoglycans are essential components of the unconventional export machinery of FGF-2. Proceedings of the National Academy of Sciences of the United States of America 103:15479–15484

954. D'Angelo G, Polishchuk E, Di Tullio G, Santoro M, Di Campli A, Godi A, West G, Bielawski J, Chuang CC, van der Spoel AC, Platt FM, Hannun YA, Polishchuk R, Mattjus P, De Matteis MA (2007) Glycosphingolipid synthesis requires FAPP2 transfer of glucosylceramide. Nature 449:62–67

955. Hasdemir B, Bunnett NW, Cottrell GS (2007) Hepatocyte growth factor-regulated tyrosine kinase substrate (HRS) mediates post-endocytic trafficking of protease-activated receptor 2 and calcitonin receptor-like receptor. Journal of Biological Chemistry 282:29646–29657

956. Babst M (2005) A protein's final ESCRT. Traffic 6:2–9

957. Merrifield CJ, Moss SE, Ballestrem C, Imhof BA, Giese G, Wunderlich I, Almers W (1999) Endocytic vesicles move at the tips of actin tails in cultured mast cells. Nature – Cell Biology 1:72–74

958. Grant BD, Donaldson JG (2009) Pathways and mechanisms of endocytic recycling. Nature Reviews – Molecular Cell Biology 10:597–608

959. Hager HA, Roberts RJ, Cross EE, Proux-Gillardeaux V, Bader D (2010) Identification of a novel Bves function: regulation of vesicular transport. EMBO Journal 29:532–545

960. Glyvuk N, Tsytsyura Y, Geumann C, D'Hooge R, Hüve J, Kratzke M, Baltes J, Böning D, Klingauf J, Schu P (2010) AP-1/σ1B-adaptin mediates endosomal synaptic vesicle recycling, learning and memory. EMBO Journal 29:1318–1330

961. Johannes L, Lamaze C (2002) Clathrin-dependent or not: is it still the question? Traffic 3:443–451

962. van Kerkhof P, Lee J, McCormick L, Tetrault E, Lu W, Schoenfish M, Oorschot V, Strous GJ, Klumperman J, Bu G (2005) Sorting nexin 17 facilitates LRP recycling in the early endosome. EMBO Journal 24:2851–2861

963. Naga Prasad SV, Jayatilleke A, Madamanchi A, Rockman HA (2005) Protein kinase activity of phosphoinositide 3-kinase regulates β-adrenergic receptor endocytosis. Nature – Cell Biology 7:785–796

964. Haglund K, Dikic I (2005) Ubiquitylation and cell signaling. EMBO Journal 24:3353–3359

965. Jozic D, Cárdenes N, Deribe YL, Moncalián G, Hoeller D, Groemping Y, Dikic Y, Rittinger K, Bravo J (2005) Cbl promotes clustering of endocytic adaptor proteins. Nature – Structural and Molecular Biology 12:972–979

966. Pelkmans L, Fava E, Grabner H, Hannus M, Habermann B, Krausz E, Zerial M (2005) Genome-wide analysis of human kinases in clathrin- and caveolae/raft-mediated endocytosis. Nature 436:78–86

967. Sabharanjak S, Sharma P, Parton RG, Mayor S (2002) GPI-anchored proteins are delivered to recycling endosomes via a distinct cdc42-regulated, clathrin-independent pinocytic pathway. Developmental Cell 2:411–423

968. Bonazzi M, Spanò S, Turacchio G, Cericola C, Valente C, Colanzi A, Kweon HS, Hsu VW, Polishchuck EV, Polishchuck RS, Sallese M, Pulvirenti T, Corda D, Luini A (2005) CtBP3/BARS drives membrane fission in dynamin-independent transport pathways. Nature – Cell Biology 7:570–580

969. Smythe E, Ayscough KR (2003) The Ark1/Prk1 family of protein kinases. Regulators of endocytosis and the actin cytoskeleton. EMBO Reports 4:246–251

970. Traub LM (2009) Tickets to ride: selecting cargo for clathrin-regulated internalization. Nature Reviews – Molecular Cell Biology 10:583–596

971. Fotin A, Cheng Y, Sliz P, Grigorieff N, Harrison SC, Kirchhausen T, Walz T (2004) Molecular model for a complete clathrin lattice from electron cryomicroscopy. Nature 432:573–579

972. Hinrichsen L, Meyerholz A, Groos S, Ungewickell EJ (2006) Bending a membrane: How clathrin affects budding. Proceedings of the National Academy of Sciences of the United States of America 103:8715-8720

973. Merrifield CJ, Perrais D, Zenisek D (2005) Coupling between clathrin-coated-pit invagination, cortactin recruitment, and membrane scission observed in live cells. Cell 121:593–606

974. Toshima J, Toshima JY, Martin AC, Drubin DG (2005) Phosphoregulation of Arp2/3-dependent actin assembly during receptor-mediated endocytosis. Nature – Cell Biology 7:246–254

975. Liu J, Kaksonen M, Drubin DG, Oster G (2006) Endocytic vesicle scission by lipid phase boundary forces. Proceedings of the National Academy of Sciences of the United States of America 103:10277–10282

976. Massol RH, Boll W, Griffin AM, Kirchhausen T (2006) A burst of auxilin recruitment determines the onset of clathrin-coated vesicle uncoating. Proceedings of the National Academy of Sciences of the United States of America 103:10265–10270

977. Brodsky FM, Chen CY, Knuehl C, Towler MC, Wakeham DE (2001) Biological basket weaving: formation and function of clathrin-coated vesicles. Annual Review of Cell and Developmental Biology 17:517–568 .

978. Owen DJ, Collins BM, Evans PR (2004) Adaptors for clathrin coats: structure and function. Annual Review of Cell and Developmental Biology 20:153–191

979. Edeling MA, Smith C, Owen D (2006) Life of a clathrin coat: insights from clathrin and AP structures. Nature Reviews – Molecular Cell Biology 7:32–44

980. Fotin A, Cheng Y, Grigorieff N, Walz T, Harrison SC, Kirchhausen T (2004) Structure of an auxilin-bound clathrin coat and its implications for the mechanism of uncoating. Nature 432:649–653

981. Irie F, Okuno M, Pasquale EB, Yamaguchi Y (2005) EphrinB-EphB signalling regulates clathrin-mediated endocytosis through tyrosine phosphorylation of synaptojanin 1. Nature – Cell Biology 7:501–509

982. Rothberg KG, Heuser JE, Donzell WC, Ying YS, Glenney JR, Anderson RGW (1992) Caveolin, a protein component of caveolae membrane coats. Cell 68:673–682

983. Thompsen P, Roepstorff K, Stahlhut M, van Deurs B (2002) Caveolae are highly immobile plasma membrane microdomains, which are not involved in constitutive endocytis trafficking. Molecular Biology of the Cell 13:238–250.

984. Parton RG, Simons K (2007) The multiple faces of caveolae. Nature Reviews – Molecular Cell Biology 8:185–194

985. Le PU, Nabi IR (2003) Distinct caveolae-mediated endocytic pathways target the Golgi apparatus and the endoplasmic reticulum. Journal of Cell Science 116:1059–1071

986. Hansen CG, Bright NA, Howard G, Nichols BJ (2009) SDPR induces membrane curvature and functions in the formation of caveolae. Nature – Cell Biology 11:807–814

987. Nabi IR (2009) Cavin fever: regulating caveolae. Nature – Cell Biology 11:789–791

988. Choudhury A, Marks DL, Proctor KM, Gould GW, Pagano RE (2006) Regulation of caveolar endocytosis by syntaxin 6-dependent delivery of membrane components to the cell surface. Nature – Cell Biology 8:317–328

989. Roux A, Uyhazi K, Frost A, de Camilli P (2006) GTP-dependent twisting of dynamin implicates constriction and tension in membrane fission. Nature 441:528–531

990. Mellman I, Nelson WJ (2008) Coordinated protein sorting, targeting and distribution in polarized cells. Nature Reviews – Molecular Cell Biology 9:833–845

991. Harada A (2010) Molecular mechanism of polarized transport. Journal of Biochemistry 147:619–624

992. Iden S, Collard JG (2008) Crosstalk between small GTPases and polarity proteins in cell polarization. Nature Reviews – Molecular Cell Biology 9:846–859

993. Cresawn KO, Potter BA, Oztan A, Guerriero CJ, Ihrke G, Goldenring JR, Apodaca G, Weisz OA (2007) Differential involvement of endocytic compartments in the biosynthetic traffic of apical proteins. EMBO Journal 26:3737–3748

994. Murphy JE, Padilla BE, Hasdemir B, Cottrell GS, Bunnett NW (2009) Endosomes: a legitimate platform for the signaling train. Proceedings of the National Academy of Sciences of the United States of America 106:17615–17622

995. Sancak Y, Bar-Peled L, Zoncu R, Markhard AL, Nada S, Sabatini DM (2010) Ragulator-Rag complex targets mTORC1 to the lysosomal surface and is necessary for its activation by amino acids. Cell 141:290–303

996. Nada S, Hondo A, Kasai A, Koike M, Saito K, Uchiyama Y, Okada M (2009) The novel lipid raft adaptor p18 controls endosome dynamics by anchoring the MEK-ERK pathway to late endosomes. EMBO Journal 28:477–489

997. Teis D, Wunderlich W, Huber LA (2002) Localization of the MP1–MAPK scaffold complex to endosomes is mediated by p14 and required for signal transduction. Developmental Cell 3:803–814

998. McDonald PH, Chow CW, Miller WE, Laporte SA, Field ME, Lin FT, Davis RJ, Lefkowitz RJ (2000) η-Arrestin 2: a receptor-regulated MAPK scaffold for the activation of JNK3. Science 290:1574–1577

999. Pegtel DM, Cosmopoulos K, Thorley-Lawson DA, van Eijndhoven MA, Hopmans ES, Lindenberg JL, de Gruijl TD, Würdinger T, Middeldorp JM (2010) Functional delivery of viral miRNAs via exosomes. Proceedings of the National Academy of Sciences of the United States of America 107:6328–6333

1000. Amzallag N, Passer BJ, Allanic D, Segura E, Théry C, Goud B, Amson R, Telerman A (2004) TSAP6 facilitates the secretion of translationally controlled tumor protein/histamine-releasing factor via a nonclassical pathway. Journal of Biological Chemistry 279:46104–46112

1001. Chipot M, Hastings S, Kinderlehrer D (2004) Transport in a molecular motor system. ESAIM: Mathematical Modelling and Numerical Analysis 38:1011–1034

1002. Perthame B, Souganidis PE (2010) personal communication

A

Notations: Aliases and Symbols

A.1 Aliases for Molecules

Aliases include all written variants, i.e., any abbreviation, in particular acronyms.[1] *Acronym* corresponds to a word made from the initial letters or syllables of nouns that are pronounceable as a word. Acronyms are generally written with all letters in upper case. Yet, some acronyms are treated as words and written in lower case (e.g., laser [originally LASER] is an acronym for light amplification by stimulated emission of radiation, sonar [originally SONAR] for sound navigation and ranging). A substance name can derive from its chemical name (e.g., amphetamine: α-methylphenethylamine).

Acronyms can give rise to molecule names by adding a scientific suffix such as "-in", a common ending of a molecule name (e.g., sirtuin stands for silent information regulator-2 [alias SIRT]). Other scientific prefixes and suffixes can be frequently detected throughout the present text. Their meaning is given in Notations, particularly for readers from Asia. Many prefixes are used to specify position, configuration and behavior, quantity, direction and motion, structure, timing, frequency, and speed.

Initialisms are abbreviations that are formed from initial letters of a single long noun or several nouns and, instead of being pronounced like an ordinary word, are read letter-by-letter (e.g., DNA that stands deoxyribonucleic acid).

Some abbreviations can give rise to alphabetisms that are written as new words (e.g., Rho-associated, coiled-coil-containing protein kinase (RoCK) that is also called Rho kinase). In biochemistry, multiple-letter abbreviations can also be formed from a single word that can be long (e.g., Cam stands for

---

[1] In general, abbreviations exclude the initials of short function words, such as "and", "or", "of", or "to". However they are sometimes included in acronyms to make them pronounceable (e.g., radar [originally RADAR] for radio detection and ranging). These letters are often written in lower case. In addition, both cardinal (size, molecular weight, etc.) and ordinal (isoform discovery order) numbers in names are represented by digits.

calmodulin that is itself a portmanteau word, Trx for thioredoxin, etc.) as well as short (e.g., Ttn for titin, etc.). In addition, single-letter symbols of amino acids are often used to define a molecule alias (e.g., tyrosine can be abbreviated as Tyr or Y, hence SYK stands for spleen tyrosine kinase).

A *portmanteau* is a word that combines initials and some inner letters of at least 2words (e.g., calmodulin stands for calcium modulated protein; caspase, a cysteine-dependent aspartate-specific protease; chanzyme, an ion channel and enzyme; chemokine, a chemoattractant cytokine;[2] emilin, an elastin microfibril interfacer; porin, a pore-forming protein; restin, a Reed-Steinberg cell-expressed intermediate filament-associated protein, an alias for cytoplasmic linker protein CLiP1 (or CLiP170); serpin, a serine protease inhibitor; siglec, a sialic acid-binding Ig-like lectin; sirtuin, a silent information regulator-2 (two); transceptor, a transporter-related receptor; and Prompt, a promoter upstream transcript).[3]

Aliases use, in general, capital letters and can include hyphens and dots. Yet, as a given protein can represent a proto-oncogene[4] encoded by a gene that can give rise to an oncogene (tumor promoter) after gain- or loss-of-function mutations,[5] the same acronym represents 3 different entities.[6]

---

[2] Cytokines are peptidic, proteic, or glycoproteic regulators that are secreted by cells of the immune system. These immunomodulating agents serve as auto- or paracrine signals.

[3] The upper case initial P in Prompt is used to avoid confusion with command-line interpreter prompt or prompt book to direct precise timing of actions on theater stage.

[4] In 1911, P. Rous isolated a virus that was capable of generating tumors of connective tissue (sarcomas) in chicken. Proteins were afterward identified, the activity of which, when uncontrolled, can provoke cancer, hence the name oncogene given to genes that encode these proteins. Most of these proteins are enzymes, more precisely kinases. The first oncogene was isolated from the avian Rous virus by D. Stéhelin and called Src (from sarcoma). This investigator demonstrated that the abnormal functioning of the Src protein resulted from mutation of a normal gene, or proto-oncogene, which is involved in cell division.

[5] Loss-of-function mutations cause complete or partial loss of function of gene products that operate as tumor suppressors, whereas gain-of-function mutations generate gene products with new or abnormal function that can then act as oncogenes. Typical tumor-inducing agents are enzymes, mostly regulatory kinases and small guanosine triphosphatases, that favor proliferation of cells, which do normally need to be activated to exert their activities. Once their genes are mutated, these enzymes become constitutively active. Other oncogenes include growth factors (a.k.a. mitogens) and transcription factors. Mutations can also disturb signaling axis regulation, thereby raising protein expression. Last, but not least, chromosomal translocation can also provoke the expression of a constitutively active hybrid protein.

[6] Like Latin-derived shortened expressions – as well as foreign words – that are currently written in italics, genes can be italicized. However, this usage is not required in scientific textbooks published by Springer. Italic characters are then used to highlight words within a text to easily target them. Proteins are currently

Besides, a given abbreviation can designate distinct molecules without necessarily erroneous consequence in a given context (e.g., PAR: polyADPribose or protease-activated receptor and GCK: germinal center kinases or glucokinase; in the latter case, the glucokinase abbreviation should be written as GcK or, better, GK).

In addition, a large number of aliases that designate a single molecule results from the fact that molecules have been discovered independently several times with possibly updated functions. Some biochemists uppercase the name of a given molecule, whereas others lowercase (e.g., cell division cycle guanosine triphosphatase of the Rho family CDC42 or Cdc42, adaptor growth factor receptor-bound protein GRB2 or Grb2, chicken tumor virus regulator of kinase CRK or Crk, guanine-nucleotide exchange factor Son-of-sevenless SOS or Sos, etc.). Acronyms are then not always capitalized. Printing style of aliases should not only avoid confusion, but also help one in remembering alias meaning.

In the present textbook, choice of lower and upper case letters in molecule aliases is dictated by the following criteria. (1) An upper case letter is used for initials of molecules constituting nouns (e.g., receptor tyrosine kinase RTK). An alias of any compound takes into account added atoms or molecules (e.g., PI: phosphoinositide and PIP: phosphoinositide phosphate) as well as their number (e.g., PIP2: phosphatidylinositol bisphosphate and DAG: diacylglycerol).

(2) A lower case letter is used when a single letter denotes a subfamily or an isoform when it is preceded by a capital letter (e.g., PTPRe: protein tyrosine phosphatase receptor-like type-E). Nevertheless, an upper case letter is used in an alias after a single or several lower case letters to distinguish the isoform type (e.g., RhoA isoform and DNA-repair protein RecA for recombination protein-A), but OSM stands for oncostatin-M, not osmole Osm[7] to optimize molecule identification.

These criteria enable to use differently written aliases with the same sequence of letters for distinct molecules (e.g., CLIP for corticotropin-like intermediate peptide, CLiP: cytoplasmic CAP-Gly domain-containing linker protein, and iCliP: intramembrane-cleaving protease).

As the exception proves the rule, current aliases, such as PKA and PLA that designate protein kinase-A and phospholipase-A, respectively, have been kept. Preceded by only 2 upper case letters, a lower case letter that should be used to specify an isoform can bring confusion with acronyms of other protein types (e.g., phospholamban alias PLb).

---

romanized (ordinary print), but with a capital initial. Nevertheless, names (not aliases) of chemical species are entirely lowercased in most – if not all – scientific articles, except to avoid confusion with a usual word (e.g., hedgehog animal vs. Hedgehog protein).

[7] Osmole: the amount of osmotically active particles that exerts an osmotic pressure of 1 atm when dissolved in 22.4 l of solvent at 0 C.

Nouns (e.g., hormone-like fibroblast growth factor [hFGF] and urokinase-type plasminogen activator [uPA]) or adjectives (e.g., intracellular FGF isoform [iFGF]) that categorize a subtype of a given molecule correspond to lower case letter to emphasize the molecule species. Hence, an upper case letter with a commonly used hyphen (e.g., I[R]-SMAD that stands for inhibitory [receptor-regulated] SMAD; V-ATPase for vacuolar adenosine triphosphatase; MT1-MMP for membrane type-1 matrix metalloproteinase; and T[V]-SNARE for target [vesicle-associated] soluble ^{N}ethylmaleimide-sensitive factor-attachment protein receptor) is then replaced by a lower case letter (e.g., i[r]SMAD, vATPase, mt1MMP, and t[v]SNARE), as is usual for RNA subtypes (mRNA, rRNA, snRNA, and tRNA for messenger, ribosomal, small nuclear, and transfer RNA). Similarly, membrane-bound and secreted forms of receptors and coreceptors that can derive from alternative mRNA splicing are defined by a lower case letter (e.g., sFGFR for secreted extracellular FGFR form and sFRP for soluble Frizzled-related protein), as well as eukaryotic translation elongation (eEF) and initiation (eIF) factors.

(3) Although l, r, and t can stand for molecule-like, -related, and -type, respectively, when a chemical is related to another one, in general, upper case letters are used for the sake of homogenity and to clearly distinguish between the letter L and numeral 1 (e.g., KLF: Krüppel-like factor, CTK: C-terminal Src kinase (CSK)-type kinase, and SLA: Src-like adaptor).

(4) An upper case letter is most often used for initials of adjectives contained in the molecule name (e.g., AIP: actin-interacting protein; BAX: BCL2-associated X protein; HIF: hypoxia-inducible factor; KHC: kinesin heavy chain; LAB: linker of activated B lymphocytes; MAPK: mitogen-activated protein kinase; and SNAP: soluble N-ethylmaleimide-sensitive factor-attachment protein);

(5) Lower case letters are used when alias letters do not correspond to initials (e.g., Fox – not fox –: forkhead box), except for portmanteau words that are entirely written in minuscules (e.g., gadkin: γ1-adaptin and kinesin interactor).

This rule applies, whether alias letters do correspond to successive noun letters (e.g., Par: partitioning defective protein and Pax: paxillin, as well as BrK: breast tumor kinase and ChK: checkpoint kinase, whereas CHK denotes C-terminal Src kinase [CSK]-homologous kinase) or not (e.g., Fz: Frizzled and HhIP: Hedgehog-interacting protein),[8] except for composite chemical species (e.g., DAG: diacylglycerol). However, some current usages have been kept for

---

[8] The Hedgehog gene was originally identified in the fruit fly Drosophila melanogaster. It encodes a protein involved in the determination of segmental polarity and intercellular signaling during morphogenesis. Homologous gene and protein exist in various vertebrate species. The name of the mammal hedgehog comes from hecg and hegge (dense row of shrubs or low trees), as it resides in hedgerows, and hogg and hogge, due to its pig-like, long projecting nose (snout). The word Hedgehog hence is considered as a seamless whole.

short aliases of chemical species name (e.g., Rho for Ras homolog rather than RHo).

In any case, molecule family names are written in capital letters as well as their members, such as the IGSF (IGSFi; immunoglobulin-), KIF (KIFi; kinesin-), SLC (SLCi; solute carrier-), TNFSF (TNFSFi; tumor-necrosis factor-), and TNFRSF (TNFRSFi; tumor-necrosis factor receptor-) superfamily.

Gene names are also written with majuscules when the corresponding protein name contains at least one minuscule, otherwise only the gene name initial is written with an upper case letter that is then followed by lower case letters.

To highlight its function, substrate aliases (e.g., ARF GTPases) contained in a molecule alias are partly written with lower case letters (e.g., ArfRP, ArfGEF, ArfGAP stand for ARF-related protein, ARF guanine-nucleotide exchange factor, and ARF GTPase-activating protein, respectively).

Last, but not least, heavy and pedantic designation of protein isoforms based on roman numerals have been avoided and replaced by usual arabic numerals (e.g., angiotensin-2 rather than angiotensin-II), except for coagulation (or clotting) factors. Moreover, character I can mean either letter I or number 1 without obvious discrimination at first glance (e.g., GAPI that stands for Ras GTPase-activating protein GAP1, but can be used to designate a growth-associated protein inhibitor).

Unnecessary hyphenation in aliases of substances (between an upper case letter, which can define the molecule function, and the chemical alias, or between it and assigned isotype number) has been avoided. In any case, the Notation section serves not only to define aliases, but also, in some instances, as disambiguation pages.

A.2 Symbols for Physical Variables

Unlike substances aliases, symbols for physical quantities are most often represented by a single letter of the Latin or Greek alphabet (i: current; J: flux; L: length; m: mass; p: pressure; P: power; T: temperature; t: time; u: displacement; v: velocity; x: space; λ: wavelength; μ: dynamic viscosity; ρ: mass density; etc.). These symbols are specified using sub- and superscripts (c_p and c_v: heat capacity at constant pressure and volume, respectively; α_T: thermal diffusivity; λ_T: thermal conductivity; etc.).

A physical quantity associated with a given point in space at a given time can be: (1) a scalar uniquely defined by its magnitude; (2) a vector characterized by a magnitude, a support, and a direction represented by an oriented line segment defined by a unit vector; and (3) a tensor specified by a magnitude and a few directions. To ensure a straightforward meaning of symbols used for scalar, vectorial, and tensorial quantities, bold face upper ($\mathbf{T}$) and lower ($\mathbf{v}$) case letters are used to denote a tensor and a vector, respectively, whereas both roman (plain, upright)-style upper and lower case letters designate a scalar.

Notations — Prefixes and Suffixes

Prefixes (localization)

"ab-" (Latin) and "apo-" (Greek: απο): away from or off (abluminal: endothelial edge opposite to wetted surface; apolipoproteins: lipid carriers that cause egress [also ingress] from cells; aponeurosis (απονευρωσις; νευρον: sinew, tendon) muscle sheath that limits radial motion and enhances axial contraction; and apoptosis: separation ["-ptosis": fall (πτωσις): as leaves fall away from a tree], a type of programmed cell death)

"acr-" (variant "acro-" [ακρος]): top or apex

"ad-" (adfecto: to reach; adfio: to blow toward; adfluo: to flow toward): toward (ad- becomes "ac-" before c, k, or q; "af-" before f [afferent]; "ag-" before g [agglutination]; "al-" before l; "ap-" before p [approximation]; "as-" before s; and "at-" before t)

"cis-", "juxta-", and "para-" (παρα): near, beside, or alongside

"contra-": opposite side; "ipsi-" (ipse): same side; "latero-": side;

"ecto-" (εκτος), "exo-" (εξο), and "extra-": outside, outer, external, or beyond (exogenous chemicals produced by an external source, or xenobiotics ["xeno-": foreigner])

"endo-" (ενδον) and "intra-": inside (endogenous substances synthesized by the body's cells)

"ep-" (variant "eph-", or "epi-" [επι]): upon (epigenetics refers to the inheritance ("-genetic": ability to procreate [γεννητικος]) of variations in gene expression beyond ("epi-": on, upon, above, close to, beside, near, toward, against, among, beyond, and also) change in the DNA sequence.

"front-" and "pre-": anterior or in front of

"post-": behind

"infra-" and "sub-": under or below

"super-" and "supra-": above

"inter-": between or among

"peri-" (περι): around

"tele-" (τελε): remote

"trans-": across

Prefixes (composition)

"an-" and "aniso-" (ανισος): unequal, uneven, heterogeneous

"iso-" (ισος): equal, alike (isomer [μερος: part, portion]

"mono-" (μονος) and "uni-" (unicus): single

"oligo-" (ολιγος): few, little, small

"multi-" (multus), 'pluri-" (plus, plures), and "poly-" (πολυς): many, much

"ultra-": in excess.

Prefixes (quantity)

"demi-" (dimidius) and "hemi-" (ημι): half

"sesqui-": one and a half (half more)

"di-" or "dis-" (δυο; δις) as well as "bi-" or "bis-": 2, twice

"tri" (τρεις, τρι-; tres, tria): 3

"tetra-" (τετρα), "quadri-" (variant: "quadr-" and "quadru-"): 4

"penta-" (πεντας; pentas), "quinqu-", and "quint-": 5

"hexa-" (εξ) and "sexa-": 6

"hepta-" (επτα): 7

"octa-" (οκτα): 8

"nona-" (εννεα): 9 (ninth part)

"deca-" (δεκα): 10

"quadra-" (quadragenarius): 40 (elements)

"quinqua-" (quinquagenarius): 50

"sexa-" (sexagenarius [sex: 6]: 60

"septua-" (septuagenarius [septem: 7]): 70

"nona-" (nonagenarius): 90

Prefixes (motion and direction)

"af-": toward the center (single master object); e.g., nerve and vascular afferents (ferre: to carry) to brain and heart, respectively, rather than toward any slave, supplied tissue from the set of body's organs; also affector, i.e., chemical messenger that brings a signal to the cell considered as the object of interest, this exploration focus being virtually excised from the organism with its central command system, except received signals

"ef-" (effero: to take away): from the center (efferent; effector, i.e., chemical transmitter recruited by the previous mediator of a signaling cascade at a given locus to possibly translocate to another subcellular compartment)

"antero-" (anterior): before, in front of, facing, or forward

"retro-": behind or backward

"tropo-" (τροπος): duct direction; (tropa: rotation; celestial revolution); e.g., tropomyosin (μυς, musculus: muscle; μυο-: refers to muscle [μυοτρωτος: injured at a muscle])

Prefixes (structure and size)

"macro-" (μακρος): large, long, or big

"mega-" (μεγας): great, large

"meso-" (μεσος): middle

"micro-" (μικρος): small

"nano-" (νανος): dwarf, tiny

"homo-" (ομο-): same (ομολογος: agreeing, corroborating; variant: "homeo-" [homeostasis])

"hetero-" (ετερο-): other

Prefixes (timing)

"ana-" (ανα): culminating (anaphase of the cell division cycle), up, above (ανοδος: a way up, anode [positive electrode; οδος; way, path, road, track])

"ante-": before

"circa-": approximately, around (circadian: approximately one day)

"infra-": below, shorter (infradian: rhythm with lower frequency than that of circadian rhythm, not smaller period)

"inter-": among, between, during

"meta-" (μετα): after, beyond, behind, later; in the middle of (metaphase of the cell division cycle); as well as connected to, but with a change of state (metabolism) and about (metadata)

"post-": after

"pre-": earlier

"pro-" (προ): preceding, first, before (prophase of the cell division cycle)

"telo-" (τελος): end, completion

"ultra-": beyond, longer (ultradian: period smaller than that of

24–28-hour cycle, i.e., frequency greater than that of the circadian rhythm)

Prefixes (functioning modality)

"auto-" (αυτος): same, self

"brady-" (βραδυς): slow (decelerate)

"tachy-" (ταχος): rapid (accelerate)

"amphi-" (αμφι): both (amphiphilic substances are both hydrophilic and lipophilic; amphisomes are generated by both autophago-somes and endosomes)

"ana-" : upward (anabolism) or against (anaphylaxis)

"cata-" (κατα): downward (catabolism, cathode [negative electrode; οδος; way, path, road, track])

"anti-" (αντι): against

"pro-": favoring

"co-" (coaccedo: add itself to): together

"contra-": adverse, against, beside, next to, opposite

"de-": remove, reduce, separation after association (Latin de; e.g., deoxy-)

"dys-" (δυς): abnormal (δυσαης: ill-blowing)

"equi-" (æque): equal or alike

"hem-" or "hemat-" (αιμα: blood): related to blood

"hypo-" (υπο): under, beneath, and low

"hyper-" (υπερ): above, beyond, and large

"per-": through (e.g., percutaneous) and during (e.g., peroperative)

"pseudo-" (ψευδο): pretended, false

"re-"; again

Scientific suffixes

"-ase": enzyme (synthase, lipase, etc.)

"-ate": salt of a base

"-cyte" (κυτος): cell (erythro- [ερυθρος: red], leuko- [λευκος: light, bright, clear, white], thrombo- [θρομβος: lump, clot], adipo- [adeps: fat; adipalis, adipatus, adipeus, adipinus: fatty], fibro- [fibra: fiber,

filament], myo- [μυς: muscle, mouse, mussel], myocardiocyte [κραδια: heart; cardiacus: related to heart, stomach; to have heart trouble, stomach trouble], etc.);

"-crine" (κρινω): to decide, to separate, and to secrete (e.g., endocrine regulator) (ευκρινεω: keep in order)

"-elle": small (organelle in a cell [like an organ in a body])

"-ium", "-ion", "-isk", and "-iscus": little ("-ium": tissue interface and envelop, such as endothelium and pericardium)

"-phil" (φιλια): attracted (αφιλια: want of friends)

"-phob" (φοβια): repulsed (υδροφοβια, hydrophobia [Latin]: horror of water)

"-phore" (φερω): carrier (αμφερω: to bring up)

"-yl" denotes a radical (molecules with unpaired electrons)

"-ploid" (πλοω): double, fold (diploid, twofold; διπλοω: to double; διαπλοω: unfold)

"-emia": in relation to flow (ανεμια: flatulence; ευηνεμια: fair wind), particularly blood condition

"-genesis" (γενεσις): cause, generation, life source, origin, productive force

"-iasis": for diseased condition

"-itis": inflammation

"-lemma" (λεμμα: skin): sheath

"-ole" and "-ule": small (arteriole and venule (variant "-ula" [blastula] and "-ulum")

"-plasma" (πλασμα): anything moulded (plasma: creature generated from silt of earth)

"-plasia" (πλασια): formation, moulding

"-podium" (ποδος: foot; podium [Latin]: small knoll, small protuberance): protrusion

"-poiesis" (ποιεω): production

"-soma" (σωμα): body

"-sclerosis" (σκλημα): hardness, induration

"-stasis" (στασις): stabilization (αποκαταστασις: restoration; ανυποστασις: migration)

"-stomosis" (στομα: mouth): equipped with an outlet

"-taxy/tactic" (ταχυ: rapid; τακτικος: to maneuver): related to motion (also prefix, i.e., ταχυκινησις: quick motion; ταχυνω: to accelerate; and ταχυπνοια: short breath; not [δια]ταξις: disposition, arrangement)

"-trophy/trophic" (τροφις: well fed): related to growth

"-oma": tumor of

"-pathy" (παθος, παθεια): disease of

"-tomy" (τομια) and "-ectomy": surgical removal (απλοτομια: simple incision; φαûrhουγγοτομια: laryngotomy)

Biochemical, Medical, and Physical Aliases

A

$\mathcal{A}$: Avogadro number

$\mathcal{A}(p)$: area-pressure relation

A: Almansi strain tensor

A: cross sectional area

A: actin-binding site

a: acceleration

a: major semi-axis

AA: arachidonic acid

AAA: ATPase associated with diverse cellular activities

AAA: abdominal aortic aneurism

AAAP: aneurism-associated antigenic protein

AAK: adaptin-associated kinase

ABC: ATP-binding cassette transporter (transfer ATPase)

AbI: Abelson kinase interactor

Abl: Abelson leukemia viral proto-oncogene product (NRTK)

ABLIM: actin-binding LIM domain-containing protein

ABP: actin-binding protein

AC: atrial contraction

ACAP: ArfGAP with coiled-coil, ankyrin repeat, PH domains

ACase: adenylyl cyclase

ACi: adenylyl cyclase isoform i

ACAT: acylCoA–cholesterol acyltransferase

ACC: acetyl coenzyme-A carboxylase

ACE: angiotensin converting enzyme

ACh: acetylcholine

mAChR: acetylcholine muscarinic receptor (metabotropic; GPCR)

nAChR: acetylcholine nicotinic receptor (ionotropic; LGIC)

ACK: activated CDC42-associated kinase

ACTH: adrenocorticotropic hormone

Cav-actin: caveolin-associated F actin

^{F}actin: filamentous actin

^{G}actin: monomeric globular actin

AcvR: activin receptor (TGFβ receptor superfamily)

Ad: adrenaline

ADAM: a disintegrin and metalloprotease (adamalysin)

ADAMTS: a disintegrin and metalloprotease with thrombospondin

ADAP: adhesion and degranulation-promoting adaptor protein

ADAP: ArfGAP with dual PH domains

ADF: actin depolymerizing factor (cofilin-related destrin)

ADH: antidiuretic hormone (vasopressin)

ADMA: asymmetric dimethylarginine

ADP: adenosine diphosphate

AE: anion exchanger

AEA: N-arachidonoyl ethanolamine (anandamide)

AF: atrial fibrillation

AFAP: ArfGAP with phosphoinositide-binding and PH domains

AGAP: ArfGAP with GTPAse, ankyrin repeat, and PH domains

AGF: autocrine growth factor

AGFG: ArfGAP with FG repeats

Ago: Argonaute protein

AGS: activator of G-protein signaling

AHR: aryl hydrocarbon receptor

AIF: apoptosis-inducing factor

AIP: actin-interacting protein

AIRe: autoimmune regulator

AKAP: A-kinase anchoring protein (for protein kinase A)

ALE: arbitrary Eulerian Lagrangian

ALIX: apoptosis-linked gene-2-interacting protein-X

ALK: anaplastic lymphoma kinase

ALKi: type-i activin receptor-like kinase (TGFβ receptor superfamily)

ALOx5: arachidonate 5-lipoxygenase

ALOx5AP: arachidonate 5-lipoxygenase activation protein

ALP: actinin-associated LIM protein (PDLIM3)

alsin: amyotrophic lateral sclerosis protein

ALX: adaptor in lymphocytes of unknown function X

AMAP: A multidomain ArfGAP protein

AMBRA: activating molecule in beclin-1-regulated autophagy protein

AMHR: anti-Mullerian hormone receptor (TGFβ receptor superfamily)

AMIS: apical membrane initiation site (lumenogenesis)

AMPAR: α-amino 3-hydroxy 5-methyl 4-isoxazole propionic acid receptor

AMPK: AMP-activated protein kinase

AMSH: associated molecule with SH3 domain (deubiquitinase)

AmyR: amylin receptor

Ang: angiopoietin

AngL: angiopoietin-like molecule

Ank: ankyrin

ANP: atrial natriuretic peptide

ANPR (NP$_1$): atrial natriuretic peptide receptor (guanylyl cyclase)

ARNT: aryl hydrocarbon nuclear receptor translocator

ANS: autonomic nervous system

ANT: adenine nucleotide transporter

Anx: annexin

AOC: amine oxidase copper-containing protein

AoV: aortic valve

AP: (clathrin-associated) adaptor proteic complex

AP: Activator protein (transcription factor)

AP: activating enhancer-binding protein

AP4A: diadenosine tetraphosphate

APAF: apoptotic protease-activating factor

APAP: ArfGAP with PIx- and paxillin-binding domains

APC: antigen-presenting cell

APC: adenomatous polyposis coli protein (Ub ligase)

APC/C: anaphase-promoting complex (or cyclosome; Ub ligase)

APH: anterior pharynx defective phenotype homolog

APl: action potential

Apn: adiponectin

Apo: apolipoprotein

ApoER: apolipoprotein-E receptor

APPL: adaptor containing phospho-Tyr interaction, PH domain, and Leu zipper

APS: adaptor with a PH- and SH2 domain

Aqp: aquaporin

AR: adrenergic receptor (adrenoceptor)

AR: androgen receptor (nuclear receptor NR3c4; transcription factor)

AR: area ratio

ARAP: ArfGAP with RhoGAP, ankyrin repeat, PH domains

Areg: amphiregulin (EGF superfamily member)

ARF: ADP-ribosylation factor

ArfRP: ARF-related protein

ARFTS: CKI2A-locus alternate reading frame tumor suppressor (ARF or p14^{ARF})

ARH: autosomal recessive hypercholesterolemia adaptor (low-density lipoprotein receptor adaptor)

ARH: aplysia Ras-related homolog

ArhGEF: RhoGEF

ARL: ADP-ribosylation factor-like protein

ARNO: Arf nucleotide site opener

ARP: absolute refractory period

ARP: actin-related protein

ARPP: cAMP-regulated phosphoprotein

ART: arrestin-related transport adaptor (α-arrestin)

Artn: artemin

ARVCF: armadillo repeat gene deleted in velocardiofacial syndrome

ARVD: arrythmogenic right ventricular dystrophy

AS: Akt (PKB) substrate

ASAP: artery-specific antigenic protein

ASAP: ArfGAP with SH3, ankyrin repeat, PH domains

ASIC: acid-sensing ion channel

ASP: actin-severing protein

ASK: apoptosis signal-regulating kinase

AT: antithrombin

ATAA: ascending thoracic aortic aneurism

ATF: activating transcription factor

AtG: autophagy-related gene product

ATMK: ataxia telangiectasia mutated kinase

ATn: angiotensin

ATng: angiotensinogen

ATP: adenosine triphosphate

ATPase: adenosine triphosphatase

ATR ($AT_{1/2}$): angiotensin receptor

ATRK: ataxia telangiectasia and Rad3-related kinase

AVN: atrioventricular node

AVV: atrioventricular valves

AW: analysis window

B

B: Biot-Finger strain tensor

B: bulk modulus

$\mathcal{B}$: bilinear form

b: minor semi-axis

b: body force

$\hat{\mathbf{b}}$: unit binormal

BACE: β-amyloid precursor protein-converting enzyme

BAD: BCL2 antagonist of cell death

BAF: barrier-to-auto-integration factor

BAG: BCL2-associated athanogene (chaperone regulator)

BAI: brain-specific angiogenesis inhibitor (adhesion-GPCR)

BAIAP: brain-specific angiogenesis inhibitor-1-associated protein (insulin receptor substrate)

BAK: BCL2-antagonist–killer

(i)BALT: (inducible) bronchus-associated lymphoid tissu

BAnk: B-cell scaffold with ankyrin repeats

Barkor: beclin-1-associated autophagy-related key regulator

BAT: brown adipose tissue

BATF: basic leucine zipper ATF-like transcription factor (B-cell-activating transcription factor)

BAX: BCL2-associated X protein

BBB: blood–brain barrier

BC: boundary condition

bCAM: basal cell adhesion molecule (Lutheran blood group glycoprotein)

BCAP: B-cell adaptor for phosphatidyl-inositol 3-kinase

BCAR: Breast cancer anti-estrogen resistance docking protein

BCL: B-cell lymphoma (leukemia) protein

BCLxL: B-cell lymphoma extra-large protein

BCR: B-cell receptor

BCR: breakpoint cluster region protein

Bdk: bradykinin

BDNF: brain-derived neurotrophic factor

Becn, beclin: BCL2-interacting protein

BEM: boundary element method

Best: bestrophin

BFUe: burst forming unit erythroid

BFUmeg: burst forming unit megakaryocyte

BGT: betaine–GABA transporter

BID: BH3-interacting domain death agonist

BIG: brefeldin-A-inhibited GEFs for ARFs

BIK: BCL2-interacting killer

BIM: BH3-containing protein BCL2-like 11 (BCL2L11)

BK: high-conductance, Ca^{++}-activated, voltage-gated K^+ channel

BLK: B-lymphoid tyrosine kinase

Blm: Bloom syndrome, RecQ DNA helicase-like protein

BLnk: B-cell linker protein

BM: basement membrane

BMAL: brain and muscle ARNT-like protein (gene Bmal)

BMAT: bone-marrow adipose tissue

BMF: BCL2 modifying factor

BMP: bone morphogenetic protein (TGFβ superfamily)

BMPR: bone morphogenetic protein receptor

BNIP: BCL2/adenovirus E1B 19-kDa protein-interacting protein

BNP: B-type natriuretic peptide

BMX: bone marrow Tyr kinase gene in chromosome-X product

BOC: brother of CDO

BOK: BCL2-related ovarian killer

BORG: binder of Rho GTPase

BRAG: brefeldin-resistant ArfGEF

BrCa: breast cancer-associated (susceptibility) protein (tumor suppressor; DNA damage repair; a.k.a. FancD1)

BrD: bromodomain-containing protein

BrK: breast tumor kinase

BrSK: brain-selective kinase

BSEP: bile salt export pump

BTF: basic transcription factor

BTK: Bruton Tyr kinase

BUB: budding uninhibited by benzimidazoles

C

C: stress tensor

C: compliance

C: heat capacity

C: chronotropy

Cx: type-x chemokine C (γ)

C_D: drag coefficient

C_f: friction coefficient

C_L: lift coefficient

C_p: pressure coefficient

c: stress vector

c_τ: shear

c_w: wall shear stress

c: concentration

$c(p)$: wave speed

c_p: isobar heat capacity

c_v: isochor heat capacity

C1P: ceramide 1-phosphate

C-terminus: carboxy (carboxyl group COOH)-terminus

C/EBP: CCAAT/enhancer-binding protein

CA: computed angiography

CAi: carbonic anhydrase isoform i

Ca: calcium

Ca_V: voltage-gated Ca^{++} channel

$Ca_V1.x$: L-type high-voltage-gated Ca^{++} channel

$Ca_V2.x$: P/Q/R-type Ca^{++} channel

$Ca_V3.x$: T-type low-voltage-gated Ca^{++} channel

CAAT: cationic amino acid transporter

CABG: coronary artery bypass grafting

Cables: CDK5 and Abl enzyme substrate

CAK: CDK-activating kinase (pseudokinase)

Cam: calmodulin (calcium-modulated protein)

CamK: calmodulin-dependent kinase

cAMP: cyclic adenosine monophosphate

CAP: adenylyl cyclase-associated protein

CAP: carboxyalkylpyrrole protein adduct

CAP: chromosome-associated protein (BrD4)

CaPON: carboxy-terminal PDZ ligand of NOS1

CAPN: calpain gene

CAR: constitutive androstane receptor (NR1i3)

CaR: calcium-sensing receptor

CARP: cell division cycle and apoptosis regulatory protein

CAS: cellular apoptosis susceptibility protein

CAS: CRK-associated substrate (or P130CAS and BCAR1)

CAs: cadherin-associated protein

CASK: calcium–calmodulin-dependent serine kinase (pseudokinase)

CASL: CRK-associated substrate-related protein (CAS2)

CASP: cytohesin-associated scaffold protein

caspase: cysteine-dependent aspartate-specific protease

Cav: caveolin

CBF: coronary blood flow

CBF: core-binding factor

CBL: Casitas B-lineage lymphoma adaptor and Ub ligase

CBLb: CBL-related adaptor

CBP: cap-binding protein

CBP: CREB-binding protein

CBP: C-terminal Src kinase-binding protein

CBS: cystathionine β-synthase (H_2S production)

CCDC: coiled-coil domain-containing protein

CCICR: calcium channel-induced Ca^{++} release

CCK4: colon carcinoma kinase 4 (PTK7)

CCL: chemokine CC-motif ligand

Ccn: cyclin

Ccnx–CDKi: type-x cyclin–type-i cyclin-dependent kinase dimer

CCPg: cell-cycle progression protein

CCT: chaperonin containing T-complex protein

CCx: type-x chemokine CC (β)

CCR: chemokine CC motif receptor

CD: cluster determinant protein (cluster of differentiation)

CDase: ceramidase

CDC: cell-division-cycle protein

CDH: CDC20 homolog

Cdh: cadherin

CDK: cyclin-dependent kinase

Cdm: caldesmon

CDO: cell adhesion molecule-related/ downregulated by oncogenes

CE (CsE): cholesteryl esters

CEC: circulating endothelial cell

CELSR: cadherin, EGF-like, LAG-like, and seven-pass receptor

CenP: centromere protein

CEP: carboxyethylpyrrole

CeP: centrosomal protein

CEPC: circulating endothelial progenitor cell

Cer: ceramide

CerK: ceramide kinase

Cer:

CeRT: ceramide-transfer protein

CETP: cholesterol ester transfer protein

CFD: computational fluid dynamics

CFLAR: caspase-8 and FADD-like apoptosis regulator

CFTR: cystic fibrosis transmembrane conductance regulator

CFU: colony-forming unit

CFUb: CFU basophil (basophil-committed stem cells)

CFUc: CFU in culture (granulocyte precursors, i.e., CFUgm)

CFUe: CFU erythroid

CFUeo: CFU eosinophil

CFUg: CFU granulocyte

CFUgm: CFU granulocyte–macrophage

CFUgemm: CFU granulocyte–erythroid–macrophage–megakaryocyte

CFUm: CFU macrophage

CFUmeg: CFU megakaryocyte

CFUs: colony-forming unit spleen (pluripotent stem cells)

CG: chromogranin

cGK: cGMP-dependent protein kinase (protein kinase G)

cGMP: cyclic guanosine monophosphate

CGN: cis-Golgi network

CGRP: calcitonin gene-related peptide

chanzyme: ion channel and enzyme

chemokine: chemoattractant cytokine

CHIP: C-terminus heat shock cognate-70-interacting protein

ChK: checkpoint kinase

CHK: CSK homologous kinase

CHOP: CCAAT/enhancer-binding protein homologous protein

CHREBP: carbohydrate-responsive element-binding protein

CI: cardiac index

CICR: calcium-induced calcium release

Cin: chronophin

CIP: CDC42-interacting protein

CIP2a: cancerous inhibitor of protein phosphatase-2A

CIPC: CLOCK-interacting protein, circadian

CIS: cytokine-inducible SH2-containing protein

CK: creatine kinase

CK: casein kinase

CKI: cyclin-dependent kinase inhibitor

CLAsP: CLiP-associated protein (microtubule binder)

ClASP: clathrin-associated sorting protein

CLC: cardiotrophin-like cytokine

ClC: voltage-gated chloride channel

ClCa: calcium-activated chloride channel

ClIC: chloride intracellular channel

CLINT: clathrin-interacting protein located in the trans-Golgi network

CLIP: corticotropin-like intermediate peptide

CLiP: cytoplasmic CAP-Gly domain-containing linker protein

iCliP: intramembrane-cleaving protease (that clips)

CLK: CDC-like kinase

ClNS: Cl$^-$ channel nucleotide-sensitive

CLOCK: circadian locomotor output cycles kaput

CLP: common lymphoid progenitor

CLS: ciliary localization signal

CMLP: common myeloid–lymphoid progenitor

CMP: common myeloid progenitor

CMC: cardiomyocyte

Cmi: chylomicron

Col: collagen

CoLec: collectin

ColF: collagen fiber

CNG: cyclic nucleotide-gated channel

CNS: central nervous system

CNT: connecting tubule

CNTi: concentrative nucleoside transporter (SLC28ai)

CNTF: ciliary neurotrophic factor

CntnAP: contactin-associated protein

CO: cardiac output

CoBl: Cordon-bleu homolog (actin nucleator)

COLD: chronic obstructive lung disease

COOL: Cloned out of library (RhoGEF6/7)

COx: cyclooxygenase

COx17: cytochrome-C oxidase copper chaperone

CoP: coat protein

CoP: constitutive photomorphogenic protein (Ub ligase)

COPD: chronic obstructive pulmonary disease

COUP-TF: chicken ovalbumin upstream promoter-transcription factor (NR2f1/2)

CP4H: collagen prolyl 4-hydroxylase

CPC: chromosomal passenger complex

CpG: cytidine-phosphate–guanosine oligodeoxynucleotide (motif)

Cpx: complexin

CR: complement component receptor

Cr: creatine

CRABP: cellular retinoic acid-binding protein

CRAC: Ca^{++} release-activated Ca^{++} channel

CRACR: CRAC regulator

Crb: Crumbs homolog polarity complex

CRE: cAMP-responsive element

CREB: cAMP-responsive element-binding protein

CRF: corticotropin-releasing factor (family)

CRH: corticotropin-releasing hormone

CRIB: CDC42/Rac interactive-binding protein

CRIK: citron Rho-interacting, serine–threonine kinase (STK21)

CRK: CT10 regulator of kinase

CRK: chicken tumor virus regulator of kinase

CRKL: V-CRK avian sarcoma virus CT10 homolog-like

CRL4: cullin-4A RING E3 ubiquitin ligase

CRLR: calcitonin receptor-like receptor

CRP: C-reactive protein

CRTC: CREB-regulated transcription coactivator

Cry: cryptochrome

Cs: cholesterol

CSBP: cytokine-suppressive anti-inflammatory drug-binding protein

CSE: cystathionine γ-lyase (H_2S production)

CSF: cerebrospinal fluid

CSF: colony-stimulating factor

CSF1: macrophage colony-stimulating factor (mCSF)

CSF2: granulocyte macrophage colony-stimulating factors (gmCSF and sargramostim)

CSF3: granulocyte colony-stimulating factors (gCSF and filgrastim)

CSK: C-terminal Src kinase

Csk: cytoskeleton

Csq: calsequestrin

CSS: candidate sphingomyelin synthase

CT: cardiotrophin

CT: computed tomography

CTBP: C-terminal-binding protein

CTen: C-terminal tensin-like protein

CTF: C-terminal fragment

CTGF: connective tissue growth factor

CTL: cytotoxic T lymphocyte

CTLA: cytotoxic T-lymphocyte-associated protein

Ctn: catenin

CtR: calcitonin receptor

CTRC: CREB-regulated transcription coactivator

Cul: cullin

CUT: cryptic unstable transcript

CVI: chronic venous insufficiency

CVLM: caudal ventrolateral medulla

CVP: central venous pressure

CVS: cardiovascular system

Cx: connexin

CXCLi: type-i CXC (C-X-C motif; α) chemokine ligand

CXCRi: type-i CXC (C-X-C motif; α) chemokine receptor

CX3CLi: type-i CX3C (δ) chemokine ligand

CX3CRi: type-i CX3C (δ) chemokine receptor

cyCK: cytosolic creatine kinase

Cyld: cylindromatosis tumor suppressor protein (deubiquitinase)

CyP: member of the cytochrome-P450 superfamily

C3G: Crk SH3-binding GEF

D

D: dromotropy

D: vessel distensibility

$\mathcal{D}$: diffusion coefficient

D: deformation rate tensor

d: displacement vector

D: flexural rigidity

D: demobilisation function (from proliferation to quiescence)

d: death rate, degradation rate

d: duration

Dab: Disabled homolog

DAD: delayed afterdepolarization

DAG: diacylglycerol

DAPC: dystrophin-associated protein complex

DAPK: death-associated protein kinase

DARC: Duffy antigen receptor for chemokine

DAT: dopamine active transporter

DAX: dosage-sensitive sex reversal, adrenal hypoplasia critical region on chromosome X (NR0b1)

DBC: deleted in breast cancer protein

DBP: albumin D-element binding protein (PAR/b–ZIP family)

DC: dendritic cell

cDC: classical dendritic cell

lpDC: lamina propria dendritic cell

mDC: myeloid dendritic cell

pDC: plasmacytoid dendritic cell

pre-cDC: pre-classical dendritic cell

DCA: directional coronary atherectomy

DCAF: DDB1- and Cul4-associated factor

DCC: deleted in colorectal carcinoma (netrin receptor)

DCT: distal convoluted tubule

Dctn: dynactin

DDAH: dimethylarginine dimethylaminohydrolase

DDB: damage-specific DNA-binding protein

DDEF: development and differentiation-enhancing factor (ArfGAP)

DDR: discoidin domain receptor

De: Dean number

DEC: differentially expressed in chondrocytes (DEC1 and DEC2 are a.k.a bHLHe40 and bHLHe41, bHLHb2 and bHLHb3, or HRT2 and HRT1)

DEC: deleted in esophageal cancer

DETC: dendritic epidermal $\gamma\delta$ T cell

DH: Dbl homology

DHET: dihydroxyeicosatrienoic acid

DHh: desert Hedgehog

Dia: Diaphanous

DICOM: digital imaging and communication for medicine

DICR: depolarization-induced Ca^{++} release

DISC: death-inducing signaling complex

Dkk: Dickkopf

DLg: Disc large homolog

DLL: Delta-like (Notch) ligand

DLx: distal-less homeobox protein

DM: double minute

DMM: DNA methylation modulator

DMPK: myotonic dystrophy-associated protein kinase

DN1: double-negative-1 cell

DN2: double-negative-2 cell

DN3: double-negative-3 cell

DNA: deoxyribonucleic acid

DNAPK: DNA-dependent protein kinase

DoC2: double C2-like domain-containing protein

DOCK: dedicator of cytokinesis (GEF)

DOK: downstream of tyrosine kinase docking protein

DOR: δ-opioid receptor

DPG: diphosphoglyceric acid

DRAM: damage-regulated modulator of autophagy

DRF: Diaphanous-related formin (for GTPase-triggered actin rearrangement)

Drl: Derailed

Dsc: desmocollin

Dsg: desmoglein

Dsh: Disheveled (Wnt-signaling mediator)

DSK: dual-specificity kinase

Dst: dystonin

DUb: deubiquitinase

DUS: Doppler ultrasound

DUSP: dual-specificity phosphatase

DV: dead space volume

Dvl: Disheveled (cytoplasmic phospho-protein; other alias Dsh)

DVT: deep-vein thrombosis

dynactin: dynein activator

DYRK: dual-specificity Tyr (Y) phosphorylation-regulated kinase

E

$\mathbf{E}$: strain tensor

$\mathbf{E}$: electric field

E: elastic modulus

E: elastance

$\mathcal{E}$: energy

$\{\hat{\mathbf{e}}_i\}_{i=1}^3$: basis

$\mathbf{e}$: strain vector

e: specific free energy

E-box: enhancer box sequence of DNA (e.g., see circadian control)

E2: E2 ubiquitin-conjugase

E3: E3 ubiquitin-ligase

EAAT: excitatory amino acid (glutamate) transporter

EAD: early afterdepolarization

EAR: V-erbA-related nuclear receptor (NR2f6)

EB: end-binding protein

EBCT: electron beam CT

EBF: early B-cell factor

EC: endothelial cell

ECA: external carotid artery

ECF: extracellular fluid

ECG: electrocardiogram

ECM: extracellular matrix

ED1L: EGF-like repeat- and discoidin-1-like domain-containing protein

EDGR: endothelial differentiation gene receptor

EDHF: endothelial-derived hyperpolarizing factor

EDIL: EGF-like repeats and discoidin-1 (I)-like domain-containing protein

EDV: end-diastolic volume

EEA: early endosomal antigen

eEF: eukaryotic translation elongation factor

EEL: external elastic lamina

EET: epoxyeicosatrienoic acid

EFA6: exchange factor for ARF6 (ArfGEF)

EF-Tu: elongation factor Tu

EGF: epidermal growth factor

EGFL: EGF-like domain-containing protein

EGFR: epidermal growth factor receptor

EGR: early growth response transcription factor

EHD: C-terminal EGFR substrate-15 homology domain-containing protein

eIF: eukaryotic translation initiation factor

EL: endothelial lipase

ELAM: endothelial–leukocyte adhesion molecules

ELCA: excimer laser coronary angioplasty

ELK: ETS-like transcription factor (ternary complex factor [TCF] subfamily)

ElMo: engulfment and cell motility adaptor

ELP: early lymphoid progenitor

EMI: early mitotic inhibitor

EMR: EGF-like module containing, mucin-like, hormone receptor-like protein

EMT: epithelial–mesenchymal transition

Eln: elastin

EnaH: Enabled homolog

ENA–VASP: Enabled homolog and vasoactive (vasodilator)-stimulated phosphoprotein family

ENaC: epithelial Na^+ channel

ElnF: elastin fiber

ENPP: ectonucleotide pyrophosphatase–phosphodiesterase

Ens: endosulfine

ENT: equilibrative nucleoside transporter

ENTPD: ectonucleoside triphosphate diphosphohydrolase

EPAC: exchange protein-activated by cAMP

EPAS: endothelial PAS domain protein

EPC: endothelial progenitor cell

EPCR: endothelial protein-C receptor

EPDC: epicardial-derived cell

Epgn: epigen (EGF superfamily member)

Eph: erythropoietin-producing hepatocyte kinase or pseudokinase (EphA10 and EphB6)

Ephrin: Eph receptor interactor

Epo: erythropoietin

EPS: epidermal growth factor receptor pathway substrate

ER: endoplasmic reticulum

ER: estrogen receptor

ErbB: erythroblastoma viral gene product B (HER)

ERE: estrogen response element (DNA sequence)

Ereg: epiregulin (EGF superfamily member)

eRF: eukaryotic release factor

ERGIC: endoplasmic reticulum–Golgi intermediate compartment

ERK: extracellular signal-regulated protein kinase

ERK1/2: usually refers to ERK1 and ERK2

ERM: ezrin–radixin–moesin

ERMES: endoplasmic reticulum–
   mitochondrion encounter
   structure
ERP: effective refractory period
ERR: estrogen-related receptor
   (NR3b1–NR3b3)
ESCRT: endosomal sorting complex
   required for transport
ESL: E-selectin ligand
ESRP: epithelial splicing regulatory
   protein
ESV: end-systolic volume
ET: endothelin
ETP: early thymocyte progenitor
ETR ($ET_{A/B}$): endothelin receptor
ETS: E-twenty six (transcription
   factor; erythroblastosis virus E26
   proto-oncogene product homolog)
ETV: ETS-related translocation variant
EVAR: endovascular aneurism repair
Exo: exocyst subunit
Ext: exostosin (glycosyltransferase)

F

F: transformation gradient tensor
F: function fraction of proliferating cells
F: erythrocytic rouleau fragmentation
   rate
f: surface force
f̂: fiber direction unit vector
f: binding frequency
f_C: cardiac frequency
f_R: breathing frequency
f: friction shape factor
f_v: head loss per unit length
f_i: molar fraction of gas component i
FA: fatty acid
FABP: fatty acid-binding protein
FABP: filamentous actin-binding
   protein
FACAP: F-actin complex-associated
   protein
FAD: flavine adenine dinucleotide
FADD: Fas receptor-associated death
   domain
FAK: focal adhesion kinase
Fanc: Fanconi anemia protein

FAN: Fanconi anemia-associated
   nuclease
FAPP: phosphatidylinositol four-
   phosphate adapter protein
Fas: death receptor (TNFRSF6a)
FasL: death ligand (TNFSF6)
FAST: Forkhead activin signal
   transducer
FB: fibroblast
Fbln (Fibl): fibulin
Fbn: fibrillin
FBS: F-box, Sec7 protein (ArfGEF)
FBX: F-box only protein (ArfGEF)
FC: fibrocyte
FCHO: FCH domain only protein
FcαR: Fc receptor for IgA
FcγR: Fc receptor for IgG
FcεR: Fc receptor for IgE
FDM: finite difference method
FEM: finite element method
FERM: four point-1, ezrin, radixin, and
   moesin domain
Fer: Fes-related Tyr kinase
Fes: feline sarcoma kinase
FFA: free fatty acid
FGF: fibroblast growth factor
aFGF: acidic fibroblast growth factor
   (FGF1)
bFGF: basic fibroblast growth factor
   (FGF2)
FGFR: fibroblast growth factor receptor
FGR: viral feline Gardner-Rasheed
   sarcoma oncogene homolog kinase
FHL: four-and-a-half LIM-only protein
FHoD: formin homology domain-
   containing protein (FmnL)
FIH: factor inhibiting HIF1α (as-
   paraginyl hydroxylase)
FIP: focal adhesion kinase family-
   interacting protein
FIT: Fat-inducing transcript
FKBP: FK506-binding protein
FlIP: flice-inhibitory protein
FLK: fetal liver kinase
fMLP: N-formyl methionyl-leucyl-
   phenylalanine
FN: fibronectin
Fn: fibrin
Fng: fibrinogen

Fos: Finkel Biskis Jinkins murine osteosarcoma virus sarcoma proto-oncogene product

Fox: forkhead box transcription factor

Fpn: ferroportin

FR: flow ratio

FRK: Fyn-related kinase

FrmD: FERM domain-containing adaptor

FRNK: FAK-related non-kinase

FSH: follicle-stimulating hormone

FSI: fluid–structure interaction

FVM: finite volume method

FXR: farnesoid X receptor (NR1h4)

Fz: Frizzled (Wnt GPCR)

sFRP: secreted Frizzled-related protein

G

G: Green-Lagrange strain tensor

G: shear modulus

G': storage modulus

G'': loss modulus

$\mathcal{G}$: Gibbs function

G: conductance

G_p: pressure gradient

G_e: electrical conductivity

G_h: hydraulic conductivity

G_T: thermal conductivity

$\mathbf{g}$: gravity acceleration

$\mathbf{g}$: physical quantity

g: detachment frequency

g: free enthalpy

G protein: guanine nucleotide-binding protein ($G\alpha\beta\gamma$ trimer)

$G\alpha$: α subunit (signaling mediator) of G protein

$G\alpha_i$ (Gi): inhibitory $G\alpha$ subunit

$G\alpha_s$ (Gs): stimulatory $G\alpha$ subunit

$G\alpha_t$ (Gt): transducin, $G\alpha$ subunit of rhodopsin

Gs_{XL}: extra-large Gs protein

$G\alpha_{i/o}$ (Gi/o): $G\alpha$ subunit class

$G\alpha_{q/11}$ (Gq/11): $G\alpha$ subunit class

$G\alpha_{12/13}$ (G12/13): $G\alpha$ subunit class

$G\beta\gamma$: dimeric subunit (signaling effector) of G protein

G_{gust}: gustducin, G protein α subunit (Gi/o) of taste receptor

G_{olf}: G protein α subunit (Gs) of olfactory receptor

GAB: GRB2-associated binder

GABA: γ-aminobutyric acid

$GABA_A$: GABA ionotropic receptor (Cl^- channel)

$GABA_B$: GABA metabotropic receptor (GPCR)

GABARAP: $GABA_A$ receptor-associated protein

GaBP: globular actin-binding protein

GADD: growth arrest and DNA damage-induced protein

gadkin: γ1-adaptin and kinesin interactor

GAG: glycosaminoglycan

GAK: cyclin G-associated kinase

Gal: galanin

GAP: GTPase-activating protein

GAPDH: glyceraldehyde 3-phosphate dehydrogenase

GAS: growth arrest-specific gene product

GAT: γ-aminobutyric acid transporter

GATA: DNA sequence GATA-binding protein (TF)

GBF: Golgi-associated brefeldin-A-resistant guanine nucleotide exchange factor

GCAP: guanylyl cyclase-activating protein

GCC: Golgi coiled-coil domain-containing protein

GCK: germinal center kinase

GCKR: GCK-related kinase

GCNF: germ cell nuclear factor (NR6a1)

GCN2: general control non-derepressible 2 (pseudokinase)

gCSF: granulocyte colony-stimulating factor (G-CSF; CSF3)

GD: disialoganglioside

GDP: guanosine diphosphate

GDF: growth differentiation factor

GDF: (Rab)GDI displacement (dissociation) factor

GDI: guanine nucleotide-dissociation inhibitor

GDNF: glial cell line-derived neurotrophic factor

GEF: guanine nucleotide (GDP-to-
GTP)-exchange factor

GF: growth factor

GFAP: glial fibrillary acidic protein
(intermediate filament)

GFL: GDNF family of ligands

GFP: geodesic front propagation

GGA: Golgi-localized γ-adaptin
ear-containing Arf-binding protein

Ggust: (G protein) Gα subunit
gustducin

GH: growth hormone

GHR: growth hormone receptor

GHRH: growth hormone-releasing
hormone

GIP: GPCR-interacting protein

GIRK: Gβγ-regulated inwardly
rectifying K^+ channel

GIT: GPCR kinase-interacting protein

GKAP: guanylyl kinase-associated
protein

GLK: GCK-like kinase

mGl: metabotropic glutamate receptor

GluK: ionotropic glutamate receptor
(kainate)

GluN: ionotropic glutamate receptor
(NMDA)

GluR: ionotropic glutamate receptor
(AMPA)

GluT: glucose transporter

GlyCAM: glycosylation-dependent cell
adhesion molecule

GlyR: glycine receptor (channel)

GlyT: glycine transporter

GM: monosialoganglioside

gmCSF: granulocyte–monocyte colony-
stimulating factor (GM-CSF;
CSF2)

GMP: granulocyte–monocyte progenitor

GMP: guanosine monophosphate

CNG: cyclic nucleotide-gated channel

GnRH: gonadotropin-releasing hormone

GP: glycoprotein

Gpc: glypican

GPI: glycosyl-phosphatidylinositol
anchor

gpiAP: GPI-anchored protein

GPCR: G-protein-coupled receptor

GPx: glutathione peroxidase

GQ: quadrisialoganglioside

GR: glucocorticoid receptor (NR3c1)

GRAP: GRB2-related adaptor protein
(or GAds)

GRB: growth factor receptor-bound
protein

GRE: glucocorticoid response element
(DNA sequence)

GRK: G-protein-coupled receptor
kinase

GRP: G-protein-coupled receptor
phosphatase

GSK: glycogen synthase kinase

GT: trisialoganglioside

GTF: general transcription factor

GTP: guanosine triphosphate

GTPase: guanosine triphosphatase

GuCy: guanylyl cyclase (CyG)

GWAS: genome-wide association study

H

H: height

$\mathcal{H}$: history function

$\mathtt{H}$: dissipation

$\mathbf{h}$: head loss

h: thickness

h: specific enthalpy

h_T: heat transfer coefficient

h_m: mass transfer coefficient

HA: hyaluronic acid

HAD: haloacid dehalogenase

HAP: huntingtin-associated protein

HAT: histone acetyltransferase

HAAT: heterodimeric amino acid
transporter

HAND: heart and neural crest
derivatives expressed protein

Hb: hemoglobin

Hb^{SNO}: S-nitrosohemoglobin

HBEGF: heparin-binding epidermal
growth factor

HCK: hematopoietic cell kinase

HCLS: hematopoietic lineage cell-
specific Lyn substrate protein

HCN: hyperpolarization-activated, cyclic nucleotide-gated K$^+$ channel

HCT: helical CT

HDAC: histone deacetylase complex

HCNP: hippocampal cholinergic neurostimulatory peptide

HDL: high-density lipoprotein

HDL–C: HDL–cholesterol

HDL–CE: HDL–cholesteryl ester

HDM: human double minute (E3 ubiquitin ligase)

HEET: hydroxyepoxyeicosatrienoic

HERG: human ether-a-go-go related gene

HER: human epidermal growth factor receptor (HER3: pseudokinase)

HES: Hairy enhancer of split

HETE: hydroxyeicosatrienoic acid

HEV: high endothelial venule

HGF: hepatocyte growth factor

HGFR: hepatocyte growth factor receptor

HGS: HGF-regulated Tyr kinase substrate (HRS)

HhIP: Hedgehog-interacting protein

HIF: hypoxia-inducible factor

HIP: huntingtin-interacting protein

HIP1R: HIP1-related protein

His: histamine

Hjv: hemojuvelin

HK: hexokinase

HL: hepatic lipase

HMG: high mobility group protein

HMGB: high mobility group box protein

HMGCoAR: hydroxy methylglutaryl coenzyme-A reductase

HMT: histone methyl transferase

HMWK: high-molecular-weight kininogen

HNF: hepatocyte nuclear factor (NR2a1/2)

HNP: human neutrophil peptide

HODE: hydroxy octadecadienoic acid

HOP: HSP70–HSP90 complex-organizing protein

HotAIR: HOX antisense intergenic RNA (large intergenic non-coding RNA)

HOx: heme oxygenase

Hox: homeobox DNA sequence (encodes homeodomain-containing morphogens)

HPK: hematopoietic progenitor kinase (MAP4K)

HRE: hormone response element (DNA sequence)

HRM: hypoxia-regulated microRNA

HRS: hepatocyte growth factor-regulated tyrosine kinase substrate

HRT: Hairy and enhancer of Split-related transcription factor

HS: heparan sulfate

HSC: hematopoietic stem cell

HSC: heat shock cognate

HSER: heat stable enterotoxin receptor (guanylyl cyclase 2C)

HSP: heat shock protein (chaperone)

HSPG: heparan sulfate proteoglycan

Ht: hematocrit

HTR: high temperature requirement endoprotease

I

I: identity tensor

i: current

I: inotropy

IAP: inhibitor of apoptosis protein

IBABP: intestinal bile acid-binding protein

IC: isovolumetric contraction

ICA: internal carotid artery

ICAM: intercellular adhesion molecule (IgCAM member)

IgCAM: immunoglobulin-like adhesion molecule

ICF: intracellular fluid

ICliP: intramembrane-cleaving protease

ID: inhibitor of DNA binding

IDL: intermediate-density lipoprotein

IDOL: inducible degrader of LDL receptor (E3-Ub ligase)

IEL: internal elastic lamina

IEL: intra-epithelial lymphocyte

IfIH: interferon-induced with helicase-C domain-containing protein

Ifn: interferon

IFT: intraflagellar transport complex
Ig: immunoglobulin
IgCAM: immunoglobulin (class)-cell
adhesion molecule
IGF: insulin-like growth factor
IGFBP: IGF-binding protein
IgHC: immunoglobulin heavy chain
IgLC: immunoglobulin light chain
iGluR: ionotropic glutamate receptor
IH: intimal hyperplasia
IHh: indian Hedgehog
IK: intermediate-conductance Ca^{++}-
activated K^+ channel
IκB: inhibitor of NFκB
IKK: IκB kinase
IL: interleukin
ILC: innate lymphoid cell
ILK: integrin-linked (pseudo)kinase
IMP: Impedes mitogenic signal
propagation
INADl: inactivation no after-potential D
protein
InCenP: inner centromere protein
InF: inverted formin
IMM: inner mitochondrial membrane
INPP: inositol polyphosphate 5-
phosphatase
InsIG: insulin-induced gene product
(ER anchor)
InsL: insulin-like peptide
InsRR: insulin receptor-related receptor
IP: inositol phosphate
IP_3: inositol (1,4,5)-triphosphate
IP_3R: IP_3 receptor (IP_3-sensitive
Ca^{++}-release channel)
IP_4: inositol (1,3,4,5)-tetrakisphosphate
IP_5: inositol pentakisphosphate
IP_6: inositol hexakisphosphate
IPCEF: interaction protein for cytohesin
exchange factor
IPOD: perivacuolar insoluble protein
deposit
IPP: ILK–PINCH–parvin complex
IPSC: induced pluripotent stem cell
IQGAP: IQ motif-containing GTPase-
activating protein (IQ: first
2 amino acids of the motif:
isoleucine [I; commonly] and
glutamine [Q; invariably]).

IR (InsR): insulin receptor
IR: isovolumetric relaxation
IRAK: IL1 receptor-associated kinase
(IRAK2: pseudokinase)
IRE: irreversible electroporation
IRES: internal ribosome entry site
IRF: interferon-regulatory protein
(transcription factor)
IRFF: interferon-regulatory factor
family
IRS: insulin receptor substrate
ISA: intracranial saccular aneurism
ISG: interferon-stimulated gene product
ITAM: immunoreceptor tyrosine-based
activation motif
Itch: Itchy homolog (ubiquitin ligase)
ITIM: immunoreceptor tyrosine-based
inhibitory motif
ITK: interleukin-2-inducible T-cell
kinase
ITPK: inositol trisphosphate kinase
IVC: inferior vena cava
IVP: initial value problem
IVUS: intravascular ultrasound

J

J: flux
J_m: cell surface current density
JAM: junctional adhesion molecule
JaK: Janus (pseudo)kinase
JMy: junction-mediating and regulatory
protein
JIP: JNK-interacting protein
(MAPK8IP1 and -2)
JNK: Jun N-terminal kinase (MAPK8–
MAPK10)
JNKK: JNK kinase
JNKBP: JNK-binding protein;
JSAP: JNK/SAPK-associated protein
Jun: avian sarcoma virus-17 proto-
oncogene product (Japanese
juunana: 17; TF)
JUNQ: juxtanuclear quality-control
compartment

K

K: conductivity tensor

K: bending stiffness
K: reflection coefficient
K_d: dissociation constant
K_M: Michaelis constant (chemical reaction kinetics)
K_m: material compressibility
k: cross section ellipticity
k_{ATP}: myosin ATPasic rate
k_B: Boltzmann constant $(1.38 \times 10^{-23}$ J/K)
k_c: spring stiffness
k_m: mass-transfer coefficient
k_P: Planck constant
K_R: resistance coefficient
KaP: karyopherin
K_{ATP}: ATP-sensitive K^+ channel
$K_{Ca}1.x$: BK channel
$K_{Ca}2/3/4.x$: SK channel
$K_{Ca}5.x$: IK channel
K_{IR}: inwardly rectifying K^+ channel
K_V: voltage-gated K^+ channel
KAP: kinesin (KIF)-associated protein
Kap: karyopherin
KAT: lysine (K) acetyltransferase
KCC: K^+–Cl^- cotransporter
KChAP: K^+ channel-associated protein
KChIP: K_V channel-interacting protein
KDELR: KDEL (Lys–Asp–Glu–Leu) endoplasmic reticulum retention receptor
KDR: kinase insert domain receptor
KHC: kinesin heavy chain
KIF: kinesin family
KIR: killer-cell immunoglobulin-like receptor
KIT: cellular kinase in tyrosine (SCFR)
Kk: kallikrein
KLC: kinesin light chain
KLF: Krüppel-like factor
KLR: killer cell lectin-like receptor
Kn: Knudsen number
KOR: κ-opioid receptor
Krt: keratin
KSR: kinase suppressor of Ras (adaptor; pseudokinase)

L

L: velocity gradient tensor

L: inertance
L: length
LA: left atrium
LAB: linker of activated B lymphocyte
LAd: LcK-associated adaptor
LANP: long-acting natriuretic peptide
LAP: leucine-rich repeat and PDZ domain-containing protein (4-member family)
LAP: latency-associated peptide (4 isoforms LAP1–LAP4)
LAP: nuclear lamina-associated polypeptide
LAR: leukocyte common-antigen-related receptor (PTPRF)
LAT: linker of activated T lymphocytes
LaTS: large tumor suppressor
LAX: linker of activated X cells (both B and T cells)
LBR: lamin-B receptor
LCA: left coronary artery
LCAT: lysolecithin cholesterol acyltransferase
LCC: left coronary cusp
LCK: leukocyte-specific cytosolic (non-receptor) Tyr kinase
LCP: lymphocyte cytosolic protein (adaptor SLP76)
LDL: low-density lipoprotein
LDLR: low-density lipoprotein receptor
LDV: laser Doppler velocimetry
Le: entry length
LEF: lymphoid enhancer-binding transcription factor
LGalS: lectin, galactoside-binding, soluble cell-adhesion molecule
LGIC: ligand-gated ion channel
LGL: lethal giant larva protein
LH: luteinizing hormone
LIF: leukemia-inhibitory factor
LIFR: leukemia-inhibitory factor receptor
LIMA: LIM domain and actin-binding protein
LIME: LcK-interacting molecule
LIMK: Lin1, Isl1, and Mec3 kinase
LIMS: LIM and senescent cell antigen-like-containing domain protein

LiNC: linker of nucleoskeleton and
    cytoskeleton
LipC: hepatic lipase
LipD: lipoprotein lipase
LipE: hormone-sensitive lipase
LipG: endothelial lipase
LipH: lipase-H
liprin: LAR PTP-interacting protein
LIR: leukocyte immunoglobulin-like
    receptor
LIS: lissencephaly protein
LKB: liver kinase-B
LKLF: lung Kruppel-like factor
LLTC: large latent TGFβ complex
LMan: lectin, mannose-binding
LMO: LIM domain-only-7 protein
Lmod: leiomodin (actin nucleator)
LMPP: lymphoid-primed multipotent
    progenitor
LMR: laser myocardial revascularization
Ln: laminin
LOx: lipoxygenase
LP: lipoprotein
LPA: lysophosphatidic acid
Lphn: latrophilin (adhesion-GPCR)
LPL: lysophospholipid
LPLase: lysophospholipase
LPase: lipoprotein lipase
LPP: lipid phosphate phosphatase
LPR: lipid phosphatase-related protein
LPS: lipopolysaccharide
LQTS: long-QT syndrome
LRH: liver receptor homolog (NR5a2)
LRO: lysosome-related organelle
LRP: LDL receptor-related protein
LRRTM: leucine-rich repeat-containing
    transmembrane protein
LSK: Lin−, SCA1+, KIT+ cell
LST: lethal with Sec-thirteen
LSV: long saphenous vein
LT (Lkt): leukotriene
LTBP: latent TGFβ-binding protein
LTCC: L-type Ca^{++} channel ($Ca_V 1$)
LTK: leukocyte tyrosine kinase
LUbAC: Linear ubiquitin chain
    assembly complex
LV: left ventricle
LVAD: left ventricular assist device
LX: lipoxin

LXR: liver X receptor (NR1h2/3)
LyVE: lymphatic vessel endothelial
    hyaluronan receptor

M

M: molar mass
$\mathcal{M}$: moment
m: mass
Ma: Mach number
MACF: microtubule-actin crosslinking
    factor
MAD: mothers against decapentaplegic
    homolog
MAD: mitotic arrest-deficient protein
MAdCAM: mucosal vascular addressin
    cell adhesion molecule
MAF: V-maf musculoaponeurotic
    fibrosarcoma oncogene homolog
    (TF)
MAGI: membrane-associated guanylate
    kinase-related protein with
    inverted domain organization
MAGP: microfibril-associated glycopro-
    tein
MAGuK: membrane-associated
    guanylyl kinase
MAIT: mucosal-associated invariant
    T lymphocyte
MALT: mucosa-associated lymphoid
    tissue
MAO: monoamine oxidase
MAP: microtubule-associated protein
MAP1LC3: microtubule-associated
    protein-1 light chain-3 (LC3)
mAP: mean arterial pressure
MAPK: mitogen-activated protein
    kinase
MAP2K: mitogen-activated protein
    kinase kinase
MAP3K: MAP kinase kinase kinase
MAPKAPK: MAPK-activated protein
    kinase
MaRCo: macrophage receptor with
    collagenous structure (ScaRa2)
MARCKS: myristoylated alanine-rich
    C kinase substrate
MARK: microtubule affinity-regulating
    kinase

MASTL: microtubule-associated Ser/ Thr kinase-like protein

MAT: ménage à trois

MATK: megacaryocyte-associated Tyr kinase

MAVS: mitochondrial antiviral signaling protein

MBP: myosin-binding protein

MBP: myeloid–B-cell progenitor

MBTPSi: membrane-bound transcription factor peptidase site i

MCAK: mitotic centromere-associated kinesin

MCAM: melanoma cell adhesion molecule

MCL1: BCL2-related myeloid cell leukemia sequence protein-1

MCLC: stretch-gated Mid1-related chloride channel

MCM: minichromosome maintenance protein

MCP: monocyte chemoattractant protein

mCSF: macrophage colony-stimulating factor (M-CSF; CSF1)

MCT: monocarboxylate–proton cotransporter

MDM: mitochondrial distribution and morphology protein

MEF: myocyte enhancer factor

megCSF: megakaryocyte colony-stimulating factor

MEJ: myoendothelial junction

MELK: maternal embryonic leucine zipper kinase

MEP: megakaryocyte erythroid progenitor

MEP: myeloid–erythroid progenitor

MET: mesenchymal–epithelial transition factor (proto-oncogene; HGFR)

METC: mitochondrial electron transport chain

MGP: matrix Gla protein

mGluR: metabotropic glutamate receptor

MHC: major histocompatibility complex

MHC: myosin heavy chain

MyHC or MYH: myosin heavy chain gene

miCK: mitochondrial creatine kinase

MinK: misshapen-like kinase

MiRP: MinK-related peptide

MIRR: multichain immune-recognition receptor

MIS: Müllerian inhibiting substance

MIS: mini-invasive surgery

MIS: mitochondrial intermembrane space

MIST: mast-cell immunoreceptor signal transducer

MIT: mini-invasive therapy

MiV: mitral valve

MIZ: Myc-interacting zinc-finger protein

MJD: Machado-Joseph disease protein domain-containing protease (DUb)

MKL: megakaryoblastic leukemia-1 fusion coactivator

MKP: mitogen-activated protein kinase phosphatase

MLC: myosin light chain

MyLC or MYL: myosin light chain gene

MLCK: myosin light-chain kinase

MLCP: myosin light-chain phosphatase

MLK: mixed lineage kinase

MLKL: mixed lineage kinase-like pseudokinase

MLP: muscle LIM protein

MLL: mixed-lineage [myeloid–lymphoid] leukemia factor

MLLT: mixed-lineage leukemia translocated protein

mmCK: myofibrillar creatine kinase

MME: membrane metalloendopeptidase

MMM: maintenance of mitochondrial morphology protein

MMP: matrix metalloproteinase

mtMMP: membrane-type MMP (mt$i$MMP: type-i mtMMP)

MO: mouse protein

Mo: monocyte

MOMP: mitochondrial outer membrane permeabilization

MOR: μ-opioid receptor

MP: MAPK partner

MPF: mitosis (maturation)-promoting factor (CcnB–CDK1 complex)

MPG: N-methylpurine (N-methyladenine)-DNA glycosylase

MPO: median preoptic nucleus

Mpo: myeloperoxidase

MP_P: membrane protein, palmitoylated

MPP: multipotent progenitor

MR: mineralocorticoid receptor (NR3c2)

MRCK: myotonic dystrophy kinase-related CDC42-binding kinase

MRI: (nuclear) magnetic resonance imaging

MRTF: myocardin-related transcription factor

MSC: mesenchymal stem cell

MSH: melanocyte-stimulating hormone

MSSCT: multi-slice spiral CT

MST: mammalian sterile-twenty-like kinase

MST1R: macrophage-stimulating-1 factor receptor

MT: metallothionein

MTM: myotubularin (myotubular myopathy-associated gene product)

MTMR: myotubularin-related phosphatase

MTOC: microtubule organizing center

MTP: myeloid–T-cell progenitor

MTP: microsomal triglyceride transfer protein

MuRF: muscle-specific RING finger (Ub ligase)

MuSK: muscle-specific kinase

MVB: multivesicular body

MVE: multivesicular endosome (MVB)

MVO2: myocardial oxygen consumption

MWSS: maximal wall shear stress

MyB: V-myb myeloblastosis viral oncogene homolog (TF)

MyC: V-myc myelocytomatosis viral oncogene homolog (TF)

MyHC: myosin heavy chain

MyPT: myosin phosphatase targeting subunit

MyT: myelin transcription factor

N

N: sarcomere number

$\hat{n}$: unit normal vector

n: mole number

n: PAM density with elongation x

$\mathbf{n}$: myosin head density

$\mathcal{N}_A$: Avogadro number

N-terminus: amino (amine group NH_2)-terminus

NAADP: nicotinic acid adenine dinucleotide phosphate

NAD: nicotine adenine dinucleotide

NADPH: reduced form of nicotinamide adenine dinucleotide phosphate.

NAd: noradrenaline

NAF: nutrient-deprivation autophagy factor

NALT: nasal-associated lymphoid tissu

NAmPT: nicotinamide phosphoribosyl-transferase

Nanog: ever young (Gaelic)

NAP: NCK-associated protein (NCKAP)

NAT: noradrenaline transporter

Na_V voltage-gated Na^+ channel

NPAS: neuronal PAS domain-containing transcription factor

NBC: Na^+–HCO_3^- cotransporters

NCC: non-coronary cusp

NCC: Na^+–Cl^- cotransporter

Ncdn: neurochondrin

NCK: non-catalytic region of tyrosine kinase adaptor

NCoA: nuclear receptor coactivator

NCoR: nuclear receptor corepressor

NCR: natural cytotoxicity-triggering receptor

NCS: neuronal calcium sensor

NCX: Na^+–Ca^{++} exchanger

NCKX: Na^+–Ca^{++}–K^+ exchanger

NCLX: Na^+–Ca^{++}–Li^+ exchanger

NDCBE: Na^+-dependent Cl^-–HCO_3^- exchanger

NecL: nectin-like molecule

NEDD: neural precursor cell expressed, developmentally downregulated

NeK: never in mitosis gene-A (NIMA)-related kinase

NES: nuclear export signal

NESK: NIK-like embryo-specific kinase

nesprin: nuclear envelope spectrin repeat protein

NeuroD: neurogenic differentiation protein

NF: neurofilament protein (intermediate filament)

NFH: neurofilament, heavy polypeptide

NFL: neurofilament, light polypeptide

NFM: neurofilament, medium polypeptide

NF: neurofibromin (RasGAP)

NFκB: nuclear factor κ light chain-enhancer of activated B cells

NFAT: nuclear factor of activated T cells

NFe2: erythroid-derived nuclear factor-2

NGAL: neutrophil gelatinase-associated lipocalin

NGF: nerve growth factor

Ngn: neogenin (netrin receptor)

NHA: Na^+–H^+ antiporter

NHE: sodium–hydrogen exchanger

NHERF: NHE regulatory factor

NHR: nuclear hormone receptor

NIc: nucleoporin-interacting protein

NIK: NFκB-inducing kinase

NIK: NCK-interacting kinase

NIP: neointimal proliferation

NK: natural killer cell

NKCC: Na^+–Ka^+–$2Cl^-$ cotransporter

NKG: NK receptor group

NKT: natural killer T cell

NKx2: NK2 transcription factor-related homeobox protein

NLR: NOD-like receptor (nucleotide-binding oligomerization domain, Leu-rich repeat-containing)

NLS: nuclear localization signal

NMDAR: N-methyl D-aspartate receptor

NmU: neuromedin-U

NO: nitric oxide (nitrogen monoxide)

NonO: non-POU domain-containing octamer-binding protein

NOR: neuron-derived orphan receptor (NR4a3)

NOS: nitric oxide synthase

NOS1: neuronal NOS

NOS2: inducible NOS

NOS3: endothelial NOS

NOx: NAD(P)H oxidase

Noxa: damage (Latin)

NPC: Niemann-Pick disease type-C protein

NPC1L: Niemann-Pick protein-C1-like

NPC: nuclear pore complex

NPY: neuropeptide Y

NR: nuclear receptor

NRAP: nebulin-related actinin-binding protein

NRBP: nuclear receptor-binding protein

NRF: nuclear factor erythroid-derived-2 (NF-E2)-related factor

NRF1: nuclear respiratory factor-1

Nrg: neuregulin (EGF superfamily member)

Nrgn: neuroligin

Nrp: neuropilin (VEGF-binding molecule; VEGFR coreceptor)

NRSTK: non-receptor serine/threonine kinase

NRTK: non-receptor tyrosine kinase

Nrxn: neurexin

NSCLC: non-small-cell lung cancer

NSF: N-ethylmaleimide-sensitive factor

NSLTP: non-specific lipid-transfer protein

NST: nucleus of the solitary tract

NT: neurotrophin

NT5E: ecto-5′-nucleotidase

NTCP: sodium–taurocholate cotransporter polypeptide

NTF: N-terminal fragment

NTRK: neurotrophic tyrosine receptor kinase (TRK)

NTP: nucleoside triphosphate

NTPase: nucleoside triphosphate hydrolase superfamily member

Nu: Nusselt number

NuAK: nuclear AMPK-related kinase

NuP: nucleoporin (nuclear pore complex protein)

NuRD: nucleosome remodeling and histone deacetylase

NuRR: nuclear receptor-related factor (NR4a2)

O

^{O}Glc^{N}Ac: β^{N}acetyl ^{D}glucosamine

OSBP: oxysterol-binding protein

OCRL: oculocerebrorenal syndrome of Lowe phosphatase

Oct: octamer-binding transcription factor

ODE: ordinary differential equation

OGA: ^{O}Glc^{N}Acase (β^{N}acetylglucosaminidase)

OMM: outer mitochondrial membrane

ORC: origin recognition complex

ORF: open reading frame

ORP: OSBP-related protein

OSA: obstructive sleep apnea

OSI: oscillatory shear index

OSM: oncostatin M

OSMR: oncostatin M receptor

OSR (OxSR): oxidative stress-responsive kinase

OTK: off-track (pseudo)kinase

OTU: ovarian tumor superfamily protease (deubiquitinase)

OTUB: otubain (Ub thioesterase of the OTU superfamily)

OVLT: organum vasculosum lamina terminalis

P

$\mathcal{P}$: permeability

P: power

P: cell division rate

p: pressure

p_i: partial pressure of gas component i

PA: phosphatidic acid

PAAT: proton–amino acid transporter

PAF: platelet-activating factor

PAFAH: platelet-activating factor acetylhydrolase

PAG: phosphoprotein associated with glycosphingolipid-enriched microdomains

PAH: polycyclic aromatic hydrocarbon

PAH: pulmonary arterial hypertension

PAI: plasminogen activator inhibitor

PAK: P21-activated kinase

PAMP: pathogen-associated molecular pattern

PAMP: proadrenomedullin peptide

PAR: protease-activated receptor

PAR: polyADPribose

PAR: promoter-associated, non-coding RNA

Par: partitioning defective protein

PALR: promoter-associated long RNA

PALS: protein associated with Lin-7

PARP: polyADPribose polymerase

PASR: promoter-associated short RNA

PATJ: protein (PALS1) associated to tight junctions

Pax: paxillin

Pax$i$: paired box protein-i (transcription regulator)

PBC: pre-Bötzinger complex (ventilation frequency)

PBIP: Polo box-interacting protein

PC: protein C

PCMRV: phase-contrast MR velocimetry

PCr: phosphocreatine

PCT: proximal convoluted tubule

PCTP: phosphatidylcholine-transfer protein

PD: pharmacodynamics

PdCD: programmed cell death protein

PdCD6IP: PdCD 6-interacting protein

PdCD1Lg: programmed cell death-1 ligand

PDE: phosphodiesterase

PDE: partial differential equation

PDGF: platelet-derived growth factor

PDGFR: platelet-derived growth factor receptor

PDI: protein disulfide isomerase

PDK: phosphoinositide-dependent kinase

Pe: Péclet number

PE: pulmonary embolism

PEBP: phosphatidylethanolamine-binding protein

PECAM: platelet–endothelial cell adhesion molecule

PEDF: pigment epithelium-derived factor (serpin F1)

PEn2: presenilin enhancer-2

PEO: proepicardial organ

Per: Period homolog

PERK: protein kinase-like endoplasmic reticulum kinase

PERP: P53 apoptosis effector related to peripheral myelin protein PMP22

PET: positron emission tomography

Pex: peroxin

PF: platelet factor

PFK: phosphofructokinase

pFRG: parafacial respiratory group

PG: prostaglandin

PGG: prostaglandin glycerol ester

PGEA: prostaglandin ethanolamide

PGC: PPARγ coactivator

pGC: particulate guanylyl cyclase

PGi2: prostacyclin

PGF: paracrine growth factor

PGP: permeability glycoprotein

PH: pleckstrin homology domain

PHD: prolyl hydroxylase

PHLPP: PH domain and Leu-rich repeat protein phosphatase

PI: phosphoinositide (phosphorylated phosphatidylinositol)

PI(4)P: phosphatidylinositol 4-phosphate

PI(i)PiK: phosphatidylinositol i-phosphate i-kinase

PI(i,j)P$_2$: phosphatidylinositol (i,j)-bisphosphate (PIP$_2$)

PI(3,4,5)P$_3$: phosphatidylinositol (3,4,5)-trisphosphate (PIP$_3$)

PI3K: phosphatidylinositol 3-kinase

PIiK: phosphatidylinositol i-kinase

PIAS: protein inhibitor of activated STAT (SUMo E3 ligase)

PIC: pre-initiation complex

PICK: protein that interacts with C-kinase

PIDD: P53-induced protein with a death domain

PIKE: phosphoinositide 3-kinase enhancer (GTPase; ArfGAP)

PIKK: phosphatidylinositol 3-kinase-related kinase (pseudokinase)

PIK3AP: PI3K adaptor protein

PIP: phosphoinositide monophosphate

PIPiK: phosphatidylinositol phosphate i-kinase

PIP$_2$: phosphatidylinositol bisphosphate

PIP$_3$: phosphatidylinositol triphosphate

PIM: provirus insertion of Molony murine leukemia virus gene product

PIN: protein peptidyl prolyl isomerase NIMA-interacting

PINCH: particularly interesting new Cys–His protein (or LIMS1)

PInK: PTen-induced kinase

PIPP: proline-rich inositol polyphosphate 5-phosphatase

PIR: paired immunoglobulin-like receptor

PITP: phosphatidylinositol-transfer protein

Pitx: pituitary (or paired-like) homeobox transcription factor

PIV: particle image velocimetry

PIX: P21-activated kinase (PAK)-interacting exchange factor (Rho(Arh)GEF6/7)

PK: pharmacokinetics

PK: protein kinase

PKA: protein kinase A

PKB: protein kinase B

PKC: protein kinase C

aPKC: atypical protein kinase C

cPKC: conventional protein kinase C

nPKC: novel protein kinase C

PKD: protein kinase D

PKG: protein kinase G

PKL: paxillin kinase linker

PKMYT (MYT): membrane-associated Tyr–Thr protein kinase

PKN: protein kinase novel

Pkp: plakophilin

PL: phospholipase

PLA2: phospholipase A2

PLC: phospholipase C

PLD: phospholipase D

PLb: phospholamban

PLd: phospholipid

PlGF: placenta growth factor

PLK: Polo-like kinase

PLTP: phospholipid transfer protein

PMCA: plasma membrane Ca^{++} ATPase

PML: promyelocytic leukemia protein

PMR: percutaneous (laser) myocardial revascularization

PMRT: protein arginine methyltransferase

Pn: plasmin

Png: plasminogen

PoG: proteoglycan

PoM: pore membrane protein

Pon: paraoxonase

POPx: partner of PIX

POSH: scaffold plenty of SH3 domains

PP: protein phosphatase

PP3: protein phosphatase 3 (PP2b or calcineurin)

PPAR: peroxisome proliferator-activated receptor (NR1c1–3)

PPG: photoplethysmography

PPId: peptidyl prolyl isomerase-D

PPIP: monopyrophosphorylated inositol phosphate

$(PP)_2IP$: bispyrophosphorylated inositol phosphate

PPK: PIP kinase

PPM: protein phosphatase (magnesium-dependent)

PPR: pathogen-recognition receptor

PPRE: PPAR response element (DNA sequence)

PR: progesterone receptor (NR3c3)

PRC: protein regulator of cytokinesis

PRC: Polycomb repressive complex

Prdx: peroxiredoxin

preBotC: preBötzinger complex

preKk: prekallikrein

PREx: PIP_3-dependent Rac exchanger (RacGEF)

PRG: plasticity-related gene product

PRH: prolactin-releasing hormone

Prl: prolactin

PrlR: prolactin receptor

PRMT: protein arginine (R) N-methyltransferase

Prompt: promoter upstream transcript

Protor: protein observed with Rictor

PROX: prospero homeobox gene

Prox: PROX gene product (transcription factor)

PrP: processing protein

PRPK: P53-related protein kinase

PRR: pattern recognition receptor

PRR: prorenin and renin receptor

PS: presenilin

PS: protein S

PSC: pluripotent stem cell

iPSC: induced pluripotent stem cell

PSD: post-synaptic density adaptor

PsD: post-synaptic density

PSEF: pseudo-strain energy function

PSer: phosphatidylserine

PSGL: P-selectin glycoprotein ligand

PSKh: protein serine kinase H

Psm: proteasome subunit

PSTPIP: Pro–Ser–Thr phosphatase-interacting protein

PTA: plasma thromboplastin antecedent

Ptc: Patched receptor (Hedgehog signaling)

PtcH: Patched Hedgehog receptor

PTCA: percutaneous transluminal coronary angioplasty

PTCRA: PTC rotational burr atherectomy

PtdCho: phosphatidylcholine

PtdEtn: phosphatidylethanolamine

PtdSer: phosphatidylserine

PtdIns: phosphatidylinositol

PTen: phosphatase and tensin homolog deleted on chromosome ten (phosphatidylinositol 3-phosphatase)

PTFE: polytetrafluoroethylene

PTH: parathyroid hormone

PTHRP: parathyroid hormone-related protein

PTK: protein tyrosine kinase

PTK7: pseudokinase (RTK)

PTP: protein tyrosine phosphatase

PTPni: protein Tyr phosphatase non-receptor type i

PTPR: protein tyrosine phosphatase receptor

PTRF: RNA polymerase-1 and transcript release factor

PUFA: polyunsaturated fatty acid

PUMA: P53-upregulated modulator of apoptosis

PuV: pulmonary valve

PVF: PDGF- and VEGF-related factor

PVNH: paraventricular nucleus of
hypothalamus
PVR: pulmonary vascular resistance
PWS: pulse wave speed
Px: pannexin
PXR: pregnane X receptor (NR1i2)
PYK: proline-rich tyrosine kinase
P2X: purinergic ligand-gated channel
P53AIP: P53-regulated apoptosis-
inducing protein
p75NtR: pan-neurotrophin receptor

Q

Q: material quantity
Q_e: electric current density
Q_T: thermal energy (heat)
q_T: transfer rate of thermal energy
(power)
q: flow rate

R

R: resistance
$\mathcal{R}$: local reaction term
R_h: hydraulic radius
R_g: gas constant
R_R: respiratory quotient
R: recruitment function (from quies-
cence to proliferation)
r: cell renewal rate
r: radial coordinate
RA: right atrium
RAAS: renin–angiotensin–aldosterone
system
Rab: Ras from brain
Rab11FIP: Rab11 family-interacting
protein
Rac: Ras-related C3-botulinum toxin
substrate
RACC: receptor-activated cation
channel
RACK: receptor for activated C-kinase
RAD: recombination protein-A
(RecA)-homolog DNA-repair
protein
Rad: radiation sensitivity protein
Rag: Ras-related GTP-binding protein
Ral: Ras-related protein

RalGDS: Ral guanine nucleotide
dissociation stimulator
RAMP: receptor activity-modifying
protein
Ran: Ras-related nuclear protein
RANTES: regulated upon activation,
normal T-cell expressed, and
secreted product (CCL5)
RAP: receptor-associated protein
Rap: Ras-related protein
Raptor: regulatory associated protein of
TOR
RAR: retinoic acid receptor (NR1b2/3)
Ras: rat sarcoma virus homolog (small
GTPase)
RasA: Ras p21 protein activator
RASSF: Ras interaction/interference
protein RIN1, afadin, and Ras
association domain-containing
family protein
RB: retinoblastoma protein
RBC: red blood cell (erythrocyte)
RBP: retinol-binding protein
RC: ryanodine calcium channel (RyR)
RCA: right coronary artery
RCan: regulator of calcineurin
RCC: right coronary cusp
RCC: regulator of chromosome
condensation
Re: Reynolds number
REDD: regulated in development
and DNA-damage response gene
product
Rel: reticuloendotheliosis proto-
oncogene product (TF; member of
NFκB)
REP: Rab escort protein
ReR: renin receptor (PRR)
restin: Reed-Steinberg cell-expressed
intermediate filament-associated
protein (CLiP1)
ReT: rearranged during transfection
(receptor Tyr kinase)
RevRE: reverse (Rev)-ErbA (NR1d1/2)
response element (DNA sequence)
RFA: radiofrequency ablation
RGL: Ral guanine nucleotide disso-
ciation stimulator-like protein
(GEF)

RGS: regulator of G-protein signaling

RHEB: Ras homolog enriched in brain

RHS: equation right hand side

Rho: Ras homologous

RIAM: Rap1-GTP-interacting adapter

RIBP: RLK- and ITK-binding protein

RICH: RhoGAP interacting with CIP4 homolog

RICK: receptor for inactive C-kinase

Rictor: rapamycin-insensitive companion of TOR

RIF: Rho in filopodium

RIn: Ras and Rab interactor (RabGEF)

RIN: Ras-like protein expressed in neurons (GTPase)

RIP: regulated intramembrane proteolysis

RIPK: receptor-interacting protein kinase

RISC: RNA-induced silencing complex

RIT: Ras-like protein expressed in many tissues

RKIP: Raf kinase inhibitor protein

RlBP: retinaldehyde-binding protein

RLC: RISC-loading complex

RLK: resting lymphocyte kinase (TXK)

RNA: ribonucleic acid

dsRNA: double-stranded RNA

endo-siRNA: endogenous small interfering RNA

hpRNA: long hairpin RNA

lincRNA: large intergenic non-coding RNA

mRNA: messenger RNA

miR: microRNA

ncRNA: non-coding RNA

piRNA: P-element-induced wimpy testis-interacting (PIWI) RNA

pre-miR: precursor microRNA

pri-miR: primary microRNA

rRNA: ribosomal RNA

rasiRNA: repeat-associated small interfering RNA (PIWI)

shRNA: small or short hairpin RNA

siRNA: small interfering RNA

snRNA: small nuclear RNA

snoRNA: small nucleolar RNA

ssRNA: single-stranded RNA

tRNA: transfer RNA

tiRNA: transcription initiation RNA

tsRNA: tRNA-derived small RNA

tssaRNA: transcription start site-associated RNA

RNABP: RNA-binding protein

RNase: ribonuclease

RnBP: renin-binding protein

RnF2: ring finger protein-2 (E3 ubiquitin ligase)

RNP: ribonucleoprotein

hRNP: heterogeneous ribonucleoprotein

hnRNP: heterogeneous nuclear ribonucleoprotein

mRNP: messenger ribonucleoprotein

miRNP: microribonucleoprotein

snRNP: small nuclear ribonucleoprotein

snoRNP: small nucleolar ribonucleoprotein

Robo: roundabout

ROC: receptor-operated channel

RoCK: Rho-associated, coiled-coil-containing protein kinase

ROI: region of interest

ROMK: renal outer medullary potassium channel

ROR: RAR-related orphan receptor (NR1f1–NR1f3)

ROR(RTK): receptor Tyr kinase-like orphan receptor

ROS: reactive oxygen species

Ros: V-ros UR2 sarcoma virus proto-oncogene product (RTK)

RPIP: Rap2-interacting protein

RPS6: ribosomal protein S6

RPTP: receptor protein tyrosine phosphatase

RSA: respiratory sinus arrhythmia

RSE: rapid systolic ejection

RSK: P90 ribosomal S6 kinase

RSKL: ribosomal protein S6 kinase-like (pseudokinase)

RSMCS: robot-supported medical and surgical system

RSpo: R-spondin

RSTK: receptor serine/threonine kinase

RTK: receptor tyrosine kinase

RTN: retrotrapezoid nucleus

Rubicon: RUN domain and Cys-rich
domain-containing, beclin-1-
interacting protein
Runx: Runt-related transcription factor
RV: right ventricle
RVF: rapid ventricular filling
RVLM: rostral ventrolateral medulla
RVMM: rostral ventromedial medulla
RXR: retinoid X receptor (NR2b1–
NR2b3)
RYK: receptor-like tyrosine (Y) kinase
(pseudokinase)
RyR: ryanodine receptor (ryanodine-
sensitive Ca^{++}-release channel)

S

S: Cauchy-Green deformation tensor
s: entropy
s: sarcomere length
s: evolution speed
SAA: serum amyloid A
SAC: stretch-activated channel
SAc: suppressor of actin domain-
containing 5-phosphatase
sAC: soluble adenylyl cyclase
SACM1L: suppressor of actin mutations
1-like
SAH: subarachnoid hemorrhage
SAN: sinoatrial node
SAP: SLAM-associated protein
SAP: stress-activated protein
SAPi: synapse-associated protein i
SAPK: stress-activated protein kinase
(MAPK)
SAR: secretion-associated and
Ras-related protein
Sc: Schmidt number
SCA: stem cell antigen
SCAMP: secretory carrier membrane
protein
SCAP: SREBP cleavage-activating
protein (SREBP escort)
SCAR: suppressor of cAMP receptor
(WAVe)
ScaR: scavenger receptor
SCF: SKP1–Cul1–F-box Ub-ligase
complex
SCF: stem cell factor

SCFR: stem cell factor receptor (KIT)
Scgb: secretoglobin
SCLC: small-cell lung cancer
scLC: squamous-cell lung cancer
(NSCLC subtype)
SCN: suprachiasmatic nucleus
SCO: synthesis of cytochrome-C oxidase
Scp: stresscopin (urocortin 3)
Scrib: Scribble polarity protein
Sdc: syndecan
SDF: stromal cell-derived factor
SDPR: serum deprivation protein
response
SE: systolic ejection
SEF: strain-energy function
SEF: similar expression to FGF genes
(inhibitor of RTK signaling)
SEK: SAPK/ERK kinase
Sema: semaphorin (Sema-, Ig-,
transmembrane-, and short
cytoplasmic domain)
SERCA: sarco(endo)plamic reticulum
calcium ATPase
serpin: serine protease inhibitor
SerT: serotonin transporter
SF: steroidogenic factor (NR5a1)
SFK: SRC-family kinase
SFO: subfornical organ
sGC: soluble guanylyl cyclase
SGK: serum- and glucocorticoid-
regulated kinase
SGlT: Na^+–glucose cotransporter
(SLC5a)
Sgo: shugoshin (japanese: guardian
spirit)
SH: Src homology domain
Sh: Sherwood number
SH3P: Src homology-3 domain-
containing adaptor protein
Shank: SH3- and multiple ankyrin
repeat domain-containing protein
SHAX: SNF7 (VSP32) homolog
associated with ALIX
SHB: Src homology-2 domain-containing
adaptor
SHC: Src-homologous and collagen-like
substrate
SHC: Src homology-2 domain-
containing transforming protein

SHh: sonic Hedgehog

SHIP: SH-containing inositol phosphatase

SHP: SH-containing protein tyrosine
phosphatase (PTPn6/11)

SHP: small heterodimer partner
(NR0b2)

SIAH: Seven in absentia homolog
(E3 ubiquitin ligase)

siglec: sialic acid-binding Ig-like lectin

SIK: salt-inducible kinase

SIn: stress-activated protein kinase-
interacting protein

SIP: steroid receptor coactivator-
interacting protein

SiRP: signal-regulatory protein

SIRT: sirtuin (silent information
regulator-2 [two]; histone
deacetylase)

SIT: SHP2-interacting transmembrane
adaptor

SKi: sphingosine kinase-i

SK: small conductance Ca^{++}-activated
K^+ channel

SKIP: sphingosine kinase-1-interacting
protein

SKIP: skeletal muscle and kidney-
enriched inositol phosphatase

SKP: S-phase kinase-associated protein

SLA: Src-like adaptor

SLAM: signaling lymphocytic activation
molecule

SLAMF: SLAM family member

SLAP: Src-like adaptor protein

SLC: solute carrier class member

SLK: Ste20-like kinase

Sln: sarcolipin

SLTC: small latent TGFβ complex

SM: sphingomyelin

SMA: smooth muscle actin

SMAD: small mothers against
decapentaplegic homolog

coSMAD: common-mediator SMAD
(SMAD4)

iSMAD: inhibitory SMAD (SMAD6 or
SMAD7)

rSMAD: receptor-regulated SMAD
(SMAD1–SMAD3, SMAD5, and
SMAD9)

SmAP: Small ArfGAP protein

SMase: sphingomyelinase

SMC: smooth muscle cell

aSMC: airway smooth muscle cell

vSMC: vascular smooth muscle cell

Smo: Smoothened

SMPD: sphingomyelin phosphodiesterase

SMRT: silencing mediator of retinoic
acid and thyroid hormone receptor

SMS: sphingomyelin synthase

SMURF: SMAD ubiquitination
regulatory factor

SNAAT: sodium-coupled neutral amino
acid transporter

SNAP: soluble N-ethylmaleimide-
sensitive factor-attachment
protein

SnAP: synaptosomal-associated protein

SNARE: SNAP receptor

tSNARE: target SNARE

vSNARE: vesicle SNARE

SNF7: sucrose non-fermenting (VPS32)

SNIP: SMAD nuclear-interacting
protein

SNP: single-nucleotide polymorphism

SNx: sorting nexin

SOC: store-operated Ca^{++} channel

SOCE: store-operated Ca^{++} entry

SOCS: suppressor of cytokine signaling
protein

SOD: superoxide dismutase

SorbS: sorbin and SH3 domain-
containing adaptor

SOS: Son of sevenless (GEF)

Sost: sclerostin

SostDC: sclerostin domain-containing
protein

SOX: sex-determining region Y
(SRY)-box gene

Sox: SOX gene product (transcription
factor)

SftP: surfactant protein

SP1: specificity protein (transcription
factor)

SPARC: secreted protein acidic and rich
in cysteine

SPC: sphingosylphosphorylcholine

SPCA: secretory pathway Ca^{++} ATPase

SPECT: single photon emission CT

SPI: spleen focus forming virus (SFFV) proviral integration proto-oncogene product (transcription factor)

Sph: sphingosine

SphK: sphingosine kinase

SPN: supernormal period

SPP: sphingosine phosphate phosphatase

SQTS: short-QT syndrome

SR: sarcoplasmic reticulum

SR: Arg/Ser domain-containing protein (alternative splicing)

SRA: steroid receptor RNA activator

SRC: steroid receptor coactivator

Src: **sarc**oma-associated (Schmidt-Ruppin A2 viral oncogene homolog) kinase

SREBP: sterol regulatory element-binding protein

SpRED: Sprouty-related protein with an EVH1 domain

SRF: serum response factor

SRM/SMRS: Src-related kinase lacking regulatory and myristylation sites

SRP: stresscopin-related peptide (urocortin 2)

SRPK: splicing factor RS domain-containing protein kinase

SPURT: secretory protein in upper respiratory tract

SRY: sex determining region Y

SSAC: shear stress-activated channel

SSE: slow systolic ejection

SSH: slingshot homolog protein

SSI: STAT-induced STAT inhibitor

Sst: somatostatin

SSV: short saphenous vein

St: Strouhal number

STAM: signal-transducing adaptor molecule

StAR: steroidogenic acute regulatory protein

StART: StAR-related lipid transfer protein

STAT: signal transducer and activator of transduction

STEAP: six transmembrane epithelial antigen of the prostate

STICK: substrate that interacts with C-kinase

StIM: stromal interaction molecule

STK: protein Ser/Thr kinase

STK1: stem-cell protein Tyr kinase receptor

STLK: Ser/Thr kinase-like (pseudo)kinase

Sto: Stokes number

StRAd: STe20-related adaptor

StRAP: stress-responsive activator of P300

Stx: syntaxin ($SNARE^Q$)

SUMo: small ubiquitin-related modifier

SUn: Sad1 and Unc84 homology protein

SUR: sulfonylurea receptor

SUT: stable unannotated transcript

SV: stroke volume

SVC: superior vena cava

SVF: slow ventricular filling

SVP: synaptic vesicle precursor

SVR: systemic vascular resistance

SW: stroke work

SwAP70: 70-kDa switch-associated protein (RacGEF)

SYK: spleen tyrosine kinase

Synj: synaptojanin

Syp: synaptophysin

Syt: synaptotagmin

S1P: sphingosine 1-phosphate

S6K: P70 ribosomal S6 kinase ($p70^{RSK}$)

T

$\mathbf{T}$: extrastress tensor

T: transition rate from a cell cycle phase to the next

T: temperature

T_L: transfer capacity

T_C: cytotoxic T lymphocyte (CD8+ effector T cell; CTL)

T_{C1}: type-1 cytotoxic T lymphocyte

T_{C2}: type-2 cytotoxic T lymphocyte

T_{CM}: central memory T lymphocyte

T_{Conv}: conventional T lymphocyte

T_{Eff}: effector T lymphocyte

T_{EM}: effector memory T lymphocyte

T_{FH}: follicular helper T lymphocyte

T_H: helper T lymphocyte (CD4+ effector T cell)

T_{Hi}: type-i helper T lymphocyte ($i = 1/2/9/17/22$)

T_{H3}: TGFβ-secreting T_{Reg} lymphocyte

T_L: lung transfer capacity (alveolocapillary membrane)

T_{R1}: type-1, IL10-secreting, regulatory T lymphocyte

T_{Reg}: regulatory T lymphocyte

aT_{Reg}: CD45RA−, FoxP3^{hi}, activated T_{Reg} cell

iT_{Reg}: inducible T_{Reg} lymphocyte

nT_{Reg}: naturally occurring (natural) T_{Reg} lymphocyte

rT_{Reg}: CD45RA+, FoxP3^{low}, resting T_{Reg} cell

$\hat{t}$: unit tangent

t: time

TβR(1/2): TGFβ receptor

TAA: thoracic aortic aneurism

TAB: TAK1-binding protein

TACE: tumor necrosis factorα-converting enzyme

TACE: transarterial chemoembolization

TAF: TBP-associated factor

TAK: TGFβ-activated kinase (MAP3K7)

TALK: TWIK-related alkaline pH-activated K^+ channel

TANK: TRAF family member-associated NFκB activator

TASK: TWIK-related acid-sensitive K^+ channel

TASR: terminus-associated short RNA

TAP: transporter associated with antigen processing

Taz: taffazin

TBC1D: Tre2 (or USP6), BUB2, CDC16 domain-containg RabGAP

TBCK: tubulin-binding cofactor kinase (pseudokinase)

TBK: TANK-binding kinase

TBP: TATA box-binding protein (subclass-4F transcription factor)

TBx: T-box transcription factor

TC: thrombocyte (platelet)

TCF: T-cell factor

TCF: ternary complex factor

TcFi: type-i transcription factor

TCR: T-cell receptor

TCA: tricarboxylic acid cycle

TCP: T-complex protein

TEA: transluminal extraction atherectomy

TEC: tyrosine kinase expressed in hepatocellular carcinoma

TEF: thyrotroph embryonic factor (PAR/b–ZIP family)

TEK: Tyr endothelial kinase

TEM: transendothelial migration

Ten: tenascin

TF: transcription factor

TFPI: tissue factor pathway inhibitor

TG: triglyceride (triacylglycerol)

TGF: transforming growth factor

TGFBR: TGFβ receptor gene

TGN: trans-Golgi network

THETE: trihydroxyeicosatrienoic acid

THIK: tandem pore-domain halothane-inhibited K^+ channel

TIAM: T-lymphoma invasion and metastasis-inducing protein (RacGEF)

TICE: transintestinal cholesterol efflux

TIE: Tyr kinase with Ig and EGF homology domains (angiopoietin receptor)

TIEG: TGFβ-inducible early gene product

TIGAR: TP53-inducible glycolysis and apoptosis regulator

TIM: T-cell immunoglobulin and mucin domain-containing protein

Tim: timeless homolog

TIMM: translocase of inner mitochondrial membrane

TIMP: tissue inhibitor of metalloproteinase

TIRAP: Toll–IL1R domain-containing adaptor protein

TJ: tight junction

TKR: tyrosine kinase receptor

TLC: total lung capacity

TLR: Toll-like receptor

TLT: TREM-like transcript

TLX: tailless receptor (NR2e1)

TM: thrombomodulin
TMC: twisting magnetocytometry
TMePAI: transmembrane prostate
    androgen-induced protein
TMy: tropomyosin
Tnn (TN): troponin
Tn: thrombin
TNF: tumor-necrosis factor
TNFαIP: tumor necrosis factor-α-
    induced protein
TNFR: tumor-necrosis factor receptor
TNFSF: tumor-necrosis factor
    superfamily member
TNFRSF: tumor-necrosis factor
    receptor superfamily member
TNK: Tyr kinase inhitor of NFκB
Tns: tensin
TOR: target of rapamycin
TORC: target of rapamycin complex
TORC: transducer of regulated CREB
    activity (a.k.a. CRTC)
TP: thromboxane-A2 Gq/11-coupled
    receptor
TP53I: tumor protein P53-inducible
    protein
tPA: tissue plasminogen activator
Tpo: thrombopoietin
TPPP: tubulin polymerization-
    promoting protein
TPST: tyrosylprotein sulftotransferase
TR: testicular receptor (NR2c1/2)
T(H)R: thyroid hormone receptor
    (NR1a1/2)
TRAAK: TWIK-related arachidonic
    acid-stimulated K$^+$ channel
TRADD: tumor-necrosis factor
    receptor-associated death domain
    adaptor
TRAF: tumor-necrosis factor receptor-
    associated factor
TRAM: TRIF-related adaptor molecule
transceptor: transporter-related
    receptor
TraPP: transport protein particle
TRAT: T-cell receptor-associated
    transmembrane adaptor
Trb: Tribbles homolog (pseudokinase)
TRE: trapped in endoderm
TREK: TWIK-related K$^+$ channel

TREM: triggering receptor expressed
    on myeloid cells
TRESK: TWIK-related spinal cord K$^+$
    channel
TRF: TBP-related factor
TRH: thyrotropin-releasing hormone
TRIF: Toll–IL1R domain-containing
    adaptor inducing Ifnβ
TRIM: T-cell receptor interacting
    molecule
TRK: tropomyosin receptor kinase
    (NTRK)
TRP: transient receptor potential
    channel
TRPA: ankyrin-like transient receptor
    potential channel
TRPC: canonical transient receptor
    potential channel
TRPM: melastatin-related transient
    receptor potential channel
TRPML: mucolipin-related transient
    receptor potential channel
TRPN: no mechanoreceptor potential C
TRPP: polycystin-related transient
    receptor potential channel
TRPV: vanilloid transient receptor
    potential channel
TrrAP: transactivation (transfor-
    mation)/transcription domain-
    associated protein (pseudokinase)
TrV: tricuspid valve
TRx: thioredoxin
TRxIP: thioredoxin-interacting protein
TSC: tuberous sclerosis complex
TSH: thyroid-stimulating hormone
TSLP: thymic stromal lymphopoietin
Tsp: thrombospondin
Tspan: tetraspanin
TsPO: translocator protein of the outer
    mitochondrial membrane
Ttn: titin (pseudokinase)
TUT: terminal uridine transferase
TWIK: tandem of P domains in a weak
    inwardly rectifying K$^+$ channel
TxA2: thromboxane A2 (thromboxane)
TxB2: thromboxane B2 (thromboxane
    metabolite)
TXK: tyrosine kinase mutated in
    X-linked agammaglobulinemia

TyK: tyrosine kinase
T_3: tri-iodothyronine
T_4: thyroxine
$^+$TP: plus-end-tracking proteins

U

U: right stretch tensor
u: displacement vector
u: electrochemical command
u: specific internal energy
Ub: ubiquitin
UbC: ubiquitin-conjugating enzyme
UbE2: E2 ubiquitin-conjugase
UbE3: E3 ubiquitin-ligase
UbL: ubiquitin-like protein
UCH: ubiquitin C-terminal hydrolase
    (DUb)
Ucn: urocortin
UCP: uncoupling protein
UDP: uridine diphosphate-glucose
UK: urokinase
ULK: uncoordinated-51-like kinase
    (pseudokinase)
Unc: uncoordinated receptor
uPA: urokinase-type plasminogen
    activator (urokinase)
uPAR: uPA receptor
uPARAP: uPAR-associated protein
    (CLec13e)
UPR: unfolded protein response
UPS: ubiquitin-proteasome system
UP4A: uridine adenosine tetraphos-
    phate
Uro: urodilatin
US: ultrasound
USC: unipotential stem cell
USF: upstream stimulatory factor
USI: ultrasound imaging
USP: ubiquitin-specific protease
    (deubiquitinase)
UTP: uridine triphosphate
UTR: untranslated region
UVRAG: ultraviolet wave resistance-
    associated gene product

V

V: left stretch tensor

V: volume
V_q: cross-sectional average velocity
V_s: specific volume
v: velocity vector
v: recovery variable
V1(2)R: type-1(2) vomeronasal receptor
$V_{1A/1B/2}$: type-1a/1b/2 arginine
    vasopressin receptor
VAAC: volume-activated anion channel
VACamKL: vesicle-associated CamK-
    like (pseudokinase)
VAMP: vesicle-associated membrane
    protein (synaptobrevin)
VanGL: Van Gogh (Strabismus)-like
    protein
VAP: VAMP-associated protein
VASP: vasoactive stimulatory phospho-
    protein
VAT: vesicular amine transporter
vATPase: vesicular-type H^+ ATPase
VAV: ventriculoarterial valve
Vav: GEF named from Hebrew sixth
    letter
VC: vital capacity
VCAM: vascular cell adhesion molecule
VCt: vasoconstriction
VDAC: voltage-dependent anion
    channel (porin)
VDCC: voltage-dependent calcium
    channel
VDP: vesicle docking protein
VDt: vasodilation
VEGF: vascular endothelial growth
    factor
VEGFR: vascular endothelial growth
    factor receptor
VF: ventricular fibrillation
VF: ventricular filling
VGAT: vesicular GABA transporter
VGC: voltage-gated channel
VgL: Vestigial-like protein
VGluT: vesicular glutamate transporter
VHL: von Hippel-Lindau protein
    (E3 ubiquitin ligase)
VIP: vasoactive intestinal peptide
VLDL: very low-density lipoprotein
VLDLR: very low-density lipoprotein
    receptor

VMAT: vesicular monoamine transporter

VN: vitronectin

VPS: vacuolar protein sorting-associated kinase

VR: venous return

VRAC: volume-regulated anion channel

VRC: ventral respiratory column

VRK: vaccinia-related kinase

VS: vasostatin

vSNARE: vesicular SNAP receptor

VSOR: volume-sensitive outwardly rectifying anion channel

VSP: voltage-sensing phosphatase

VVO: vesiculo-vacuolar organelle

vWF: von Willenbrand factor

W

$\mathbf{W}$: vorticity tensor

$\mathcal{W}$: strain energy density

W: work, deformation energy

$\mathbf{w}$: weight

$\mathbf{w}$: grid velocity

WASH: WASP and SCAR homolog

WASP: Wiskott-Aldrich syndrome protein

nWASP: neuronal WASP

WAT: white adipose tissue

WAVe: WASP-family verprolin homolog

WBC: white blood cell

WDR: WD repeat-containing protein

Wee: small (Scottish)

WHAMM: WASP homolog associated with actin, membranes, and microtubules

WIP: WASP-interacting protein

WIPF: WASP-interacting protein family protein

WIPI: WD repeat domain-containing phosphoinositide-interacting protein

WNK: with no K (Lys) kinase

Wnt: wingless-type

WPWS: Wolff-Parkinson-White syndrome

WSB: WD-repeat and SOCS box-containing protein (Ub ligase)

WSS: wall shear stress

WSSTG: WSS transverse gradient

WWTR: WW domain-containing transcription regulator

X

$\mathcal{X}$: trajectory

X: reactance

$\mathbf{X}$: Lagrangian position vector

$\mathbf{x}$: position vector

$\{x, y, z\}$: Cartesian coordinates

XBP: X-box-binding protein (transcription factor)

XIAP: X-linked inhibitor of apoptosis (Ub ligase)

Y

Y: admittance coefficient

YAP: Yes-associated protein

YBP: Y-box-binding protein (transcription factor)

YY: yin yang (transcriptional repressor)

Z

Z: impedance

ZAP70: ζ-associated protein 70

ZBTB: zinc finger and BTB (Broad complex, Tramtrack, and bric-à-brac) domain-containing transcription factor

ZnF: zinc finger protein

ZO: zonula occludens

Miscellaneous

2AG: 2-arachidonyl glycerol

3DR: three-dimensional reconstruction

3BP2: Abl Src homology-3 domain-binding adaptor

4eBP1: inhibitory eIF4e-binding protein

5HT: serotonin

7TMR: 7-transmembrane receptor (GPCR)

Mathematical Symbols, Molecules, and Physical Quantities

Greek Letters

α: volumic fraction

α: convergence/divergence angle

α: attenuation coefficient

α_k: kinetic energy coefficient

α_m: momentum coefficient

α_T: thermal diffusivity

β: inclination angle

$\{\beta_i\}_1^2$: myocyte parameters

β_T: coefficient of thermal expansion

Γ: domain boundary

Γ_L: local reflection coefficient

Γ_G: global reflection coefficient

γ: heat capacity ratio

γ: activation factor

γ_g: amplitude ratio (modulation rate) of **g**

γ_s: surface tension

$\dot{\gamma}$: shear rate

δ: boundary layer thickness

ϵ_T: emissivity (thermal energy radiation)

ϵ_e: electric permittivity

ϵ: strain

ε: small quantity

ζ: singular head loss coefficient

ζ: transmural coordinate

$\{\zeta_j\}_1^3$: local coordinate

η: azimuthal spheroidal coordinate

θ: circonferential polar coordinate

θ: $(\hat{\mathbf{e}}_x, \hat{\mathbf{t}})$ angle

κ: wall curvature

κ_c: curvature ratio

κ_d: drag coefficient

κ_h: hindrance coefficient

κ_o: osmotic coefficient

κ_s: size ratio

$\{\kappa_k\}_{k=1}^9$: tube law coefficients

κ_e: correction factor

Λ: head loss coefficient

λ_L: Lamé coefficient

λ: stretch ratio

λ: wavelength

λ_A: area ratio

λ_a: acceleration ratio

λ_L: length ratio

λ_q: flow rate ratio

λ_T: thermal conductivity

λ_t: time ratio

λ_v: velocity ratio

μ: dynamic viscosity

μ_L: Lamé coefficient

ν: kinematic viscosity

ν_P: Poisson ratio

Π: osmotic pressure

ρ: mass density

τ: time constant

Φ: potential

$\phi(t)$: creep function

φ: phase

χ: Lagrangian label

chi_i: molar fraction of species i

χ_i: wetted perimeter

$\psi(t)$: relaxation function

Ψ: porosity

ω: angular frequency
Ω: computational domain

Dual Notations

Bφ: basophil
Eφ: eosinophil
Lφ: lymphocyte
Mφ: macrophage
aaMφ: alternatively activated
    macrophage
caMφ: classically activated macrophage
Nφ: neutrophil
Σc: sympathetic
pΣc: parasympathetic

Subscripts

$_A$: alveolar, atrial
$_{Ao}$: aortic
$_a$: arterial
$_{app}$: apparent
$_{atm}$: atmospheric
$_b$: blood
$_\mathbf{c}$: contractile
$_c$: center
$_c$: point-contact
$_D$: Darcy (filtration)
$_d$: diastolic
$_{dyn}$: dynamic
$_E$: expiration, Eulerian
$_e$: external
$_\mathbf{e}$: extremum
$_{eff}$: effective
$_f$: fluid
$_g$: grid
$_I$: inspiration
$_i$: internal
$_{inc}$: incremental
$_L$: Lagrangian
$_l$: limit
$_\ell$: line-contact
$_M$: macroscopic
$_m$: mean
$_{max}$: maximum
$_m$: muscular, mouth
$_{met}$: metabolic
$_\mu$: microscopic
$_P$: pulmonary

$_\mathbf{p}$: parallel
$_p$: particle
$_q$: quasi-ovalization
$_r$: radial
$_{rel}$: relative
$_S$: systemic
$_s$: solute
$_\mathbf{s}$: serial
$_s$: systolic
$_t$: stream division
$_T$: total
$_t$: turbulence
$_t$: time derivative of order 1
$_{tt}$: time derivative of order 2
$_{tis}$: tissue
$_V$: ventricular
$_v$: venous
$_w$: wall
$_\mathbf{w}$: water (solvent)
$_\Gamma$: boundary
$_\theta$: azimuthal
$_+$: positive command
$_-$: negative command
$_*$: at interface
$_0$: reference state ($_{\cdot 0}$: unstressed or low
    shear rate)
$_\infty$: high shear rate

Superscripts

^{a}: active state
^{e}: elastic
^{f}: fluid
^{h}: hypertensive
^{n}: normotensive
^{P}: passive state
^{P}: power
^{s}: solid
^{T}: transpose
^{v}: viscoelastic
*: scale
*: complex variable
$^\cdot{}'$: first component of complex elastic
    and shear moduli
$^\cdot{}''$: second component of complex elastic
    and shear moduli
$^\natural$: static, stationary, steady variable

Mathematical Notations

T: bold face capital letter means tensor

v: bold face lower case letter means vector

S, s: upper or lower case letter means scalar

$\Delta\bullet$: difference

$\delta\bullet$: increment

$d\bullet/dt$: time gradient

∂_t: first-order time partial derivative

∂_{tt}: second-order time partial derivative

∂_i: first-order space partial derivative with respect to spatial coordinate x_i

∇: gradient operator

$\nabla\mathbf{u}$: displacement gradient tensor

$\nabla\mathbf{v}$: velocity gradient tensor

$\nabla\cdot$: divergence operator

∇^2: Laplace operator

$|\ |_+$: positive part

$|\ |_-$: negative part

$\dot{\bullet}$: time derivative

$\bar{\bullet}$: time mean

$\breve{\bullet}$: space averaged

$\langle\bullet\rangle$: ensemble averaged

$\tilde{\bullet}$: dimensionless

$\bullet^+$: normalized ($\in [0,1]$)

$\hat{\bullet}$: peak value

$\bullet_\sim$: modulation amplitude

$\det(\bullet)$: determinant

$\operatorname{cof}(\bullet)$: cofactor

$\operatorname{tr}(\bullet)$: trace

Cranial Nerves

I: olfactory nerve (sensory)

II: optic nerve (sensory)

III: oculomotor nerve (mainly motor)

IV: trochlear nerve (mainly motor)

V: trigeminal nerve (sensory and motor)

VI: abducens nerve (mainly motor)

VII: facial nerve (sensory and motor)

VIII: vestibulocochlear (auditory-vestibular) nerve (mainly sensory)

IX: glossopharyngeal nerve (sensory and motor)

X: vagus nerve (sensory and motor)

XI: cranial accessory nerve (mainly motor)

XII: hypoglossal nerve (mainly motor)

Chemical Notations

[**X**]: concentration of **X** species

X: upper and lower case letters correspond to gene and corresponding protein or conversely (i.e., Fes, FES, and fes designate protein, a proto-oncogene product that acts as a kinase, and corresponding gene and oncogene product, respectively)

$\mathbf{X}_i$: receptor isoform i of ligand **X** (i: integer)

XRi: receptor isoform i of ligand **X** (i: integer)

X+: molecule **X** expressed (**X**-positive)

$\mathbf{X}^+$: cation; also intermediate product **X** of oxidation (loss of electron) from a reductant (or reducer) by an oxidant (electron acceptor that removes electrons from a reductant)

X−: molecule **X** absent (**X**-negative)

$\mathbf{X}^-$: anion; also intermediate product **X** of reduction (gain of electron) from an oxidant (or oxidizer) by a reductant (electron donor that transfers electrons to an oxidant)

small GTPase$^{\mathrm{GTP(GDP)}}$: active (inactive) form of small (monomeric), regulatory guanosine triphosphatase

$\mathbf{X}^{\mathrm{GTP(GDP)}}$: GTP (GDP)-loaded molecule **X**

$\mathbf{X}^{\mathrm{M}}$: methylated molecule **X**

$\mathbf{X}^{\mathrm{P}}$: phosphorylated molecule **X**

p**AA**: phosphorylated amino acid (pSer, pThr, and pTyr)

$\mathbf{X}^S$: soluble form

$\mathbf{X}^{\mathrm{SNO}}$: ^{S}nitrosylated molecule **X**

$\mathbf{X}^{\mathrm{U}}$: ubiquitinated protein **X**

$\mathbf{X}_{\mathrm{alt}}$: alternative splice variant

$\mathbf{X}_{\mathrm{h(l)}}$: high (low)-molecular weight isotype

$\mathbf{X}_{\mathrm{L(S)}}$: long (short) isoform (splice variants)

$\mathbf{X}_{\mathrm{C}}$: C-terminal fragment (after proteolytic cleavage)

$\mathbf{X}_{\mathrm{c}}$: catalytic subunit

X_N: N-terminal fragment

X_P: palmitoylated molecule X

X_i: number of molecule or atom (i: integer, often 2 or 3)

$(X_1-X_2)_i$: oligomer made of i complexes constituted of molecules X_1 and X_2 (e.g., histones)

$^{D(L)}X$: D (L)-stereoisomer of amino acids and carbohydrates (chirality prefixes for dextro- [dexter: right] and levorotation [lævus: left]), i.e., dextro(levo)rotatory enantiomer

$^{F(G)}$actin: polymeric, filamentous (monomeric, globular) actin

$_tX$: truncated isoform

a, c, nX: atypical, conventional, novel molecule X (e.g., PKC)

acX: acetylated molecule X (e.g., acLDL)

al, ac, nX: alkaline, acidic, neutral molecule X (e.g., sphingomyelinase)

asX: alternatively spliced molecule X (e.g., asTF)

cX: cellular, cytosolic, constitutive (e.g., cNOS), or cyclic (e.g., cAMP and cGMP) molecule X

caX: cardiomyocyte isoform (e.g., caMLCK)

dX: deoxyX

eX: endothelial isoform (e.g., eNOS and eMLCK)

hX: human form (ortholog); heart type (e.g., hFABP); hormone-like isoform (FGF)

iX: inhibitory mediator (e.g., iSMAD) or intracellular (e.g., iFGF) or inducible (e.g., iNOS) isoform

kX: renal type (kidney) molecule X

ksX: kidney-specific isoform of molecule X

lX: lysosomal molecule X

l,acX: lysosomal, acidic molecule X

mX: mammalian species or membrane-associated molecule X (e.g., mTGFβ)

mtX: mitochondrial type of molecule X

nX: neutral X; neuronal type (e.g., nWASP)

oxX: oxidized molecule X (e.g., oxLDL)

plX: plasmalemmal type of molecule X

rX: receptor-associated mediator or receptor-like enzyme; also regulatory type of molecular species (e.g., rSMAD)

sX: secreted, soluble form of molecule X

s,acX: secreted, acidic molecule X

skX: skeletal myocyte isoform (e.g., skMLCK)

smcX: smooth muscle cell isoform (e.g., smcMLCK)

tX: target type of X (e.g., tSNARE); tissue type (e.g., tPA)

vX: vesicle-associated (e.g., vSNARE) or vacuolar (e.g., vATPase) type of X

GPX: glycoprotein (X: molecule abbreviation or assigned numeral)

Xx: (x: single letter) splice variants

X1: human form (ortholog)

Xi: isoform type i (paralog or splice variant; i: integer)

Xi/j: (i,j: integers) refers to either both isoforms (i.e., Xi and Xj, such as ERK1/2) or heterodimer (i.e., Xi-Xj, such as ARP2/3)

X1/X2: molecular homologs or commonly used aliases (e.g., contactin-1/F3)

PI(i)P, PI(i,j)P_2, PI(i,j,k)P_3: i,j,k (integers): position(s) of phosphorylated OH groups of the inositol ring of phosphatidylinositol mono-, bis-, and trisphosphates

Amino Acids

Ala (A): alanine

Arg (R): arginine

Asn (N): asparagine

Asp (D): aspartic acid

CysH (C): cysteine

Cys: cystine

Gln (Q): glutamine

Glu (E): glutamic acid

Gly (G): glycine

His (H): histidine

Iso, Ile (I): isoleucine

Leu (L): leucine

Lys (K): lysine

Met (M): methionine

Phe (F): phenylalanine
Pro (P): proline
Ser (S): serine
Thr (T): threonine
Trp (W): tryptophan
Tyr (Y): tyrosine
Val (V): valine

Ions

Asp^-: aspartate (carboxylate anion of aspartic acid)
ATP^{4-}: ATP anion
Ca^{++}: calcium cation
Cl^-: chloride anion
Co^{++}: cobalt cation
Cu^+: copper monovalent cation
Cu^{++}: copper divalent cation
Fe^{++}: ferrous iron cation
Fe^{3+}: ferric iron cation
Glu^-: glutamate (carboxylate anion of glutamic acid)
H^+: hydrogen cation (proton)
H_3O^+: hydronium (oxonium or hydroxonium) cation
HCO_3^-: bicarbonate anion
HPO_4^{2-}: hydrogen phosphate anion
K^+: potassium cation
Mg^{++}: magnesium cation
$MgATP^{2-}$: ATP anion
Mn^{++}: manganese cation
Na^+: sodium cation
Ni^{++}: nickel cation (common oxidation state)
OH^-: hydroxide anion
PO_4^{3-}: phosphate anion

SO_4^{2-}: sulfate anion
Zn^{++}: zinc cation (common oxidation state)

Inhaled and Signaling Gas

CO: carbon monoxide (or carbonic oxide; signaling gas and pollutant)
CO_2: carbon dioxide (cell waste)
H_2S: hydrogen sulfide (signaling gas)
He: helium (inert monatomic gas)
N_2: nitrogen (inert diatomic gas)
NO: nitric oxide (or nitrogen monoxide; signaling gas and pollutant)
NO_2: nitrogen dioxide (air pollutant)
O_2: oxygen (cell energy producer)
SO_2: sulfur dioxide (air pollutant)

Nitric Oxide Derivatives

$NO^\star$: free radical form
NO^+: nitrosonium (nitrosyl) cation
NO^-: nitroxyl or hyponitrite anion (inodilator)
HNO: nitroxyl (protonated nitroxyl anion)
NO_2^-: nitrite anion
NO_3^-: nitrate anion

Reactive Oxygen Species

H_2O_2: hydrogen peroxide
O_2^-: superoxide
OH^-: hydroxyl radical, hydroxide
$ONOO^-$: peroxynitrite

Index

M